International Cryogenics Monograph Series

Series Editors

The International Cryogenics Monograph Series was established in the early 1960s to present an opportunity for active researchers in various areas associated with cryogenic engineering to cover their area of expertise by thoroughly covering its past development and its present status. These high level reviews assist young researchers to initiate research programs of their own in these key areas of cryogenic engineering without an extensive search of literature.

More information about this series at http://www.springer.com/series/6086

Guy Gistau Baguer

Cryogenic Helium Refrigeration for Middle and Large Powers

Springer

Guy Gistau Baguer
Biviers, France

ISSN 0538-7051 ISSN 2199-3084 (electronic)
International Cryogenics Monograph Series
ISBN 978-3-030-51676-5 ISBN 978-3-030-51677-2 (eBook)
https://doi.org/10.1007/978-3-030-51677-2

This Springer imprint is published by the registered company Springer Nature Switzerland AG
The registered company address is: Gewerbestrasse 11, 6330 Cham, Switzerland

Preface

I would like this book to be used by today's cryogenic community, mainly by operators of cryogenic systems but also by designers, even if not all detailed information is made available for the latter's activity.

This book is a result of my personal journey into cryogenics and helium refrigeration, which is presented in logical progression through main topics such as cycle design, technology of components, system control, commissioning, operation and maintenance. At every occasion, I emphasise the importance of "feeling" the behaviour of a cryogenic system.

The book is intended to provide, in a single document, most of the material that is necessary to deal with helium refrigeration.

Biviers, France — Guy Gistau Baguer

Acknowledgements

The very first person I think about is my father, Joaquín Gistau Baguer, who was a self-educated man. He almost never went to school, did not like working in the field with his family and learnt blacksmithing and the trade of mechanics by himself. He wanted to "understand". He gave me the desire to understand and build machines. He would be very proud to see a book written by his son. I must also thank the French education system that allowed me to pursue engineering studies in comfortable conditions.

I would like to thank Anna, my wife, who supported me all along this time-consuming and stressful job. I promised her not to write another book!

Thanks to the large number of people I have had the opportunity to meet and interact with during my career in various places around the world. I will not try to cite all names lest I forget to mention some of them.

A special thanks to my friends who dared to review the material of this book and helped me on various specific issues: M. Bonneton, B. Bradu, P. Briend, K. Brodzinski, L. Delprat, V. Gahier, D. Grillot, Ph. Lebrun, E. Monneret, A. Ravex, J.C. Villard, U. Wagner, J. Weisend and G. Zick.

In addition to the reviewers, I would like to acknowledge the important contributions of H. Quack and U. Wagner, who brought more rigour in the thermodynamic aspects, and J. C. Villard and F. Delcayre, who made private communications on cryogenic expansion turbines and cryogenic centrifugal compressors.

Finally, I add a deep thank you to the Springer team who has "golden eyes" to hunt and find all kinds of mistakes I made in the manuscript.

About the Book

Why Such a Book?

A lot of very good books dealing with various cryogenic topics are available, but none, to my knowledge, are really and totally dedicated to practical aspect of helium refrigeration.

Among the general information on cryogenics, the most important, at least for me, had been the well-known "*Cryogenic Engineering*" by Russell B. Scott (see Fig. 1). For many people from my generation interested in cryogenics, it has been "The Reference", because at that time very little literature concerning cryogenics could be found. The first edition of this book was issued in March 1959, more than 60 years ago! At that time, I did not even know the word "cryogenics". Time has flown, and cryogenics has much developed and improved, both in quality, size and, most importantly, reliability and efficiency. However, today, I would not be able to indicate an existing up-to-date one-off book that could comfortably lead newcomers to the world of helium refrigeration. Most of them deal with, for example, introduction to general thermodynamic cycles for helium refrigeration or components (heat exchangers). Some others deal with conventional machines that are incorporated into helium refrigerators as compressors. Very few deal with specific cryogenic components such as cryogenic expansion turbines or cryogenic compressors; therefore, one must "dig" into various books, as the information provided in a particular book may not be really necessary for this job.

If one looks at what is necessary to deal with helium refrigeration, the range is, as often in an engineering activity, much extended:

- Theoretical background in mechanics, physics and, especially, thermodynamics
- Thermodynamic process design aiming at smart high-efficiency thermodynamic cycles often operating in off-design conditions

Fig. 1 My very old personal exemplar of Cryogenic Engineering by R. B. Scott (D. Van Nostrand Company)

- Technology of conventional machines such as compressors from a few tens of kilowatts up to a few megawatts (not to forget electric motors), heat exchangers, valves and vacuum systems
- Technology of special equipment to be incorporated into conventional machines as oil removal system
- Technology of very special machines like cryogenic expansion turbines, cryogenic centrifugal compressors and circulators
- Specific knowledge like high vacuum technology, high-performance thermal insulation and leak checking
- Construction: selection of structural materials, designing and sizing pressure vessels
- Quality assurance procedures
- Rather sophisticated process control procedures
- A little bit of chemistry for gas and oil analysis
- Testing and operating complex systems
- General maintenance, but also maintenance specifically dedicated to cryogenic systems

One cannot be an expert on all such topics, but one must know a little bit of each one, and, of course, the “little bit” that is of interest in the domain of helium refrigeration. This is the way the book is written: information on each topic is not extensive but, hopefully, sufficient.

In this book, I aim to provide the reader with the maximum useful information, saving him or her the trouble of looking for relevant information in multiple documents. Even if this book is not intended to teach the design of refrigerating systems, though some information in this field is provided.

It means that there are a lot of pragmatic technics (tricks) or remarks that should help the reader. This information comes mainly from either personal experience gained through my career or from what I have learned from people operating cryogenic systems, for example, CERN operators at Geneva, Switzerland, which is till date the only place in the world where the highest operating experience in large cryogenic systems is concentrated.

It is also obvious that even if I tried to be accurate, mistakes and/or inaccuracies might exist in the text. The reader should not hesitate to point these out if he or she finds some of them.

Machine manufacturers, especially of compressors, are not all very cooperative with an individual like me who has retired and is no more a potential customer and who tries to write on the equipment they make. This also can lead to inaccuracies by lack of information.

The Philosophy of the Book

I would like this book to be used by today's cryogenic community in the same manner as I had used *Cryogenic Engineering* by Russell B. Scott, but, to be fair, I can only help in the field I know best, which is helium refrigeration.

Thermodynamics is a necessary tool to understand the operation and get into the process calculations. In this book, contrary to a lot of others, which I anyway advise to have a look at, I have tried my best to avoid deep discussion on thermodynamics. Very few thermodynamic equations are used in the book, the bare minimum. Therefore, only a few elements out of the wonderful thermodynamic tool box will be used. The emphasis is on the way a system operates and behaves. A good cryogenic system designer or operator should ideally be able to think as a kind of a "system process simulator". By these words, I mean that one should be able to guess what would be the consequence of an external perturbation or a voluntary action. This is what I call "feeling the system".

A long time ago, the French philosopher René Descartes (1596–1650) had said:

- Split the problem into simpler questions.
- Begin with the simplest question, not by laziness, but to obtain the necessary elements to solve the more complex questions.
- Move forward of a stage only when the previous one contains no more ambiguity.

This way of behaving is still up to date. Therefore, in the experience that I have gained conducting educational sessions, I always start from very basic issues and then move on to more complicated cases. The readers might have varied technical knowledge; therefore, some of them could feel that the starting level is too low, but I believe that this is the only way to be sure that at a given point, each reader has the same background. Nobody should be left behind.

The detailed level of information that I provide can be very different according to the topics that are discussed: This is because my knowledge is not at the same level in all matters!

The Content of the Book

This book is the result of a kind of journey (my personal trip into cryogenics and helium refrigeration) made in a logical way into helium refrigeration engineering: design, building, commissioning, operation and maintenance.

In order that all readers start with a common amount of information, even if they are supposed to already know most of the basic principles of thermodynamics and cryogenics, some general points about physics and cryogenics are revised, for example, physics principles that are not common and therefore might have been partially (or totally!) forgotten by the reader. The formulae that are used along the book are generally not demonstrated at the place they appear in order not to make the text awkward. However, if the reader wants to know about their origin, a digest in thermodynamics has been added at the end of the book, in Chap. 17.

A "light" theory of heat exchangers allows the reader to understand the importance of heat exchanger role in the various thermodynamic cycle structures that are discussed later.

Along the book, every time where a calculation could help, a copy of the spreadsheet is displayed.

The thermodynamic cycles are considered from the simplest (Joule Thomson) to the most complicated for the very large refrigeration plants and, finally, those operating at temperatures lower than 4.5 K. Liquid nitrogen pre-cooling and cooling of thermal shields are also considered.

When the cycle organisation is decided, one must size and combine the main components. It is of interest to know how the main components operate and what their capabilities and limits are. This is not a course on machine design: how to design a cryogenic expansion turbine will not be taught, only the general principles that are involved in this process are mentioned. However, specificities of room temperature or cryogenic machines such as room temperature cycle screw compressors, heat exchangers, cryogenic expansion turbines, cryogenic centrifugal compressors and circulators will be pointed out. A few specific characteristic digests (models) summarise the behaviour of each component.

In the above cycle description, the location where the thermal load is dissipated is a part of the sketch. This is for simplification. In reality, the thermal load is always located at a distance, either a liquid helium dewar or one or more cryostats. The most conventional among the various ways to connect the refrigerator cold box to the thermal load are described and related to existing plants.

In most of the cases, a plant does not exactly operate in the regime for which it has been designed or it operates at partial load due to a safety factor that has been considered in writing the specification; it is of interest to understand how the system behaves. This is explained in the introduction to the "Off-Design Operation" chapter.

When the refrigerator has been "constructed", it is to be operated. The book goes through the basic principles of process control and discusses more particularly the operating situations of helium plants. Emphasis is laid on keeping the plant efficiency at the highest level, whatever the operating regime.

But one should not forget that there is a paramount condition to be fulfilled: the cycle helium must be pure! Therefore, the book addresses what to do in order to be sure all components are filled with pure helium prior to starting: after a reminder on the helium fluid, one learns why helium is polluted, how to purify it and, also, how to analyse its purity.

The correct operation of a plant means that its availability is permanently kept at the highest level. To reach this target, permanent checking of the plant is necessary, mainly performed by the control system but also by the operator. And, of course, correct maintenance action is necessary.

The description of a selection of typical or characteristic existing plants allows illustrating the material that has been provided along the former chapters.

To help the reader when he is to set up a project, some advice concerning a Request for Quotation are given. Guide lines for commissioning tests of large systems are also provided.

Important calculations that are performed along the former chapters are detailed in the Cryo Tool Box. Even if this book is not specifically planned to design thermodynamic cycles, it is of interest, both as a designer and as an operator of a cryogenic system, to be in a position to perform some simple calculations to get an idea of how components or systems behave. Such calculations are generally performed using the well-known Microsoft® Excel software associated with the REFPROP® software. Other software such as Gaspak® or Hepak® can be used too. In order not to redo the same job (with the ever-present risk of making new mistakes!), such simple calculations are saved to a file that can be called the "Cryogenic Tool Box".

Examples of proposed Cryogenic Tool Box calculations:

- Thermodynamic properties of helium (how to use them comfortably)
- Room temperature compressor
- Behaviour of a simple circuit dealing with heat loads and pressure drops
- Operation of a heat exchanger
- Operation of a cryogenic expander
- Operation of a cryogenic compressor or a circulator
- Equivalent power at 4.5 K

The present helium plant operators are, obviously, younger than me. They are used to run systems that are very sophisticated and do not, generally, have a clear knowledge about "old" former systems. I thought that it would be interesting to write a few words on what I used to call the "cryogenic refrigeration saga" along more than one century.

Finally, a digest in thermodynamics provides explanations on most of the formulae that have been used throughout the book.

As the material that is dealt with in the book is rather important, it is possible, when going deep into the book, that the reader have forgotten an issue that may have been explained a number of pages before. To help him locate this, a lot of references, both to paragraphs and figures, are given along the text.

The reader might be surprised to find so few references on books or publications. As I collected information over a long period of time, I have often forgotten where I got this from or from whom.

Help from Readers

The author would be very much interested in getting remarks from readers about the book. Some may find that a topic is not clearly explained or some points have been forgotten; therefore, do not hesitate to send your positive remarks or suggestions at: guy@cryoguy.com. They would be taken into account in case of a re-edition.

Disclaimer

Several systems are described and discussed in this book. They must be considered as examples and not as unique solutions or structures. As usually in life, a problem might have many solutions that cope with the same requirement. Not all of them are optimal but all can fulfil the duty.

Contents

About the Author

When I was studying mechanical engineering at "École Nationale Supérieure des Arts et Métiers", I do not remember ever having heard this curious word: cryogenics.

My first and unique contact with this world, even if I did not know the word, was in 1957. I was visiting a mine in the northeast of France (Lorraine) where I met miners who were using rather special explosives made out of wood sawdust cartridges they had soaked in liquid oxygen. In case the cartridge did not explode, such stuff had the advantage of being inexplosive after the oxygen had been vaporised. Therefore, there was no more a risk of explosion when miners recovered the material after a blast. These miners explained to me that soaking the sawdust cartridge in oil before soaking it in liquid oxygen improved the effect strongly! It was only years later, at the time I was operating an air separation plant, that I realised how dangerous such a procedure had been.

I have spent most of my time dealing with helium refrigeration and liquefaction. I was hired by Air Liquide in 1965, my first job in the "*Centre d'Etudes Cryogéniques*" in Sassenage, in the southeast of France. I was in charge of building

7 L/h helium liquefiers. My second position was to help operate an air separation plant and its distribution system (a fleet of trucks that were delivering gas and liquid) during the 3 years I spent in the Paris area. Later, I spent another 3 years selling gases, both industrial bulk and speciality gases, including gases for the electronic industry (very pure gases, doping and epitaxy gases) and liquid helium. My next job was at the Air Liquide Head Office, Paris, where I coordinated the activities of electronic gases and liquid helium. Finally, in 1975, I had the opportunity to come back to the Sassenage site where I dedicated myself again to helium refrigeration. As the team was very small, I had the opportunity (or, more exactly, the necessity!) to deal with all steps of a project: sales, design, construction, erection, tests, after sales, troubleshooting and, simultaneously, development. That was an incredibly efficient way to become familiar, or even intimate, with all aspects of helium refrigeration, at a time when there was no possibility to get such information from literature.

Among various projects and machines designed and built, I would like to emphasise a few of them. In the beginning of the 1980s, our team gave birth to the first fully automatic liquefier/refrigerator: HELIAL. The expertise that was acquired at this occasion, especially in automation of the process, allowed, later, to deal with more complicated systems. Such an opportunity occurred with the controlled thermonuclear fusion project Tore Supra that needed 300 W at 1.8 K. A few plants of required size had already been built, incorporating a lot of room temperature Roots machines. If fusion was to develop, it was obvious that such a technology would rapidly hit limits. Therefore, it was decided to develop centrifugal cryogenic compressors that were integrated into the refrigerator, making it the first one operating on such a technology and opening the door to the modern large systems operating at temperatures lower than 4.5 K. I should also narrate the saga of starting up the cryogenic centrifugal train of four machines that was integrated into the CEBAF (Continuous Electron Beam Accelerating Facility) refrigerator. And, finally, when CERN required refrigerators for high power at 1.8 K, I had the opportunity of putting together most of my knowledge in helium refrigeration into the design of refrigerators for this project.

In 2011, the largest helium liquefier, built by my younger colleagues, delivered 7000 L/h! Therefore, during my lifetime, I had the opportunity to see the helium liquefier's size increase by three orders of magnitude.

During this period, I had the opportunity to write around 50 publications, mainly about the machines that were built but also on topics dealing with helium refrigeration. A few patents on helium refrigeration were filed too, some of which are referred to, later in the book.

Attending the usual cryogenic conferences led me to become a member of the International Cryogenic Engineering Committee for 16 years and eventually its chairman for another 12 years.

When time of retirement came, in 2000, I felt guilty to let such an amount of self-gained experience (mistakes, thinking, trying again, success sometimes) be lost. Therefore, I decided to try and transfer this knowledge to other people who would need it, especially young researchers. I knew that transmitting experience is a very difficult task, with a low efficiency, but anyways, I thought that if I did not try, nothing would happen.

In 1997, I started, by chance, giving notions from scratch on helium refrigeration to a customer who had just stepped into the field. I liked my ideas of starting an education action and began improving my teaching material and method, switching from transparent to animated PowerPoint. Every time somebody did not understand what I was explaining in my courses, I questioned myself: did he (or she) not simply understand or did I not explain correctly? For almost each new teaching session, I had introduced what I expected to be improvements. For each session, a set of prints of the PowerPoint slides was provided to each school participant. Up till now (2020), I have held more than 60 sessions on cryogenics, most of them dealing with helium refrigeration. Such an experience made me understand that explanations must be clear, sometimes starting from a rather low level to be sure that everybody follows.

In 2012, I wrote an article in the French journal *Techniques de l'ingénieur* about helium refrigeration. The number of pages was limited to 16, making it a challenge to fit the subject into such a compact format. This experience gave me the idea to write down the talks I give at each school, which in turn gave birth to this book.

In 2007, I was presented the Charles Tellier Medal from the *Association Française du Froid.*

In 2020, I was the recipient of the Mendelssohn Award, which should have been presented to me at the International Cryogenic Engineering Conference that was to be held in Hangzhou, China, in August 2020, but that has been shifted to 2022 due to the Covid-19 pandemic.

By the way this book is kind of my "cryogenic testament"!

Abbreviations

API	American Petroleum Institute
BOS	Bulk Oil Separator
CAPEX	CAPital EXpenditure
CB	Cold Box
CCC	Centrifugal Cryogenic Compressor
CEA	*Commissariat à l'Energie Atomique (France)*
CEBAF	Continuous Electron Beam Accelerating Facility (Newport News, USA)
CERN	European Organisation for Nuclear Research (Geneva, Switzerland)
CICC	Cable In Conduct Conductor
CL	Control Loop
CNS	Cold Neutron Source
COP	Coefficient of Performance
CTB	Cryogenic Tool Box
DP	Dew Point
e	Error
ESS	European Spallation Source (Lund, Sweden)
FT	Flow Transmitter
GD_2	Gaseous Deuterium
GH_2	Gaseous Hydrogen
GHe	Gaseous Helium
GN_2	Gaseous Nitrogen
GRAFCET	*Graphe Fonctionnel de Commande des Etapes et Transitions*
HEPAK	Computer Program to Calculate Thermophysical Properties of Helium
HP	High Pressure
HTR	Heater
HX	Heat Exchanger
IHI	Isikawajima Heavy Industry (Japan)

ITER	International Thermonuclear Experimental Reactor (Cadarache, France)
JT	Joule Thomson
LD_2	Liquid Deuterium
LEP	Large Electron Positron accelerator (CERN, Geneva, Switzerland)
LH_2	Liquid Hydrogen
LHC	Large Hadron Collider (CERN, Geneva, Switzerland)
LHe	Liquid Helium
LMTD	Logarithmic Mean Temperature Difference
LN_2	Liquid Nitrogen
LP	Low-Pressure
LSH	Level Switch High
LSL	Level Switch Low
MV	Measured Variable
MTD	Mean Temperature Difference
NTU	Number of Transfer Units
OPEX	OPerational EXpenditure
ORS	Oil Removal System
PCV	Pressure Control Valve
PID	Proportional Integrative Derivative
PT	Pressure Transmitter
PTFE	Polytetrafluoroethylene
PV	Process Variable
REFPROP	REFerence Fluid PROperties
SP	Set Point
ST	Speed Transmitter
SV	Solenoid Valve
TBT	Très Basses Températures (a former subsidiary of Air Liquide)
TJLAB	Thomas Jefferson Lab (Newport News, USA)
T-s	Temperature-Entropy (diagram)
TT	Temperature Transmitter
VFD	Variable Frequency Drive
WEST	Wolfram Environment in Steady-state Tokamak (Cadarache, France)
X	Normalised dimensionless flow

List of Symbols

A Area (m^2)
c_p Specific isobar heat capacity ($J.g^{-1}.K^{-1}$)
c_v Specific isochoric heat capacity ($J.g^{-1}.K^{-1}$)
E Exergy (J)
f_c Flow coefficient (−)
h Mass specific enthalpy of the fluid ($J.g^{-1}$)
K Global coefficient of thermal transmission ($W.K^{-1}.m^{-2}$)
K Kinetic energy (J)
$\dot{l}$ Liquefaction mass flow rate ($g.s^{-1}$)
M Molar mass (helium: 4,0026) (g)
m Mass (g)
$\dot{m}_1$ Mass flow rate at point 1 ($g.s^{-1}$)
Nm^3 Normal cubic meter (m^3)
P_1 Pressure at point 1 (bar)
P_{dis} Discharge pressure (bar)
ppM V part per Million expressed by Volume (10^{-6})
ppM W part per Million expressed by Weight (10^{-6})
P_{suc} Suction pressure (bar)
$\dot{Q}$ Thermal power (W)
$\dot{Q}_p$ Parasitic thermal power (W)
$\dot{Q}_{T_c}$ Thermal power absorbed at cold temperature T_c (W)
$\dot{Q}_{T_w}$ Thermal power absorbed at warm temperature T_w (W)
$\dot{Q}_u$ Useful thermal power (W)
R Thermal resistance ($K.W^{-1}$)
T_0 Room temperature (300 K) (K)
T_1 Temperature at point 1 (K)
T_c Cold temperature (K)
T_{dis} Discharge temperature (K)
T_{suc} Suction temperature (K)

T_w	Warm temperature (K)
U	Global heat exchange coefficient (W.K^{-1})
U	Internal energy (J)
V	Volume (m^3)
$\dot{W}$	Mechanical power (W)
Δh	Mass specific enthalpy difference of the fluid (J.g^{-1})
Δh_{actu}	Actual specific mass enthalpy drop (J.g^{-1})
Δh_{is}	Isentropic specific mass enthalpy drop (J.g^{-1})
Δs_{actu}	Actual specific mass entropy drop (J.g^{-1})
Δs_{is}	Isentropic specific mass entropy drop (J.g^{-1})
ΔT_c	Cold temperature difference (K)
ΔT_w	Warm temperature difference (K)
ε	Efficiency (–)
η	Efficiency (–)
ρ	Density (kg.m^{-3})

Chapter 1
Some Reminders About Cryogenics and Physics

Abstract This chapter is only a reminder (not a course) about a few important basics that are needed to deal with helium refrigeration like thermodynamics, cryogenic fluid properties, materials that are used in cryogenics and principle of process calculation.

1.1 Introduction

In order to keep the main text light, formulae that are used are not demonstrated in this chapter. However, details in some thermodynamics demonstrations are given in Chap. 17.

1.2 The Typical Structure of a Helium Cryogenic System

For newcomers in the field of helium refrigeration, it might be good to name the main components of a helium cryogenic system:

- A cycle compression station composed of:
  - Cycle compressor(s)
  - Inter- and aftercoolers
  - Oil removal system
  - Helium management valves
  - Cycle buffer capacity
- A so-called cold box composed of:
  - A vacuum enclosure that houses the cryogenic components such as:
    - A set of heat exchangers
    - Cryogenic machines (turbines, compressors)
    - Cycle adsorbers

G. Gistau Baguer, *Cryogenic Helium Refrigeration for Middle and Large Powers*, International Cryogenics Monograph Series,
https://doi.org/10.1007/978-3-030-51677-2_1

    - Phase separators
    - Various piping and valves
  - Instrumentation
  - A vacuum set
- A liquid helium storage or a cryostat that houses the components to be cooled
- Cryogenic transfer lines that connect the cold box to the liquid helium storage or the cryostat (the component in which the object to be cooled, for example, a coil or a resonant cavity is housed)
- A helium recovery and purification system

This book deals with only the two first items that are parts of the refrigerator: the compression station and the cold box.

Such systems can be pure liquefiers or mixed duty refrigerators (the difference is explained later, in Sect. 3.1). In the book, except in specific situations, the wording "refrigeration plant or system" includes the refrigeration or liquefaction systems.

1.3 Specific Operating Conditions of a Cryogenic System

The operation of a cryogenic system happens roughly in the reverse way compared to most of other current-life systems that one is used to deal with: when the system is started up, temperatures decrease instead of increasing.

Any component of a cryogenic system must be able to operate in very different operating conditions from the time of start-up to the time where the system is in operational cold steady state. In some large systems, such a transient operation might last for a few weeks, for example, during the cool-down process of a large system.

1.4 The Cryogenic Fluids

Compared to conventional refrigeration, the number of cryogenic fluids is small. Their names and their boiling temperatures at atmospheric pressure are shown on Fig. 1.1. By *convention*, cryogenics deals with temperatures that are lower than 120 K.

One should note that, except for hydrogen (explanation is given in Sect. 1.4.4.1), latent heats decrease with temperature.

Among those fluids, only helium, nitrogen and hydrogen are considered in this book.

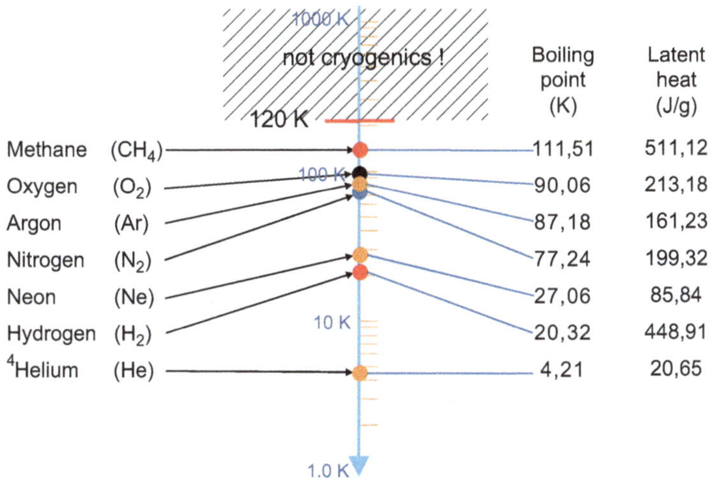

Fig. 1.1 The usual cryogenic fluids: boiling points and latent heat at atmospheric pressure (REFPROP)

1.4.1 Properties of Cryogenic Fluids

1.4.1.1 The Pressure-Temperature (P-T) Diagram

The pressure-temperature diagram is a well-known tool, as shown for nitrogen in Fig. 1.2.

The liquid-vapour saturation curve separates the liquid and vapour domains. Towards high pressure and temperature, it ends on the critical point where liquid no longer exists (the liquid-vapour separation interface disappears). Towards low pressure and temperature, the liquid-vapour line ends on the triple point where the three states of matter (vapour, liquid and gas) exist simultaneously. Vapour can be superheated if its temperature is higher than the equilibrium temperature or its pressure is lower. Similarly, liquid can be subcooled if its temperature is lower than the equilibrium temperature or its pressure is higher (see Fig. 1.2). Liquid and solid are separated by the melting line leaving from the triple point. The melting temperature is nearly independent on the pressure. At pressure or/and temperature slightly higher than the critical point, gas is in a so-called supercritical state, the properties of which are not very different from that of liquid.

Remark It happens that such a diagram is plotted with logarithmic scale(s), either on one or both coordinates. That changes heavily the shape of the curve (see Fig. 1.2, right, where the pressure scale is logarithmic).

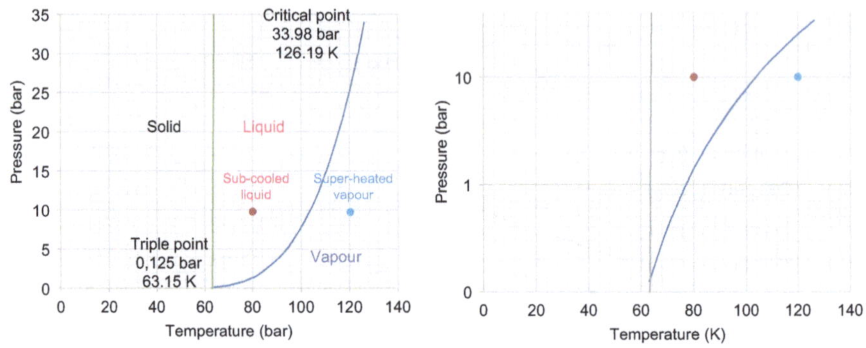

Fig. 1.2 The nitrogen PT diagram (according to REFPROP values) with arithmetic and logarithmic pressure scale

1.4.1.2 Thermal Properties of Fluids

Thermal properties change with both pressure and temperature. To get exact property values for a real gas, one has to use one among various tools (tables, diagrams, etc.), but the easiest way the latter is to use such specific software as REFPROP® or GASPAK® or HEPAK® the latter being specific for helium (see Sect. 1.4.2.1).

Specific heat (or heat capacity): c_p or c_v
The specific heat is the amount of thermal energy that is needed to cause a temperature-specific variation of one unit of a given body or system of one mass unit (under specified conditions). It is assumed that during the process, no phase change occurs. The conditions usually are isobaric-specific heat capacity (at constant pressure), c_p, or isochoric-specific heat capacity (at constant volume), c_v.

In ISO units, the specific heat capacity is expressed in joule per kilogram per kelvin: **J/kg.K**

Physics reminder
For monatomic ideal gases:

$$c\_(p\ ) = R \times 7/2 = 29100\ \mathrm{J/kmol.K} \tag{1.1}$$

$$c\_(v) = R \times 5/2 = 20790\ \mathrm{J/kmol.K} \tag{1.2}$$

Example: At atmospheric pressure and at temperatures higher than 100 K, the isobaric specific heat of helium c_p is approximately 5193 J/kg.K, and the isochoric is 3116 J/kg.K (REFPROP). (Be careful! at lower temperatures, especially near the liquefaction temperature, c_p is no more constant.)

Adiabatic coefficient γ (gamma)
The adiabatic coefficient is the ratio c_p/c_v.

At atmospheric pressure and 300 K, the adiabatic coefficient of helium is 1.6665 (REFPROP).

Specific latent heat (or heat of vaporisation): L
The latent heat is the amount of energy (enthalpy) that is necessary to vaporise one unit of mass of liquid at equilibrium conditions (or turn one unit of mass of liquid at equilibrium into saturated vapour).

In ISO units, the latent heat is expressed in joule per kilogram: **J/kg**.

Example: At atmospheric pressure, the latent heat of helium is 20 790 J/kg (REFPROP).

Note that the latent heat of helium is the lowest among all fluids.

Specific sensible heat
The sensible heat is the heat supplied to or extracted from a body or substance that is associated with a change in temperature and is not accompanied by a change of phase.

The sensible heat is the amount of energy that is necessary to warm up from saturated temperature up to room temperature one unit of mass of saturated vapour.

In ISO units, the sensible heat is expressed in joule per kilogram: **J/kg**.

Example: At atmospheric pressure, the sensible heat of helium from saturated vapour at 1.0 bar up to 300 K is 1,542,674 J/kg (REFPROP).

Specific enthalpy h: c_p
The absolute value of enthalpy is never used alone; it is the enthalpy difference between two states that is considered.

Tables or software that return enthalpy may have different origin for enthalpy values. Therefore, a calculation must ALL be performed using the *same table or software*.

Remember The enthalpy of an *ideal gas* is only related to its temperature, not to its pressure!

$$dh = c_p \times dT \tag{1.3}$$

or

$$h = c_p \times T + Cte \tag{1.4}$$

In ISO units, the specific enthalpy **h** is expressed in joule per kilogram: **J/kg**. For example, at atmospheric pressure and 300 K, the specific enthalpy of helium is: 1563320 J/kg

Specific entropy: s
As for enthalpy, the absolute value of entropy is never used alone; it is the entropy difference between two states that is considered. As for enthalpy, calculation must ALL be performed using the same table or software.

In ISO units, the specific enthalpy **s** is expressed in joule per kilogram per kelvin: **J/kg.K**.

Example: At atmospheric pressure and 300 K, the specific entropy of helium is: 28010 J/kg.K

Sound velocity: c
The sound velocity for an *ideal gas* is:

$$c_{\text{ideal}} = \sqrt{\gamma RT/M} \tag{1.5}$$

γ: Adiabatic coefficient c_p/c_v
R: universal ideal gas constant 8.314462 J/kg. Mole
M: molar weight

For real gases, properties are generally not very different from ideal gases as soon as pressure is low and the conditions are sufficiently far away from liquefaction. See how sound velocity changes for helium in Fig. 1.16.

Example: At atmospheric pressure and 300 K, the sound velocity in helium is 1019.58 m/s (REFPROP).

Behaviour of a sonic orifice
When the ratio of inlet versus discharge pressure of an orifice is higher than two, the gas velocity in the throat is equal to sound velocity.

The volume flow through an orifice is:

$$\dot{V} = A\, v \tag{1.6}$$

A: cross section of the orifice
v: velocity of gas in the throat
The mass flow is:

$$\dot{m} = \rho \times \dot{V} \tag{1.7}$$

and, as ρ is roughly proportional to P,

$$\dot{m} = \propto \frac{P}{\sqrt{T}} \tag{1.8}$$

Consequently, through a sonic orifice, the mass flow rate is proportional to the inlet pressure and inversely proportional to the square root of the inlet temperature.

See how helium mass flow rate changes in Fig. 1.17.

1.4.1.3 The Temperature-Entropy (T-s) Diagram

The temperature-entropy diagram (T-s diagram) (see Fig. 1.3) is a very interesting tool for studying thermodynamic cycles. It eases the understanding of the fluid behaviours and thermodynamic cycles.

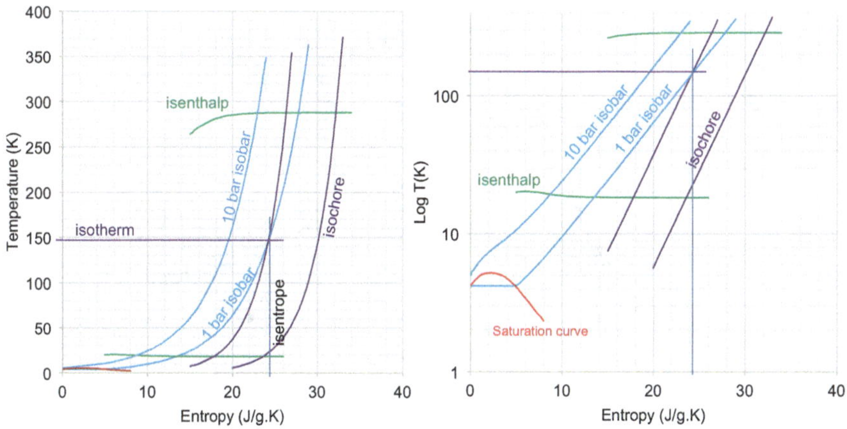

Fig. 1.3 A T-s diagram displaying same curves, with an arithmetic (left) or a logarithmic (right) temperature scale

On the T-s diagram, one can plot various thermodynamic evolutions of a gas:

- An isothermal transformation follows a horizontal straight line (isotherm).
- An isentropic transformation follows a vertical strait line (isentrope).
- An isobaric (constant pressure) transformation follows an arc of exponential (isobar) if c_p is constant. When it is a two-phase mixture, it is a horizontal straight line. For an ideal gas, isobars are arcs of exponentials that are shifted horizontally according to pressure (see Fig. 1.3, left).
- An isochore (constant volume) is an arc of exponential if c_v is constant. At a given point, the slope of the isochore is larger than that of an isobar. For an ideal gas, isochores are arcs of exponentials that are shifted horizontally according to pressure.

When the T-s diagram is plotted in a semi-log scale, isobars are parallel straight lines (see Fig. 1.3, right) in the ideal gas region. It is the same for isochores.

In Fig. 1.4, the red curve is the saturation curve that encloses the two-phase domain where the iso-quality curves split the two-phase segments proportionally to the quality.

It is a common practice to use a simplified log (T)-s diagram as shown in Fig. 1.5, where one can follow the evolutions of the gas in a *qualitative way*, in order to "feel" how the system behaves.

The compression of a gas shown on the T-s diagram

The first phase of a cooling cycle is the compression of the cycle gas. Let us take the opportunity to describe such a process on the T-s diagram. As usually, it is of interest to follow the gas transformations both on the T-s diagram as in Fig. 1.6, left and the flow diagram Fig. 1.6, right.

Gas at low pressure (1) is compressed up to high pressure. An ideal isentropic compression would lead to point 2_{is}: at high pressure and suction temperature. The

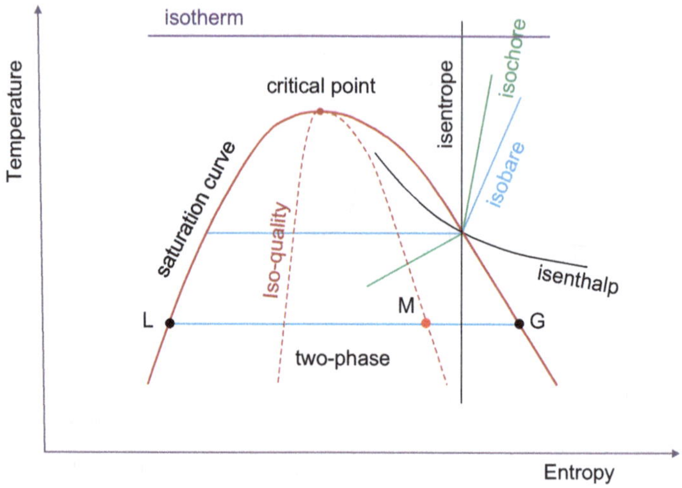

Fig. 1.4 The cold part of a T-s diagram

Fig. 1.5 The simplified helium Log(T)-s chart

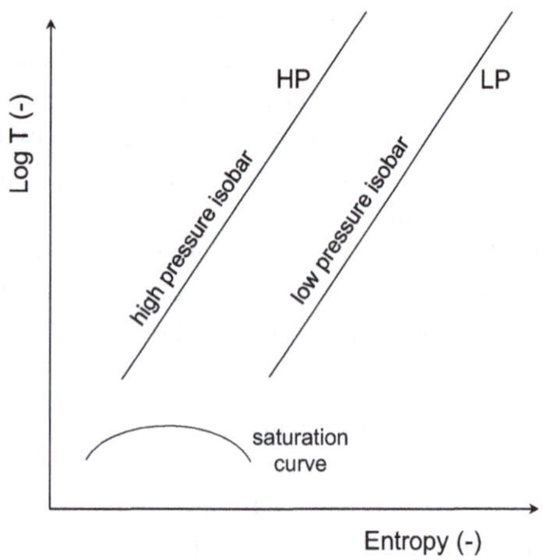

compressor follows an actual process (such phenomena will be described in Sect. 6.2.1) to reach point 2. The gas temperature increases. As this gas is to be used to reach low temperature, it is obvious that it must be returned as close as possible to the suction temperature (room temperature) prior to be sent into the cooling process (3). Cooling is performed at constant pressure, by exchanging heat against water or air, both at around room temperature, into a so-called cooler that is a water/helium or air/helium heat exchanger.

For simplification of the next T-s diagrams, the compression process will be simply described as almost isothermal (from 1 to 3).

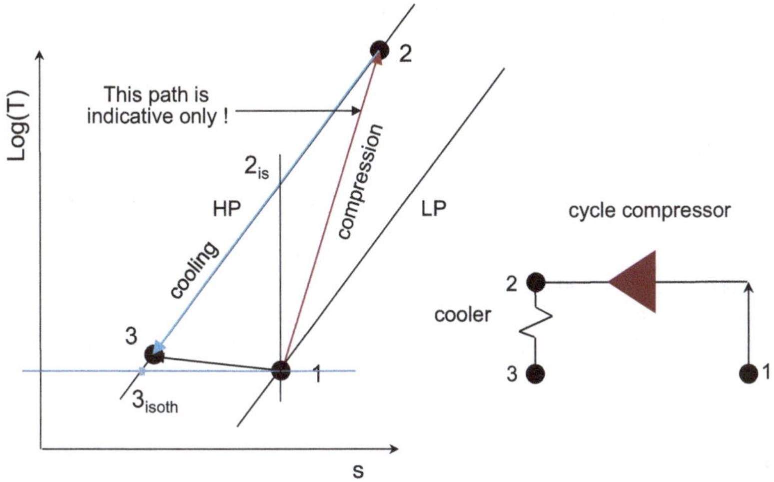

Fig. 1.6 Compression of a gas on the Log(T)-s diagram

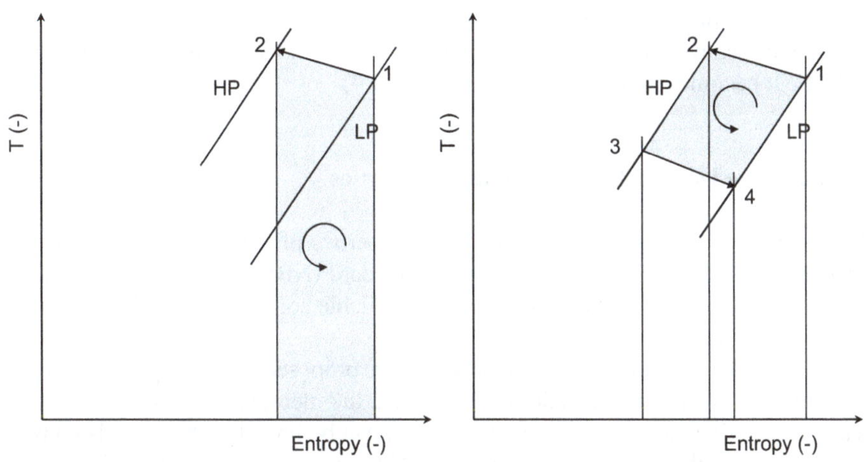

Fig. 1.7 Work exchanged on the T-s diagram

Heat and work that are exchanged

During a process from point 1 to point 2 in Fig. 1.7, left, the area that is located under the transformation line 1–2 is proportional to the work that is exchanged. In this case, as the trip is made counterclockwise, the work that is getting into the system is positive. It is what happens for the compression of a gas.

When a closed cycle is described, the work that is absorbed or released is proportional to the area of the cycle. An example for a refrigeration cycle is shown in Fig. 1.7, right.

In refrigeration cycles, the T-s diagrams are circulated counterclockwise.

Table 1.1 Main characteristics of helium (REFPROP)

Helium	P	T	ρ_{gas}	ρ_{liq}	Δh			
	(bar)	(K)	(kg/m^3)	(kg/m^3)	(J/g)	Conc. in air	5.2 10^{-6}	
Gas	1.00	300.00	0.16			$\rho_{liq}/\rho_{gas\ 300\ K}$	779	
Normal boiling point	1.00	4.21	16.67	124.94		$\rho_{liq}/\rho_{sat.\ vap.}$	7.49	
Latent heat					20.65	To vaporise 1 L/h	0.72	W
Sensible heat					1542.75	Sens./Lat. heat	74.73	
Critical point	*2.28*	*5.20*						
Triple point	*0.05*	*2.18*						
Safety:	Helium is a neutral gas: there is a risk of anoxia if it displaces oxygen							

1.4.2 Helium

Helium is the gas used for reaching the lowest "natural" temperature of a liquid. Information on this fluid can be found later, in Sect. 9.1. The main characteristics of helium are displayed in Table 1.1.

Helium behaves in a very special way as soon as its temperature is lower than 2.17 K. It becomes "superfluid" (see Sect. 1.4.2.4).

1.4.2.1 The Helium Thermophysical Properties

In the old days (60s!), the thermophysical properties of helium were to be copied manually from the US National Bureau of Standard (NBS) Technical Note 631and interpolated (see Fig. 15.1). Today, several available software return thermal properties of helium.

Along this book, some Excel spread sheets are proposed, each of them related to a situation that is often met in helium refrigeration. In order that the reader can perform same calculations on his own, a software, that can be linked with Excel, is considered: REFPROP[1] from NIST. However, HEPAK[2] from Cryodata can also be used in a similar way. At temperatures lower than 3.0 K, it is advised to use preferably HEPAK and lower than 2.18 K, using HEPAK is mandatory.

The most important of these spread sheets are explained in Chap. 15. They are to be saved into a large Excel workbook that can be called "the Cryo Tool Box". Such a tool box is an everyday tool that can be used for rather frequent calculations: instead

[1]Lemmon, E.W., Bell, I.H., Huber, M.L., McLinden, M.O. NIST Standard Reference Database 23: Reference Fluid Thermodynamic and Transport Properties-REFPROP, Version 10.0, National Institute of Standards and Technology, Standard Reference Data Program, Gaithersburg, 2018

[2]HEPAK is a registered trademark of Cryodata Inc., USA

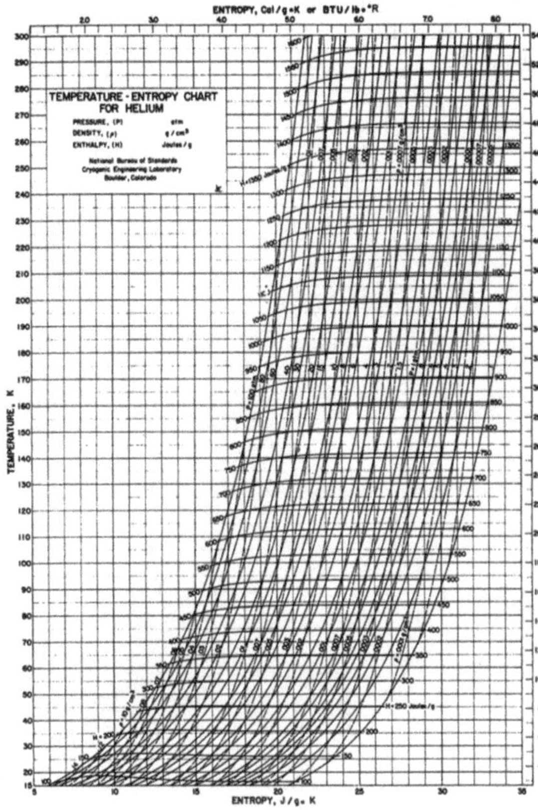

Fig. 1.8 The warm part of the helium T-s diagram (NBS). *T-s* chart for helium. *P* is in atm, density ρ is in g/cm^3, temperature *T* is in K or °R, enthalpy *H* is in J/g, and entropy *S* is in Cal/g$^-$ K or Btu/lb$^-$ °R. (From National Bureau of Standards Cryogenic Engineering Laboratory, Boulder, CO)

of rewriting a spread sheet, it is only necessary to copy and paste the sheet in the working document. It saves time and, also, errors.

When the helium T-s diagram from room temperature down to liquid is displayed in a one-off picture, it is difficult to use. It is generally split into two parts: the warm part (Fig. 1.8) and the cold part (Fig. 1.9). By the way, the T-s diagrams that are displayed in Figs. 1.8 and 1.9 and Figs. 1.24 and 1.25 are historical documents that were established a long time ago by the NBS.

It is possible to plot a T-s or a Log(T)-s diagram with REFPROP, showing only the parameters of interest, for example, in Figs. 1.10 and 1.11, only pressures 1, 4, 20 and 1000 bar are displayed.

Remark When looking at the helium T-s diagram, it is interesting to notice that for low pressures, helium behaves like an ideal gas:

- The isobars are straight lines (in Fig. 1.11, with a logarithmic temperature scale).
- The specific heat is constant (isenthalps are horizontal).

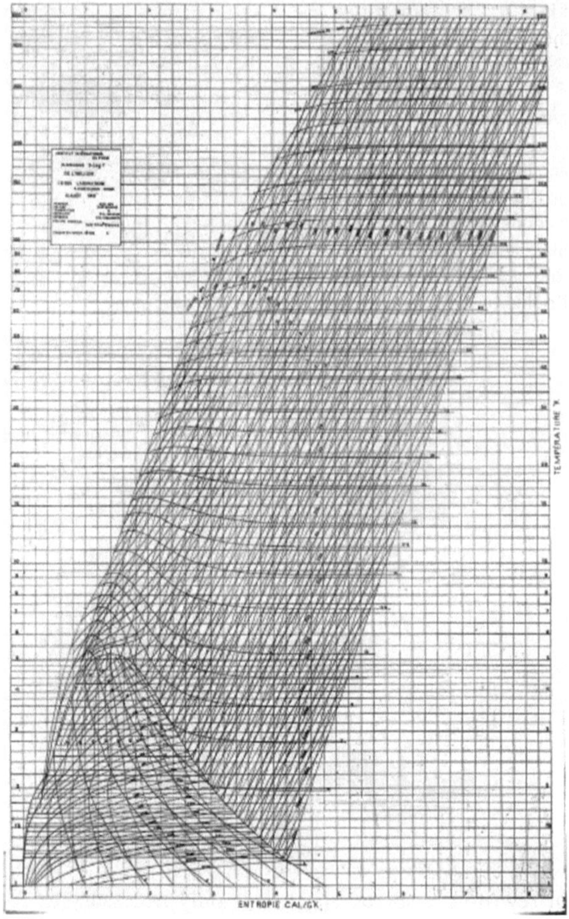

Fig. 1.9 The historical cold part of the helium T-s diagram (IIR)

However, for low entropy values, isenthalps have a positive derivative, for "high" temperatures the slope comes to zero or near zero, and for "cold" temperatures, the derivative comes to negative values.

1.4.2.2 The Simple or Isenthalpic Expansion of Helium

The so-called simple expansion is the "natural" expansion that takes place in an orifice or a valve. As no work is performed by the expanded gas, the enthalpy does not change: it is an isenthalpic expansion.

Simple or isenthalpic expansion of helium is somewhat special. In Sect. 1.6, one notices that the expansion of helium from 14 bar, 300 K down to 1 bar, generates an *increase of temperature* of the gas. This is rather uncommon; in most of the cases,

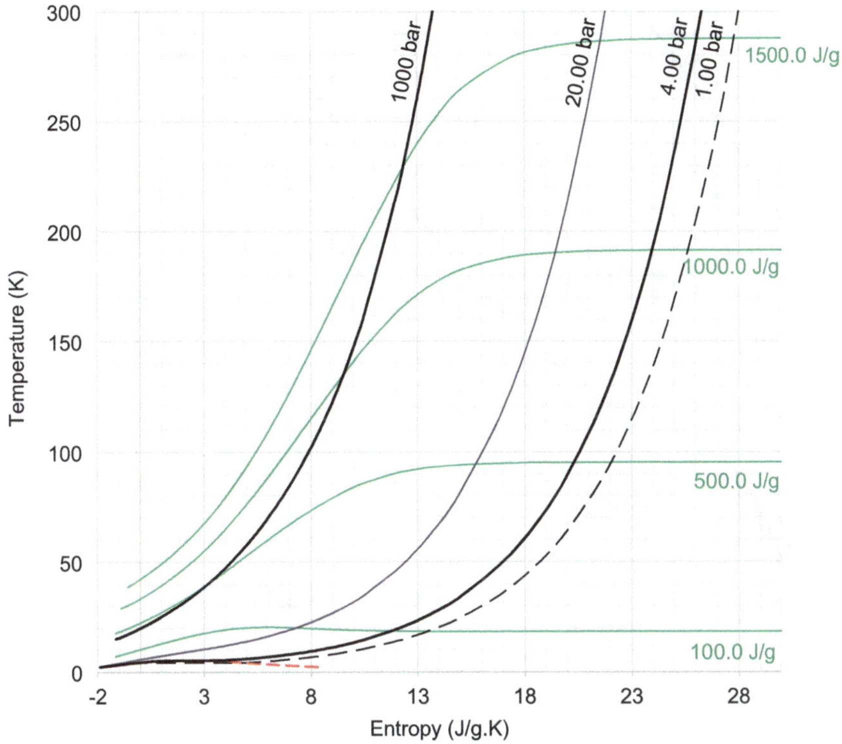

Fig. 1.10 The warm part of the helium T-s diagram from 0 to 300 K, for 1.00, 4.00, 20.00 and 1000.00 bar (REFPROP)

the common thinking is that the expansion of a gas cools it. What happens with helium?

Let us calculate the expansion of helium from various pressures down to 1 bar, changing the upstream temperature. For example, from 25 bar, 300 K, the expansion generates a 1.5 K increase of temperature (see Fig. 1.12). Such temperature increase is almost constant down to 150 K, then starts to decrease and reaches zero at around 40 K. Lower than 40 K, the temperature change is negative.

The behaviour is similar for other pressures (20, 15, 10 and 5 bar, see Fig. 1.12). In all cases, the temperature change is zero at around 40 K. Well, this is the "inversion temperature" of helium!

The calculation of the helium inversion temperature is proposed in the CTB, Sect. 15.8.2.

The conclusion of this calculation could seem surprising for some readers who have experienced an iced part on a valve that is expanding helium at room temperature. Is such an observation contrary to the calculation?

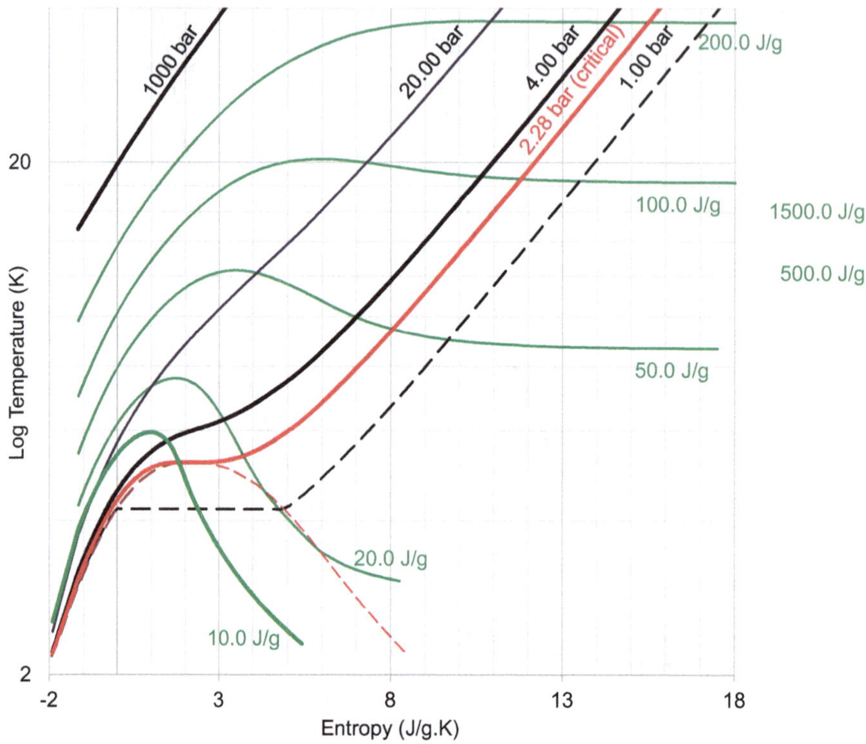

Fig. 1.11 The cold part of the helium Log(T)-s diagram up to 40 K, for 1.00, 2.28 (critical), 4.00, 20.00 and 1000.00 bar (REFPROP)

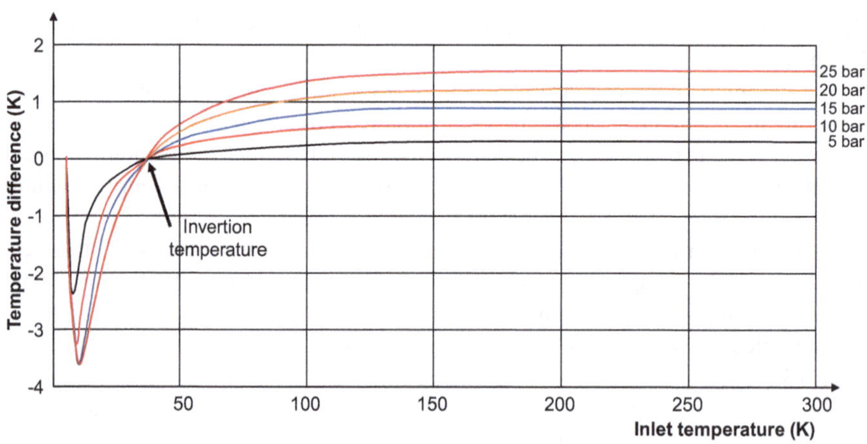

Fig. 1.12 Expansion of helium: the inversion temperature

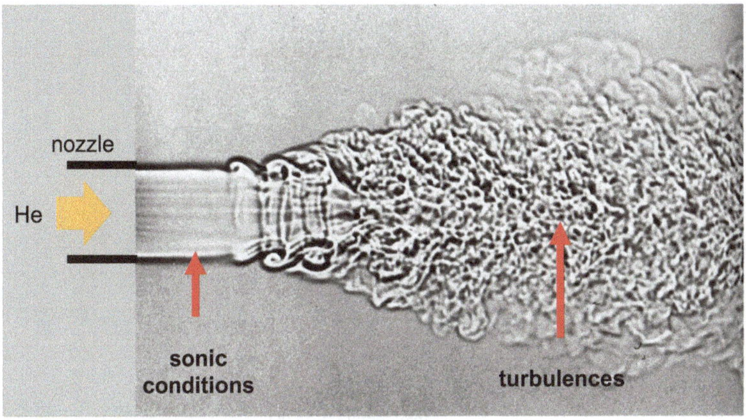

Fig. 1.13 Free expansion of a gas (F. Landys and A. Shapiro)

If one looks at a very nice picture of gas expanding through a nozzle in a sonic regime, which means that the expansion ratio higher than 2, one can see two regions (see Fig. 1.13):

- A "clean" flow, perfectly laminar. In this place, the flow is generally sonic (when the pressure ratio is higher than 2/1). The helium pressure has been turned into velocity, and the temperature has decreased down to 200 K (−73 °C)[3]. This is why the valve is coated with ice!
- A turbulence region. In this place, vortexes are generated, which produce big friction between the gas molecules that results in heat production, and the resulting temperature is 301 K.

Is really helium a special gas because it warms up during expansion? Let us perform same isenthalpic expansion with two other gases: nitrogen and hydrogen.

On Fig. 1.14, one can notice that for these gases, above the inversion temperature, the temperature change is positive too; therefore helium behaves as any other gas. In the case of nitrogen, as the inversion temperature is higher than room temperature (603 K for 14 bar), expansion from room temperature provides a cooling effect (see calculation in CTB, Sect. 15.8.2). However, for hydrogen, the Joule Thomson cooling effect can only be seen if the temperature is lower than 193 K.

This behaviour can also be observed on the hydrogen T-s diagram.

In Fig. 1.15, an isenthalp curve is plotted from 14.0 to 1.0 bar, at 300.00 K, 43.12 K and 15.00 K. One can see that:

---

[3]According to $v = \sqrt{2}\Delta H$ Eq. 17.18, one can calculate this temperature.
v is the sound velocity: 1025 m/s according to REFPROP.
$\Delta h = 0.5 \times 1025^2 = 525{,}697$ J/kg = 526 J/g
h at 14 bar, 300 K = 1567 J/g
h at 1 bar = 1567–526 = 1042 J/g
T at 1 bar, 1042 J/g = 200 K

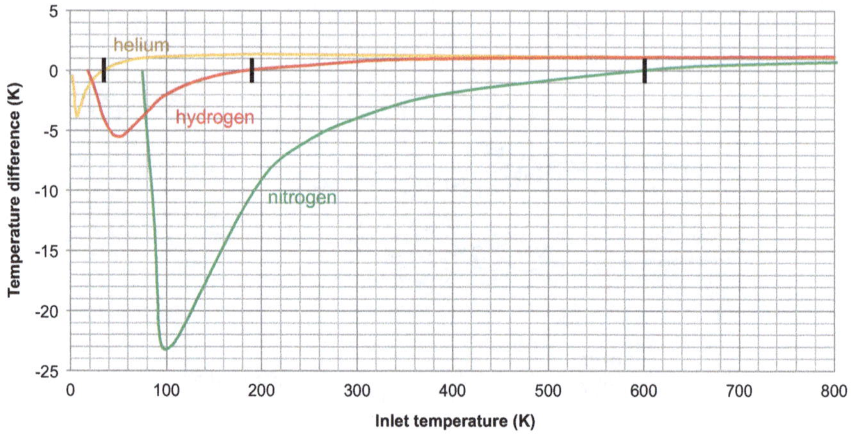

Fig. 1.14 Isenthalpic expansion of various gases: temperature change versus temperature

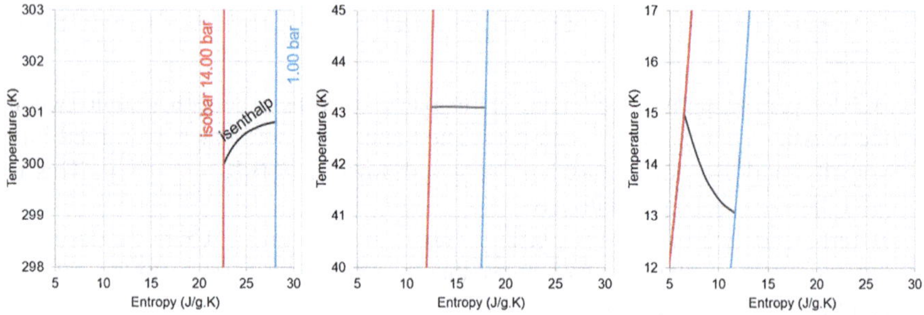

Fig. 1.15 An enlargement on the temperature-enthalpy diagram of helium, showing the shape of isenthalps from 14.00 to 1.00 bar, at, respectively, 300.00 K, 43.12 K (inversion temperature) and 15.00 K (REFPROP)

- At 300.00 K the temperature after expansion is higher.
- At 15.00 K the temperature after expansion is lower.
- At 43.12 K the temperature after expansion is equal: 43.15 K is the inversion temperature for an expansion from 14 to 1 bar.

1.4.2.3 Evolution of a Few Properties of Helium

It is interesting to see how some properties vary with temperature and pressure.

Enthalpy and Entropy

In Fig. 1.16, one can see that enthalpy is exactly linear to temperature down to less than 10 K at 1 bar, but discrepancy happens as soon as 50 K at 20 bar.

c_p and γ

Similarly, in Fig. 1.17, c_p and γ values are almost constant down to 100 K, whatever the pressure is.

Thermal conductivity

The helium thermal conductivity decreases with temperature and is little dependent on pressure, except for low temperatures (Fig. 1.18).

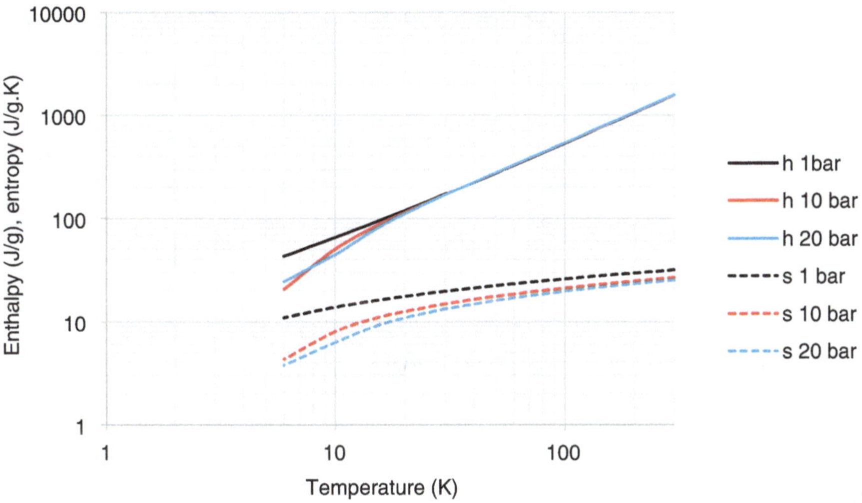

Fig. 1.16 Helium enthalpy and entropy versus temperature (REFPROP)

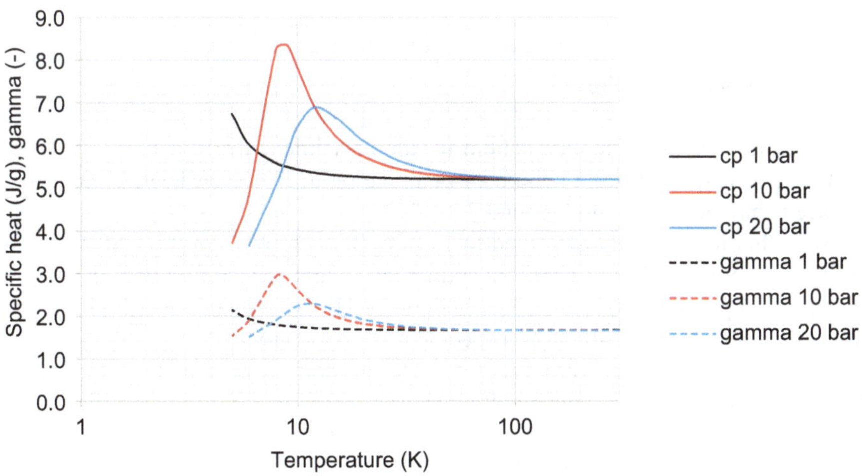

Fig. 1.17 Helium c_p and γ versus temperature (REFPROP)

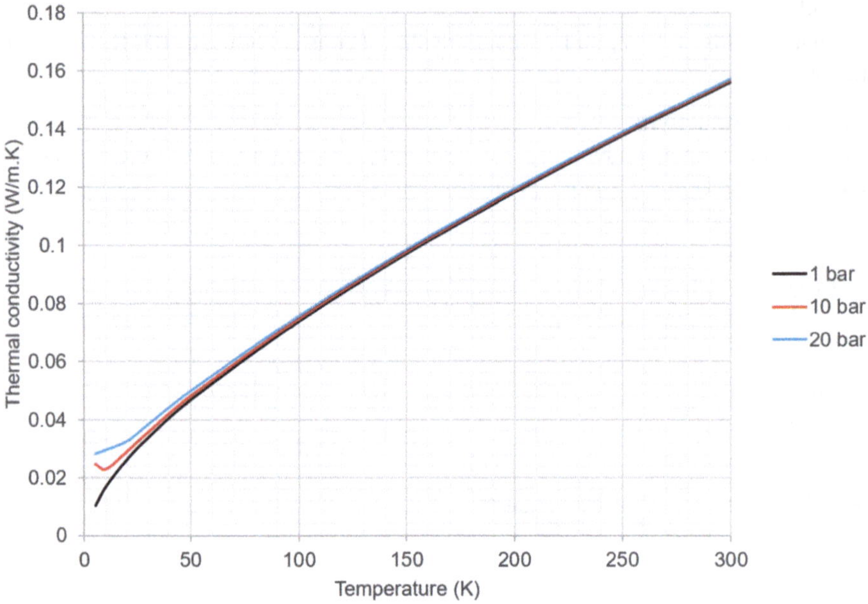

Fig. 1.18 Thermal conductivity of helium (REFPROP)

Sound velocity
The sound velocity versus temperature is shown on Fig. 1.19, left, according to the upstream pressure. The way the mass flow varies though a sonic orifice versus temperature is shown on Fig. 1.19, right (flow is unit at 300 K).

1.4.2.4 Superfluid Helium

At the time K. Onnes liquefied helium for the first time (July 10th 1908), he tried, without success, to get solid helium by pumping on it. However, later, he noticed a bizarre behaviour of helium "around 2.2 K".

Superfluidity was only identified in 1938 by Piotr Kapitza in the USSR. When helium is cooled below 2.17 K, by pumping at less than 0.050 bar, the liquid undergoes a phase transition: normal helium (or He-I) becomes superfluid helium, also called helium II (He-II). It is a specific property of helium.

Superfluid helium has zero viscosity. The proportion of superfluid helium in He-II increases when the temperature decreases. This property induces disturbing behaviour in relation to habits: He-II flows through media considered as leak tight for normal liquids and remains stationary when the vessel that contains it is driven by a rotary motion. Even more amazing, He-II climbs up to the walls of the vessel that contains it: it is a combination of its surface tension and no viscosity (see Fig. 1.20)!

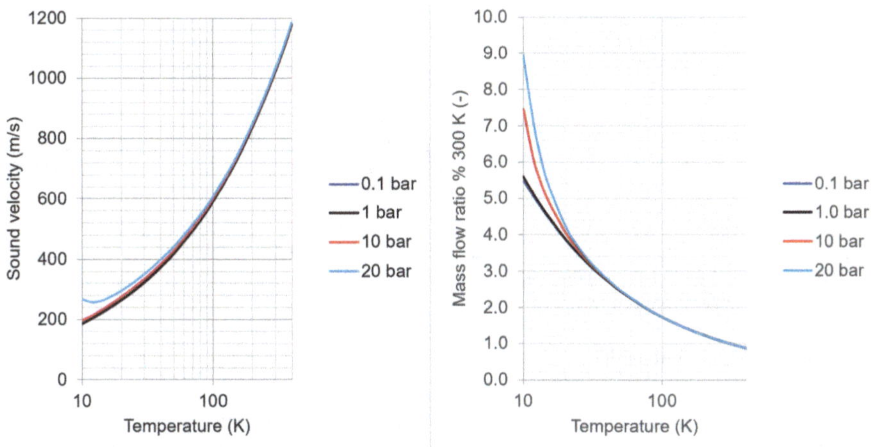

Fig. 1.19 Sound velocity and helium mass flow rate change in a sonic orifice versus temperature (REFPROP)

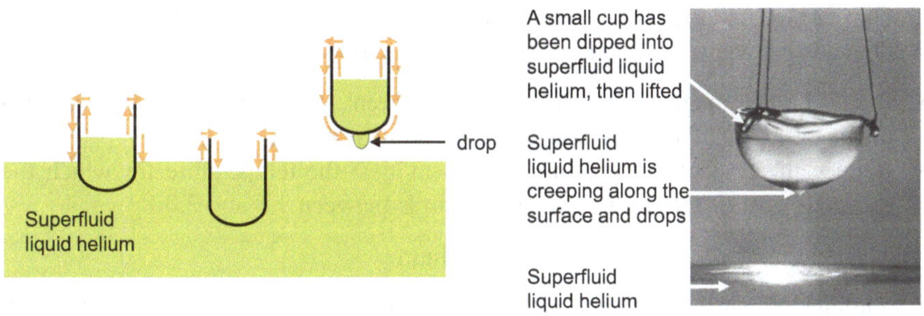

Fig. 1.20 A curious behaviour of superfluid helium

He-II has an enormous thermal conductivity (around 1000 times more than OFHC copper); it can be considered a "superconductive heat flow", and the slightest temperature difference is spread throughout the body almost instantaneously: He-II transfers heat without neither circulation nor bubbles!

This behaviour is explained by the two-fluid model in Fig. 1.21.

At temperatures colder than that of lambda point (2.17 K), the proportion of superfluid He/normal He varies according to the curves in Fig. 1.21, left: over 2.17 K all helium is normal, lower than 1.0 K, almost all helium is superfluid. In Fig. 1.21, right, one can see how heat is transferred in a pipe filled with He-II: superfluid helium that is created on the cold side (right) moves towards the warm end (left); conversely, normal helium flows towards the cold side, transporting heat, and is converted into superfluid. So, looking in detail, heat is actually transported, not by heat conduction but by convection. This kind of heat transport is very effective, so the thermal conductivity of He-II is very much better than the best materials.

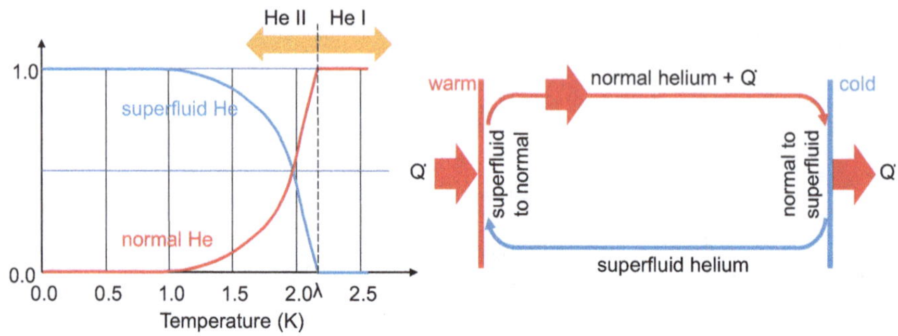

Fig. 1.21 The two-fluid model

Another surprising behaviour is the so-called fountain effect.

The bottom of the U-tube in Fig. 1.22, that is immersed in He-II, is filled with a very fine powder. When the heater is energised, it warms liquid helium in the U-tube where some superfluid helium is turned into normal helium. As the concentration of superfluid helium decreases, superfluid helium is flowing through the fine powder and pushes out the normal helium that gushes at the top of the U-tube.

One can located the various helium phase in the PT diagram in Fig. 1.23.

Operating at superfluid helium temperatures allows to:

- Reach higher magnetic fields: for same current, Nb-Ti at 1.9 K can reach a magnetic field 3 T higher than when operating at 4.2 K.
- Reduce the cost of refrigeration of RF cavities: the temperature for which the energetic cost of refrigeration is minimum is between 1.5 and 2.0 K.

There are two ways to use superfluid helium:

- Saturated superfluid helium, for situations where there is no risk to meet a high voltage difference in the device such as RF cavities (at the saturated superfluid helium pressure, the Paschen curve is almost at its minimum).
- Static pressurised superfluid helium when there is a risk that a high voltage difference happens, for example, during the quench of a superconducting coil. Furthermore, there are no risk to get atmospheric air into the circuit by pressure difference. Such technology has been developed in *Commissariat à l'Energie Atomique* (CEA), Grenoble, France, under the guidance of Gérard Claudet, in the 1970s.

Examples of cryogenic systems using superfluid helium are given in Sect. 5.3.4. A few large cryogenic systems operating with super fluid helium are described in Sect. 12.8.

Detailed info on superfluid helium can be found in S. W. Van Sciver, *Helium Cryogenics,* (Plenum Press, New York, 1986).

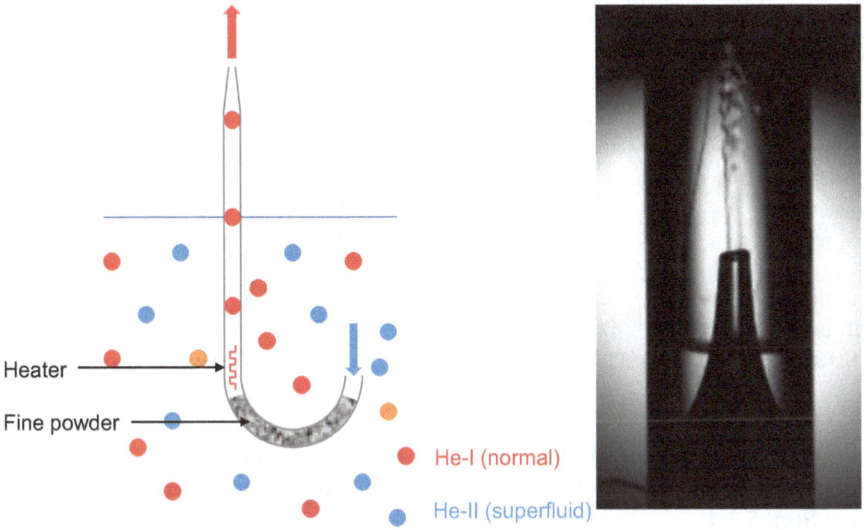

Fig. 1.22 The fountain effect

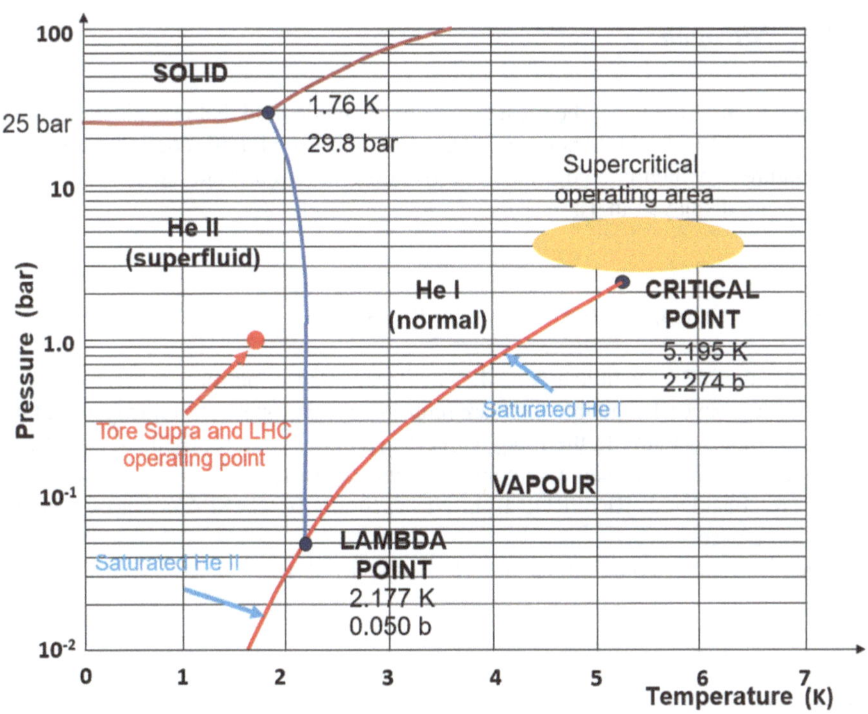

Fig. 1.23 The helium PT diagram

Table 1.2 Main characteristics of nitrogen (REFPROP)

Nitrogen	P	T	ρ_{gas}	ρ_{liq}	Δh			
	(bar)	(K)	(kg/m^3)	(kg/m^3)	(J/g)	Conc. in air	0.78	
Gas	1.00	300.00	1.12			$\rho_{liq.}/\rho_{gas\ 300\ K}$	718	
Normal boiling point	1.00	77.24	4.56	806.59		$\rho_{liq.}/\rho_{sat.\ vap.}$	177.02	
Latent heat					199.32	To vap 1 L/h	44.66	W
Sensible heat					234.12	Sens./Lat. heat	1.17	
Critical point	33.96	126.19						
Triple point	0.13	63.15						
Safety:	Nitrogen is a neutral gas, there is a risk of anoxia if it displaces oxygen							

Remark He-II is generally located inside the cryostat; the refrigerator cold box does not deal directly with He-II but only with gaseous helium at very low pressure (see Sect. 5.3.4).

1.4.3 Nitrogen

Nitrogen is, after methane, the most common cryogenic fluid that is available almost everywhere in the world. In helium refrigeration, it is used for pre-cooling cycles, the temperature of which is colder than that of liquid nitrogen. Main characteristics of nitrogen are displayed in Table 1.2.

1.4.3.1 The Nitrogen Thermophysical Properties

Formerly, the NBS Technical Note 129 has been used to perform process calculations. Today, several available software return thermal properties of nitrogen. In this book, examples of simple thermodynamic calculations are proposed. In order that the reader can perform same calculations, only software that can be linked with Excel are considered: REFPROP from NIST. Similarly, GASPAK from Cryodata can also be used.

The historical T-s diagram of nitrogen is displayed in Fig. 1.24.

1.4.4 Hydrogen

This book deals with helium refrigeration, but there are a few topics that are related to hydrogen; therefore, some information on hydrogen is of interest.

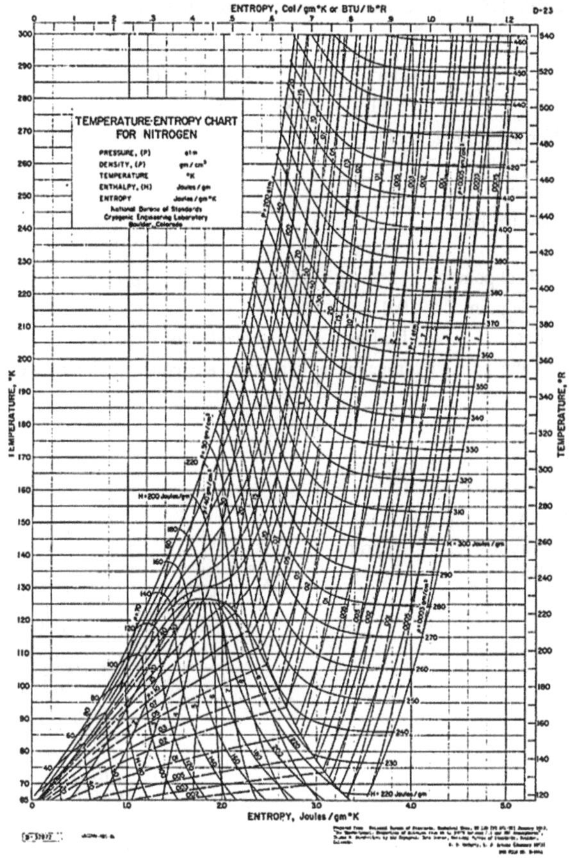

Fig. 1.24 The historical nitrogen T-s diagram (NBS)

1.4.4.1 The Hydrogen Thermophysical Properties

Main characteristics of hydrogen are displayed in Table 1.3.

As for nitrogen, thermodynamic properties can be calculated with either the REFPROP or GASPAK software.

One of the historical NBS T-s diagrams for parahydrogen is displayed in Fig. 1.25.

A peculiarity of hydrogen: the ortho-para conversion

Hydrogen coexists in two isomeric forms: orthohydrogen, where the nuclear spins are aligned in same direction, and parahydrogen, where the spins are aligned, but in opposite directions. The relative proportions depend only on the temperature. The

Table 1.3 Main characteristics of normal hydrogen (REFPROP)

Hydrogen	P	T	ρ_{gas}	ρ_{liq}	Δh			
	(bar)	(K)	(kg/m^3)	(kg/m^3)	(J/g)	Conc. In air	5.0 10^{-7}	
Gas	1.00	300.00	0.08			$\rho_{liq.}/\rho_{gas\ 300\ K}$	878	
Normal boiling point	1.00	20.32	1.32	70.90		$\rho_{liq.}/\rho_{sat.\ vap.}$	53.86	
Latent heat					448.91	To vap 1 L/h	8.84	W
Sensible heat					3509.80	Sens./Lat. heat	7.82	
Critical point	12.96	33.15						
Triple point	0.07	13.96						
Safety:	Hydrogen is a flammable gas; there are risks of burning or explosion							
Combustion of hydrogen in air happens at any concentration between 4.0% and 74.5%								

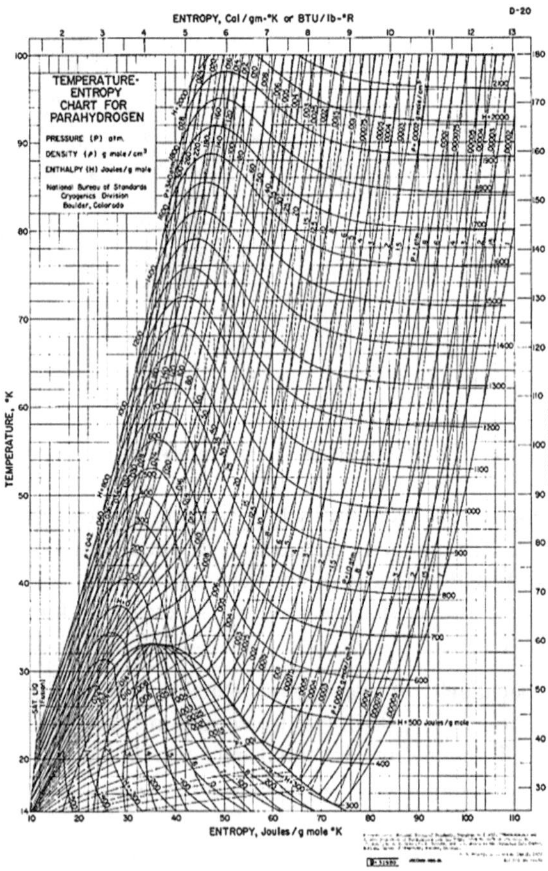

Fig. 1.25 The historical parahydrogen T-s diagram (NBS)

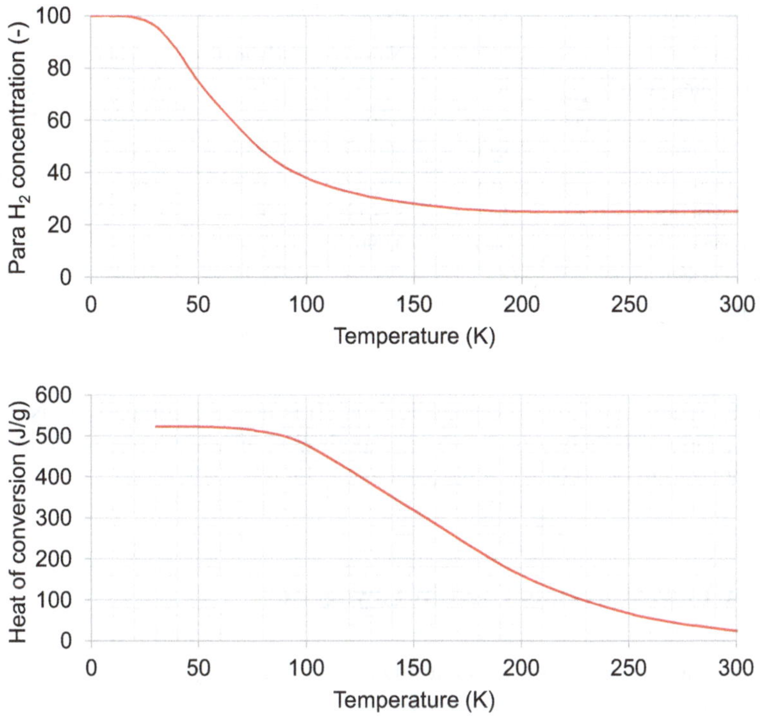

Fig. 1.26 Ortho-para conversion of hydrogen, parahydrogen concentration, top; heat of conversion, bottom

so-called normal hydrogen, at room temperature and thermal equilibrium, is 75% ortho and 25% para. At equilibrium, at boiling temperature and atmospheric pressure, most of hydrogen is turned into parahydrogen as shown in Fig. 1.26, top. Such transition is exothermic (527 J/g) (see Fig. 1.26, bottom). It takes place naturally over a few days. As the conversion heat is about the same order of magnitude as the latent heat (446 J/g), when the equilibrium is reached, more than half the mass of liquid has been vaporised. In order to fight against this inconvenience, the ortho-para conversion is accelerated by a converter, generally an iron oxide, during liquefaction (see Sect. 4.5.3).

1.4.5 Comparison of Helium, Hydrogen and Nitrogen Properties

Comparison is displayed in Table 1.4.

Table 1.4 Comparison of some fluid properties (REFPROP)

At 1,00 bar		Nitrogen	Hydrogen	Helium	Water
Normal boiling point	(K)	77.24	20.23	4.21	372.76
Density of saturated vapour	(kg/m^3)	4.56	1.32	16.64	0.59
Density of liquid	(kg/m^3)	806.59	70.88	125.25	958.63
Heat of vaporisation	(J/g)	199.32	448.91	20.72	2257.44
Sensible heat, up to 300 K	(J/g)	234.12	3512.45	1542.77	-2562.29
Critical pressure	(bar)	33.96	12.86	2.28	220.64
Critical temperature	(K)	126.19	32.94	5.20	5.20
Triple point temperature	(bar)	0.13	0.07	0.05	0.01
Triple point pressure	(K)	63.15	13.96	2.18	273.16

Remark Water that is not a cryogenic fluid (!) is here for comparison. Note that there is a ratio of 10 between the heat of vaporisation of water compared to nitrogen and another ratio of 10 between nitrogen and helium.

1.5 A Few Materials Used in Cryogenics

Here, only a few properties of some materials used in cryogenics are dealt with. More details can be found in specialised books.

1.5.1 Specific Heat (or Heat Capacity)

In a general way, the specific heat of materials decreases with the temperature as shown in Fig. 1.27, left. One must be careful when dealing with logarithmic scales as can be seen in Fig. 1.27, right. Remark the higher value of the helium and even nitrogen-specific heat compared to that of other materials (Figs. 1.27 and 1.28).

A rather surprising situation: at temperatures lower than 20 K, one notices a surprising phenomenon: the *volume* specific heat of helium is very much higher than the one of almost any material (see Fig. 1.29)!

This important phenomenon means that at low temperatures, almost all the thermal inertia of a system sits in the helium it contains! Such peculiarity must be taken into consideration when designing regenerators (see Sect. 16.8.1).

1.5.2 Thermal Conductivity

The thermal conductivity of a few materials is shown in Fig. 1.30, the thermal conductivity integral from 5 K is shown in Fig. 1.31. Among the usual materials, one

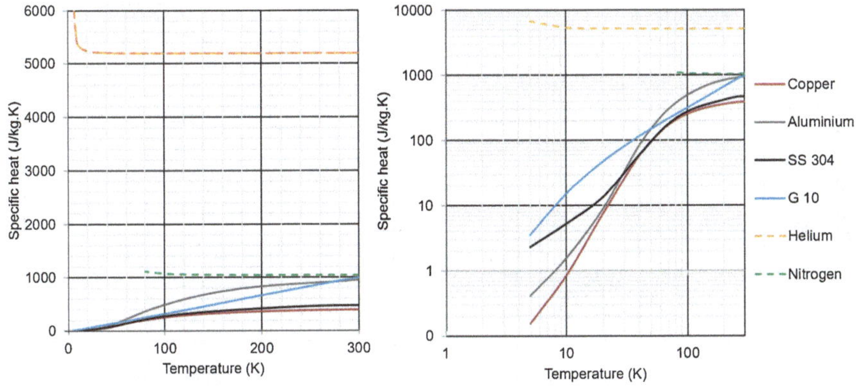

Fig. 1.27 Specific heat of some materials (NIST, REFPROP)

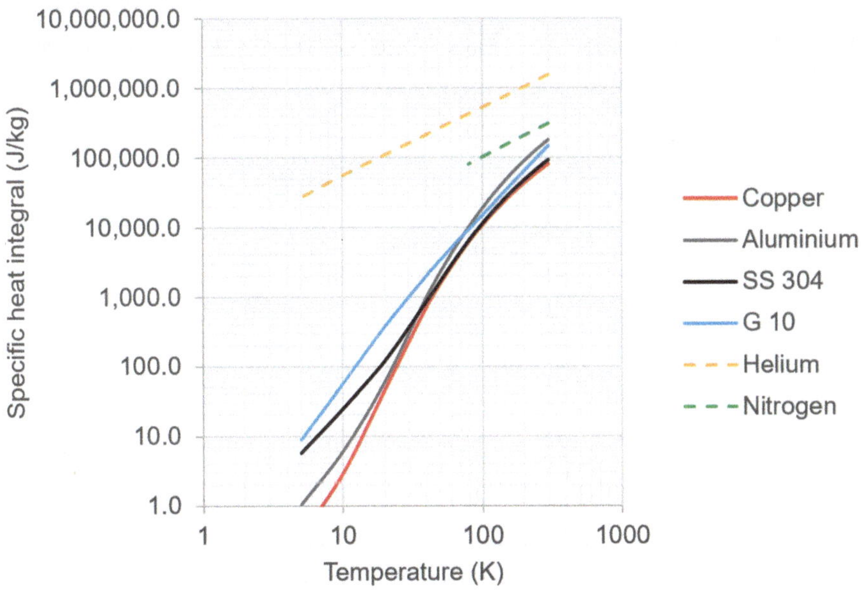

Fig. 1.28 Integral from 1 K, of the specific heat (NIST, REFPROP)

should keep in mind that copper is a good heat conductor, stainless steel is a bad one and composite materials have very low heat conductivities.

Remark Thermal conductivity of two gases, helium and nitrogen, has been added for comparison.

To be kept in mind, thermal conductivity of a metal is higher than that of a liquid that is higher than that of a gas.

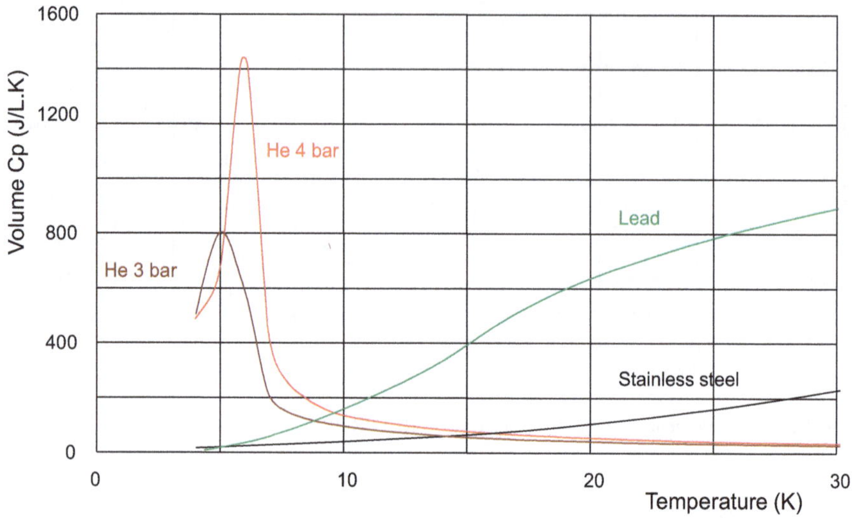

Fig. 1.29 Volume c_p variation versus temperature of stainless steel, lead and helium (NIST, REFPROP)

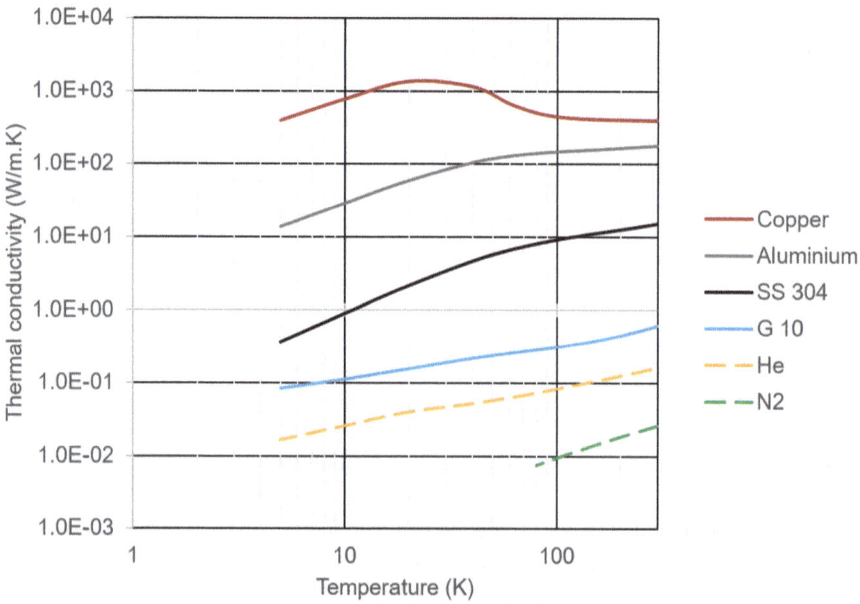

Fig. 1.30 Thermal conductivity of some materials (NIST, REFPROP)

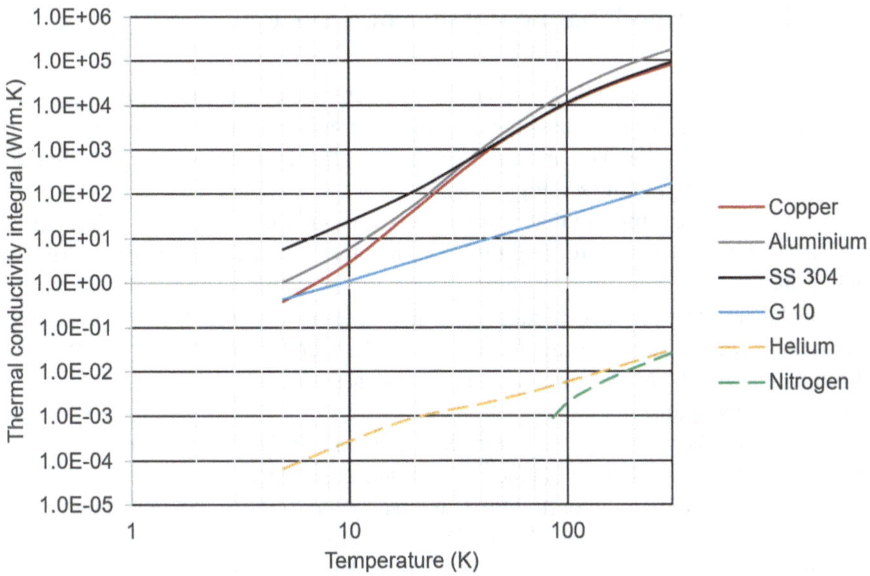

Fig. 1.31 Integral from 5 K of thermal conductivity (NIST, REFPROP)

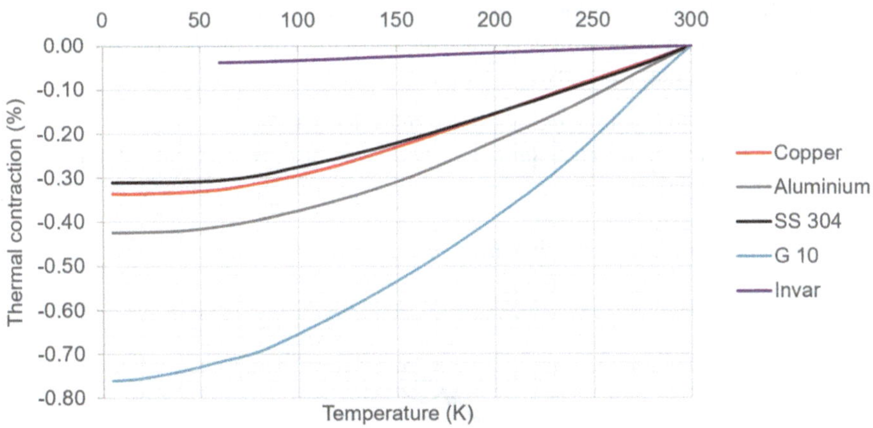

Fig. 1.32 Thermal contraction of some materials (NIST)

1.5.3 Thermal Contraction

The thermal contraction of a few materials is shown in Fig. 1.32. Note the special behaviour of Invar.

1.6 The Thermodynamic Balance of a System

Process calculations that allow to size the refrigeration cycles in a steady state are based upon two balances: a mass balance and a thermal balance. Each balance, for a *steady state*, is null: the mass and energy quantities that enter the system are equal to the quantities that get out (see Fig. 1.33).

An important assumption that is sometimes forgotten is that the gas velocities at the limit of the systems are *negligible* or, in other words, the kinetic energy is negligible.

For the simplest process calculation example, let us consider the operation of a valve that expands helium from pressure P_1 and temperature T_1 down to P_2. What is the temperature T_2 in the discharge pipe, downstream the valve, where the gas velocity is low again?

Let us take the opportunity to introduce the way of dealing with the mass and thermal (or thermodynamic) balance of a system. The system (here, the valve) is isolated from the remaining part of the Universe by the rounded corner rectangle, the "border" (see Fig. 1.33, right). When circulating along this border, the ingoing and outgoing quantities are identified and calculated. By convention, ingoing quantities are positive, and the outgoing quantities are negative.

- At point 1 the mass flow $\dot{m}$ enters the system with a specific enthalpy h_1: therefore, the quantity of entering power is $\dot{m} \times h_1$. (enthalpy h_1 is calculated with REFPROP® according to pressure P_1 and temperature T_1).
- At point 2 the mass flow $\dot{m}$ gets out with a specific enthalpy h_2. The quantity of outgoing power is $- \dot{m} \times h_2$.

The mass and thermal balance can be written: $\dot{m} \times \mathrm{h}_1 - \dot{m} \times h_2$.

As the system is in a *steady state and fluid velocities are negligible*, the balance equals to 0. One writes:

$$\dot{m} \times h_1 - \dot{m} \times h_2 = 0 \tag{1.9}$$

It comes:

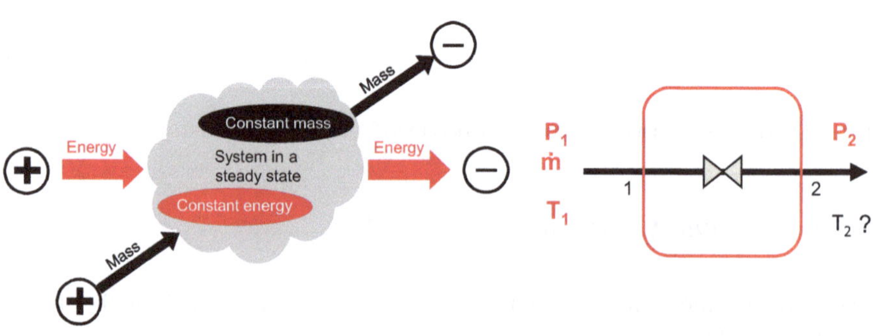

Fig. 1.33 Thermodynamic balance of a system

$$h_1 = h_2 \tag{1.10}$$

The temperature T_2 is calculated with REFPROP®, according to pressure P_2 and mass enthalpy h_2.

Numerical application
Process inputs:

- $P_1 = 14.0$ bar
- $T_1 = 300.0$ K
- $P_2 = 1.0$ bar

Calculation (again, assuming that gas velocities are negligible at the limit of the system):

- P_1, T_1 and REFPROP return $h_1 = 1567.68$ J/g
- P_2, h_1 and REFPROP return $T_2 = 300.82$ K

This very simple exercise allows to see that, sometimes, during an expansion process, the discharge temperature T_2 can be *higher* than the inlet temperature T_1! See Sect. 1.4.2.2.

Such a calculation is performed in Sect. 15.8.2 Cryogenic Tool Box, using an Excel spread sheet.

As an example, let us now study the same process using the helium T-s diagram.

Here above, we learned that enthalpy stays constant during the expansion; therefore, let us follow the iso-enthalpy (isenthalp) curve passing at 14 bar and 300 K on Fig. 1.34 until 1 bar.

One can see that the isenthalp curve issued from 14.00 bar, 300.00 K, crosses the 1.00 bar isobar at a temperature that is higher than 300 K.

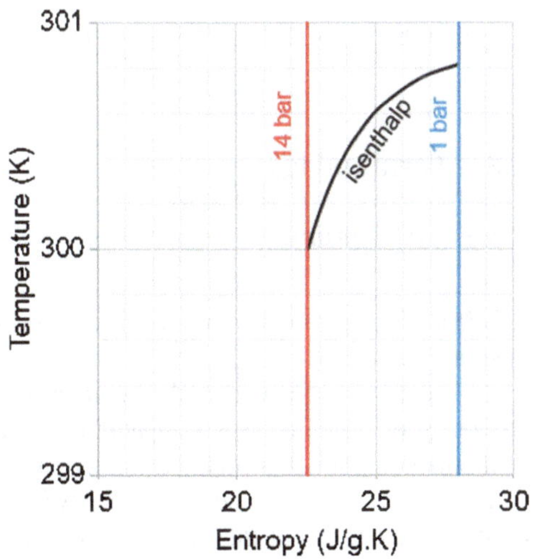

Fig. 1.34 Isenthalpic expansion of helium on a blow-up of the T-s diagram (REFPROP)

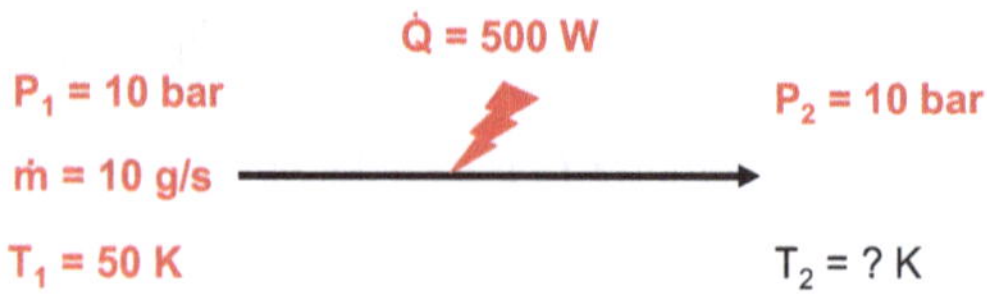

Fig. 1.35 Power into helium

1.7 Thermal Energy and Gas

To change the temperature of a mass of gas m from T_1 to T_2, it is necessary to bring or remove a quantity of energy that is:

$$Q = m \times (h_2 - h_1) \tag{1.11}$$

or, if c_p is constant:

$$Q = m \times (T_2 - T_1) \times c_p \tag{1.12}$$

$$\dot{Q} = \dot{m} \times (T_2 - T_1) \times c_p \tag{1.13}$$

Remark A dot on the symbol means that it is a derivative according to time: $\dot{m}$ is a mass flow rate expressed in g/s, and Q is a thermal power expressed in W.

Example A power of 500 W is dissipated into a flow of 10 g/s helium at 10 bar, 50 K (see Fig. 1.35). What is the downstream temperature?

Specific enthalpy at inlet: for 10.00 bar, 50.00 K, REFPROP returns $h_1 = 265.22$ J/g

Enthalpy at inlet: $H_1 = 265.22 \times 10 = 2652.2$ J

Enthalpy at discharge: $H_2 = 2652.2 + 500 = 3152.2$ J

Specific enthalpy at discharge: 10.00 bar, 3152.2 J: $h_2 = 3152.2/10 = 315.22$ J/g

For 10 bar, 315.22 J/g, REFPROP returns $T_2 = 59.51$ K

1.8 Terminology

1.8.1 Efficiency

Efficiency is the ratio of what a single real machine can perform versus what an ideal machine could.

Efficiency is to be expressed by comparison with an ideal identified process. The efficiency of a compressor is generally calculated by comparison with the ideal isothermal compression process: isothermal power/actual absorbed power.

For an expansion turbine the efficiency is calculated by comparison with the ideal isentropic process, but the ratio is inversed to get a figure lower than one: actual extracted power/isentropic power.

1.8.2 Yield

Yield is the efficiency of a machine (a refrigerator) integrating various components (compressor, expander).

Comparison of an actual refrigerator to an ideal (Carnot, see Sect. 3.2.1) one.

Yield of a refrigerator operating at an isothermal-duty regime = Carnot power/actual power absorbed.

1.8.3 Coefficient of Performance (COP)

For a refrigerator, the COP is the ratio of thermal power that is extracted from the cold source versus mechanical power input into the machine.

$$\mathrm{COP} = \dot{Q}_{\mathrm{cold}}/\dot{W}_{300\mathrm{K}} (\text{no dimension})$$

1.8.4 Specific Power

The inverse of COP, also called specific power:

$$1/\mathrm{COP} = W_{300\mathrm{K}}/\dot{Q}_{\mathrm{cold}}$$

is a practical way to compare efficiencies of various refrigerators. For example, a refrigerator needs 230 W at 300 K per W at 4.5 K.

1.9 Digest

In order to perform simple rough thumb calculations, one should keep in mind the rounded figures at right in Table 1.5.

In an isolated system that is at steady state, the mass and thermal balances are equal to zero.

Table 1.5 A few helium values to be kept in mind

	Gaseous helium	***To be kept in mind***
Cp = 5.19 J/g.K		*(~ 5 J/g.K)*
1 g/s	≥20.4 Nm^3/h	*(~ 20 Nm^3/h)*
	Liquid helium at 1.2 bar (4.4 K)	
1 g/s LHe	≥29.82 L/h	*(~ 30 L/h)*
1 g/s LHe	≥19.33 W	*(~ 20 W)*

Chapter 2
A Light Theory of Heat Exchangers for Cryogenic Use

Abstract This chapter is *not a course* on heat exchangers. It does not explain how to size a heat exchanger. It only contains a few reminders, the knowledge of them being necessary to understand how heat exchangers work and how do they behave in the various operating situations occurring in a helium refrigeration or liquefaction system.

2.1 Introduction

Compared to heat exchangers operating around room temperature, the heat exchangers that are used for cryogenic purposes have some specificities that are related to the high-level duty they have to fulfil in a cryogenic system and to the important variations of gas and material (see Sect. 1.5) properties according to the temperature along the heat exchanger (see Sect. 1.5.2). When calculating simple thermodynamic cycles, one will notice in Sect. 3.3.2.6.2 the direct incidence of heat exchanger efficiency (or small temperature differences) on the system efficiency, making necessary to aim towards high heat exchanger efficiency.

Deeper information can be found into specialised books.[1]

In the chapter, the results of simple calculations, performed with Excel© and REFPROP© (see Sect. 1.2), are displayed. Their principle is explained in Sect. 15.8.3.2. The author advises the reader to try to re-perform these calculations by himself (or herself). This allows a better comprehension of the system. Such spread sheets become also components of the "Cryo Tool Box" that is described in Chap. 15.

[1]Cryogenic heat transfer, R. F. Barron

G. Gistau Baguer, *Cryogenic Helium Refrigeration for Middle and Large Powers*, International Cryogenics Monograph Series,
https://doi.org/10.1007/978-3-030-51677-2_2

2.2 Duty of a Heat Exchanger

A heat exchanger transfers energy, as heat, from a fluid (warm) to a colder fluid. An "intuitive" heat exchanger can be easily built by arranging two pipes of different diameters, one inserted into the other (see Fig. 2.1). The internal small diameter pipe is circulated, for example, by the warm entering fluid (from right to left), the space between the large diameter pipe, and the small one is circulated by the cold entering fluid (from left to right).

The fluid that enters warm at the right end gets out cold at the left end and conversely. As fluids circulate in opposite directions, it is a pure *countercurrent* heat exchanger.

Such a heat exchange process is performed without mixing the fluids; therefore, a wall must separate the two fluids. This wall *is* the heat exchanger. A pipe in room temperature air in which a fluid which temperature is different from room temperature circulates is also a heat exchanger. In such a case, it is generally a *parasitic* heat exchanger; according to the temperature of the circulating fluid, the later warms up or cools down.

Here, we deal only with countercurrent heat exchangers, almost the only kind that is used in helium refrigeration.

An *ideal* heat exchanger would transfer energy with a temperature difference that would be *zero*. Obviously, such a heat exchanger cannot exist because, without a temperature difference, no heat transfer can be achieved! In Fig. 2.2 that is one usual way a heat exchanger is shown (a rectangle with two lines representing each fluid circuit), one can see where such minimum temperature difference could be located: at the warm end, the cold end or even, anywhere inside the heat exchanger, where one cannot measure it.

Therefore, in a real heat exchanger there is always a finite temperature difference, somewhere, between the processed fluids.

The temperature difference $T_1 - T_2$, along the heat exchanger is the temperature gradient: grad T.

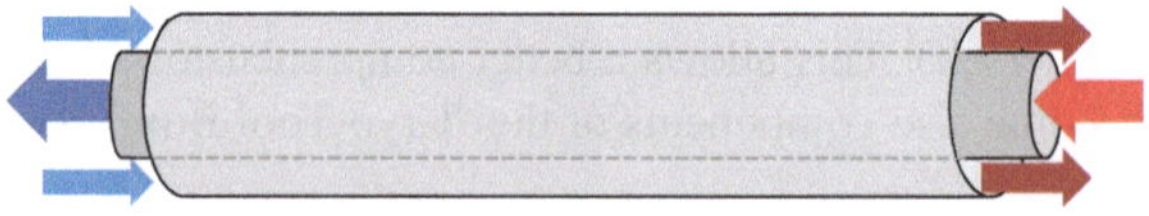

Fig. 2.1 An "intuitive" heat exchanger

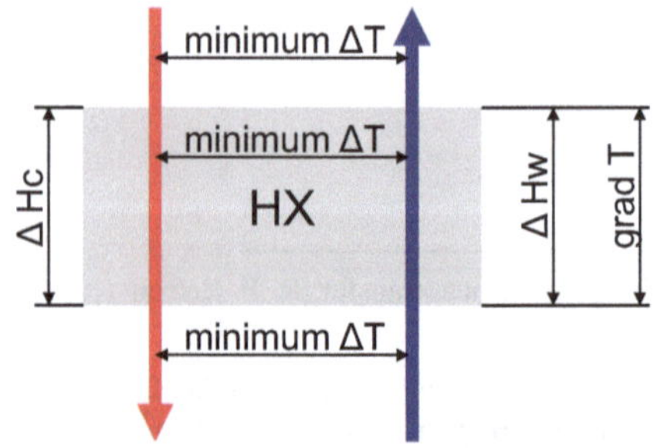

Fig. 2.2 Possible locations of the minimum temperature difference in a heat exchanger

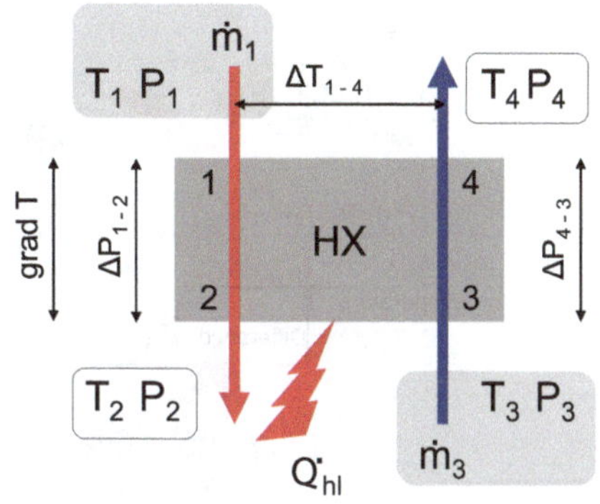

Fig. 2.3 A heat exchanger considered from "outdoors"

2.2.1 *Operation of a Heat Exchanger (Considered from "Outdoors")*

A heat exchanger can be functionally represented as in Fig. 2.3. As a convention, let us identify the heat exchanger connections by markers 1, 2, 3 and 4: odd numbers are in, even numbers are out.

The heat exchanger operating parameters can be sorted in three groups:

- The *process inputs* (in the grey rounded corner rectangles): they depend on the duty to be fulfilled by the heat exchanger. They are:
  - Pressure P_1, temperature T_1 and mass flow $\dot{m}_1$ of the fluid entering the heat exchanger at the warm end
  - Pressure P_3, temperature T_3 and mass flow $\dot{m}_3$ of the fluid entering the heat exchanger at the cold end
  - Heat leaks $\dot{Q}_{hl}$ that are the result of a non-ideal thermal insulation
- The *design parameters*:
  - Pressure drops in each circuit $\Delta P_{1\text{-}2}$ and $\Delta P_{3\text{-}4}$
  - Allowed temperature difference $\Delta T_{1\text{-}4}$

  Such parameters are selected according to experience. For example, a reasonable value for the temperature difference is 1% of the temperature of the warm end of the heat exchanger.
- The *parameters that depend on the operation* of the heat exchanger (into the transparent rounded corner rectangles):
  - Pressures P_2 and P_4 and temperatures T_2 and T_4 of the fluids exiting the heat exchanger

They are the result of the operation of the heat exchanger.

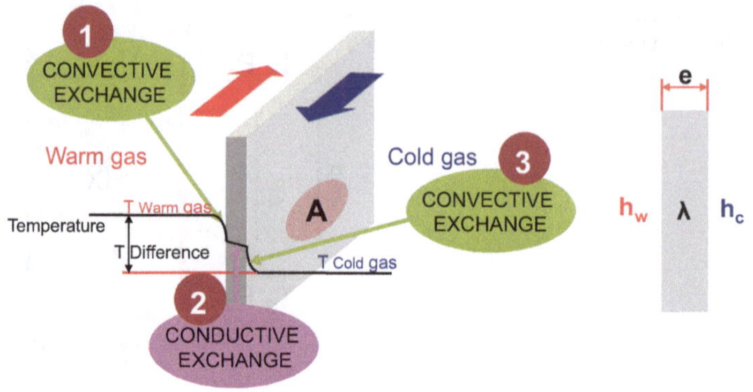

Fig. 2.4 The heat exchange process

2.2.2 *Operation of a Heat Exchanger (Considered from "Indoors")*

It is of interest to understand how the heat exchanger is operating "inside".

The wall, the area of which is A (see Fig. 2.4, left), separates warm and cold fluids that circulate in opposite directions (countercurrent arrangement). If, with a "magic" punctual thermometer, we could measure temperatures into the system by moving the thermometer from left to right into the fluid vein, we could register a constant temperature that *sharply decreases* as soon as the probe nears the wall. When the probe crosses the wall (as it is "magic"!), the temperature *decreases linearly* at a constant rate. Finally, when the probe is being dipped in the cold fluid, its temperature *decreases sharply* first and then more smoothly until it stays constant. The temperature difference between warm and cold gas is noted: ΔT.

Heat exchange proceeds in three stages:

- *Convective* exchange between warm fluid and the wall
- *Conductive* exchange through the wall
- *Convective* exchange between the wall and the warm fluid

The power that is transferred is:

$$\dot{Q} = U \times A \times \Delta T \tag{2.1}$$

with U, the heat exchange coefficient (W/m^2.K); A, the area of the wall (m^2); and ΔT, the temperature difference between fluids (K).

2.2.2.1 Heat Exchange Coefficients

The temperature difference between the two fluids generates a surface heat flow proportional to the difference and the reciprocal of a surface thermal resistance.

When:

- e is the thickness of the wall (m)
- λ the thermal conductivity of the all material (W/m.K)
- h_w the warm fluid heat exchange coefficient (W/m^2.K)
- h_c the cold fluid heat exchange coefficient (W/m^2.K)

(see Fig. 2.4 right)

The thermal resistance R is, by analogy with an electrical circuit:

$$R = 1/h_w + e/\lambda + 1/h_c = 1/U \quad (2.2)$$

The heat exchange coefficients are calculated by the heat exchanger designer. They depend on the thermophysical properties of the fluids and the heat exchanger configuration among others, the gas velocity.

Remark In Sect. 7.3.1, one can see how the heat exchange coefficients change with flow variation, during off-design regimes.

2.2.2.2 Incidence of the Wall

Let us consider an example: What are the heat transfer coefficients for a sheet of copper of 1 mm thickness with gas flowing in forced convection on both sides?

For a gas in forced convection: $30 < h < 300$ W/m^2.K. For simplification, let us select same average value for both fluids: 150 W/m^2.K

For copper: $\lambda \cong 390$ W/m.K

$$1/U = 1/h_w + e/\lambda + 1/h_c$$
$$1/U = 1/150 + 0.001/390 + 1/150$$
$$0.00666667 + 0.00000256 + 0.00666667 = 0.013333$$

As from Eq. [2.1]:

$$\Delta T = 1/U \times \dot{Q}/A \quad (2.3)$$

normalising the above calculation for a 1 K temperature difference gives the temperature difference repartition between the three steps: warm fluid convective transfer, conductive transfer and cold fluid convective transfer.

$$1 = (0,499999 + 0,000192 + 0,499999) \times \dot{Q}/A.\Delta T$$

These values are visualised in Fig. 2.5a.

Similar calculation with a 1 mm thick bad thermal conductor stainless steel sheet ($\lambda_{SS} = 26$ W/m. K) gives:

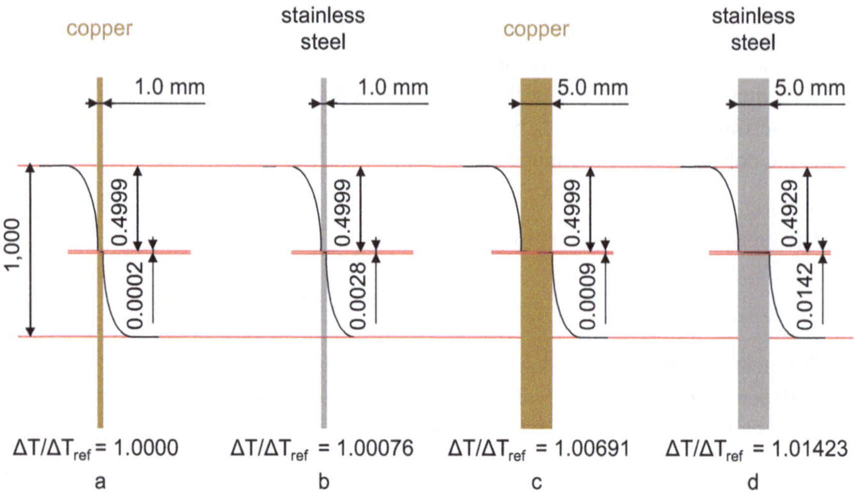

Fig. 2.5 Impact of material and thickness of the wall

$$1 = (0,499999 + 0,000002 + 0,499999) \times \dot{Q}/A.\Delta T$$

Other calculation results with a 5 mm thickness of both copper and stainless steel are shown on Fig. 2.5c, d.

From this simple exercise, one notices that the temperature difference through the wall is *generally* negligible in the type of heat exchangers that are being used for general cryogenic purposes and there is no noticeable difference between a good thermal conductivity material as copper and a bad one as stainless steel. Only a thicker wall with stainless steel shows a minor difference.

However, when dealing with temperatures around liquid helium, one must be careful because thermal conductivity of metals may decrease drastically (remember Fig. 1.30).

2.3 Thermodynamic Balance of a Heat Exchanger

Remark Such thermodynamic calculation is explained in detail in Sect. 15.8.3.

After having plotted the "border" (remember Sect. 1.6) that isolates a system from the universe (see Fig. 2.6), let us write its thermodynamic balance (identification of connections as said in Sect. 2.2.1:

- At point 1: (In, positive) $+\dot{m}_1 \times h_1$
- At point 2: (Out, negative) $-\dot{m}_1 \times h_2$

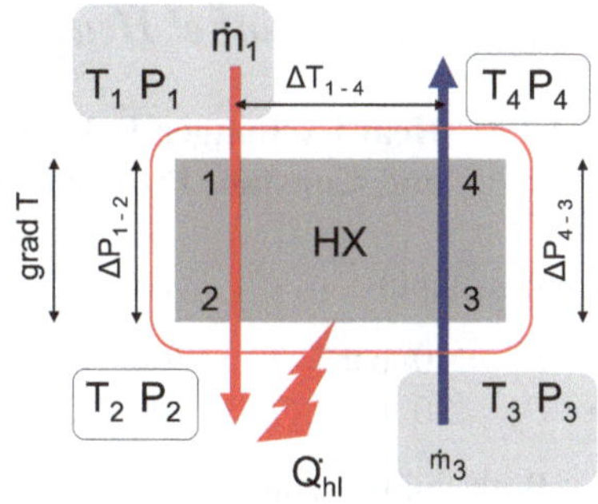

Fig. 2.6 Operating parameters of a heat exchanger

- At point 3: (In, positive) $+\dot{m}_3 \times h_3$
- At point 4: (Out, negative) $-\dot{m}_3 \times h_4$

and, *as the system is in a steady state and the gas velocity at the limits is negligible*, the sum equals to zero:

$$\dot{m}_1 \times h_1 - \dot{m}_1 \times h_2 + \dot{m}_3 \times h_3 - \dot{m}_3 \times h_4 = 0 \tag{2.4}$$

As we know conditions at points 1, 3 and 4, we can calculate the corresponding enthalpies and write:

$$h_2 = [\dot{m}_3 \times (h_3 - h_4) + \dot{m}_1 \times h_1]/\dot{m}_1 \tag{2.5}$$

Knowing P_2 and h_2, T_2 is calculated using REFPROP®, as an example.

2.4 A Few Operating Situations Illustrated by Simple Heat Exchanger Calculations

In this paragraph, various heat exchangers are considered, shifting from a very simple one to an almost real one, each situation being more complicated than the former.

In order to simplify calculations, some assumptions are made:

- No pressure drops
- No heat leaks
- No longitudinal heat transfer by solid conduction

Important Remark Each of the calculations, here under in this chapter, refers to a one-off heat exchanger. None of these calculations refer to the operation of a one-off heat exchanger *in a different operating condition*. Operating a one-off heat exchanger in various operating conditions is discussed in Sect. 7.2.

2.4.1 *Two-Channel Heat Exchangers*

2.4.1.1 Heat Exchanger 1: Very Simple, Flow-Balanced, Same and Constant Properties for Both Fluids

Process inputs:

- $P_1 = 20\ \text{bar},\quad T_1 = 300\ \text{K},\quad \dot{m}_1 = 50\ \text{g/s}$
- $P_3 = 20\ \text{bar},\quad T_3 = 200\ \text{K},\quad \dot{m}_3 = 50\ \text{g/s}$

Both fluids are at same pressure: they have constant and same properties.
Design parameters:

- $\Delta(T_1 - T_4) = 3.0\ \text{K}$
- $\Delta(P_1 - P_2) = 0$ (assumption for simplification of calculation)
- $\Delta(P_4 - P_3) = 0$ (assumption for simplification of calculation)

Using $h_2 = [\dot{m}_3 \times (h_3 - h_4) + \dot{m}_1 \times h_1]/\dot{m}_1$ Eq. 2.5, the calculation result is:

$$h_2 = 1065.54\ \text{J/g}$$

with $P_2 = 20.0$ bar and $h_2 = 1065.54$ J/g, $T_2 = 203.00$ K (see Fig. 2.7).

The calculation can easily be performed using an Excel® spread sheet, as shown in Fig. 2.7, top, that computes the mass and thermal balance using the REFPROP® (NIST) software that returns the helium thermodynamic properties (remember Sect. 1.4.2.1.). Such calculations are also detailed in the CTB, Sect. 15.8.3.

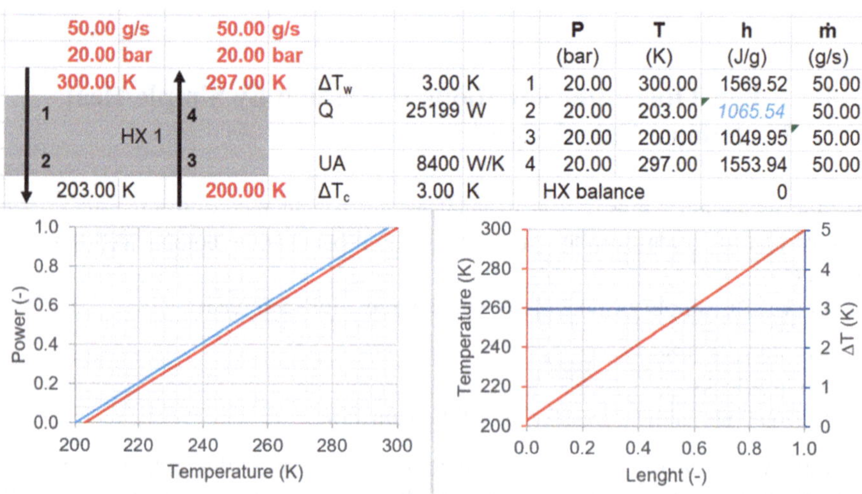

Fig. 2.7 A very simple balanced heat exchanger calculation performed with Excel and REFPROP

The "HX balance" cell checks that the power that is provided by the warm fluid is transferred to the cold one. It is a check of the heat exchanger calculation.

One can notice that the warm end temperature difference is *equal* to the one at the cold end. (This is related to constant helium properties in the temperature domain.)

2.4.1.2 Characteristics of a Heat Exchanger

2.4.1.2.1 The Temperature Difference

As it has been introduced in Sect. 2.2.2, the temperature difference relates to the warm and cold fluids. In the above example, such temperature difference is constant along the heat exchanger.

Remark: in this example, the *warm end* temperature difference has been set at 3.0 K. It is a wise behaviour to set the warm end temperature difference of a heat exchanger at a value that is *1% of the warmest temperature*. Here 3.0 K for 300 K. Furthermore, it is advised *not to select a temperature difference lower than 0.20 K, whatever the temperature is.*

2.4.1.2.2 The Heat Exchanger Diagram

Plotting the normalised energy (or power) that is transferred from one fluid to the other versus the temperature for each fluid of the heat exchanger (see Fig. 2.8) gives the so-called heat exchanger diagram. The warm fluid is represented by the line linking T_1 to T_2 and the cold fluid by line linking T_3 to T_4.

As, in the temperature domain in which this heat exchanger operates (300 to 200 K), there is almost no variation of the helium thermodynamic properties of both fluids, particularly the c_p, such lines are *straight lines*. Their slope, which is:

$$d\dot{Q}/dT = \dot{m} \times c_p/dT,$$

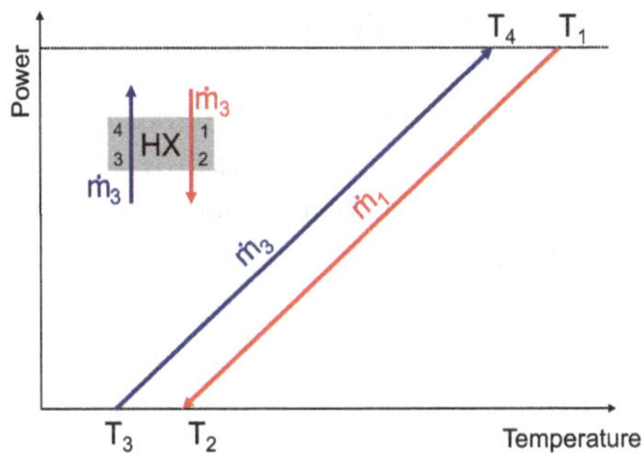

Fig. 2.8 The diagram of a heat exchanger

is proportional to the mass flow and to the isobaric specific heat c_p of the fluid. As the warm and cold mass flows are equal, these straight lines are *parallels*.

Such kind of operation is typical of a "refrigeration" regime where all liquid that is produced by the refrigerator is vaporised by the thermal load; therefore the gas mass flows into both the heat exchanger channels are equal.

On Fig. 2.7, bottom right, both the temperature of the warm fluid and the temperature difference are plotted versus the length of the heat exchanger. As the temperature difference is constant, the energy that is transferred in each unit of length is constant; therefore, the temperature changes linearly along the heat exchanger.

2.4.1.2.3 The "UA"

The power of the heat exchanger is the power that is transferred from one fluid to the other:

$$\dot{Q} = \dot{m}_1 \times (h_1 - h_2) = \dot{m}_2 \times (h_4 - h_3) \tag{2.6}$$

In the example displayed in Fig. 2.7:

$$\dot{Q} = 50 \times (1569.52 - 1065.54) = 50 \times (1553.94 - 1049.95) = 25199 \text{ W}$$

The product UA, (U = global heat exchange coefficient, A = heat exchange area) that is a characteristic of the heat exchanger, is:

$$UA = \dot{Q}/\Delta T \tag{2.7}$$

here:

$$UA = 8400 \text{ W/K}$$

2.4.1.2.4 The Efficiency

An ideal heat exchanger would transfer *all the energy* from the warm fluid to the cold fluid, with a zero-temperature difference; in other words, in Fig. 2.2 left, T_2 would be equal to T_3 (when keeping the same numbering for the four connections of a heat exchanger as in Fig. 2.3).

$$\dot{Q} = \dot{m}_1 \times \left(h_1 - h_{(P_2,T_3)}\right) = \dot{m}_3 \times \left(h_3 - h_{(P_4,T_1)}\right) \tag{2.8}$$

An actual heat exchanger transfers power from the warm fluid to the cold fluid, with a *finite temperature difference*:

$$\dot{Q}_{act} = \dot{m}_1 \times (h_1 - h_2) = \dot{m}_3 \times (h_3 - h_4) \tag{2.9}$$

The efficiency of the heat exchanger is the ratio of actual versus ideal transferred power:

$$\varepsilon_{hp} = \dot{m}_1 \times (h_1 - h_2)/\dot{m}_1 \times \left(h_1 - h_{(P_2,T_3)}\right) \tag{2.10}$$

$$\varepsilon_{hp} = (h_1 - h_2)/\left(h_1 - h_{(P_2,T_3)}\right) \tag{2.11}$$

Example:
Calculate the efficiency of the heat exchanger in Fig. 2.7:

$$P_2, T_3 \text{ and REFPROP} \rightarrow h_{(P_2,T_3)} = h_{(20.0,200.0)} = 1050.08 \ \mathrm{J/g}$$
$$\varepsilon_{hp} = (1569.52 - 1065.54)/(1569.52 - 1050.08) = 0.970$$

The calculation can also be performed on the cold fluid side:

$$\varepsilon_{lp} = (h_3 - h_4)/\left(h_3 - h_{(P_4,T_1)}\right)$$
$$P_4, T_1 \ \text{ and REFPROP} \rightarrow h_{(P_4,T_1)} = h(1.2,\ \ 300.0) = 1563.49 \ \mathrm{J/g}$$
$$\varepsilon_{lp} = (1049.95 - 1553.94)/(1049.95 - 1569.64) = 0.970$$

Remark On a balanced heat exchanger, the efficiency that is calculated on the warm fluid side is the same as the one on the cold fluid side.

2.4.1.2.5 The Number of Transfer Units (NTU)

One transfer unit corresponds to an elementary heat exchanger where the temperature gradient grad T along it equals the temperature difference between the two fluids: $\Delta T_{w\ -\ c}$

$$\mathrm{NTU} = \text{grad}\ \ T/\Delta T_{w-c} = 1$$

NTU can be expressed as the temperature gradient over a stream divided by the mean temperature difference across the heat exchange. In Fig. 2.9, it is:

$$\mathrm{NTU} = (T_1 - T_2)/(T_2 - T_3)$$

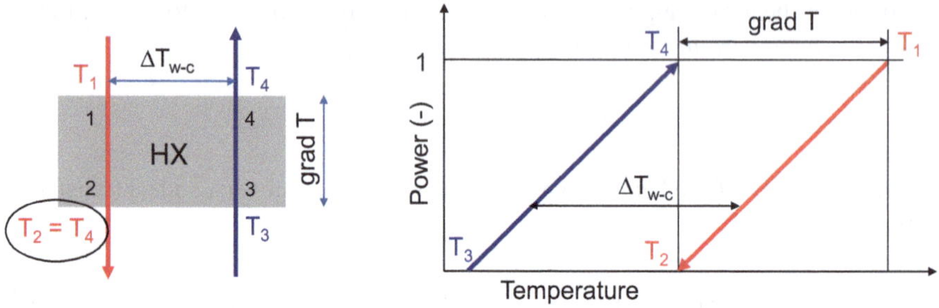

Fig. 2.9 Visualisation of one transfer unit

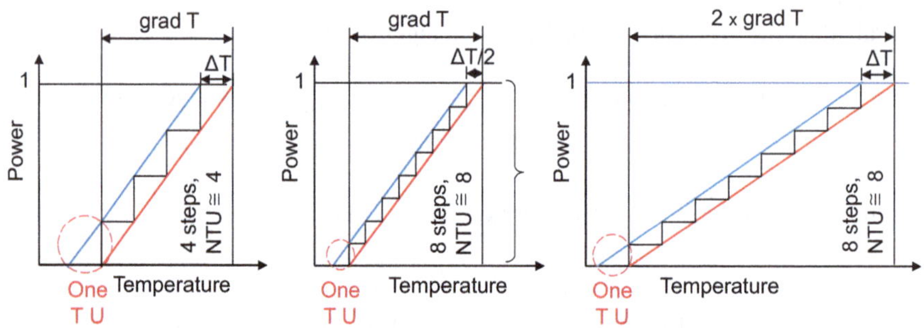

Fig. 2.10 A way to visualise the NTU on the heat exchanger diagram

On a heat exchanger diagram where the exchanged power is normalised to unit, the transfer units can also be visualised as shown in Fig. 2.10 where each "stair step" is one thermal unit.

$$\dot{Q} = \dot{m} \times c_{\mathrm{p}} \times \mathrm{grad}\ \ T$$

The NTU of a heat exchanger is an indicator of the actual heat transfer area or physical size of the heat exchanger. The larger the value of NTU, the closer the heat exchanger is to its thermodynamic limit.

The heat capacity rate of one fluid is:

$$C_i = \dot{m}_i \times \Delta h_i / \Delta T_i \tag{2.12}$$

The number of thermal units (NTU) of a heat exchanger is strictly defined as:

$$\mathrm{NTU} = \mathrm{UA}/C_{\min} \tag{2.13}$$

with $C_{\min}$ = the smallest of C_{w} or C_{c}

For the heat exchanger in Fig. 2.7:

$$C_w = 50 \times (1569.64 - 1065.66)/(300.00 - 203.00) = 259.78$$
$$C_c = 50 \times (1554.06 - 1050.08)/(297.00 - 200.00) = 259.78$$
$$C_{min} = 259.78$$
$$\text{NTU} = 3369/259.78 = 32.35$$

Remark As for the efficiency, here above, heat capacity rates are equal. This is the special situation that happens only for a balanced heat exchanger.

In such a balanced heat exchanger operating in a region where the gas properties are constant, the calculation is simple:

$$\text{NTU} = \text{grad}\, T/\Delta T \qquad (2.14)$$

NTU increases when:

- The temperature difference decreases
- The temperature gradient increase

2.4.1.2.6 Relation Between NTU and Efficiency

For a balanced heat exchanger, there is a relation between efficiency and NTU:

$$\varepsilon = \text{NTU}/(1 + \text{NTU})$$

or

$$\text{NTU} = \varepsilon/(\varepsilon - 1)$$

that is plotted in Fig. 2.11. Here, one can see that the NTU varies in a way that is more "representative" than efficiency.

Remark As the reasonable rule related to the minimum temperature difference (1% of the absolute temperature), there is a reasonable rule for the maximum NTU: around 50. Again, as for the warm end temperature difference for which it is possible to go to lower values, it is possible to go to higher values, but one must make sure that the heat exchanger supplier is able to fit to the requirement.

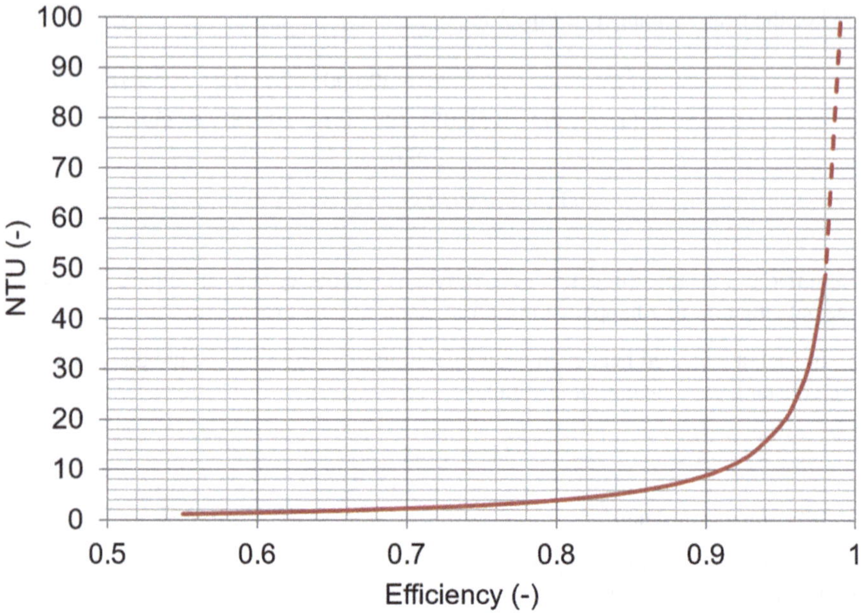

Fig. 2.11 Relation between efficiency and NTU for a balanced heat exchanger

2.4.1.3 Heat Exchanger 2: A Simple Heat Exchanger, Flow-Unbalanced, Different, But Constant, Fluid Properties

2.4.1.3.1 A Heat Exchanger Operating in a "Liquefier" Regime

Process inputs:

- $P_1 = 20$ bar, $T_1 = 300$ K, $\dot{m}_1 = 50\ \text{g/s}$
- $P_3 = 1.20$ bar, $T_3 = 200$ K, $\dot{m}_3 = 45\ \text{g/s}$

The differences with heat exchanger in Fig. 2.7 are:

- The cold flow is 45 instead of 50 g/s. It is a so-called unbalanced heat exchanger.
- The cold flow pressure is 1.20 bar: warm and cold fluid properties are different.

Design parameters:

- $\Delta(T_1 - T_4) = 3.0\ \text{K}$
- $\Delta(P_1 - P_2) = 0$ (assumption for simplification of calculation)
- $\Delta(P_4 - P_3) = 0$ (assumption for simplification of calculation)

Using $h_2 = [\dot{m}_3 \times (h_3 - h_4) + \dot{m}_1 \times h_1]/\dot{m}_1$ Eq. 2.5, the calculation result is:

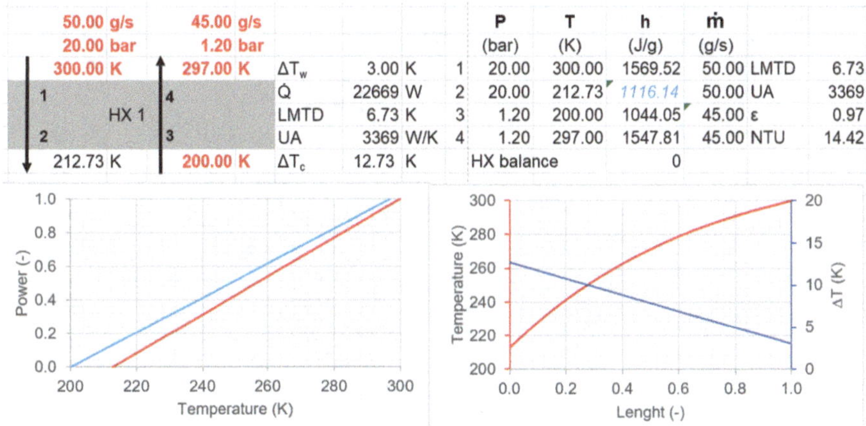

Fig. 2.12 Calculation of a flow-unbalanced heat exchanger operating in a "liquefier" mode

$$h_2 = 1116.14 \text{ J/g}$$

with $P_2 = 20.0$ bar and $h_2 = 1116.14$ J/g, $T_2 = 212.73$ K

Such regime is characteristic of a liquefier where part of the high-pressure flow is deposited as liquid and, therefore, reduces the low-pressure flow.

The HP temperature calculation along the heat exchanger that is displayed in Fig. 2.12, bottom right, is explained in Sect. 15.8.3.2.

The Temperature Difference

Here, the temperature difference is not the same at each end of the heat exchanger.

As the temperature difference between the two fluids is not constant as shown in Fig. 2.12, bottom right, as in the former example, it is no longer possible to use equation (2.1):

$$\dot{Q} = U \times A \times \Delta T \tag{2.1}$$

As fluid properties and U are considered as constant, one must use the "logarithmic mean temperature difference" (LMTD) instead of ΔT, the value of which is:

$$\text{LMTD} = (\Delta T_\text{w} - \Delta T_\text{c})/\text{Ln}(\Delta T_\text{w}/\Delta T_\text{c}) \tag{2.15}$$

or,

$$\text{LMTD} = ((T_1 - T_4) - (T_2 - T_3))/\text{Ln}\left(\frac{(T_1 - T_4)}{(T_2 - T_3)}\right) \tag{2.16}$$

here:

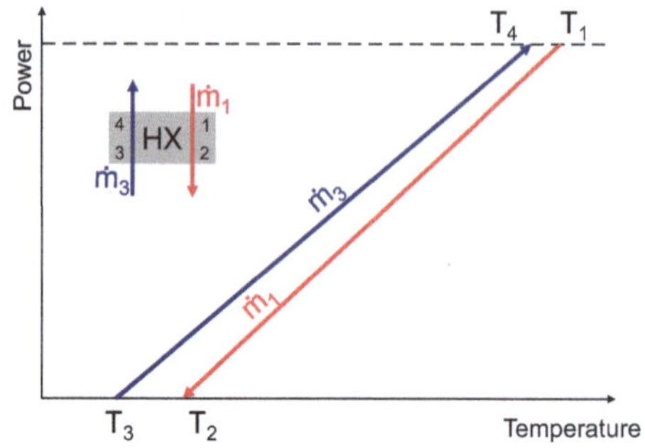

Fig. 2.13 The exchange diagram of an unbalanced heat exchanger

$$\text{LMTD} = 6.73\ \text{K}.$$

A demonstration of LMTD, taken from Wikipedia, is given in Sect. 17.2.

The plot of HP temperature along the length of the heat exchanger is explained in Sect. 15.8.3.2.

The Heat Exchanger Diagram

Here, the heat exchanger diagram is constituted of two straight lines (because c_p is constant) in Fig. 2.13, but, as the mass flow rates are different, the respective slopes are different. Such a kind of operation is characteristic of a "liquefaction" regime for which the cold flow is lower than the warm flow because some part of the fluid that has been liquefied stays in the phase separator, at the cold end of the system.

In Fig. 2.12, bottom right, one can see that for a larger temperature difference each unit of area transfers a larger quantity of energy. Therefore, the temperature changes faster than when the temperature difference is smaller. For this reason, the temperature profile along the heat exchanger is no longer linear. In the configuration of "liquefaction" regime where the warm mass flow is higher than the cold and, consequently, the heat exchanger is pinched at the warm end, the temperature variation rate (the slope of the curve) is smaller at the warm end and larger at the cold end.

Efficiency

Calculate the efficiency of the heat exchanger in Fig. 2.12.

On the warm side:

$$P_2, T_3 \text{ and REFPROP} \rightarrow h_{(P_2,T_3)} = h_{(20.0,200.0)} = 1050.08\ \text{J/g}$$

Using $\varepsilon_{hp} = (h_1 - h_2)/\left(h_1 - h_{(P_2,T_3)}\right)$ Eq. 2.11,

$$\varepsilon_{hp} = (1569.52 - 1116.14)/(1569.52 - 1050.08) = 0.873$$

On the cold side:

$$P_4, T_1 \text{ and REFPROP} \rightarrow h_{(P_4,T_1)} = h(1.2,\ 300.0) = 1563.49\ \text{J/g}$$
$$\varepsilon_{lp} = (1547.81 - 1044.05)/(1547.81 - 1563.49) = 0.970$$

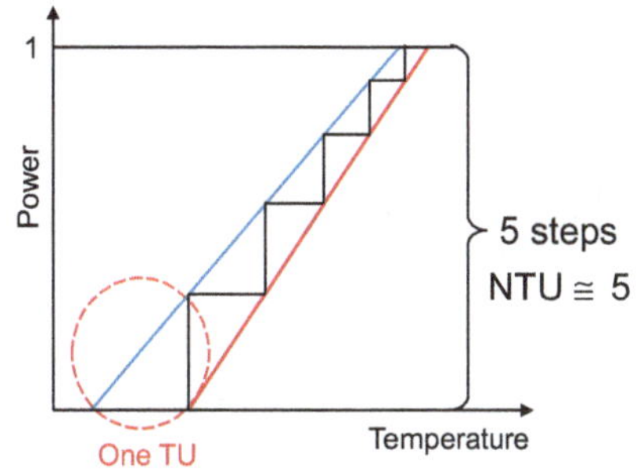

Fig. 2.14 NTU visualisation for an unbalanced heat exchanger

The efficiency that is calculated on the warm fluid side is not the same as for the one on the cold fluid side! For an unbalanced heat exchanger, the figure to be considered is *the highest*: here 0.970.

NTU

Calculate the NTU of the heat exchanger in Fig. 2.12 ($C_i = \dot{m}_i \times \Delta h_i / \Delta T_i$ Eq. 2.12):

$$C_w = 50 \times (1569.52 - 1116.14)/(300.00 - 212.73) = 259.76$$
$$C_c = 45 \times (1547.81 - 1044.05)/(297.00 - 200.00) = 233.70$$
$$C_{min} = 233.70$$
$$\text{NTU} = 3369/233.70 = 14.42$$

2.4.1.3.2 A Heat Exchanger Operating in an "Economiser" Regime

When the flows are unbalanced in the reverse way (cold flow higher than warm one as shown on Fig. 2.15), the heat exchanger diagram is "pinched" at the cold end.

Such a kind of operation is characteristic of an "economiser" regime (see Sect. 3.3.2.4.5).

The HP temperature calculation along the heat exchanger that is displayed in Fig. 2.12, bottom right, is explained in Sect. 15.8.3.2. One can see, on the heat exchanger diagram in Fig. 2.15, that, conversely, in an "economiser" regime, the temperature variation rate is larger at the warm end.

2.4.1.4 Heat Exchanger 3: Flow-Unbalanced, Non-constant Fluid Properties

This heat exchanger is more complicated to calculate. It is operating in a domain where the thermodynamic properties of helium vary because it is not far away from the liquid-gas domain.

Process inputs:

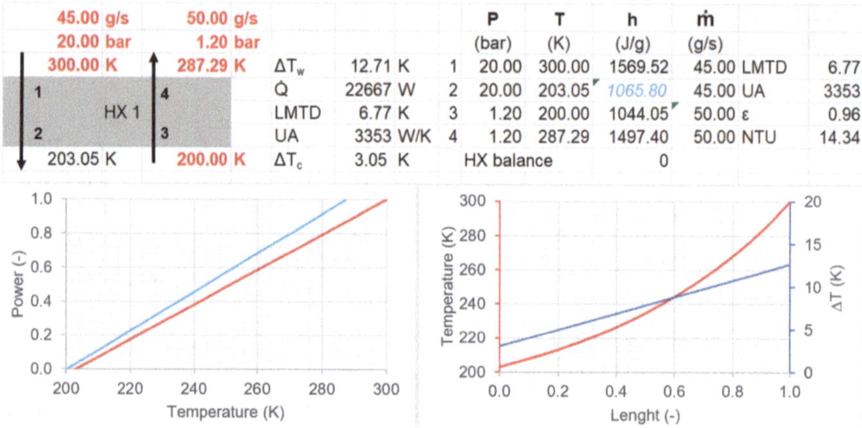

Fig. 2.15 Calculation of a flow-unbalanced heat exchanger operating in an "economiser" mode

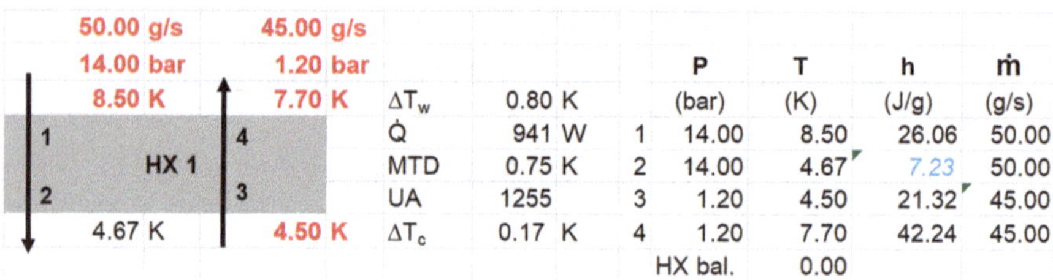

Fig. 2.16 Calculation of a heat exchanger in which the fluid properties are not constant

- $P_1 = 14.0$ bar, $T_1 = 8.5$ K, $\dot{m}_1 = 50$ g/s
- $P_3 = 1.2$ bar, $T_3 = 4.5$ K, $\dot{m}_3 = 45$ g/s

Design parameters:

- $(T_1 - T_4) = 0.80$ K
- $(P_1 - P_2) = 0$ *(assumption for simplification of calculation)*
- $(P_4 - P_3) = 0$ *(assumption for simplification of calculation)*

Using $h_2 = [\dot{m}_3 \times (h_3 - h_4) + \dot{m}_1 \times h_1]/\dot{m}_1$ Eq. 2.5, the calculation result is:

$$h_2 = 7.23 \text{ J/g}$$

with $P_2 = 14.0$ bar and $h_2 = 7.23$ J/g, $T_2 = 4.70$ K (see Fig. 2.16).

The Temperature Difference

Like in the unbalanced heat exchanger, the temperature difference is not the same at both ends. As U and the fluid properties are no more constant, one must use the MTD (mean temperature difference) instead of the LMTD.

Here, the MTD is:

$$941/1255 = 0.75 \text{ K}$$

If we had calculated the logarithmic mean temperature difference with $h_2 = [\dot{m}_3 \times (h_3 - h_4) + \dot{m}_1 \times h_1]/\dot{m}_1$ Eq. 2.5 that takes into account only the fluids conditions at inlets and outlets, we would have considered a heat exchanger the characteristics of which are represented by the doted straight lines in Fig. 2.18, bottom, left. Such calculation would have given a wrong value: 0.40.

Therefore, if this heat exchanger would have been specified with an MTD of 0.40 K instead of 0.75 K, its area would have been oversized by a factor of 1.86!

The Heat Exchanger Diagram

The calculation, based on the general balance at the heat exchanger inlets and outlets, is exact. However, in this example as the operating conditions are *near the helium liquid domain where properties change, especially* c_p, the curves representing the two fluids in the diagram cannot be straight lines anymore because the slope, that is, $\dot{m} \times c_p/dT$ (remember Sect. 2.4.1.1), is not constant.

In such a situation, one has to split the heat exchanger into equal parts (slices), as shown in Fig. 2.17, each of them transferring the *same quantity* of energy and within which helium thermodynamic properties are *considered as constant*.

In each elementary heat exchanger (slice), the diagram is linear (dotted lines in Fig. 2.17). The actual heat exchanger is the sum of all the elementary heat exchangers.

The heat exchanger is split into ten slices of equal transferred power. Its diagram is plotted in Fig. 2.18. The lines represent the actual evolution of the fluids. The hyphens separate each of the ten elementary heat exchangers.

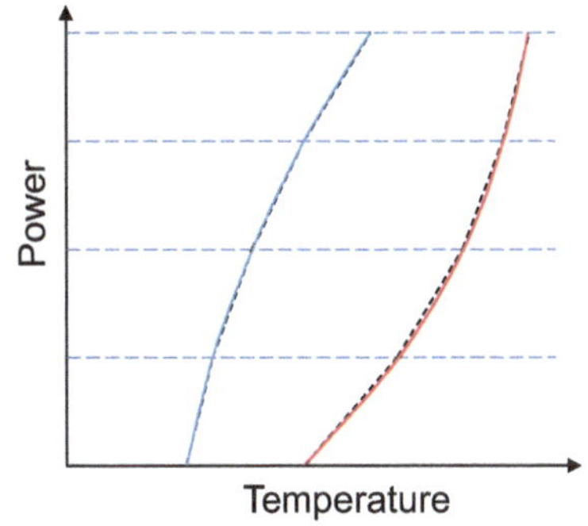

Fig. 2.17 Splitting the heat exchanger into four slices

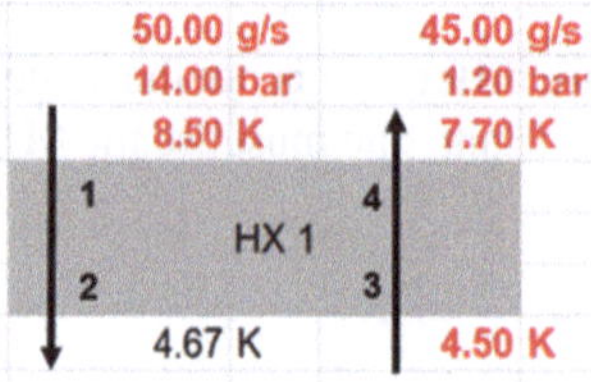

			P (bar)	T (K)	h (J/g)	ṁ (g/s)
ΔT_w	0.80 K	1	14.00	8.50	26.06	50.00
$\dot{Q}$	941 W	2	14.00	4.67	7.23	50.00
MTD	0.75 K	3	1.20	4.50	21.32	45.00
UA	1255 W/K	4	1.20	7.70	42.24	45.00
ΔT_c	0.17 K	HX balance			0	

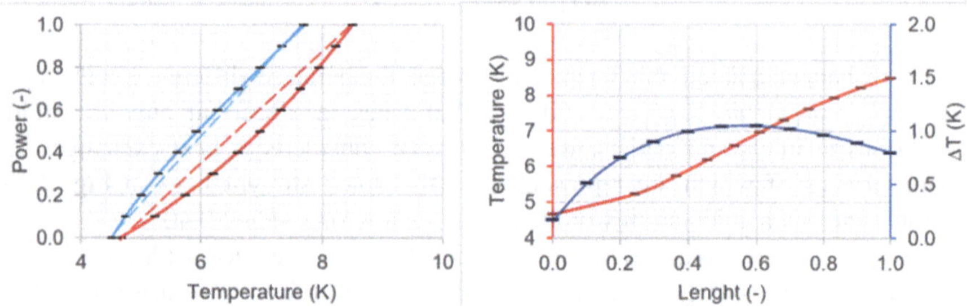

HP P	HP T	HP h	LP P	LP T	LP h	ΔT	% $\dot{Q}$	UA elem.	UA integr.	%UA
14.00	8.50	26.06	1.20	7.70	42.24	0.80	1.00		1255	1.00
14.00	8.23	24.18	1.20	7.34	40.15	0.89	0.90	111	1143	0.91
14.00	7.94	22.29	1.20	6.97	38.06	0.97	0.80	101	1042	0.83
14.00	7.64	20.41	1.20	6.62	35.96	1.02	0.70	95	947	0.75
14.00	7.32	18.53	1.20	6.27	33.87	1.05	0.60	91	856	0.68
14.00	6.98	16.64	1.20	5.94	31.78	1.04	0.50	90	766	0.61
14.00	6.61	14.76	1.20	5.61	29.69	1.00	0.40	92	674	0.54
14.00	6.20	12.88	1.20	5.30	27.60	0.90	0.30	99	575	0.46
14.00	5.75	11.00	1.20	5.00	25.50	0.75	0.20	115	460	0.37
14.00	5.24	9.11	1.20	4.73	23.41	0.51	0.10	152	308	0.25
14.00	4.67	7.23	1.20	4.50	21.32	0.17	0.00	308	0	0.00

Fig. 2.18 Calculation of an unbalanced heat exchanger with variable helium properties. Plots of the diagrams

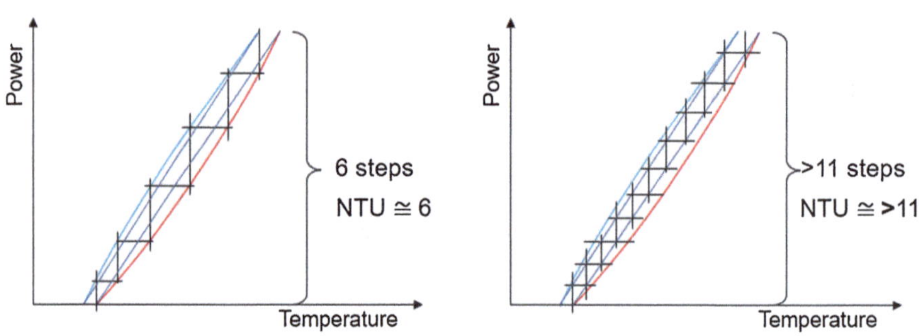

Fig. 2.19 The NTU "steps" of a heat exchanger with variable fluid properties

Calculation of the heat exchanger total UA is done by adding the UA values of each elementary heat exchanger, here: 1255 W/K. The detailed way of calculation is shown in Sect. 15.8.3.2.

Note the shape of the temperature curve along the heat exchanger length, in Fig. 2.18, bottom right: its slope changes according to the temperature difference.

NTU

Another consequence of the property variations is that NTU must also be calculated in splitting the heat exchanger into slices. Figure 2.19 shows what would be the kind of error done if NTU were calculated from the values of fluid properties at inlets and outlets. For this heat exchanger, the wrong value would have been 4.78 instead of 5.50.

Remark When operating temperatures are lower than 3.0 K, better use the HEPAK software!

2.4.1.5 Heat Exchanger 4: Non-constant Fluid Properties – A Trap

Mind! Even is the thermal balance of a heat exchanger is just at its interfaces, it is wise to check if *heat flows from warm to cold everywhere along the heat exchanger*.

Consider this example.

Process inputs:

- $P_1 = 20.0$ bar, $T_1 = 8.5$ K, $\dot{m}_1 = 50\ \text{g/s}$
- $P_3 = 1.2$ bar, $T_3 = 4.6$ K, $\dot{m}_3 = 45\ \text{g/s}$

Design parameters:

- $(T_1 - T_4) = 0.5\ \text{K}$
- $(P_1 - P_2) = 0$ (assumption for simplification of calculation)
- $(P_4 - P_3) = 0$ (assumption for simplification of calculation)

The calculation results are displayed in Fig. 2.20

The heat exchanger thermal balance is correct according to $h_2 = [\dot{m}_3 \times (h_3 - h_4) + \dot{m}_1 \times h_1]/\dot{m}_1$ Eq. 2.5; however, the calculation gives $T_2 = 2.45$ K. This temperature is lower than the one of the cold fluid ($T_3 = 4.5$ K)! This is, of course, impossible. In the heat exchanger people jargon, one says about this impossible heat exchanger that it "crosses" (see the diagram in Fig. 2.20 where, obviously, the temperature along the heat exchanger cannot be plotted and the MTD and UA cannot be calculated).

One must specify the temperature difference at the warm end in order that the cold end temperature difference becomes realistic. Here, with a temperature difference of 1.40 K at the warm end, the calculation gives $T_2 = 4.70$ K for the cold end (see Fig. 2.21).

Remark When operating at rather low temperatures where thermophysical properties change, one must always cut the heat exchangers in slices!

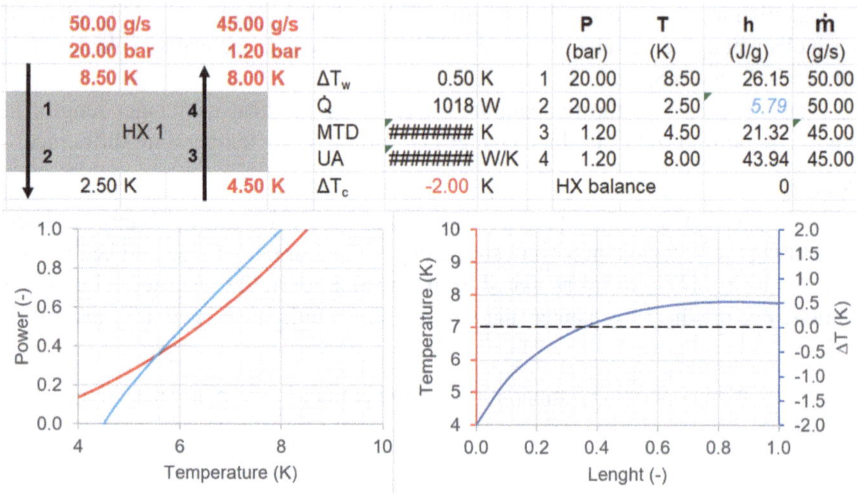

Fig. 2.20 A trap

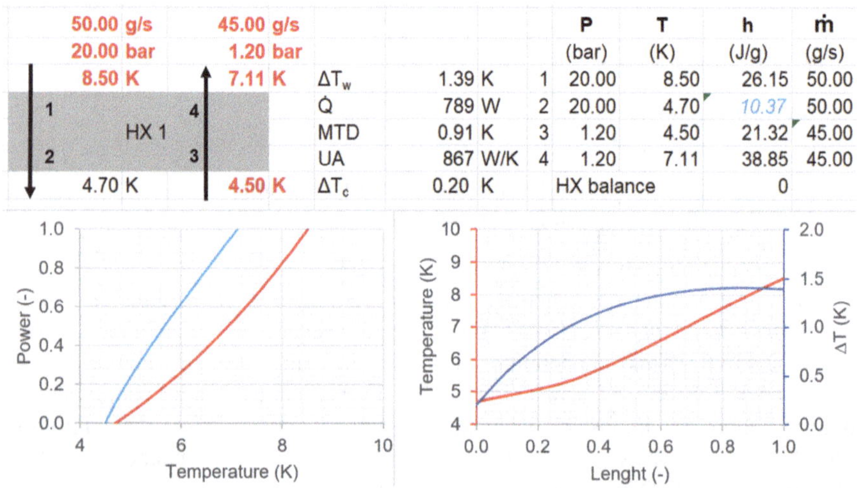

Fig. 2.21 Correction of the trap of Fig. 2.20

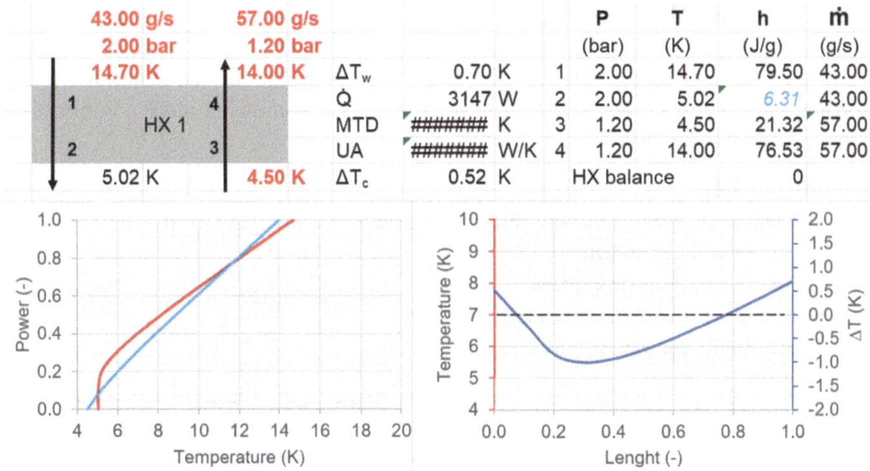

Fig. 2.22 Calculation and diagram of an impossible heat exchanger that "crosses" inside

2.4.1.6 Heat Exchanger 5: Non-constant Fluid Properties – Another Trap

Here is another example, shown in Fig. 2.22, where the trap is much more difficult to detect.

Process inputs:

- $P_1 = 2.0$ bar, $T_1 = 14.70$ K, $\dot{m}_1 = 43$ g/s
- $P_3 = 1.2$ bar, $T_3 = 4.5$ K, $\dot{m}_3 = 57$ g/s

Design parameters:

- $(T_1 - T_4) = 0, 70$ K
- $(P_1 - P_2) = 0$ (assumption for simplification of calculation)
- $(P_4 - P_3) = 0$ (assumption for simplification of calculation)

The calculation, performed as here above, gives $T_2 = 5.03$ K. This temperature is higher than the one of the cold fluid: everything seems correct.

However, if we split the heat exchanger into elementary parts as we did in Sect. 2.4.1.4 and we plot its diagram (Fig. 2.22), we see that, although if the balance calculation is correct, heat is supposed to flow from cold to warm inside the core of the heat exchanger! As in the former example, one must reconsider the hypothesis of warm end temperature difference in order that, in the whole heat exchanger, heat circulates from warm to cold. With a warm end temperature difference of 2.00 K, the calculation gives 0.53 K at the cold end, and the minimum temperature pinch is 0.14 K.

2.4.1.7 A Real Heat Exchanger

Assumptions have been made to perform easily some simple calculations in the former paragraphs. However, a real heat exchanger is somewhat more complicated. One has to take care of:

- Pressure drops
- Heat leaks from environment due to imperfect thermal insulation
- Change in fluid properties
- Change in material properties
- Change in heat exchange coefficient
- Longitudinal heat conduction

Only the simplest issues, heat leaks and pressure drop consequences are considered in the following paragraphs.

2.4.1.7.1 The Incidence of Pressure Drops and Heat Leaks

The pressure drops sit on all channels.

The heat leaks enter the heat exchanger by conduction through the supporting elements, by radiation and by gas conduction/convection through the external surface. In $\dot{m}_1 \times h_1 - \dot{m}_1 \times h_2 + \dot{m}_3 \times h_3 - \dot{m}_3 \times h_4 = 0$ Eq. 2.4, there are no heat leaks. To take care of them, one has to write:

$$\dot{m}_1 \times h_1 - \dot{m}_1 \times h_2 + \dot{m}_3 \times h_3 - \dot{m}_3 \times h_4 + \dot{Q}_{\mathrm{hl}} = 0$$

so,

$$h_2 = [\dot{m}_3 \times (h_3 - h_4) + \dot{m}_1 \times h_1]/\dot{m}_1 \tag{2.5}$$

Remark that when writing the heat exchanger balance, there is no special place where the heat leaks "fall".

Let us compare four different heat exchangers, one with neither pressure drops nor heat leaks (Fig. 2.23, top) with a second one having only pressure drops a third one having only heat leaks and a fourth one with both heat leaks and pressure drops.

Pressure drops affect only slightly the results because enthalpy is mainly related to temperature, not pressure, remember Sect. 1.4.1.2. The MTD is almost unchanged: 3.02 K instead of 3.01 K.

Heat leaks result in a higher MTD: 4.02 K instead of 3.01 K. According to the input conditions (the warm temperature difference is fixed), the result is a higher temperature at point 2: 205.26.

Both pressure drops and heat leaks result in a higher MTD 4.03 K instead of 3.01 K and a higher temperature at point 2: 205.28 K.

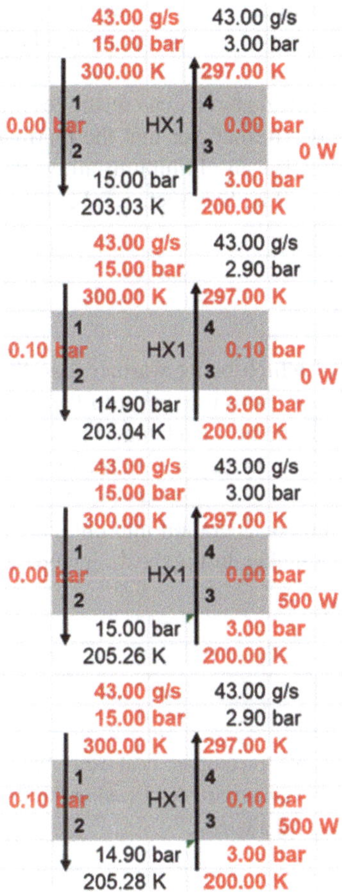

			P (bar)	T (K)	h (J/g)	ṁ (g/s)
ΔT_w	3.00 K	1	15.00	300.00	1567.89	43.00
$\dot{Q}$	21662 W	2	15.00	203.03	1064.12	43.00
MTD	3.01 K	3	3.00	200.00	1044.62	43.00
UA	7198 W/K	4	3.00	297.00	1548.39	43.00
ΔT_c	3.03 K	HX Balance			0.00	

			P (bar)	T (K)	h (J/g)	ṁ (g/s)
ΔT_w	3.00 K	1	15.00	300.00	1567.89	43.00
$\dot{Q}$	21661 W	2	14.90	203.04	1064.15	43.00
MTD	3.02 K	3	3.00	200.00	1044.62	43.00
UA	7183 W/K	4	2.90	297.00	1548.36	43.00
ΔT_c	3.04 K	HX Balance			0.00	

			P (bar)	T (K)	h (J/g)	ṁ (g/s)
ΔT_w	3.00 K	1	15.00	300.00	1567.89	43.00
$\dot{Q}$	21162 W	2	15.00	205.26	1075.74	43.00
MTD	4.02 K	3	3.00	200.00	1044.62	43.00
UA	5260 W/K	4	3.00	297.00	1548.39	43.00
ΔT_c	5.26 K	HX Balance			0.00	

			P (bar)	T (K)	h (J/g)	ṁ (g/s)
ΔT_w	3.00 K	1	15.00	300.00	1567.89	43.00
$\dot{Q}$	21161 W	2	14.90	205.28	1075.78	43.00
MTD	4.03 K	3	3.00	200.00	1044.62	43.00
UA	5253 W/K	4	2.90	297.00	1548.36	43.00
ΔT_c	5.28 K	HX Balance			0.00	

Fig. 2.23 Incidence of pressure drops and heat leaks

Note that the above calculations deal with four one-off different heat exchangers. The behaviour of a one-off heat exchanger in different operating conditions is discussed in Sect. 7.3.1.

2.4.1.7.2 Change of Properties of Fluids and Materials

There are two other issues not to be forgotten when considering a heat exchanger: the change in properties of gas and material.

On top of the variation of the gas thermodynamic properties according to temperature and pressure, transport properties of the gas vary also along the heat exchanger.

One should not forget the variation of properties of the material that constitutes the heat exchanger, in particular its thermal conductivity. (Remember that the thermal conductivity of a metal varies with the temperature; for an aluminium alloy, the thermal conductivity at 10 K is about 1/10 of that at 300 K. See Sect. 1.5.2.)

2.4.1.7.3 The Longitudinal Conduction

As a heat exchanger is a piece of metal, one end of which being warmer than the other, heat is naturally transferred by solid conduction from the warm to the cold end. This phenomenon "fights" again the operation of the heat exchanger. It is kind of a parasitic "thermal by-pass", the incidence of which increases with the temperature gradient along the heat exchanger.

Of course, all such issues, which are not taken into consideration into our simple calculations, must be considered by the heat exchanger maker during the design phase.

2.4.2 *Comparison of Heat Exchangers*

With the parameters that have been introduced, it is possible to compare heat exchangers. In column REF of Table 2.1, the parameters of a heat exchanger are displayed in the top part. In the bottom part, $\dot{Q}$, LMTD, UA, NTU and ε are calculated.

Table 2.1 Comparing various heat exchangers

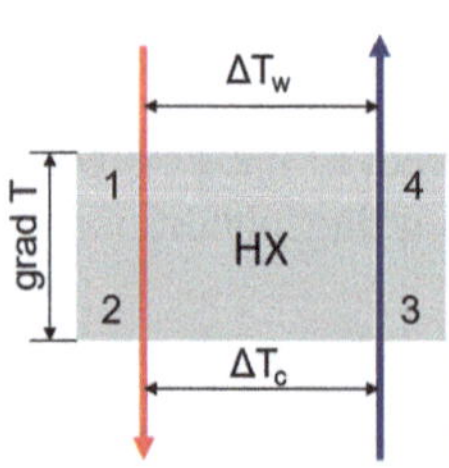

		REF	2 x $\dot{m}$	2 x grad T	2 x ΔT	Unbal
$\dot{m}_{HP}$	(g/s)	50.00	100.00	50.00	50.00	50.00
P_{HP}	(bar)	20.00	20.00	20.00	20.00	20.00
T_1	(K)	300.00	300.00	300.00	300.00	300.00
T_2	(K)	253.00	253.01	203.04	256.01	257.71
$\dot{m}_{LP}$	(g/s)	50.00	100.00	50.00	50.00	45.00
P_{LP}	(bar)	1.20	1.20	1.20	1.20	1.20
T_4	(K)	297.00	297.00	297.00	294.00	297.00
T_3	(K)	250.00	250.00	200.00	250.00	250.00
ΔT_w	(K)	3.00	3.00	3.00	6.00	3.00
ΔT_c	(K)	3.01	3.01	3.04	6.01	7.71
$\dot{Q}$	(W)	12204	24408	25188	11425	10984
LMTD	(K)	3.00	3.00	3.01	6.00	4.99
UA	(W/K)	4062	8124	8356	1903	2202
NTU	(-)	15.64	15.64	32.18	7.33	9.42
ε	(-)	0,940	0.940	0.970	0.880	0.846
			cause			consequence

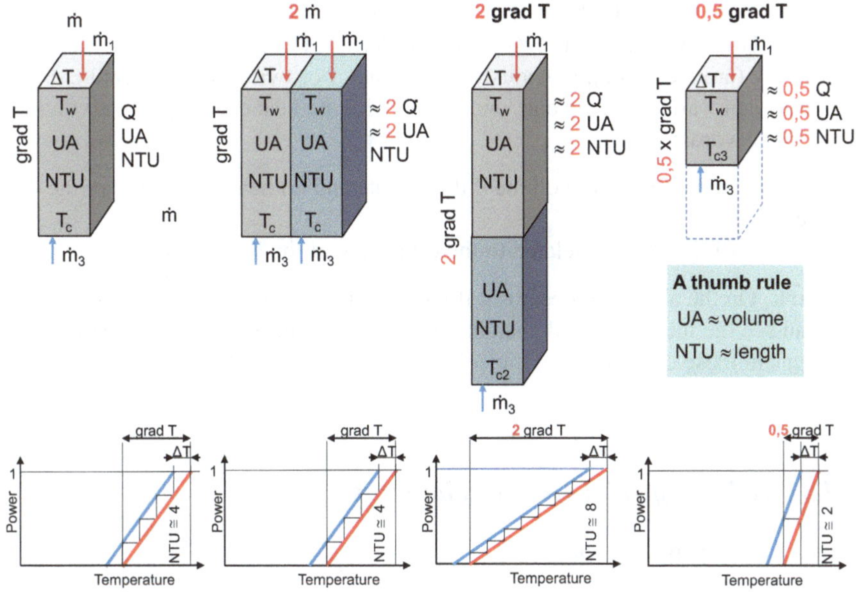

Fig. 2.24 A way to visualise UA and NTU (with help of *U*. Wagner)

In column "$2 \times \dot{m}$", the mass flow is doubled. $\dot{Q}$ and UA double. LMTD, NTU and ε stay constant. For twice the mass flow rate, one needs twice the heat exchange area (therefore, in case of plate and fins heat exchanger, twice the volume). See Fig. 2.24 for a visualisation of these parameters.

In column "2 × grad *T*", the T_3 temperature is lower for same warm end temperature difference. $\dot{Q}$ is doubled due to the increased temperature gradient (300.00 − 203.04 = 96.96)) instead of (300.00 − 253.01 = 46.99). LMTD stays constant. UA doubled because $\dot{Q}$ doubled. As LMTD decreases, NTU increases conversely (around twice). Obviously, ε is higher. For twice the temperature difference between warm and cold end, one needs twice the heat exchange area (volume).

In column "$2 \times \Delta T$", the temperature difference is doubled for same mass flow rate as reference. $\dot{Q}$ stays almost constant, LMTD doubles, but UA is half, as NTU. Obviously, ε is lower. For twice the temperature difference, one needs half of the heat exchange area (volume).

In column "Unbal", the LP mass flow rate is lower (45 instead of 50 g/s) for same warm end temperature difference (3.00 K). $\dot{Q}$ is slightly reduced due to the reduced temperature gradient (300.00 − 257.71 = 42.29) instead of (300.00 − 256.01 = 43.99). LMTD increases because the cold end temperature difference increases (3.01 to 7.71). As LMTD increases, UA and NTU decrease conversely (around half). Obviously, ε is lower. For an unbalanced heat exchanger, the heat exchanger area is reduced because the temperature difference is larger.

In column "2 × grad *T*", the T_3 temperature is lower for same warm end temperature difference. $\dot{Q}$ is doubled due to the increased temperature gradient

(300.00 − 203.04 = 96.96) instead of (300.00 − 253.01 = 46.99). LMTD stays constant. UA doubled because $\dot{Q}$ doubled. As LMTD decreases, NTU increases conversely (around twice). Obviously, ε is higher. For twice the temperature difference between warm and cold end, one needs twice the heat exchange area (volume).

In other words:

- UA is related to the size (volume if it is a plate and fin heat exchanger) of the heat exchanger.
- LMTD, NTU and ε are related to the efficiency of the heat exchanger.

Remark On the above figures, even if the operating temperatures are rather warm, one can see the incidence of the slight change of helium properties compared to an ideal gas in these conditions on the cold end temperature difference.

2.4.3 A Few Special Heat Exchangers

2.4.3.1 Two Different Fluids

2.4.3.1.1 Gas-Gas

One may have to operate heat exchangers with different fluids. Such a situation happens, for example, when helium is to be cooled down with nitrogen, using a two-fluid heat exchanger as shown on Fig. 2.25. This could be part of a liquid nitrogen pre-cooling system for a liquefier like it is shown in Sect. 3.6.

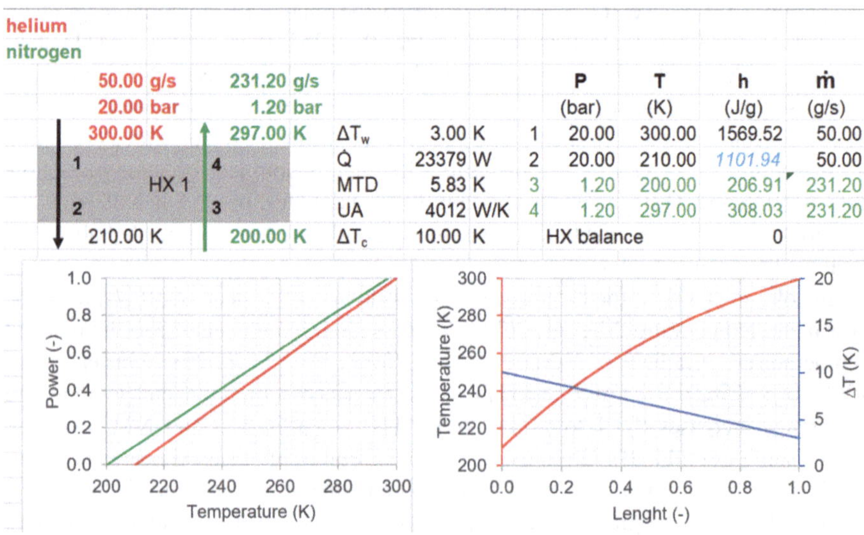

Fig. 2.25 Heat exchanger operating with two different fluids (helium and nitrogen)

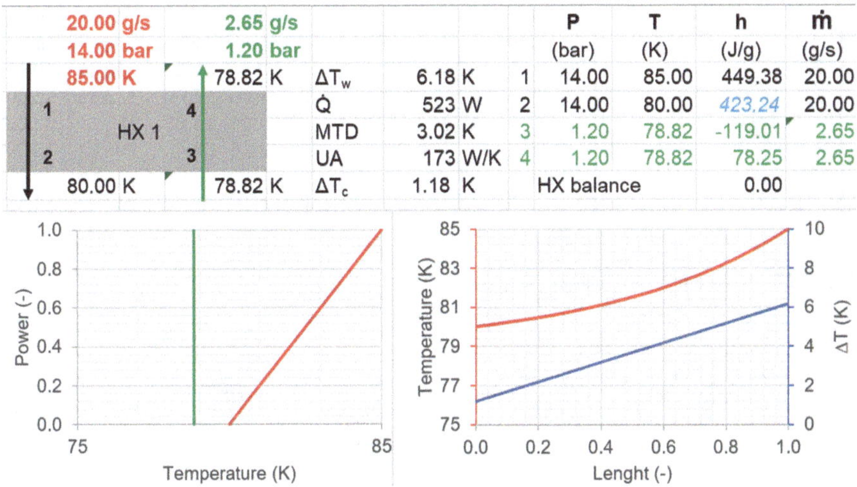

Fig. 2.26 The liquid nitrogen vaporiser

The only change in the calculation is that properties at points 3 and 4 have to be calculated according to the other fluid, here nitrogen.

Details on such a calculation are given in CTB, Sect. 15.8.3.3.

Remark Other special cases of non-cryogenic heat exchangers operating with different fluids are the oil and helium coolers that are incorporated into the compression station (see Sect. 6.3.7).

2.4.3.1.2 Gas-Liquid (Vaporiser)

Helium might also be cooled down to the nearest temperature of liquid nitrogen. This is performed in a "vaporiser" (a liquid nitrogen vaporiser), which can be a simple coiled pipe or a plate and fin heat exchanger that is dipped into the liquid nitrogen or heat exchanger operating as a thermal syphon (see Sect. 3.6.5). It is a two-channel heat exchanger that has a peculiarity: as power is dumped into liquid, the liquid side stays at constant temperature (see Fig. 2.26). At point 3 there is pure liquid and at point 4, saturated vapour.

A liquid nitrogen vaporiser (or helium cooler with liquid nitrogen) is often part of a pre-cooling system, at the warm end of a liquefier or refrigerator (see Sect. 3.6).

Remark Such calculation can also be performed for a supercritical helium sub-cooler in which supercritical helium is cooled with boiling liquid helium (see Sect. 5.3.3).

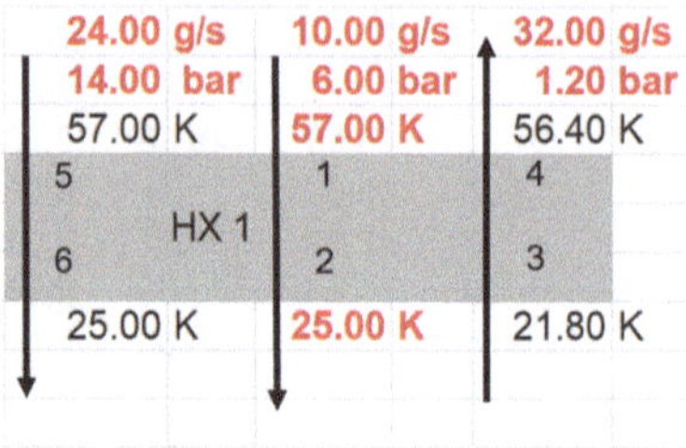

ΔTw = 0.60 K
Q̇ = 1720
MTD = 1.55 K
ΔTc = 3.20 K

	P (bar)	T (K)	h (J/g)	ṁ (g/s)
1	14.00	57.00	302.49	10.00
2	14.00	25.00	130.49	10.00
3	1.20	21.80	117.66	32.00
4	1.20	56.40	298.04	32.00
5	6.00	57.00	301.62	24.00
6	6.00	25.00	132.79	24.10
HX balance			0.00	

Fig. 2.27 A three-pressure heat exchanger

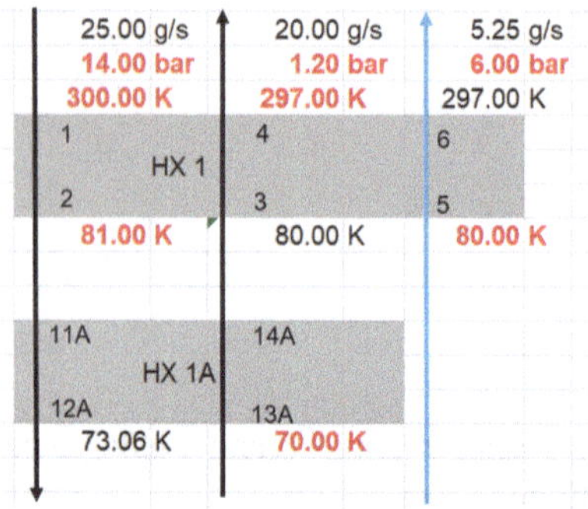

HX 1: ΔTw = 3.00 K; Q̇ = 28478 W; LMTD = 1.82 K; ΔTc = 1.00 K

HX 1A: Q̇ = 1039 W; LMTD = 1.84 K; ΔTc = 3.06 K

HX 1	P	T	h	ṁ
1	14.00	300.00	1567.56	25.00
2	14.00	81.00	428.46	25.00
3	1.20	80.00	420.72	20.00
4	1.20	297.00	1547.81	20.00
5	6.00	80.00	421.65	5.26
6	6.00	297.00	1549.37	5.26
HX 1 balance			0.00	
HX 1A				
11A	14.00	81.00	428.46	25.00
12A	14.00	73.06	386.88	25.00
13A	1.20	70.00	368.75	20.00
14A	1.20	80.00	420.72	20.00
HX 1A balance			0.00	
System balance			0.00	

Fig. 2.28 Fluids at different temperatures

2.4.3.2 More Than Two Fluids

When a heat exchanger is operating with more than two fluids, the assumption to be made is that all the fluids entering the heat exchanger are at the same temperature. Similarly, all the exiting fluids are at same temperature. The mass and thermal balance is calculated in the usual way. Such a heat exchanger is shown on Fig. 2.27.

But sometimes fluids are at different temperatures: one must split the heat exchanger into elementary heat exchangers in which all in-going fluids have the same temperature and all outgoing fluids have also same temperature, as shown in Fig. 2.28.

A similar arrangement can be found in Sect. 3.5.3.3, where HX3 and HX4 are located between the two expansion turbines of a series arranged turbine Claude cycle.

Of course, such a three-channel heat exchanger can operate with different gases (see Sect. 2.4.3.3).

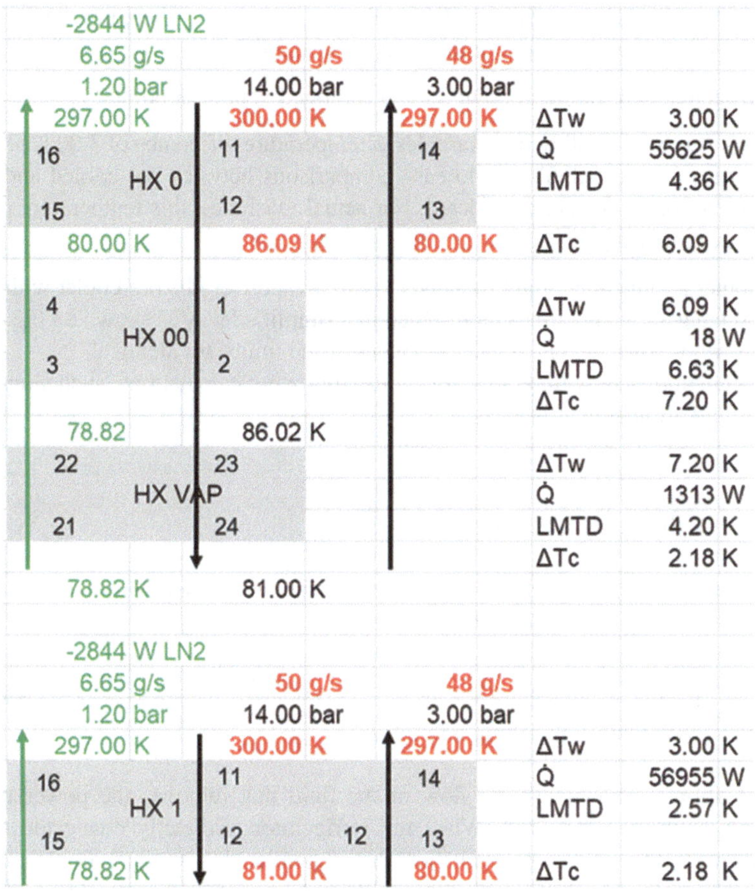

Fig. 2.29 *Top* the full structure of a liquid nitrogen pre-cooler calculation bottom, simplified pre-cooling heat exchanger for a refrigerator or a liquefier

2.4.3.3 The Liquid Nitrogen Pre-cooler

To pre-cool a cycle (see Sect. 3.6), either a refrigerator or a liquefier, one must add, at the warm end of the cold box, a set of heat exchangers to cool helium down to a temperature that is generally near that of liquid nitrogen. Nitrogen is provided in a liquid phase that is firstly vaporised and then warmed up to room temperature.

For an easy understanding, one must discuss the duty of each heat exchanger:

- First, a gas-gas heat exchanger, HX 0, in Fig. 2.29, top, where high-pressure helium is cooled down against both low-pressure helium and gaseous nitrogen
- Then, a second gas-gas heat exchanger, HX 00, where high-pressure helium is cooled down against gaseous nitrogen

- Finally, a third heat exchanger HX VAP, the "vaporiser", where helium is cooled down against boiling liquid nitrogen (see Sect. 2.4.3.1.2)

  Nitrogen is pure liquid at point 23 and saturated vapour at point 24.

Remark On all the following examples, a temperature difference of 3.00 K at room temperature is selected. This is to ease comparisons between pre-cooled and non-pre-cooled cycles that are considered. For actual machines, this temperature difference is generally higher.

Another arrangement that fills same duty as a vaporiser: the pre-cooler with three heat exchangers is sometimes represented in a simplified way as shown on Fig. 2.29, bottom, where one can check that all values at the limits are identical.

To get simpler flow diagrams, for all the following calculations, such simplified pre-cooling arrangement is used.

2.4.3.4 Dividing Heat Exchangers

In some situations, the load power is to be absorbed within a rather narrow temperature difference. Such situation happens in two cases.

2.4.3.4.1 Avoiding Solidifying a Fluid

One solution is to increase the flow of the fluid that removes the power that is released by the thermal load. When the refrigerator is directly connected to the thermal load, it is a part of the cycle flow that is circulated (the turbine flow in case of a Brayton cycle, the JT flow in case of a Claude cycle). If the circulating flow is increased, the cycle might not operate at its optimal conditions. Furthermore, such a simple solution has drawbacks: higher flow means larger components (compressor, heat exchangers, etc.) and a higher heat input at the warm end of the heat exchanger, therefore a higher power consumption.

In such a case, it is of interest to use dividing heat exchangers, as shown in Fig. 2.30 where hydrogen (red) is cooled between 21 and 18 K with helium (black) between 14 and 20 K without any risk to solidify hydrogen (triple point is 13.96 K).

2.4.3.4.2 Cooling a Fragile Component

Cooling heavy (and rather fragile) loads such as superconducting coils is to be performed with a helium temperature difference that is typically limited to 50 K. When the cold source is at a much lower temperature, for example, 80 K in case of using liquid nitrogen, and if it is possible to have various cooling channels, an arrangement as shown on Fig. 2.31 is of interest because instead of being submitted

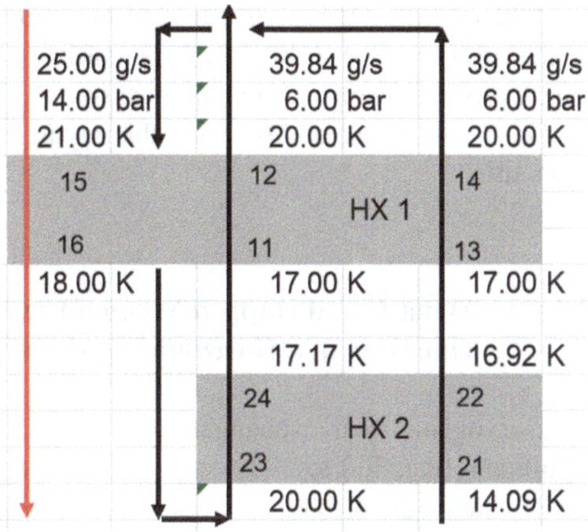

Fig. 2.30 Dividing heat exchanger to avoid solidification of hydrogen

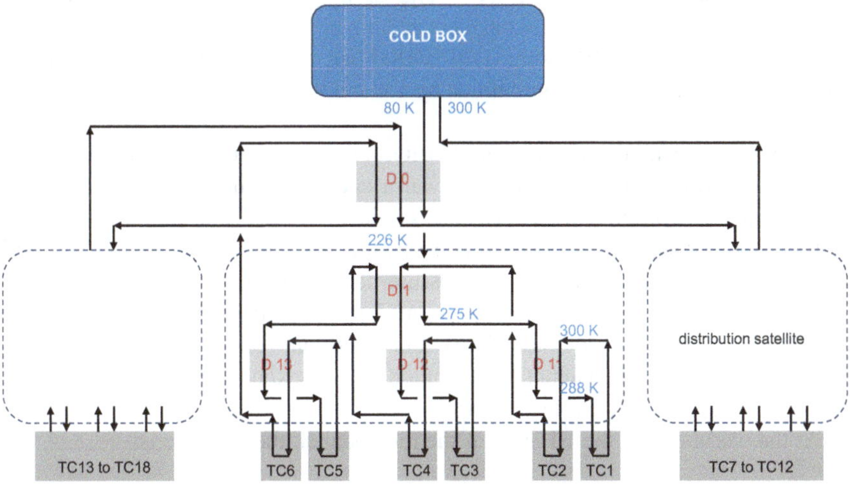

Fig. 2.31 Dividing heat exchangers for cooling a heavy and fragile load

to a 300 − 80 = 220 K temperature difference, the actual temperature difference is reduced to 12 K, which is very comfortable for cooldown of a fragile component.

This arrangement has been implemented for cooling of the thick casings (TC), in the Tore Supra system (see Fig. 12.13).

2.4.4 *About Heat Exchangers Operating Horizontally*

Cryogenics, as any other activity, is sometimes disturbed by curious sayings that are not questioned by people who repeat them. It is of interest to find why such sayings exist. Generally, it is because, at one time, a mistake has been made and not correctly analysed. Therefore, it becomes the starting point of a rumour. One such saying is that "cryogenic heat exchangers must operate in a vertical position".

This section aims at discussing this saying and answers the question that is asked in its title.

2.4.4.1 What Could Happen When a Cryogenic Heat Exchanger Is Operated in a Horizontal Position?

In a cryogenic heat exchanger, the temperature gradient along the core can be significant (e.g. 300 to 20 K), and, as a consequence, the physical properties of the fluids change drastically.

Heat exchanger manufacturers have very sophisticated tools to size heat exchangers, but here, the goal is to understand the behaviour using tools available for everyone such as Microsoft Excel™ and Cryodata Hepak™.

As a working example, let us consider a helium heat exchanger operating between 300 and 20 K. The high pressure (HP) is 18.0 bar; the low pressure (LP) is 4 bar. Along the core, the density is multiplied by 15, and the viscosity is divided by 5 (see Fig. 2.32)!

If one considers the change in density, the hydrostatic pressure difference between the top and the bottom at the warm end and at the cold end of the core differ as the ratio of the density, that is to say in a ratio of 15 to 1!

In a first approach, if one does not consider the incidence of the gravity, Fig. 2.33 gives a graphic representation of the pressure field in the core.

To make thinking easier, let us consider three identical elementary heat exchangers: a "top" one, a "middle" one and a "bottom" one (see Fig. 2.34).

We consider the HP channel of the "middle" heat exchanger operating between pressures HP_{in} and HP_{out}, ($HP_{in} - HP_{out} = \Delta P_{HP}$) as a reference.

At the warm end of the HP core of the top channel, the pressure is:

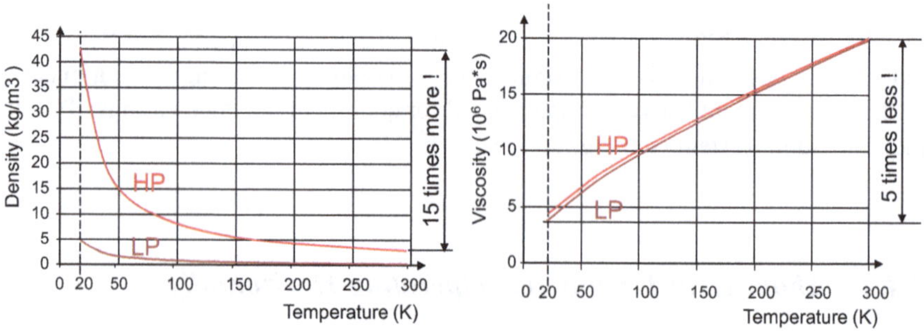

Fig. 2.32 Variation of helium properties with temperature

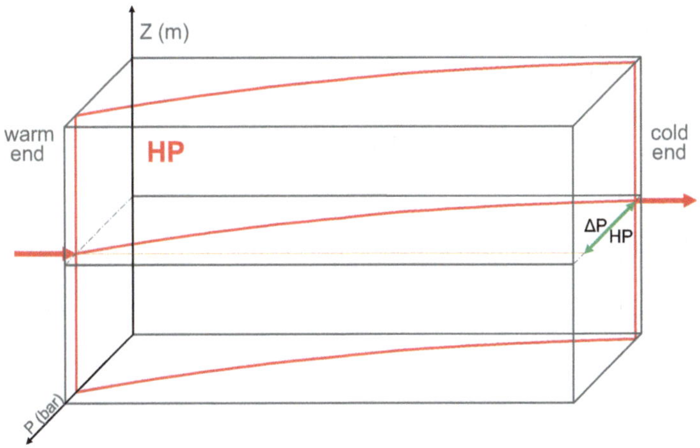

Fig. 2.33 Pressure field in the core, no incidence of gravity

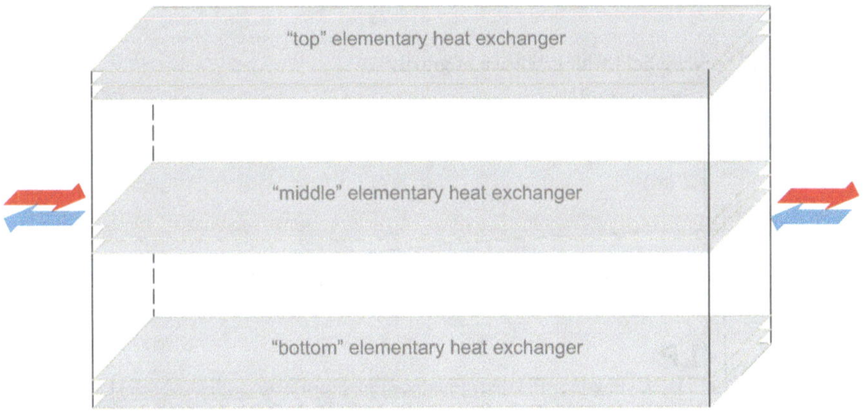

Fig. 2.34 The three elementary heat exchangers

$$\text{HP In Top} = \text{HP}_{\text{in}} - (\rho \text{HP}_{\text{w}} . g . h)/2$$

At the cold end of the HP core of the top channel, the pressure is:

$$\text{HP In Bot} = \text{HP}_{\text{in}} - \text{HP}_{\text{dp}} - (\rho \text{HP}_{\text{c}} . g . h)/2$$

Therefore, the driving pressures in the different channels are (Figs. 2.35 and 2.36):

Top: $\Delta P_{\text{HP}}\text{Top} = \Delta P_{\text{HP}} + ((\rho \text{HP}_{\text{cold}} - \rho \text{HP}_{\text{warm}}).\ g.\ h)/2$
Middle: ΔP_{HP}

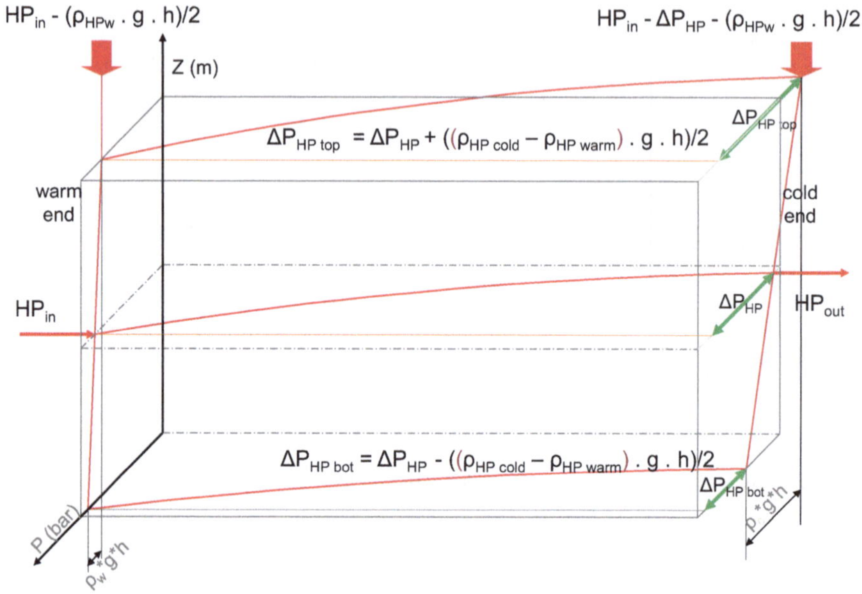

Fig. 2.35 HP pressure field with incidence of gravity

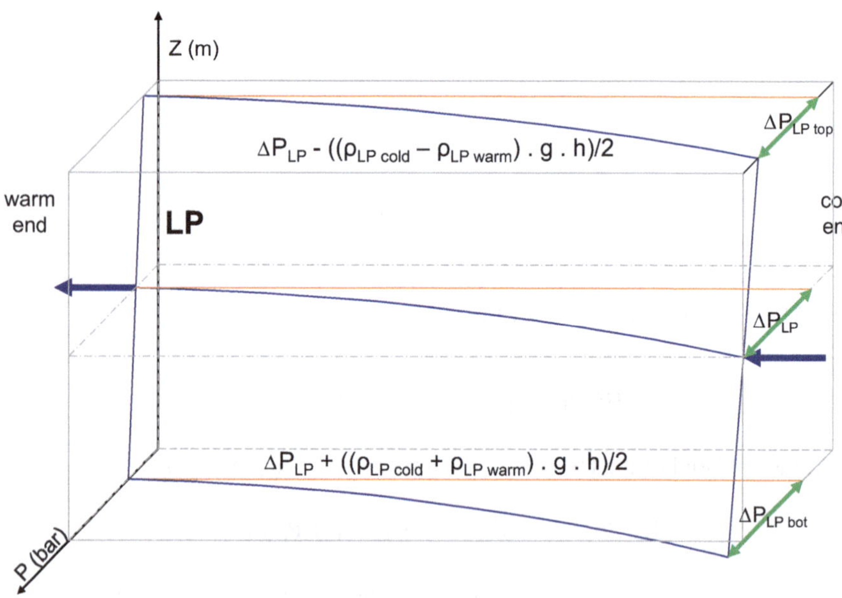

Fig. 2.36 LP pressure field with incidence of gravity

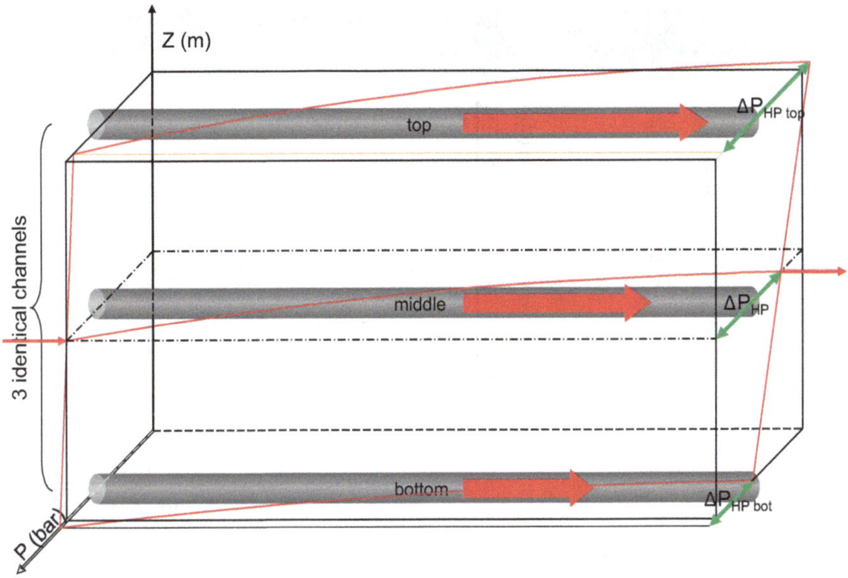

Fig. 2.37 Flow unbalance due to gravity

Bottom: $\Delta P_{\mathrm{HP}}\mathrm{Bot} = \Delta P_{\mathrm{HP}} - ((\rho \mathrm{HP}_{\mathrm{cold}} - \rho \mathrm{HP}_{\mathrm{warm}}).\ g.\ h)/2$

where h = height of the core
$\rho \mathrm{HP}_{\mathrm{cold}}$ = density of HP helium at the cold end temperature
$\rho \mathrm{HP}_{\mathrm{warm}}$ = density of HP helium at the warm end temperature

In the HP channels, the driving force at the bottom is then lower than the driving force at the top. Consequently, the top HP channel has more mass flow than the bottom channel. A similar calculation involving the LP channels shows that the top channel mass flow is lower than the bottom one.

As a result of these remarks, it happens that the top and bottom elementary heat exchangers are both unbalanced, but in an opposite way: the top one has more high-pressure flow than low pressure, and the bottom one has less high-pressure flow than low pressure.

Let us check the behaviour of these elementary heat exchangers and compare it to the elementary middle heat exchanger. The middle elementary balanced heat exchanger is modelled by two pipes; the diameter and length of them are consistent with the specified pressure drop. The top and bottom elementary heat exchangers are now calculated with the same channel conductance (same pipes as for the middle elementary heat exchanger); as a consequence, the mass flow, thus the unbalancing, is established according to the driving pressure in each channel (see Fig. 2.37).

For the calculations we will consider specified core pressure drops as 0.020 and 0.014 bar, respectively, for the HP and LP channels. The height of the core is 0.60 m. Figures 2.38, 2.39 and 2.40 show the calculation of respectively the middle, top and bottom elementary heat exchangers.

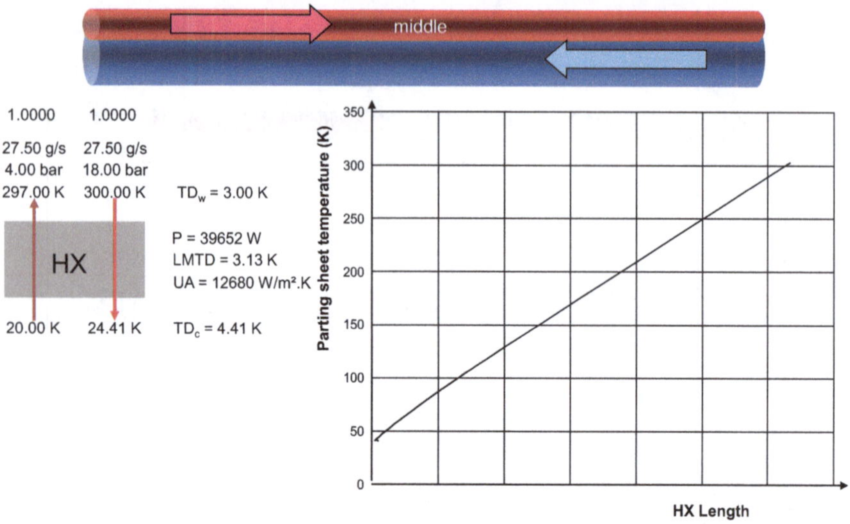

Fig. 2.38 The "middle" elementary heat exchanger

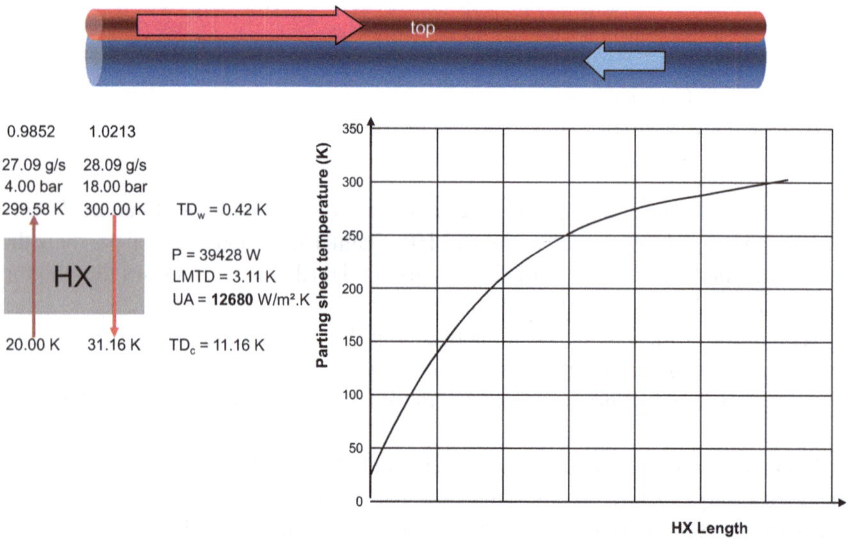

Fig. 2.39 The "top" elementary heat exchanger

Now, if one mixes the helium flows getting out of the top and bottom elementary heat exchangers, one can set an equivalent average heat exchanger having the same performances. Such "average" elementary heat exchanger UA is only 40% of that of the middle heat exchanger (see Fig. 2.41)!

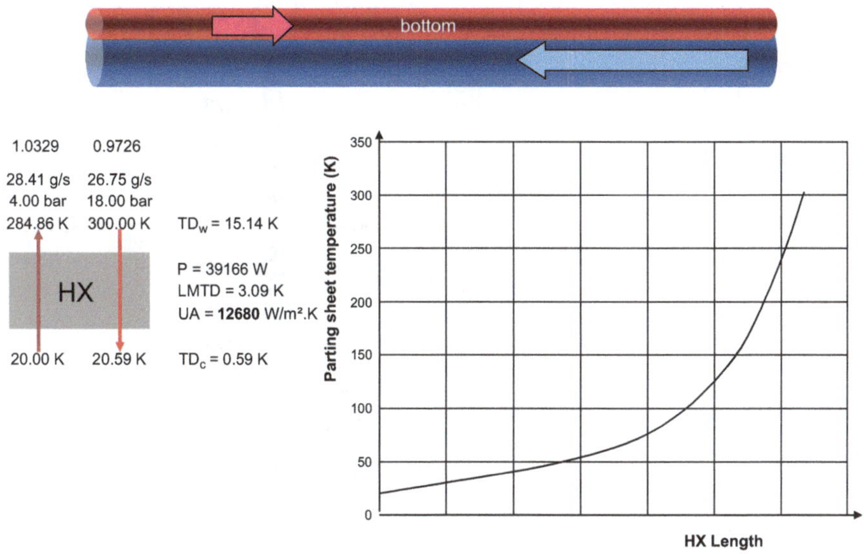

Fig. 2.40 The "bottom" elementary heat exchanger

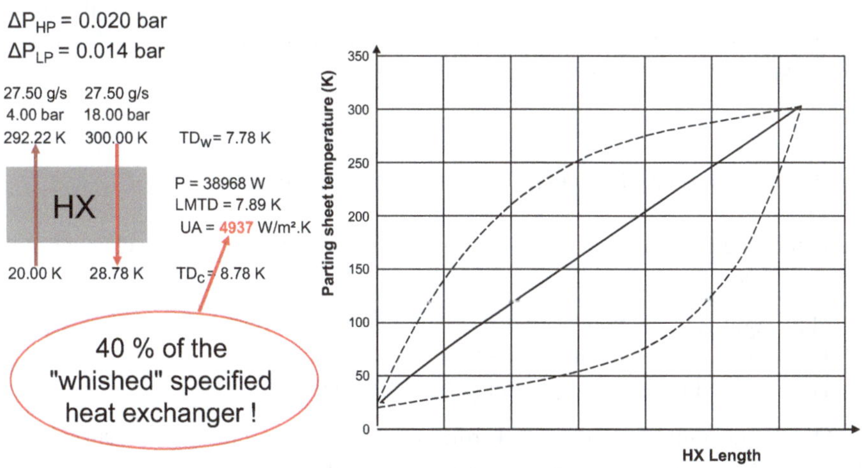

Fig. 2.41 The "average" elementary heat exchanger

If one models the whole heat exchanger by putting together all elementary heat exchangers, the UA of each changing linearly with their position in the core, the UA of the equivalent heat exchanger comes to only 66% of the "expected" specified heat exchanger (see Fig. 2.42)!

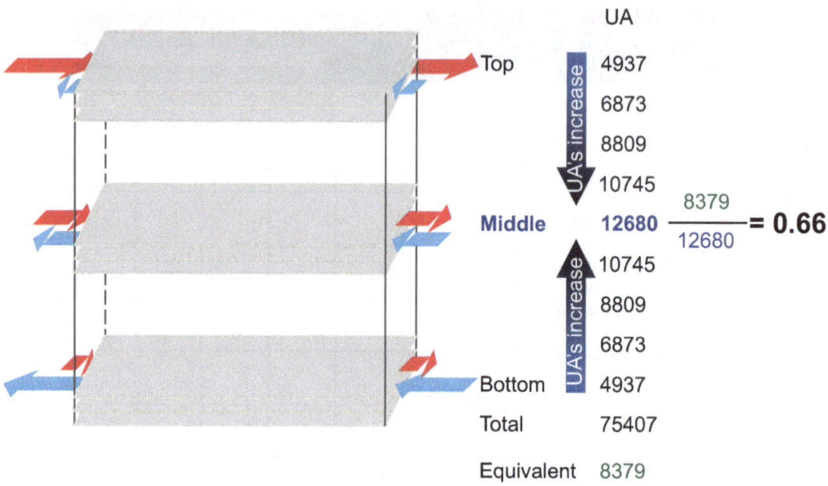

Fig. 2.42 The equivalent heat exchanger

2.4.4.2 Sensitivity of the Specified Pressure Drops in the Heat Exchange Zone

For the same core, operating in similar conditions, when the pressure drop specification values are higher (0.040 and 0.027 bar, respectively, for HP and LP), the UA of the equivalent heat exchanger comes to 86%.

2.4.4.3 A Correctly Designed Horizontal Heat Exchanger

To reduce the effect of gravity on the operation of a horizontal heat exchanger, it is obvious that:

- The height is to be "low",
- The temperature gradient is to be "not too high"
- The specified pressure drops should be "sufficient".

One could complain with such wordings as "low", "not too high" or "sufficient". For a given heat exchanger, these parameters will differ, and it is not possible to give absolute values. However, knowing the phenomenon, it is now possible by thorough thinking to put figures on that.

Let us consider another heat exchanger operating in the same conditions, but with some slightly different parameters:

- The height of the core is 0.234 m (instead of 0.60 m).
- There are two cores: the temperature gradient in each core is 140 K (instead of 280 K) (see Fig. 2.43).

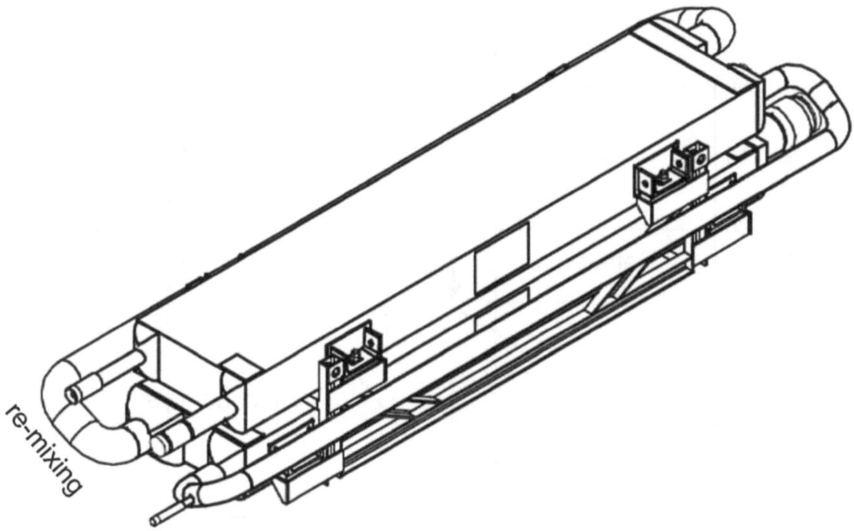

Fig. 2.43 A correctly designed horizontal heat exchanger

The specified pressure drops in the heat exchange zone are, respectively, in the warm core (300 to 140 K), 0.060 for HP and 0.140 bar for LP (instead of 0.020 and 0.014 bar), and in the cold core (140 to 20 K), 0.020 and 0.045 bar, respectively, for HP and LP, therefore, a total of 0.080 and 0.185 bar, respectively, for HP and LP (instead of 0.020 and 0.014 bar).

For the warm core, the equivalent heat exchanger UA is 99.9% of that of the specified heat exchanger and 98% for the cold core. Therefore, it is possible to operate correctly horizontal cryogenic heat exchangers, provided they are correctly designed.

Remark that, furthermore, there is a re-mixing of helium in the piping between the two cores.

2.4.4.4 Conclusions

When there are no constraints, it is more comfortable to operate a cryogenic heat exchanger in a vertical position. However, when there are some constraints (available room, cost, etc.), it is possible to operate a cryogenic heat exchanger in a horizontal position provided that:

- The operating temperatures are higher than 20 K
- The temperature gradient is reasonable (not too high)
- The height of the core of the heat exchanger is reasonable (not too high)
- The specified pressure drops are not too low

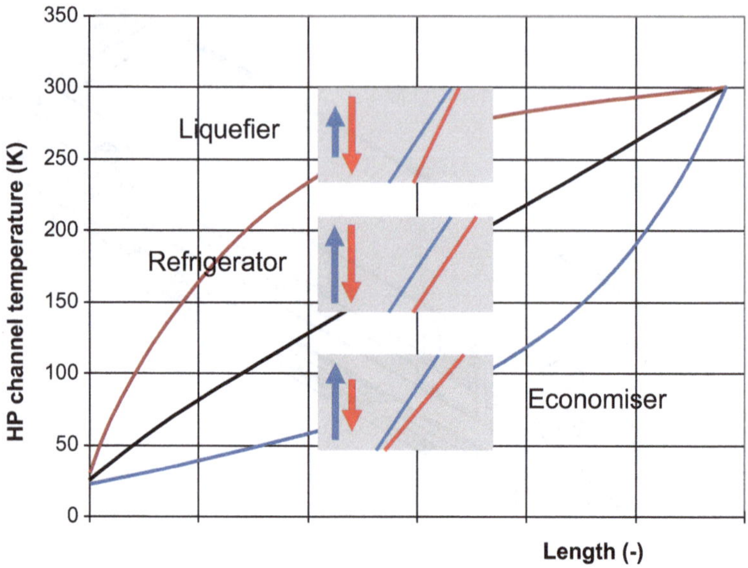

Fig. 2.44 Reminder: temperature profile along a heat exchanger

2.4.5 *Heat Exchanger Digest*

Do not forget that heat flows from warm to cold!

All the energy of the warm fluid is transferred to cold fluid (2.17):

$$\dot{m}_1 \times (h_1 - h_2) = \dot{m}_3 \times (h_4 - h_3) \tag{2.17}$$

Visualisation of the operation of a heat exchanger is made with diagrams (remember Fig. 2.8).

UA gives an idea about the size of the heat exchanger.

NTU increases when the temperature difference decreases and the temperature gradient increases. It gives an idea of the length of the heat exchanger.

"Reasonable" design rules:

- Minimum temperature difference: 1% of the absolute temperature and no less than 0.2 K.
- NTU: maximum 50.
- At temperatures lower than around 20 K, calculation must be performed in "slices".
- At temperatures lower than around 20 K, heat exchangers should be vertical.

The heat exchanger material has low incidence on its operation, except at temperatures lower than 10 K.

The temperature profile along the heat exchanger is related to the flow unbalance, as remembered in Fig. 2.44.

Chapter 3
Basic Thermodynamic Cycles

Abstract After making clear the difference between isothermal-duty and non-isothermal-duty regimes, the basic cycles, Joule Thomson, Brayton and Claude, are considered and simply calculated. Emphasis is put on cycle optimisation.

3.1 Introduction

One knows that refrigerators and liquefiers have similar structures that combine same components: compressor(s), heat exchanger(s) and expander(s). Such similarity has for consequence a possible confusion between liquefaction and refrigeration regimes.

Therefore, first, one must make this difference be a clear one before going farther.

Then, it is of interest to know the theoretical quantity of energy that is necessary to obtain refrigeration. What is the ideal (the minimum) amount of power to be spent at room temperature (that is to say, generally, the power taken on the electrical net by the cycle compression system) in order to absorb one unit of refrigeration power at temperatures lower than ambient? Various situations are considered.

Finally, the basic cycles (Joule Thomson and Brayton) are visited in detail.

Along the chapter, the results of simple calculations, performed with Excel© and REFPROP© (see Sect. 1.4.2.1), are displayed. The author advises the reader to try to perform these calculations by himself (or herself). This allows a better comprehension of the system. This becomes also possible elements of the "Cryo toolbox" that is described in Chap. 15.

G. Gistau Baguer, *Cryogenic Helium Refrigeration for Middle and Large Powers*, International Cryogenics Monograph Series,
https://doi.org/10.1007/978-3-030-51677-2_3

3.2 The Various Operating Regimes of a Refrigerator

When dealing with a refrigerator operating at liquid helium temperature or a liquefier, the pressure of the bath is generally around atmospheric pressure, except special machines operating at lower temperatures (see Sect. 4.3) where the bath pressure can be much lower.

If a cryostat is cooled from a dewar and helium is recovered at room temperature into a gas holder, the bath pressure is 1.0 bar, and the temperature is 4.2 K.

If liquid is produced in a refrigerator or a liquefier, its pressure is the result of:

- The pressure drops in all components that are circulated (heat exchangers, valves, pipes…) from the liquid bath up to the suction flange of the cycle compressor: generally, around 0.150 bar
- A small safety margin on the suction pressure of the compressor in order it is higher than the local atmospheric pressure to avoid possible air leak getting into the helium circuits: generally, around 0.05 bar

Consequently, the pressure on the liquid helium bath into a liquefier or a refrigerator is usually about 1.2 bar, corresponding to 4.41 K. This is the typical bath pressure that is used in the examples in this book.

3.2.1 The Isothermal-Duty Regime

If the power that is released by the user (a superconducting coil, a superconducting resonant cavity, etc.) is dissipated into a liquid bath boiling at constant pressure, the bath temperature that is only related to the bath pressure stays constant, whatever the power that is dumped into the liquid: the process is *isothermal*.

The only duty to be fulfilled is the re-condensation of the saturated vapour, i.e. transformation of the saturated vapour at the equilibrium temperature into liquid boiling at the same temperature. One must extract the amount of energy to condense saturated vapour (extract the latent heat): 19.33 J/g in Table 3.1.

Between 300 and 4.41 K (temperature of saturated vapour at 1.2 bar), there is, theoretically, "nothing to do" in a thermodynamic sense, but, as the surrounding world is at room temperature, a transition is needed between it and the liquid helium temperature. This temperature transition is taken care by the cold box with its heat exchangers (see Fig. 3.1).

Table 3.1 (REFPROP)

	Pressure	Temperature	Enthalpy
	P	T	h
	(bar)	(K)	(J/g)
Vapour	***1.20***	4.41	20.37
Liquid	1.20	4.41	1.04
			19.33

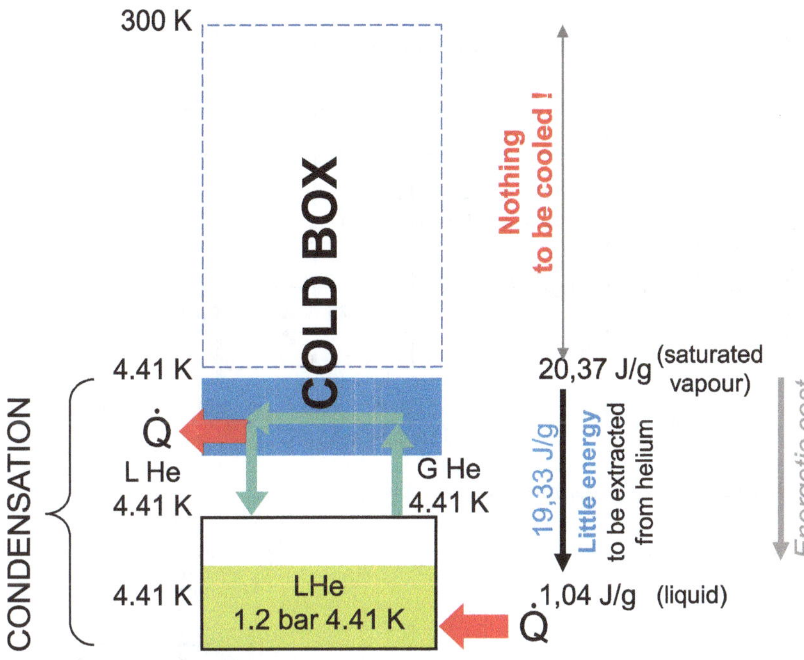

Fig. 3.1 Pure isothermal-duty refrigeration regime (REFPROP)

In order to calculate the ideal (minimum) amount of power to be spent at room temperature, one has to use the Carnot formula (see Sect. 17.3.2.1. for demonstration) that gives the ratio between the mechanical power to be spent at room temperature and the thermal refrigeration power that is to be absorbed at cold temperature:

$$\dot{W}/\dot{Q}_c = (Tw/T_c - 1) \tag{3.1}$$

Example: mechanical power to be spent at room temperature to absorb 1 Watt at the temperature of liquid helium boiling at 1.2 bar (4.41 K):

$$\dot{W}/\dot{Q}_c = (300/4.41 - 1) = 67.03$$

The mechanical power to be spent at ambient temperature is almost inversely proportional to the cold temperature: the lower the cold temperature, the higher the energetic cost is.

Figure 3.2 is a graphic representation of $\dot{W}/\dot{Q}_c = (Tw/T_c - 1)$. It shows the large variation of the energetic cost versus cold temperature. Such a remark will heavily influence the designer thinking when studying and optimising thermodynamic cycles.

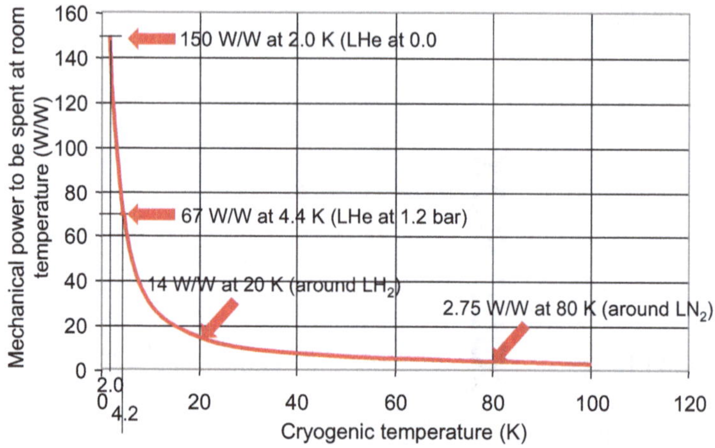

Fig. 3.2 Energetical cost of isothermal cryogenic power

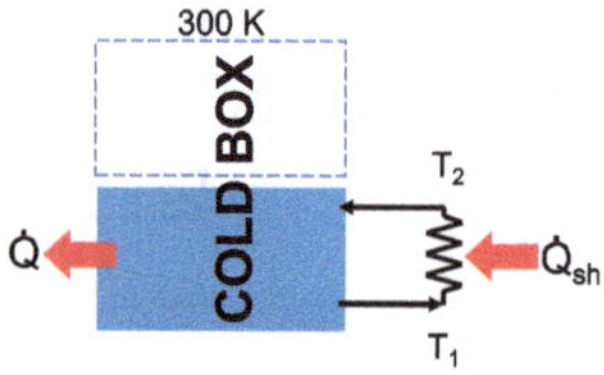

Fig. 3.3 Non-isothermal-duty refrigeration for thermal shields

The operation in pure isothermal-duty refrigeration is defined by a *given refrigeration power* that is absorbed at a *given temperature*. Example: a refrigerator absorbs 300 W at 4.5 K.

3.2.2 The Non-isothermal-Duty Regime

If heat is dissipated into some gas, the gas temperature varies along the process: the operation is not isothermal.

Such a situation happens in various cases:

- When a gas is circulated into a system where energy is dumped in.
  Examples: thermal shields or circulation in CICC (see Sect. 4.2.4.1). The gas exits the refrigerator at a temperature T_1 and returns to it at a higher temperature T_2 (see Fig. 3.3).
- When a gas is liquefied, it enters the system (liquefier) generally at room temperature and exits in a liquid form at the boiling temperature corresponding to a pressure slightly above the atmospheric pressure, generally 1.2 bar (see Fig. 3.4). In such a process, one has, first, to withdraw a large quantity of energy in order to cool helium from room temperature down to the condensation

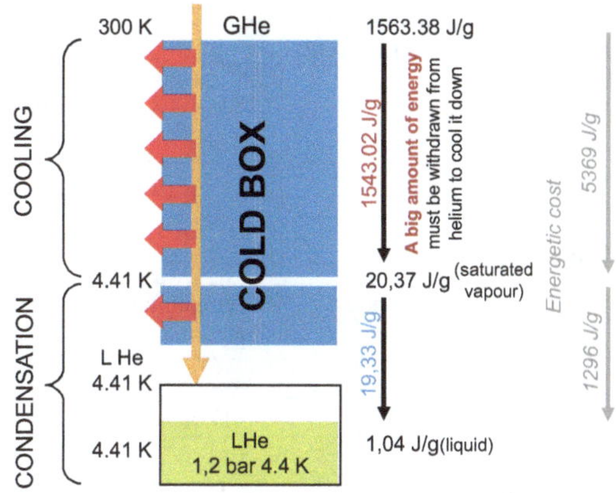

Fig. 3.4 Pure liquefaction regime (REFPROP)

Table 3.2 Liquefaction of helium (REFPROP)

Helium	Pressure	Temperature	Enthalpy
	P	T	h
	(bar)	(K)	(J/g)
Gas	1.20	300.00	1563.38
Saturated vapour	1.20	4.41	20.37
Sensible heat	1.20		1543.02
Liquid	1.20	4.41	1.04
Latent heat			19.33

temperature (to withdraw the *sensible* heat, 1543.02 J/g) and secondly to turn the saturated vapour into liquid helium at the same temperature (to withdraw the *latent (or vaporisation)* heat, 19.33 J/g).

In non-isothermal-duty refrigeration, thermal energy is withdrawn at an infinite number of temperature levels between warm and cold temperatures. As its energetic cost varies according to the temperature, it is no more possible to use the simple isothermal Carnot formula; one must use the exergy difference expression (minimum work to be brought to the system). (See Sect. 17.3.2.2.2 for demonstration.)

$$\dot{W}_{T0} = \dot{m} \times [(T_0 \times \Delta s) - \Delta h] \tag{3.2}$$

T_0 is the reference temperature (room temperature), and Δs and Δh are, respectively, the fluid specific entropy and enthalpy differences between the maxi and mini temperatures of the gas that is circulated into the thermal load.

Example 1 Mechanical power to be spent at room temperature to absorb 1 Watt on thermal shields circulated with helium entering at 18 bar, 70 K, and exiting at 17.5 bar, 80 K (Table 3.3).

Mass flow rate to be circulated to absorb 1 Watt:

Table 3.3 Non-isothermal duty refrigeration (REFPROP)

P	*T*	*h*	*s*
(bar)	(K)	(J/g)	(J/g.K)
18.00	70.00	371.53	14.42
17.50	80.00	423.93	15.18

Table 3.4 (REFPROP)

	Pressure	Temperature	Enthalpy	Entropy	
	P	T	h	s	
	(bar)	(K)	(J/g)	(J/g.K)	(W/W)
	1.20	300.00	1563.38	27.63	
Vapour	1.20	4.41	20.37	4.59	5369
Liquid	1.20	4.41	1.04	0.20	1296
			1562.35		6665

$$\dot{m} = 1/(423.93 - 371.53) = 0.0191 \;\; \text{g/s}$$

Mechanical power:

$$\dot{W}_{\text{T0}} = 0.0191 \times [(300 \times (15.18 - 14.42)) - (423.93 - 371.53)] = 3.35 \;\; \text{W}$$

Note that 3.35 W is same order of magnitude as 3.0 W to absorb 1 W isothermal at 75 K.

Operation in a non-isothermal-duty regime is expressed by the absorbed power between two temperatures: a refrigerator absorbs 1000 W between 60 K and 80 K. However, in order not to put a constraint on the helium flow, one can also say: 100 W at less than 80 K.

Example 2 Power to be spent at room temperature to liquefy 1 g/s of helium (approximately 30 L/h; remember Sect. 1.6). This means that helium is cooled from room temperature (300 K) to 4.41 K and then condensed at 4.41 K.

With (Table 3.4):

$$\dot{W}_0 = 1 \times [(300 \times (27.63 - 0.2)) - (1563.38 - 1.04)] = 6665 \;\; \text{W}$$

Remark Among the 6665 W, to cool helium from 300 K down to 4.4 K (extracting the 1543.02 W of sensible heat):

$$1 \times [(300 \times (27.63 - 4.59)) - (1563.38 - 20.37)] = 5369 \;\; \text{W}$$

are necessary,
and to condense the saturated vapour (extracting the 19.33 W of latent heat):

$$1 \times [(300 \times (4.59 - 40.20)) - (20.37 - 1.04)] = 1296 \ \text{W}$$

are necessary.

Operation in a pure liquefaction regime is expressed by the helium quantity that is liquefied within one unit of time, gram per second (g/s), even if the usual practical expression is litre per hour (L/h), a liquefier for 3 g/s or 90 L/h.

Remark In cycles where the coldest temperature is lower than 4.5 K (see Sect. 5.2), the helium flow that is processed by the cryogenic compressors is a non-isothermal load for the refrigerator.

3.2.3 Mixed-Duty Regimes

A superconducting magnet is fed through electrical current leads; a cavity is supplied by high frequency waves through couplers, the extremities of which are, respectively, at room and liquid helium temperatures (Fig. 3.5).

Along these devices, generally made out of good electrical conductor materials, and therefore good thermal conductors too, as copper, heat is transferred by thermal conduction from ambient to liquid helium temperature.

In order to fight against such a parasitic heat load, a small gaseous helium mass flow rate issued from liquid helium or saturated vapour is circulated along the current lead of a coil or the coupler of a resonant cavity. The heat exchange between helium and the device allows most of the heat to be transferred by convection from the device to helium that is exiting the system at approximately room temperature. The high thermal load that would have been brought by conduction along the device (47 W for 1 kA) is compensated by an equivalent lower thermal load for the re-liquefaction of the circulated helium. A thumb rule for conventional current leads is 0.05 g/s for 1 kA, which is roughly equivalent to 1 W.

In Fig. 3.6, one can see that the thermal load of the refrigerator is the sum of:

- An isothermal load at liquid helium temperature: $\dot{Q}_u$
- A non-isothermal load between T_1 and T_2 for cooling the thermal shields: $\dot{Q}_{sh}$
- A liquefaction load for the helium flow that cools the current lead

An operation in mixed-duty regime is expressed by:

- Isothermal-duty power
- Non-isothermal-duty power
- And flow to be re-liquefied

Example: a refrigerator absorbs 300 W at 4.5 K + 800 W at 80 K and liquefies 0.5 g/s of He.

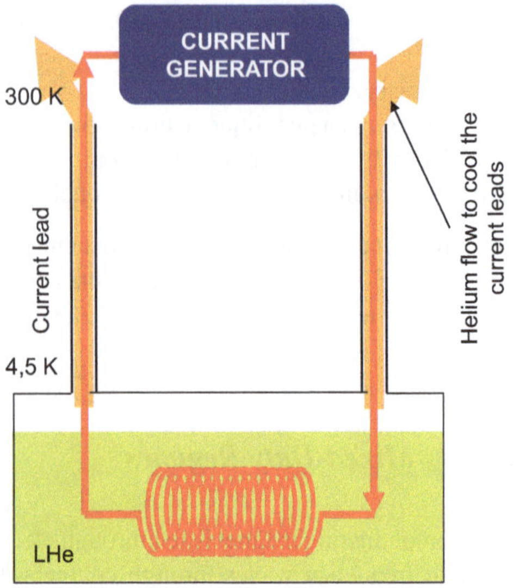

Fig. 3.5 Cooling of the current leads of a superconducting coil

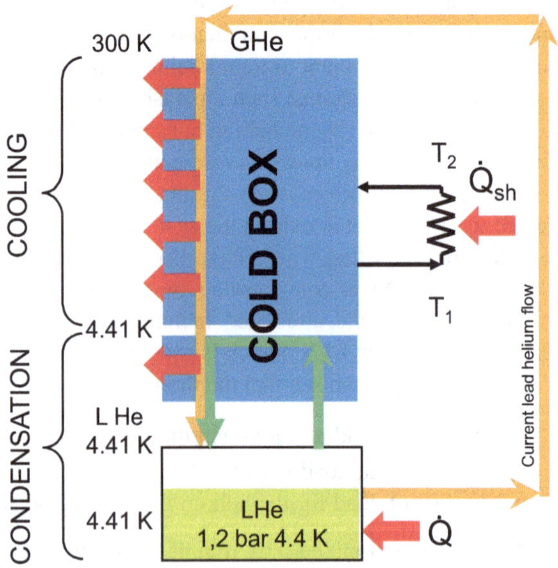

Fig. 3.6 A mixed-duty regime

3.2.4 *Easy Comparison of the Results of Cycle Calculations*

In order to be in a position to compare the results of the various cycle calculations in this chapter, three regimes will be used: 100 W for a pure liquefier, 1 g/s for a pure liquefier and 50 W plus 0.5 g/s for a mixed operation. In addition, the high pressure is set at 14.0 bar and the efficiency of any expander is set at 0.75.

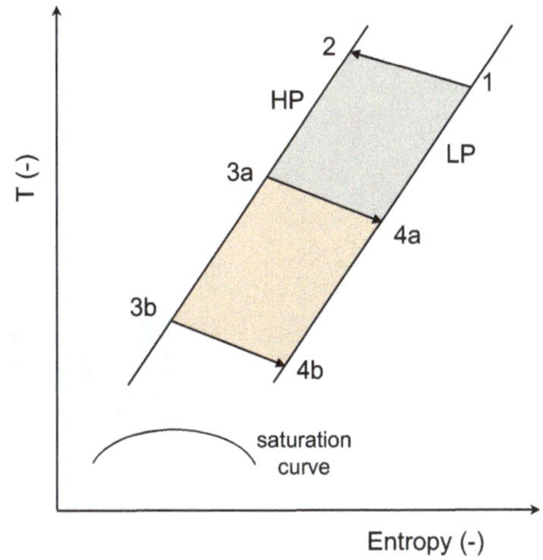

Fig. 3.7 The work that is related to a thermodynamic cycle

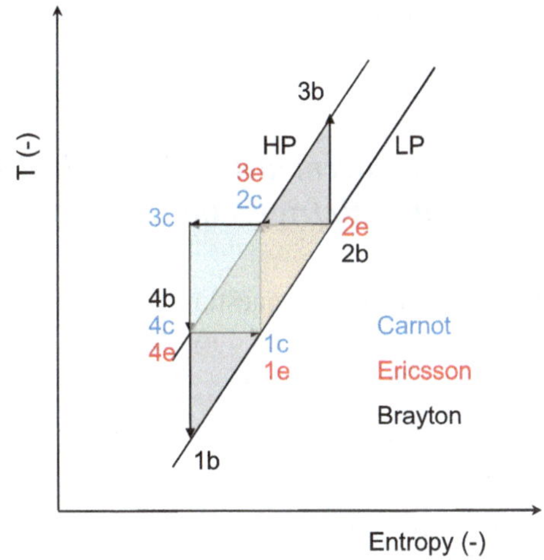

Fig. 3.8 Comparing various cycles on the T-s diagram

3.2.5 An Interest of the T-s Diagram

In Sect. 1.4.1.3, it has been said that the work that is related to a cycle is proportional to the area that is enclosed in the transformation lines of the cycle. Let us consider two cycles: 1, 2, 3, 4 and 1, 2, 3a, 4a displayed in Fig. 3.7 (the transformation lines 1–2, 3–4 and 3a–4a are typical; they do not represent an identified transformation).

The T-s diagram shows that the area of cycle 1, 2, 3a, 4a is smaller than the area of the cycle 1, 2, 3b, 4b because the latter is operating at a lower temperature; therefore cycle 1, 2, 3b, 4b has a lower COP than cycle 1, 2, 3a, 4a.

With such a tool, it is rather easy to show that the Carnot cycle is the most efficient. In Fig. 3.8, one can see that Carnot and Ericsson cycles have the same area

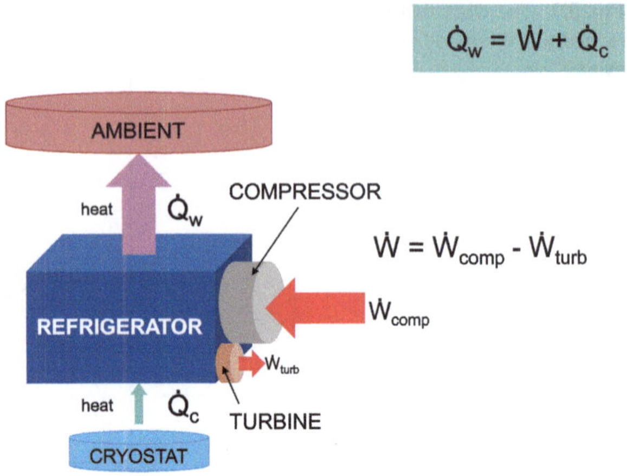

Fig. 3.9 Theoretical energy balance of an ideal refrigerator

in order to fulfil the same duty. They have same efficiency. However, the Brayton cycle has a larger area: it is less efficient.

3.2.6 A Thermodynamic Equivalence Between Liquefaction and Refrigeration Regimes

One often-recurring question concerns the comparison of pure refrigeration duty versus pure liquefaction duty. As equivalent duty means equivalent exergy, a simple exergy calculation provides the answer.

Remark Here and in all following instances, liquefaction is understood to be "constant level liquefaction".

The two values that have been calculated here above, 67.03 W to get 1 Watt at 4.41 K and 6665 W to liquefy 1 g/s of helium, can be compared. Indeed, if one spends 6665 W to liquefy 1 g/s of helium, one could either expect:

$$6665/67.03 = 99.43 \text{ W at } 4.2 \text{ K}$$

in a refrigerator mode.

This is a *pure thermodynamic* comparison. It refers to the theoretical algebraic sum of mechanical powers $\dot{W}$ that are spend and recovered: $\dot{W}_{\text{comp}}$ that is input into the compressor(s) and $\dot{W}_{\text{turb}}$ that is extracted by the turbine(s) (see Fig. 3.9).

$$\dot{W} = \dot{W}_{\text{comp}} - \dot{W}_{\text{turb}} \tag{3.3}$$

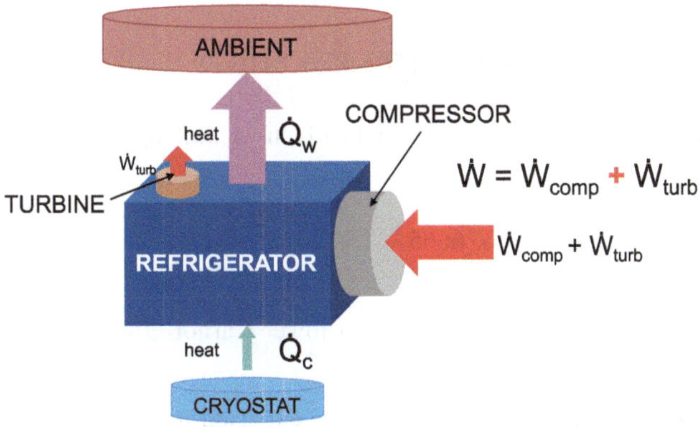

Fig. 3.10 Energy balance of an actual refrigerator

Remark It is the situation in the turbo Brayton machine that is described in Sect. 4. 6.1.1 where the energy that is extracted by the turbine is re-injected into the cycle compressor.

However, in most of the actual helium refrigeration plants, the energy that is extracted by the turbine(s) is not recovered because, for example, it is difficult to recover energy from a high rotational speed rotor. Furthermore, the quantity of energy is small. Therefore, instead of recovering such an energy, one must provide more energy to the system (see Fig. 3.10):

$$\dot{W} = \dot{W}_{\text{comp}} + \dot{W}_{\text{turb}} \tag{3.4}$$

The energy that could be recovered is:

- For an isothermal-duty regime: 1 W at room temperature for each Watt at cold temperature
- For a pure liquefaction regime: 1562.35 W for each gram that is liquefied

Consequence 1 The minimum mechanical power to be spent at room temperature in an *actual refrigerator* to get 1 W at 4.2 K is:

$$67.03 + 1 = 68.03 \text{ W}$$

The maximum Carnot yield of an *actual refrigerator where expander energy is not recovered* is:

$$67.03/68.03 = 98.5\%$$

Consequence 2 The minimum mechanical power to be spent at room temperature in an *actual liquefier* to liquefy 1 g/s at 4.2 K is:

$$6665 + 1562.35 = 8416 \text{ W}$$

Therefore, the maximum Carnot yield of an *actual liquefier where expander energy is not recovered* is:

$$6665/(6665 + 1562.35) = 81.0\%$$

Remark an actual liquefier has always a lower Carnot efficiency than an actual refrigerator.

Consequence 3 The equivalence that can be expected for *actual machines* that do not recover the energy that is extracted by the expanders is:

$$(6665 + 1562.35)/(67.03 + 1) = 120.9 \text{ W at } 4.4 \text{ K}$$

A simple rule of equivalence between productions in regimes of non-liquid nitrogen pre-cooled refrigeration or liquefaction of an actual machine where energy extracted by the expanders is not recovered: when the operating point of a refrigeration/liquefaction machine shifts from a refrigeration regime towards a liquefaction regime, a reduction of approximately 120 W at the liquid temperature is approximately compensated by the liquefaction of an extra 1 g/s and conversely.

Such situation is more accurately considered in Chap. 7 (Off-design).

Mind! This equivalence is to be used with caution and only for non-liquid nitrogen pre-cooled machines. It can be used to evaluate the performance of a machine the operating regime of which shifts from refrigeration to liquefaction and conversely, in a proportion not exceeding a *few tens of percent* of the equivalent total power. Farther, the equivalence is less and less accurate.

A liquefier has to extract a large quantity of energy from the gas to be liquefied: its expansion turbines are powerful. A refrigerator does not have to extract a big quantity of energy from the gas that is to be re-condensed: its turbines are not so powerful. For an equivalent exergy level, the heat exchangers of a refrigerator have a larger heat exchange area than the ones of a liquefier. These points are further discussed in Sect. 3.5.4.

3.2.7 The Efficiency of a Thermodynamic Cycle, the Carnot Equivalent Power

The efficiency of a thermodynamic cycle is the ratio of the power that is absorbed by the actual refrigerator (mainly compressors) compared to the power that would be necessary to feed an ideal machine operating on a Carnot cycle.

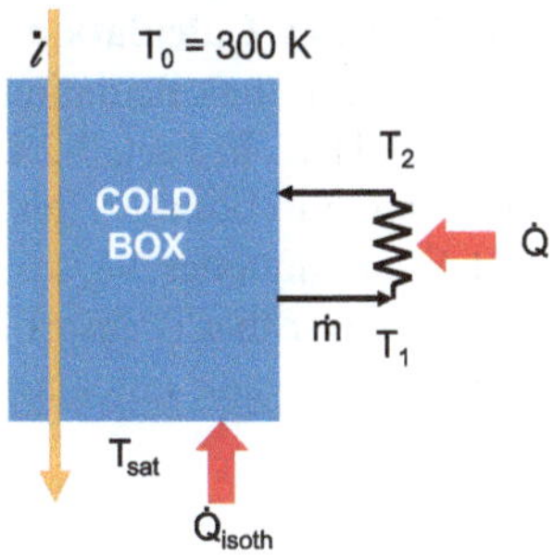

Fig. 3.11 Duties of a mixed refrigerator

$$\eta_{\text{Carnot}} = \frac{\text{Carnot power}}{\text{Absorbed power}}$$

For an *isothermal-duty* cycle operating at T_c temperature:

$$\eta_{\text{Carnot}} = \frac{(T_w - T_c)}{T_c \times \text{Absorbed power}}$$

with T_0 = room temperature = 300 K

For a *non-isothermal-duty* cycle operating between T_{c1} and T_{c2} temperatures:

$$\eta_{\text{Carnot}} = \frac{\dot{m} \times [(T_0 \times (s_2 - s_1)) - (h_2 - h_1)]}{\text{Absorbed power}}$$

with $\dot{m}$ = helium mass flow rate that is circulated between T_{c1} and T_{c2}

For a *mixed cycle* as shown in Fig. 3.11, it is interesting to calculate a Carnot equivalent power which is the total Carnot powers of all duties compared to the Carnot power at a reference temperature, generally 4.5 K.

$$\text{Carnot eq.power} = \frac{\Sigma\ \text{Isothermal powers} + \Sigma\ \text{Non isothermal powers}}{\text{Carnot power at 4.5 K}}$$

$$\begin{aligned}\text{Carnot eq.power} = &\left(\dot{Q}_{\text{isoth}} \times \frac{(T_0 - T_{\text{isoth}})}{T_{\text{isoth}}}\right)\\ &+ [\dot{m} \times (T_0 \times (s_2 - s_1)) - (h_2 - h_1)]\\ &+ \left[\dot{l} \times (T_0 \times (s_{\text{sat}} - s_0)) - (h_{\text{sat}} - h_0)\right]\end{aligned}$$

$$\eta_{\text{Carnot}} = \frac{\text{Carnot equivalent power}}{\text{Absorbed power}}$$

The calculation of the equivalent power and the Carnot efficiency is performed in Sect. 15.8.7.

An Efficiency Calculation

In Sect. 3.2.1, the calculation of Carnot power gave 67.05 W at room temperature for 1 W at 4.41 K. In Sect. 3.2.2, the calculation of Carnot power gave 6665 W at room temperature to liquefy 1 g/s. Therefore, an ideal refrigerator that absorbs 100 W at 4.41 K and liquefies 1 g/s would need: $100 \times 67.05 + 6665 = 13370$ W.

The power that is absorbed by the actual refrigerator is 125 kW.

The refrigerator efficiency is:

$$13370/125000 = 0.11$$

3.3 The Joule Thomson Cycle

The Joule Thomson cycle is a thermodynamic cycle named after James Prescott Joule (1818–1889), an English physicist, and William Thomson (1824–1907), knighted as Lord Kelvin in 1866, an Irish-Scottish mathematical physicist and engineer.

The Joule Thomson cycle is based on the simple isenthalpic expansion.

3.3.1 An Important Remark

Chapters 3 and 4 deal mainly with *simplified* calculations of various helium refrigeration cycles. Each calculation deals with a *one-off machine*. When systems that are liquid nitrogen pre-cooled or not are compared, there are different machines and *not a same actual machine* that is operated without or with liquid nitrogen pre-cooling.

It is only in Chap. 7 (Off-design) that the behaviour of an actual one-off machine with its compression station, heat exchangers and expanders is discussed.

3.3.2 "Re-discovering" the Joule Thomson Cycle

Now, one knows that, when helium is expanded, it cools, *provided expansion takes place at a temperature that is lower than the inversion one*. Therefore, one can rise a question: isn't it possible to liquefy helium by a simple expansion?

A simple calculation shows that expanding helium from 20 bar, 20 K, down to 1.2 bar, leads to a discharge temperature of 18.46 K, only (see Fig. 3.12, left).

Therefore, and unfortunately, it is not possible to liquefy helium by a simple Joule Thomson expansion, from 20 K, but as the gas temperature has a little bit decreased, why not use this lower temperature to cool the "next" gas to be

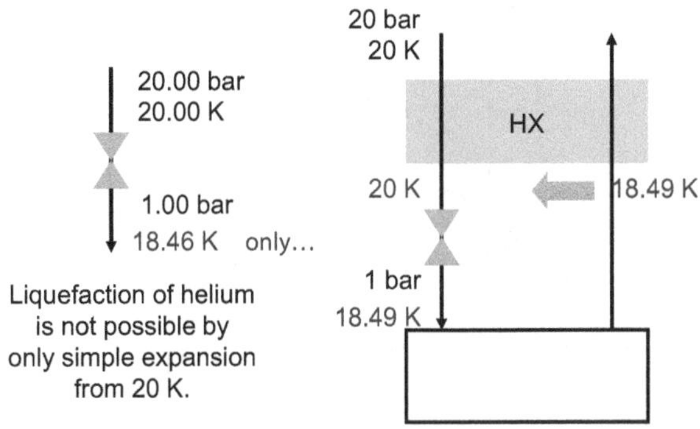

Fig. 3.12 Re-discovering the Joule Thomson cycle

expanded? This duty is fulfilled by a heat exchanger. The arrangement is shown in Fig. 3.12, right. After the "first" expansion down to 1.2 bar into the valve, helium is cooled down to 18.49 K.[1] This gas is sent into the heat exchanger where its enthalpy is recovered and compressed and is used to cool the high-pressure flow.

If one makes the assumptions that the heat exchanger is perfect (no temperature difference between high- and low-pressure fluids) and that there is no thermal inertia in the system, helium exiting the heat exchanger is at 18.49 K, upstream the expansion valve, at point 3. When it is expanded, its temperature reaches 16.72 K at point 4. This gas cools the "next" expanded gas at 14.61 K and so on. One understands that the temperature decreases progressively until it reaches the boiling temperature of helium. During this process, one can notice that the temperature difference that is generated by expansion is increasing while the temperature is decreasing. This can be seen on the helium T-s diagram in Fig. 3.12, right. The "latest" isenthalpic expansion produces a two-phase mixture at 4.41 K. One can also notice the surprising behaviour of helium that warms up prior to cool-down during the "latest" expansion.

3.3.2.1 Description and Representation of the Joule Thomson Cycle on the Temperature – Entropy (T-s) Diagram.

It is of interest to use a simplified T-s diagram, as introduced in Sect. 1.4.1.3, on which one plots the evolution of the cycle fluid in a qualitative way, the target being to "feel" how the cycle operates.

On the simplified T-s diagram in Fig. 3.13, the isobar curves are schematically shown as inclined straight lines, the condensation curve being assimilated to an arc of ellipse. The axes do not bear any scale.

- From point 4 to point 1, gas is compressed.

[1] Such a calculation can be performed in Sect. 15.8.2.

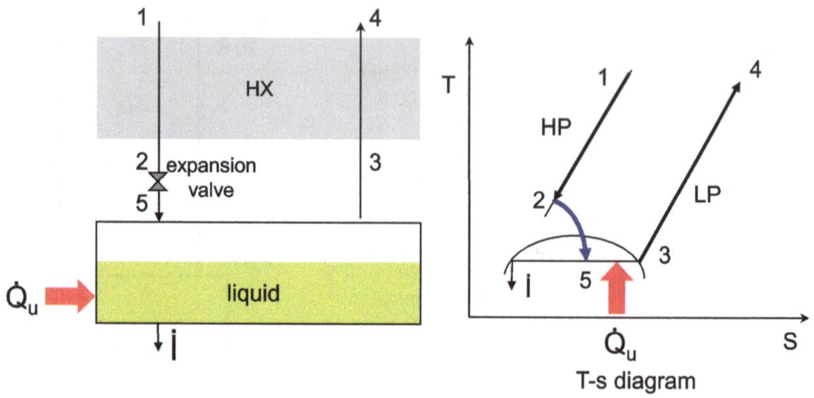

Fig. 3.13 The Joule Thomson cycle

- From point 1 to point 2, gas is cooled down into the heat exchanger, along the high-pressure isobar by counter current against the low-pressure gas.
- From points 2 to 5, the Joule Thomson expansion takes place, turning part of the gas into liquid. The liquid deposits into the phase separator. The gas that has been expanded but not liquefied is saturated vapour.
- The thermal power $\dot{Q}$ that is injected into the liquid vaporises *exactly all the liquid that is produced.*
- The part of gas that has not been liquefied and the vapour produced by the liquid that has been vaporised (that is to say all the cycle mass flow rate) are warmed along the low-pressure isobar, by flowing into the heat exchanger low-pressure circuit from point 3 to point 4 towards the cycle compressor.

The cycle gas runs through the *four characteristic stages* of a refrigeration cycle:

- Compression (4 to 1)
- Cooling by counter current heat exchange (1 to 2)
- Expansion (2 to 5)
- Warming by counter current heat exchange (3 to 4)

Remark When the liquid level in the phase separator of a liquefier/refrigerator stays constant (*provided no liquid is withdrawn*), it means that all liquid that is generated is vaporised: the machine is in a *thermal steady state.*

In Fig. 3.14, one could get an idea about the cool-down process, assuming that the heat exchanger is ideal and that materials have no heat capacity:

- Expansion from 14.00 bar, 15.0 K, to 1.2 bar cools helium to 13.80 K.
- Expansion from 14.00 bar, 13.80 K, to 1.2 bar cools helium to 10.76 K.

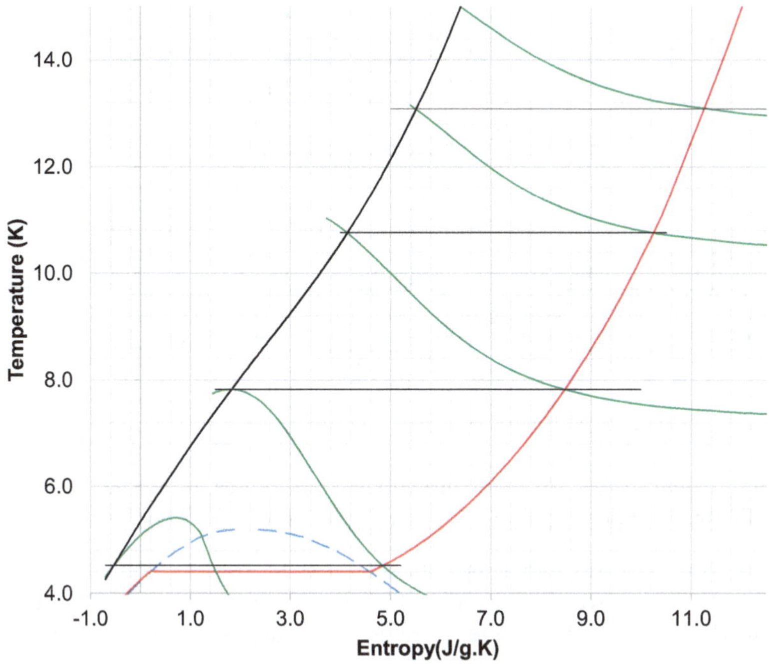

Fig. 3.14 Cooling down kinematics of a JT cycle (REFPROP)

- Expansion from 14.00 bar, 10.76 K, to 1.2 bar cools helium to 7.82 K.
- Expansion from 14.00 bar, 7.82 K, to 1.2 bar cools helium to 4.52 K.
- And, finally, expansion from 14.00 bar, 4.52 K, to 1.2 bar turns helium into two-phase at 4.41 K.

3.3.2.2 The Thermodynamic Balance of a Helium Joule Thomson Cycle

Let us consider the limits of a very simplified Joule Thomson refrigerator cycle as shown in Fig. 3.15 top. One can write the mass and thermal balance as:

$$\dot{m} \times h_1 - \dot{m} \times h_2 + \dot{Q}_u = 0 \tag{3.5}$$

by considering masses and energy that cross the red border (rounded corner rectangle),

or:

$$\dot{Q}_u = \dot{m} \times (h_2 - h_1) \tag{3.6}$$

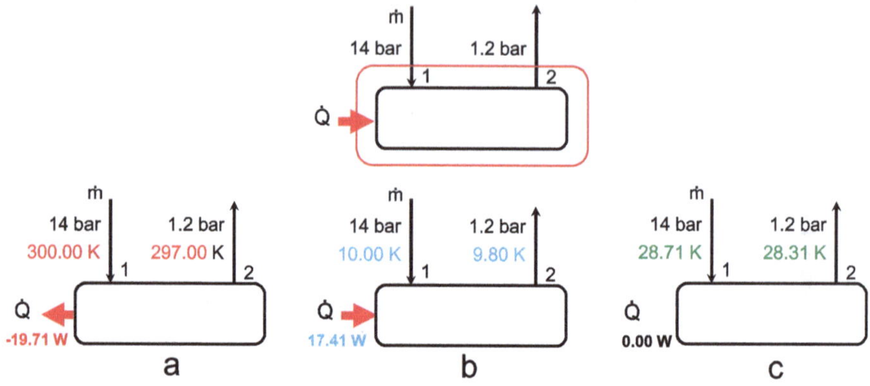

Fig. 3.15 Simplified Joule Thomson cycle

Let us calculate $\dot{m} \times (h_2 - h_1)$ at various temperatures, using the "1 % temperature difference" rule (Sect. 2.4.1.2.1) for 1 g/s (see Fig. 3.15 bottom).

Case a With a warm end temperature at 300 K, the calculation gives $\dot{Q}_u = -19.71\ \text{W}$. The power is negative, which means that heat is getting out of the system. This is obviously not a refrigerator that is supposed to absorb heat!

Case b With a warm end temperature at 10 K, the calculation gives $\dot{Q}_u = 17.41\ \text{W}$. The power is positive, so it is a machine that *absorbs* heat, therefore, a refrigerator.

What is the warm end temperature for which the cycle does not exchange energy with the rest of the universe? On C, one can see that it is 28.71 K (for 14.00 bar!).

Such a simple exercise shows that the warm end of a helium Joule Thomson cycle must be lower than 30 K.

Remark Such temperature is lower than the inversion temperature (~40 K) due to the heat exchanger warm end temperature difference.

3.3.2.3 Calculation of a Helium Joule Thomson Cycle

Let us isolate the system by plotting its "border" (rounded corner rectangle in Fig. 3.16) and write the heat exchange balance along the border. Remember that, by convention, the ingoing quantities are positive and the outgoing quantities are negative.

- At point 1 (in) $+\dot{m}_1 \times h_1 + \dot{l} \times h_1$
- At point 6 (out) $-\dot{l} \times h_L$
- At point 7 (in) $+\dot{Q}_u$
- At point 4 (out) $-\dot{m}_1 \times h_4$

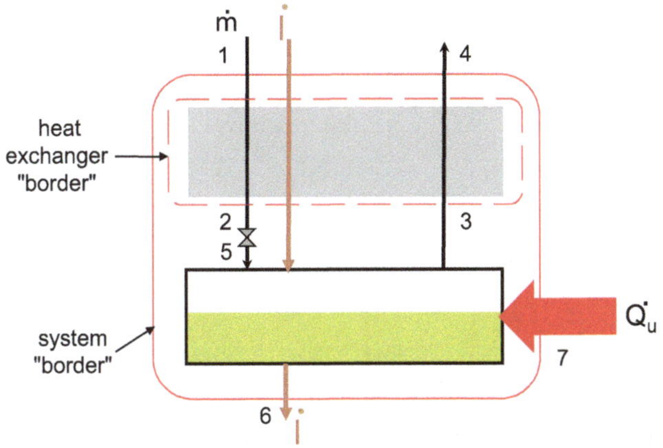

Fig. 3.16 Flow circulation through the Joule Thomson cycle

And, as *the liquid level stays constant*, the system is in a *steady state*, and the sum equals 0.

$$\dot{m} \times h_1 + \dot{l} \times h_1 - \dot{l} \times h_L + \dot{Q}_u - \dot{m} \times h_4 = 0 \tag{3.7}$$

This allows to calculate ṁ:

$$\dot{m} = \left(\dot{l} \times (h_L - h_1) - \dot{Q}_u\right)/(h_1 - h_4) \tag{3.8}$$

This equation can also be written:

$$\dot{Q}_u + \dot{l} \times (h_1 - h_L) = \dot{m}_1 \times (h_4 - h_1) \tag{3.9}$$

This equation can be interpreted as:

- The left member represents the power that enters the system: the useful power $(\dot{Q}_u)+$ the power to be withdrawn from the helium mass flow rate that is to be liquefied in order to cool and condense it $\left(\dot{l} \times (h_1 - h_L)\right)$.
- The right member represents the power that gets out of the system. One can see that all the power that enters the system gets out by the warm end of the heat exchanger (points 1 and 4).

The Joule Thomson cycle transports all the entering energy, *whatever the level it enters into the system,* at the temperature level of the warm end of the heat exchanger.

By using, as an example, the REFPROP® software (see Sect. 2.4.2.) that computes the thermodynamic properties of helium, this equation can easily be programmed into an Excel® spread sheet as shown in Fig. 3.17.

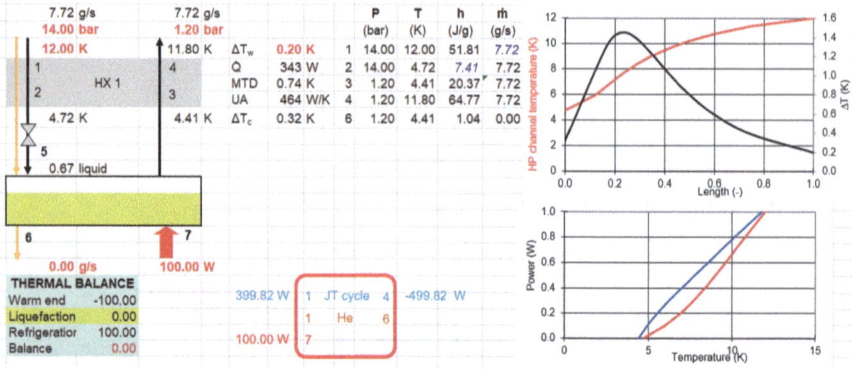

Fig. 3.17 Joule Thomson helium cycle operating in pure refrigerator. In bottom, centre, circulation of energy

3.3.2.4 Various Operating Conditions of a Joule Thomson Cycle

Depending on the way the heat loads are injected into the cycle, one can sort various operations of the Joule Thomson cycle.

3.3.2.4.1 Pure Refrigerator Operation

Problem How much cycle helium flow rate is necessary in order to absorb 100 W at 4.4 K?

Process input

$$\dot{Q}_{\mathrm{u}} = 100 \mathrm{~W}$$

Design parameters (for which the technological limitations of the components must be considered):

- $P_1 = 14.00$ bar (high pressure depends on the cycle compressor).
- $T_1 = 12.00$ K (remind: T_1 must be <30 K!).
- $P_3 = 1.2$ bar (because a pressure drop of about 0.150 bar sits in the low-pressure circuit and the suction pressure of the cycle compressor must not be lower than atmospheric pressure).
- $T_3 =$ saturation temperature for pressure P_3: 4.41 K, according to REFPROP.
- $P_4 = 1.2$ bar (for simplification, pressure drop $= 0$ between 3 and 4).
- $T_1 - T_4 = 0.20$ K (temperature difference, a characteristic of the heat exchanger Sect. 2.4.1.2.1)

  In Fig. 3.17, the calculation returns the mass flow rate at point 1: $\dot{m} = 7.72 \;\; \mathrm{g/s}$.

In a second step, the heat exchanger is calculated as in Sect. 2.4.1.4, which leads to $T_2 = 4.72$ K. Knowing h_2 and P_3, one can calculate the liquid concentration after expansion: 0.67.

At the bottom of Fig. 3.17 where the border of the system is shown as a rounded-corner rectangle, one can see the energy circulation.

Important remark In $\dot{Q}_u + \dot{l} \times (h_1 - h_L) = \dot{m}_1 \times (h_4 - h_1)$ Eq. 3.9, $\dot{Q}_u$ stands for the power to be absorbed by the cycle. When it has been introduced into the equation, it meant the useful power. However, if the system is not perfectly insulated from the rest of the universe, which is the general situation because the thermal insulation of a system cannot be perfect, any other heat input must be evacuated from the cycle as well as the useful power, whatever the place it enters the system.

Therefore, the real calculation must take care of the heat leaks and the equation to be used is:

$$\dot{Q}_{\text{useful}} + \dot{Q}_{\text{leaks}} = \dot{m}_1 \times (h_4 - h_1) \tag{3.10}$$

Any heat getting into the cycle *whatever it is either useful or parasitic* is to be extracted by the cycle.

3.3.2.4.2 Pure Liquefier Operating at Constant Level

Problem How much cycle helium flow rate is necessary in order to liquefy 1.0 g/s at constant level?

Process input

$$\dot{l} = 1.0 \text{ g/s}$$

Design parameters:

- $P_1 = 14.00$ bar,
- $T_1 = 12.00$ K
- $P_3 = 1.20$ bar
- $T_3 = 4.41$ K,
- $P_4 = 1.20$ bar
- $T_1 - T_4 = 0.20$ K

Calculation is performed in Fig. 3.18.

Remark There are two ways to deal with the mass flow to be liquefied. It can be supplied at the low- or the high-pressure side of the compressor. In all the following examples, helium gas to be liquefied is *supplied at the high-pressure* side. Therefore, the compressor flow rate is calculated at point 4, here 3.92 g/s.

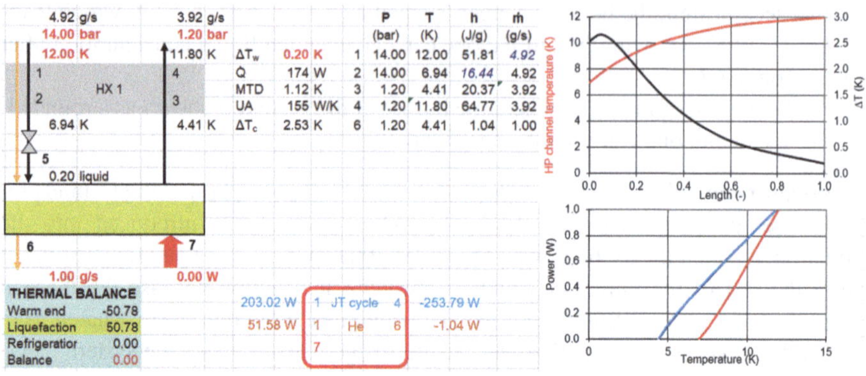

Fig. 3.18 Joule Thomson helium cycle operating in pure liquefier. In bottom, centre, circulation of energy

3.3.2.4.3 Pure Liquefier Operating at Rising Level

The calculation that is performed in Sect. 3.3.2.4.2 follows the assumption that the liquid helium level is constant, which means that any liquid quantity that is produced is permanently withdrawn. (This is what happens during a mixed operated refrigerator because the helium that is re-liquefied is permanently withdrawn to feed the current leads.)

However, in a liquefaction centre, during night, when nobody withdraws liquid helium, the liquid level rises in the stationary storage. Each quantity of liquid that is deposited into the reservoir displaces an equal volume of saturated vapour (Fig. 3.19) that circulates through the low-pressure circuits of the cold box, helping the pre-cooling of the high-pressure helium upstream the Joule Thomson valve. Consequently, the concentration of liquid at the discharge side of the JT valve is higher: the liquefaction rate is apparently "increased".

In order to perform the calculation in such operating conditions, one has to make the system a closed one with at constant mass balance to be in a steady state. For that, one considers that a flow of saturated vapour $\dot{v}$ that is displaced by the liquid phase is added at point 8.

Balance (Fig. 3.19):

$$\dot{m}\,x\,(h_1 - h_4) + \dot{l}\,x\,(h_1 - h_l) + \dot{v}\,x\,(h_v - h_4) = 0 \tag{3.11}$$

One volume of liquid displaces the same volume of saturated vapour. Therefore, when a liquid mass flow rate $\dot{l}$ is stored in the dewar, a mass flow rate: $\dot{v} = x\,\rho_v / \rho_l$ of saturated vapour is displaced from the dewar to the LP circuit.

with:

ρ_l = the liquid density,

ρ_v = the saturated vapour density,

this allows to calculate $\dot{m}$:

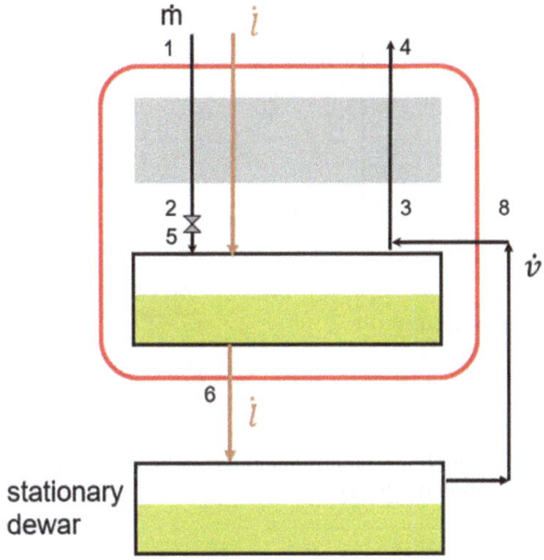

Fig. 3.19 Balance rising level

$$\dot{m} = \dot{l}\,x\left(h_l - h_4 - \left(\frac{\rho_v}{\rho_l}\right)\right)x(h_v - h_4)/(h_1 - h_4) \quad (3.12)$$

Problem Which quantity of liquid helium can be deposited into the dewar when using the same cycle mass flow rate as a 1.0 g/s liquefier operating at constant level?

Design parameters:

- $P_1 = 14.00$ bar,
- $T_1 = 12.00$ K
- $P_3 = 1.20$ bar
- $T_3 = 4.41$ K,
- $P_4 = 1.20$ bar
- $T_1 - T_4 = 0.20$ K

Calculation is performed in Fig. 3.20.

When processing the same mass flow rate into the compressor (3.98 g/s), 1.13 g/s of liquid helium can be deposited into the dewar. Note, also, that the temperature upstream the JT valve is decreased from 6.94 K to 6.84 K, due to the heat exchanger larger unbalance and the liquid concentration after expansion has increased from 0.20 to 0.23.

3.3.2.4.4 Mixed Duty Operation

Problem How much helium flow rate is necessary in order to simultaneously absorb 50 W at 4.4 K and, simultaneously, liquefy 0.5 g/s at constant level?

Remark These values, 50 W at 4.4 K + 0.5 g/s, will also be used for most of the other cycle calculations that are performed here under, in order to allow making easy comparisons.

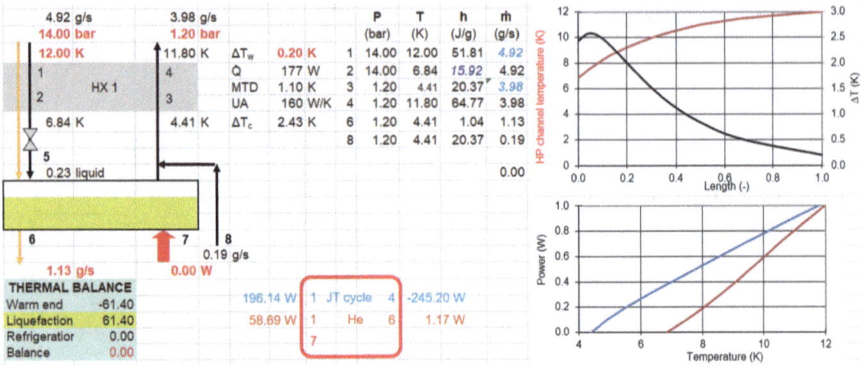

Fig. 3.20 Joule Thomson helium cycle operating in pure liquefier at rising level. In bottom, centre, circulation of energy

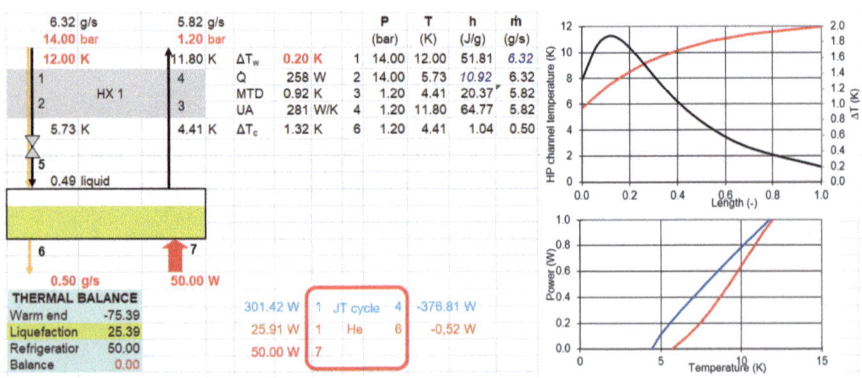

Fig. 3.21 Joule Thomson helium mixed duty cycle calculation

Process inputs

- $\dot{Q}_u = 50\ \text{W}$
- $\dot{l} = 0.5\ \text{g/s}$

Design parameters:

- $P_1 = 14.00$ bar,
- $T_1 = 12.00$ K
- $P_3 = 1.20$ bar
- $T_3 = 4.41$ K,
- $P_4 = 1.20$ bar
- $T_1 - T_4 = 0.20$ K

Calculation is performed in Fig. 3.21

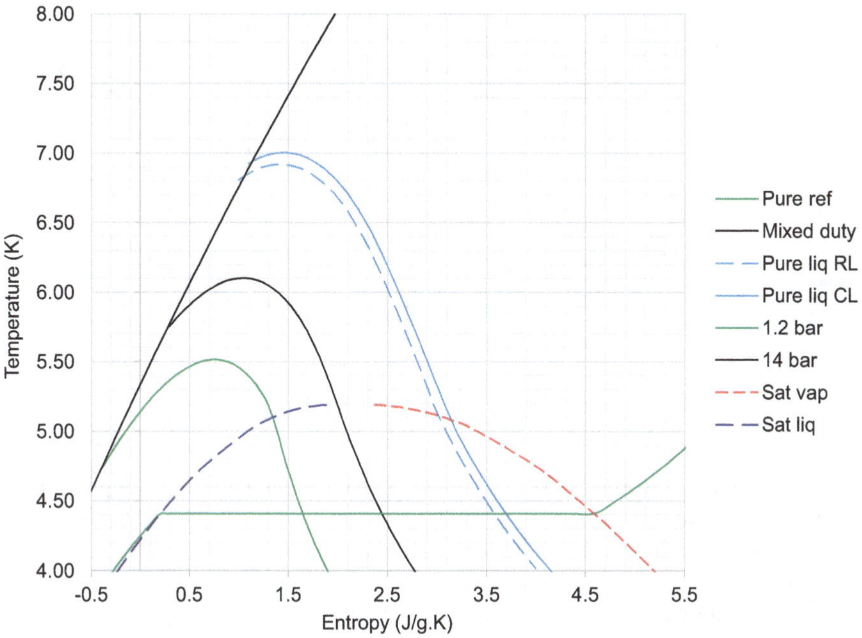

Fig. 3.22 Visualisation of the JT expansions in the various regimes from 14.0 bar to 1.2 bar (REFPROP)

The *design parameters* (high pressure, warm end temperature, temperature difference at the warm end) are identical as for Sect. 3.3.2.4.1 (Fig. 3.21).

One can notice that the temperature upstream the JT valve has changed: 4.72 K for pure refrigerator, 5.73 K for mixed-duty regime and 6.94 K for pure liquefier. Such temperature increases from pure refrigerator to pure liquefier regime because the LP flow decreases and, consequently, the temperature upstream the JT increases and the liquid concentration after expansion decreases too (Fig. 3.22).

3.3.2.4.5 Pure "Economiser" Operation

The economiser cycle (see Fig. 3.23) is kind of a reciprocal liquefier cycle: instead of producing some liquid, the economiser cycle vaporises liquid helium that is supplied at the cold end and then warms up the corresponding vapour up to room temperature in order to recover the enthalpy.

Interest of the Economiser Cycle

If one simply vaporises 2.0 g/s of liquid helium, the power that can be absorbed is only around 40 W (latent heat). Here, the economiser cycle recovers into the heat exchanger all the sensible heat of helium between saturated vapour and room temperature. Such sensible heat (enthalpy) is used to cool the HP cycle flow upstream the Joule Thomson expansion. As a result, for the example above, instead

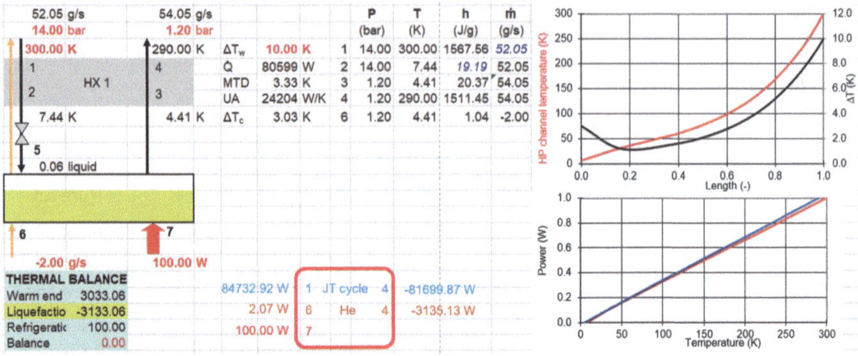

Fig. 3.23 Joule Thomson "economiser" cycle calculation

of 40 W, one can absorb 100 W. Of course, a heat exchanger and a compressor are necessary. But an economiser is a much lower investment than a full refrigerator (no expansion machines, very simple structure).

Such a solution can be of interest in case one needs a rather high power for a limited time, for example, the test of a component. Of course, adequate storages, to provide liquid and recover gas, are necessary.

See, in Sect. 12.7.3, a special example of a system that operated with quasi-economisers at FERMILAB.

See, in Chap. 7, where it is said that such operating conditions might happen in some circumstances, in the actual operation on any refrigerator.

3.3.2.5 The First Drop of Liquid

On the helium T-s diagram, the first drop of liquid appears at the right end of the line that represents the two-phase mixture at the bath pressure (see Fig. 3.24) at 4.4 K, on the 1 20 bar isobar. If one "circulates" along the isenthalp that crosses this point, "climbing up" the pressures, one can see that the highest temperature that is reached is approximately 7.7 K.

This means that a Joule Thomson expansion down to 1.20 bar can only produce liquid if the temperature upstream the valve is *lower than 7.7 K, whatever the upstream pressure is.*

3.3.2.6 Joule Thomson Cycle Analysis: Incidence of Some Parameters

When the former cycles have been calculated, only one parameter that is related to helium properties is imposed: the liquid bath pressure and, consequently, the temperature, because one wants the coldest possible liquid.

However, various values are possible for other design parameters like cycle high pressure, temperature at the warm end of the heat exchanger and temperature difference at the warm end of the heat exchanger. The consequence of the

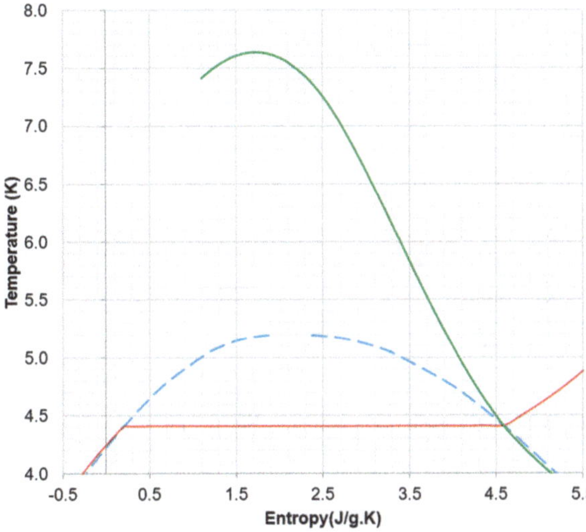

Fig. 3.24 Apparition of the first drop of liquid (REFPROP)

combination of these parameters has an incidence on the cycle mass flow rate and, consequently, on the heat exchanger size.

Let us calculate the incidence of the "free" parameters on the cycle mass flow rate in order to cope with the same thermal duty. To do it, only one parameter at a time is modified, keeping the two others constant.

Comparisons are made with the mixed duty reference cycle that has been formerly calculated in Sect. 3.3.2.4.4 and is reminded here in Fig. 3.25.

3.3.2.6.1 Changing the Heat Exchanger Warm End Temperature

Instead of 12 K, let us decrease the heat exchanger warm end temperature down to 10 K (see Fig. 3.26).

What are the differences?

- The liquefaction thermal load is lower, 18.08 instead of 25.39 W, because the mass flow rate to be liquefied is cooled from a lower temperature.
- The temperature upstream the Joule Thomson valve is lower, 4.69 instead of 5.73 K, because the temperature difference at the warm end of the heat exchanger is smaller.
- The liquid fraction is higher, 0.68 instead of 0.49, because the temperature upstream the JT valve is lower.
- The compressor mass flow rate is lower: 4.07 instead of 5.82 g/s.
- The heat exchanger power is lower, 137 instead of 258 W, because the mass flow and the warm end temperature differences are lower.
- The MTD is lower: 0.69 instead of 0.92 K.

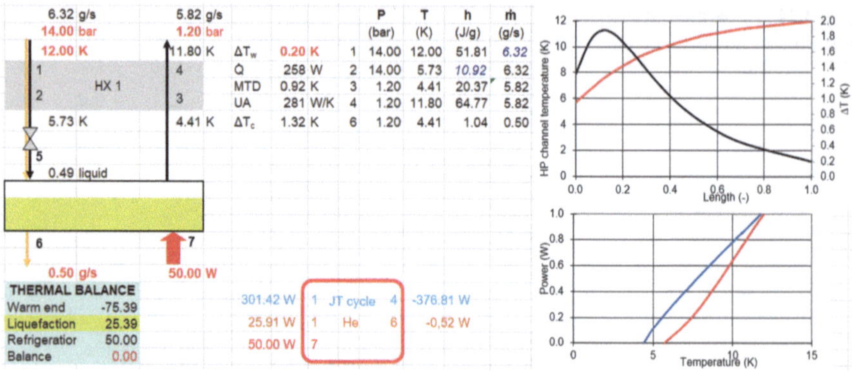

Fig. 3.25 Reference cycle (same as Fig. 3.21)

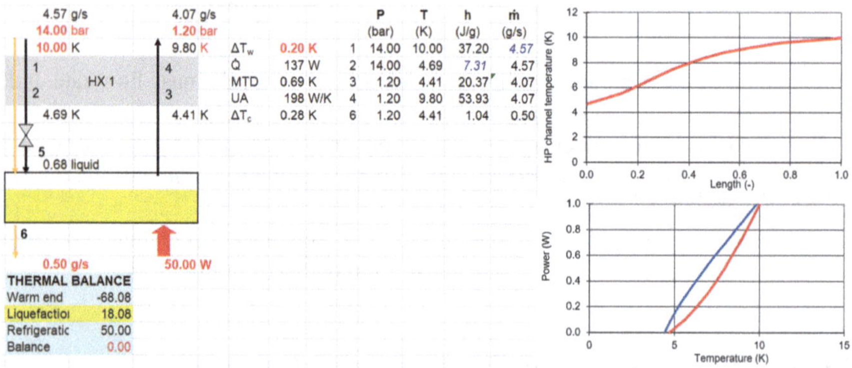

Fig. 3.26 Lower warm-end temperature

Other calculations, performed with either higher or lower warm end temperature, show that the cycle mass flow rate respectively increases or decreases.

Therefore, when the heat exchanger warm end temperature increases, the cycle mass flow rate increases. This evolution is shown on Fig. 3.27 left.

3.3.2.6.2 Changing the Heat Exchanger Warm End Temperature Difference

Instead of 0.2 K, let us increase the heat exchanger warm end temperature difference up to 0.4 K (see Fig. 3.28).

One can make same kind of observations on change of parameter values as in Sect. 3.3.2.6.1. Here, the cycle compressor mass flow rate has increased from 5.82 to 6.34 g/s.

Another calculation, performed with a smaller warm end temperature difference, shows that the cycle mass flow rate decreases.

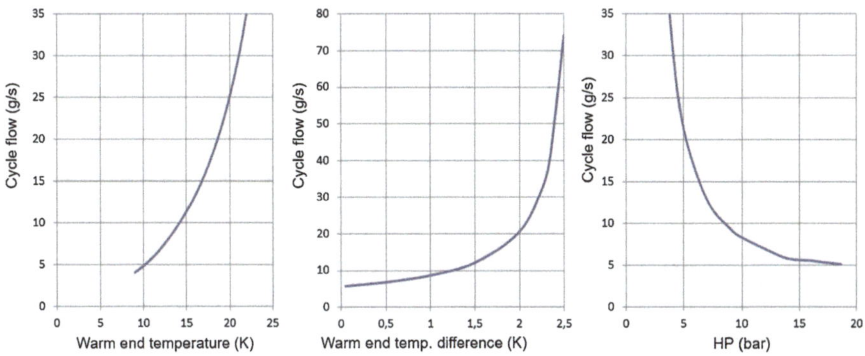

Fig. 3.27 Incidence of various parameters on the cycle mass flow rate

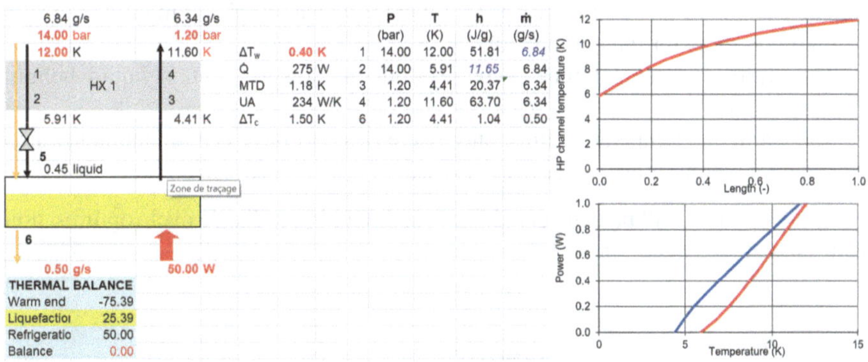

			P (bar)	T (K)	h (J/g)	ṁ (g/s)
ΔT_w	0.40 K	1	14.00	12.00	51.81	6.84
$\dot{Q}$	275 W	2	14.00	5.91	11.65	6.84
MTD	1.18 K	3	1.20	4.41	20.37	6.34
UA	234 W/K	4	1.20	11.60	63.70	6.34
ΔT_c	1.50 K	6	1.20	4.41	1.04	0.50

Fig. 3.28 Larger warm end temperature difference

Therefore, when the warm end temperature difference increases, the cycle mass flow rate increases. This evolution is shown on Fig. 3.27 centre.

3.3.2.6.3 Changing the Cycle High Pressure

Instead of 14 bar, let us decrease the cycle high pressure down to 12 bar (see Fig. 3.29).

One can make same kind of observations on changes happening for parameter values as in Sect. 3.3.2.6.1. Here, the cycle compressor mass flow rate has increased from 5.82 to 6.62 g/s.

Other calculations, performed with higher or lower high pressure, show that the cycle mass flow rate respectively decreases or increases.

Therefore, when the warm end temperature difference increases, the cycle mass flow rate increases. This evolution is shown on Fig. 3.27 right.

The Joule Thomson cycle mass flow rate decreases when:

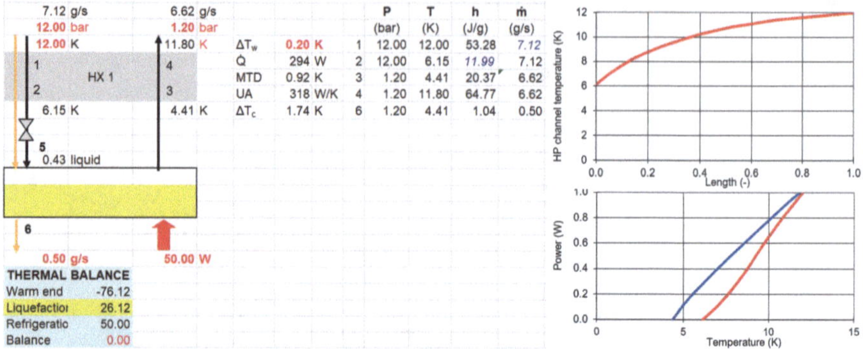

Fig. 3.29 Lower high pressure

- The cycle high pressure increases (but there is a technological limitation towards high pressures for the compressor).
- The warm end temperature decreases (but the lowest limit is liquid helium temperature)
- The warm end temperature difference decreases (but there is a limitation towards small temperature difference).

These remarks will be considered, later, at the time the Claude cycle optimisation is discussed (see Sect. 3.5.3.1).

3.3.3 The Double JT Expansion

For a given set of conditions at the warm end of the JT heat exchanger, the power is limited by the warm end temperature difference. In Fig. 3.30, the minimum warm end temperature difference is 1.66 K because the designer does not want to have a temperature difference lower than 0.2 K (remember the "reasonable" value, Sect. 2. 4.1.2.1) at the heat exchanger cold end. One could notice that, even if this "reasonable" rule is not implemented, the warm end temperature difference cannot be reduced significantly whatever the size of the heat exchanger would be. As a consequence, the cycle compressor mass flow rate is 8.87 g/s. If one selects two heat exchangers HX 11 and HX 12 with another expansion valve between them (see Fig. 3.31), it is possible to decrease the warm end temperature difference of HX 1 down to 0.60 K, keeping all temperature differences higher or equal to the reasonable value: 0.20 K. The reduction of the warm end temperature difference makes possible to absorb 155 W instead of 100 W *with the same cycle compressor mass flow rate* (8.87 g/s) at same warm end temperature: 8.0 K.

One should notice that the total UA of HX 11 and HX 12 405 W/K is higher than that of HX 1 130 W/K.

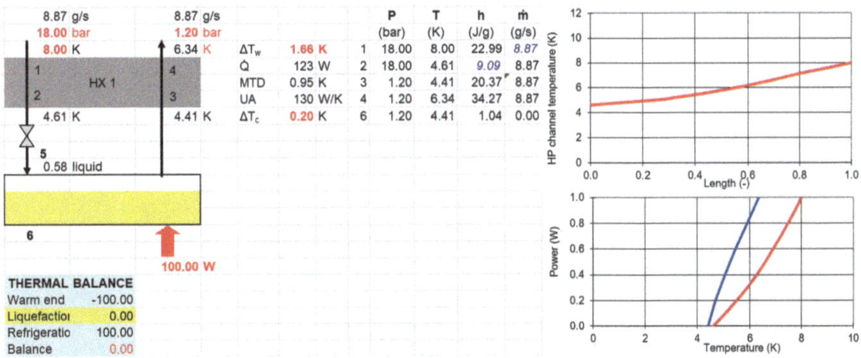

Fig. 3.30 Reference cycle

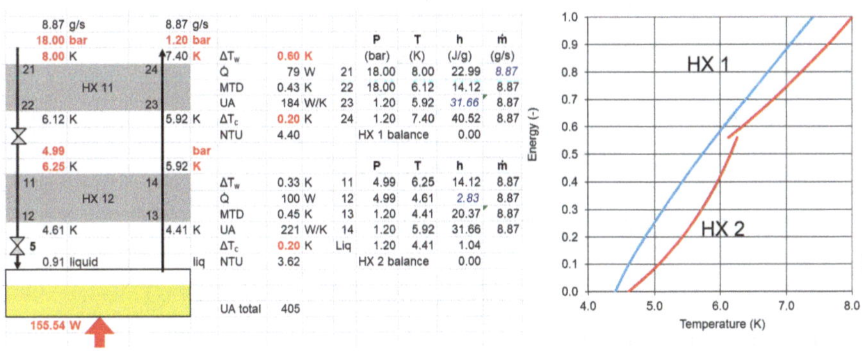

Fig. 3.31 The double JT expansion

When looking at the heat exchanger diagrams, one can see that the temperature difference along the double heat exchanger arrangement is lower and rather constant than in the one-off heat exchanger.

It is interesting to follow and compare the two processes on the T-s diagram, in Fig. 3.32, where one can see that, when following the blue curves from point 1, isobar at 18.00 bar to point 22, isenthalp to point 11, isobar at 4.99 bar to point 12, and finally isenthalp to point 50, the liquid concentration is higher than when expanding from point 1, isobar at 18.00 bar to point 2, and isenthalp to point 5.

One could also think of a one-off heat exchanger, HX 111, in which the HP circuit has, for example, a linearly distributed pressure drop so as the pressure at point 2 is 2,0 bar. The calculation of this JT cycle is displayed in Fig. 3.33.

The diagram shows a rather constant and small temperature difference. When processing the same cycle compressor flow rate from the same temperature, the available power is now 161 W because the warm heat exchanger temperature

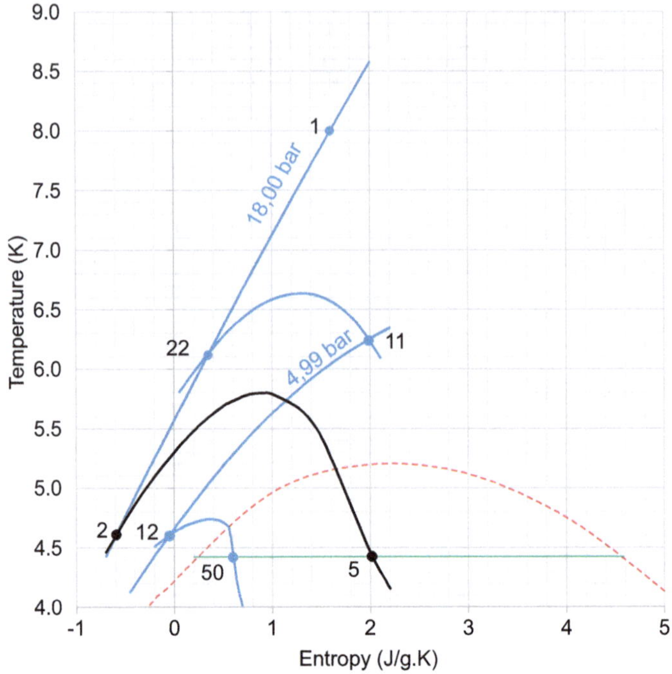

Fig. 3.32 Comparing simple and double JT expansions

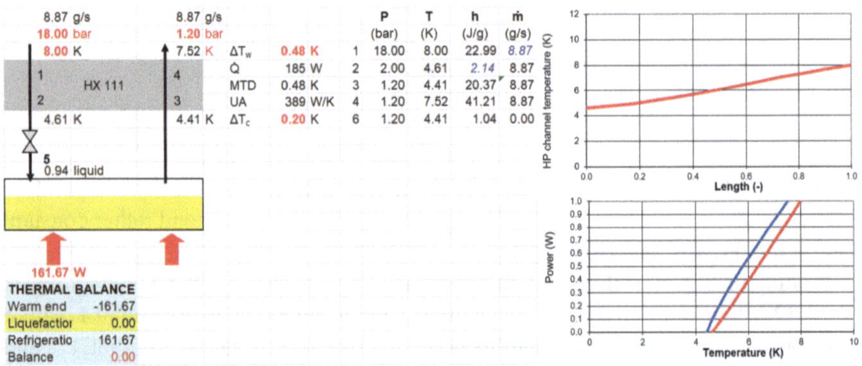

Fig. 3.33 Pressure drops in the JT HX

difference has been again decreased. The UA is 389 W/K, a little bit lower than the total of HX 11 and HX 12; therefore, the double expansion arrangement is not far from the optimum solution and, consequently, very interesting.

Such a double expansion process is of interest only for refrigerators, not for a liquefier, because the warm end heat exchanger temperature difference is, naturally, small. It cannot be reduced.

3.3.4 A Digest About the Joule Thomson Cycle

The Joule Thomson cycle rejects at its warm end any heat getting anywhere into the cycle, whatever it is *useful* or *parasitic*. Therefore, any such heat, whatever its origin, has the same thermal "cost".

The Joule Thomson cycle mass flow rate decreases when:

- The cycle high pressure increases
- The warm end temperature decreases
- The warm end temperature difference decreases

There is no cooling effect if the warm end temperature is higher than 30 K.

There is no liquid production if the temperature upstream the JT valve is higher than 7.5 K.

3.4 The Brayton Cycle

The Brayton cycle is a thermodynamic cycle named after George Brayton (1830–1892), an American engineer that described the operation of a heat engine in order to generate mechanical work. In cryogenic refrigeration, one inputs energy into the system, and heat is moved from the cold to the warm source according to the same cycle structure. The correct wording is "reverse Brayton cycle"; however, in the book the wording is shortened in "Brayton cycle".

3.4.1 The Expansion with Work Extraction (or Quasi-isentropic Expansion)

As it is said in the title, such expansion generates work. Such work (or energy or enthalpy difference) is extracted from the fluid by an expander that is kind of a gas engine. The enthalpy of a fluid, which is mainly connected to its temperature (ref Sect. 1.4.1.2), decreases during expansion. On the simplified T-s diagram (Fig. 3.34), the fluid that is ideally expanded stays at a constant entropy (isentropic expansion). It follows a vertical straight line from h_1 to h_{is}.

During an expansion process with external work (or quasi isentropic expansion), the temperature decreases *whatever the value it is upstream the expander*.

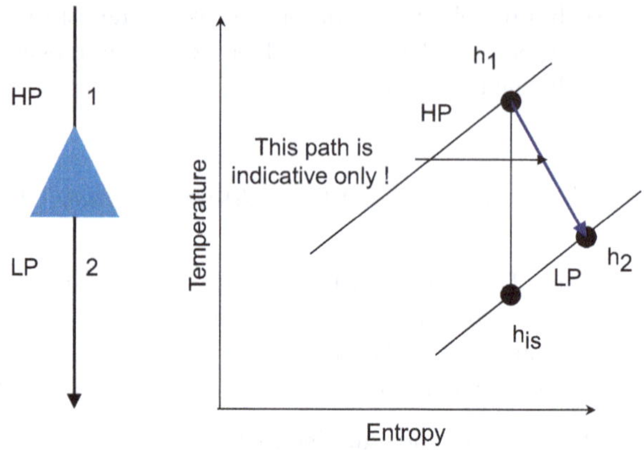

Fig. 3.34 Quasi-isentropic expansion

As the expander is not ideal, its discharge temperature is at same low pressure P_2 (this is easy!) but at a higher temperature. Expansion starts from point 1 down to point 2 (the straight line on Fig. 3.34 is an indicative path only).

One can compare the enthalpy drop that is generated by the expander ($h_1 - h_2$) to the available enthalpy drop ($h_1 - h_{is}$); the ratio of these two values is the *isentropic efficiency* of the expander:

$$\eta_{is} = (h_1 - h_2)/(h_1 - h_{is}) \tag{3.13}$$

Remark Cryogenic expansion of a gas can be performed with any kind of reciprocating or rotating expander ("engine"). Presently, almost all these expanders are turbines (see Sect. 6.6). In this book, only turbines are considered as cryogenic expanders.

Calculation of a helium expansion with work extraction

Problem: What are the discharge temperature and the power of a turbine operating with such conditions:

- Inlet pressure 15.0 bar
- Inlet temperature 20.0 K
- Discharge pressure 1.2 bar
- Mass flow rate 50.0 g/s
- Efficiency 0.75

Calculate the isentropic expansion that would give the enthalpy difference (or enthalpy drop) Δh_{is} (Table 3.5).

$$\Delta h_{is} = h_1 - h_{is} = 62.31 \text{ J/g}$$

Table 3.5 (REFPROP)

	P (bar)	T (K)	h (J/g)	s (J/g.K)	η (–)
1	15.00	20.00	101.55	7.98	
is	1.20	7.18	39.24	7.98	
	1.20		62.31		
	1.20		46.73		0.75
2	1.20	9.96	54.81		

The isentropic enthalpy drop times the expander isentropic efficiency gives the actual enthalpy drop.

$$\Delta h_{\text{act}} = 62.31 \times 0.75 = 46.73 \ \text{J/g}$$

Calculate enthalpy h_2 and then temperature T_2.

$$h_2 = 101.55 - 46.73 = 54.81 \ \text{J/g}, \ \ T_2 = 9.96 \ \text{K}$$

The expander power is:

$$\Delta h_{\text{act}} \times \text{Mass flow rate} = 2336.65 \ \text{W}$$

On the T-s diagram of Fig. 3.35, it is of interest to compare both expansions: whatever the temperature, the isentropic expansion cools the gas.

Let us have a glance at the enthalpy drop of an expander with an isentropic efficiency of 1 (ideal machine) that is expanding helium from a given pressure down to 1 bar (a similar calculation is performed in Sect. 6.6.3.7.1). If one plots the temperature drop versus the upstream temperature of the expander, the result is an almost straight line in the diagram of Fig. 3.36, because, for an ideal gas, the enthalpy is proportional to the upstream temperature (remember Sect. 1.4.1.2, $h = c_p \times T + \text{Cte}$ Eq. 1.4). Helium behaviour is not very different from an ideal gas.

One can see that the enthalpy drop (enthalpy difference between inlet and discharge of the expander) is:

- Proportional to the inlet temperature
- Not very sensitive to the expansion ratio

To be kept in mind: for same high and low pressures, the enthalpy drop is proportional to the turbine inlet temperature.

3.4.2 "Re-discovering" the Brayton Cycle

The Brayton cycle uses the quasi-isentropic or so-called expansion with work extraction.

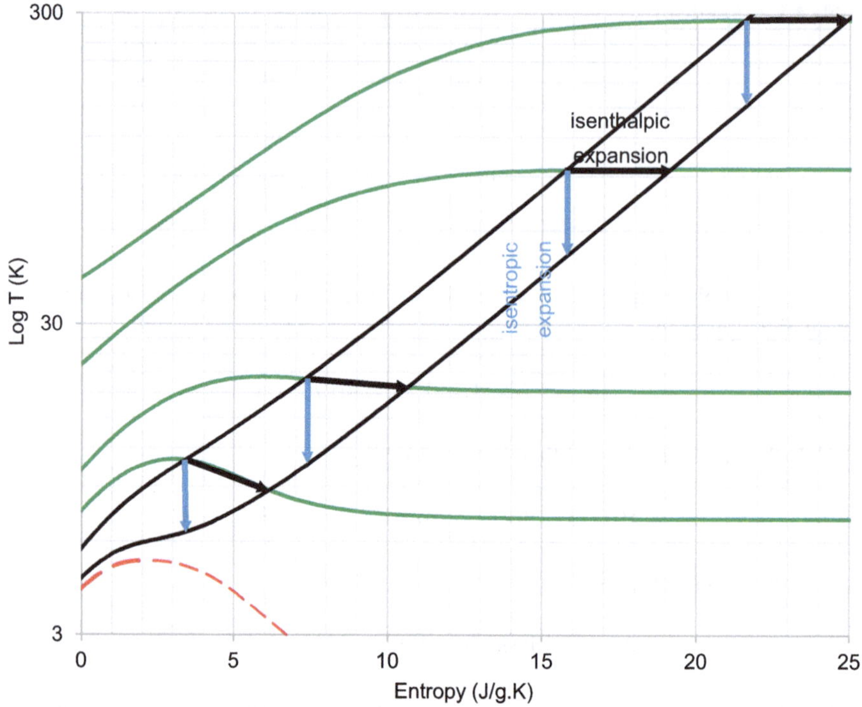

Fig. 3.35 Comparison between isenthalpic and isentropic expansions (REFPROP)

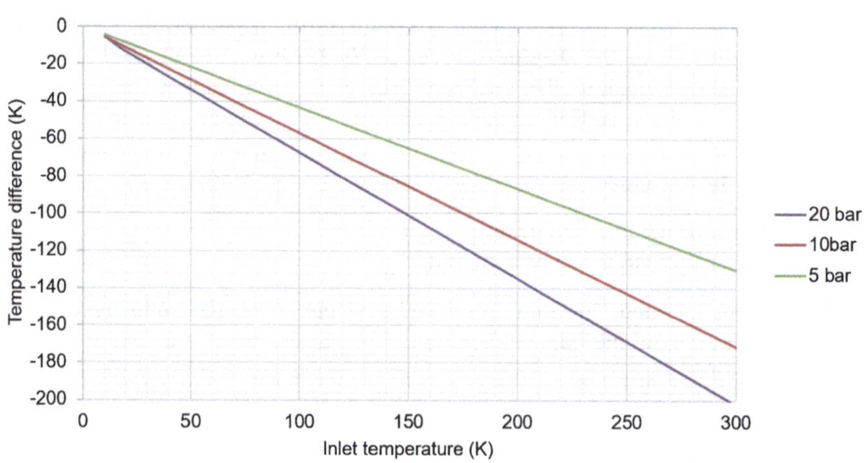

Fig. 3.36 Temperature drop versus temperature for an isentropic expansion (REFPROP)

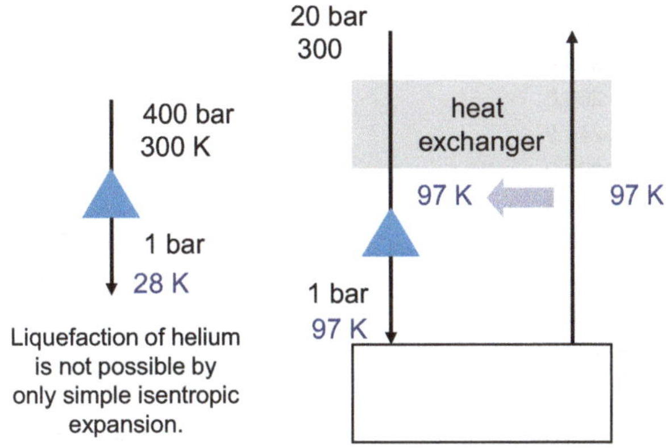

Fig. 3.37 Re-discovering the Brayton cycle

As helium that is expanded in such a way cools whatever the inlet temperature is, could it be possible to liquefy it with a simple isentropic expansion?

A simple calculation shows that expanding helium from a very high pressure, 400 bar, 300 K, down to 1 bar, with an ideal expander (isentropic efficiency 100%) leads to a discharge temperature of 28 K, only (see Fig. 3.37, left).[2]

Therefore, and unfortunately, it is not possible to liquefy helium by a simple isentropic expansion from room temperature, but as the gas temperature has strongly decreased, why not use this lower temperature to cool the "next" gas to be expanded? (remember what has been done on the JT cycle in Sect. 3.3.2). This duty is fulfilled by a heat exchanger. The arrangement is shown in Fig. 3.37, right. After the "first" expansion from 20 bar, 300 K, helium is cooled down to 97 K. This gas is sent into the heat exchanger where its enthalpy is recovered and is used to cool the incoming high-pressure flow. If one makes the assumptions that the heat exchanger is perfect (no temperature difference between high- and low-pressure fluids) and there is no thermal inertia in the system, helium exiting the heat exchanger is at 97 K upstream the expander.

In Fig. 3.38, the cool-down kinematics is described with same assumptions (perfect heat exchanger, no heat capacity) for expansion from 18.0 bar to 3.0 bar with an expander the efficiency of which is 0.75.

- Expansion from 18.00 bar, 300.0 K, to 3.0 bar cools helium to 185.16 K.
- Expansion from 18.00 bar, 185.16 K, to 3.0 bar cools helium to 114.35 K.
- Expansion from 18.00 bar, 114.35 K, to 3.0 bar cools helium to 70.76 K.
- Expansion from 18.00 bar, 70.76 K, to 3.0 bar cools helium to 43.63 K.
- Expansion from 18.00 bar, 43.63 K, to 3.0 bar cools helium to 26.82 K.
- Expansion from 18.00 bar, 26.82 K, to 3.0 bar cools helium to 16.23 K.

[2]Such calculation is performed in Sect. 15.8.3.2.

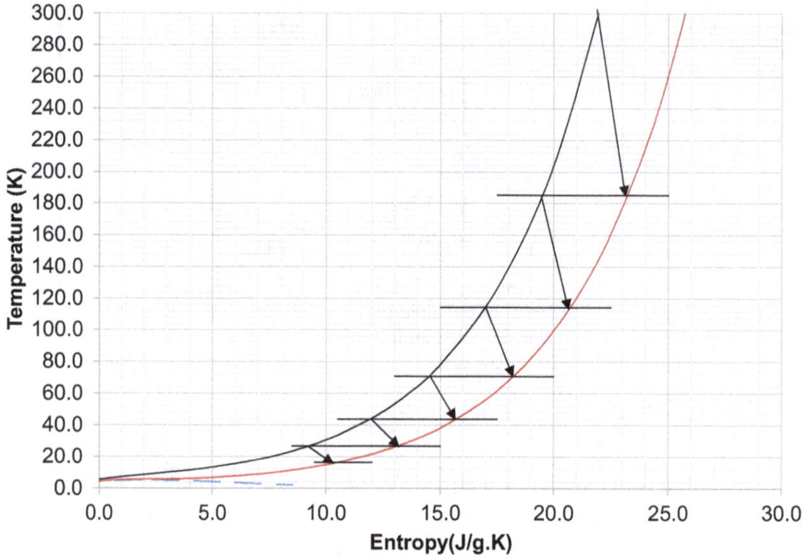

Fig. 3.38 Cool-down kinematics from 18 to 3 bar, expander efficiency 0.75 (REFPROP)

During this process, one can notice that the temperature difference that is generated by isentropic expansion is decreasing with the temperature (remember Fig. 3.36).

3.4.2.1 Description and Representation of the Brayton Cycle on the Temperature-Entropy (T-s) Diagram

The Brayton cycle structure is shown in Fig. 3.39, left. It looks like pretty much to the Joule Thomson one; the only differences are:

- The heat exchanger warm end temperature can be up to room temperature.
- The expansion valve is replaced by an expansion machine.

In most of the uses of the Brayton cycle, one does not look for obtaining liquid. Indeed, the creation of liquid in the expansion can be problematic (see Sect. 6.6.3.1).

Remark When the heat load balances exactly the cryogenic power that is extracted by the cycle, all cold temperatures reach a stable value: the cycle is in a steady state. (Analogy with Sect. 3.3.2.1 where it is said that the stability of a JT cycle is demonstrated by a *constant liquid level*!)

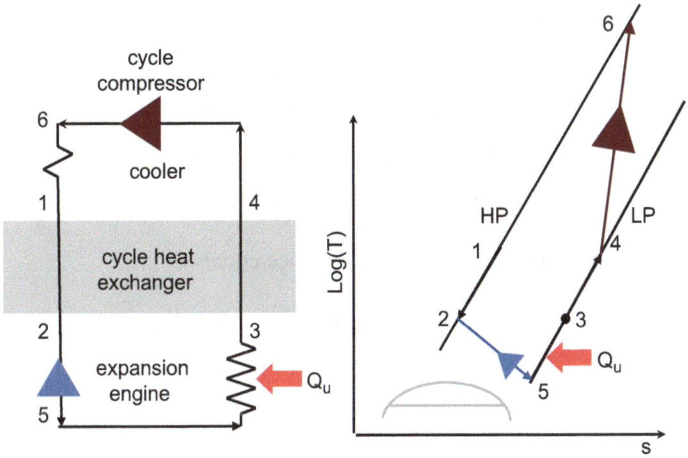

Fig. 3.39 The Brayton cycle

Fig. 3.40 Energy circulation in the Brayton cycle

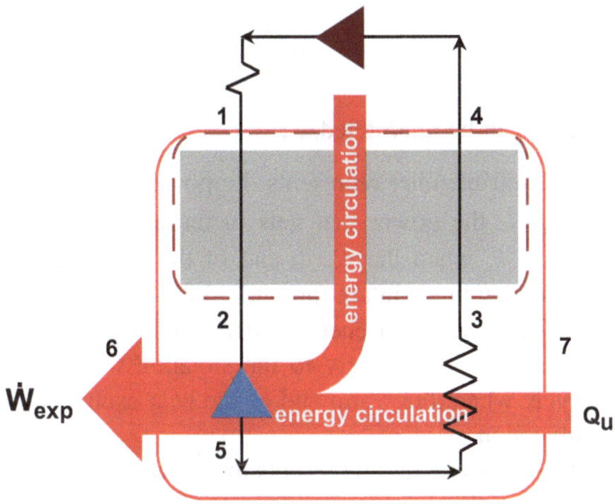

3.4.3 Calculation of a Brayton Refrigerator Cycle

An assumption for calculations Every time the calculation of a turbine will be performed, an efficiency of 0.75 is selected, whatever the operating conditions of the turbine. This is for reasons of simplicity and possibility to compare various cycles.

Let us isolate the system by plotting its "border" (see Fig. 3.40) and write the heat exchange balance along the border. As usually, by convention, the ingoing quantities are positive, and the outgoing quantities are negative.

Along the red "border":

- At point 1 (in, positive) $+\dot{m}_1 \times h_1$
- At point 6 (out, negative) $-\dot{W}_{\text{exp}}$
- At point 7 (in, positive) $+\dot{Q}_{\text{u}}$
- At point 4 (out, negative) $-\dot{m}_1 \times h_4$

and as the system is in a steady state, the balance equals 0.

$$\dot{m}_1 \times h_1 - \dot{W}_{\text{exp}} + \dot{Q}_{\text{u}} - \dot{m}_1 \times h_4 = 0 \tag{3.14}$$

and (see Sect. 3.4.1):

$$\dot{W}_{\text{exp}} = \dot{m}_1 \times (h_2 - h_5) \tag{3.15}$$

(h_2–h_5) depends on the turbine efficiency

One can write:

$$\dot{Q}_{\text{u}} + \dot{m}_1 \times (h_4 - h_1) = \dot{W}_{\text{exp}} \tag{3.16}$$

That can be interpreted as:

- The left member represents the power that enters the system: the useful power $(\dot{Q}_{\text{u}})+$ the power that gets in through the warm end of the heat exchanger. *Remark:* when the warm end of the heat exchanger temperature is above the inversion temperature (around 40 K), the enthalpy difference $h_4 - h_1$ is positive, which means that energy *enters the system through the warm end of the heat exchanger*, conversely to the situation formerly met with the Joule Thomson cycle where the warm end of the heat exchanger is at a lower temperature than inversion.
- The right member represents the power that gets out of the system.

One can see that all the power that enters the system gets out through the expander (see Fig. 3.40).

Problem How much helium flow rate is it necessary to process in order to absorb 1000 W at less than 20.00 K (20 K is the warmest temperature)?

Process inputs:

- $\dot{Q}_{\text{u}} = 1000$ W
- $T_3 = 20$ K

Sizing parameters:

- $P_1 = 18.0$ bar,

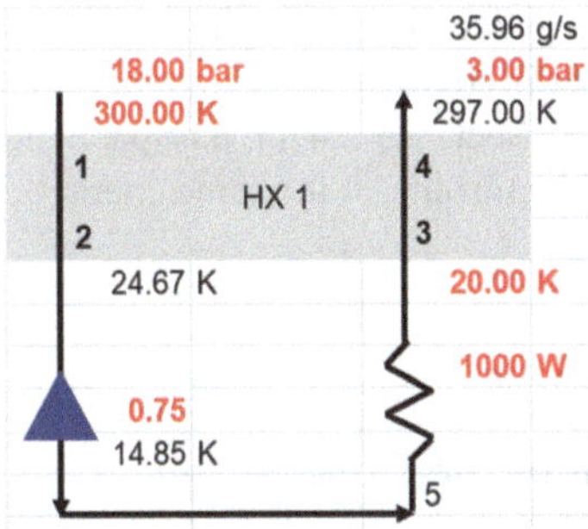

ΔT_w	3.00 K
$\dot{Q}$	51821 W
LMTD	3.21 K
UA	16146 W/K
ΔT_c	4.67 K
$\dot{W}_{exp}$	-1736 W

	P (bar)	T (K)	h (J/g)	$\dot{m}$ (g/s)
1	18.00	300.00	1568.87	35.96
2	18.00	24.67	127.70	35.96
3	3.00	20.00	107.22	35.96
4	3.00	297.00	1548.39	35.96
2'	18.00	24.67	127.70	35.96
5'	3.00	14.85	79.41	35.96

THERMAL BALANCE	
Warm end	736.26
Turbine	-1736.26
Refrigeration	1000.00
Balance	0.00

Fig. 3.41 Calculation of a Brayton pure refrigeration cycle

- $T_1 = 300.00$ K
- $P_4 = 3.00$ bar (this pressure is not connected to the boiling temperature of a liquid as in case of a Joule Thomson; it is free)
- $P_5 = P_3 = P_4$ (no pressure drop)
- $\Delta T_1 - T_4 = 3.00$ K (1% of 300 K, remember Sect. 2.4.1.2.1)
- Isentropic efficiency of the expander = 0.75

  Using $h_2 = [\dot{m}_3 \times (h_3 - h_4) + \dot{m}_1 \times h_1]/\dot{m}_1$ See Eq. 2.5, calculate h_2:

$$h_2 = h_1 + h_3 - h_4$$

Calculate the expander discharge enthalpy h_5 as in Sect. 3.4.1.
Calculate the cycle mass flow rate:

$$\dot{m}_1 = \dot{Q}_u/(h_3 - h_5) \tag{3.17}$$

Calculation results are displayed in Fig. 3.41.
Comments about the cycle calculation results.

The heat exchanger is flow balanced: the exchange diagram shows two parallel (almost) straight lines. As the temperature domain is rather far away from the liquid area, the evolution of temperature along the heat exchanger is rather linear, except at temperatures lower than 50 K: on HP temperature versus length, where one can see a slight curvature.

The thermal balance of the cycle shows that the power that enters the cycle through the warm end of the heat exchanger is important 736 W compared to the useful power: 1000 W. Such a power is almost proportional to the warm end temperature difference. With such a remark, one can understand the high importance of the heat exchanger efficiency (small temperature difference).

As already explained for the JT cycle (remember Sect. 3.3.2.4.1), any parasitic heat load getting into the cycle is to be extracted by the turbine, at the same energetic cost as useful power.

3.4.4 *Brayton Cycle Analysis: Incidence of the Cold Temperature*

Now, let us perform the calculation with same cryogenic power, all other conditions being identical, except that the coldest temperature is now 15 K instead of 20 K.

What can be expected?

- When the requested temperature decreases from 20 K to 15 K, the turbine temperature decreases in the same ratio; therefore, its power is around 0.75.
- In order to extract the same power, the turbine mass flow rate should be increased by 1/0.75 = 1.33.
- But, with a higher mass flow rate, the heat load at the warm end of the heat exchanger increases too, and finally the turbine mass flow ratio is higher than 1.33.

Calculation results are displayed in Fig. 3.42.

Here, the cycle mass flow rate is higher, and one can see that the power that enters at the warm end of the heat exchanger, 1178 W, is even higher than the useful power!

Remark Another calculation performed for same power but for a higher temperature would give a lower cycle mass flow rate.

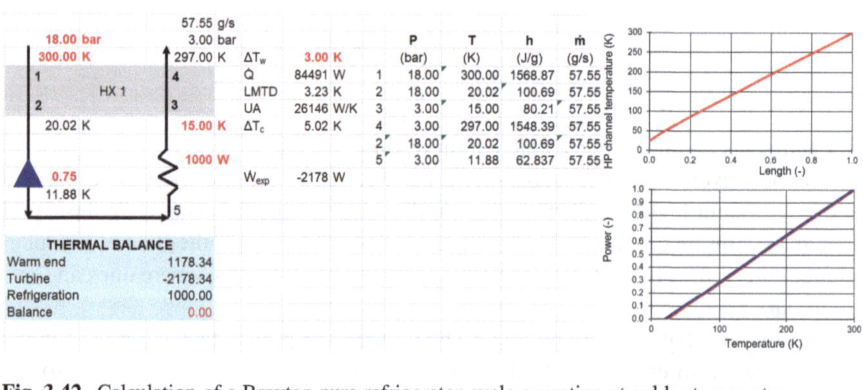

Fig. 3.42 Calculation of a Brayton pure refrigerator cycle operating at colder temperatures

3.4.5 Brayton Cycles with Multiple Turbines

It might be of interest to integrate a warmer Brayton cycle in order to pre-cool the cold Brayton cycle, as shown in Fig. 3.43.

The warm Brayton brings the warm end of the cold one down to 80 K. To comply with the temperature difference rule of 1%, the warm end temperature difference of the cold Brayton $T_{31} - T_{34}$ is set at 1 K.

The consequence is that the HX3 warm end heat load decreased by $736 - 259 = 478$ W, a power that is "spared" for the turbine that, consequently, needs less mass flow rate in about same proportion, $36 \times 1258/1736 = 26$ g/s, not far from the real flow 28.70 g/s. (By the way, the difference is related to the operating

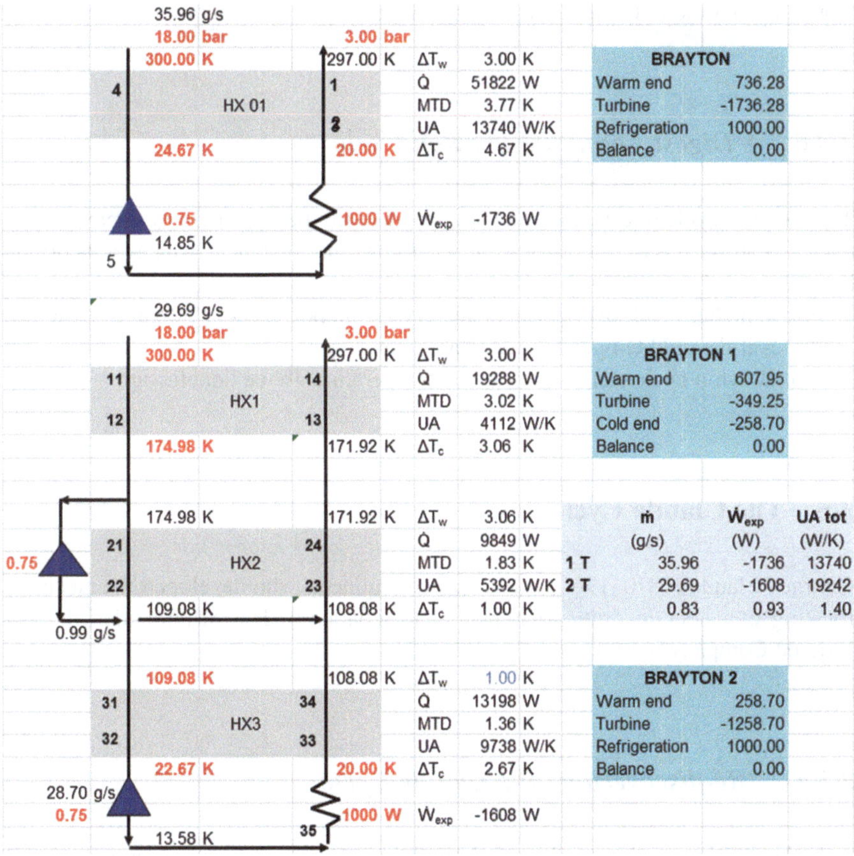

Fig. 3.43 Incidence of the number of turbines

temperature of the turbine that decreased slightly; remember Fig. 3.36. As the enthalpy drop is lower, the mass flow is higher to get same power.)

One can see at the right of Fig. 3.43 that, for same power (1000 W at less than 20 K), the cycle mass flow rate is 0.83 but the heat exchanger UA increased in a ratio of 1.4.

Such a solution is of interest only for high powers, when the increase in capital cost (smaller compressor but second turbine and larger heat exchangers) is compensated by a lower operational cost.

Remark Would it be interesting to add a third pre-cooling Brayton stage? The warm end heat load of the coldest Brayton could be reduced in proportion of the warm end temperature, say 0.7. The warm end heat load would decrease by around 180 W, which would reduce the coldest turbine flow rate by 29 × (1608 − 180)/1608 = 26 g/s. But as above, as the operating temperature is lower, the mass flow is slightly larger. By the way, calculation shows that the cycle mass flow would be 28.46 g/s. In other words, the addition of a third pre-cooling Brayton cycle has little interest.

3.4.6 A Digest About the Brayton Cycle

The Brayton cycle rejects through the expander, any heat getting anywhere into the cycle, whatever it is useful or parasitic. Therefore, any such heat, whatever its origin, has the same thermal "cost".

For same high and low pressures and cryogenic power, the cycle mass flow rate increases if the cold temperature decreases.

The Brayton cycle can operate at any temperature above liquefaction.

3.5 The Claude Cycle

Georges Claude (1870–1960) was a French engineer, who developed the air reciprocating expander in order to improve air separation. He then founded the Air Liquide Company in 1902.

3.5.1 "Re-discovering" the Claude Cycle

A Joule Thomson cycle can liquefy helium, provided that its warm temperature is lower than 30 K (see Sect. 3.3.2.3). A Brayton cycle can absorb cryogenic power at low temperature without any maximum temperature constraint at the warm end (see Sect. 3.4.2).

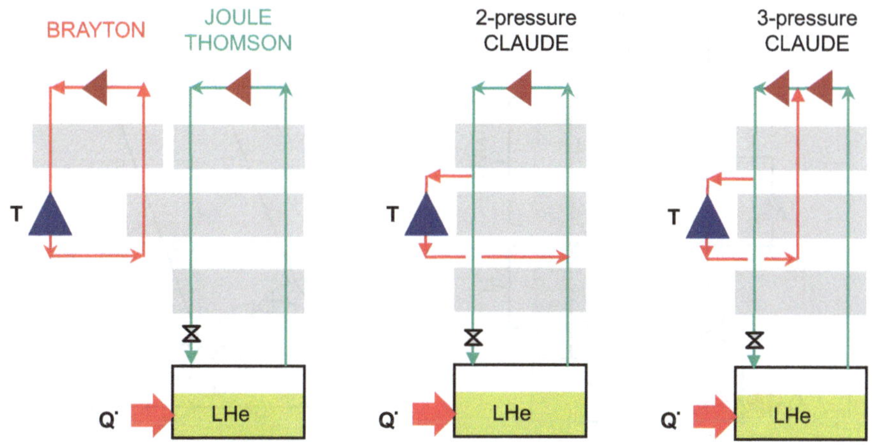

Fig. 3.44 Combining a Joule Thomson and a Brayton cycle to get a Claude cycle

Therefore, why not use a Brayton cycle in order to "pre-cool" a Joule Thomson one?

This could be done, as shown in Fig. 3.44 left, where the Brayton cycle pre-cools the Joule Thomson one. But, obviously, it makes a rather complicated machine with two cycle compressors and four heat exchangers. There is a smarter way to combine both cycles as indicated in Fig. 3.44 centre: one-off compressor processes both flow rates of Joule Thomson and Brayton cycles, and the helium flow is split between Brayton (red) and Joule Thomson (green) cycles. By the way, this is the Claude cycle! Another structure is shown in Fig. 3.44 right, so-called a three-pressure one. The interest of this arrangement is discussed later in the book (Sect. 8.7.4).

We are, now, in a position to liquefy helium or absorb power at liquid helium temperature starting from room temperature!

3.5.2 Description and Representation of the Claude Cycle on the Temperature-Entropy (T-s) Diagram

In Fig. 3.45, the flow diagram is on the left and the T-s diagram is on the right. On the T-s diagram, the isobar curves are schematically shown as inclined straight lines, the condensation curve being assimilated to an arc of ellipse. Both axes do not have any scale (Sect. 1.4.1.3).

- From point 9 to point 1, helium is compressed and then cooled into a cooler, in order that the temperature is returned to the ambient.
- From point 1 to point 2, helium is cooled down along the high-pressure isobar by countercurrent against the low-pressure gas.

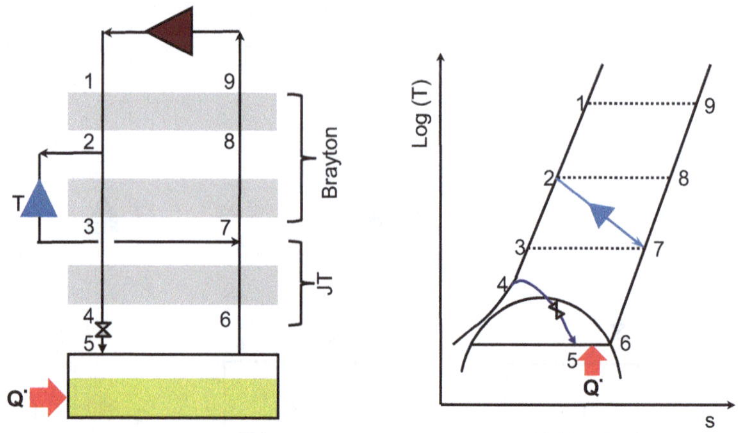

Fig. 3.45 Claude cycle with one turbine

- From points 2 to point 7, the expansion into the turbine takes place.
- From point 2 to point 4, helium is cooled down along the high-pressure isobar by countercurrent against the low-pressure gas.
- From point 4 to point 5, the Joule Thomson expansion takes place, turning part of the gas into liquid. The liquid deposits into the phase separator. The gas that has been expanded but not liquefied is saturated vapour.
- A thermal power $\dot{Q}$ is injected into the liquid: its amount is so that exactly all the liquid that is produced is vaporised.
- The part of gas that has not been liquefied and the saturated vapour produced by the liquid that has been vaporised (that is to say all the cycle mass flow rate) are warmed along the low-pressure isobar.
- JT and turbine flows that are at same temperature mix up at point 7 and are warmed up to room temperature by flowing into the heat exchanger low-pressure circuit from point 7 to point 9 towards the cycle compressor.

3.5.3 Various Arrangements for Claude Cycles

3.5.3.1 Claude Cycle with One Turbine

3.5.3.1.1 Calculation of the Thermodynamic Balance

See Fig. 3.46.

Problem What is the cycle mass flow rate that is needed to absorb 50 W at 4.41 K and, simultaneously, liquefy 0.5 g/s of helium at constant level (liquid is withdrawn as soon as it is produced)?

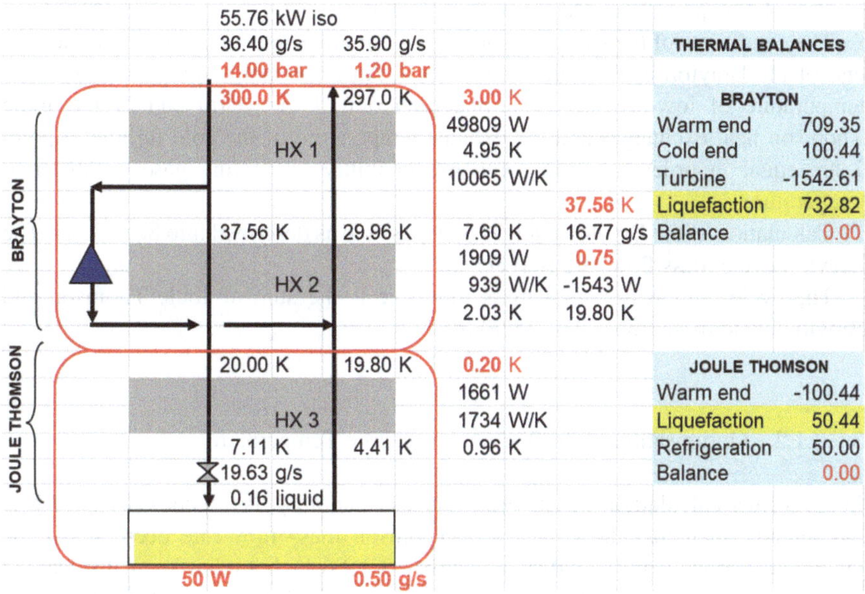

Fig. 3.46 Calculation of the Claude cycle with one turbine

Process inputs:

- $\dot{Q}_u = 50\ \text{W}$
- $\dot{l} = 0{,}5\ \text{g/s}$

Design parameters:

- $P_1 = 14.0\ \text{bar}$
- $T_1 = 300{,}\ 00\ \text{K}$
- $P_9 = 1{,}\ 2\ \text{bar}$ (such pressure is connected to the saturation temperature)
- $T_1 - T_9 = 3{,}\ 00\ \text{K}$ (the 1% of 300 K reasonable rule)
- $T_3 - T_7 = 0{,}\ 20\ \text{K}$
- Efficiency of the expander = 0,75
- $T_7 = 20.00\ \text{K}$ (initial selection)
- Assumption: the turbine discharge temperature is same as the low-pressure warm end of HX 3.

Calculation of the Claude cycle is performed from the cold end towards the warm end. As usually, before performing calculations, let us isolate the system from the rest of the universe by plotting the border, first around the Joule Thomson cycle.

The Joule-Thomson cycle calculation (see Sect. 3.5.3.1.1), for which assumptions of temperature at the warm end (here $T_3 = 20.00$ K) and temperature difference at the hot end (0.2 K) are made, allows to know the mass flow rate (19.63 g/s).

Now, let us plot the border around the Brayton cycle.

The conditions of the warm end of the cycle Joule-Thomson are those of the cold end of the Brayton cycle which pre-cools it. An important assumption is that the temperature of low-pressure helium getting out of the warm end of the Joule Thomson heat exchanger and the exhaust temperature of the cold turbine at point 7 are equal in order to avoid some irreversibility by mixing gases at different temperatures.

Calculation of the Brayton cycle is now possible as described into Sect. 3.4.3. The Brayton cycle mass flow rate is 16.77 g/s.

The cycle compressor total mass flow rate is the sum of Joule Thomson and Brayton cycle mass flow rates: 35.90 g/s.

3.5.3.1.2 Optimisation of the Turbine Discharge Temperature

In the former calculation, T_3 has been selected randomly. Is it a correct value? One has already seen that the cycle Joule-Thomson mass flow rate decreases if the temperature at the warm end decreases (Sect. 3.3.2.6.1), but it has been also learned that the mass flow rate of the Brayton cycle increases if the temperature at the cold end decreases (Sect. 3.4.4). Figure 3.47 shows these changes.

In the Claude cycle, the temperature at the warm end of the Joule-Thomson cycle is the same as the temperature at the cold end of the Brayton cycle! What is the optimal temperature such that the sum of the mass flow rates of Joule Thomson and Brayton cycles is minimum? This duty is fulfilled by a solver like the one that is part of the Excel tools.

In Fig. 3.48, for the optimized temperature, the turbine flow has increased from 16.77 to 20.68 g/s, but the Joule Thomson flow has decreased from 19.63 to 10.01 g/s. The cycle compressor mass flow rate is now minimum at 30.19 g/s, compared to 35,90 g/s. Consequently, each component (cycle compressor, heat exchangers, turbines) is smaller, and the cycle compressor absorbs less power.

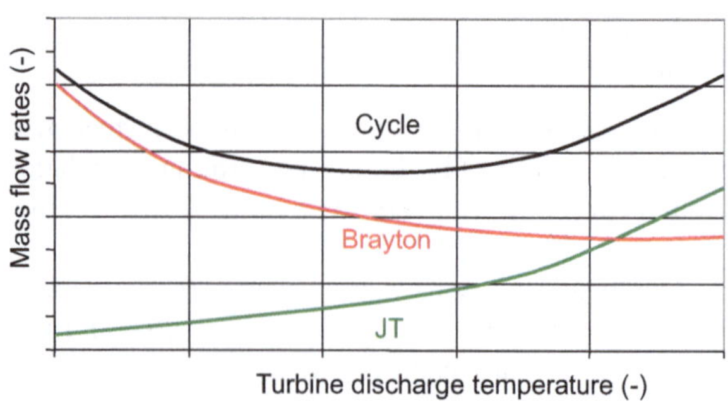

Fig. 3.47 Mass flow rates of JT, turbine and cycle

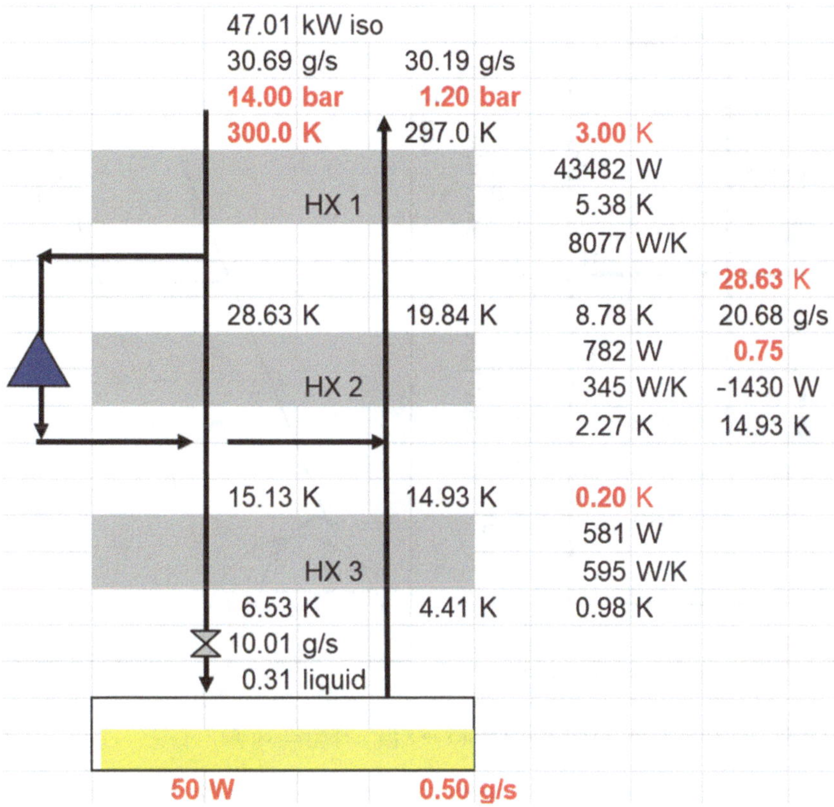

Fig. 3.48 Optimised turbine discharge temperature

3.5.3.2 Claude Cycle with Two Turbines in a Parallel Arrangement

Joule Thomson cycle may be pre-cooled by another "warm" Brayton cycle (see Fig. 3.49). It is the addition of one stage of cooling, the Brayton cycle, "on top" (or at the warm end) of the one-turbine Claude cycle. Such pre-cooling can also be performed with liquid nitrogen (see Sect. 4.5)

This arrangement has been implemented in the Collins liquefier shown in Sect. 16.6.1 fitted with reciprocating expanders.

3.5.3.2.1 Calculation of the Thermodynamic Balance

Problem What is the cycle mass flow rate that is needed to absorb 50 W at 4.41 K and simultaneously liquefy 0.5 g/s at constant level?

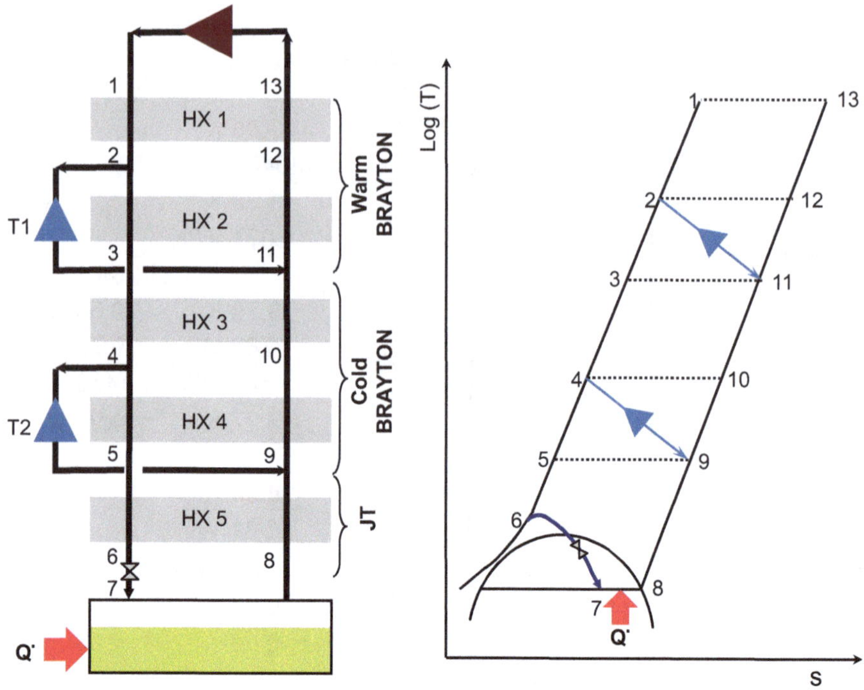

Fig. 3.49 The Claude cycle with two turbines in a parallel arrangement

Process inputs:

- $Q_u = 50$ W
- $\dot{l} = 0{,}5$ g/s

Design parameters:

- $P_1 = 14.0$ bar
- $T_1 = 300{,}00$ K
- $P_8 = 1{,}2$ bar (such pressure is connected to the saturation temperature)
- $\Delta_{T1-T13} = 3{,}00$ K (1% of 300 K)
- $\Delta_{T3-T11} = 1.00$ K
- $\Delta_{T5-T9} = 0{,}20$ K
- Efficiency of each expander = 0,75
- $T_{11} = 80$ K (initial selection)
- $T_9 = 30$ K (initial selection)

Calculation is performed from the cold end towards the hot end. The Joule-Thomson cycle calculation (see Sect. 3.3.2.3), for which assumptions of temperature

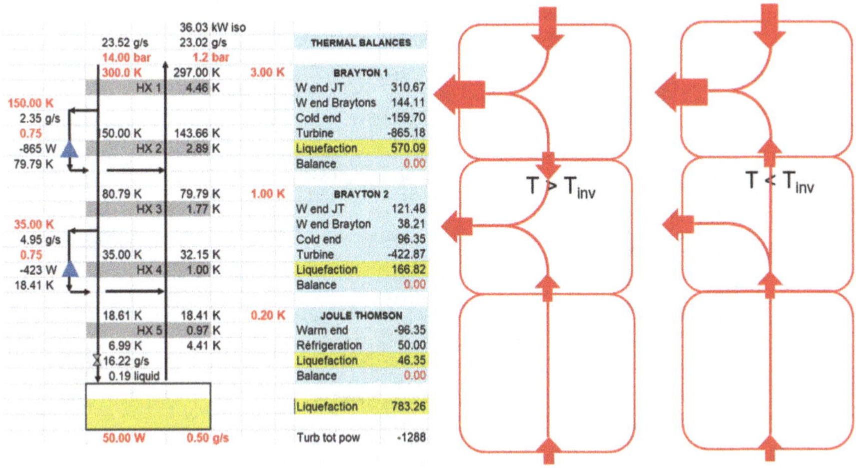

Fig. 3.50 Claude cycle with two turbines in a parallel arrangement. Energy circulation

at the warm end (here 15.2 K) and temperature difference at the warm end (by 0.2 K) are to be made, allows to know the mass flow rate (16.22 g/s).

The conditions of the warm end of the cycle Joule-Thomson are those of the cold end of the Brayton cycle which pre-cools it. The same assumption as made earlier for the one-turbine cycle is repeated: the temperature of low-pressure helium getting out of the warm end of the Joule Thomson (HX5) and HX3 heat exchangers and the exhaust temperature of the cold and warm turbines are equal, respectively.

Here, details of the calculation are similar to those made in previous calculations. The results are shown in Fig. 3.50.

Energy Circulation into the Cycle

Note how the energy flows through the system (see Fig. 3.50, centre):

- Joule Thomson cycle: all the energy gets out at the warm end.
- Cold Brayton cycle: energy from the warm end of the Joule-Thomson cycle and the cold end of the warm Brayton cycle is extracted by the cold turbine.
- Warm Brayton cycle: energy getting in at the room temperature end of the heat exchanger is extracted mainly by the warm turbine and partially by the cold Brayton cycle.

However, if the temperature break between Brayton cycles were below the inversion temperature (about 40 K), as shown in Fig. 3.50, right, energy would flow in the opposite direction: from the cold Brayton cycle to the warm one.

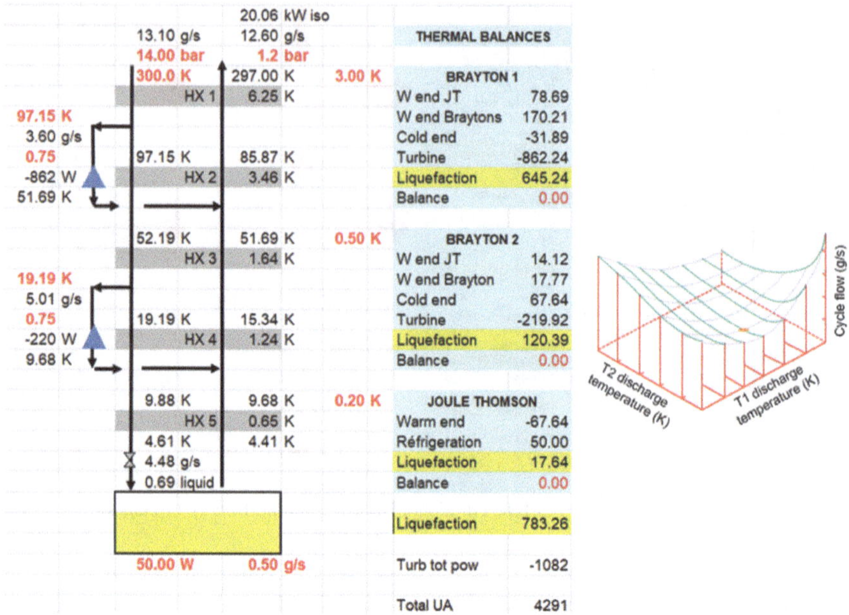

Fig. 3.51 The Claude cycle with two parallel arranged turbines, optimisation of both turbine temperatures

3.5.3.2.2 Optimisation of Both Turbine Discharge Temperatures

In case of a Claude cycle with one turbine, it has been shown that there is one optimum turbine discharge temperature corresponding to the minimum cycle mass flow rate (see Sect. 3.5.3.1.2). When the cycle is fitted with two turbines, same question arises for each turbine. Therefore, the turbine optimum temperature must be found for each turbine. Again, the duty can be fulfilled by a solver (Fig. 3.51).

One finds here the advantages of seeking optimum temperatures: the cycle compressor mass flow rate has decreased from 23.02 to 12.60 g/s.

3.5.3.3 Claude Cycle with Two Turbines in a Series Arrangement

In the examples discussed above, the turbines operate between the cycle high and low pressures: they are connected in parallel. They can also be connected in series, as shown in Fig. 3.52.

Calculation results of the optimised turbine series arrangement cycle are shown in Fig. 3.53, right.

It is interesting to compare the series and the parallel arrangements (remark that in order to make a significant comparison, the process conditions are similar in each

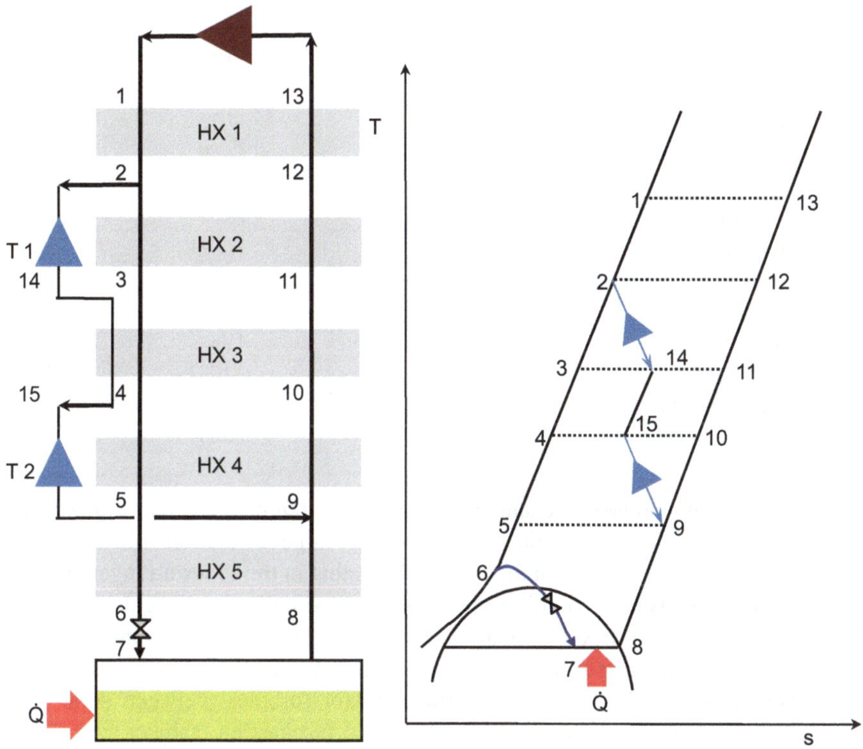

Fig. 3.52 Claude cycle with two turbines in a series arrangement

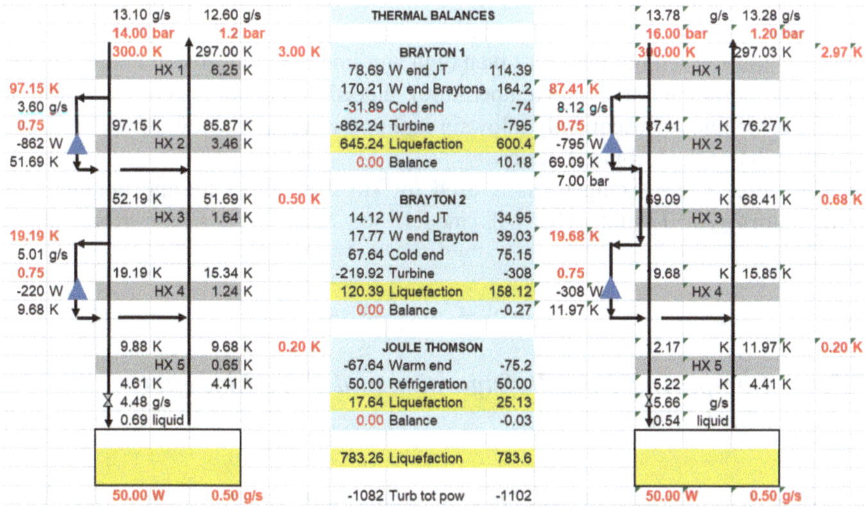

Fig. 3.53 Claude cycle with two series-arranged turbines

Table 3.6 Comparing parallel and series turbine arrangements

		Parallel	Series
Heat load at warm end HX 3 (JT)	(W)	14.12	34.95
Heat load at warm end HX 3 (Brayton)	(W)	17.77	39.03
Total heat load at warm end HX 3	*(W)*	*31.89*	*73.98*
Cold turbine power	(W)	−220	−308
Heat load at warm end HX 1 (JT)	(W)	78.69	114.39
Heat load at warm end HX 1 (Brayton)	(W)	170.21	164.19
Total heat load at warm end HX 1	*(W)*	*248.90*	*278.58*
Warm turbine power	(W)	−862	−795
JT mass flow rate	(g/s)	4.48	5.66
Total turbine power	(W)	−1082	−1102
Turbine mass flow rate	(g/s)	8.61	8.12
Compressor mass flow rate	(g/s)	12.60	13.28

cycle, same warm end heat exchanger temperature differences and same *unrealistic* turbine efficiencies). At a first glance, there are not large differences; see Table 3.6.

The main difference between the two cycles is that in the series arrangement, the warm turbine flow is circulated in the coldest Brayton cycle, bringing an extra heat load at the warm end of HX 3, which is to be extracted by the cold turbine. The final result is that the compressor mas flow rate is higher for a series arrangement.

However, the condition of equal efficiency for turbines, between series and parallel, is not realistic. Indeed, parallel arranged turbines are subject to a higher expansion ratio, their enthalpy drop is higher than the one that is extracted by turbines connected in series, and, consequently, they have a lower efficiency (see Sect. 6.7.3.1.2). If the calculations are re-performed with realistic values of efficiency, the cycle with turbines in series happens to be the most effective. The turbine arrangement in series is used when the turbines face limitations due to excessive enthalpy drops that are related to high expansion ratio or high operating temperatures (remember that, for given high and low pressures, the enthalpy drop is almost proportional to the inlet temperature). Indeed, it reduces the enthalpy drop of each machine and consequently allows a higher turbine efficiency. In addition, the construction of the turbine is easier.

The series cycle arrangement is used for standard small turbine liquefiers or refrigerators (Air Liquide HELIAL model, see Sect. 12.4, Linde Kryotechnik L-series).

3.5.4 Comparing Pure Refrigerator and Pure Liquefier Machines

In all the following calculations, the cycle optimisation is performed.

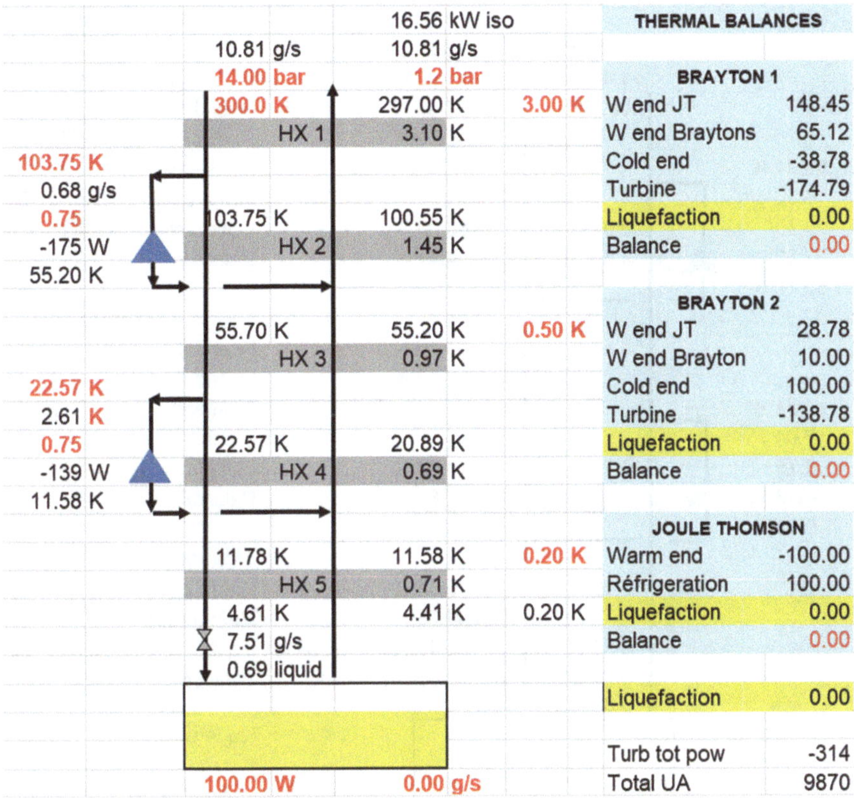

Fig. 3.54 An optimised 100 W pure refrigerator

Let us compare two roughly equivalent machines: a 100 W refrigerator and a 1 g/s liquefier at constant level (remember Sect. 3.2.4), the *optimised* calculation of both of them being shown in Figs. 3.54 and 3.55.

It is of interest to compare the sizes of the main components in both machines, as shown in Table 3.7.

One can easily see that:

- Compared to a liquefier, a refrigerator has:
  - Smaller compressor
  - Larger heat exchangers
  - Less powerful turbines

and, conversely,

- Compared to a refrigerator, a liquefier has:

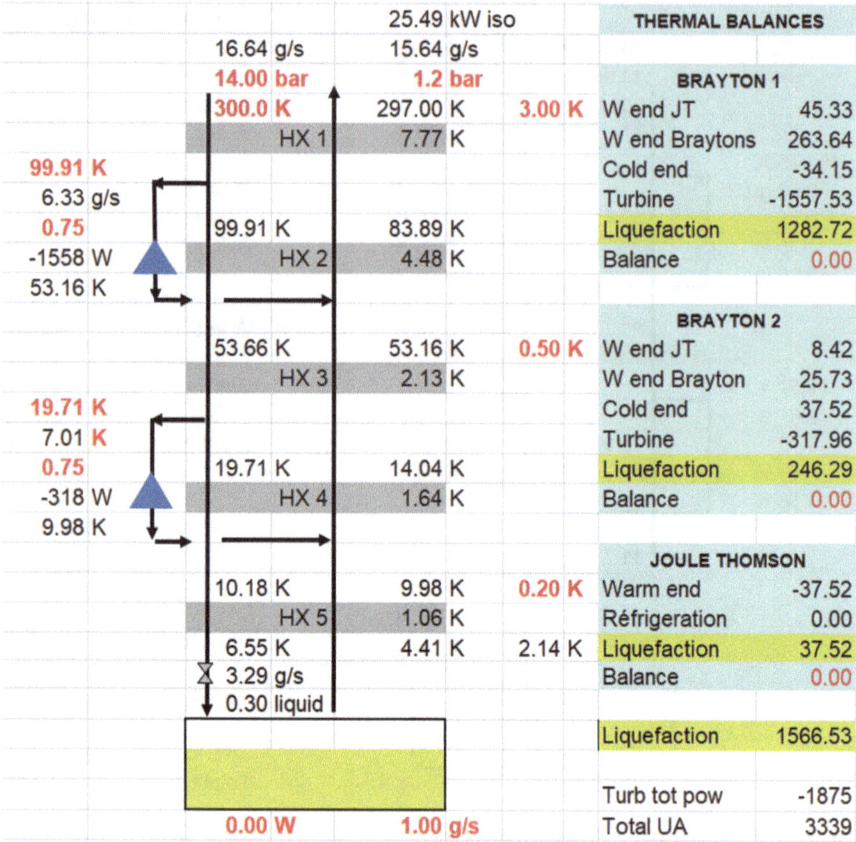

Fig. 3.55 An optimised 1 g/s pure liquefier at constant level

Table 3.7 Comparing the sizes of components

	Cycle compressor mass flow rate	Heat exchanger total UA	Turbine total power
	(g/s)	(W/K)	(W)
Pure refrigerator	10.81	9870	−314
Pure liquefier	15.64	3339	−1875
Ratio ref/liq	*0.69*	*2.96*	*0.17*
A refrigerator has	**"Smaller" compressor**	**"Large" HX**	**"Weak" turbines**
A liquefier has	**"Smaller" compressor**	**"Smaller" HX**	**Powerful turbines**

- Larger compressor
- Smaller heat exchangers
- Powerful turbines

This information makes easier to see the difference between two unknown machines, by simply looking at the size of their components.

3.5.5 *About Efficiency*

3.5.5.1 Efficiency of Existing Refrigerators

In the time period 1966 to 1974, Tom Strobridge had the idea to plot the Carnot efficiency (remember Sect. 3.2.7) versus the cryogenic power of existing helium refrigeration plants that are *not pre-cooled with liquid nitrogen.*

He noticed that the Carnot efficiency of a plant in only related to its cryogenic power, not to the operating temperature. He could plot a regression curve. This job was followed by the actions of R. Byrns in 1997 and, finally, M. Green in 2008 whose most recent diagram is displayed in Fig. 3.56.[3] Such a curve is an easy tool that can be used in order to evaluate the compressor power when projecting a plant. However, the author has re-calculated the regression equation, taking into account only the figures of refrigerators built after 1997, the power of which is higher than 100 W. For these values, he proposes the regression equation:

$$\eta\ (\%) = 16.7 * R^{0.2}$$

An approximation of the absorbed power at room temperature can be calculated by using the equation:

$$\dot{W}\ (\text{kW}) = 393 \times \dot{Q}^{0.8} \tag{3.18}$$

with $\dot{Q}$(kW) the equivalent cryogenic load at 4.5 K.

3.5.5.2 A Refrigerator Built with Ideal Components?

What would be the efficiency of a refrigerator that would incorporate ideal components:

- Ideal compressors would have an isothermal efficiency of 1.00.

[3]The cost of helium refrigerators and coolers for superconducting devices as a function of cooling at 4 K. M. A. Green, Advances in cryogenic engineering: Transactions of the Cryogenic Engineering Conference - CEC, Vol 53

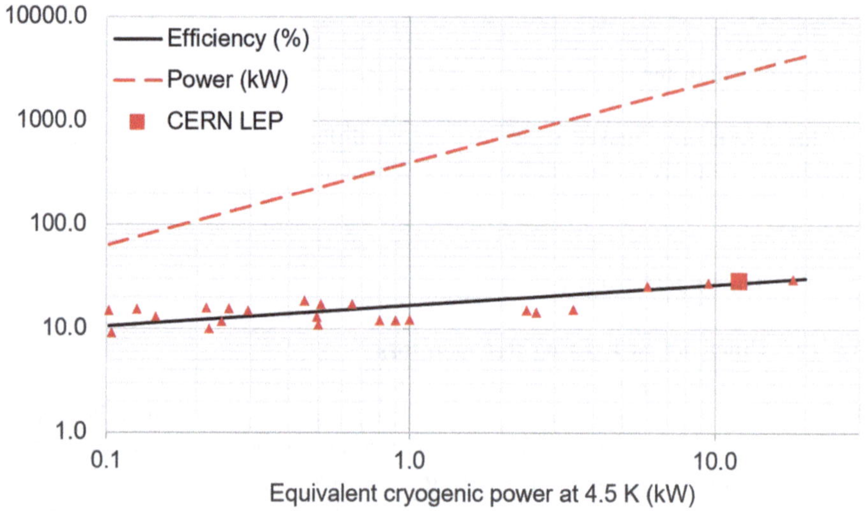

Fig. 3.56 Strobridge, Byrns and Green diagram for efficiency (black) and compressor power (dotted red)

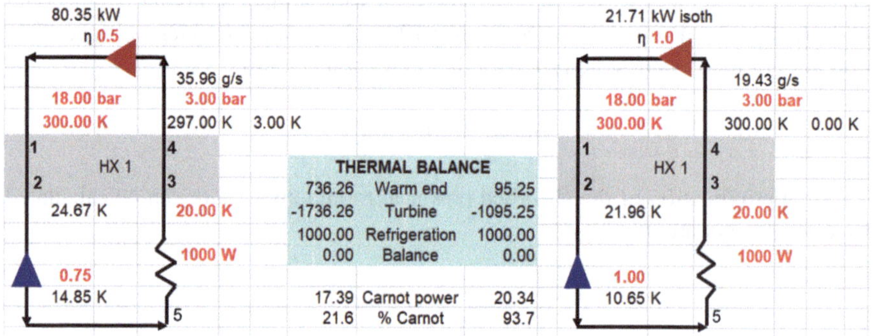

Fig. 3.57 An ideal Brayton cycle

- Ideal heat exchangers would have a warm end temperature difference of 0.00 K.
- Ideal expanders would have an isentropic efficiency of 1.00.
- In addition, there would be neither pressure drops nor heat leaks.

The simple Brayton cycle of Sect. 3.4.3 in the left part of Fig. 3.57 has realistic efficient components:

- 0.50 for isothermal efficiency of the cycle compressor
- 3.00 K for warm end temperature difference of the heat exchanger
- 0.75 for isentropic efficiency of the expansion turbine

The cycle compressor absorbs 80.35 kW, which makes 21.6% of the Carnot power for the refrigerator.

The cycle compressor of the ideal refrigerator in Fig. 3.57 right needs only 21.71 kW, which increases up to 93.7% of the Carnot power for the refrigerator.

3.5.6 Replacing the JT Valve by an Expander?

In the 1980s, J. Smith from MIT replaced the JT valve by a reciprocating expander into a Collins liquefier as a demonstrator. A reciprocating expander was also incorporated into an 800 L/h liquefier at the Otis, Kansas, helium plant to enhance the production. CERN has also operated a KPS liquefier fitted with a two-phase reciprocating expander.

This arrangement is discussed in Sect. 4.2.2 for large machines. As shown in Fig. 4.8, one sees that the liquid helium output can be increased by around 25%.

However, for small machines, the liquid turbine that could be fitted would be difficult to make. On top of that, its extra cost might make the liquefier price no more competitive.

3.5.7 A Digest About the Claude Cycle

The Claude cycle is a Joule Thomson cycle that is pre-cooled by one or more Brayton cycles. Each Brayton cycle gets power at its cold end from the heat exchanger that is at a lower temperature, except if the temperature between them is lower than the inversion one.

The operating temperature(s) of the turbine(s) must be optimized at the time of design.

A cycle with two turbines in a parallel arrangement is more efficient than a series arrangement one when turbines have same efficiency. However, as turbines in series have an actual higher efficiency than parallel, an actual series arrangement cycle is more efficient.

Liquefier operation at rising level increases the apparent liquefaction rate.

3.6 Liquid Nitrogen Pre-cooling of Thermodynamic Cycles

In Sect. 3.5.3.2, a Brayton cycle is used to pre-cool a Claude cycle. Such pre-cooling can also be performed by using liquid nitrogen that is vaporised and then warmed up to room temperature.

Liquid nitrogen is an industrial product easily available that is produced in large quantities. It is a powerful and economically attractive source of refrigeration. Pre-cooling is performed by both vaporizing, heating and then venting of liquid nitrogen or by connecting to the helium cycle a nitrogen re-condensation cycle. Remember that 1 g of liquid nitrogen at atmospheric pressure provides 199 J during vaporisation (latent heat, see Sect. 1.4.1.2) at constant temperature and 234 J when saturated vapour is warmed up to room temperature (sensible heat). Cooling performed by liquid nitrogen reduces the thermal load of turbines: their power and so the flow they process are reduced, leading to a decrease in component sizes and, therefore, in investment and consumption of electrical energy, but one must not forget the incidence of liquid nitrogen cost.

An economic assessment considering parameters such as investment, cost of electricity, cost of liquid nitrogen, local availability of liquid nitrogen and run times allows selecting the operation with or without commercially available liquid nitrogen pre-cooling that is the best adapted to local conditions.

3.6.1 Refrigeration Regimes

As there is "nothing to cool" between 300 K and 4.5 K (see Sect. 3.1), liquid nitrogen that is boiling at about 80 K hardly brings help. However, as the warm end of the cycle that is cooled by the turbines is colder (80 K instead of 300 K, about one third), it is possible to set a smaller warm end temperature difference, according to the "1 % temperature" rule. Consequently, the reduction of the warm end temperature difference results in a lower heat load getting in at the warm end of the cycle.

3.6.2 Liquefaction Regimes

The situation is very different in the case of helium liquefaction, because cooling the helium to be liquefied from room temperature down to 80 K corresponds to 1143 J/g (or 73%) of the energy to be extracted from it. This duty is fulfilled by nitrogen, and, consequently, the turbine duty is lower, as shown in Fig. 3.58.

In addition, as for refrigeration, the warm end of the part of the cycle that is cooled by the turbines is colder, it is possible to reduce the warm end temperature difference (the 1% rule) resulting in a lower heat load getting in at the warm end of the cycle. Therefore, one can expect that liquid nitrogen pre-cooling can have a higher incidence on a liquefaction cycle.

Remark To facilitate comprehension and comparison, on the following paragraph, both non-pre-cooled and pre-cooled calculations are displayed on the same figure.

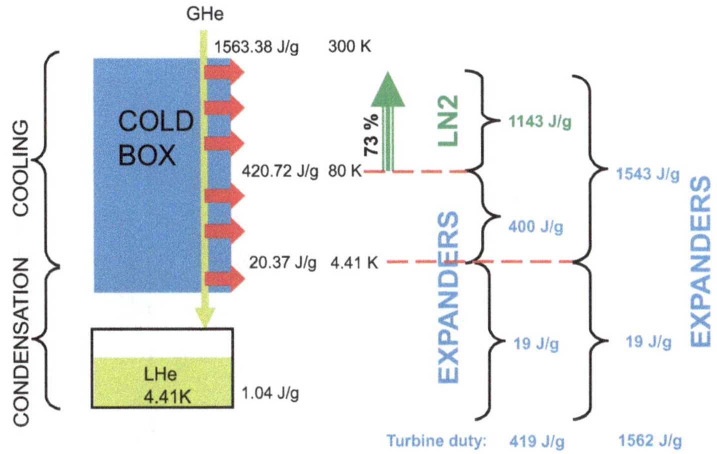

Fig. 3.58 Pre-cooling a helium liquefier with liquid nitrogen

3.6.3 *Liquid Nitrogen Pre-cooled Brayton Cycle*

Let us compare two cycles performing the same duty, 1000 W at 20.00 K, with same high and low pressures and same expander efficiency, one of them being liquid nitrogen pre-cooled.

The reference cycle in Fig. 3.59, left, needs a cycle flow of 35.96 g/s.

The nitrogen pre-cooled cycle in Fig. 3.59, right, needs a lower cycle flow of 28.30 g/s (79% of the non-pre-cooled cycle). Note that this is explained by the lower heat load getting in at the warm end of the part of the cycle that is cooled by the turbine that decreases from 736.26 to 232.32 W due to the smaller warm end temperature difference (applying the "1 % rule"!). Consequently, the turbine power is reduced accordingly from 1736 to 1232 W. Another consequence is that the turbine discharge temperature is decreased from 14.85 to 11.69 K, due to the cycle flow reduction.

3.6.4 *Liquid Nitrogen Pre-cooled Claude Cycle with Two Turbines in a Parallel Arrangement*

3.6.4.1 Pure Refrigerator (Fig. 3.60)

One can make the same comparison between the two cycles; the cycle flow decreases from 10.81 down to 9.94 g/s, which is not important. Remember Fig. 3.1, where it is said that "nothing to be cooled" from room temperature down to saturated vapour! Therefore, it is obvious that liquid nitrogen pre-cooling cannot help so much.

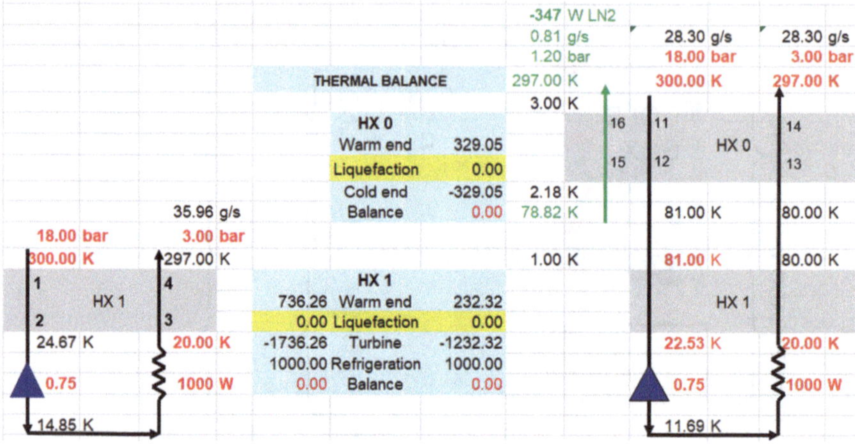

Fig. 3.59 Left, Brayton reference cycle, not pre-cooled, right, Brayton liquid nitrogen pre-cooled cycle

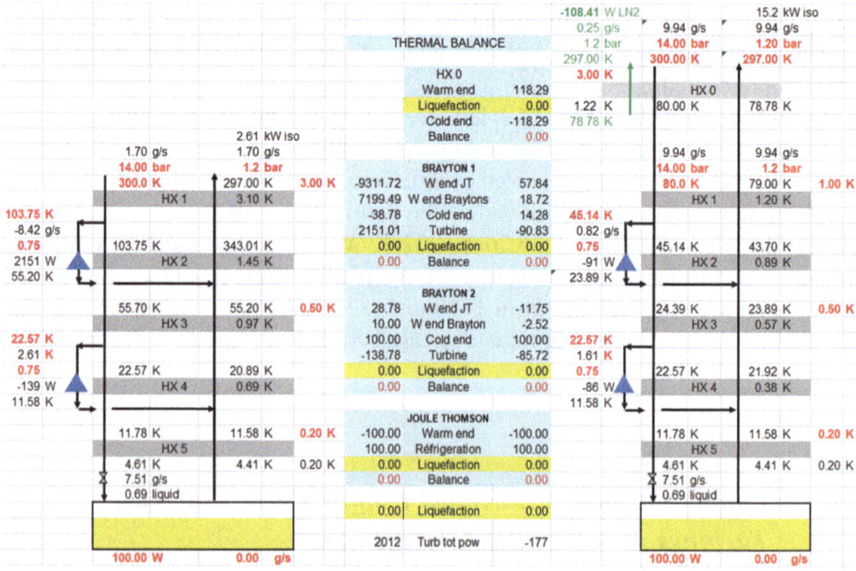

Fig. 3.60 Left, pure Claude refrigerator reference cycle with two turbines in parallel and optimised turbine temperatures, right, liquid nitrogen pre-cooled

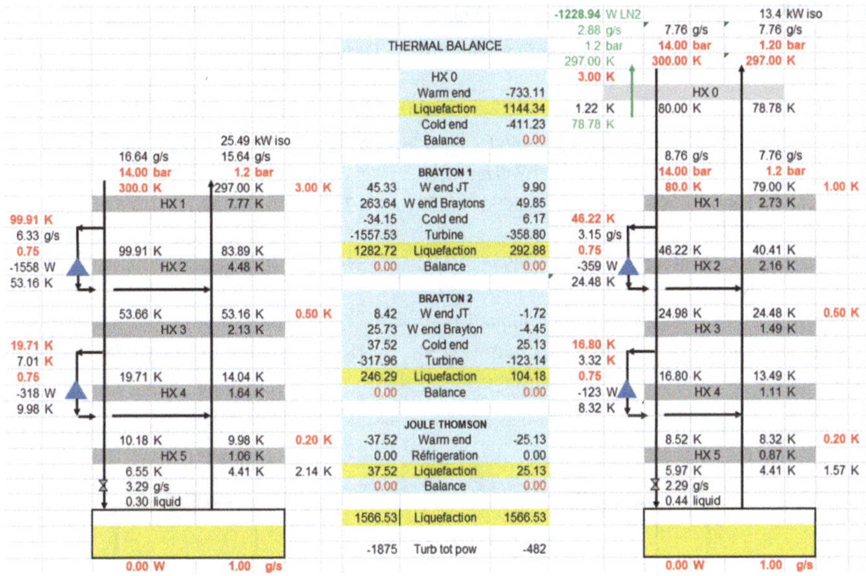

Fig. 3.61 Left, Claude pure liquefier reference cycle with two turbines in parallel and optimised turbine temperatures, right, liquid nitrogen pre-cooled

3.6.4.2 Pure Liquefier (Fig. 3.61)

For a liquefier, the cycle mass flow difference is larger: from 15.64 to 7.76 g/s.

3.6.4.3 Comparison of the Liquid Nitrogen Pre-cooling Incidence on Various Cycle Arrangements

For an equivalent duty (100 W or 1.00 g/s), same high pressure, same heat exchanger temperature differences and same turbine efficiency, the cycle flow rates vary as shown in Table 3.8.

Remark Liquid nitrogen cooling power does not compensate the turbine power reduction because, as mass flow rates are lower, the heat load on the warm end of the cycle is lower.

One can notice that liquid nitrogen pre-cooling is of a higher interest for liquefiers where the power that is extracted by the turbines is high.

Note The author has noticed that liquid nitrogen pre-cooling responds also to local habits. In fact, even in the case of a refrigerator, this method is often used in the USA and Japan, but rather rarely in Europe.

Table 3.8 Comparison of mass flow rates and turbine powers for various cycles

		Cycle compressor flow			Turbine power			*Nitrogen*	
		No LN2	LN2	Ratio	No LN2	LN2	Ratio	*Flow*	*Power*
		(g/s)	(g/s)	(–)	(W)	(W)	(–)	*(g/s)*	*(W)*
			Refrigerator						
Brayton	1000 W at 20 K	35.96	28.30	0.79	−1736	−1232	0.71	*0.81*	*−347*
Claude 2 T	100 W at 4.5 K	10.81	9.94	0.92	−314	−177	0.56	*0.25*	*−108*
			Liquefier						
Claude 2 T	1 g/s	15.64	7.76	0.50	−1875	−482	0.26	*2.88*	*−1229*

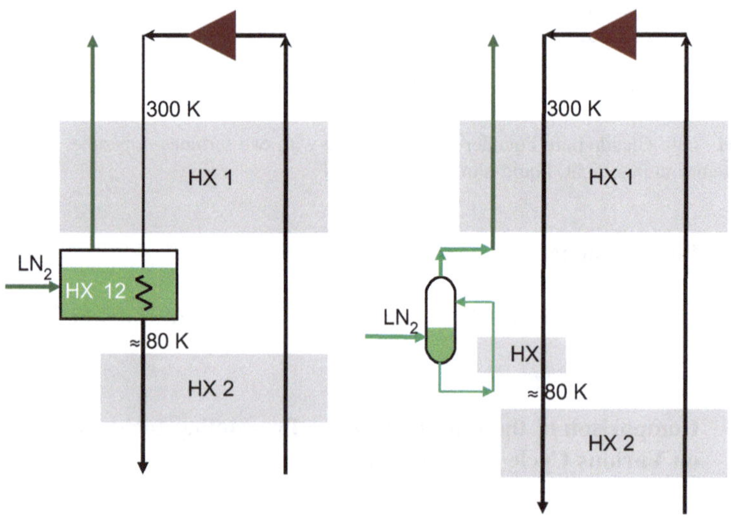

Fig. 3.62 Heat exchangers for liquid nitrogen pre-cooling

3.6.5 Liquid Nitrogen Pre-cooling Arrangements

The conventional way to pre-cool a cycle is to add a liquid nitrogen vaporiser and to recover the nitrogen-sensible heat into the warmest cycle heat exchanger, as shown on Fig. 3.62, left. The heat exchanger HX 12 is generally dipped into liquid nitrogen. For small systems, the arrangement shown in Fig. 3.63, left, can be implemented. It allows sparing the vaporiser as a separate component: the lower part of HX1 is used as the vaporiser. The liquid nitrogen level is not positively controlled; it establishes automatically at a height corresponding to the requested temperature at inlet of HX 2.

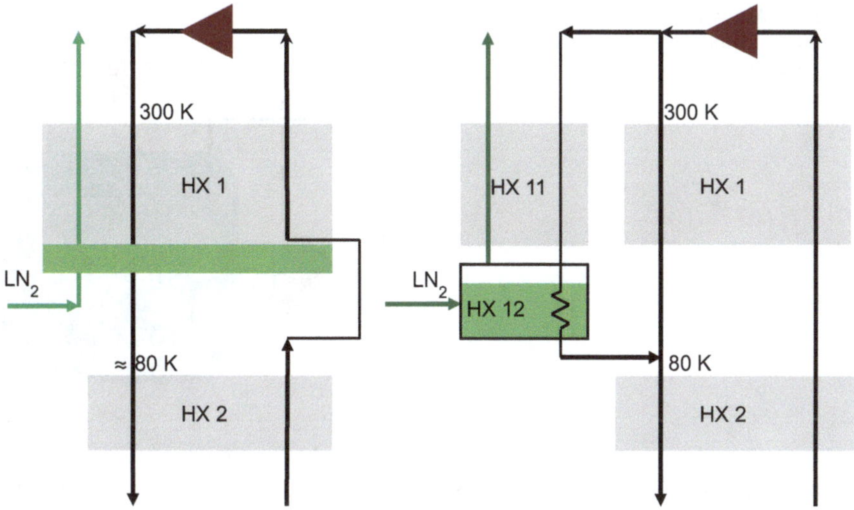

Fig. 3.63 Integrated, left, and separated, right, liquid nitrogen vaporiser

Another possibility is to use a thermal syphon as shown in Fig. 3.62, right, where two-phase that is generated in the heat exchanger decreases the density of the fluid and initiates a natural nitrogen circulation. The higher the power that is dissipated into the heat exchanger, the higher the nitrogen flow is.

The above arrangements might be dangerous if, for example, during a quench, a big amount of cold low-pressure helium is circulated from the cold to the warm end: gaseous nitrogen can be solidified into the heat exchanger channels. Nitrogen solidification is not dangerous as solid nitrogen has not a lower density as liquid (this is only right for water). However, at the time of warming up, it is not possible to control the process: if a volume of solid that is warmed up and turned into liquid and later into gas is trapped between two parts that are still full of solid, the heat exchanger cannot resist to such a pressure increase and is destroyed.

Therefore, for systems where such risk exists, the arrangement shown in Fig. 3.63, right, is better suited.

3.6.6 Interest of Liquid Nitrogen Pre-cooling

On top of sparing cycle mass flow rate and, consequently, dealing with smaller components, liquid nitrogen pre-cooling is of interest as a protection of the expansion turbine because the cycle adsorber (see Sect. 9.4.2.2) operates at a constant temperature.

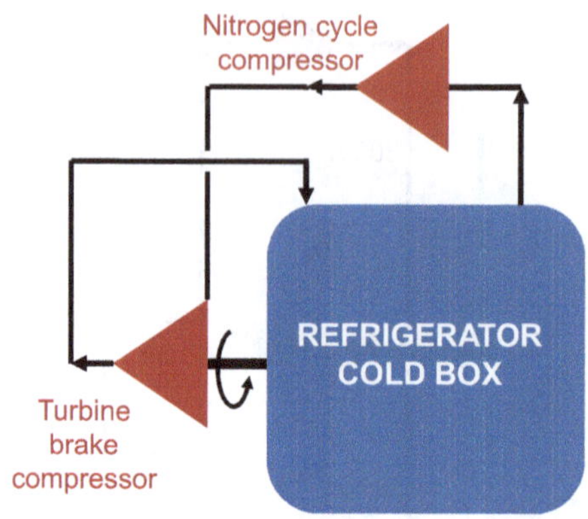

Fig. 3.64 A way to recover the power of a turbine on a nitrogen cycle

In case a refrigerator has to cool down a heavy load as a magnet, it is interesting to "help" the cooling phase down to 100 K by vaporising liquid nitrogen (see Sect. 10.3.3).

3.6.7 Nitrogen Re-condensation Cycles

For large refrigeration plants, it is of interest to pre-cool the helium cycle with nitrogen that is re-condensed by a dedicated refrigerator (re-condenser). The nitrogen re-condenser incorporates usual features for conventional air separation plants as:

- The nitrogen compression is performed with a centrifugal compressor the isothermal efficiency of which can reach up to 75% (compared to a 55% for a helium screw compressor).
- The power that is extracted by the expansion turbine is recovered, either by driving a hypersynchronous generator or a brake compressor that is arranged in series with the cycle compressor, as shown in Fig. 3.64.

Consequently, the re-condenser efficiency is higher than that of a helium cycle. Such a solution allows also keeping the system at a temperature lower than 100 K when the helium refrigerator for temperatures lower than 80 K is stopped, which is of interest in case of a long shut-down of a large system that needs a long cool-down time or that is delicate to cool down.

A new option is to integrate a turbo Brayton cycle as described in Sect. 4.6.1.1 to pre-cool the helium cycle.

Remark Such cycle structure suits exactly to the Carnot theoretical machine where the energy input is the algebraic sum of energy that is taken by the compressor and rejected by the turbine (see Sect. 3.2.4).

3.6.8 A Digest About Liquid Nitrogen Pre-cooling

Liquid nitrogen pre-cooling allows both a reduction of the size of components and energy consumption. Pre-cooling is of a higher interest for liquefiers than for refrigerators because the big duty to be fulfilled by the turbines to cool the helium to be liquefied is much reduced by nitrogen vaporisation and warming. A thumb rule: for same cycle flow, liquid nitrogen pre-cooling doubles the liquid helium production.

An economical simulation and logistic consideration should be performed prior to making the choice between pre-cooled or not system.

3.7 Cooling of Thermal Shields

3.7.1 Interest of a Thermal Shield

Reminder Heat is transmitted according to three processes:

- Solid conduction along supports that are generally made out of low-conductivity material and have a maximal length versus cross section ratio.
- Gaseous conduction and convection that are almost cut by insulation vacuum
- Radiation that is the most important process

The energy that is transmitted by radiation can be calculated according to the Stefan-Boltzmann's law:

$$\dot{Q} = kA\left(T_{\mathrm{w}}{}^{4} - T_{\mathrm{c}}{}^{4}\right) \tag{3.19}$$

where the energy that is transferred by radiation is proportional to:

- The area A of opposing surfaces times a coefficient k that depends on the shape and emissivity of the surfaces
- And, most important, to the difference of the *fourth power* of temperatures.

It is not common to deal with the fourth power of a number; therefore, let us make a simple calculation of the heat that is transferred in order to "feel" the sensitivity of this parameter with two examples, assuming, for simplification, that $k = 1$.

These two systems are supposedly made out of same materials having same shapes and emissivity (see Fig. 3.65).

System a with no thermal shield (Fig. 3.65, left)
Radiation is directly transmitted from a surface at 300 K to a surface at 4.5 K.

$$T_{\mathrm{w}} = 300\ \mathrm{K} \quad 300^4 = 8.10 \times 10^9$$

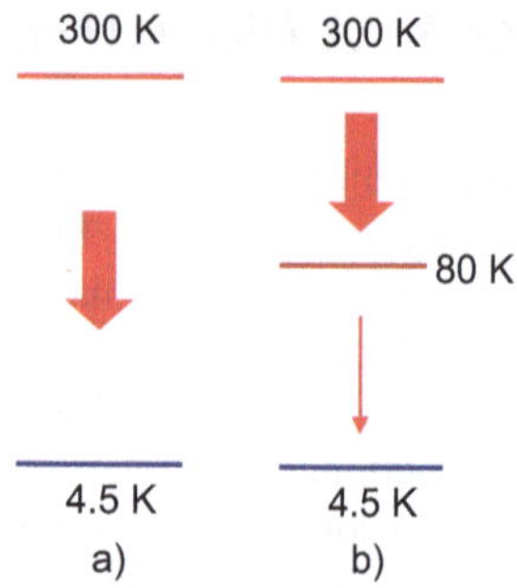

Fig. 3.65 Energy transmitted by radiation

$$T_c = 4.5\ \text{K} \quad 4.5^4 = 410$$

$$T_w{}^4 - T_c{}^4 = 8.10 \times 10^9$$

410 is negligible compared to 8.10×10^9.

The energy that is transmitted by radiation is proportional to 8.10×10^9.

System b with a thermal shield (Fig. 3.65, right)

Radiation is transmitted from a surface at 300 K to a surface at 80 K and then to a surface at 4 K.

First step:

$$T_w = 300\ \text{K} \quad 300^4 = 8.10 \times 10^9$$
$$T_c = 80\ \text{K} \quad 80^4 = 4.09 \times 10^7$$

$$T_w{}^4 - T_c{}^4 = 8.06 \times 10^9$$

8.06×10^9 is not really different from 8.10×10^9.

Second step:

$$T_w = 80\ \text{K}$$
$$T_c = 4.5\ \text{K} \quad 4.5^4 = 410$$

$$T_w{}^4 - T_c{}^4 = 4.10 \times 10^7$$

410 is negligible compared to 4.10×10^7.

The energy that is transmitted by radiation is proportional to 4.10×10^7.

The ratio of energy that is transmitted in each example is:

$$8.10 \times 10^9 / 4.10 \times 10^7 = \sim 198$$

According to this very simple example, one can draw two consequences:

1. The amount of energy that is transferred by radiation between two walls is almost only depending on the temperature of the *warmer* surface.
2. A thermal shield kept at 80 K decreases the amount of energy transferred to the colder surface by a ratio of approximately 200/1.

What is the energetic interest of a thermal shield?

It is nice to see that the amount of energy transmitted to the colder surface is divided by approximately 200, but one should not forget the cost of cooling of the thermal shield. Let us make full balances of the two systems in order to compare their thermodynamic cost of refrigeration.

System a (no shield)

1 Watt of refrigeration at 4.5 K costs 65.67 Watt at 300 K. The cost of energy at 300 K to reject the energy that hits the 4.5 K surface is proportional to:

$$8.10 \times 10^9 \times 65.67 = 5.32 \times 10^{11}$$

System b (shield)

1 Watt of refrigeration at 80 K costs 2.75 Watt at 300 K. The cost of energy at 300 K to reject the energy that hits the 80 K surface is proportional to:

$$8.06 \times 10^9 \times 2.75 = 2.22 \times 10^{10}.$$

The cost of energy at 80 K to reject the energy that hits the 4.5 K surface is proportional to:

$$4.10 \times 10^7 \times 65.67 = 2.69 \times 10^9.$$

The total cost is proportional to:

$$2.22 \times 10^{10} + 2.69 \times 10^9 = 2.49 \times 10^{10}.$$

The total amount of energy to be spent at 300 K to reject the energy that hits the 80 K and the 4.5 K surfaces is proportional to:

$$2.22 \times 10^{10} + 2.49 \times 10^{10} = 4.71 \times 10^{10}.$$

It is reduced by a ratio of:

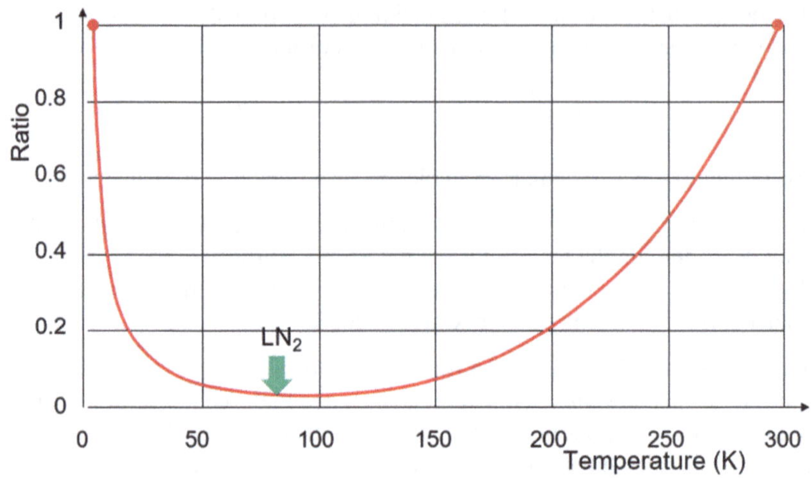

Fig. 3.66 Relative refrigeration versus shield operating temperature

$$5.32 \times 10^{11}/2.49 \times 10^{10} = 21.4/1.$$

Why is 80 K generally selected as a shield temperature? Almost certainly, because we are so much used to deal with the "liquid nitrogen" temperature and also because liquid nitrogen is, today, a fluid that is available almost anywhere. But is it wise to select such a temperature? In other words, what is the optimum operating temperature of an active cooled shield energy wise? Keeping the same simplifying assumptions as above, let us make a parametric study and plot the ratio (cost of energy with a thermal shield versus cost of energy without a thermal shield) versus shield operating temperature (see Fig. 3.66).

By some chance (that happens sometimes!), we see that the optimum shield operating temperature is approximately 100 K. So, 80 K is a correct choice and we are lucky to have liquid nitrogen. Nevertheless, in some special cases, it happens that a simulation study shows that a colder shield temperature is mandatory in order to decrease even more the heat load at LHe temperatures.

3.7.2 Various Ways to Cool Thermal Shields

The helium that cools the thermal shields is circulated in pipes that are in good thermal contact with the shield or in conducts made in extruded aluminium alloy profiles. To avoid too high pressure drops that would decrease the efficiency of the system, it is advised to circulate helium at the highest possible pressure. Such helium is generally part of the cycle high pressure flow.

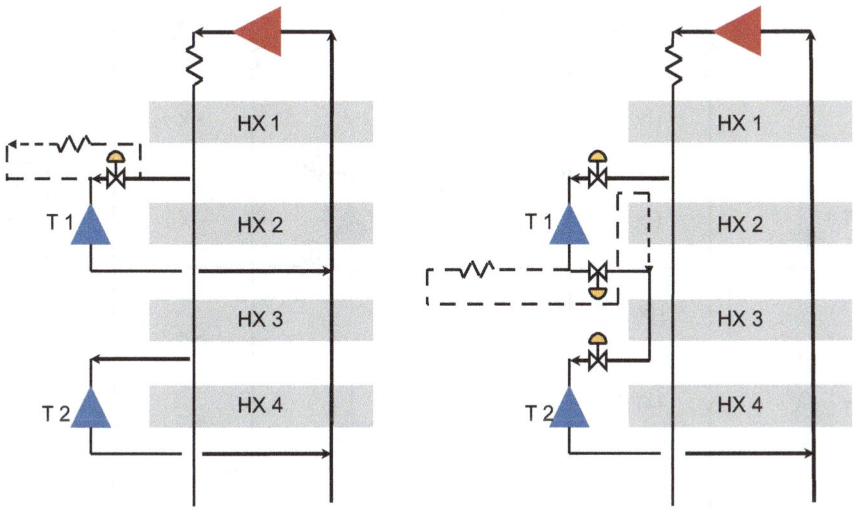

Fig. 3.67 Shield cooling 2 and 3

The operating temperatures and the pressure drop into the heat shields are defined by the system designer (the "final user"). This is a constraint for the designer of the refrigerator.

Here under, various shield cooling configurations are described. Not all of them are satisfactory, but they are examples that can be witnessed in actual machines.

3.7.2.1 When the Refrigerator Is Not Liquid Nitrogen Pre-cooled

In Fig. 3.67, left, helium is withdrawn upstream the turbine valve, then circulated in the shields and then expanded into the turbine. The shield heat load is absorbed by the turbine. Control of the shield return temperature is performed by the turbine upstream valve.

In Fig. 3.67, right, helium is withdrawn at the discharge of the turbine and circulated in the shields. A valve that is located at the discharge of the turbine allows controlling the flow in the shields. In order that the thermal load is taken care by the warm turbine instead of the cold one, helium is circulated again into heat exchanger HX 2 before it is injected in the turbine circuit. The operating pressure is lower than the high pressure: cross section of the shield circuits must be correctly sized in order to avoid too high pressure drops that would decrease the extracted power of the cold turbine. The amount of gas that is stored in the shields might be rather large. If a dangerous situation happens for the cold turbine, the time to empty such gas through the turbine might be fatal. An extra valve must be installed at the cold turbine inlet.

In Fig. 3.68 left, the operating pressure is not high, and helium that is circulated in the shields is not expanded in the turbine.

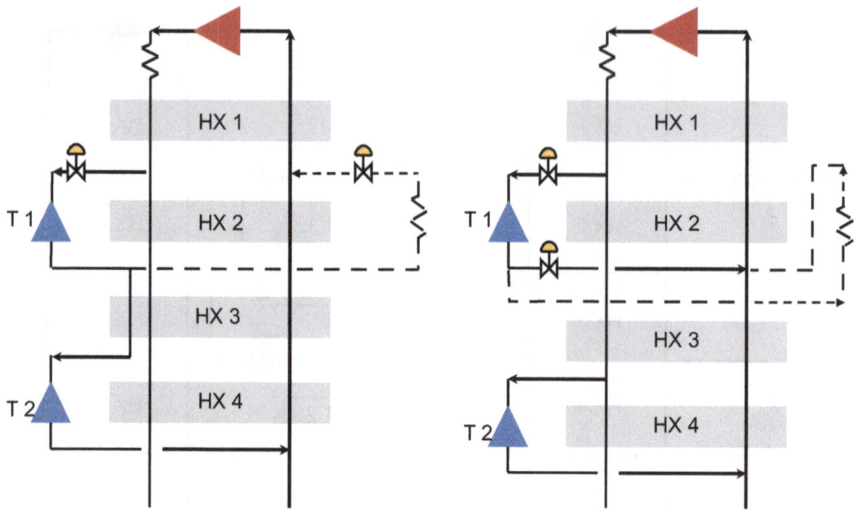

Fig. 3.68 Shield cooling 4 and 5

Fig. 3.69 Best arrangement for shield cooling

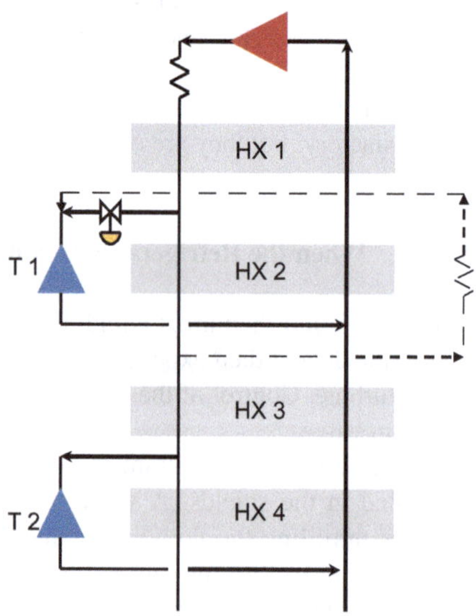

In Fig. 3.69 high-pressure helium is withdrawn in between heat exchangers HX 1 and HX 2, circulated in the shields, and then expanded into the turbine. The shield heat load is totally taken by the turbine that operates at almost the same temperatures as thermal shields, ensuring the best efficiency of the system. Control of the shield return temperature is performed by the turbine upstream valve.

This is the most efficient arrangement. It is very often used in large refrigerators.

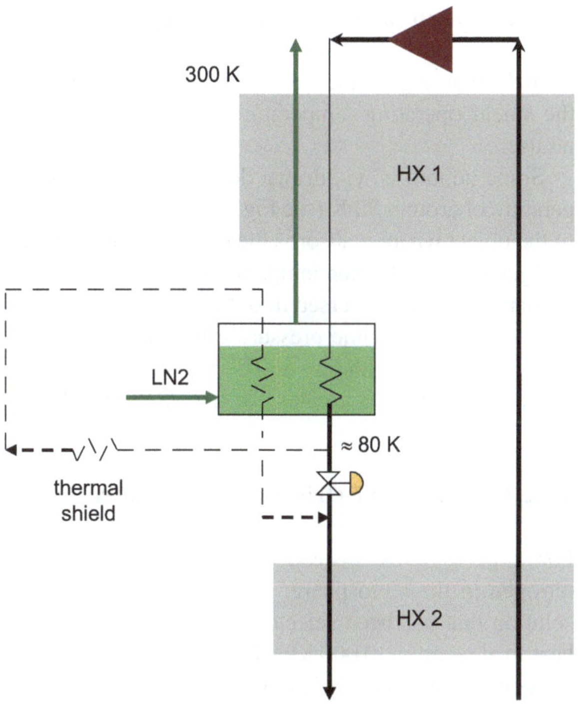

Fig. 3.70 Shield cooling when the refrigerator is LN2 pre-cooled

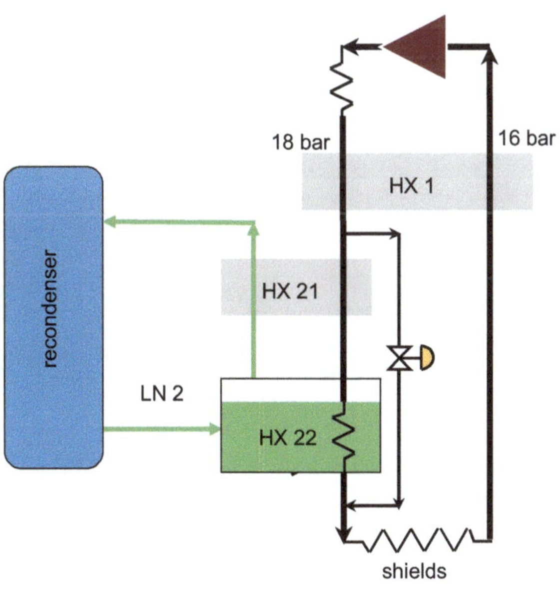

Fig. 3.71 The shield cooling system of ITER

3.7.2.2 When the Refrigerator Is Liquid Nitrogen Pre-cooled

In such an arrangement, the minimum shield temperature is, obviously, 80 K. When the shield operating temperature must be lower than 80 K, liquid nitrogen is of no use.

Some helium is withdrawn downstream the liquid nitrogen vaporiser, at a temperature of around 80 K (see Fig. 3.70). It is circulated into the circuits of the shields of the object where it absorbs the energy deposited thereon; then, it is returned to the refrigerator at a higher temperature. Helium is then circulated again into a heat exchanger that is immersed into the liquid nitrogen of the vaporiser. In order that the flow is driven by some pressure difference, a valve is fitted downstream the point where helium is withdrawn. The valve adjusts the flow rate through the thermal shields: if the flow is greater, the temperature difference is reduced.

3.7.2.3 When the Refrigerator Has a Nitrogen Re-condenser

For large systems, instead of vaporising liquid nitrogen and rejecting gaseous nitrogen in the atmosphere, a re-condenser can be an interesting feature. It is the solution that has been selected for the ITER refrigeration system where the shield heat load is around 1000 kW. For such a high power, achieving a high efficiency is imperative. (In Sect. 3.6.7, it has been explained why a nitrogen system has a higher efficiency than a helium one). The ITER nitrogen re-condensation cycle is incorporated as shown in Fig. 3.71.

Chapter 4
Special Thermodynamic Cycles

Abstract Combining the Joule Thomson and Brayton basic cycles allows designing special cycles for high powers or operating at temperatures lower than 4.5 K. The electrical power that is taken on the net increases so much that it is mandatory to find most efficient solutions.

4.1 Introduction

The cryogenic power of refrigerator units has tremendously increased in the latest decades. Figure 16.35 shows how, in the 1990s, an important step has been passed at the occasion CERN built the LEP accelerator (see Sect. 12.7.2). Here, let us assume as a convention that the wording "high power" refers to cryogenic powers higher than 1 kW equivalent at 4.5 K.

4.2 Cycles for High Cryogenic Powers

After what has been learned about basic refrigeration cycles, how an efficient cycle for a high power should be designed?

Let us use what has been learned on the former chapters.

- First, follow Mr. Carnot precepts: have a large number of cooling stages.
- Expand, in an *almost isentropic* way, *all* helium that is available, down to the *lowest possible pressure* (not any expansion into valves where no energy is extracted).
- Have the smallest, but reasonable, temperature differences in the heat exchangers.
- Operate each machine at its highest efficiency, if possible.
- But, as there could be antagonist situations between different kinds of machines (compressor, turbine), find a global operation that leads to the highest global efficiency of the system.
- Fight against any pressure drop:

G. Gistau Baguer, *Cryogenic Helium Refrigeration for Middle and Large Powers*, International Cryogenics Monograph Series,
https://doi.org/10.1007/978-3-030-51677-2_4

- In the refrigerator and distribution system warm and cold circuits
- In the final user circuits where helium that is issued from the refrigerator is circulated

In the following paragraphs, we will first concentrate on the way to improve the Brayton cycle arrangements, and then we will consider same target with the final expansion in the Joule Thomson cycle.

Remark In this chapter, cooling of thermal shields is not discussed; see Sect. 3.7.

4.2.1 Improving the Brayton Cycle Arrangements

Remember Sect. 3.4.5 where it has been shown that adding a pre-cooling Brayton cycle allows decreasing the cycle mass flow rate. What can be done in following this simple arrangement?

4.2.1.1 Piling Up, Temperature Wise, the Maximum Number of Expanders

Let us follow Mr. Carnot advice: try to introduce the maximum number of cooling stages into a cycle.

Design parameters:

- High pressure: 20.0 bar
- Low pressure: 1.0 bar
- Efficiency of expanders: 0.75

The expanders operate between high and low pressure; they are arranged in series, temperature wise (but in parallel, pressure wise), the inlet temperature of one is the discharge temperature of the warmer, as shown in Fig. 4.1 left. Due to the temperature difference that is generated by each expander, the maximum number of expanders that can be "piled up" into this cycle is 5. The final "JT" expansion is also performed with a sixth expander that can recover energy, instead of a JT valve.

Without going farther, a major impossibility occurs: the enthalpy drop of the two warmest expanders is not acceptable for actual expansion turbines (refer to Sect. 6.6. 3.1 about the enthalpy drop limitation of an expansion turbine). The high pressure can be reduced. A trial down to 15 bar (Fig. 4.1 centre) allows adding one expansion stage, but the same impossibility occurs with the two warmest turbines. Another trial at 10 bar (Fig. 4.1 right) shows that there is still an impossibility related to the warmest expander's enthalpy drops.

There are also other possible problems if the actual expanders have performances that differ from design (efficiency, mass flow rate); there will be temperature mismatches.

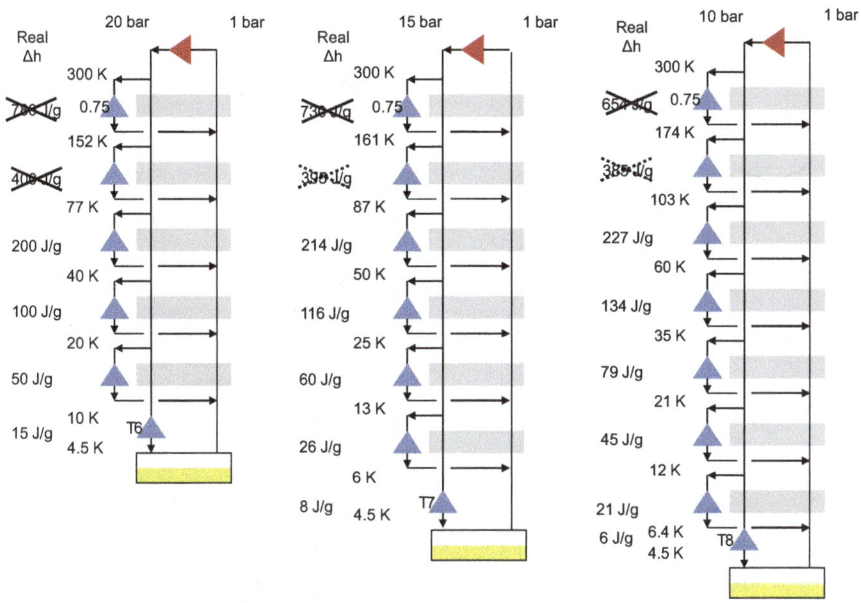

Fig. 4.1 Maximum number of expanders that can be thermally "piled up"

4.2.1.2 Inserting Heat Exchangers Between Expanders

In order to better accommodate the temperature position of the expanders, a heat exchanger can be inserted between each cooling stage.

- The first one decreases the operating temperature of the warmest expander and therefore its enthalpy drop.
- The others avoid the possible temperature mismatches.

In Fig. 4.2, left, the warmest turbine has still a high enthalpy drop. It is possible to decrease the enthalpy drop of the warmest turbines by reducing their expansion ratio in a series arrangement, as shown in Fig. 4.2, right, where the warmest turbine has, now, an acceptable, even still rather high, enthalpy drop. In Fig. 4.3, two other ways to decrease the enthalpy drop of the warmest turbines are shown. In Fig. 4.3 left these turbines expand helium from high pressure to middle pressure (4.00 bar). In Fig. 4.3 right, the warmest turbines operate in a series arrangement from high pressure to middle pressure, reducing even more their enthalpy drop.

4.2.1.3 Arranging Several Expanders in a Pure Series

When an expansion turbine has to deal with too high an expansion ratio that leads to too high an enthalpy drop, it is possible to install two expansion turbines in a pure

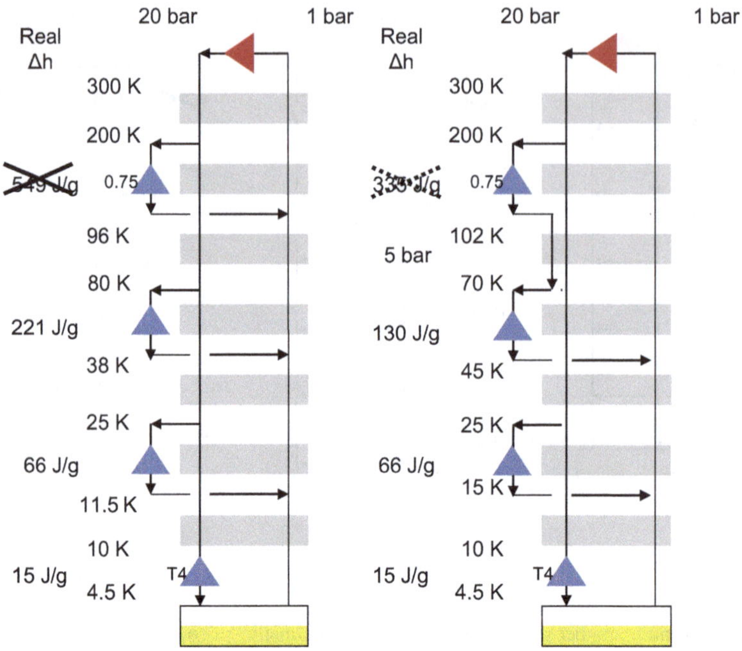

Fig. 4.2 Inserting heat exchangers between turbines

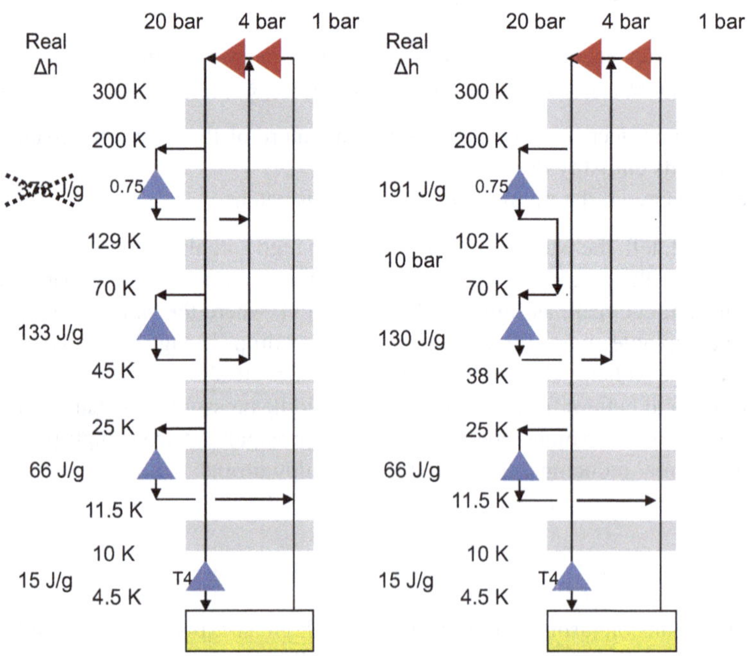

Fig. 4.3 Other ways to decrease the enthalpy drop

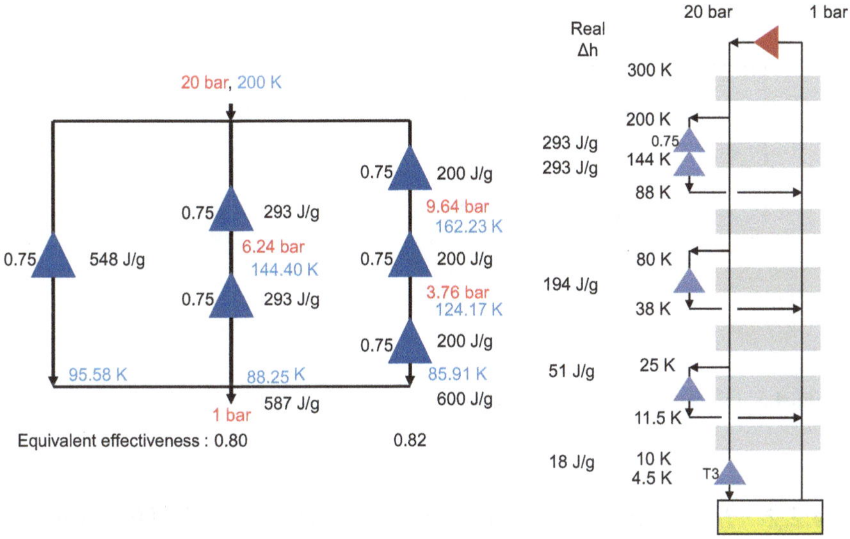

Fig. 4.4 Comparison of turbines in pure series arrangements and incorporation into a cycle

temperature and flow series arrangement: the enthalpy drop of each turbine is therefore reduced. In Fig. 4.4, three turbine arrangements are compared: one turbine alone and two or three turbines in pure series. To ease understanding, let us consider turbines having the same isentropic efficiency: 0.75. When two turbines are arranged in series, the enthalpy drop is equally shared between them; here, they have same enthalpy drop: 2×293 J/g. The calculation shows that the equivalent efficiency of the train is increased up to 0.80 and even to 0.82 for a three-turbine arrangement.

But, when the enthalpy drop is reduced, it is generally possible to get a more efficient turbine! Therefore, the equivalent efficiency of the turbine train is again increased.

On the right part of Fig. 4.4, one can see the incorporation of twin-series turbines instead of turbine T1, in order to avoid the unrealistic too high enthalpy drop that was required on Fig. 4.2, left.

4.2.1.4 General Rules for Efficient-Cycle Turbine Arrangements

Turbines must be combined in order they operate with a reasonable enthalpy drop and, therefore, a higher efficiency:

- When the enthalpy drop is high, turbines must be arranged in series, with or without heat exchangers in between. They should be spread temperature wise (at a theoretical equal constant temperature ratio).
- As soon as the enthalpy drop becomes reasonable, the parallel arrangement must be selected because it is more efficient (remember Sect. 3.5.3.3).

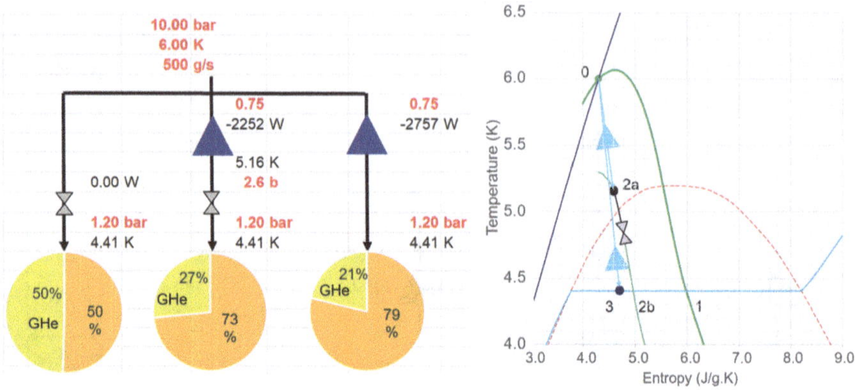

Fig. 4.5 Effect of the type of final expansion

4.2.2 *Improving the Final Expansion of the Joule Thomson Cycle*

For high efficiency cycles, it is obvious that any free expansion (e.g. in the Joule Thomson valve) must be avoided because it does not extract energy from the gas that is expanded.

Let us compare various ways to expand the "Joule Thomson" flow in Fig. 4.5.

- Through a valve (left),
- Through a turbine expanding into the supercritical region, followed by a valve (centre)
- Through a turbine expanding directly into the two-phase area (right), from 10 bar, 6.00 K (point 0 in Fig. 4.5, right) down to 1.0 bar

On the T-s diagram, in Fig. 4.5, expansion through the valve takes place along an isenthalp that crosses the 1.2 bar isobar at point 1. The second type of expansion takes place first into the turbine (0,75 efficiency) down to point 2a that is above the liquid-vapour line (therefore super-critical) and then through the valve, along an isenthalp down to point 2 b that is located at the left of point 1, which means that the concentration of liquid in the resulting two-phase is higher. The third expansion takes place through the two-phase turbine (0,75 efficiency) down to point 3 that is located at the left of point 2b, meaning that the liquid concentration is even higher. The liquid proportion that is reached by means of each expansion is also indicated on a T-s diagram in Fig. 4.5, right. The corresponding calculations are displayed in Table 4.1.

Let us consider the thermal and mass balance of the cold end (dotted rounded corner rectangle in Fig. 4.6, left). As already explained, the useful power Q_u that gets into liquid helium is transferred to the cold Brayton cycle (remember Sect. 3.3.2.3).

Table 4.1 Calculation of various kinds of expansion to generate liquid

	P	*T*	*h*	$\dot{m}$	Power	Liquid
	(bar)	(K)	(J/g)	(g/s)	(W)	(–)
	10.00	***6.00***	20.75	***500***		
Valve	1.20	4.41	20.75		0	0.50
Turbine	2.60	5.16	16.25			
+ valve	1.20	4.41	16.25		−2252	0.73
Turbine	1.20	4.41	15.24		−2757	0.79

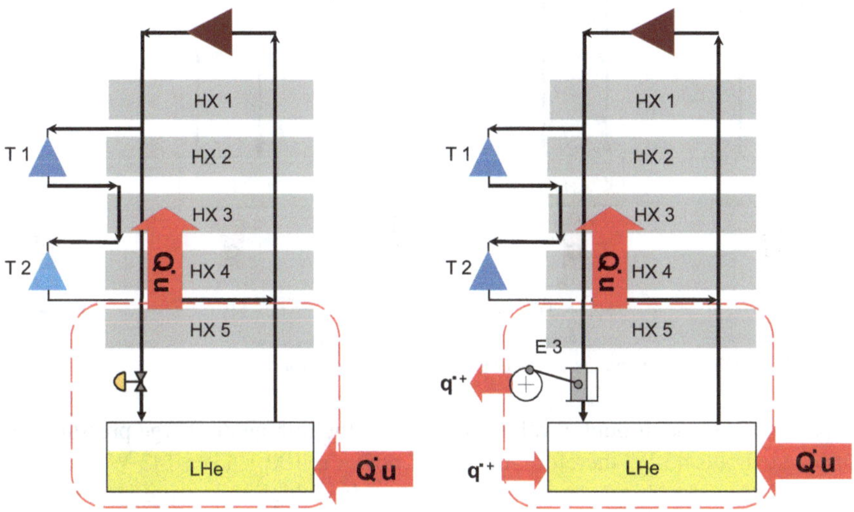

Fig. 4.6 Addition of the expander power to the cycle power

When the expander extracts a power q̇, same quantity must be injected into liquid helium to keep the thermal balance correct (Fig. 4.6, right). The refrigerator power is now Q + q̇. In other words, any power that is extracted by an expander in the cold end is extra useful power for the refrigerator.

4.2.2.1 Possible Arrangements of the Cold End for a Refrigerator

There are various ways to arrange the cold end. In Fig. 4.7, five arrangements are compared. All of them are connected to the same set of two Brayton cycles fed by same compressor processing 10.73 g/s. Remember that helium that is liquefied is added at high pressure to the cycle compressor mass flow rate.

The left cycle, SE (single expansion), is taken as a reference: its cryogenic power is 100 W.

In HT + JT, a supercritical turbine (HT) that discharges in the supercritical region, thus avoiding liquid production, performs the first part of the expansion. Final

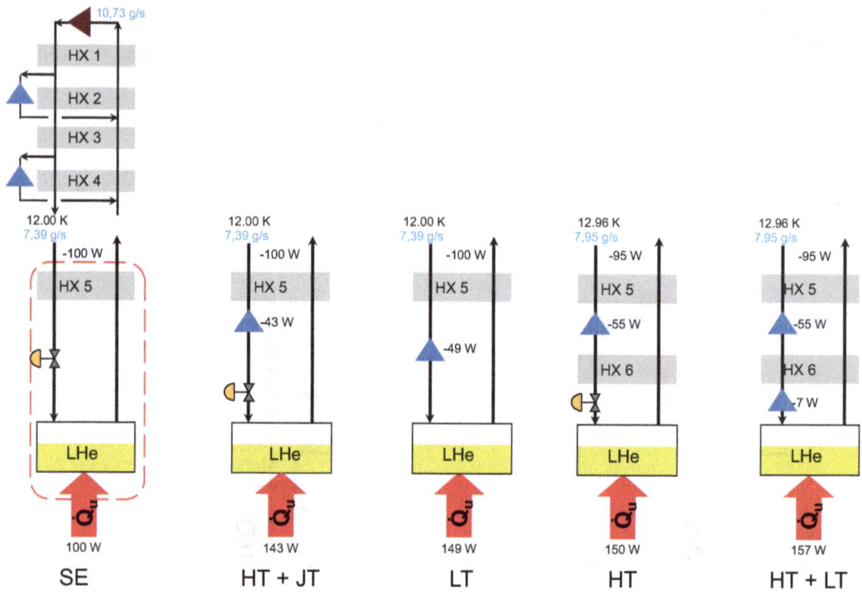

Fig. 4.7 Possible arrangements of the cold end for a refrigerator

expansion is made through a valve that controls the turbine discharge pressure. The turbine extracts 43 W; therefore, the useful power is 100 + 43 = 143 W.

In LT (liquid turbine), an expander operates from high to low pressure. There is two-phase at the discharge side. Such an expansion can be performed by either a reciprocating expander or a turbine. The limit of size of a reciprocating expander brings a limitation towards high mass flow rates. Furthermore, such machines do not have the high reliability that is required for large refrigerators. Such a solution has been used formerly because, for not very clear reasons, it was believed that a cryogenic expansion turbine could not discharge in two-phase. By the way, it is possible to operate a cryogenic expansion turbine in conditions where liquid appears at the discharge side; therefore, one can use the arrangement that is shown in LT where expansion is performed by a turbine that extracts 49 W; therefore, the useful power is 149 W. (The author selected such an arrangement that is shown in Fig. 12.17, where turbine T8 expands in two-phase.)

Introducing a heat exchanger between the turbine and the JT valve (HT) (see Fig. 4.7) increases a little bit the operating temperature of the turbine. Consequently, the discharge temperature of the Brayton cycle increases from 12.00 to 12.96 K, which allows the JT flow to be slightly increased from 7.39 to 7.95 g/s, because the Brayton cycles that operate at a higher temperature need less flow; therefore, the new useful power is 150 W.

And, finally, replacing the JT valve by another turbine (HT + LT) gives 157 W. The later arrangement has been patented by the author: N° 5,205,134, April 27, 1993.

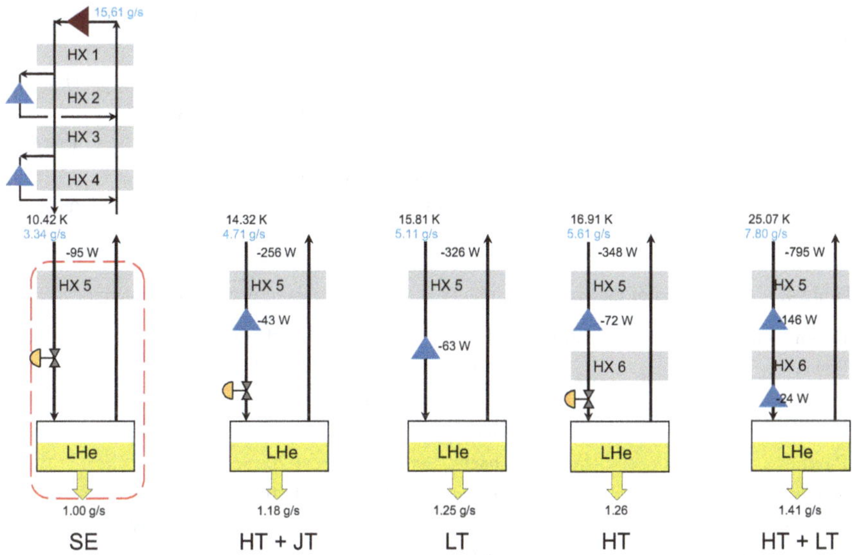

Fig. 4.8 Possible arrangements of the cold end for a liquefier

Other examples of large refrigerators incorporating such cold end arrangements are shown in Chap. 12.

4.2.2.2 Possible Arrangements of the Cold End for a Liquefier

For same cold end arrangements as for a refrigerator, one can look at similar situations in Fig. 4.8. All of them are connected to the same set of two Brayton cycles fed by same compressor processing 15.61 g/s. The helium to be liquefied is provided at the HP side of the cycle.

The left cycle, SE, is taken as a reference: it liquefies 1.0 g/s.

4.2.3 Comparing Various Refrigerators: The Equivalent Power

Another often-recurring question is how to compare refrigerators that fulfil different duties at different temperatures? According to the definition of exergy, two equivalent refrigerators would need same amount of energy to fulfil their duties.

Let us calculate the exergy of a refrigerator that can:

- Absorb 3000 W between 18.00 bar at 70 K and 17.50 bar at 80 K (thermal shield heat load)
- 10,000 W at LHe boiling temperature at 1.20 bar

Table 4.2 Thermal shields (REFPROP)

	Pressure	Temperature	Enthalpy	Entropy
	P	*T*	*h*	*s*
	(bar)	(K)	(J/g)	(J/g.K)
	1.20	300.00	1563.38	27.63
	18.00	70.00	371.53	14.42
	17.50	80.00	423.93	15.18
Liquid	1.20	4.41	1.04	0.20

Table 4.3 Calculation of the equivalent power at 4.5 K of a mixed operation refrigerator

Room temperature	***300***	(K)	Shields	Isothermal	Liquid	
Power		(W)	***3000***	***10000***		
P in		(bar)	***18.00***		***1.20***	
T in		(K)	***70.00***	4.41	300	
P out		(bar)	***17.50***	1.20	1.20	
T out		(K)	***80.00***	4.4087	4.4087	
Mass flow		(g/s)	57.25		***13.00***	
Carnot power		(W/W)	10055	670479	86643	*767177*
Carnot power at	4.5	K				65.67
Equivalent at	***4.5***	K	153	10210	1319	*11683*
Absorbed power		(kW)				***2665***
Specific power	(W_{300K}/W_{cold}					228
Carnot yield		(–)				0.288
Estimated absorbed power		(kW)				2761

- Liquefy 13 g/s at 1.20 bar (Table 4.2)

  The calculation is displayed in Table 4.3

Thermal shields

Mass flow rate to be circulated in the shields:

$$\dot{m}_{ts} = (423.93 - 371.53)/3000 = 57.25\ \mathrm{g/s}$$

$$\dot{W}_0 = 57.25 \times [(300 \times (15.18 - 14.42)) - (423.93 - 371.53)] = 10054.69\ \mathrm{W}$$

Isothermal power

$$\dot{W}_0 = 10000 \times (300 - 4.41)/4.41 = 670478.96\ \mathrm{W}$$

Liquefaction

$$\dot{W}_0 = 13.00 \times [(300 \times (27.63 - 0.20)) - (1563.38 - 1.04)] = 88643.06\ \mathrm{W}$$

Total exergy

$$\dot{W}_0 = 10054.69 + 670478.96 + 88643.06 = 767176.71\ \text{W}$$

In order to compare various refrigerators, it is interesting to evaluate their equivalent power to a reference temperature, generally 4.5 K.

At 4.5 K, $\dot{W}_0 = (300 - 4.5)/4.5 = 65.67\ \text{W/W}$

Equivalent power at 4.5 K:

$$768654.53/65.67 = 11682.89\ \text{W}$$

$$\text{The factor of merit} = 2665/11.683 = 228\ \text{W/W}$$

$$\text{The Carnot efficiency} = 767/2665 = 0.288$$

Finally, one can estimate the compressor power by using:

$$\dot{W}\ (\text{kW}) = 393 \times \dot{Q}^{0.8} \tag{3.18}$$

The equivalent power is an extremely useful tool in order to estimate the refrigerator size, efficiency and compression electrical power during the design phase of a cryogenic project.

Such a calculation is detailed in Sect. 15.8.6.

4.2.4 The Specific Duties of High-Power Refrigeration for Thermonuclear Controlled Fusion

Thermonuclear controlled fusion begins to be an important user of high-power refrigeration. It is interesting to have a glance at some specificities.

4.2.4.1 The Cable in Conduct Conductor (CICC)

Fusion machines have two main specificities that are to be considered when designing the refrigeration system:

- The coils that are parts of a fusion tokamak are generally constituted of cable in conduct conductor (CICC), circulated by supercritical helium at around 5.0 K.
- The operation of a tokamak generates a cyclic power on a short time period basis (presently, 3–4 energy pulses per hour) and on a daily basis (night and day operations).

How are these issues solved?

The CICC (Fig. 4.9, top left) is kind of a pipe (squared or circular) in which conductor strands (superconductor and stabiliser materials) are housed. Supercritical helium circulates in the remaining free cross section which is generally in the

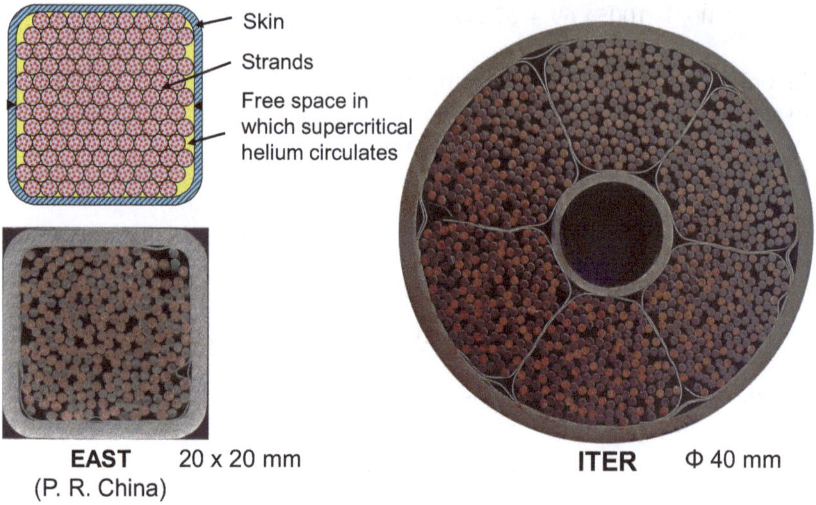

Fig. 4.9 Structure of a CICC (courtesy EAST, ITER)

order of 40% of the inner cross section of the pipe: not a very large space as can be seen in Fig. 4.9 bottom left and right. It can be noticed that in the ITER CICC, a free cross section in the centre facilitates helium circulation. However, in general, the pressure drop is a few bars.

In operation, heat leaks fall on the cable (see Fig. 4.10). As a consequence, the temperature increases along the CICC. Depending on the heat load and the supercritical helium mass flow rate, the transition temperature might be reached after a length. An acceptable temperature safety margin is found by increasing the helium mass flow rate.

4.2.4.2 Circulation of Helium Through the CICC

The thermal load of a conventional refrigerator can be cooled by circulating the "natural" coldest flow that is available for a refrigerator: the JT flow. When such flow is not sufficient, there are two possibilities:

- Circulate the available flow with intermediary re-coolers, as shown in Fig. 2.31.
- Integrate a circulator.

The first idea to integrate a circulator is shown in Fig. 4.11, left: supercritical helium is cooled into a liquid helium bath prior to be circulated through the CICC. Due to the rather high pressure drops, the power that is dissipated by the circulator is not negligible compared to the heat load on the CICC. On top left, one sees that, in this example, the circulator power is 878 W, compared to the useful load that is 3500 W.

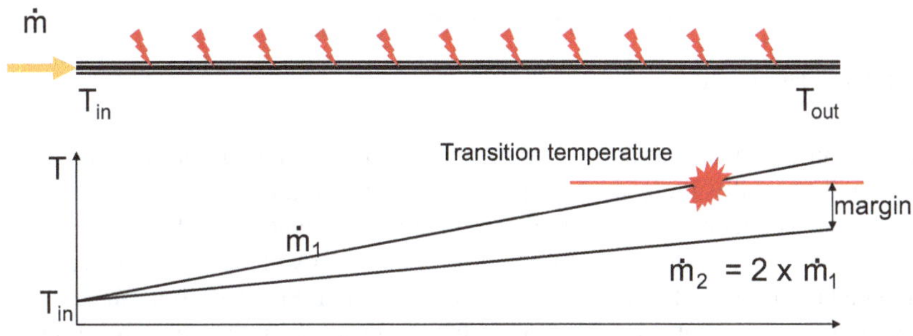

Fig. 4.10 Temperature variation along a CICC

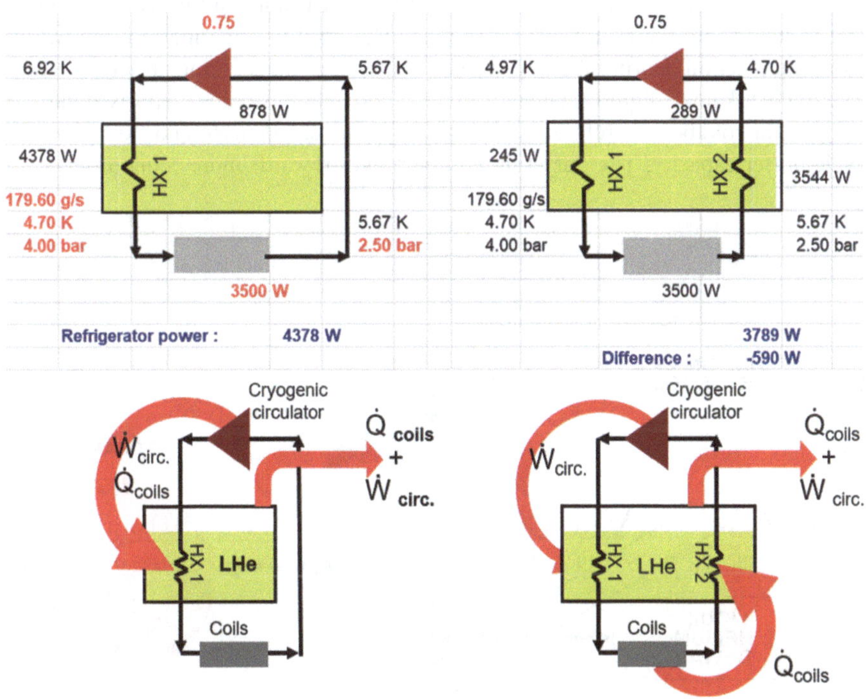

Fig. 4.11 How to integrate a cryogenic circulator

As the circulator power is very sensitive to its suction temperature, it is of interest that supercritical helium that gets into the circulator be at the lowest value: a heat exchanger, HX2, must be incorporated downstream the thermal load, so the circulator suction temperature decreases from 5.67 K to 4.70 K and the circulator power is now only 289 W. The power to be absorbed by the refrigerator is reduced by 14%!

4.2.4.3 The Power Periodic Variation Regimes

4.2.4.3.1 The Daily Regime

The present tokamaks are still experimental machines. They do not operate permanently and generally, there are no experiments during night. In order to spare energy, the tokamak is operated either at higher temperature than nominal if coils operate at a temperature lower than 4.5 K or at least at lower power because less current is circulated in the coils. The power evolution is shown in Fig. 4.12, bottom. Instead of designing the refrigerator in order to be able to absorb the maximum power and run at lower power during night, it is interesting to aim at the average power, shown by the dotted line.

During night, as the available refrigerator power is higher than that required, the system takes gaseous helium from the cycle buffer capacity and turns it into liquid that is stored into the liquid helium storage. The operating regime moves towards more liquefaction. During the day, as the refrigerator power is not sufficient to take care of the tokamak, liquid is taken from the liquid helium storage; vaporised into the refrigerator cold box, therefore, operating as a partial economiser; and stored into the cycle buffer capacity. The operating regime moves towards more economiser.

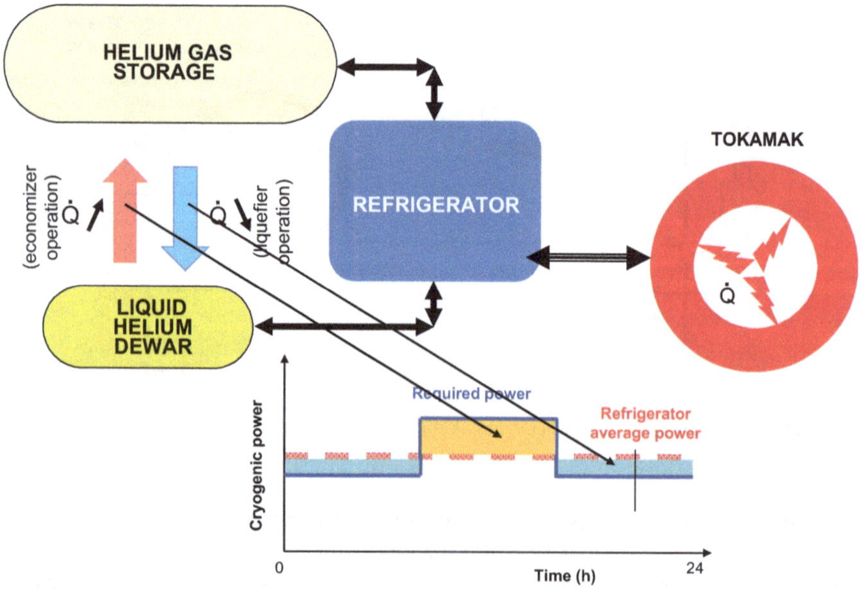

Fig. 4.12 Variation of cryogenic power along the day

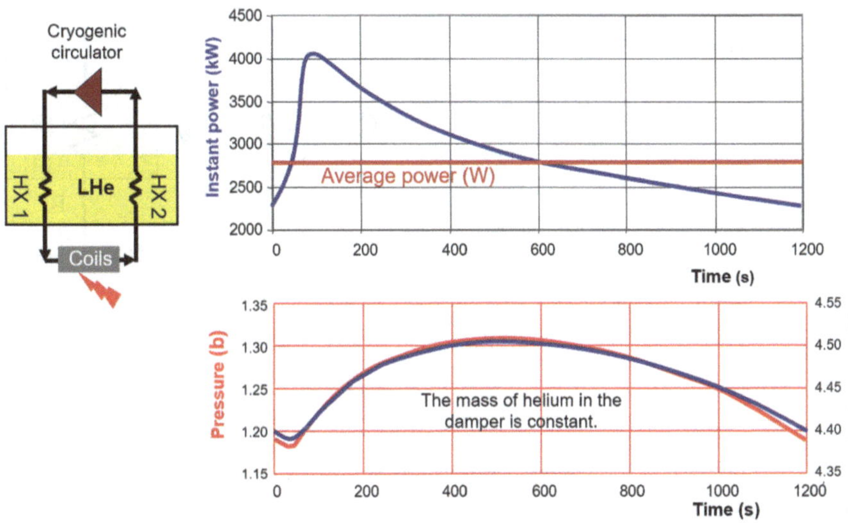

Fig. 4.13 Change in cryogenic power during one run, evolution of pressure and temperature in a constant volume thermal damper along the energy pulse

4.2.4.3.2 The Energy Pulse Regime

During the day-time, when the tokamak is operated, as the magnetic field of the toroidal coils is highly varied, energy pulses generated by eddy currents circulating in the heavy casings of the coils happen two to three times per hour. The power rises sharply and then decreases down to the initial value as shown in Fig. 4.13, top right. Again, it is interesting not to design the refrigerator for the peak power (4000 W) but for the average power (2800 W). One solution is to operate with a liquid helium bath that operates at constant volume, as shown in Fig. 4.13, left. When energy is dumped into the helium that is stored into the capacity, both pressure and temperature rise as shown in Fig. 4.13, bottom. This is a so-called constant-volume thermal damper: energy is temporarily stored into the helium quantity that is located into the damper.

Energy is to be removed from the thermal damper at constant power in order to run smoothly the refrigerator. The power that is dumped by supercritical helium circulating into the closed loop is:

$$\dot{Q} = \dot{m} \times (h_1 - h_2) \tag{4.1}$$

h_1 depends on the energy that has been dumped into the coil and casings, and there is no possible action on it; h_2 is almost constant as it depends on the temperature of the damper, and it cannot be controlled either.

The only parameter that can be controlled is the mass flow rate that is circulated in the system (Fig. 4.14).

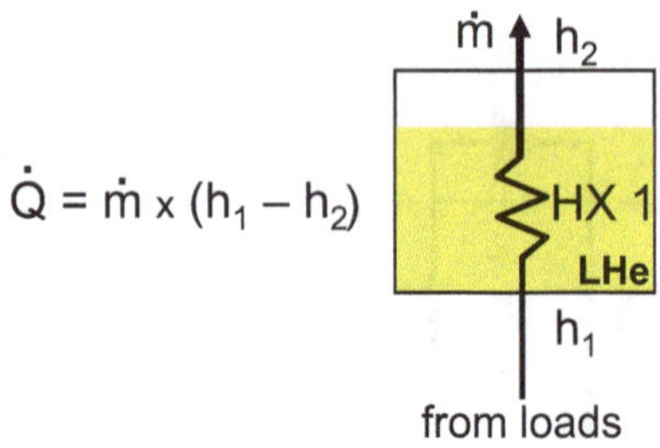

Fig. 4.14 Operation of the constant volume thermal damper

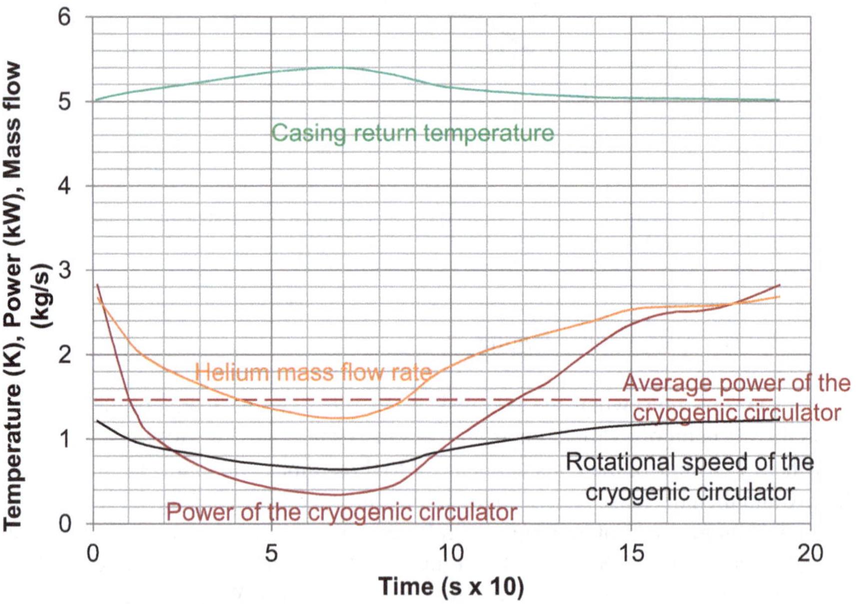

Fig. 4.15 Evolution of various parameters during the energy pulse clean

This is possible by varying the rotational speed of the circulator as shown in Fig. 4.15, where various parameters are displayed during the time of an energy pulse recovery. Another advantage is that the power that is absorbed by the circulator is reduced. In the example, the average power of the circulator is divided by two. Such spared power is available to help cooling the system.

4.3 Cycles for Temperatures Lower than 4.5 K

4.3.1 How to Reach Temperatures Lower than 4.5 K?

The temperatures lower than 4,5 K are obtained by maintaining on the liquid helium bath an adequate pressure, lower than atmospheric pressure. For example, in Fig. 4.16, one can see that the bath pressure for 1.8 K is 0.0164 bar, which corresponds to a compression ratio of 61, with regard to atmospheric pressure.

4.3.1.1 General Arrangements

The various ways to pump on the helium bath are shown in Fig. 4.17:

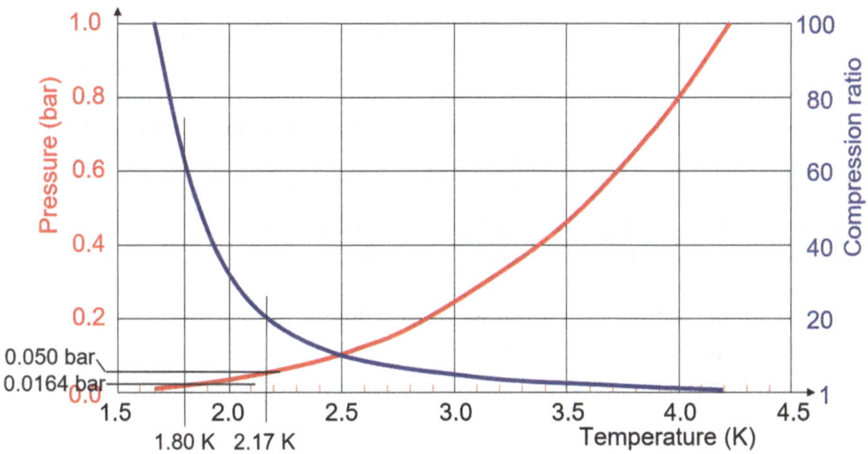

Fig. 4.16 Saturation pressure and compression ratio versus temperature

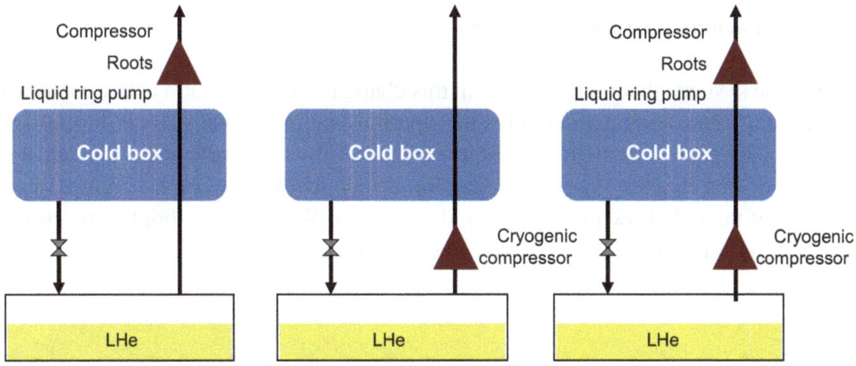

Fig. 4.17 Pumping methods

- The full compression at room temperature (left)
- The full cryogenic compression (centre)
- The mixed compression: at first cryogenic and then at room temperature (right)

The compression at room temperature can be performed with screw compressors, "Roots" (see Sect. 6.2.6.3) or, formerly, liquid ring pumps (see Sect. 6.2.6.5). The cryogenic compression is performed with cryogenic centrifugal compressors (see Sect. 6.7.3).

Of these three processes, the one who uses the compression at room temperature is, only in a pure theory, the most efficient because the heat of compression is evacuated at room temperature; therefore, it is not to be extracted by the refrigerator, as it is the case for a cryogenic compression which has an incidence on the refrigerator efficiency.

But this implementation quickly collides with a limit of size of the machines of compression: the volume flow of these machines becomes enormous according to the expected temperature and power. For example, the necessary volume flow to obtain 300 W at 1.80 K, assuming an unrealistic pressure drop of only 0.011 bar, is 60,000 m^3/h at 0.005 bar. To this issue is added another one concerning the low-pressure heat exchanger. Indeed, the pressure drop between the helium bath and the suction side of the compression system must be very low (around 0.010 bar). Yet such a heat exchanger with very low-pressure drop has a poor efficiency, i.e. a very large temperature difference. For these reasons, the current refrigerators are designed with a partially or totally cryogenic compression.

In the second and third processes, the heat of compression at cryogenic temperatures, although lower that released during a room temperature compression, must be absorbed by the refrigerator. It can be a rather high load, compared to the useful thermal load (remember Sect. 4.2.4.2).

In the first and third processes, a part of helium circuits, located between the cold box and the machines of compression, is at sub-atmospheric pressure. If the leak tightness is not perfect, air in leaks that are harmful to long-term operation can take place.

4.3.1.2 Various Possible Structures

Even if the systems that are described in this chapter allow operation at temperatures lower than 4.2 K, such temperatures or superfluid helium never exits in the refrigeration plant itself but in the cryostat where the object to be cooled sits. In the sketches of this chapter, the phase separator at such low temperatures is represented as a part of the cold bow to facilitate understanding. See Chap. 5 about connections between the refrigerator cold box and the cryostat that is housing the object to be cooled.

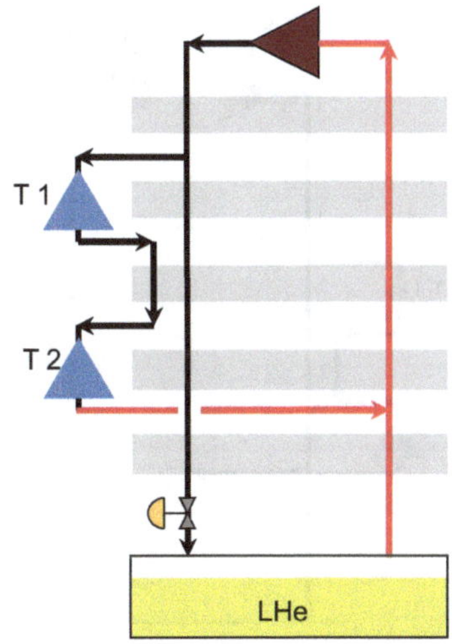

Fig. 4.18 Cycle compressor operating at sub-atmospheric suction pressure

4.3.1.2.1 Cycle Compressor Operating at Sub-atmospheric Suction Pressure

For moderate temperatures (around 4.0 K), it is possible to operate the cycle compressor at a suction pressure that is lower than atmospheric pressure (around 0.8 bar) as shown in Fig. 4.18.

The cycle compressor mass flow rate is obviously reduced; therefore, the cryogenic power is reduced too.

In the 1990s, at HERA, Hamburg, Germany, the operators wanted to run their magnets at a slightly lower temperature. They simply choose to decrease the suction pressure of the first stage cycle compressors under atmospheric pressure instead of modifying the cold box. The system operated satisfactorily for years, but one should mention that there was a full flow purifier fitted upstream the cold box.

4.3.1.2.2 Room Temperature Suppressor and Ejector

In order to avoid oversizing the cycle compressor, it is possible to process only the JT flow with a suppressor (a compressor operating at a suction pressure that is lower than atmospheric one) that feeds the cycle compressor at atmospheric pressure as shown in Fig. 4.19.

In the left part of the figure, the low-pressure helium flows into the cycle heat exchangers. The conventional heat exchangers (of the plate and fin aluminium alloy type (see Sect. 6.4.2)) hit a limitation to reach low-pressure drops. In the right part, special low-pressure heat exchangers are implemented (see Sect. 6.4.1.2).

Another way to pump on the liquid bath from room temperature is to use an ejector that is fed by a part of the high-pressure flow provided by the cycle compressor as shown in the right-top of Fig. 4.19.

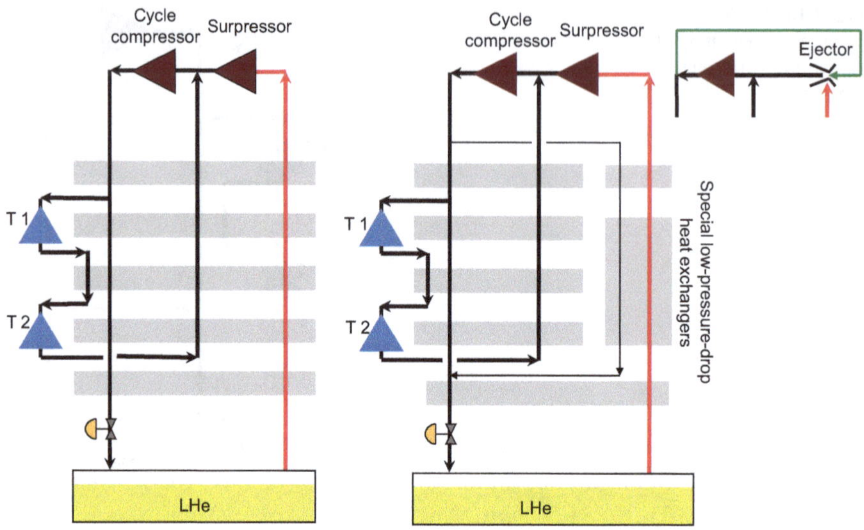

Fig. 4.19 Suppressor or ejector keeping the cycle compressor suction at atmospheric pressure

4.3.1.2.3 Cold Ejector

As helium that is expanded in a JT valve does no "work" (it is a simple isenthalpic expansion), there is a tricky way to use the JT flow by processing it through a cold ejector, in order to pump on a liquid helium bath. Consider the left part of Fig. 4.20. It is a two-turbine Claude cycle where an ejector is located in place of the JT valve (about gas ejector, see Sect. 6.2.6.1). There is not difference in operations, except that, at the suction side of the ejector, the pressure is sub-atmospheric. It is therefore possible to pump on another liquid helium bath, as shown on the right part of the figure. The second bath, the pressure of which is, for example, around 0.8 bar, is fed with pure liquid helium at 1.2 bar. The ejector processes helium vapour that is the result of vaporisation of liquid by the thermal load plus vapour that has been generated by expansion of liquid. Note that a heat exchanger is located between the two baths in order to reduce the amount of vapour generated by expansion (see Sect. 4.3.1.3).

In the 1960s, the Dutch company Philips built small helium liquefiers that incorporated a cold ejector (see Sect. 16.9).

In case current leads are to be cooled, it is necessary to use a warm ejector fed by a small flow from the cycle compressor, as shown in Fig. 4.20, right.

4.3.1.3 An Important Component: The Joule Thomson Heat Exchanger

Prior to expansion at the very low pressure, the helium can be either supercritical or liquid, at a temperature in the order of 4.5 K. In the left part of Fig. 4.21, pure liquid

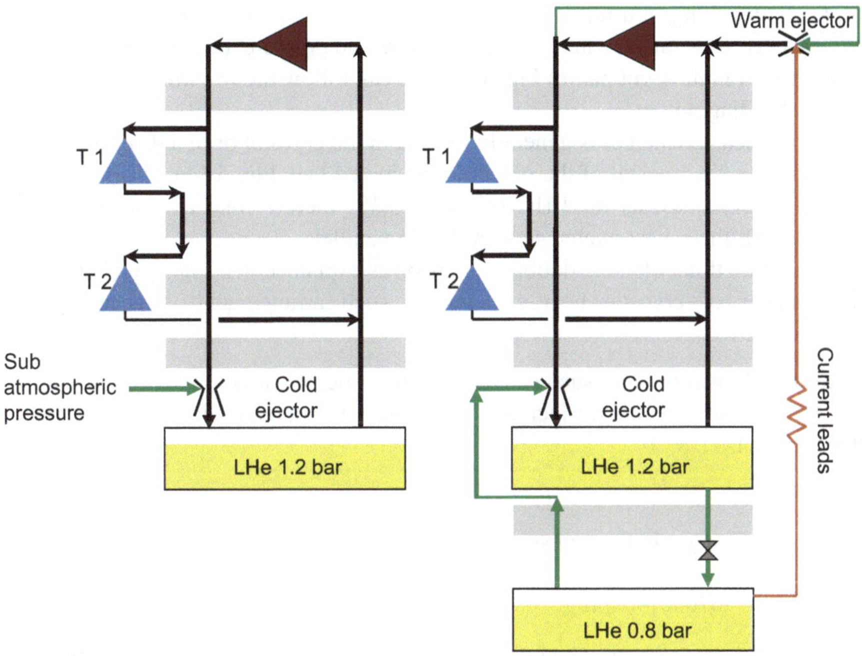

Fig. 4.20 Combination of cold and warm ejectors

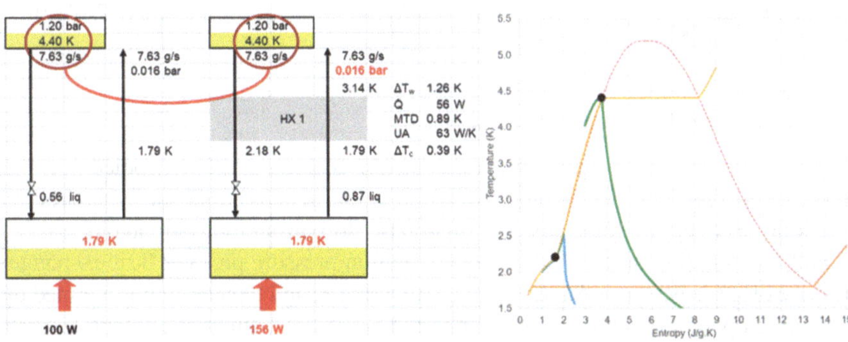

Fig. 4.21 Interest of the JT heat exchanger (HEPAK)

helium is directly expanded down to 0.016 bar (corresponding to 1.8 K). For a mass flow rate of 7.63 g/s, the cryogenic power is 100 W.

In the centre, a heat exchanger has been incorporated upstream the expansion valve. It allows to cool the JT flow against the low-pressure helium, prior to expansion. As a consequence, the temperature upstream the valve is 2.18 K instead of 4.4 K, resulting in a liquid concentration of 87% instead of 56%. For the same helium mass flow rate as same inlet temperature and pressure, the cryogenic power is now 156 W. In the right part of Fig. 4.21, one can follow the two kind of expansions on the T-s diagram.

Such a heat exchanger is generally located in the cryostat or in a distribution box located near the cryostat of the object to be cooled (see Fig. 5.11). When there are numerous cryostats, as for LHC (see Fig. 5.13) there is one exchanger in each cryostat. All these heat exchangers are fed in parallel.

One could think of using double expansion as explained in § Fig. 3.31, but, due to the number of cryostat heat exchangers, such arrangement is generally not implemented.

Remark Remember that such calculations must be performed with HEPAK only!

Descriptions of actual refrigerators operating at temperatures lower than 4.5 K are done in Sect. 12.8.

4.3.1.3.1 Cryogenic Compressor

4.3.1.3.1.1 Mixed Pumping

In order that the cryogenic compressor processes the minimum flow, expansion is performed from 1.20 bar pure liquid that is cooled against very-low-pressure flow in the JT heat exchanger that has been described in Sect. 4.3.1.3. As shown in Fig. 4.22, helium is firstly compressed in one or more cryogenic compressors and then warmed up through the heat exchangers and finally compressed up to atmospheric pressure.

4.3.1.3.1.2 Full Cryogenic Pumping

Helium that is expanded may come from supercritical that has been cooled down to about 4.5 K; see Fig. 4.22, right. It is a better solution than liquid for transportation over long distances because possible two-phase flow instabilities are avoided.

Full compression up to the compressor suction pressure plus the heat exchanger pressure drops is performed at cryogenic temperature, and then helium is warmed up to room temperature in the heat exchangers. For a same liquid helium bath temperature, there are, obviously, more compression stages. The discharge temperature of the final stage is higher; therefore, compressed helium is returned at a higher temperature into the heat exchanger battery.

4.4 Example of a Cycle T-s Diagram

In Fig. 4.23, one can see the simplified T-s diagram of a refrigerator operating mainly at 2,0 K and simultaneously cooling thermal shields.

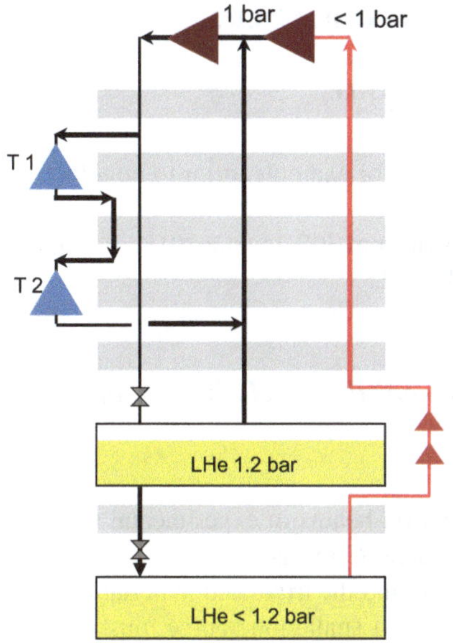

Fig. 4.22 Mixed and full cryogenic pumping

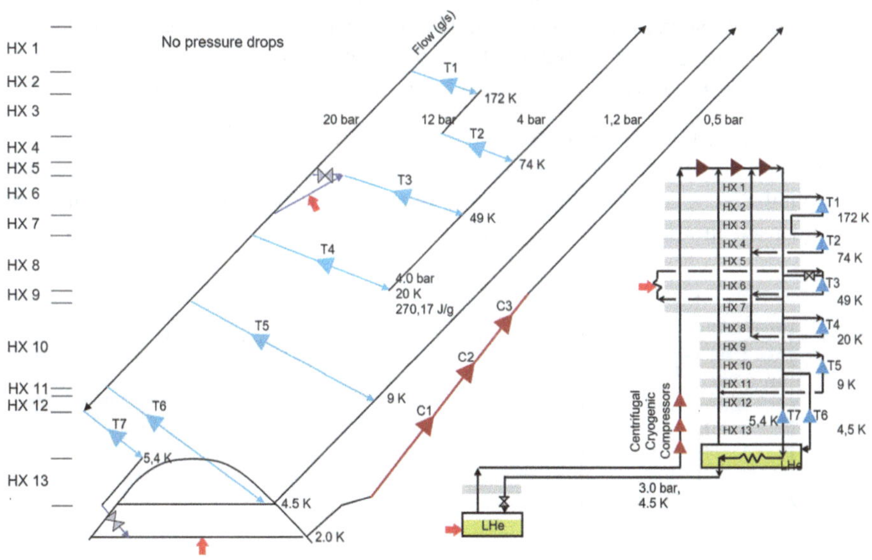

Fig. 4.23 The T-s simplified diagram of a 2 K refrigerator

At each fluid change after a transformation, pressure, temperature and enthalpy are mentioned. Mass flow rates are also indicated. At the left part, the heat exchangers are indicated.

4.5 Special Applications of Pure Brayton Cycles

4.5.1 Using a Brayton Cycle to Cool Gas

The Brayton cycle can be used for cooling gas, for example, prior to liquefaction of a gas.

In Fig. 4.24, 1 g/s of helium is cooled down from 300 down to 80 K. It is a part of a liquefier cycle that can be seen in Sect. 3.5.3.2.

4.5.2 Using a Brayton Cycle to Re-condense Hydrogen or Deuterium

The Brayton cycle is often used to cool neutron beams in experimental reactors or spallation sources where a Cold Neutron Source (CNS) is fitted.

To get better interactions between neutrons and the irradiated materials, neutrons issued either from a fission research reactor or a spallation source must be cooled down in order to reduce their energy (velocity). Slowing down is performed by making these neutrons cross some moderator material (hydrogen or deuterium) volume. In order to get the moderator fluid highest density and the neutrons to the desired energy, such process is performed at cryogenic temperatures: the moderator can be in a boiling liquid, sub-cooled liquid or supercritical state.

There are various ways to operate; three of them are described here.

Liquid hydrogen or deuterium

The saturated hydrogen (or deuterium) vapour that is generated by the neutron flux through the cell, as in Cold Neutron Sources ORPHÉE, at Saclay, France, is re-condensed in a heat exchanger that is circulated by cold helium provided by a

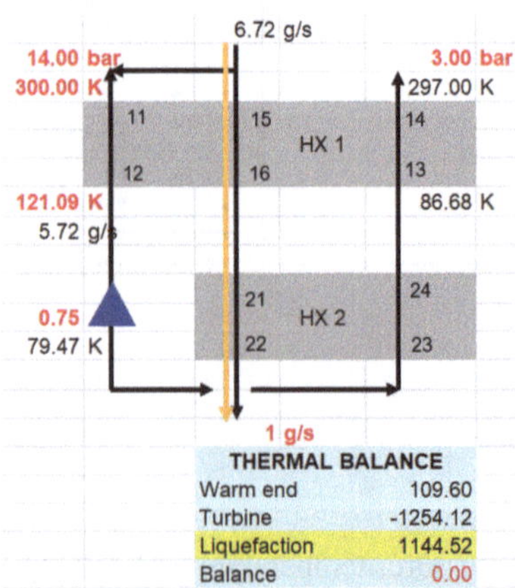

ΔT_w	3.00 K
$\dot{Q}$	5316 W
LMTD	12.85 K
UA	414 W/K
ΔT_c	34.41 K
W_T	-1254 W
ΔT_c	0.50 K

	P (bar)	T (K)	h (J/g)	ṁ (g/s)
11	14.00	300.00	1567.56	5.72
12	14.00	121.09	637.66	5.72
13	3.00	86.68	455.83	5.72
14	3.00	297.00	1548.39	5.72
15	14.00	300.00	1567.56	1.00
16	14.00	121.09	637.66	1.00
21	14.00	121.09	637.66	1.00
22	14.00	79.97	423.05	1.00
23	3.00	79.47	418.29	5.72
24	3.00	86.68	455.83	5.72
12	14.00	121.09	637.66	5.72
23	3.00	79.47	418.29	5.72

Fig. 4.24 Brayton cycle for cooling gas

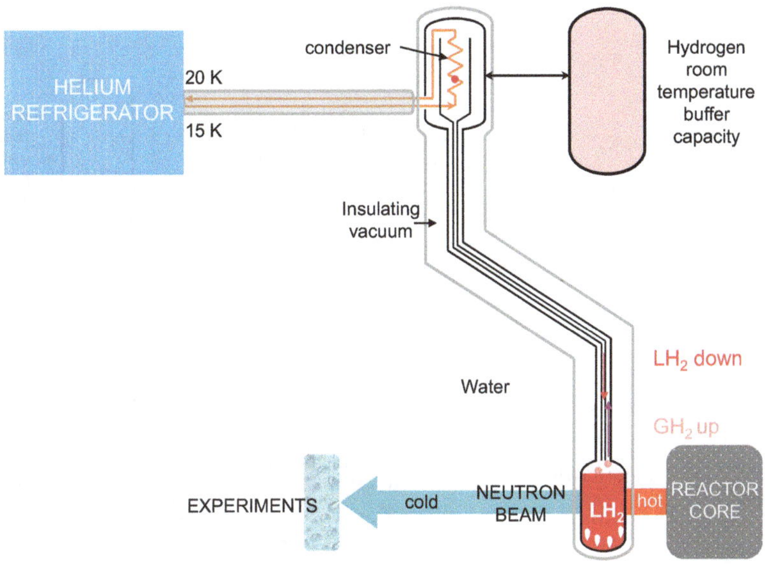

Fig. 4.25 A Cold Neutron Source cooled by liquid hydrogen or deuterium circulated by a heat pipe thermosyphon

dedicated refrigerator (see Sect. 12.1). Circulation of hydrogen in a heat pipe is "natural" by difference of density of liquid flowing down into the central pipe and vapour rising to the re-condenser.

Such a refrigerator operates according to a Brayton cycle with one expansion turbine (remember Sect. 3.4.2.1) for re-condensing the saturated vapour of either hydrogen (boiling point, 20.86 K at 1,20 bar) or deuterium (24.26 K at 1.20 bar) created by the vaporisation of the liquid under the heat load provided by the neutron radiation as in ORPHÉE, at Saclay, France (see Sect. 12.1), or in the Institute Laue Langevin (ILL), in Grenoble, France. A hydrogen or deuterium room temperature buffer capacity allows keeping the pressure at a reasonable value when the system is at room temperature (Fig. 4.25).

As, in a CNS cooling system, hydrogen or deuterium is kept cold, it is not necessary to use an ortho-para converter (remember Sect. 1.4.4.1).

Supercritical hydrogen

When hydrogen is supercritical, it is necessary to circulate it with a mechanical cryogenic circulator, generally a centrifugal machine. In case the refrigerator or cryogenic circulator is off, a room temperature circulator allows keeping the cell at almost room temperature and therefore allowing operating the reactor without melting the cell. A hydrogen buffer capacity allows keeping the pressure at a reasonable value when the system is at room temperature (Fig. 4.26).

Here, too, as hydrogen or deuterium is kept cold, it is not necessary to use an ortho-para converter.

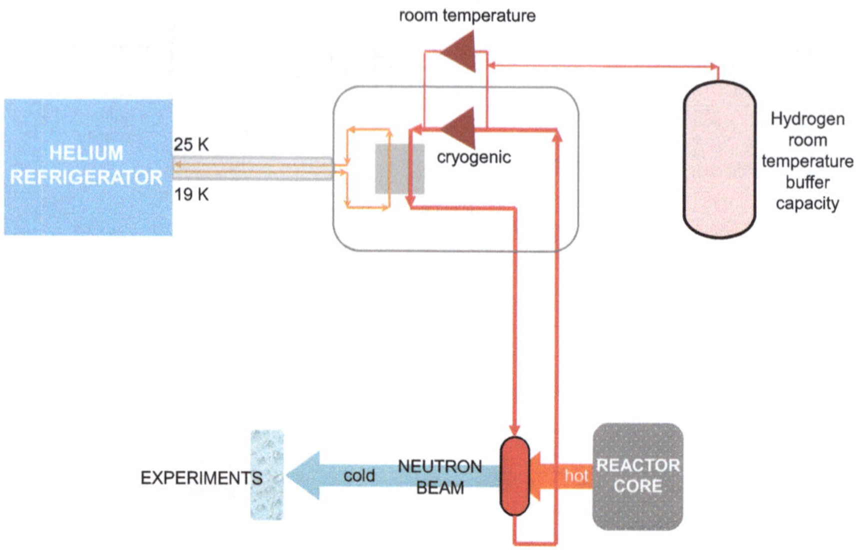

Fig. 4.26 A CNS cooled by supercritical hydrogen circulated by a mechanical circulator

Subcooled liquid deuterium

Other systems can operate with subcooled liquid deuterium as the ANSTO system, in Australia. Here it is the density difference of liquid deuterium in a thermal syphon that naturally circulates the moderator. A deuterium buffer capacity allows keeping the pressure at a reasonable value when the system is at room temperature (Fig. 4.27).

Here, too, as hydrogen or deuterium is kept cold, it is not necessary to use an ortho-para converter.

4.5.3 Using a Brayton Cycle to Liquefy Hydrogen

Liquefaction of hydrogen is easily performed with a helium Brayton cycle that cools and condenses hydrogen. Remember that an ortho-para converter must be used (Sect. 1.4.4.1).

The interest of such a cycle sits in:

- Its simplicity
- The possibility to avoid the compression of hydrogen, provided hydrogen is available at a pressure of a few bar.
- Less constraints related to hydrogen safety, only the cold box must be located in a suitable room.
- The possibility, when the liquefier is in operation, to avoid any hydrogen boil-off from the storage by feeding it with slightly subcooled hydrogen.

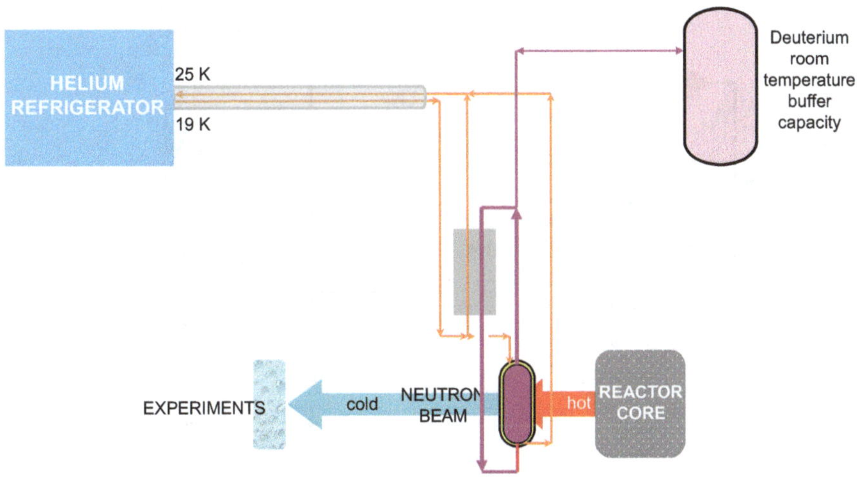

Fig. 4.27 Subcooled liquid deuterium and liquid thermosyphon

When hydrogen is liquefied, it has no time to reach a natural equilibrium, and the heat of transition from ortho to para vaporises part of the liquid (see Sect. 1.4.4.1). Therefore, it is important to force such conversion *during* the liquefaction process by using special converters as $Fe(OH)_3$. The heat of conversion is removed by the liquefier. A smart way is to "stuff" the hydrogen channels of the heat exchangers with the converter, as shown in Fig. 4.28.

If the liquefier is pre-cooled with liquid nitrogen, conversion up to 50% para is performed at liquid nitrogen temperature.

Figure 4.28, left, shows an example of a very simple hydrogen liquefier with a Brayton cycle pre-cooled with liquid nitrogen. In order to compensate the heat leaks of the storage, the liquid hydrogen is subcooled. Such a cycle can also be fitted with two turbines as shown in Fig. 4.28, right.

Such type of hydrogen liquefier is of interest up to about 2000 L/h (see Sect. 12.5).

4.6 Cycles Operating with Turbo Machinery Only

The very first helium refrigeration machines were incorporating full reciprocating components: compressor and expander (1940s). Later, the reciprocating expander was replaced by the cryogenic expansion turbine (end of the 1960s). Later on, the reciprocating compressor was replaced by the rotating screw compressor (1980s). It is, presently (2020), still the situation for almost all refrigerating machines, whatever their size.

It is a long time that refrigerator designers think about the possibility to incorporate a centrifugal compressor instead of a screw one. Two advantages were aimed at:

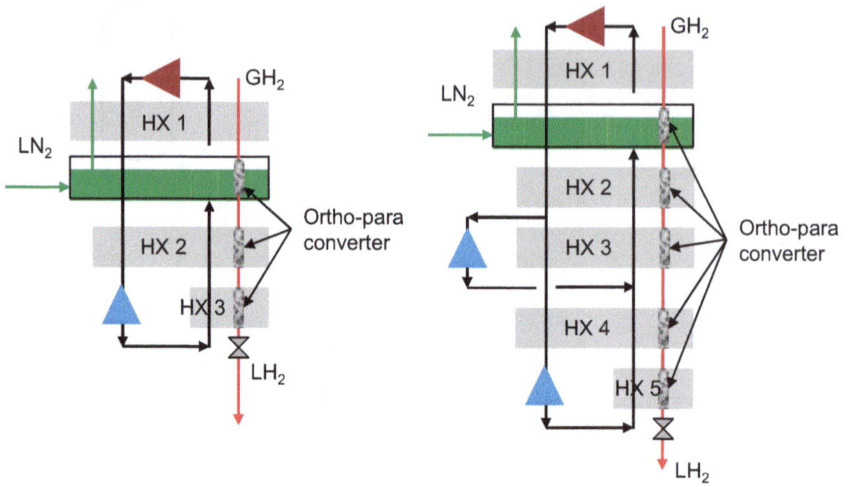

Fig. 4.28 A small capacity hydrogen liquefier incorporating a Brayton cycle

no oil removal and a higher isothermal efficiency. The difficulty sits in the low compression ratio that can be achieved in one stage, due to the low density of helium. A very high tip speed is required to reach a reasonable compression ratio, which leads to too high material characteristics. Therefore, a number of compression stages are necessary, and the investment cost becomes unacceptable. Of course, if one wants to incorporate a low-compression ratio machine into a refrigerator, it is necessary to run a trade-off procedure which, certainly, would lead to a cycle high pressure lower than usually.

The author tried a number of times to incorporate centrifugal compressors into large refrigerators. In the beginning 1990s, when he had to design the largest refrigerators for LEP at CERN (18 kW at 4.5 K equivalent power), the answer of the makers was that the necessary number of stages, due to a low compression ratio per stage, was so large that the investment was too high.

About 10 years later, at the time CERN issued the RFQ for the LHC refrigerators, a new tentative gave a better result: the number of stages had decreased, and the isothermal efficiency was around 73%. However, the combination of CAPEX and OPEX was not yet in favour of the centrifugal solution.

Later on, in 2012, ITER issued the RFQ for the biggest refrigerators (35 kW at 4.5 K). According to the progress that has been witnessed on the former 10 years, the author was almost sure that the ITER refrigerators were to incorporate cycle centrifugal compressors. Unfortunately, ITER wanted to avoid any trouble by stepping into unknown situations and decided to stay with oil lubricated compressors.

The future might be the time for cycle centrifugal compressors for large and very large machines.

4.6.1 Room Temperature Compression

4.6.1.1 A Turbo Brayton Cycle

This is a Brayton cycle operating around 80 K with nitrogen as cycle gas. Even if this machine does not operate with helium, it is of interest to have a glance at it, as a possible component for future plants.

A centrifugal compressor is a clean compressor; there is no need for an oil removal system that can be spared, as shown in Fig. 4.29, left.

In such a system where there is no liquid, it is possible to store all the gas inventory in the system, even at room temperature, by allowing the pressure to rise reasonably in all parts of the system. As the compressor can run at any suction pressure, it is not necessary to control the cycle pressures. Consequently, one can spare the cycle buffer capacity, the 3 cycle control valves and the cold box isolation valves.

The centrifugal compressor is fitted with a variable speed drive; therefore, the compressor by-pass and the turbine inlet valve can be saved also. The new sketch is in the centre of the figure.

The next step towards simplification is: why not arrange compressor and turbine on a same bearing system also incorporating the drive motor? That allows recovering the energy that is extracted by the turbine!

The solution stays in using a centrifugal compressor and an expansion turbine fitted with a dynamic magnetic bearing system incorporating the electric motor drive as shown in the right part of Fig. 4.29. Such a system is totally leak tight; therefore, one can also spare the internal adsorber and its regeneration devices. Higher-power refrigerators incorporate a number of rotating machines, either cycle compressors alone or combined compressor turbine operating in parallel as can be seen in Fig. 12.1 where there are two such assemblies. This arrangement allows reaching a high efficiency: up to 41% compared to Carnot, for 50.8 kW at 77 K.

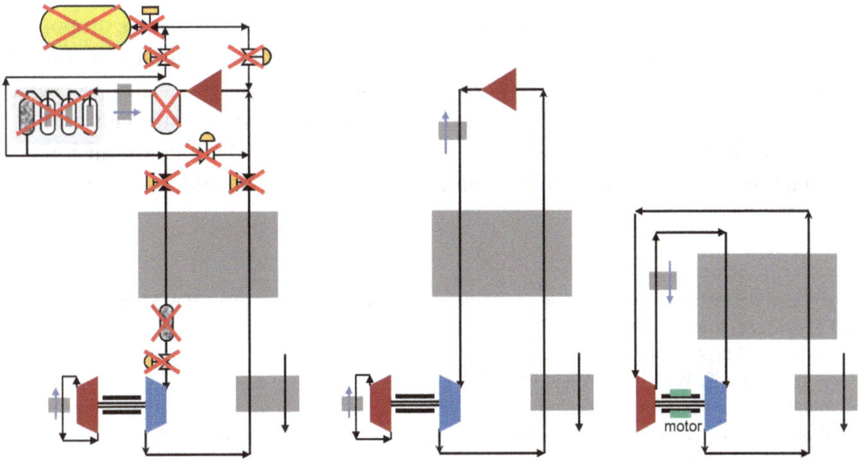

Fig. 4.29 A nitrogen turbo Brayton cycle

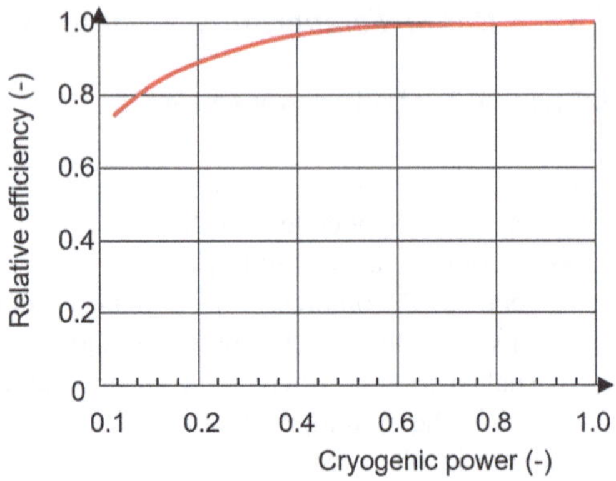

Fig. 4.30 Relative efficiency versus cryogenic power

The adjustment of the cryogenic power is performed by changing the rotational speed and, consequently, floating the cycle pressures (see Sect. 8.7.3). High efficiency is an advantage for full load but even more at partial load: at 60% partial load, the overall efficiency of the refrigerator is only decreased by 3% (see Fig. 4.30).

Such a plant has not only a high efficiency but also, as the whole rotating machine is contact-free and leak tight, a high reliability and low maintenance costs. See Sect. 12.2 for a description of this machine.

Such a refrigerator could be used as a pre-cooling stage for a helium liquefier or any other large machine, as shown in Fig. 4.31.

4.6.1.2 Other Systems

4.6.1.2.1 A Large Hydrogen Liquefier

In the frame of the European project IDEALHY (2011–2013)[1], H. Quack proposed a hydrogen liquefaction cycle based mainly on cycle centrifugal compressors operating with various gas mixtures (see Fig. 4.32) for a 50 t/day liquefier. The process can be split into four stages: pre-compression and chilling, pre-cooling with a mixed refrigerant (MR), cryogenic cooling with Brayton cycles and a final expansion and liquefaction stage. The pre-cooling and cryogenic cooling down to 80 K stages are

[1]Integrated design for demonstration of efficient liquefaction of hydrogen (IDEALHY) Fuel Cells and Hydrogen Joint Undertaking (FCH JU) Grant Agreement Number 278177 Title: Report on Modelling of Large-Scale High-Efficiency IDEALHY Hydrogen Liquefier Concept Authors/David Berstad, Harald T. Walnum, Project Partner: Petter Nekså/SINTEF Energi AS Lutz Decker/Linde Kryotechnik AG Hans Quack/Technische Universität Dresden Work Package: Component Assessment and Optimisation of Feasible Large-Scale Liquefaction Process Deliverable Number: D2.4 Date: 31 March 2013

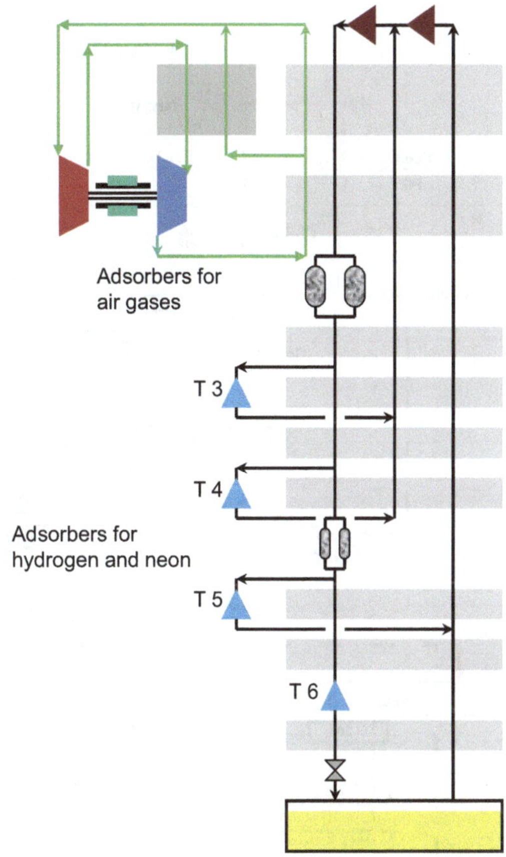

Fig. 4.31 A possible helium liquefier pre-cooled by a nitrogen turbo Brayton stage

located in one cold box, while the last cryogenic cooling stage is located in a separate cold box.

The hydrogen feed enters the liquefaction process at a pressure of 20 bar and is compressed with a piston compressor up to 82 bar. All gas streams entering the cold box are pre-chilled with a single component refrigerator down to 279 K and then down to 130 K with a single mixed refrigerant (MR) cycle operating with a mixture of nitrogen, methane, ethane, propane and n-butane.

The cooling of the hydrogen feed down to the final expansion and liquefaction stage is performed by two Brayton cycles with a common compression train. A mixture of helium and neon so-called Nelium was chosen as refrigerant to give the optimum trade-off between refrigeration efficiency and compressibility by centrifugal compressors.

From 80 bar and 26.8 K, the hydrogen feed is expanded down to 2.1 bar in a two-stage turbo expander T7 and T8 (remember Sect. 4.2.1.3). The outlet vapour quality is about 0.05 as the feed enters a flash tank. The liquid outlet from the flash

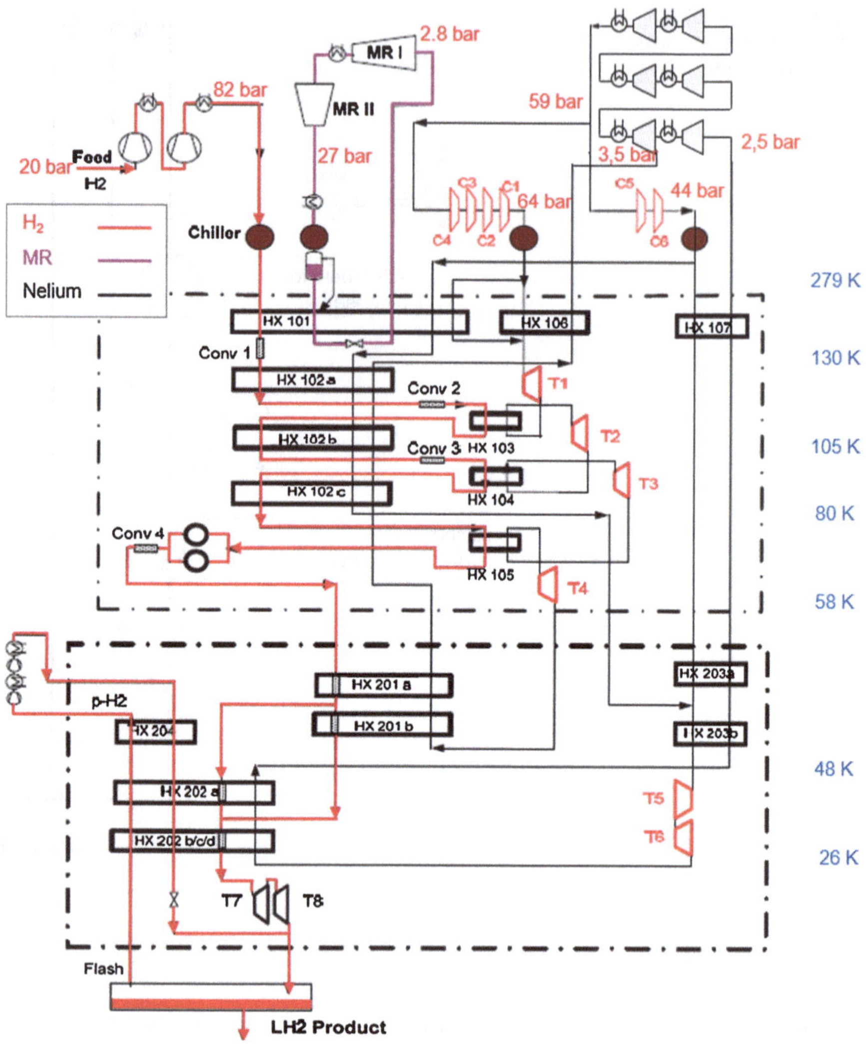

Fig. 4.32 A hydrogen liquefaction cycle concept proposed by H. Quack for the IDEALHY project. C1–C5 compressors are driven by T1–T5 turbines to recover energy

tank leads to the liquid hydrogen storage tanks. Due to the limited minimum temperature of the Nelium cycle, some flash gas is produced in the final expansion of the feed. This flash gas must be re-liquefied, and in this subprocess it is reheated up to ambient temperature and subsequently compressed to 7.4 bar in a two-stage piston compressor. The flash gas is then cooled back down to 26.8 K in parts of the heat exchanger network and throttled back to low pressure and into the feed before entering the flash tank with the main stream of liquefied hydrogen.

Such cycle should be very efficient. It would absorb about 6.3 kWh/kg of hydrogen, almost half of presently existing plants.

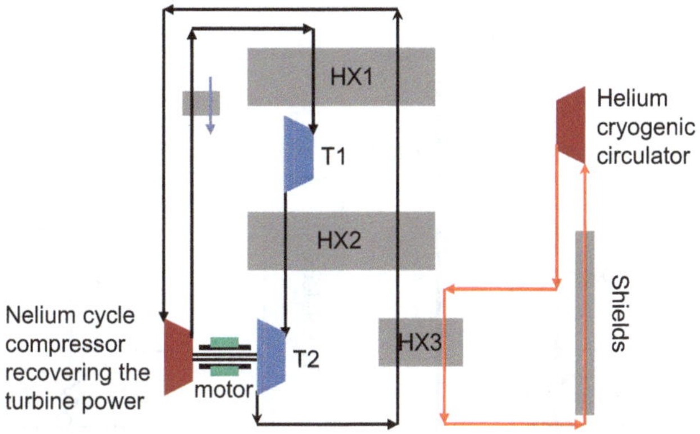

Fig. 4.33 A project for cooling thermal shields for FCC

4.6.1.2.2 Large Helium Refrigerators for Future Accelerators

CERN is working (2020) on the FCC beam screen cooling system that should absorb 570 kW between 40 and 60 K. As for the hydrogen liquefier of Sect. 4.6.1.2.1, it is planned to use cycle centrifugal compressors processing "Nelium" (Fig. 4.33).[2]

However, in order to avoid induced radioactivity, presence of neon in the shield circuits is not acceptable. Therefore, a secondary circuit operated with helium is necessary. Among various solution, one is to use a cryogenic circulator (or compressor). Both circuits are connected through the heat exchanger HX3.

4.6.2 Cryogenic Compression

In the 1990s, IHI, Japan, developed a cryogenic compressor operating at liquid nitrogen temperature.[3]

At the time of the design of the cryogenic system of ITER, circulation of 2 × 2 kg/s of helium between 16.00 and 18.00 bar was planned. The author proposed to use a cryogenic centrifugal compressor operating at 80 K instead of room temperature compression, as shown in Fig. 4.34. The energy balance comparison between a conventional room temperature compressor fitted with a large heat exchanger and the full cryogenic solution showed that the cold compressor version had an energetic cost reduced by 10%. Furthermore, the room necessary

---

[2]Neon helium mixtures as a refrigerant for the FCC beam screen cooling: comparison of cycle design Options S Kloeppel1, H Quack1, C Haberstroh1, F Holdener2

[3]Asakura H., Kato D., Saji N. and Ohya H., "80 K Centrifugal compressor for helium refrigeration system." *Advances in Cryogenic Engineering* 37B, edited by R.W. Fast, Plenum, New York, 1992, pp. 787–894

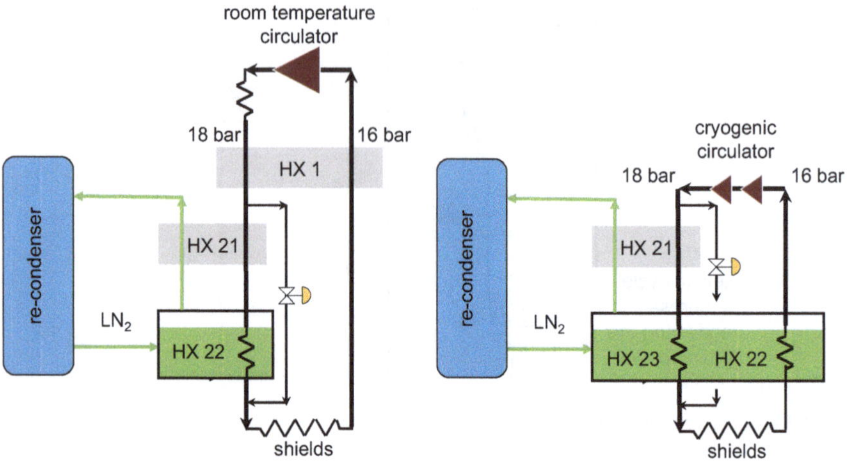

Fig. 4.34 Circulating thermal shields with a room temperature or a cryogenic circulator

Fig. 4.35 Cryogenic cycle compressor cooled with liquid methane

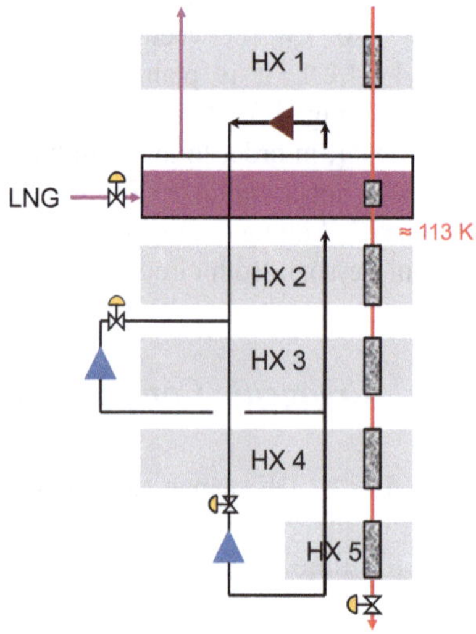

for the cryogenic compression solution is very much reduced by saving the huge 300–100 K heat exchanger.

Another idea to reduce the energetical cost of compression is to take advantage of the availability of liquid methane that is to be vaporised for injection into the gas net. Instead of spending energy, it could be possible to both use liquid methane to pre-cool the cycle and also to cool the cycle cryogenic compressors operating at around 110 K. The compression power could be around 110/300 = 0.36 that of room temperature's (see Sect. 6.2.1) (Fig. 4.35)!

Chapter 5
Various Ways to Connect the Refrigerator Cold Box to the Object to Be Cooled

Abstract Connection of the cold box to the liquid helium storage is considered for small and large liquefiers. Connection of a refrigerator depends on the way an object is to be cooled. The most common arrangements are considered.

5.1 Introduction

In the sketches of Chaps. 3 and 4, the liquid helium bath is represented as if it were a part of the liquefier or refrigerator for easiness of understanding. By the way, in almost all the cases, this is not the reality: the liquid helium bath is generally located in the liquid helium storage or in the cryostat that houses the object to be cooled. Let us consider some typical arrangements among the various ways the refrigerating system is connected to the object to be cooled. (Special circuits for cool-down or warm-up of the object are not shown (see Sects. 10.3 and 10.5).)

5.2 Connection of a Liquefier to a Dewar

5.2.1 Small or Middle Size Liquefier

A small or middle size helium liquefier is generally permanently connected to a stationary reservoir (located nearby the cold box), the volume of which being between 500 and 25,000 L, as shown in Fig. 5.1. Helium at the cold end of the JT heat exchanger is expanded in the cold box, and the two-phase is sent into the stationary dewar where separation between liquid and vapour happens and, then, the saturated vapour returns to the cold end of the JT heat exchanger. To fill mobile dewars that are later moved to the location where liquid helium is needed, helium is withdrawn by pressure difference between the stationary dewar that is about 1.2 bar and mobile dewar that is at about atmospheric pressure, through a mobile small diameter, light-weight transfer line that is inserted both into the stationary and the

G. Gistau Baguer, *Cryogenic Helium Refrigeration for Middle and Large Powers*, International Cryogenics Monograph Series,
https://doi.org/10.1007/978-3-030-51677-2_5

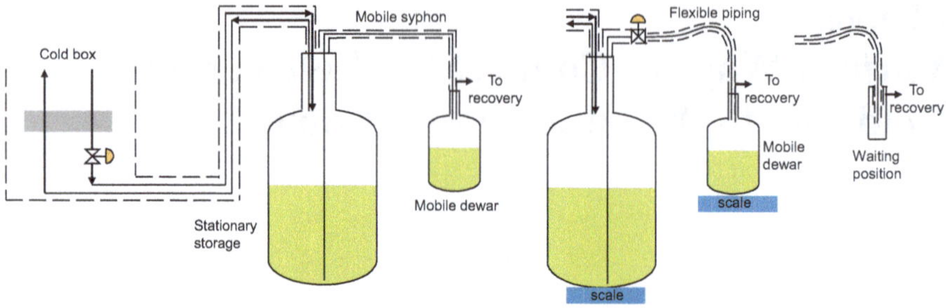

Fig. 5.1 Filling a mobile dewar

Fig. 5.2 A smart system to fill mobile dewar. On bottom, right, the display of the scale. (courtesy CNRS, Grenoble, France)

mobile dewar: the syphon, as shown Fig. 5.1, left. Another arrangement, shown in Fig. 5.2, right, incorporates a permanent connection to the stationary dewar with a valve and a flexible part. This allows not to disconnect the syphon on the stationary dewar side. Furthermore, when the flexible part is not inserted into the mobile dewar, it is connected to the recovery, avoiding air getting inside the cold piping (Fig. 5.1, right). A temperature measurement on the helium recovery pipe allows automatic stop of the transfer process.

The incoming liquid displaces gaseous helium that is in the mobile dewar. Some saturated vapour that is generated by:

- Expansion of the liquid (so-called flash)
- Vaporisation of liquid to cool-down the syphon
- Vaporisation of liquid due to heat leaks

is recovered into the gas holder of the recovery system, in addition of the vapour that is displaced by the liquid. Note that there is no recovery of the enthalpy of the cold helium.

If a mobile dewar is not at liquid helium temperature, re-cooling is done by vaporising some liquid.

Depending on its size, the stationary dewar can be partially located in a pit, as shown in Fig. 5.2. The helium quantity in the stationary dewar is generally measured by weight. It is also a good solution to measure the quantity of liquid in the mobile dewars by weight. The difference in weight between before and after filling allows the measure of liquid to be invoiced.

The helium recovery system includes a gas holder (generally a gas bag), a recovery compressor, an impure helium storage and a helium purifier (a dryer followed by a cryogenic adsorber).

5.2.2 Very Large Liquefier

A very large liquefier is permanently connected to a large storage. As an example, the large liquefier that is operated in Qatar (see Sect. 12.5) is connected to four dewars of 120 m^3 each that make 480 m^3 in total. The dewars are located at about 10 m above the mobile containers (Fig. 5.3). This allows filling the containers by gravity (see Fig. 12.6). The helium vapour that is generated by expansion and heat leaks and also displaced by liquid is sent into the cold box where its enthalpy is recovered.

If a container is not at liquid helium temperature, re-cooling is done by vaporising some liquid, but, here, the vapour enthalpy is recovered by recycling it at various temperature levels into the cold box heat exchangers as in a refrigerator (see the way vapour is recovered at various temperature levels in Sect. 10.3.5).

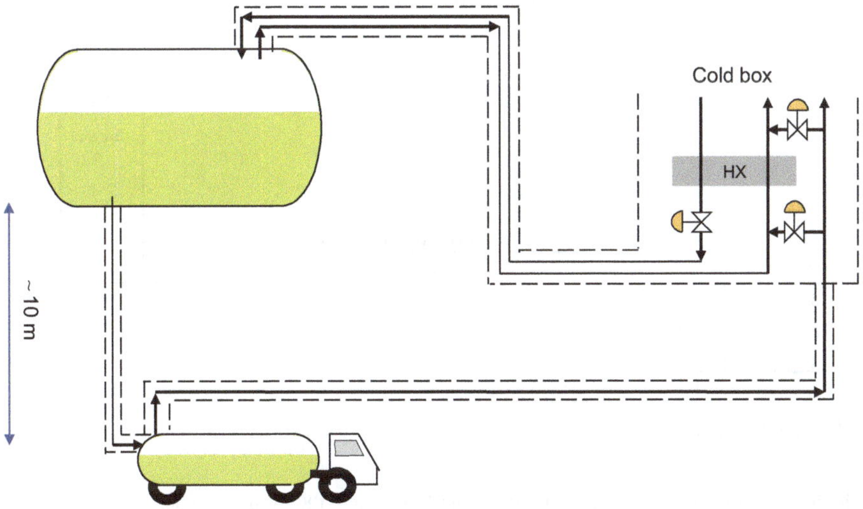

Fig. 5.3 Filling a container by gravity

5.3 Connection of a Refrigerator to the Cryostat

5.3.1 An Important Device: The Test Heater

In all the following examples, one should notice the interest of having one or more electrical heaters dipped into liquid helium for various purposes:

- At the time of plant commissioning, it allows the measurement of the available cryogenic power. In most of the cases, it is part of the test equipment.
- In current operation, one can inject, permanently, a reasonable amount of power. In case there is a sudden increase of refrigeration demand, the power that is injected into the heater can be instantaneously reduced as it is explained in Sect. 8.7, prior to adapting the turn down situation in an optimised way.
- During the lifetime of the plant, it allows, at any time, to check the availability of the cryogenic power without disconnecting physically the equipment that is cooled.

5.3.2 Cooling an Object that Is Dipped in a Liquid Bath

5.3.2.1 With Helium Coming Directly from the JT Heat Exchanger

When the cryostat is not far from the cold box, it is possible to send cold gaseous high-pressure helium towards it, where it is expanded. Heat leaks into the transfer line decrease the concentration of liquid at the time of expansion. The cryostat is directly fed with two-phase helium (Fig. 5.4).

Insulating the helium to be expanded by low-pressure return helium allows getting a higher liquid concentration after expansion.

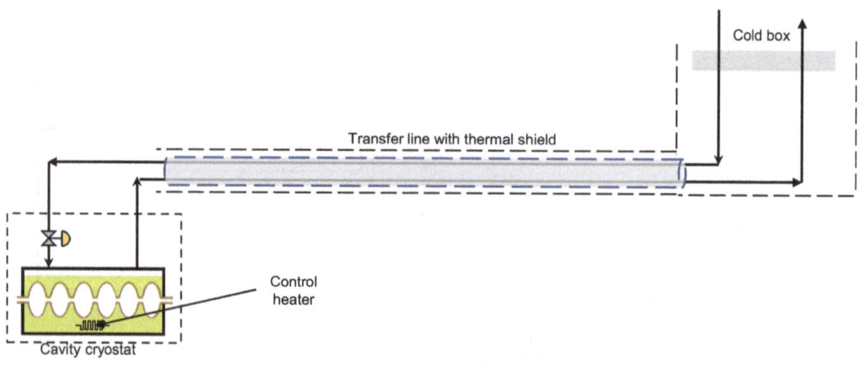

Fig. 5.4 Cooling a Super Conducting Resonant Cavity at 4.5 K from liquid helium

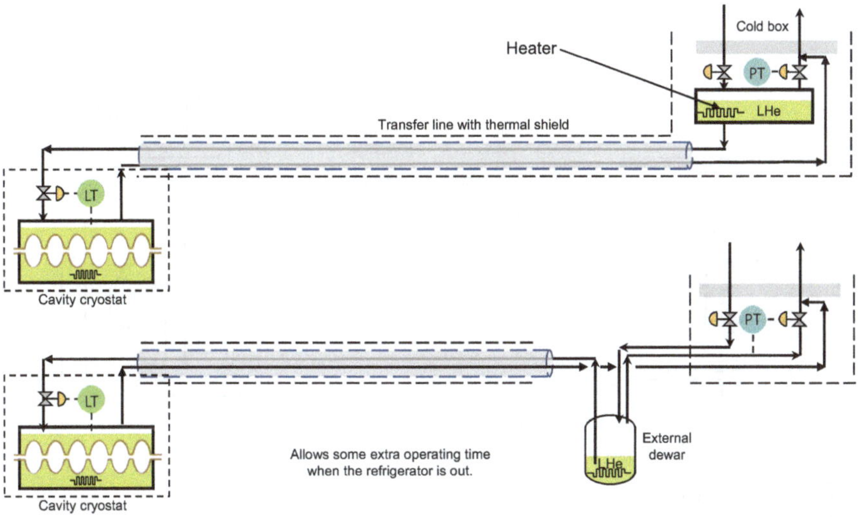

Fig. 5.5 Liquid helium fed from a phase separator located in the cold box, top, or separated, bottom

5.3.2.2 With Liquid Helium, From a Phase Separator

When the cryostat is not too far from the cold box (around 10 m), it is interesting to supply it with pure liquid helium taken from a phase separator that can be located either into the cold box as shown in Fig. 5.5, top, or separated in an external dewar, as shown in Fig. 5.5, bottom.

According to the phase separator volume, a volume of liquid helium allows keeping the cooled object at a correct temperature in order to get sufficient time to get into safe electrical standby in case of a refrigerator failure (related equipment is not shown here). Obviously, the recovery system must be in a position to recover helium during a possible power breakdown: a secure electrical supply is mandatory.

5.3.2.3 From Supercritical Helium

If the transfer line between the cold box and the cryostat is rather long, it is better to circulate supercritical helium in order to avoid possible instabilities in two-phase flow. Super critical helium is expanded into the cryostat by an expansion valve that controls the liquid helium level. A heat exchanger, located into the cryostat, allows cooling helium upstream the cryostat valve in order to get the highest liquid concentration after expansion.

Super critical helium is cooled into a heat exchanger that is dipped into the cold box phase separator as shown in Fig. 5.6, top right. There is another way that is to use a liquid-gas heat exchanger that is circulated by natural thermal syphon effect with two-phase helium (Fig. 5.6, bottom, right).

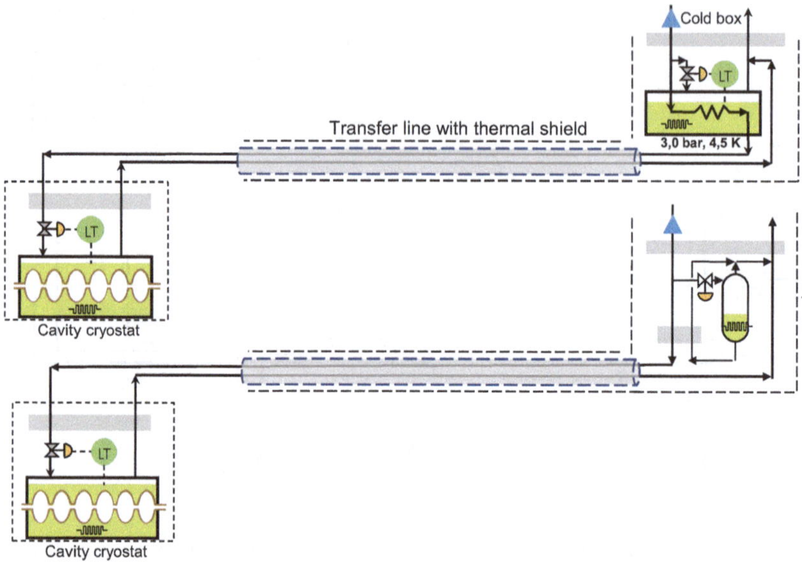

Fig. 5.6 Cooling a Super Conducting Resonant Cavity in a liquid helium bath at 4.5 K from supercritical helium

5.3.3 *Cooling an Object that Is Circulated with Supercritical Helium*

Supercritical helium cooling is generally used for cooling devices that incorporate cable-in-conduct conductor (CICC) coils. It is a common situation in fusion (remember Fig. 4.9).

5.3.3.1 Direct Circulation of the JT Flow

If the JT flow is sufficient to maintain the maximum allowable temperature difference between inlet and outlet, supercritical helium can be directly circulated into the coil after a sub-cooling in the cold box phase separator (Fig. 5.7).

After circulating in the coil, supercritical helium is expanded down to low pressure, producing liquid that is vaporised by the super critical/liquid heat exchanger that is dipped in. Some complementary liquid is produced by a dedicated JT valve that controls the level.

When the natural refrigerator JT flow is not sufficient to maintain the allowed temperature difference between inlet and outlet, it is possible to re-circulate it into sub-coolers as shown in Fig. 5.8. There is a limitation by too high-pressure drops into the objects to be cooled. (In other words, the design of the cooling channels must take this point into consideration.)

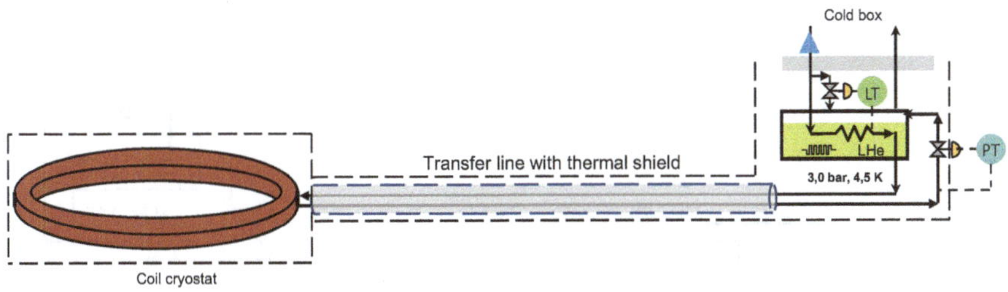

Fig. 5.7 Cooling a coil with the "natural" JT flow of the refrigerator

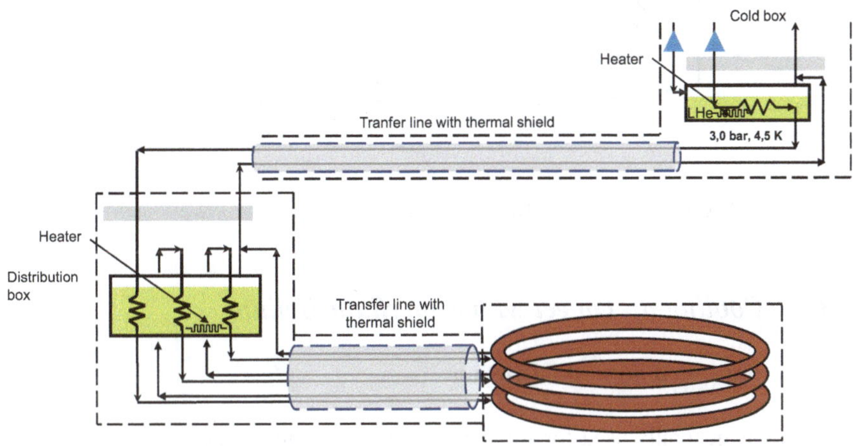

Fig. 5.8 Cooling with re-cooled JT flow

Such solution could seem complicated, but it has the advantage to not add the power of a circulator (see Sect. 5.3.3.2). It has been used on the Tore Supra plant.

5.3.3.2 Using a Circulator

When the flow to be circulated is very important (larger than the natural JT flow), one must use a circulator (remember Sect. 4.2.4.2) (Fig. 5.9).

The higher the pressure drops in the coil, the higher the power that is dissipated by the circulator!

Remark The power that is injected into the system might be not negligible: in some systems, it is as high as 25% of the refrigerator power!

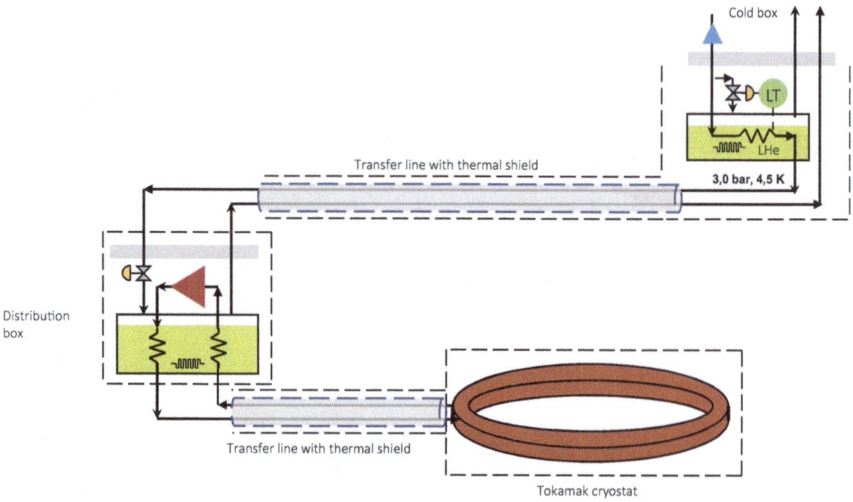

Fig. 5.9 Cooling a coil made out of CICC with a circulator

5.3.4 Cooling an Object with Superfluid Helium

5.3.4.1 Saturated Superfluid Helium

This way is generally used for superconducting resonant cavities where there is no risk to face a high voltage at the time of a quench. Superfluid saturated liquid is produced by expansion of supercritical helium, after cooling into the JT heat exchanger that is part of the cryostat, as shown in Fig. 5.10. Again, there are two ways to get supercritical helium, either by immersed heat exchanger (Fig. 5.10, top) or thermal syphon (Fig. 5.10, bottom).

5.3.4.2 Cooling an Object with Static Pressurised Superfluid Helium

This technology has been initiated by CEA, Grenoble, France, at the beginning of the Tore Supra project, at the end of the 1970s (remember Sect. 1.4.2.4).

5.3.4.2.1 Tore Supra (Now WEST)

Each coil is housed into a casing that is filled with static pressurised superfluid helium. A heat exchanger that is dipped into saturated superfluid helium cools one end of a pipe that is connected to the coil casing. For people who are not familiar with superfluid helium, the Tore Supra arrangement might be surprising on two points when looking at Fig. 5.11.

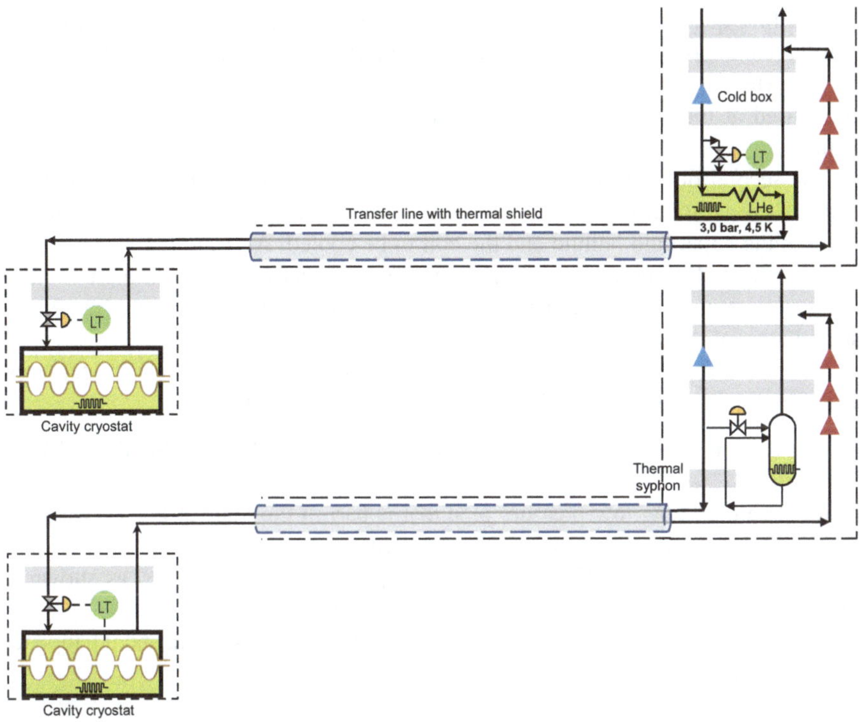

Fig. 5.10 Cooling with saturated superfluid helium

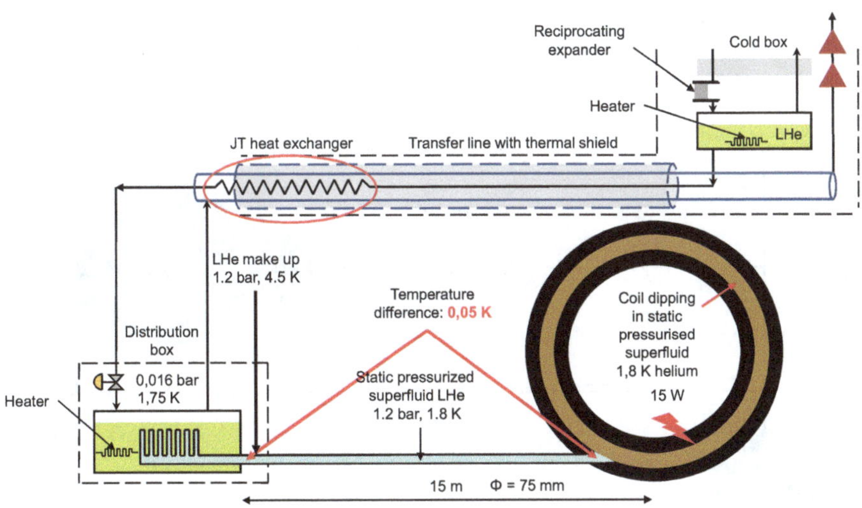

Fig. 5.11 The Tore Supra coil cooling system

The first question is: How is the coil casing filled up? The superfluid helium thermal conductivity is so high that, at the interface between He I and He II, heat leaving He I is removed by conduction in He II. The vapour phase of He I is condensed at the interface; therefore, the liquid void part of the casing fills with liquid.

The second question is: How does heat circulate from the coil to the saturated bath? There is not circulation of helium between the heat exchanger that is dipped into boiling superfluid helium and the coil to be cooled. All heat that is leaving the coil is extracted by pure conduction along the pipe that is filled with high conductivity static pressurised superfluid helium at about atmospheric pressure (remember specificities of superfluid helium in Sect. 1.4.2.4).

When cooling down from 4.5 to 1.8 K, the density of liquid helium increases; therefore liquid make-up at 1.2 bar must be fed into the coil circuit.

An Interesting Trick: Using the Transfer Line as the Housing for the JT Heat Exchanger

Inside the rather long low-pressure pipe that is part of the transfer line between the cold box and the distribution box (here called a satellite), a helically coiled pipe brings the helium to the JT valve, making it the JT heat exchanger, as shown in Fig. 5.11. The transfer line structure is shown in Fig. 5.12

Such tricky solution has been selected for the Tore Supra system, and it happened to be very successful.

5.3.4.2.2 LHC

As for Tore Supra, the LHC coils are dipped into static pressurised superfluid helium, enclosed in boxes. Cooling is performed thanks to a heat exchanger between

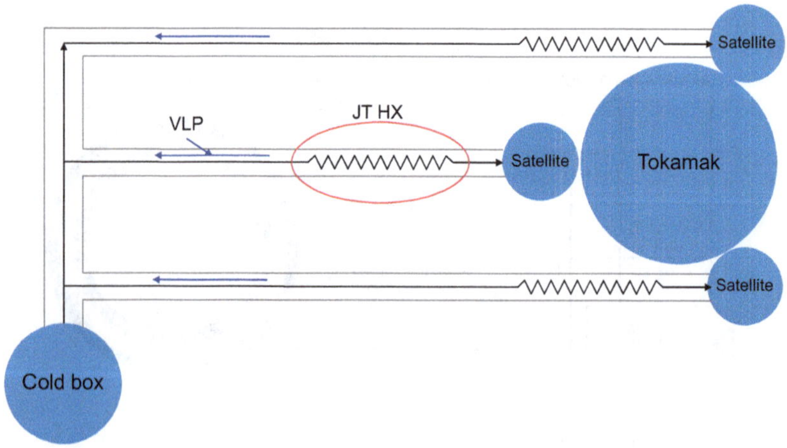

Fig. 5.12 JT heat exchangers located into the transfer lines in the Tore Supra plant

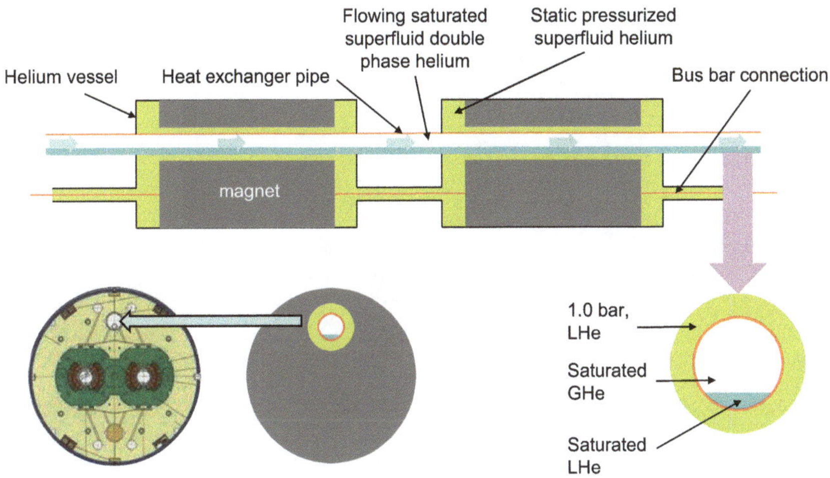

Fig. 5.13 The LHC cooling structure

saturated and pressurised helium. But the structures are very different: instead of having a few coils in a cluster like in Tora Supra, the LHC coils are lined on long distances (more than 100 m!); therefore, the cooling heat exchanger is a pipe that circulates along all coil boxes (Fig. 5.13).

Chapter 6
Technology of Components

Abstract Up to now, machines have been considered as theoretical components fulfilling a function. It is interesting to understand how they operate, what are their operating domains and what is their efficiency. Some of them have to operate in very different regimes during the cool-down and warm-up phase. This chapter considers mainly oil lubricated compressors, oil management system (including oil removal), heat exchangers, cryogenic expansion turbines and cryogenic centrifugal compressors, circulators and pumps.

6.1 Introduction

The main components of a refrigeration system are considered, here, on a technological point of view. The design of these machines is done by the maker, but the user must absolutely know how they operate, which are their peculiarities, their operating domain, their behaviour and their limitations. A few words are given about machines fulfilling some similar functions but not so frequently used as Roots, liquid ring pump, ejector, etc.

A special attention is given to the large operating conditions, especially temperature wise for expansion turbine and cryogenic compressors.

6.2 The Compression Machines

Two severe constraints are associated with helium compression for helium refrigeration:

- The compressed helium must be *perfectly clean* in order not to plug the cold box heat exchangers.
- The compression of helium, that is, a monatomic gas, dissipates more heat than a diatomic gas as nitrogen or air (see Sect. 6.2.1).

G. Gistau Baguer, *Cryogenic Helium Refrigeration for Middle and Large Powers*, International Cryogenics Monograph Series,
https://doi.org/10.1007/978-3-030-51677-2_6

Fig. 6.1 Helium oil lubricated small piston compressors formerly used in TBT helium liquefiers

Fig. 6.2 A dry reciprocating compressor with PTFE (polytetrafluoroethylene) rings

In addition, compression of same volume flow of helium from same low pressure to same high pressure needs more power than compression of nitrogen (see Sect. 15.8.1).

For these two reasons, there was not a satisfactory solution at the end of the 1970s. The conventional oil lubricated compressors (see Fig. 6.1) were rather reliable, but, unfortunately, they polluted helium, and at this time, the oil removal process was not well known. The "dry" reciprocating compressors, either being fitted with loaded plastic rings (see Fig. 6.2) or labyrinths (see Fig. 6.3), were expensive, had a not very high reliability (the author's experience on plastic ring's lifetime is between 8000 and 48 operating hours!) and were subject to very high maintenance costs.

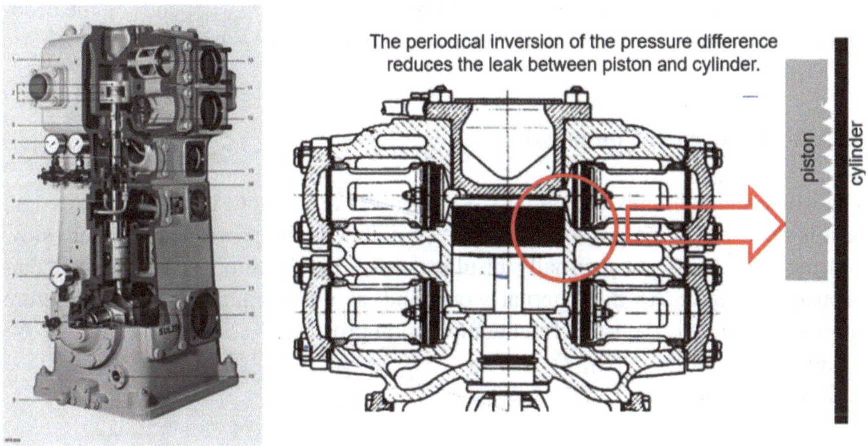

Fig. 6.3 A dry reciprocating labyrinth compressor (Sulzer)

Since the 1980s, helium compression has been almost totally made with screw compressors. Only this type of compressor is considered in this chapter.

6.2.1 Some General Reminders About Compression of Ideal Gases

In a simple and pure thermodynamic point of view, there are two processes to perform the compression of a gas: isothermal and adiabatic. In Sect. 1.4.2.1, it has been shown that helium behaves as a quasi-ideal gas; therefore it is an acceptable approximation to use ideal gas properties for such calculations.

6.2.1.1 Processed Mass Flow Rate

The mass flow rate $\dot{m}$ that is processed by a compressor is:

$$\dot{m} = \eta_{\text{vol}} \times \Delta V \times P_{\text{suc}}/T_{\text{suc}} \tag{6.1}$$

With:

ΔV = displacement
η_{vol} = volumetric efficiency,
P_{suc} = suction pressure
T_{suc} = suction temperature.

In a reciprocating compressor, the volumetric efficiency η_{vol} depends on the dead-end (top clearance) volume and the pressure difference HP – LP.

6.2.1.2 Isothermal Compression

The isothermal compression process is shown in Fig. 6.4 left. During compression, heat is released by the gas that is compressed, but, as the process is infinitely slow and the cylinder is not thermally insulated, such heat is supposed to be instantaneously transferred to the external world and, as a consequence, the temperature stays *constant*. It is the isothermal compression process.

Initial conditions are P_1, T_1, and V_1.

Final conditions are P_2, T_2, and V_2.

$$P_2 = P_1 \times V_2/V_1 \tag{6.2}$$

$$T_2 = T_1 \tag{6.3}$$

The isothermal compression power $\dot{W}$ is:

$$\dot{W} = \dot{m} \times T_1 \times R/M \times Ln(P_2/P_1) \tag{6.4}$$

R: universal ideal gas constant: 8.314462 J.mol^{-1}. K^{-1}
M: molar weight of helium: 4.0026 g.mol^{-1}

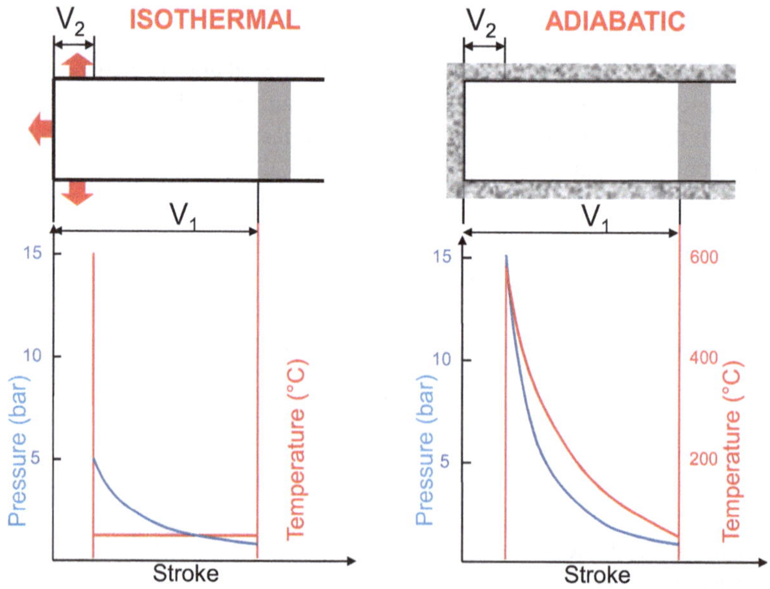

Fig. 6.4 Isothermal and adiabatic compression

At the bottom of Fig. 6.4, left, one can see the evolution of both pressure and temperature along the compression process.

6.2.1.3 Adiabatic Compression

On the contrary, the adiabatic compression happens in a way that not any heat is released to the external world because the cylinder is perfectly thermally insulated: the temperature is not constant.

$$P_2 = P_1 \times (V_2/V_1)^{\gamma} \tag{6.5}$$

$$T_2 = T_1 \times (P_{\mathrm{dis}}/P_{\mathrm{suc}})^{((\gamma-1)/\gamma)} \tag{6.6}$$

The adiabatic compression power $\dot{W}$ is:

$$\dot{W} = \dot{m} \times T_1 \times c_{\mathrm{p}} \times \left[(P_2/P_1)^{((\gamma-1)/\gamma)} - 1 \right] \tag{6.7}$$

c_{p}: helium isobaric specific heat capacity
c_{v}: helium isochoric specific heat capacity
$\gamma = c_{\mathrm{p}} / c_{\mathrm{v}} = 1.6665$ for helium at 300 K

At the bottom of Fig. 6.4, right, one can see the evolution of both pressure and temperature along the compression process. For same volume ratio V_2/V_1, note that both pressure and temperature reach higher levels during the adiabatic compression process.

Remark If the compression heat is partially released, this is called a polytropic compression process. Depending on the quantity of heat that is released, there are a number of polytropic processes. Such notion is used by compressor makers, but for our way of dealing with compressors in this book, we do not consider it.

6.2.1.4 Power Calculations

Let us perform some calculations in order to get an order of magnitude in mind.

The compressor operating conditions that we will use as an example along this entire chapter
Suction pressure: 1.05 bar
Suction temperature: 300.0 K
Discharge pressure: 12.0 bar

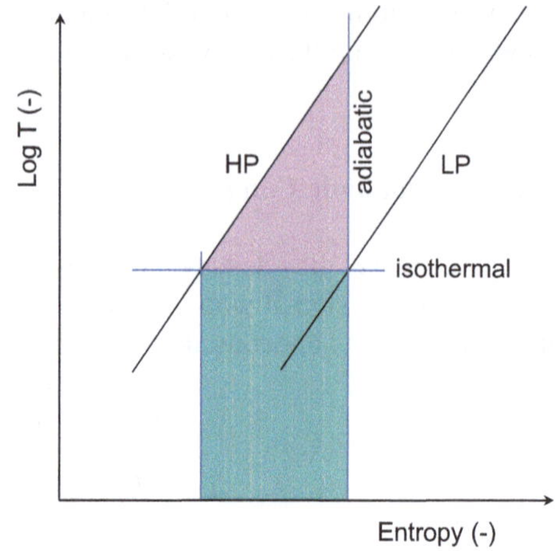

Fig. 6.5 Comparison of compression processes

- Isothermal power to compress 1 g/s of helium:

$$\dot{W} = 1 \times 300 \times (8.314/4.0026) \times Ln(12.0/1.05) = 1518 \text{ W}$$

- Adiabatic power to compress 1 g/s of helium:

$$\dot{W} = 1 \times 5.196 \times 300 \times [(12.0/1.05)((1.67 - 1)/1.67) - 1] = 2600\,\text{W}$$

Why is the isothermal power lower than the adiabatic one? When looking at Fig. 6.4, one can understand that, as pressure increases more rapidly in the adiabatic process, due to the temperature increase, than in the isothermal, the work to be provided for compression is higher. One can also compare the respective areas that are under the transformation curve of each process on a T-s diagram as in Fig. 6.5 where same compression is performed by an isothermal process or an adiabatic one (remember Fig. 1.7).

As the isothermal compression is the process that requires the minimum energy, it is used to evaluate the compressor efficiency that is the ratio of the isothermal power versus the actual compressor power.

6.2.1.5 Behaviour of a Compressor According to the Suction Temperature

What happens when the suction temperature of a compressor increases (e.g. during summer)?

If we refer to: $\dot{W} = \dot{m} \times T_1 \times R/M \times Ln(P_2/P_1)$ Eq. 6.4, the absorbed power increases proportionally to the suction temperature. However, if we refer to: $\dot{m} = \eta_{\text{vol}} \times \Delta V \times P_{\text{suc}}/T_{\text{suc}}$ Eq. 6.1, the mass flow rate decreases inversely proportionally to the suction temperature. Consequently, the power that is absorbed by the compressor is constant, but the processed flow decreases.

6.2.1.6 Comparing Compression Powers for Helium, a Monatomic Gas, and for Nitrogen, a Diatomic Gas

Let us compare a one-off compressor operating between same suction and discharge conditions, processing the same volume flow, with helium and then with nitrogen.

Isothermal Compression
Helium

$$\dot{W} = 1 \times 300 \times (8.314/4.00) \times Ln(12.0/1.05) = 1518 \text{ W}$$

Nitrogen
To get same volume flow, N2 mass flow = N2 density/ He density = 7.

$$\dot{W} = 7 \times 300 \times (8.314/28.00) \times Ln(12.0/1.05) = 1518 \text{ W}$$

In the isothermal compression process, the compressor power is same for monatomic or diatomic gases, because heat is totally rejected.

Adiabatic Compression
Helium
$\gamma = 16{,}665$ at 300 K
$c_p = 5.196$

$$\dot{W} = 1 \times 5.196 \times 300 \times [(12.0/1.05)((1.67 - 1)/1.67) - 1] = 2600\,\text{W}$$

Nitrogen
$\gamma = 1{,}40$
$c_p = 1.041$

$$\dot{W} = 7 \times 1.041 \times 300 \times [(12.0/1.05)((1.4 - 1)/1.4) - 1] = 2200\,\text{W}$$

In the adiabatic compression process, the compressor power is higher for a monatomic gas.

6.2.2 The Oil Lubricated Twin-Screw Compressor

The twin-screw compressor (Fig. 6.6) is a very simple machine: two rotors, also called screws, sustained by their journal and positioned by the thrust bearings into a

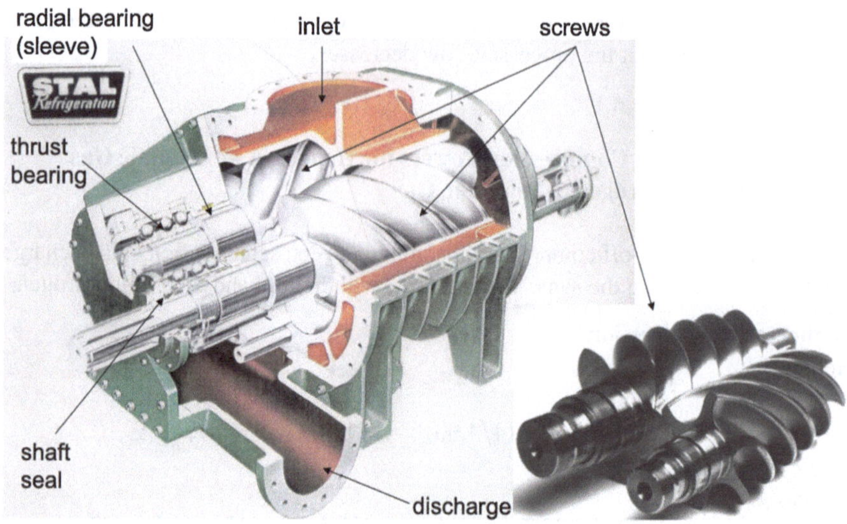

Fig. 6.6 A screw compressor (STAL)

casing. With so few moving parts, it is a very reliable machine; it can run up to 40,000 hours without any maintenance. In 2020, it feeds most of the modern helium refrigeration or liquefaction plants in the world. Later in the book, the twin-screw compressor is called a screw compressor.

A nice and very detailed literature dealing with screw compressors is available. However, to deal with helium refrigeration, it is not necessary to go so deep into the topic. In this book, it is only dealt with the most important issues that interest someone involved in helium refrigeration.

6.2.2.1 Operating Principle

The initial patent (number 4121) for a screw compressor was filed on March 24, 1878, by Heinrich Krigar, Germany. How is it possible, in 2020, to imagine such an extraordinary job with the tools that were available at this time: a pencil, a ruler and a compass? At that time, it was not possible to develop the project any further due to the lack of manufacturing technology.

One had to wait until 1935 to see the first machine that was built by Mr. Lysholm, a Swedish engineer working with Ljungströms Ångturbin AB. He developed the profile of the screw compressor and tested various configurations and rotor lobe combinations. He patented the method for accurately machining the rotors (see Fig. 6.8 right).

There are various rotor profiles (see Fig. 6.7), some of them being symmetrical (mainly for dry machines) and others being asymmetrical. Asymmetrical profiles

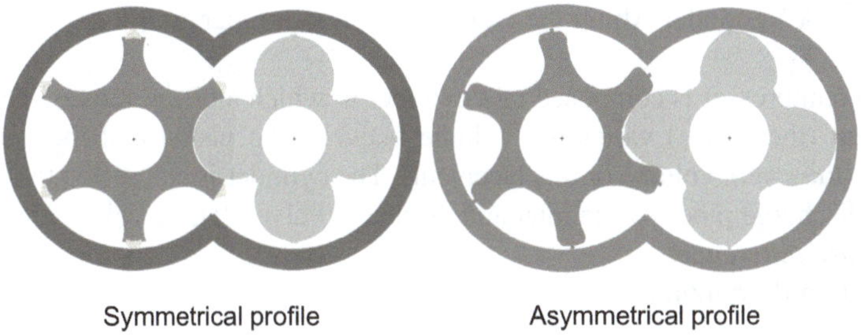

Fig. 6.7 Rotor profiles

Fig. 6.8 End view of asymmetrical rotors. Sigma and casing (patented by Kaeser), left. Screw machining, right

have a higher volumetric efficiency. Present oil lubricated screw compressors use asymmetrical profiles. Some of them even use different rotor diameters (see Fig. 6.8 left).

Ljungströms Ångturbin AB changed its name to Svenska Rotor Maskiner AB in 1951. This company is well known in the compressor world as SRM and has issued manufacturing licenses to almost every screw compressor manufacturer.

The first screw compressor was a dry machine; it needed timing gears. In 1950, oil began to be injected into the compressors. As leak tightness between rotors and rotor and casing was improved, it was possible to reach a higher-pressure difference but also allowed rotation at lower speed, which gave less noisy machines. Later, in 1957, it was understood that most of the torque from the driven rotor is transmitted to the other by the gas pressure. Consequently, the timing gears could be spared, which made cheaper and even less noisy machines.

The interest of oil injection will be discussed later, but, first, let us make clear the geometrical operation of the screw compressor. To make the explanation easier, it is of interest to start by making the comparison with the most well-known and easier-to-understand conventional reciprocating piston compressor. Then, an analogy will be made between piston and screw machines.

6.2.2.1.1 A Reminder About the Piston Compressor Operation

A piston compressor is composed of a cylinder in which a piston moves. The piston movement (the stroke) varies the volume between the piston and the end of the cylinder (the swept volume or displacement). The cylinder is connected to the intake and the discharge pipes by the inlet and discharge valves. These valves are operated by *pressure difference*.

Operation description

- The piston moves towards the end of the cylinder (towards the top on Fig. 6.9, left); the pressure into the cylinder increases as can be seen in the diagram, at right. It is the *compression* phase.
- When the pressure in the cylinder becomes slightly higher than the pressure in the discharge pipe, the spring-loaded discharge valve opens.
- The piston movement goes ahead: the gas that sits in the cylinder is pushed out at constant pressure into the discharge pipe. It is the *displacement* phase.
- The piston is at the end of its stroke; the discharge valve closes.
- There is some gas at discharge pressure that is trapped between the piston and the end of the cylinder (the clearance volume; it is small but not zero). When the piston starts moving in the opposite direction, the high-pressure gas that was remaining in the dead volume expands until its pressure equals and becomes slightly lower than that in the suction pipe. Only at this time, the spring-loaded suction valve opens, and the *suction* phase starts. That means that the volume that enters the cylinder is *lower* than the displacement (volumetric efficiency).

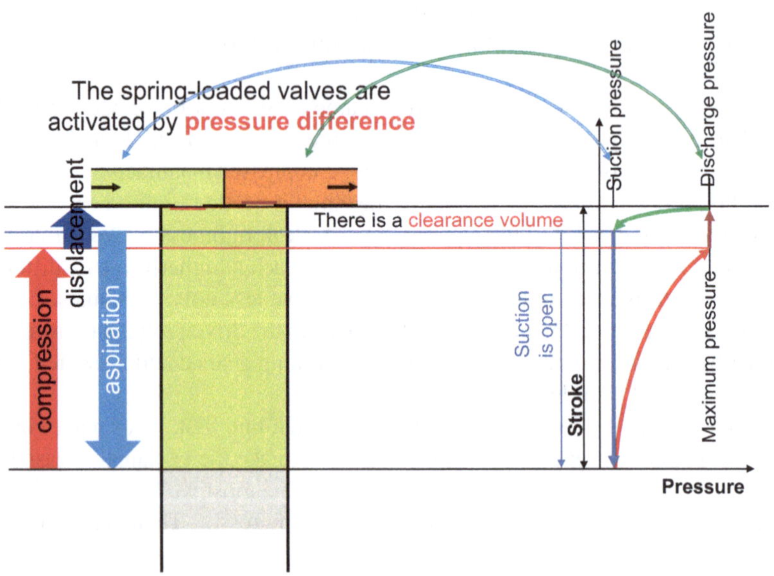

Fig. 6.9 The piston compressor

- The piston is at the farthest end of the cylinder (at the bottom, in Fig. 6.9); the volume does not increase anymore; the spring-loaded inlet valve closes.
- And so on.

Among other characteristics let us keep in mind two of them:

- The maximum pressure that is reached in the cylinder *equals and never exceeds* the pressure that is sitting in the discharge pipe (almost no pressure drop provided by the discharge valve).
- The clearance volume reduces the actual quantity of gas that is processed. This gas that has been trapped into the dead-end volume is compressed again.

One can use the piston compressor analogy in order to describe the screw compressor (Fig. 6.11).

6.2.2.1.2 The Screw Compressor Operation

Prior to start explanations, a definition is necessary: the contact line between conjugate helical surfaces separating the high-pressure compression chamber from the suction chamber is called the sealing line.

In Fig. 6.10, the author has tried to show how the screw compressor operates: the casing is drawn in light yellow, in order to see the screws by transparency, with the suction port that is visible from top and the discharge port that is visible from bottom.

- In stage 1, where the rotors are shown for upwards, low-pressure gas gets through the suction port into the cavity of each rotor (blue parts on the model). At stage 2 the sealing line moves towards the discharge side; the available volume is made larger and even larger in stage 3. In stage 4, at the time the volume reaches its maximum, the rotor cavities are no more in connection with the inlet port so that the gas is now trapped into the volume. This is the end of the *suction* phase.
- At the time the volume is disconnected from the inlet port, a new sealing line is established between rotors. The rotation of the rotors makes such sealing line to move along them, towards the discharge side.
- In stage 5 and 6, where the rotors are seen from under, the trapped volume is both reduced and moved towards the discharge side of the screws. This is the *compression* phase.
- And, finally, in stage 8, the volume in which the gas is trapped is connected to the discharge port. This is the *displacement* phase.

An analogy can be made between the equivalent parts of the two types of compressors (Table 6.1).

The sealing line is a place where leaks from high pressure to low pressure happen; therefore, it should be the shortest possible. Fortunately, its length reduces during the compression process. It is minimal at the end of the compression process. The screw compressor is a *positive displacement* volumetric machine.

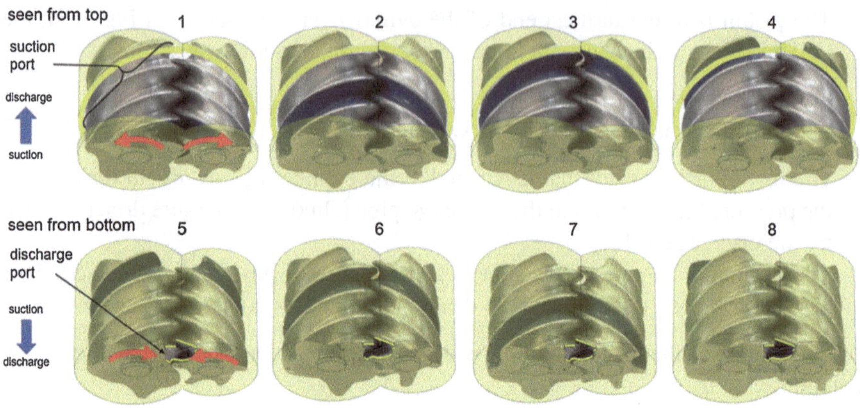

Fig. 6.10 Operating principle of the twin-screw compressor

Table 6.1 Equivalent parts between piston and screw compressors

Piston compressor	Screw compressor
Cylinder	Hollow part of the screw surrounded by the casing
Piston	Sealing line between screws
Intake valve	Suction port
Discharge valve	Discharge port

It is basically a similar operation as the piston compressor with two differences:

- The inlet and discharge valves are not actuated by pressure difference but according to the *position of the piston*, like the intake and discharge valves of a four-stoke explosion engine. The consequence is that the pressure that is reached at the end of the compression phase is *independent* of the pressure in the discharge pipe.
- There is *no dead volume*.

The screw compressor analogy becomes that of Fig. 6.11.

For an optimum efficiency, internal clearances within the screw compressor must be as small as possible. The asymmetric profiles have shorter sealing lines and smaller discharge port areas.

6.2.2.2 A Few Specific Aspects of Screw Compressors

6.2.2.2.1 The Bearings

The bearings of oil injected screw compressors that operate with high-pressure difference are heavily loaded by axial and radial forces, which are transferred to the casing (see Fig. 6.12). Rolling element bearings are generally selected for small

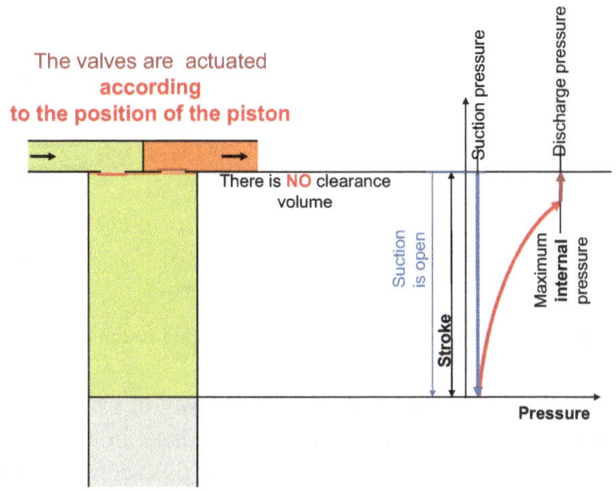

Fig. 6.11 Screw compressor analogy

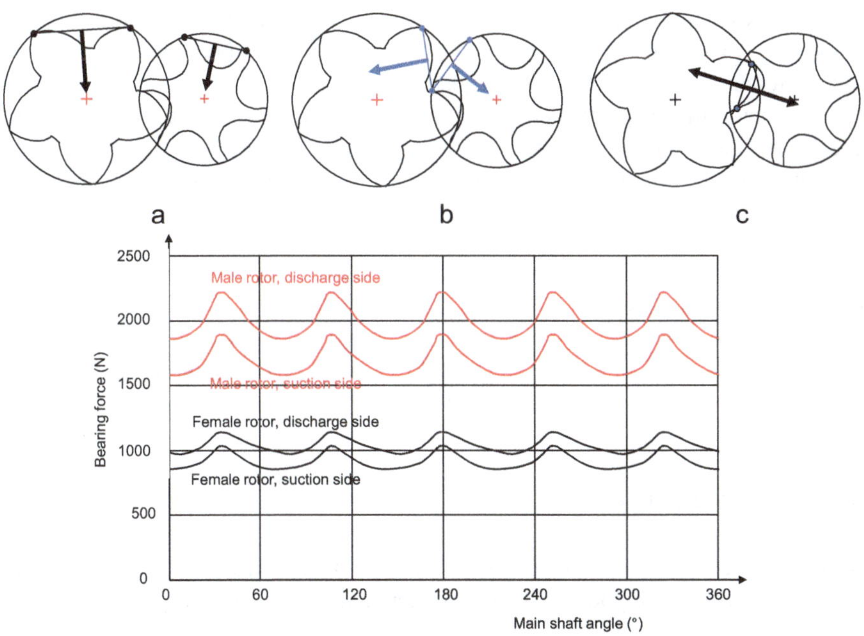

Fig. 6.12 Variation of efforts on the radial bearings

and medium machines. Usually two bearings are employed on the discharge end of the rotor shafts in order to absorb the radial and axial loads separately. For larger compressors, sleeve bearings and even tilting pad thrust bearings can be used. The contact force between the rotors is determined by the torque transferred between them and is significant where the rotors make direct contact. It is reduced when the compressor drive is through the main rotor.

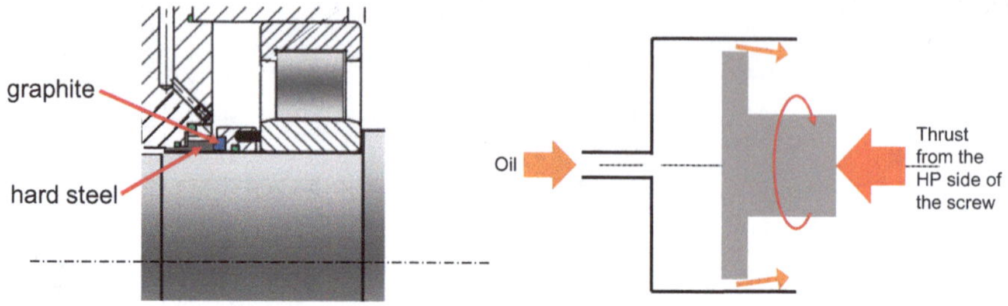

Fig. 6.13 A shaft seal, left; the thrust balancing piston principle, right

In Fig. 6.12, top, one can see the way in which efforts are combined on journal bearings: some of them are radial loads as in a, and others, as in b and c, are not radial, creating a torque. In the case of this screw profile, one can see that the higher torque is on the male screw (see Fig. 6.12, bottom): it is generally the one that is connected to the motor.

Obviously, similar periodic efforts are withstood by the thrust bearings that are helped by a piston that is connected to the low-pressure end of the screws, submitted to the pressure of oil (see Fig. 6.13, right).

The longitudinal gap between the screws and the casing is of importance because a leak can establish from high to lower pressure at the discharge end of the screws. In case the thrust bearing is fitted with tilting pads, such gap is larger, and it decreases the compressor efficiency.

6.2.2.2.2 The Shaft Seal

The shaft seal is a conventional graphite hard steel packing, as shown in Fig. 6.13, left. It allows a small oil leak that is collected and measured (see Sect. 11.5.1.1).

6.2.2.2.3 The Discharge Port

Instead of a discharge valve that operates in a piston compressor, it is the rotors that discover the discharge port; its shape is rather characteristic (see Fig. 6.14).

6.2.2.2.4 Isothermal Efficiency Versus External Compression Ratio

Another specificity of the screw compressor sits in its behaviour according to the external compression ratio (the ratio of pressures in the discharge and the suction pipes).

If one plots the isothermal efficiency of a screw compressor versus the external compression ratio, the resulting curve has a curious shape (see Fig. 6.15).

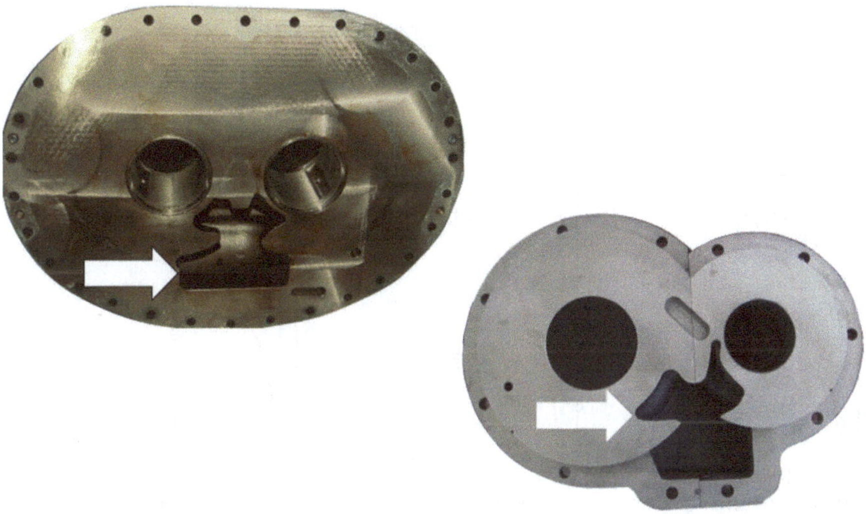

Fig. 6.14 Two typical characteristic shapes of the discharge port

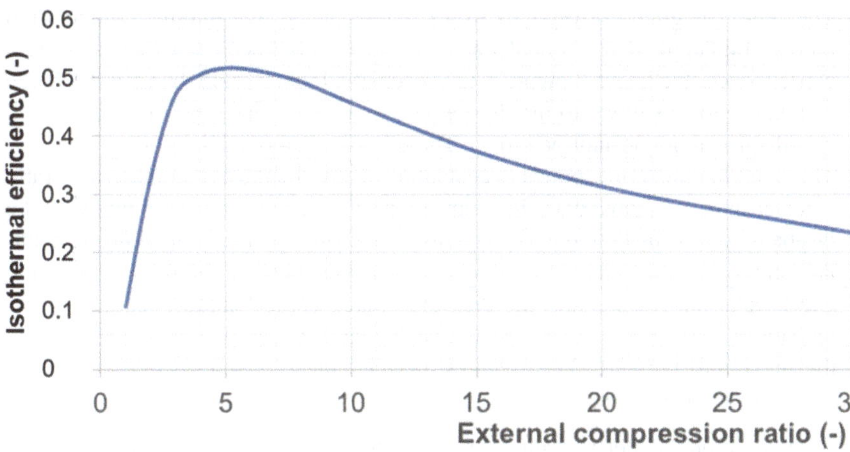

Fig. 6.15 Isothermal efficiency of a screw compressor versus the external compression ratio

What is surprising is that for very low compression ratios, there is almost no compression to do and the efficiency is very low. Efficiency reaches a maximum and then decreases again, but no so rapidly. Let us look more accurately into the compression process in order to understand such a behaviour.

In Fig. 6.16:

- The left diagram corresponds to an ideal (therefore scarcely found) situation where, by chance, the internal compression ratio is equal to the global one: the connection of the cavities of the rotors with the discharge port is made at the time both pressures are equal.

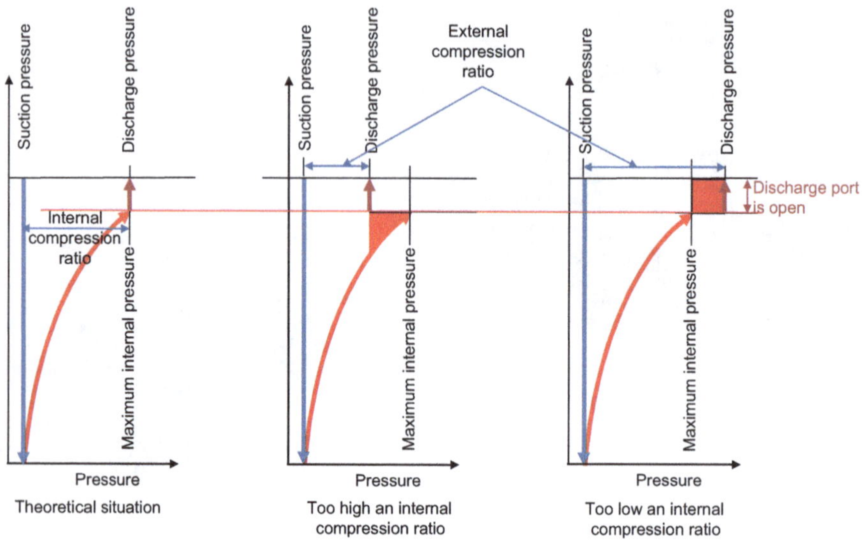

Fig. 6.16 Incorrect adaptation of the internal compression ratio to the external compression ratio

- The centre diagram corresponds to a situation where the internal compression ratio is higher than the global one: discharge port opens *too late,* and, as a consequence, the internal pressure has reached a higher value than the discharge one. Gas expands down to the discharge pressure, and later, part of it is pushed out (displaced at constant pressure). There has been too much compression.
- The right diagram corresponds to a situation where the internal compression ratio is lower than the global one: the connection of the cavities of the rotors with the discharge port is made *too early,* and, as a consequence, the internal pressure has not yet reached the value of the discharge one. Gas flows in the *reverse direction* and later is pushed out into the discharge pipe. This situation is called "Roots operation" (see Sect. 6.2.6.3). Such a gas quantity is compressed *twice* between the maximum pressure that has been reached into the screw cavities and the system discharge pressure.

Both situations, either too much or not sufficient compression, lead to a lower isothermal efficiency.

Now, it is easier to understand the shape of the curve in Fig. 6.15:

- At compression ratios, lower than the internal one, the gas is firstly too much compressed then expanded.
- At compression ratios, higher than the internal one, the gas is not sufficiently compressed. (It is a similar operation as in a "Roots" machine; see Fig. 6.28.) Such operating situation is not so ineffective than for too much compression, which explains the different slopes of the curve.

6.2.2.2.5 Control of the Gas Processed Mass Flow

Some screw compressors, generally designed for conventional refrigeration, are fitted with a system that allows controlling the gas mass flow: the slide valve.

It is a movable part that is in contact with the two screws (see Fig. 6.17, left). The slide valve establishes a low-pressure by-pass, which means that gas is not compressed but only sucked into the machine and pushed back out in the suction pipe *without having been compressed* as shown on the diagram in Fig. 6.17, right. This could be explained in another way: it is like using a shorter part of the screws, as it can be seen on Fig. 6.18.

The slide valve is generally moved by a piston that is actuated by the lubrication oil pressure. Sometimes a spring brings back the slide valve to a fully open position when the oil pressure is zero (see Sect. 6.3.6.2). This allows starting automatically the compressor at the lowest volumetric flow. The position of the slide valve can be transmitted by a system as a potentiometer (not very good, because it is often destroyed by vibrations), a magnetic inductance or capacitive system. Some other compressor builder (like Sullair, USA) moves the side valve with a screw that is actuated by an electric step motor. This later system is more accurate as for positioning the slide valve but also in transmitting its position. A local display of the slide valve position is of interest for the operator.

Figure 6.19 top shows a typical curve of the absorbed power versus the mass flow rate. It is not a perfect solution but allows saving some energy when there is no other way to reduce the processed mass flow.

When the slide valve is actuated by solenoid valves, such a solution happens to be sometimes awkward because it makes the position change by small but finite

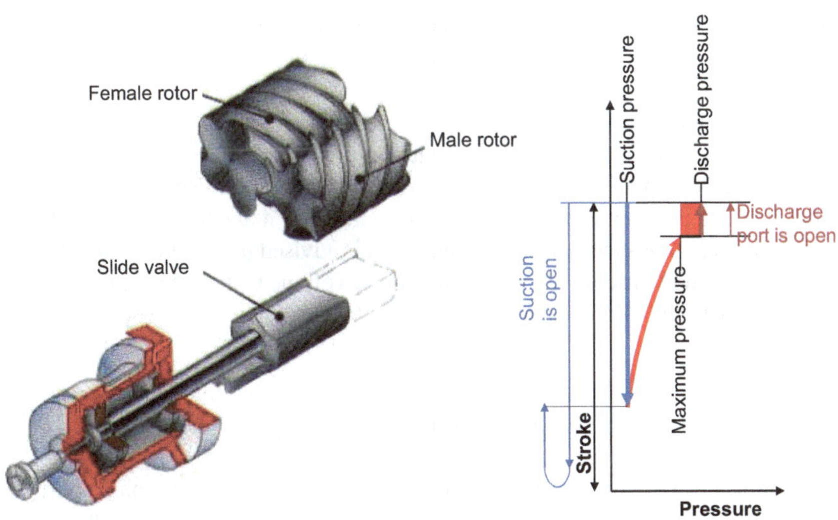

Fig. 6.17 The slide valve

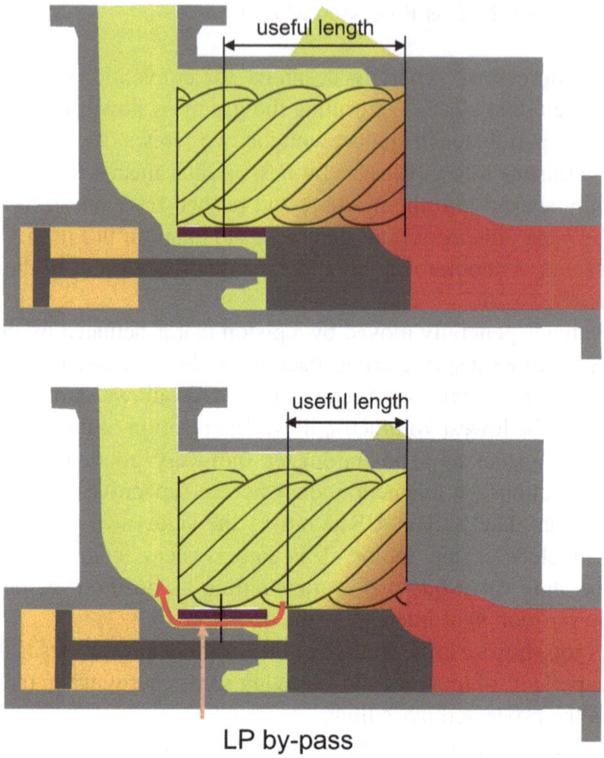

Fig. 6.18 Full flow, top; reduced flow, bottom

displacements. This is acceptable for low flow settings but becomes impracticable in the area where the flow is maximum, over 95% (see Fig. 6.19 bottom). A slide valve moved by a screw is a better suited system.

A more efficient way to reduce the processed mass flow rate is by variation of the rotational speed. For small powers a belt-and-pulley-system is acceptable but not controllable. A more flexible solution is to use a frequency converter. This allows not only to reduce the rotational speed but also to increase it above the rotational speed corresponding to the electrical net frequency. It is advised not to reduce too much the rotational speed in order to avoid too high an oil volume that can bring some trouble in relation with the oil management (see Sect. 6.2.2.3).

6.2.2.2.6 Control of the Volume Ratio V_i

The volume ratio V_i is related to the shape and dimensions of the discharge port. One can see that, in Fig. 6.20, as the discharge port is discovered later at right than at left, the resulting pressure just before opening the port is higher at right.

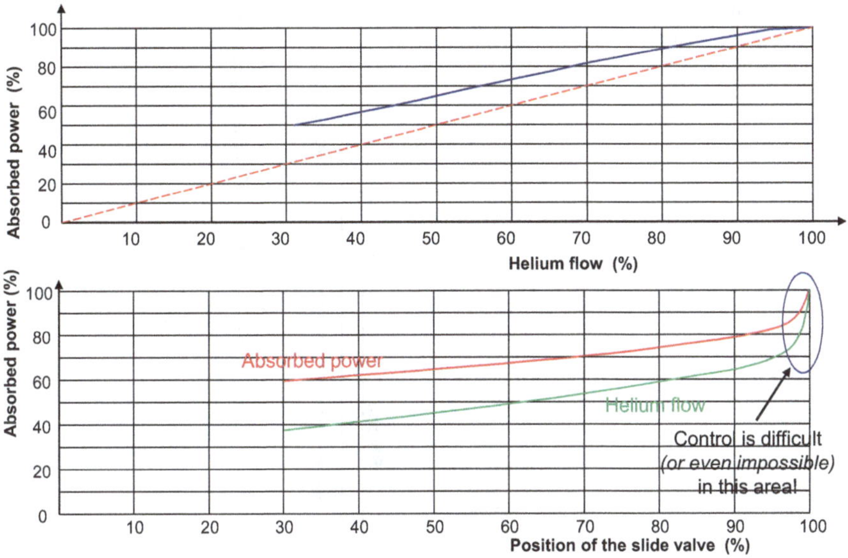

Fig. 6.19 Slide valve: mechanical absorbed power versus flow rate. Helium flow versus slide valve position

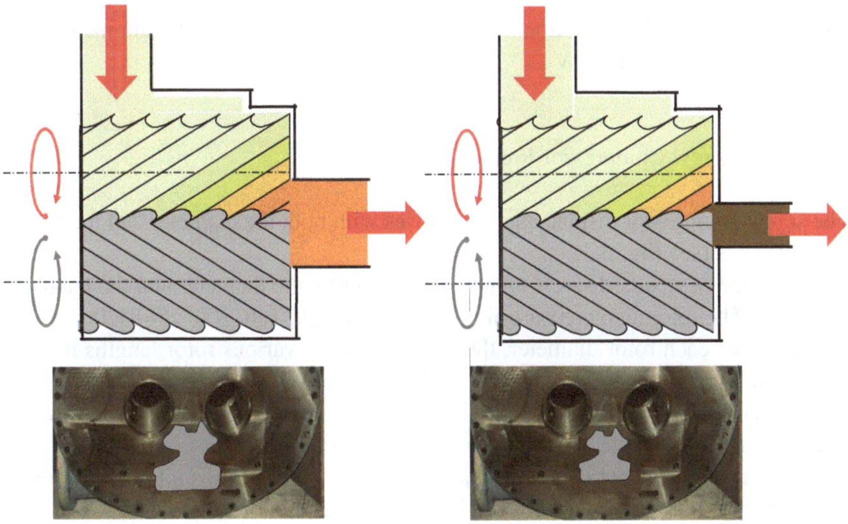

Fig. 6.20 The volume ratio is given by the shape of the discharge port

Some compressor makers have developed a way to vary the V_i during operation, either manually or automatically, using kind of a second slide valve (Fig. 6.21).

Varying both the displacement and the Vi should allow operating a screw compressor at almost maximum efficiency. However, up to now the author is not

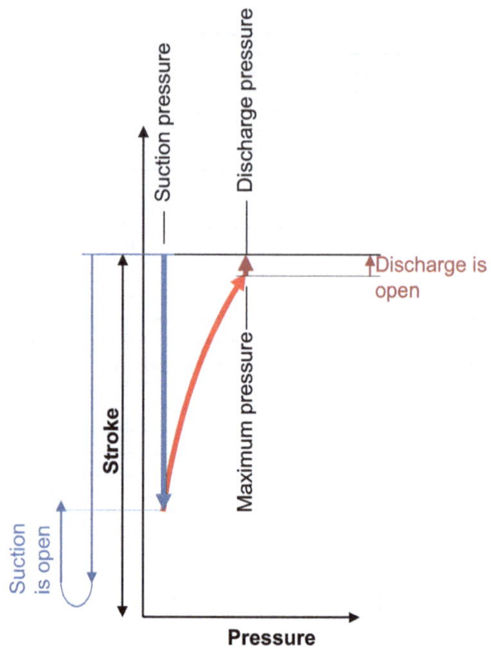

Fig. 6.21 An ideal situation where flow and Vi are correctly controlled

aware of a practical application of this system in order to operate automatically a compressor at its optimum V_i.

6.2.2.2.7 Screw Compressor Range

The screw compressor displacement increases with both the diameter and the length of the screws (see Fig. 6.22).

The range of standard screw compressors is set by the diameter and the length of the rotors. The rotor diameters are almost standard: 120, 160, 200, 250, 320, 400 and 510 mm. For each rotor diameter, there are generally various rotor lengths that are expressed by the ratio L/D: typically, 1.1, 1.6 and 2.1. The maximum displacement of a standard screw compressor is around 10,000 m^3/h (see Fig. 6.23). Therefore, for large refrigeration plants, machines are to be arranged in parallel.

There are larger machines, but, due to the small number of units that are built, their price would be unaffordable for helium refrigeration.

6.2.2.2.8 Starting Procedures

According to the arrangement of the oil circuits (see Sect. 6.3.6), compressors operate according to various starting procedures.

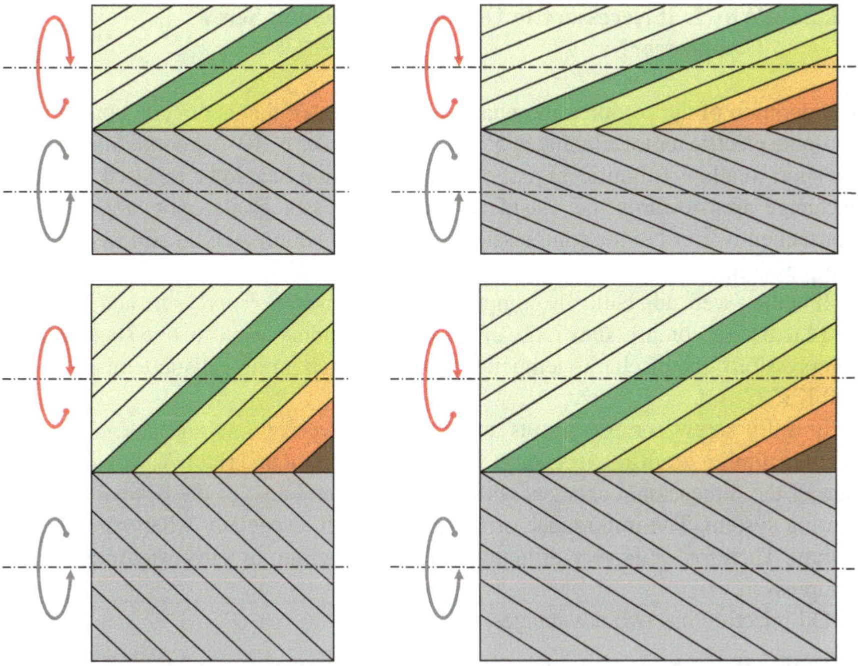

Fig. 6.22 Incidence of rotor length and diameter on the volume flow rate

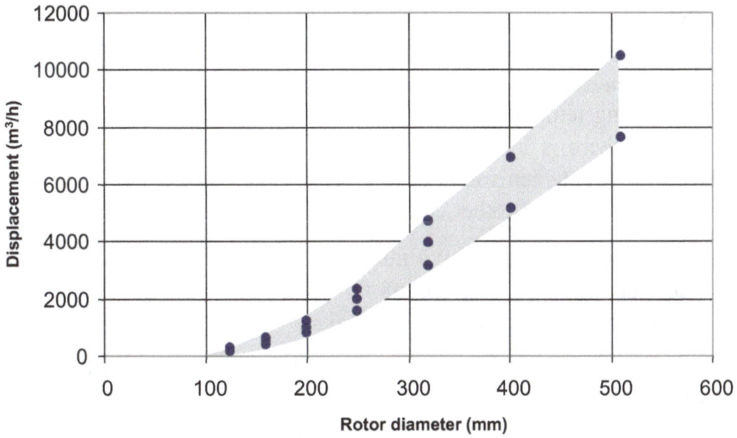

Fig. 6.23 The standard screw compressor range

6.2.2.3 Why Is It Necessary to Oil-Flood the Helium Screw Compressor?

The duration of the compression phase of a compressor that is directly coupled to a two-pole electric motor running at 50 Hz is in the order of 10^{-2} second. Such time is too short to allow a significant part of the compression heat to be released. The heat exchange area towards the outside of the compressor is also not large enough. Consequently, most of the compression heat stays in helium: it is an almost *adiabatic* compression.

If helium were adiabatically compressed from atmospheric pressure and 300 K up to 12.0 bar (a compression ratio of 12/1 is commonly used in one-stage helium compression), the discharge temperature would reach an unrealistic temperature of 800 K (527 °C).

In a dry screw compressor, as any contact between rotors and casing must be avoided, timing gears are necessary. The gaps rotor/rotor and rotors/casing must take care of the dimensional changes with temperature and also of the behaviour of the bearing system. The importance of the gaps is high, especially when compressing helium. Therefore, it is very difficult to compress helium in a dry way into a screw compressor.

Oil injection has two advantages:

- During the very short and brutal compression phase, oil is finely dispersed into helium. The size of the droplets is very small (in the order of the micrometre); consequently, the area of the interface between oil and gas is extremely large, enabling a fast and good heat transfer between gas and oil. The compression heat is distributed between helium and oil. The temperature increase is now related to the global heat capacity of the mixture of helium and oil. As a result, the *discharge temperature is drastically reduced.*
- As the operating temperatures are lower, it is easier to control and reduce the gaps. And as there is lubricant into the system, it is now possible to allow a possible contact between rotors. Therefore, oil lubrication allows an important reduction of the gaps. Another consequence is that *timing gears are spared.*

The gaps are now circulated by a mixture of helium and oil that is much more viscous than pure helium: the helium flow (or leak) is therefore substantially reduced. Let us consider the two types of leaks:

- Between casing and rotors, oil that is submitted to centrifugal forces is pushed against the casing where a layer of oil establishes. This oil layer is permanently scrapped by the edge of the rotors: the gap between rotors and casing is totally filled by oil. The helium leak is extremely low at this place. Also, at the discharge end of the rotors, a leak takes place between the end of the rotors and the casing (perpendicularly to the rotor axis) from adjacent lobes at different pressures.
- Between rotors, the helium and oil leak takes place along the sealing line.

The combination of a lower discharge temperature and a reduction of the internal helium by-passes leads to both a volumetric and isothermal higher efficiency machine at a lower cost. Another important feature of oil injection is that the oil acts as a noise damper.

Remark In the past decades, there have been a few tentative to build dry (non-oil-flooded) helium compressors in order to be rid of the oil removal issue. None of them has reached yet a satisfactory solution.

What is the quantity of oil that is necessary to reach a discharge temperature of 80 °C in conditions of the former example (compression from atmospheric pressure, 1.05 bar, 300 K, up to 12.0 bar)? The heat capacity of oil is 1.95 J/g.K. To cool helium from 527 °C down to 80 °C with oil that is injected at 45 °C, one must mix 34.1 g of oil per gram of helium! This machine is mainly processing oil.

If the concentration of oil in helium is expressed by volume at the discharge conditions, it is "only" 1.4%. Here one must remark that if the oil/helium ratio is too high, it could lead to the dangerous situation of compressing oil!

6.2.3 *Organisation of a Compression Station*

6.2.3.1 The Check Valve

If a power breakdown occurs, the compressor will rotate in the reverse direction and act as an engine. The rotational speed could reach a dangerous level and make the motor rotor explode. A check valve must be inserted either upstream or downstream the compressor, as shown in Fig. 6.24.

6.2.3.2 Compression Staging

Maximum high pressure for helium refrigeration systems reaches about 20 bar.

A compression ratio of 20/1 can easily be reached in one stage by an oil cooled screw compressor. However, its efficiency is not so good. It is common practice to arrange two compressors in series, even if the cost is higher.

Therefore, a trade-off is necessary. In a general way, for small refrigeration systems, a one-stage system is selected with an operating pressure around 12 bar. For large plants where the highest efficiency must be reached, two-stage arrangements are usual as shown in Fig. 6.25 where helium issued from the first stage is cooled down to room temperature before being processed into the second stage. This

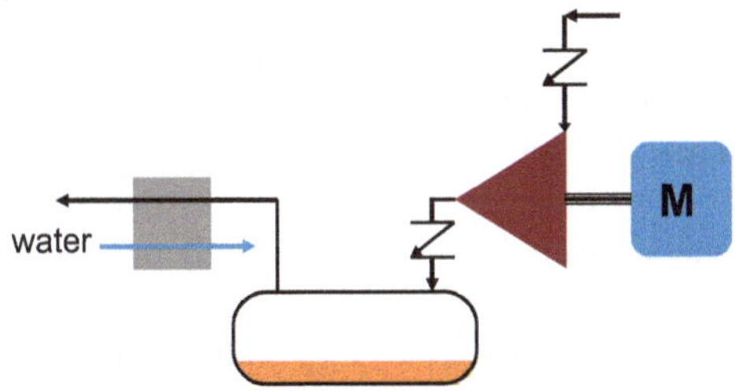

Fig. 6.24 The mandatory check valve in a screw compressor

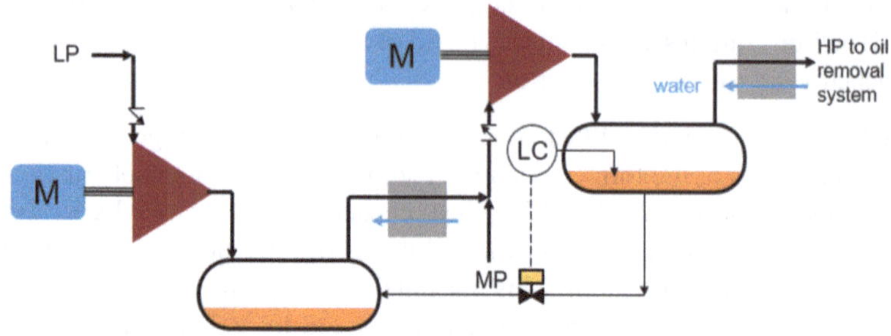

Fig. 6.25 A correct two-stage compression system

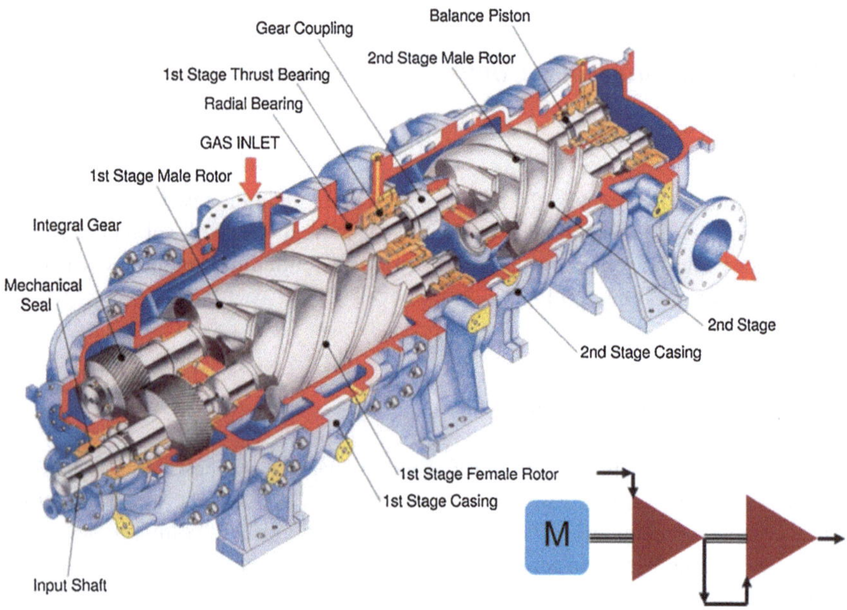

Fig. 6.26 A two-stage compound tandem compressor. (Provided by Kobelco Europe GmbH)

is an important point: remember $\dot{W} = \dot{m} \times T_1 \times R/M \times Ln(P_2/P_1)$ Eq. 6.4 where the power is proportional to the suction temperature.

If the selected cycle is a three-pressure one, the intermediate pressure return from the cold box is naturally connected between the two compression stages.

Some compressor builder build compound (or tandem) two-stage compressors. These machines are mainly designed for conventional refrigeration where a special cooling method by processed gas injection is used. This allows to avoid the gas interstage cooling. They need only one motor and are more compact.

However, in case of helium compression, the compound compressor without interstage cooling (see Fig. 6.26) should be avoided for two reasons:

- The inlet temperature of the second stage is higher; therefore, the second stage absorbs more power. As a consequence, the compressor power might be increased by around 10%.
- The pipe connecting the two stages is sized for heavy gases as freons. When processing helium (i.e. a light gas) at maximum flow, the oil flow in the pipe that connects the two stages is homogeneous and stable. However, when the helium flow is reduced, oil cannot be smoothly and continuously transferred: oil slugs appear, and bunches of oil are sent into the second stage leading to compression of oil and, consequently, faster deterioration of bearings. This can be avoided in running permanently the compressor at full flow.

6.2.4 Monitoring of a Compressor

The parameters to be monitored for one stage are indicated in Sect. 6.3.11.

6.2.5 Efficiency of Oil Cooled Compressors

The isothermal efficiency of a screw compressor, including the motor, and coupling efficiency at full load can be between 45% for small machines and 58% for large machines, for a compression of 5/1.

One should not forget that the power that is supplied to a compression system is mainly evacuated into the cooling media, water or air, as shown in Table 6.2.

This result confirms that energy in a gas is mainly related to the temperature, not to the pressure!

As nothing can be done against it, it is of interest to plan for energy recovery, for example, for heating buildings, with or without a heat pump. Such systems have been implemented in CNRS, Grenoble 100 kW, or will be at ITER (12 MW), ESS and CERN.

Table 6.2 Where does energy go? Thermal balance of a typical room temperature screw compressor

Mass flow	(g/s)	100.00	
Suction pressure	(bar)	1.05	
Suction temperature	(°C)	20.00	
Discharge pressure	(bar)	14.00	
Shaft power	(kW)	284.00	
Oil injection temperature	(°C)	45.00	
Compressor discharge temperature	(°C)	80.00	
Oil cooler power	(kW)	245.96	***0.87***
He cooler power	(kW)	25.96	***0.09***

6.2.6 Special Compression Machines

6.2.6.1 The Gas Ejector

An ejector is a device that induces a secondary fluid by momentum and energy transfer from a high velocity primary jet. A high-pressure fluid (HP) with very low velocity at the primary inlet is accelerated into a high velocity jet through a converging-diverging supersonic nozzle. The supply pressure at the inlet is partly converted into the jet momentum at the nozzle exit according to the Bernoulli's equation. The high velocity, low static pressure primary jet induces a secondary flow from the suction port (VLP) and accelerates it in the direction of the driving jet. Both flows get out at LP side. Applications of gas ejector have been explained in Sects. 6.3.9 and 4.3.1.2.3 (Fig. 6.27).

6.2.6.2 The Screw Compressor Operating at Sub-atmospheric Suction Pressure

A screw compressor can be operated at sub-atmospheric suction pressure, provided the compression ratio is acceptable. The suction pressure is often limited by the vapour pressure of water that is dissolved in oil: kind of cavitation happens at low pressures. It is important to use very well-processed oil (see Sect. 4.3.1.2.2). Experience showed that the minimum suction pressure decreases with the time that oil has been used in such a situation. Suction pressure as low as 20 mbar can be reached.

6.2.6.3 The Roots Machine

A Roots is a positive displacement machine with two lobed rotors, each looking like the figure of an 8 (see Fig. 6.28). The two rotors are mechanically linked by timing gears such that they rotate in opposite directions without contact. Their shape allows

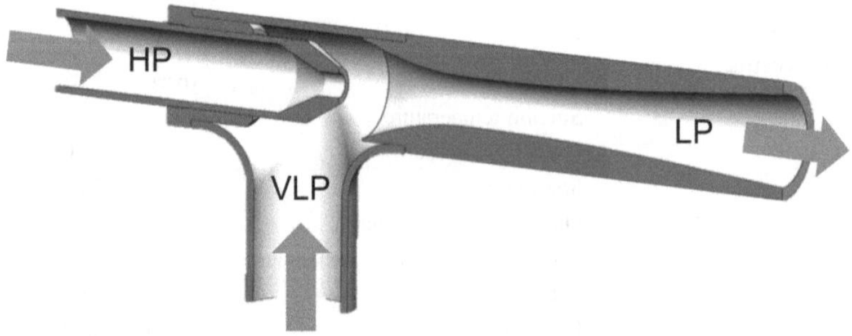

Fig. 6.27 Cross section of a gas ejector

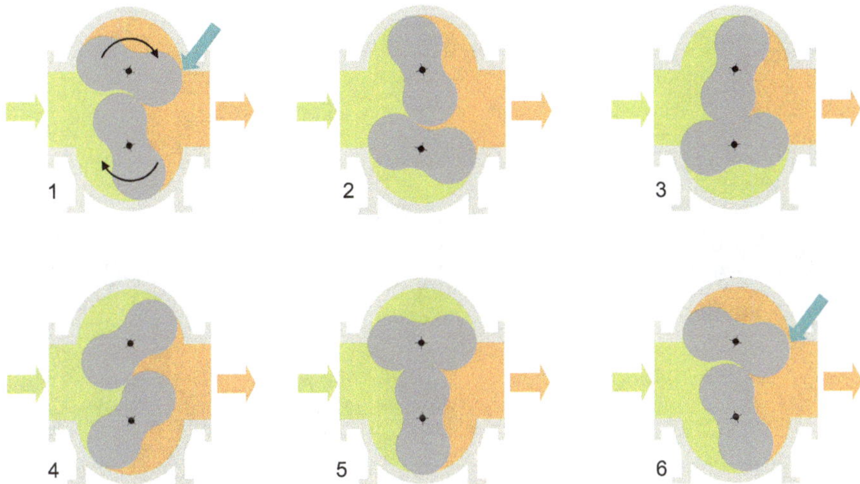

Fig. 6.28 Roots operation

them to be close to but never in contact with each other or the casing at every position during their rotation. The small gaps between the components allow the impellers to operate without a lubricant.

In state 1 and 6, one can see that the volume that is delimited by the rotor and the casing was, earlier, filled with low-pressure gas as in state 5. As soon as this volume is connected with the discharge side, the gas pressures equilibrate, and gas is *circulating backwards*! This is, obviously, not a way to reach high efficiencies. (Remember that such situation can happen in some operating conditions in screw compressors: Fig. 6.16.)

Another limitation of Roots is the rise in temperature due to gas compression: the suction side is a suction temperature, while the discharge side is a higher temperature. Consequently, the casing takes the shape of a banana, and one easily understands that contacts between rotors and casing might happen. Remember that helium, as a monatomic gas, reaches higher temperatures than air for same pressure difference! This is the reason why Roots are arranged in series, with a low-pressure difference per stage. A possibility is to inject oil during the process in order to reduce the temperature increase and reduce the gaps.

Leybold has developed a way to cool a Roots by recirculating cooled "high-pressure" gas.

6.2.6.4 The Claw Pump

The claw pump is a positive displacement machine with two rotors, having a peculiar cross section (claws), as shown in Fig. 6.29.

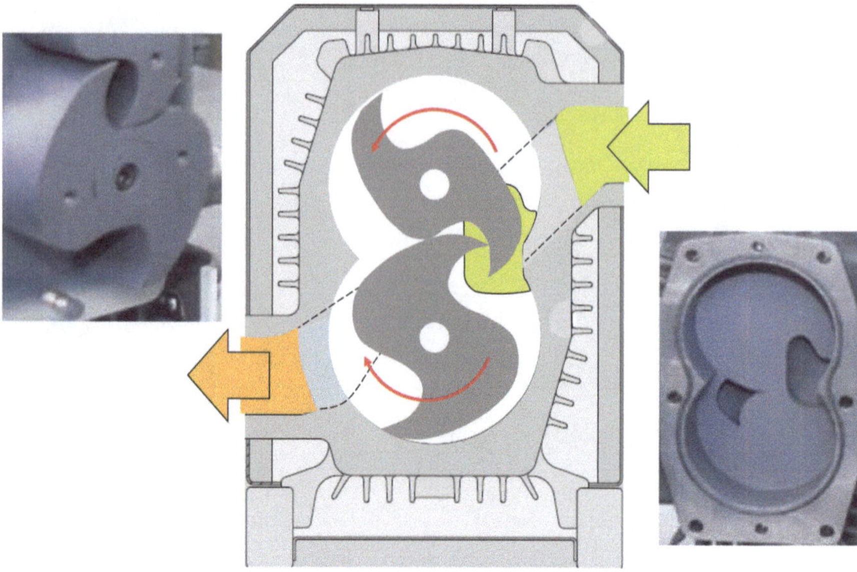

Fig. 6.29 The claw pump

It operates as a Roots, by displacing and compressing a volume of low-pressure gas to a location where it is connected to the discharge pressure. It is not lubricated.

6.2.6.5 The Liquid Ring Pump

A rotor that is fitted with wings (see Fig. 6.30) rotates on an axis that is non-concentric, compared to the axis of the casing. Oil sits in the casing. When rotating, the rotor moves oil that is pushed against the casing by centrifugal forces: it creates the oil ring. The thickness of the ring is constant, so the volumes that are trapped between the rotor wings and the oil ring:

- Increase: they are connected to the suction port.
- Then decrease: they are connected to the discharge port.
- Reach a zero volume.

The pump is fed continuously with oil that seals the clearances between inlet and discharge ports and removes the heat of compression. Oil leaves the pump together with the compressed gas and is separated in a bulk oil separator. Of course, the pump uses same oil as the compressors. With well-processed oil or oil that has been running for a long time, it is possible to reach very low suction pressure as 20 mbar.

The liquid ring pump runs at low speed and has not contact between moving parts, so it is a very reliable equipment. However, today the screw compressor is preferred.

Fig. 6.30 The liquid ring pump

6.2.6.6 The Centrifugal Compressor

The oil lubricated screw compressor is a nice machine, robust and reliable, but one has to take care of the oil that is injected in. Furthermore, its isothermal efficiency is limited to around 0.55.

However, ITER had a special problem to circulate the thermal shield cooling flow of two times 2 kg/s between 18 and 16 bar. A standard screw compressor was not acceptable, due to the very low compression ratio (1.12/1) (remember the shape of the isothermal efficiency curve in Sect. 6.2.2.2.4): for such a low compression ratio, a standard screw compressor would have had a very bad efficiency. A piston compressor was not a sufficiently reliable solution. As a consequence, it was decided to use a two-stage centrifugal compressor. But a new problem came up: due to the high rotational speed, the centrifugal compressor uses labyrinth shaft seals that are not leak tight. It happened that the helium leak would have been so high that it was not acceptable to lose so much helium. Therefore, a mixture of seal gas (nitrogen) is recovered and must be processed in order to recycle the helium quantity that has left the cycle through the labyrinth seal. Such an experience will be of very high interest for the future (Fig. 6.31).

To avoid such problem, an interesting solution is to enclose all the machine in the same casing, making it an hermetic compressor as shown in Fig. 6.32.

Remember the other possibility that is described in Sect. 4.6.2!

Fig. 6.31 Centrifugal compressor for the thermal shields of ITER (courtesy ITER)

Fig. 6.32 A possible centrifugal compressor (HOFIM™ compressor system by MAN Energy Solutions Switzerland Ltd)

6.2.7 A Compressor Digest

The mass flow rate of a compressor varies as the suction pressure and the inverse of the absolute suction temperature:

$$\dot{m} = k \times P_{\text{suc}}/T_{\text{suc}}$$

The mechanical absorbed power is proportional to the mass flow rate and the absolute suction temperature:

$$\dot{W} = k \times \dot{m} \times T_{suc}$$

If, for same suction and discharge pressures, the suction temperature changes, the mechanical absorbed *power stays constant,* but the processed *mass flow changes* as the inverse of the absolute suction temperature.

6.3 The Oil Management and Separation System

The oil removal system of a refrigeration plant is an almost static system that is generally performing correctly (provided it has been correctly designed and is correctly operated). Consequently, it happens that operators do not feel so much concerned by its operation. However, when, by no chance, oil is released by the removal system, the operation of cleaning components that have been polluted, mainly cycle heat exchangers and adsorbers, is a major trouble that can be the cause of system unavailability for a few weeks.

6.3.1 Selection of the Oil for a Helium Screw Compressor

The oil selection is made in two steps, according to:

- The viscosity: this is the choice of the compressor builder, for mechanical reasons.
- The vapour pressure: this is the choice of the cryogenic guy, for reasons of vapour separation. This topic is discussed in Sect. 6.3.9.

6.3.2 Oil Forms at the Discharge Side of the Compressor

At the discharge side of the compressor, oil stands in three various forms: droplets, aerosols (droplets <10 μm) and vapour.

Oil has to be separated into a very efficient system. Such a system operates along various specialised stages, according to the duty they have to perform, that are arranged in series.

The coarse oil separation is performed into a bulk oil separator that is generally provided by the compressor maker or the compressor integrator. Such systems are designed for general use of compressed gases, either air or refrigerant components, but the oil concentration at their discharge side is too high and not acceptable for helium refrigeration purposes. Therefore, companies dealing with helium refrigeration have to design and build the final oil separation system that takes care of aerosols and oil vapour.

6.3.3 Bulk Oil Separation for Oil Drops

At the discharge side of the compressor, the mixture of helium is moved along a pipe that is connected to the bulk oil separator. The flow conditions into this pipe should allow that the helium-oil mixture (remember Sect. 6.2.2.3, where oil occupies around 2% by volume) reaches the bulk oil separator in a stratified regime. A wise advice is that the latest bend should be located at more than 10 pipe diameters upstream the connection to the bulk oil separator. The velocity should be lower than 5.5 m/s to limit entrainment and to increase the mean size of liquid droplets, which allows an easier separation in the BOS.

Oil is collected into the bulk oil separator that has a capacity where the velocity of helium is sufficiently low in order that oil drops are simply separated by gravity. Such separator is cylindrical. It can be vertical or horizontal (see Fig. 6.33), provided its cross section allows for a sufficiently low velocity of helium in order that drops can fall down. This is a common engineering design, but experience has shown that, for cost reasons, undersizing is not scarce.

As helium is a light gas, its solubility into oil is not high. That is the reason why the retention time in the bulk oil separator is rather short, and therefore the oil quantity is not important. At the discharge end of the separator that is located on top if vertical or opposite to inlet if horizontal, there is generally a drop collector generally made out of metal wire (mesh where the separation can be as high as 99% for droplets from 5 to 10 μm) as shown on top in Fig. 6.33, left, or, better, coarse coalescing cartridges as shown in Fig. 6.33 right. Oil drops that are separated by the metal mesh fall back to the sump. In case of cartridges, oil is returned back to the suction side of the compressor by means of an automatic drainer.

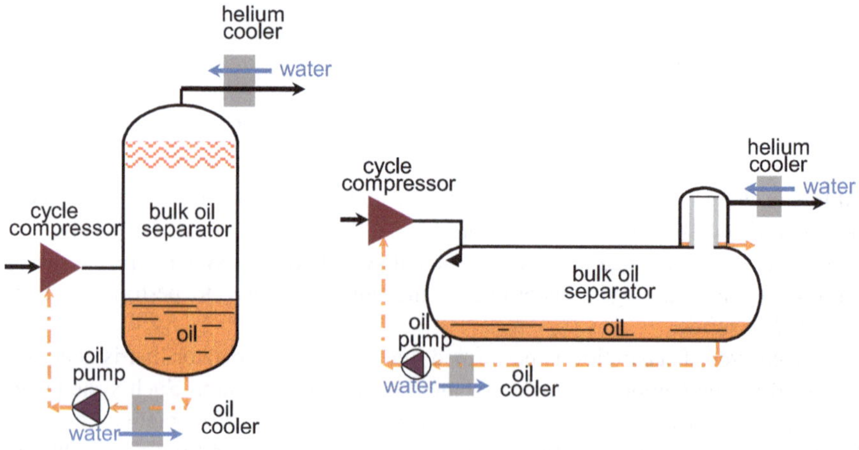

Fig. 6.33 Vertical and horizontal bulk oil separator general organisation

Remark Consider the following as a way to understand the operation of a separator, which only means how figures are changing according to parameters, but be careful with these values that should be considered as approximative!

When thinking about the operation of such a component, one can "feel" constraints as:

- Helium velocity is acceptable in order that drops are not entrained.
- Time is sufficient for such separating process happens.

6.3.4 Vertical Separator

When a flow of gas is injected into a space, its velocity decreases strongly, in the ratio of cross sections. Depending on the way the gas is injected, a lot of turbulences can be generated. However, even knowing that, the assumption is made that gas velocity in the separator is constant and isotropic.

6.3.4.1 Oil and Helium Velocity Composition

It is of interest to compare the helium and oil droplet's velocities into a vertical separator. In Fig. 6.34, one understands that if the falling velocity of a droplet is lower than the upwards velocity of helium, such droplet is entrained by the helium flow. Conversely, if the falling velocity is higher, the droplet falls and is not entrained by the helium flow.

6.3.4.2 Separation According to Droplet Sizes in a Vertical Separator

Free Fall of an Oil Droplet Using the Stokes law, one can estimate the falling velocity of drops. Figure 6.35 shows that:

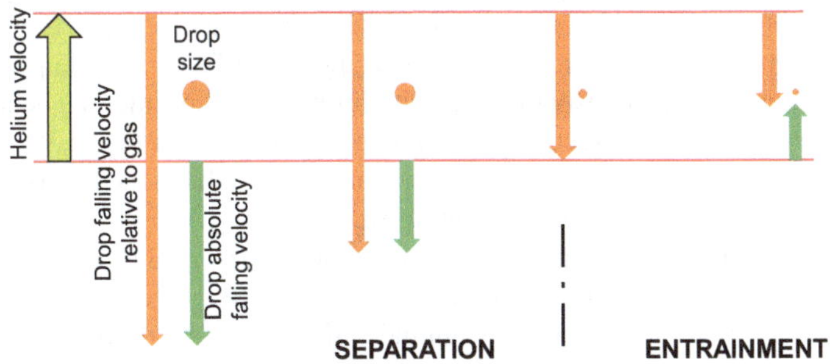

Fig. 6.34 Separation and entrainment according to the falling velocity of drops

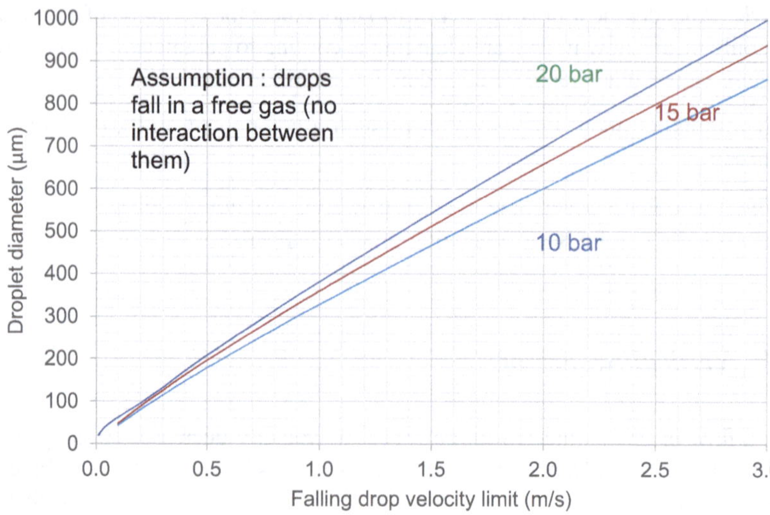

Fig. 6.35 Free falling oil droplet velocity versus diameter in helium at 80 °C

- The biggest the drop, the fastest it falls down
- Pressure has not a significant incidence on falling velocity.

In the reverse way, Fig. 6.35 can be used to show the size of the biggest drops that come out of a vertical separator, according to the helium upwards velocity: the higher the velocity, the larger the biggest drops that are released.

6.3.5 Horizontal Separator

6.3.5.1 Oil and Helium Velocity Composition

As helium flows horizontally and drops fall vertically, one has to consider the composition of velocities. Drops that are 400 µm diameter reach their falling velocity limit V_O in about 0.1 s, which is a very low value that one can neglect. For smaller drops, this time is even lower. Consequently, the simplification assumption is that the drops reach instantaneously their velocity limit. Let us consider the behaviour of droplets of various sizes issued from a same point, top left, in Fig. 6.36. Oil drops that hit the oil surface at a length lower than L are separated; others that would hit the surface at a length higher than L are entrained.

The size of the largest droplets that are released is related to the ratio h/L (h = separator diameter – oil height) and gas velocity. See Fig. 6.37.

Figure 6.37 gives an order of magnitude of the size of droplets that can be expected at the discharge side of the separator, according to the ratio h/L, without any mesh or similar equipment.

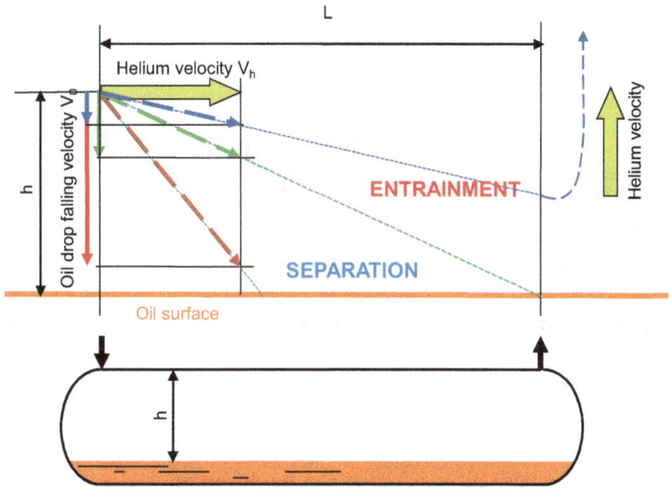

Fig. 6.36 Velocity composition in a horizontal separator

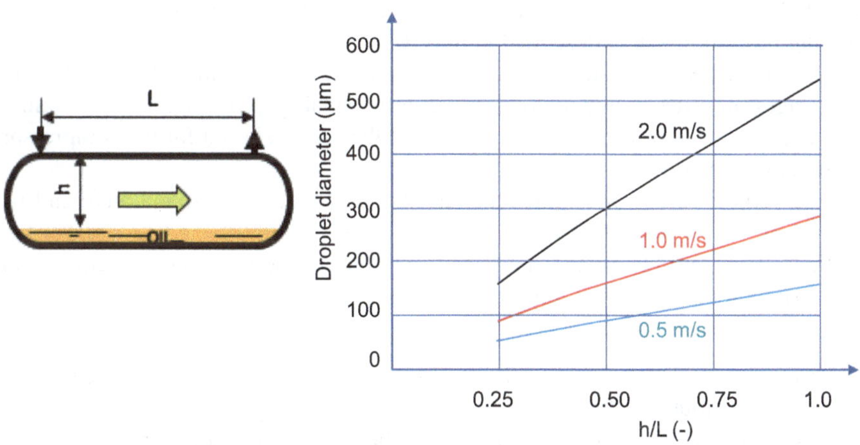

Fig. 6.37 Separation, according to h/L and gas velocity in a horizontal separator for helium at 20 bar, 80 °C

The main flow of oil that is recovered at the low part of the separator is cooled into a water or air after cooler.

Oil has to fulfil various duties:

- Cool the compressor.
- Lubricate the screws and the bearing system.
- Improve the sealing between screws and screws and casing.
- Sometimes, actuate the slide valve if any.

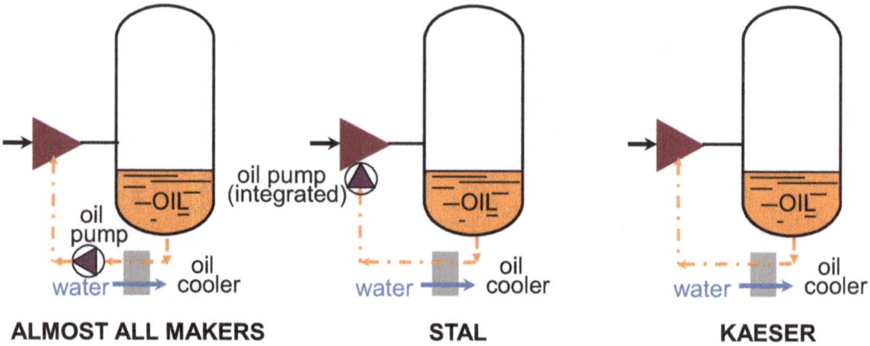

Fig. 6.38 Various ways to re-inject oil

Oil for cooling is injected into the compressor. Depending on the compressor builder, the place(s) where oil is injected can be different.

Depending also on the compressor builder, oil is processed into an oil pump or is injected by the gas pressure difference between high and low pressure (see Fig. 6.38). Such different organisations have a consequence on the compressor starting procedure that is described in Sect. 6.3.6.

Oil for the bearing system lubrication must be very finely filtered, sometimes in two stages, in order to avoid any damage in the small-gap bearing system (rolling elements or sleeve bearings). The viscosity of the oil is selected for the compressor lubrication purpose.

The vertical separator requires a smaller volume and less ground area, and its capture efficiency is not dependent of the oil level.

Such above "rough" arrangements can be improved, both in oil management and oil separation (see Fig. 6.39, right):

- A special component like a Schoepentoeter vane device (see Fig. 6.39, left) can be fitted to the inlet pipe in order to enhance the drop separation efficiency and evenly distribute the flow in the cross section of the separator,
- It is not necessary to process the totality of the oil flow into the pump. The oil flow that is used to cool the compressor can be circulated naturally by pressure difference. Only the flow that is to be delivered at a higher pressure than high-pressure helium is to be processed into the pump in order to feed the bearings and the slide valve.
- The oil quantity that is separated into the coalescing cartridges and the small amount of oil that is also separated into the helium cooler are returned to the compressor suction side by simple pressure difference.

The oil concentration in helium at the discharge side of the bulk oil separator depends on the designer. Well-designed systems reach concentrations as low as 50 ppM by weight (ppM W) but bad designed systems can release helium at concentrations as high as 10,000 ppM by weight.

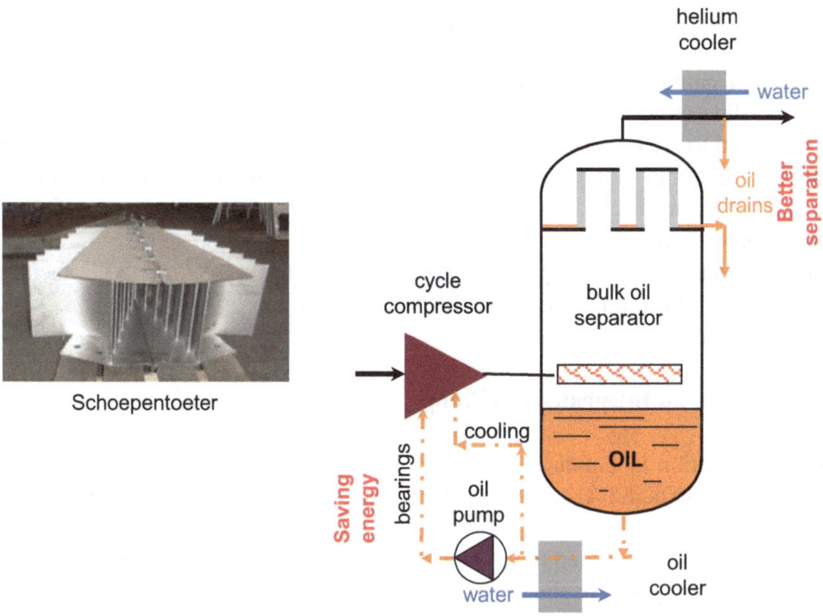

Fig. 6.39 An improved bulk oil separator. At left, a Shell Schoepentoeter TM vane device

Remark Oil concentration into helium must be expressed by a *weight ratio*, because a volume ratio would have almost no signification due to the very different densities of these two fluids. (1 ppM is one part per million: 10^{-6}.)

Such a value would be too high for a correct separation in a standard oil removal final stage (see Sect. 6.3.8.4.). As a corrective action on systems that have been incorrectly designed, an extra coalescing stage has sometime been added before entering the final system. That adds parasitic pressure drops and might not solve the problem.

An Advice It is highly advised to check the bulk oil separator efficiency at the very beginning of the commissioning tests of a compression station (see Sect. 14.4.8.1).

6.3.6 Starting and Re-starting Procedures According to the Oil Management Organisation

For compressor that is fitted with a slide valve, it is mandatory to start with the slide valve at its minimum flow position in order to avoid overloading of the electric motor.

6.3.6.1 With an External Oil Pump

Such organisation is shown in Fig. 6.38, left.

If the slide valve is not at zero position, the oil pump is to be started, the slide valve moved to zero position, and then the oil pump is stopped. During the period of time the oil pump is on, oil is obviously injected into the compressor casing; therefore, time must be allowed for oil to drip out of the compressor casing, by gravity, before starting the compressor.

If the compressor has stopped on a trip situation, this procedure to get the slide valve back to zero prior to start-up is to be applied too.

6.3.6.2 With an Integrated Oil Pump

Such organisation is shown in Fig. 6.38, centre. It has been typically the STAL, Sweden, way. (The STAL company does not exist anymore.) In the USA, Sullair also uses such a way.

The slide valve is pushed to the minimum flow position by a spring; therefore, at the time of starting up, the slide valve is always naturally at minimum flow position. If the compressor has stopped on a trip situation, the slide valve moves automatically to its minimum flow position as soon as the oil pressure is decreased.

6.3.6.3 With No Oil Pump

Such organisation is shown in Fig. 6.38, right. It is typically the Kaeser way. The arrangement is shown in Fig. 6.40.

PCV out is a FC (fail closed) valve that opens when the upstream pressure is above a set value. PCV in is a combination of a FC valve including a by-pass with a calibrated orifice. PV1 is FC; PV2 is FO (fail open).

When the compressor is stopped, all parts upstream PC out are at same pressure.

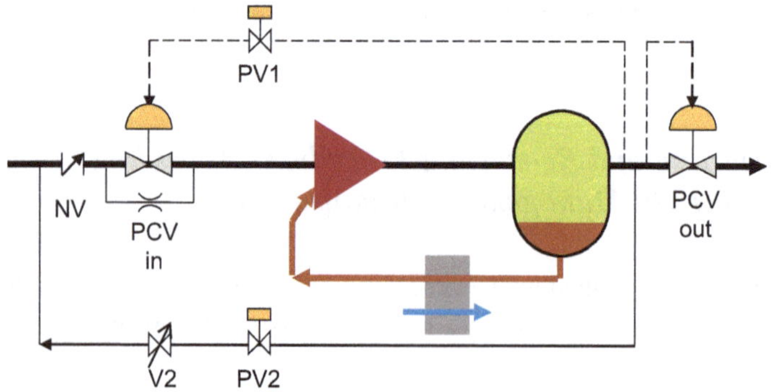

Fig. 6.40 Kaeser start-up system

When the compressor starts running, the suction pressure of the compressor decreases very fast due to the calibrated orifice. The pressure difference between the bulk oil separator and the suction side of the compressor allows quick lubrication. The small flow that is processed by the compressor is recycled through PV2 and V2. Such a situation called standby may last for a long time. When PV1 and PV2 are energised, PCV in opens, PV2 closes, and the compressor is in normal operation. PCV out opens as soon as the discharge pressure reaches the pre-set value.

6.3.7 The Coolers

The heat of compression is to be evacuated from the system into the cooling fluid. Such a duty is fulfilled by, respectively, the oil and helium coolers (remember Sect. 6.2.5). The cooling fluid can be water or air. In case it is water, the operator must be very careful about its quality.

6.3.7.1 The Oil Cooler

The oil cooler duty is to cool down oil at an acceptable temperature prior to injection into the compressor. The oil cooler can be fitted upstream or downstream the pump. In the upstream location, the pump processes colder oil, which is better for a higher reliability. It is the situation that is shown in Fig. 6.39, right, where the pump processes only the flow for bearings and slide valve. When the pump processes the whole oil flow, the oil cooler can be located downstream the pump if the relative position of the oil sump and the pump could lead to a possible too low net positive suction head. The oil injection temperature is generally around 45 °C. It is of interest to consider the duty of such a heat exchanger. Let us take, as an example, a compressor processing 50 g/s from 1.05 bar, 300 K up to 12.0 bar. In Sect. 6.2.2.3, the calculation showed that 31.4 g of oil are necessary to cool 1 g of helium. Therefore, the oil flow rate is 1704 g/s. The heat capacity of oil is 1950 J/kg.K. To cool the oil down to 45 °C, the power to be removed is:

$$1704 \times 1950 \times (80 - 45) = 116,307 \text{ W}$$

6.3.7.2 The Helium Cooler

The helium cooler duty is to cool down helium at the coldest reasonable temperature according to the cooling water temperature. Let us calculate the helium cooler power to cool down helium to 25 °C:

$$0.050 \times 5196 \times (80 - 25) = 13,780 \text{ W}$$

Remark 1 If we compare the oil and helium coolers of compressors having the same mass flow rate and discharge temperature, but various compression ratio, we notice that the oil cooler power is *related to the flow and the compression ratio,* but the helium cooler power is only related to the *processed flow*.

In any case, most of the power that is injected into the compressor is released though the oil cooler.

Remark 2 Even if the helium cooler is not designed for that, it traps a small amount of oil aerosols. It is of interest to recover this oil quantity by draining it at the lowest point (that must be at the discharge side): it reduces the amount of oil to be trapped later into the coalescers (Fig. 6.39).

6.3.7.3 Some Remarks on Oil and Helium Cooler Technologies

Tube and Shell Type
This type of cooler (see Fig. 6.41) is very common in the industry. It is advised to select, for the pipes, a kind of steel that is not sensitive to corrosion. The water pipes must be dudgeon *and welded* on the end plates. It is preferable to have straight tubes for reasons of cleaning of the water side.

Brazed Plate Type
This type of cooler, made out of stainless steel plates that are brazed together, is very compact and efficient (see Fig. 6.42). On top of that, it is a cheap component because it can be chosen on the shelf or it can easily and cheaply be assembled on demand from standard components.

As a drawback, one must be careful if it is to be used with cooling water that is not clean.

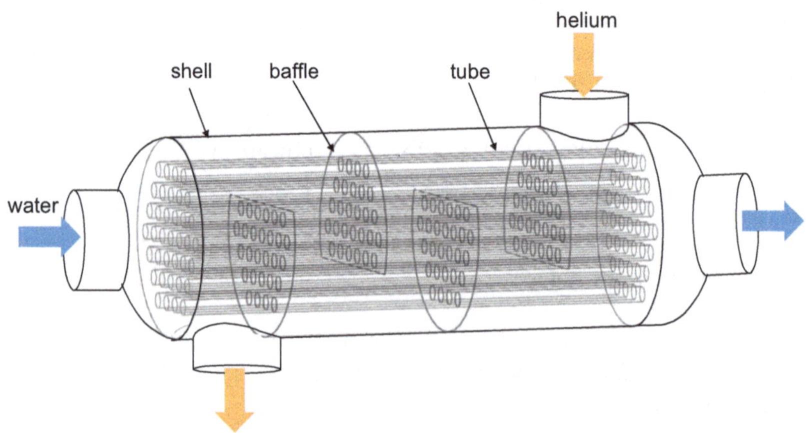

Fig. 6.41 Tube and shell after-cooler

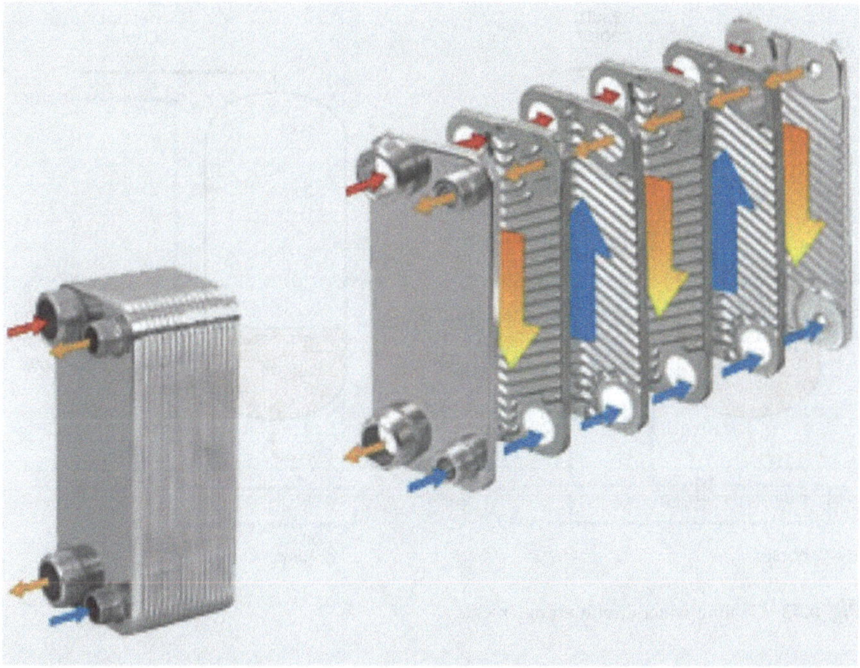

Fig. 6.42 Brazed plate cooler

6.3.7.4 The Cooling Water Circuit Organisation

Water can be circulated in parallel or in series into the oil and helium coolers (see Fig. 6.43).

In a parallel arrangement, water flow must be correctly balanced (see Sect. 11.6.4.1): the water flow in the oil cooler is to be set in order to get an oil injection temperature of around 45 °C (or any other value specified by the compressor maker). The remaining flow is used to cool helium. Experience shows that such water flow balance is rather often badly performed or not at all balanced.

In a series arrangement, all flow circulates first into the helium cooler and then into the oil cooler. Therefore, helium is naturally cooled at the coldest possible temperature, and the oil cooler is circulated by slightly warmer water, but as the oil cooling temperature is higher than the helium one, it does not make a more difficult duty for the oil cooler. The main interest is that it is no longer necessary to balance the flows.

A correct control of the oil injection temperature is important:

- Too warm (too low the oil viscosity), bearings might be damaged, or contacts might happen between rotors
- Too cold (too high the oil viscosity), could lead to damages to the bearings.

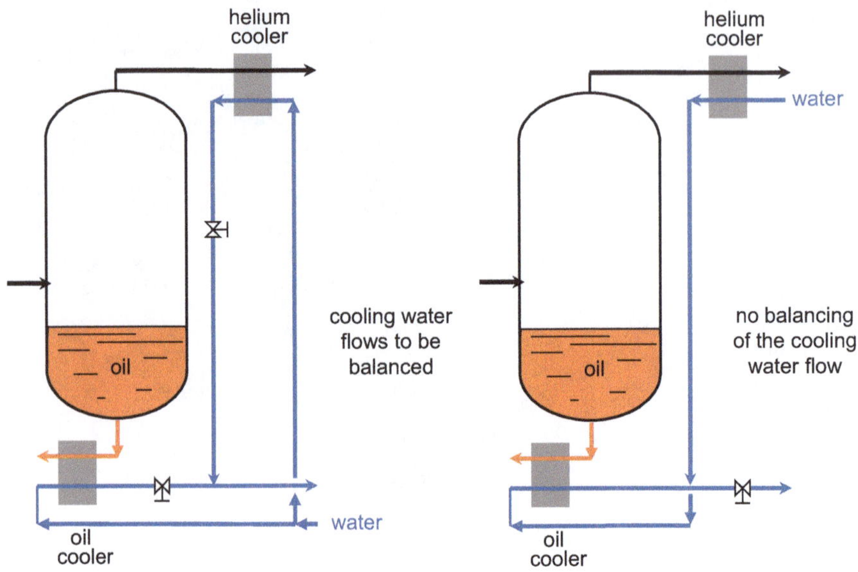

Fig. 6.43 Cooling water circuit arrangements

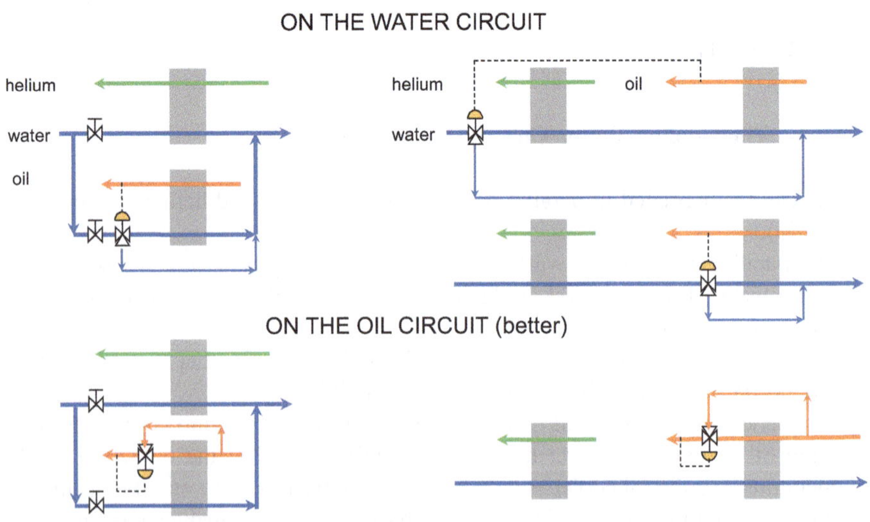

Fig. 6.44 Control of the oil injection temperature

There are various ways to control the oil injection temperature: the water flow can be controlled according to the oil temperature – the local water-cooling system might be hydraulically disturbed by the water flow changes, and when the water flow is low, deposition could happen into the water circuits; in Fig. 6.44, top, water is by–passed according to the oil temperature: there is no longer hydraulic perturbation of the local water circuit, but there is still possibility of deposition in the water circuits; in Fig. 6.44, bottom, part of the oil flow can by-pass the heat exchanger, no hydraulic perturbation of the cooling circuit and no deposition risks. This solution is the best choice.

6.3.8 *Aerosol Capture*

Aerosols are tiny droplets of a diameter around 1 micrometre. To have an idea of such aerosols, one can compare it with cigarette smoke: it is bluish and goes up. In the case of smoke, the colour is given by the particle size, and it goes up because it is warmer than ambient air. In the case of helium, the colour that is the same means that the particle sizes are identical. It goes up because helium is lighter than air. This can be easily shown by slightly opening a valve at the discharge side of the bulk oil separator: the blue "smoke" rises! This simple experiment shows that aerosols do not separate naturally: their downwards velocity is very small, due to the important friction on gas (see Table 6.3). The result of this theoretical calculation for very small diameters is certainly affected by Brownian movement of helium molecules.

A specific process must be implemented to separate them.

6.3.8.1 The Coalescing Process

Separation of aerosols is performed by the coalescing process. This process which is also called mist elimination was developed by J. Brink, working with Monsanto, back in the end of the 1950s. The original application was developed to solve air pollution issues for Monsanto's phosphoric and sulphuric acid plants. Coalescing is agglomerating tiny particles into large ones until they can be collected by simple gravity separation. This is performed by circulating aerosol-loaded helium through coalescers that are special filters made out of very thin fibres, the diameter of them being in the same order of magnitude as aerosol: approximately 1 μm. The size of void places that exist in the filtering medium is larger than the size of the particles to be trapped. Figure 6.45 gives an idea of the internal structure of the filtering medium. One should know that the filtering medium is mainly vacuum; the fibre density is very low, around 4% by volume (or 96% void!). Fibres are generally made out of

Table 6.3 Falling velocity limit of aerosols in helium at 20 bar, 80 °C

Oil in helium at 20 bar, 80 °C				
Droplet diameter	(μm)	0.016	0.12	0.91
Falling velocity	(mm/s)	0.01	0.1	1.0
Thumb rule: velocity in mm/s = diameter in μm				

Fig. 6.45 Fibres in a cartridge: mainly vacuum

glass or borosilicate. Helium is circulated at low velocity in order to stay in a laminar regime.

Figure 6.46 top, shows the way gas is circulating in a laminar regime around a cylinder (a fibre). The lines of the picture materialise the gas stream lines. According to that, one can imagine how gas is circulating through a coalescer media, as shown in Fig. 6.46 bottom.

The aerosol trapping process varies according to the particle diameter.

Big-Size Particles

The so-called "big-size" particle sizes are in the range of a few micrometres. Their inertia is rather high, and consequently, their trajectories are almost rectilinear, not at all sensitive to the direction of the gas stream lines. Consequently, they surely hit a fibre during their trip (see Fig. 6.47). It is a similar situation if one wants to run in a straight line into a forest: that will not last for long! Superficial tension of oil "sticks" the droplet to the surface of the fibre. All particles are trapped: the capture efficiency of the coalescer is 100% for big particles.

Middle-Size Particles

The middle-size particles are in the range of 0.5 to 3.0 micrometre. Their inertia is low and, consequently, they follow the gas stream lines (see Fig. 6.48). Most of them are collected on the fibres, but some of them succeed in crossing the filtering medium without having been trapped: the capture efficiency of the coalescer is not 100% for the middle-size particles.

Small-Size Particles

The small-size particles are less than 0.5 micrometre diameter. Their inertia is very low. As they are hit by gas molecules, which are submitted to Brownian movement, their trajectory is non-linear, very erratic, and so long that they have a 100%

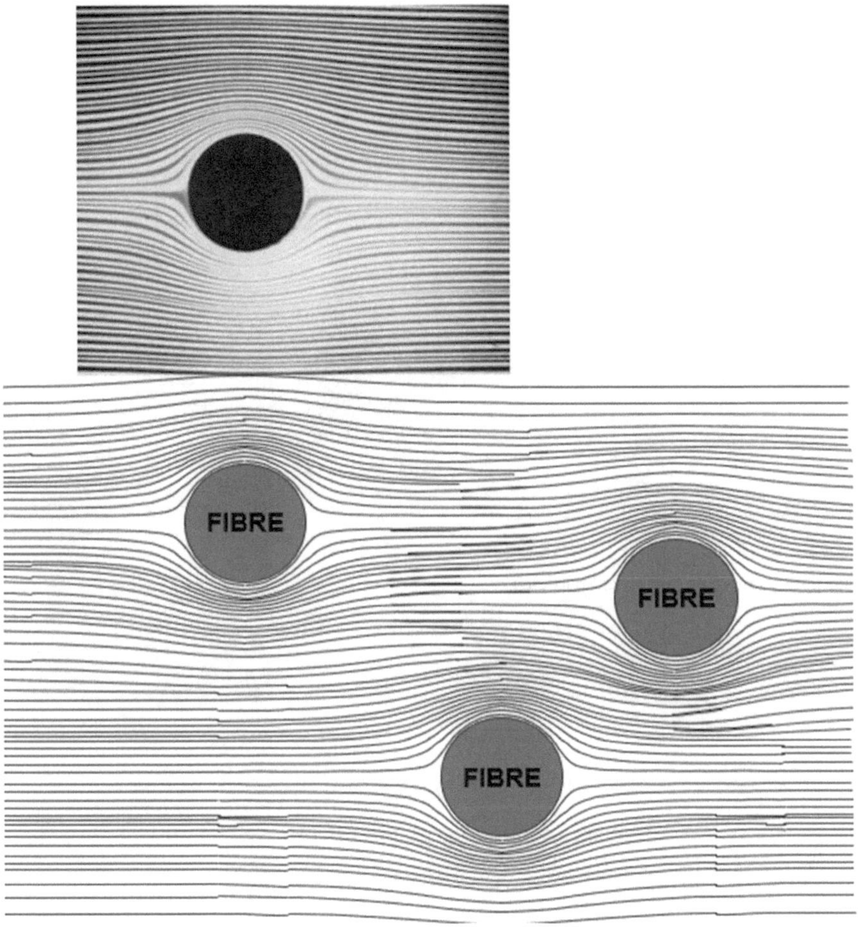

Fig. 6.46 Gas laminar flow around a cylinder (D. Peregrine), left, and fibre arrangement in a coalescer

probability to hit a fibre: the capture efficiency of the coalescer for very small particles is 100% (see Fig. 6.49).

When an aerosol droplet hits a fibre, it sticks to it; the trapping process is mainly related to surface tension. The fibre surface is generally treated in order to be "oleophilic" (see Fig. 6.50 left). Most of the oil is "stored" on the fibre nodes.

Oil droplets that are stuck on the fibres are submitted to gravity and action of the gas stream; they drip along the fibres and are finally collected at the outer lowest part of the cartridge.

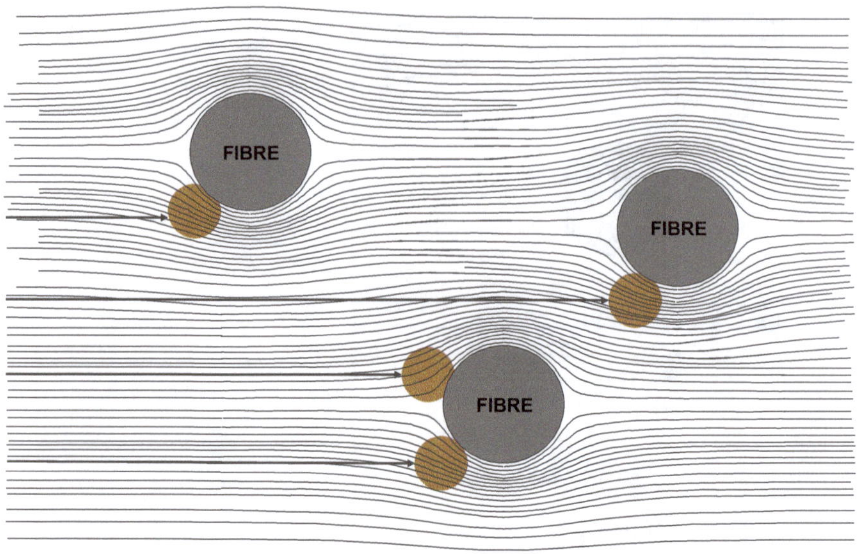

Fig. 6.47 Trapping "big-size" particles

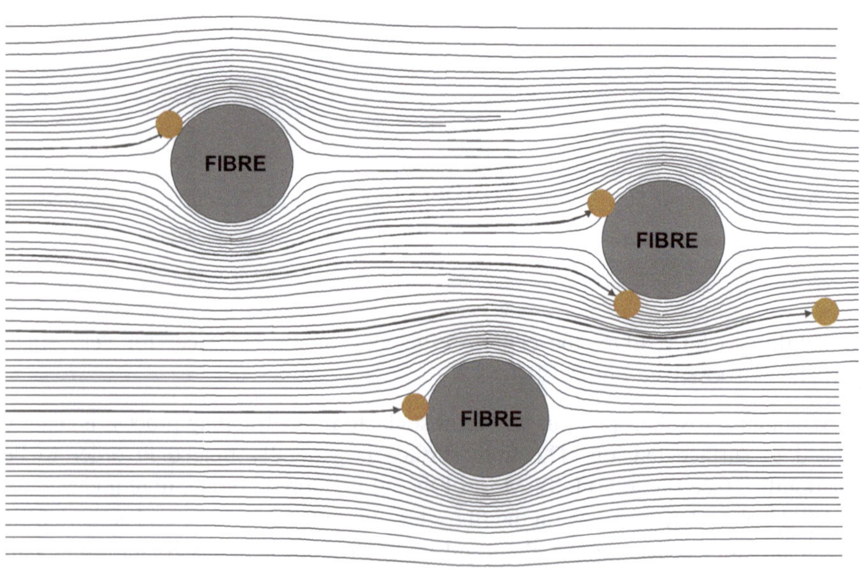

Fig. 6.48 Trapping middle-size particles

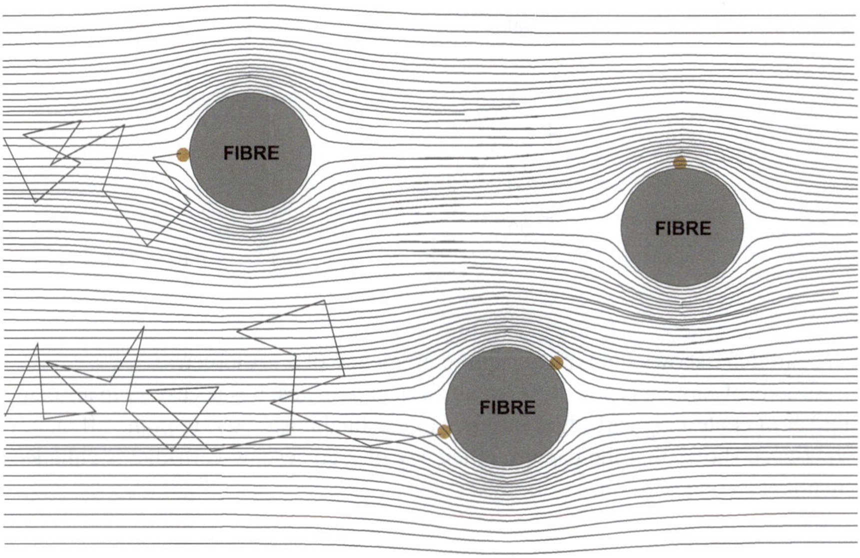

Fig. 6.49 Trapping small-size particles

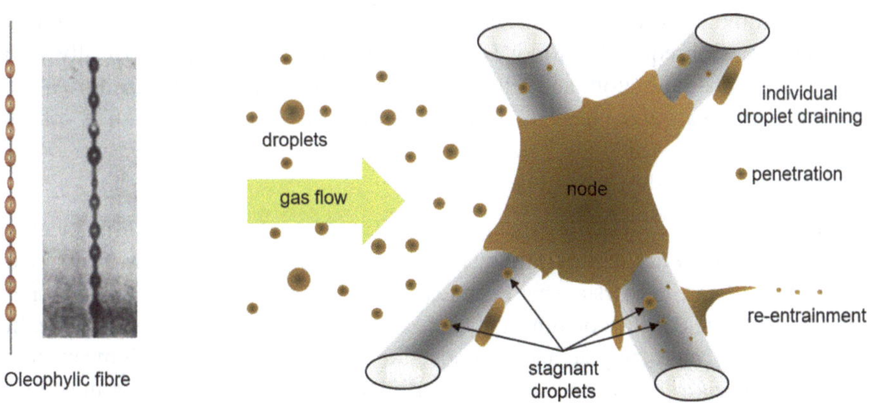

Fig. 6.50 How aerosols agglomerate on fibres

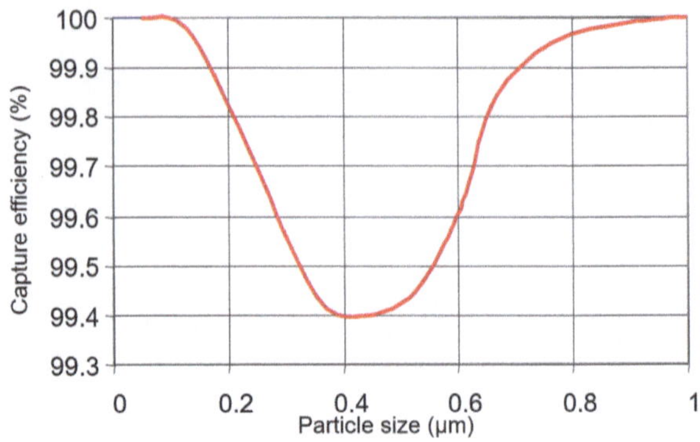

Fig. 6.51 Typical capture efficiency of a coalescer versus particle size

Figure 6.51 shows the typical capture efficiency of a coalescer.

As aerosol capture is not 100% for all sizes of particles, it is necessary to arrange a number of coalescing cartridges in series to get an almost 100% trapping efficiency. Experience shows that the capture efficiency of a coalescing cartridge is enhanced when it is wetted with oil.

6.3.8.2 The Structure of a Coalescing Cartridge

The oil loaded gas flows generally from inside to outside of the cartridge. Oil that is trapped on the coalescer fibres is submitted both to vertical gravity and horizontal gas velocity. Consequently, it moves into the filtering medium both vertically with gravity towards the bottom and laterally towards outside with gas velocity. At the bottom of the cartridge, there is a portion where the filtering medium is totally soaked by oil. It is called the "wet band". From the bottom of the wet band, drops materialize and fall down (see Fig. 6.52).

Remark Even if a coalescing cartridge looks like a filter, there is a big difference: as the separated material is only liquid, it naturally drains towards the bottom. That means that, after a steady state has been reached, where the oil quantity that is stored into the cartridge depends on time and oil concentration, there is no more loading. In other words, a coalescing element can never be "plugged". Therefore, it means that there is *no need to replace* a coalescing cartridge, except if it has been plugged by solid particles or deteriorated or any other reason (see Sect. 11.6.6.1).

For small- and middle-size cartridges, the filtering medium is kind of a blotting paper that is rolled on a perforated plate. At the outer part, another perforated plate retains the filtering medium. In Fig. 6.53, one can see a small cartridge: the white material is the filtering medium that is glued to metal parts on both ends. The inlet is fitted with an O-ring. Such O-ring is very important: if it leaks, it is kind of a parasitic by-pass of the cartridge, and therefore it jeopardises the system efficiency. One must be very careful with the selection of the seal material, because it can be deteriorated

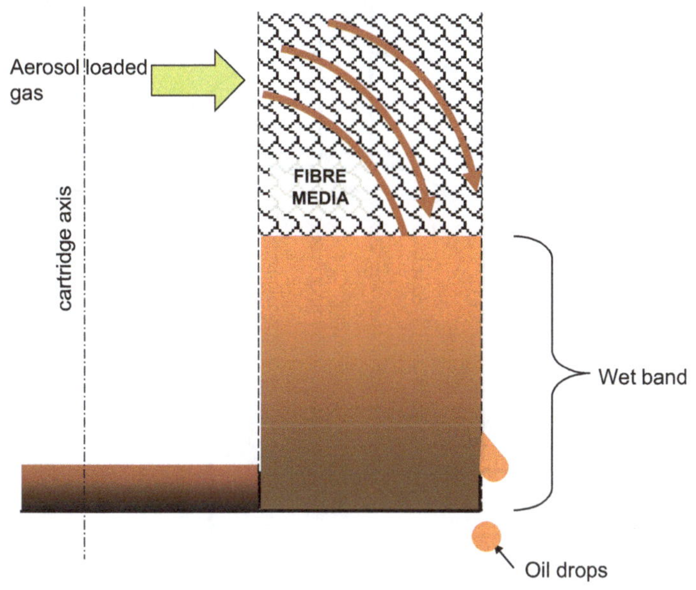

Fig. 6.52 Cross section of a coalescing cartridge

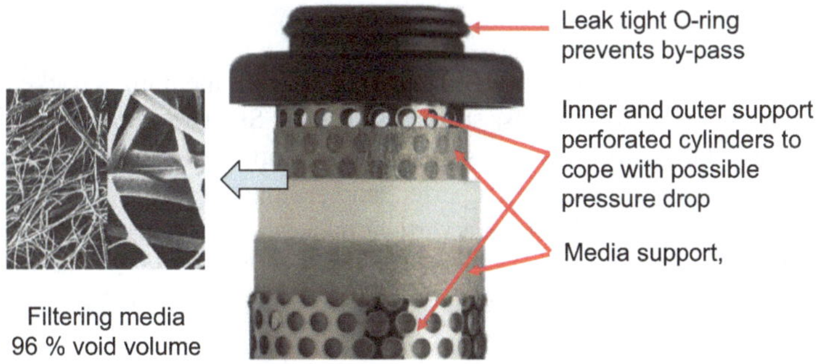

Fig. 6.53 A small coalescing cartridge (images supplied by Parker Hannifin)

by the special oil that is used for helium compression: nitrile or even better, Viton, are advised materials.

When the volume flow rate is high, it is possible to install a number of cartridges in parallel. It is a common solution in air oil removal. However, one must be aware that installing a number of cartridges must be made in a way to be sure that the connection between the cartridge and the housing is correct, which can be not easy to check, for example, as shown in Fig. 6.54, left. The right part arrangement should be preferred because it allows a visual check. If cartridges are connected by a screw system, it is wise to positively lock it in order to avoid loosening by vibrations that could develop possible undetected by-passes that could increase along time.

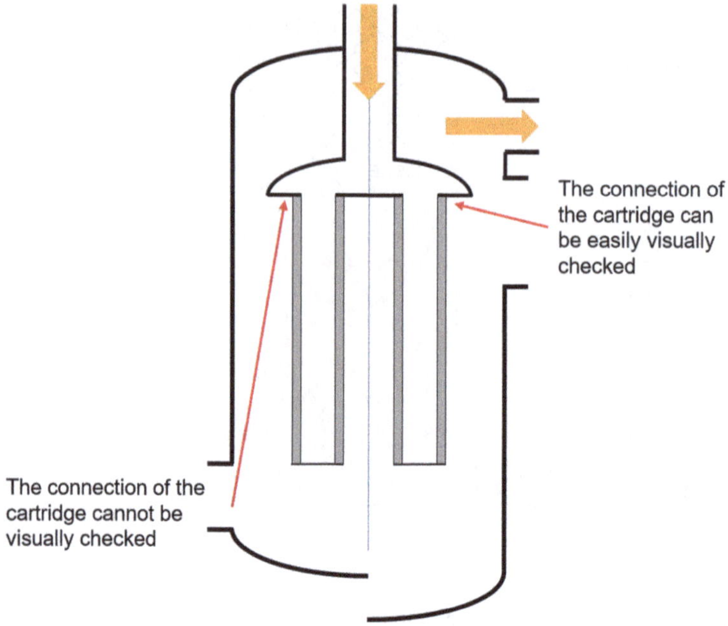

Fig. 6.54 Two ways to house cartridges

Figure 6.55 shows a large cartridge where the filtering medium is packed in kind of a large cage made out of stainless steel mesh screen. Such a solution is particularly suited for very large flows (see Fig. 6.66). The cartridge can be connected by welding to the enclosure, ensuring a perfect and permanent leak tightness.

The oil coalescing cartridge is in a vertical position. Oil loaded gas gets into the cartridge at the top and flows from inside to outside. The cartridge is housed into a capacity fitted with an oil discharge pipe at the bottom (see Fig. 6.59).

6.3.8.3 Correct Sizing of the Coalescing Cartridges

6.3.8.3.1 Front Velocity

In Fig. 6.56 top, the sizing parameters are correct:

- The front velocity (not higher than 0.10 m/s)
- The density of the medium is in relation with the oil gas load (the density of the medium is to be low if the oil load is high).

In Fig. 6.56 bottom, the front velocity is higher for a same medium. Gas velocity inside the medium is too high: some droplets are released by the “wind” from the fibres where they are stuck on and entrained again. Oil bubbles materialise at the

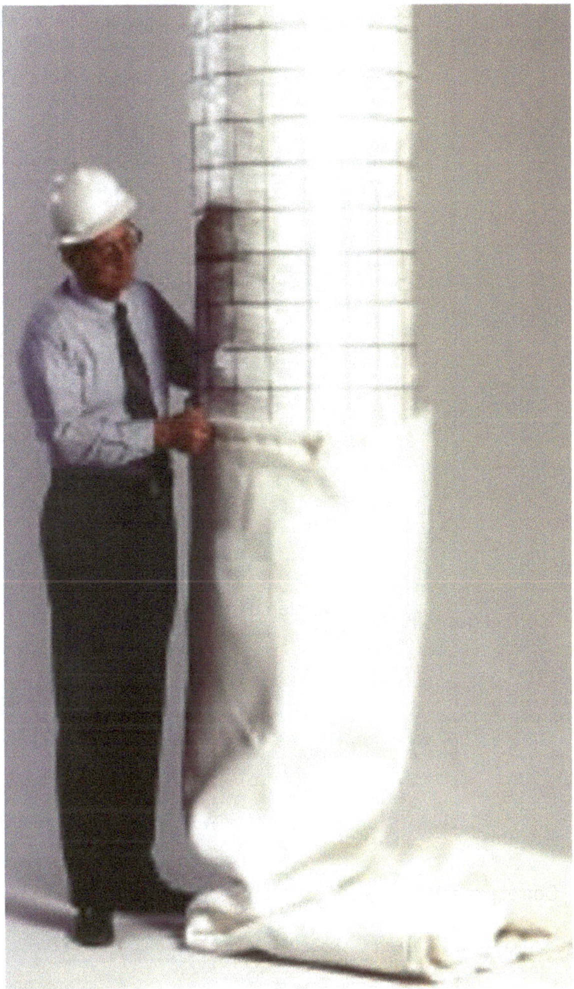

Fig. 6.55 A large coalescing cartridge (DuPont)

outer surface of the cartridge and explode, generating new aerosols, as it can be seen on the photograph in Fig. 6.57, right.

The design of the cartridge must consider the worse operating condition where the front velocity is maximum. Be careful when there are operating regimes at lower pressure than nominal.

6.3.8.3.2 Compactness of the Filtering Medium

In Fig. 6.57, top, the filtering media is much more compact (the void space is reduced). Even if the front flow velocity is correct, oil plugs partially the void space in the medium and reduces the available cross section: the local actual gas

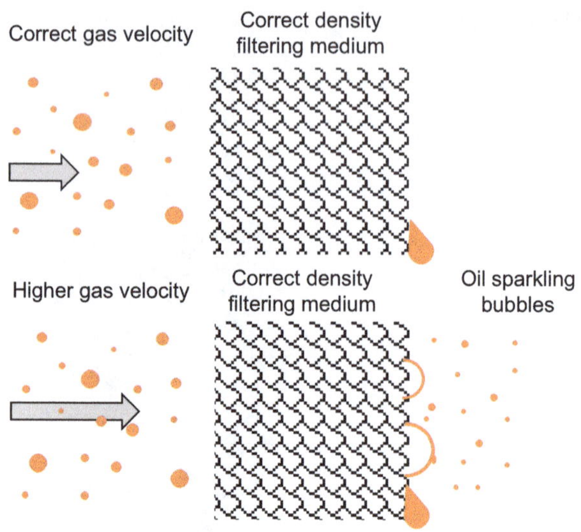

Fig. 6.56 Incidence of the front velocity

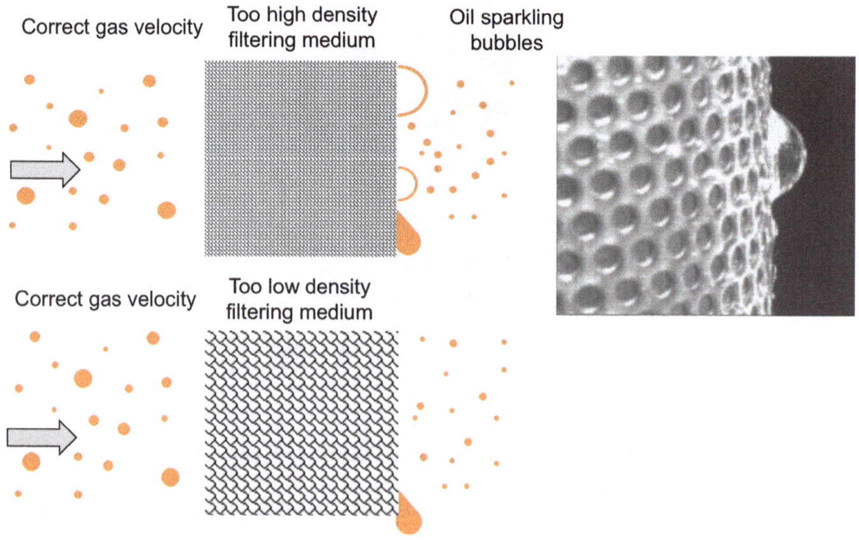

Fig. 6.57 Too compact a filtering medium

velocity is increased, and oil is pushed out of the medium where it materialises at the periphery of the cartridge in bubbles that explode, making again new aerosols. Figure 6.57, right, shows this kind of situation that is acceptable for air that is used to energise pneumatic tolls that must be slightly lubricated in order not to seize, but not correct for what is necessary for helium.

In Fig. 6.57, bottom, the filtering media is much less compact (the void space is larger). Even if the front flow velocity is correct, middle-size particles succeed in crossing the filtering media.

Remark Some coalescing cartridges are fitted with an anti-re-entrainment layer, generally made out of some plastic material foam. Such a part is designed for uses like oil removal in air where a higher oil concentration than for helium is acceptable. It is intended that the anti-re-entrainment layer reduces the oil amount that is released when the cartridge operates in conditions acceptable for air, as explained here above. In the case of helium oil removal systems, this layer is of no use because the oil concentrations are low. Moreover, some foams are made out of materials that are slowly destroyed by the special oil that is used in helium cooled compressors (cf Sect. 6.2.2.3).

6.3.8.3.3 Cartridge Housing

The cartridge housing is vertical. Oil drops that materialise at the outer surface of the cartridge are pulled downwards by gravity, fighting against the vertical upwards velocity of the gas. Such vertical velocity is zero at the bottom of the cartridge and maximum at the top. Oil drops materialise at the lower part of the cartridge if the latter has been correctly selected. If the gas vertical velocity is too high, drops might be re-entrained upwards (see Fig. 6.58).

The side velocity of helium must not be higher than 1.00 m/s (it is roughly the falling velocity of a 1 millimetre diameter drop in helium at 20 bar, 30 °C).

According to above, the discharge pipe must always be located at the top of the cartridge housing, as shown in Fig. 6.59. If the discharge pipe would have been at the bottom of the housing, it is almost sure that drops would be entrained.

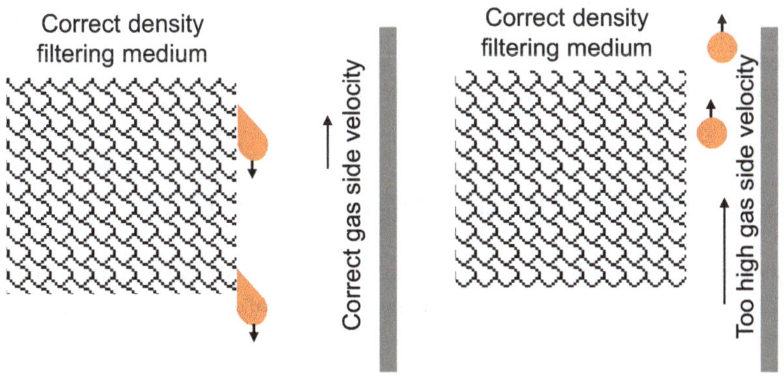

Fig. 6.58 Incidence of the side velocity

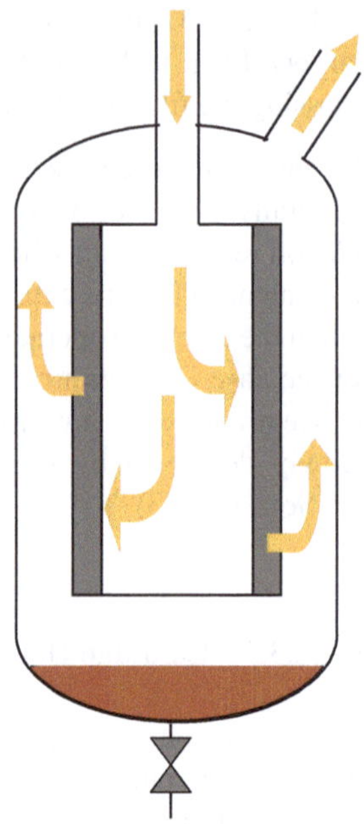

Fig. 6.59 Correct structure of a cartridge housing

6.3.8.3.4 Sizing Procedure

Such a process is performed in three steps:

- Selection of the *cartridge area* in order to get a front velocity located between 0.05 and 0.10 m/s (in order to avoid the situation shown in Sect. 6.3.8.3). Do not forget special operating conditions where front velocity can be increased!
- Selection of the *cartridge filtering media* (compactness or grade) according to the oil loading (that can be expressed as g/s.m^2). The lower the oil loading, the more compact the filtering media must be. When the oil concentration is around 100 ppM weight, one can use the finest (or highest density) available filtering media.
- Sizing of the cartridge housing *internal diameter.*

6.3.8.4 Arrangement of the Aerosol Coalescing System

As one coalescing cartridge, even if it has been correctly sized, cannot capture all aerosol particles, it is necessary to arrange a number of them in series. Assuming a capture efficiency of 99%, it is possible to estimate the concentration downstream of coalescer stage. Knowing the mass flow rate that is processed, it is possible to know the oil mass flow rate that is transported by helium. In Fig. 6.60, for 100 g/s, one can see the quantities of oil that should be drained in g/h (hour) in the first stage, in g/w (week) in the second stage and in g/y (year) oil in the third stage for various oil

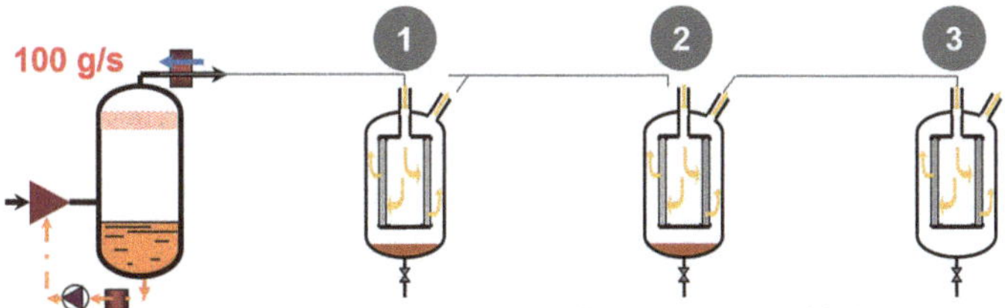

ppM W	g/h	ppM W	g/w	ppM W	g/y	ppM W	Released g per 100000 h
5000	1800	50	3024	0.5	15.76	0.005	180
1000	360	10	605	0.1	3.15	0.001	36
100	36	1	60	0.01	0.31	0.0001	3.6
10	3.6	0.1	6	0.001	0.003	0.00001	0.36

Fig. 6.60 A three-stage coalescer system

concentrations at the discharge of the bulk oil separator. The oil concentrations are also indicated.

Remark If one supposes, as an example, that the time for the first coalescer to be saturated and to release the first drop of oil is 1 hour, the second separator should deliver its first drop in 1 × 100 = 100 h and the third separator in 100 × 100 = 10,000 h. One should keep this order of magnitude in mind when performing measurements on the oil removal system!

One can see that the capture efficiency of the bulk oil separator governs the efficiency of the whole oil removal system.

As the oil concentration in helium is lower and lower, the density of the filtering medium of each stage is to be higher and higher.

What is the reasonable number of coalescing stages to be incorporated? Referring to the capture efficiency curve shown in Fig. 6.51, one can make an estimation of the quantity of oil that is released by the coalescer system (and, therefore, sent into the next equipment that is the oil vapour adsorber). Let us assume that the global capture efficiency is 99%. Results are shown in Fig. 6.61 for a 100 g/s helium mass flow, according to the oil concentration at the discharge side of the bulk oil separator and the operating time.

If the oil concentration in helium at the discharge of the bulk oil separator is lower than 100 ppM W, two stages of coalescers are sufficient. However, for reasons of safety of operation, it is good practice to add a third stage that has, theoretically, almost nothing to trap, except if some undetected trouble happens in the first or second coalescer.

A Simple Check If the oil quantity that is drained at the last coalescing stage is known, one can assume that 1/100th of this quantity is leaving the coalescer stage.

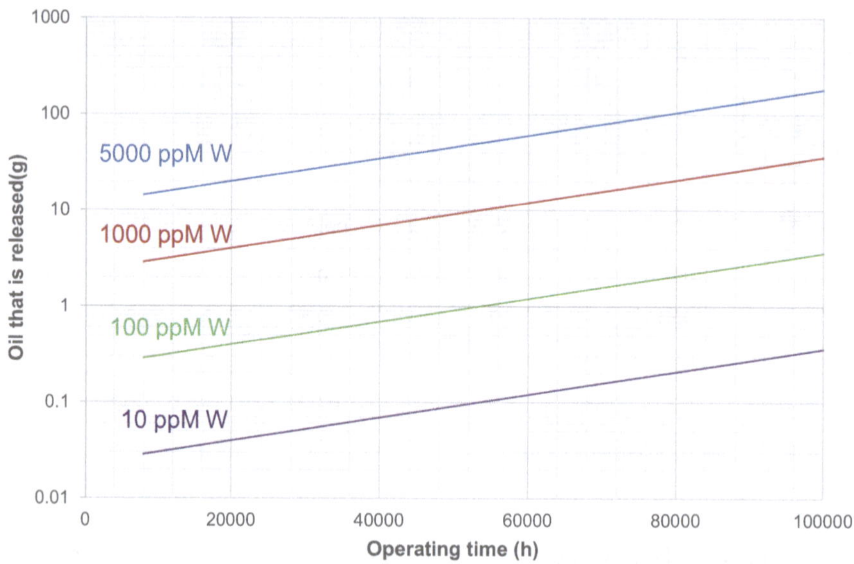

Fig. 6.61 Theoretical oil quantity that is released by the coalescer system for 100 g/s with three stages of 99% each effective coalescers

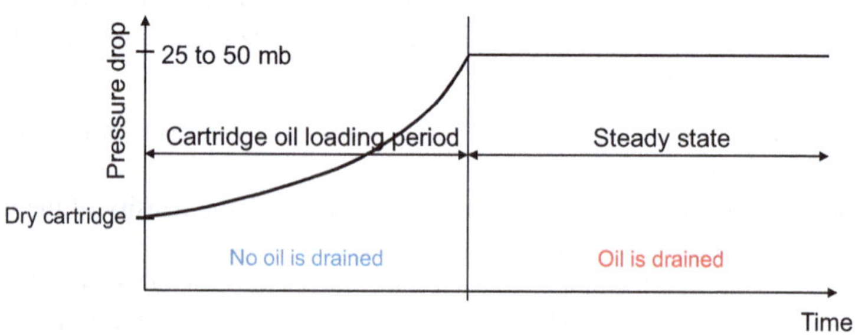

Fig. 6.62 Coalescing cartridge pressure drop evolution

Figure 6.62 shows how the pressure drop of a cartridge varies with time, at constant flow and oil load, from a low value when the cartridge is dry, up to a value where the cartridge is saturated with oil. During the oil loading period, no oil is drained out of the cartridge because it is being stored into the filtering media. Oil begins to be drained only after the cartridge is saturated. Obviously, one must wait until the cartridge is in a steady state prior to make measurements.

6.3.8.5 Managing the Coalescing System

6.3.8.5.1 Oil Draining

Oil that is separated is collected at the bottom of the cartridge enclosure. It must be drained and returned to the suction side of the compressor, by simple pressure difference. The first coalescer is separating a substantial amount of oil (see quantities in Fig. 6.60), the second one a very small amount and the third one almost nothing (by the way, the time that would be necessary to see oil dripping out of the third stage cartridge is extremely long).

Oil draining must be automatic, at least on the first stage, sometimes at the second stage. There are various solutions to fill this duty. Here are a few examples.

The Mechanical Automatic Drainer
This component that is made out of stainless steel incorporates a free-floating guided lever that opens the discharge port as soon as the oil level reaches some height inside the drainer capacity (see Fig. 6.63). Such a component, made by Armstrong, USA, is very much used for small- or middle-size plants. It is very reliable, but there is not an easy way to check or monitor its operation.

For the first stage coalescer, a sight glass can be fitted between the cartridge capacity and the oil collecting capacity (see Fig. 6.64, right). It is an interesting option for the plant operator who can, within a few seconds, check visually that draining is operating correctly, and, therefore, the first coalescer is OK.

The Controlled Draining System
For larger-size plants, it is advised to incorporate a more sophisticated assembly that allows a better control of the operation and, furthermore, an automatic way to

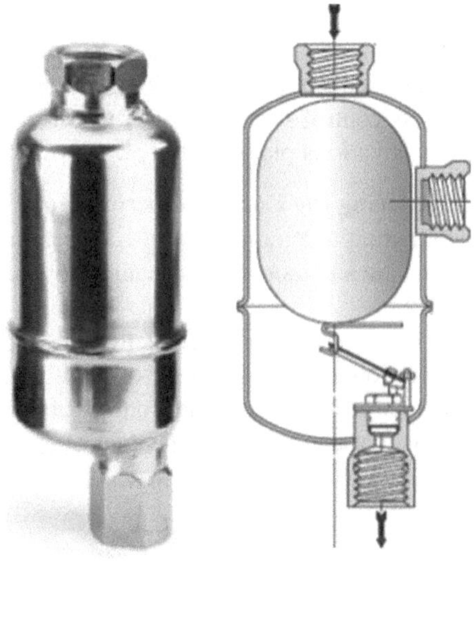

Fig. 6.63 A mechanical liquid drainer 11LD with its internal mechanism. (Courtesy Armstrong)

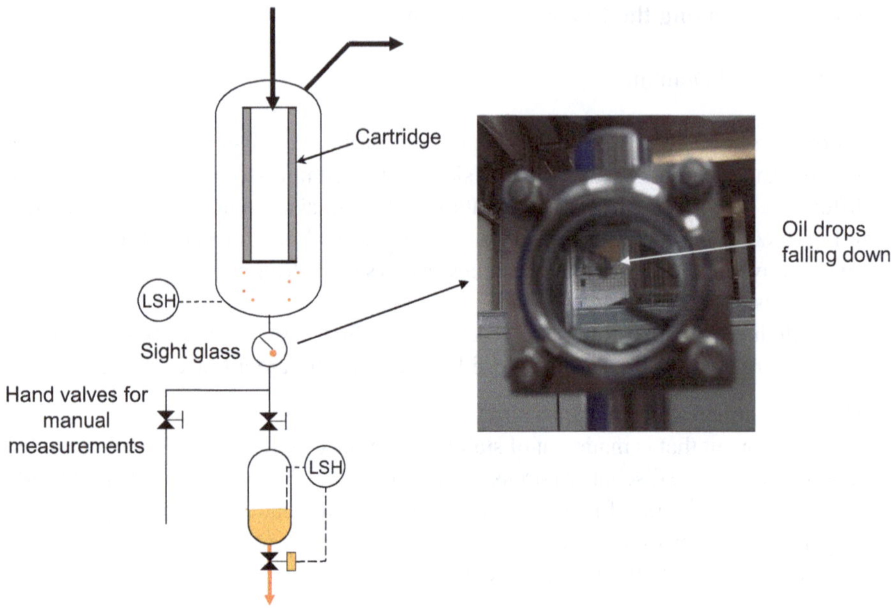

Fig. 6.64 An automatic oil draining system

measure the quantity of oil that is drained out and, therefore, estimate the oil concentration in helium upstream the coalescer that is drained. The system consists in a small capacity into which oil is collected by gravity, a level indicator and a drain valve (see Fig. 6.64, left).

There are various ways to operate such a system. Every time the level switch high (LSH) issues a signal, the valve is opened for a given time period by a timer (for a better reliability, it is advised not to use a solenoid valve but an air actuated one). A calibration of the system allows knowing the quantity of oil that is drained at each opening sequence. It is also possible to fit an oil counter delivering information to the process control system in order to perform automatic calculations of oil concentrations at various places of the oil removal system.

Monitoring the Correct Operation of a Draining System

Monitoring a draining system is of paramount importance.

In case an automatic mechanical drainer is used, it is not easy to check its correct operation.

There could also be other trouble, for example, if a leak develops at the inlet connection of the cartridge of in it. Therefore, an indirect way is to detect a maximum oil level at the bottom of the cartridge capacity: if the level rises, it means that the drainer of the enclosure stays closed (see Fig. 6.65). An alarm is to be issued by the control system.

Let us consider the situation for the three stages:

- First stage: a level switch high, located at the lowest part of the cartridge housing, gives a warning if the first stage oil drainer does not perform correctly (see Fig. 6.64, left),
- Second and third stage:
  - If the system is fitted with automatic measurement, an alarm will be issued if the oil flow that is drained is too high. There is no need of a level switch high.
  - If the coalescer is manually drained, a level switch high gives a warning prior to the manual draining action if the upstream coalescer does not fulfil its duty (leaky cartridge connection, cartridge filtering media that has been destroyed, wrong operation of the cartridges inside the bulk oil separator, badly sized cartridge).

For all these warnings, if within a reasonable period of time, no action has been performed by the operating team, the compression station must be stopped.

Remark It is of interest to have a set of isolating valves and a manual by-pass valve that allows purging manually during the time that is necessary to fix the trouble when the drainer does not operate correctly (Fig. 6.64, left). This is relevant for the first stage only.

6.3.8.5.2 Oil Concentration Measurement

First, one must remember to start such measurements only when the cartridge is saturated, operating in a steady state, i.e. when the oil flow that is drained is constant (Fig. 6.62).

As one knows that the capture efficiency of a coalescing cartridge is higher than 99%, one does not make a significant error in assuming that the oil that is collected by the cartridge is the totality of the oil that was contained into the helium that crossed the cartridge. Therefore, a very simple and, anyway, very accurate way to estimate the oil concentration in helium at the inlet of the separation stage is to calculate the ratio of the oil mass that has been trapped versus the helium mass that has been processed during a period of time.

This can be made in various ways.

Manually

Measurement of the oil quantity by collecting oil (see manual valve in Fig. 6.64) or by counting the oil drops in a sight glass, after having calibrated the mass of one drop (by measuring the volume or weight of one oil drop at about the same temperature). Estimation of the helium mass that has been processed during the sampling time. Calculation of the mass ratio oil/helium.

Example 1

The compressor processes 100 g/s of helium. The quantity of oil that has been collected for 10 hours is 350 mL.

Table 6.4 Calculation for a coalescer stage

Coalescer operation		Date
		14 March 2020
Helium mass flow	g/s	***100.00***
Oil density	g/L	***812***
Stage 1		
Drained oil quantity	(mL)	***350.00***
Time period	(h)	***8.0***
Drained oil flow	(mL/h)	43.8
Oil concentration at inlet stage 1	(ppM W)	99.3
Stage 2		
Drained oil quantity	(mL)	***350.00***
Time period	(h)	***1300.0***
Drained oil flow	(mL/h)	0.27
Oil concentration at inlet stage 2	(ppM W)	0.61
Stage 1 efficiency	(%)	99.4

For 8 hours, the helium quantity that has been processed is:

$$100 \times 8 \times 3600 = 2,880,000 \quad \text{g}$$

The oil mass is 284.2 g (density of oil = 812 g/L). Assuming that the capture efficiency of the coalescer is 100% (which is obviously not right, but does not strongly affect the accuracy of the result), the oil concentration in helium at the *inlet* of the stage is:

$$284.2/(8 \times 3600) = 0.000994 = 98.4 \ \text{ppM W}$$

A more accurate result is 98.4 + 0.61 = 99.3 ppM W, because the efficiency of the first stage coalescer is not 1.00.

Such calculation results are displayed in Table 6.4 where the efficiency of the first stage coalescer is 99.4%.

Example 2

The compressor processes 100 g/s of helium. An oil drop at 70 °C weighs 0.03 g. The operator counted 50 drops within 5 minutes. The oil mass concentration upstream the coalescer is:

$$0.03 \times 50/100 \times 5 \times 60 = 5 \times 10^{-1} \text{ or } 50 \text{ ppM W}$$

Automatically

In the system that has been described in Fig. 6.64, the time interval between two openings of the valve is an indication about the oil quantity that is separated. One could only check that the purge frequency stays constant. If a measurement of the oil quantity that is released at each purge has been done, it is possible to compare this quantity to the helium quantity that has been processed in the same time and make a permanent calculation of the oil concentration in helium at the inlet of the coalescing stage. An oil counter can also be used, but one must make sure that there is not any helium bubble into the oil. It is of interest that the control system issues warning that is turned into a safety stop after some time. The delay between warning and safety stop depends on the coalescing stage. This solution should be mandatory for any large system.

6.3.9 Oil Vapour Separation

As it has been noticed in Sect. 6.2.2.3, the interface area between oil and gas is very important during the compression process. This, combined with the temperature rise during compression, allows oil vapour to be released into helium. In order to limit such a phenomenon, a special very low vapour pressure oil is to be used. As an example, let us have a glance at a synthetic oil made out of polypropylene glycol: Breox® B 35 (Breox® oil is manufactured by BP).

The typical formula is $C_4H_9(OC_3H_6)_n - OH$, where n, that is, the number of (OC_3H_6) motives in the molecule, is distributed according to a Gaussian law with a mean value of 38.

The information provided by the maker gives a vapour pressure of 1.0×10^{-11} mbar (1.0×10^{-9} Pa) at 25 °C.

This seems to be perfect, but a finer investigation shows that:

- When $n = 1$, $C_4H_9(OC_3H_6)_1{-}OH$, the vapour pressure is 1.0×10^{-3} mbar (1.0×10^{-1} Pa) at 25 °C.
- When $n = 2$, $C_4H_9(OC_3H_6)_2{-}OH$, the vapour pressure is 6.1×10^{-5} mbar (6.1×10^{-5} Pa) at 25 °C.
- When $n = 3$, $C_4H_9(OC_3H_6)_3{-}OH$, the vapour pressure is 2.2×10^{-6} mbar (2.2×10^{-4} Pa) at 25 °C.

For a 100 g/s helium flow, the oil quantity corresponding to a 10^{-3} mbar partial pressure should be 14 g/year which is the same order of magnitude of the oil quantity released by the aerosol separation system.

It is not possible to use such an oil as it is delivered, with these high vapour pressure components. One has to be rid of these high vapour pressure components. On top of that, there are other parasitic components as:

- Some air
- Some water (such oil is very hygroscopic)

The selected oil treatment process is *heating under vacuum* into a specific equipment that warms oil under vacuum and provides a high contact area between oil and vacuum.

The difficulty of the treatment is that it is necessary, as in the compressor, to create a very large interface area between warm oil and vacuum.

The correct way is to process oil.

Remark Every time oil goes through the compressor; it is finely dispersed, and a very high interface area between oil and helium is created. It is another opportunity to release vapour into helium. Therefore, the actual operation of the compressor is kind of a very efficient oil treatment! (Such a process has been used, but it is very energy consuming.) Consequently, the *more the oil is operating* into a compressor station, the *best it is, in the sense of oil separation*. It is not advised to replace the oil of a helium compression station except if it is deteriorated, either by metal particles or any other impurity (see Sect. 11.6.1).

Even if the oil treatment is almost perfect, there might still be some possibility to release a tiny quantity of vapour. Therefore, it is usual to fit, downstream the coalescers, a capacity filled with an adsorbent material as shown in Fig. 6.65.

It is very difficult to get valuable information from the oil maker(s) in order to select the best suited adsorbent. This is the reason why activated charcoal, which has a wide adsorption spectrum (see Sect. 9.4.1.2), is mostly used for this duty. As condensing temperatures of hydrocarbons are above room temperature, it is possible to run the oil vapour adsorber at room temperature. Activated charcoal might contain a rather important amount of charcoal dust (consequence of vibration during transportation for example) and is naturally saturated with atmospheric moisture. It is to be dust cleaned and moisture regenerated prior to running the plant. To make sure the regeneration temperature is reached, the adsorbent capacity is thermally insulated. Such procedures are described in Sect. 11.6.6.2.1.

Obviously, a filter is fitted at the discharge side of the adsorber.

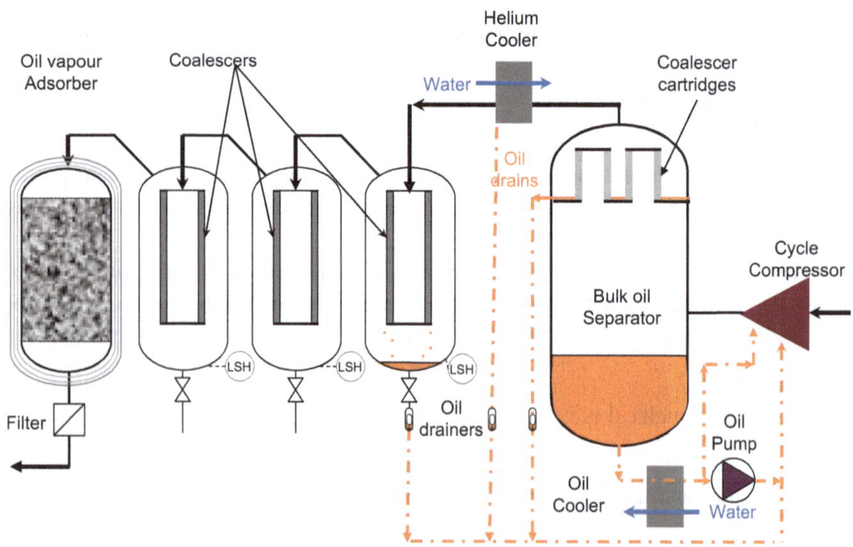

Fig. 6.65 A whole typical oil removal system

6.3.10 The Whole Oil Removal System

It consists in an arrangement in series of:

- The bulk oil separator, the helium cooler, the coalescer stages
- The oil vapour adsorber
- The filter, as shown in Figs. 6.65 and 6.66

Fig. 6.66 A final oil removal system. (Courtesy CERN). The large-cartridge coalescer capacities are behind; the thermally insulated oil vapour adsorber is on the foreground

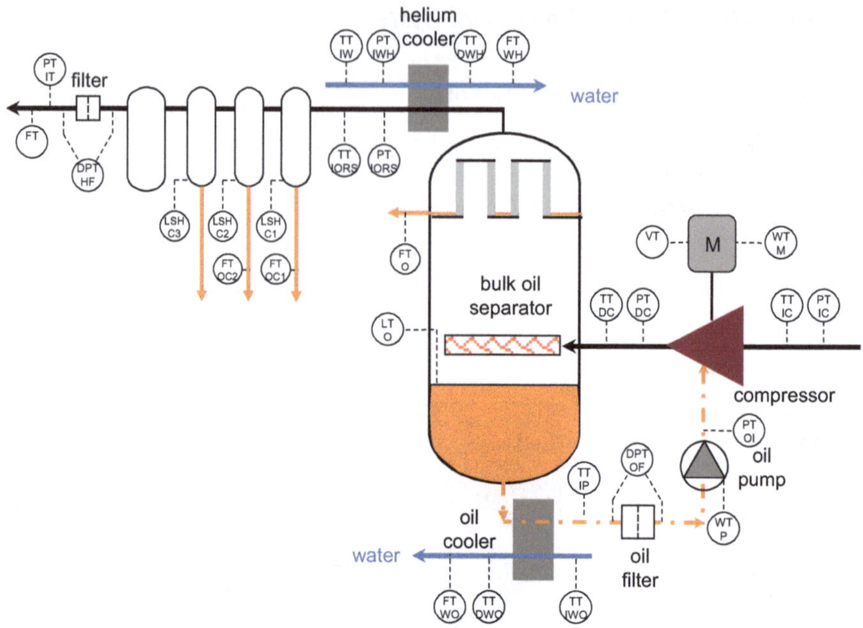

Fig. 6.67 Instrumentation around an oil lubricated screw compression station

6.3.11 Monitoring of a Compression Station

Figure 6.67 shows the possible instrumentation to monitor a one-stage cycle screw compression station. The low-switch and high-switch values should be arranged in two levels: first an alarm and then a stop action (Table 6.5).

6.3.12 A Digest on Oil Separation

The bulk oil separator efficiency governs the whole system efficiency. The oil concentration at the discharge side should be ideally around 100 ppM W.

In the coalescing cartridges, the front velocity must not be higher than 0.10 m/s; the side velocity must not be higher than 1.00 m/s.

The cartridge medium compactness must be low when the oil loading is high.

The compactness of a cartridge is higher than that of the one that is located upstream.

Oil draining must be controlled, if possible automatically, in order to check the coalescer efficiency.

Table 6.5 Monitoring a one-stage cycle screw compression station

Sensor	Low switch	High switch	Duty
PT IC		X	Helium inlet pressure at first stage
TT IC	X		Helium inlet temperature at first stage
PT DC		X	Helium discharge pressure at first stage
TT DC		X	Helium discharge temperature at first stage
FT			Helium mass flow rate (range)
WTM	X		Absorbed electrical power of the main motors
VT	X		Vibration sensors (motor and compressor, at locations indicated in the maker documents), possibly linked, with a spectral analysis system (see Sect. 11.5.1.2.1)
WT P	X		Absorbed electrical power of the oil pump
PT OI	X		Oil inlet pressure
PT OI - PT DC	X		Oil pressure difference
LTQ	X	X	Oil level
DPT OF		X	Oil pressure difference in filter
FT O	X		Drained oil from BOS
PT IWH	X		General water inlet pressure
TT IWH		X	General water inlet temperature
TT IP		X	Oil temperature after each cooler
PT IORS			Helium pressure after cooler
TT IORS		X	Helium temperature after cooler
TT DWO		X	Water discharge temperature after oil cooler
TT DWH		X	Water discharge temperature after helium cooler
FT WO	X		Water flow rate in oil cooler
FT WH	X		Water flow rate in helium cooler

6.4 The Heat Exchanger

There are a lot of much specialised books about heat exchangers. Here, it is only intended to remind some basic information related to technologies aimed at cryogenics. Emphasis is put on the most commonly used heat exchangers: the aluminium plate and fin heat exchanger.

6.4.1 Various Technologies

Heat exchangers have been designed in a very large variety of solutions, but only a few of them are used in helium refrigeration.

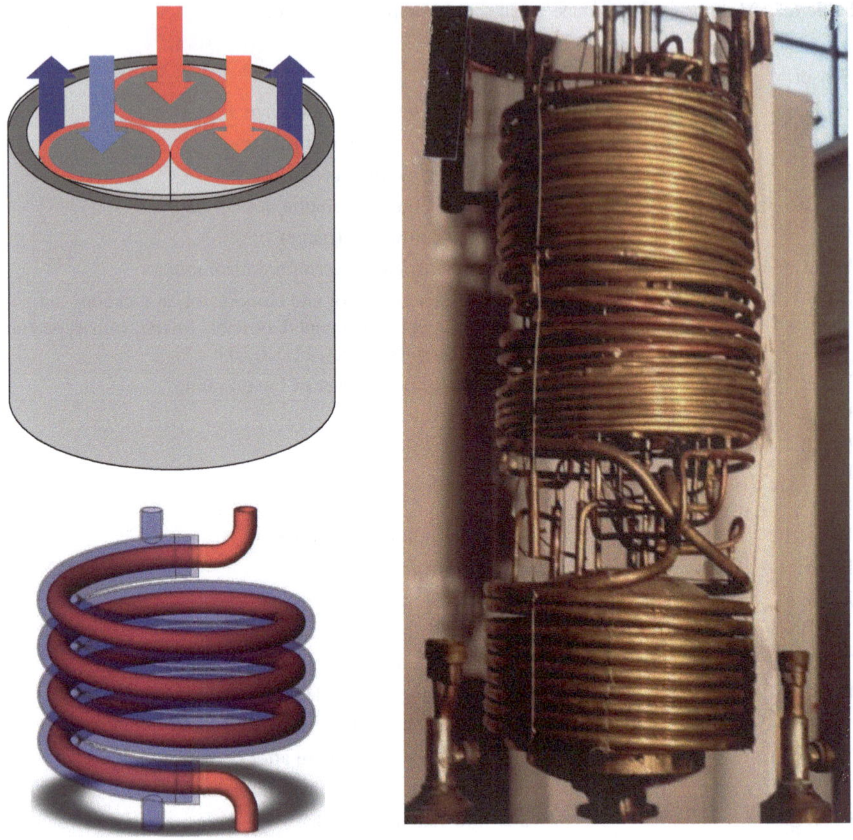

Fig. 6.68 Three-in-one pipe-in-pipe principle and a heat exchanger built in this way (TBT)

6.4.1.1 Pipe in Pipe

This is one of the very first technologies that has been initially used for small liquefiers. This kind of heat exchanger is rather easy to build, even if it needs high skills to reach a good performance (Fig. 6.68).

Characteristics of Pipe-In-Pipe Heat Exchangers

- Pressures: HP and LP more than 100 bar
- Size: very small up to a few litres
- Maximum acceptable temperature difference: no limitation

6.4.1.2 Coiled Pipes in a Shell

This is another rather easy-to-make type of heat exchanger: pipes are coiled on a mandrel (see Fig. 6.69). In order to have a correct distribution of gas in the parallel arranged pipes, the length of each pipe must be the same. Consequently, the coiling length decreases with each new layer.

The difficulty for such kind of heat exchanger is to get and keep a good leak tightness between the outer pipe layer and the external shell.

In Fig. 6.70, one can see the smart way in which coiled heat exchangers are arranged, the warmer ones, with large cross sections, thermally protecting the colder ones.

At the time refrigeration at temperatures lower than 4.5 K was obtained with room temperature compression machines, very low-pressure coiled heat exchangers have been built, as shown in Fig. 6.71.

Characteristics of Shell and Tube Heat Exchangers

- Pressures: HP more than 100 bar, LP from vacuum to a few bars
- Size: a few m^3 maximum
- Maximum acceptable temperature difference: no limitation

6.4.1.3 Plate Heat Exchanger

There are various kinds of plate heat exchangers. One example is described in Fig. 6.72. Two thin stainless steel sheets (0.5 mm thick) are joined by resistance

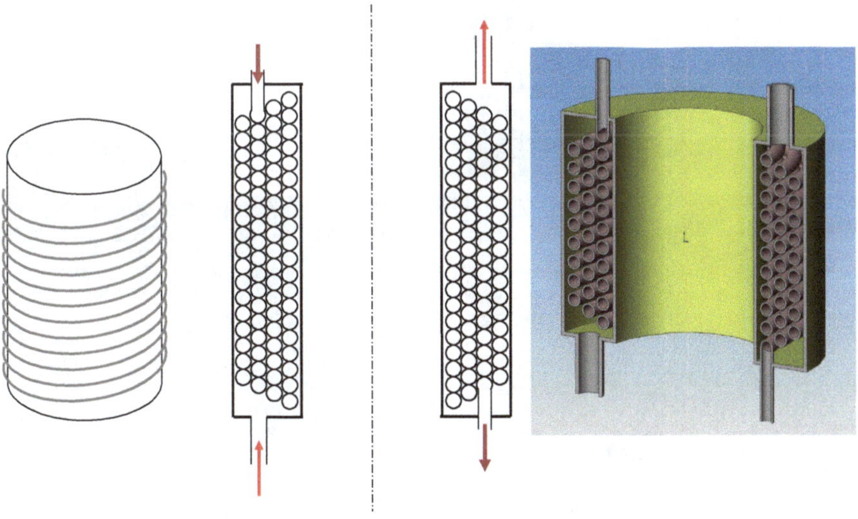

Fig. 6.69 Coiled pipe heat exchanger

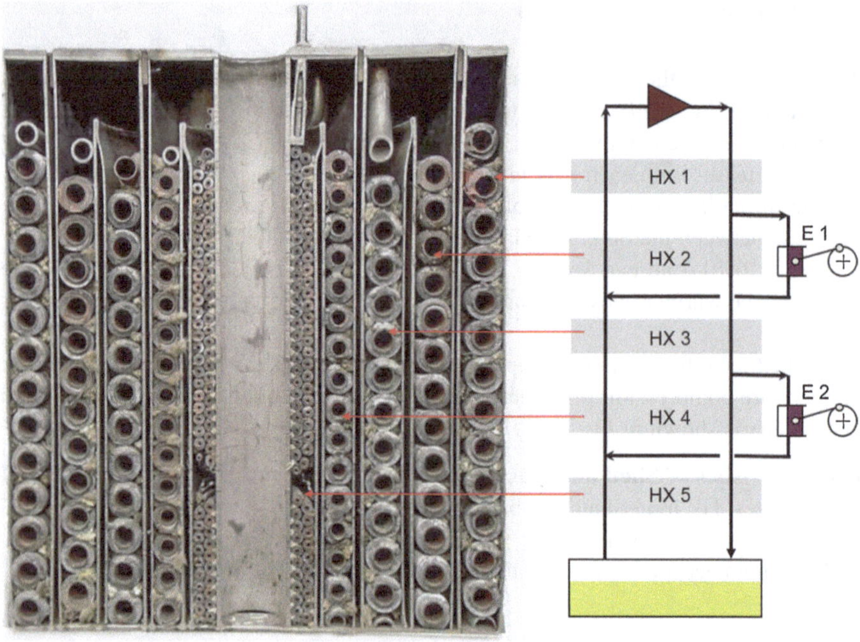

Fig. 6.70 The smart structure of a coiled heat exchanger. Collins design (courtesy Linde plc)

Fig. 6.71 Special coiled pipe, low-pressure drop heat exchangers (courtesy DESY)

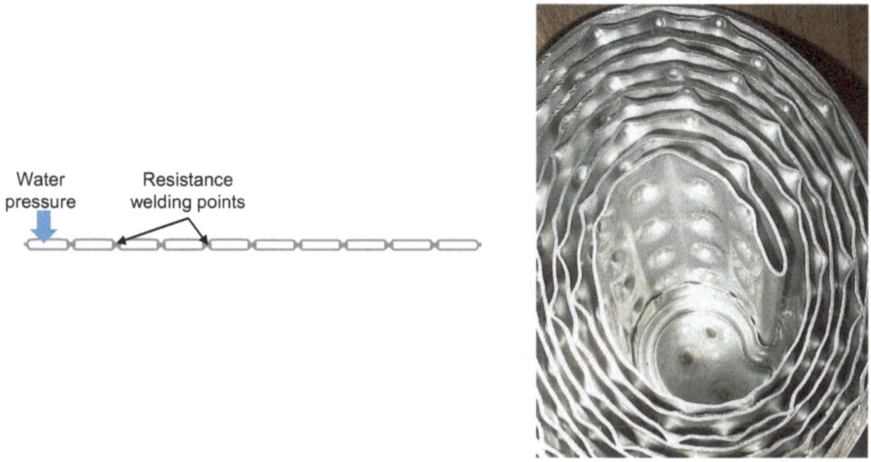

Fig. 6.72 A pressure expanded plate heat exchanger

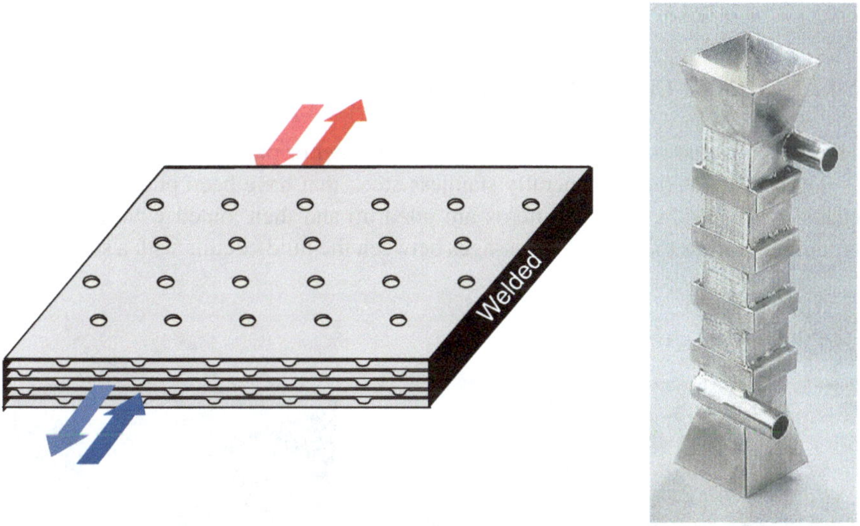

Fig. 6.73 Example of a plate heat exchanger. (Courtesy DATE)

welding points. A continuous weld closes the circuit. High-pressure water is injected into the structure that is permanently deformed, allowing a fluid to circulate. Such an element can be dipped into liquid nitrogen and is a possible structure for a nitrogen vaporiser.

In Fig. 6.73, left, 0.5 mm thick stainless steel plates are evenly deformed and then piled up, and a lateral weld creates various circuits. An example of such a heat exchanger is shown in Fig. 6.73, right: it is incorporated into the LHC coil cryostats and operates between 4.5 K and 1.8 K.

Characteristics of Plate Heat Exchangers

- Pressures: HP and LP a few bar
- Size: a few dm^3
- Maximum acceptable temperature difference: no limitation

6.4.1.4 Mesh or Perforated Plate Heat Exchanger

This kind of heat exchanger has been used in helium refrigeration by the Philips Company in the 1960s (see Sect. 16.9). It is a very compact and effective device, shown on Fig. 6.74.

A later evolution incorporated perforated copper sheets with stainless steel intercalates brazed together.

Characteristics of Perforated Sheet Heat Exchanger

- Pressures: HP and LP, 20 bar maximum
- Size: small (a few dm^3)
- Maximum acceptable temperature difference: more than 50 K

6.4.1.5 Printed Circuit Heat Exchangers

These heat exchangers are also called micro-channel or platelet.

Thin sheets of metal, generally stainless steel, that have been photo-chemically etched with specific design patterns are piled up and then joined with a diffusion bounding process that seals the passages between the fluid circuits with a solid metal

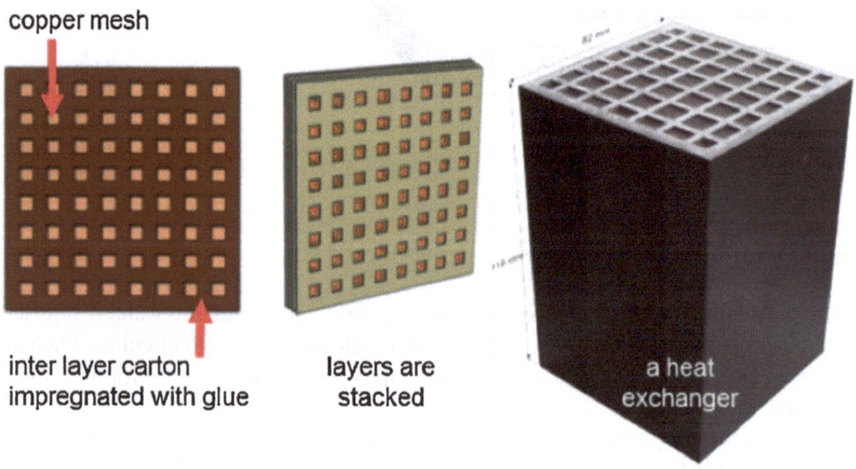

Fig. 6.74 Mesh heat exchanger. (Courtesy Philips)

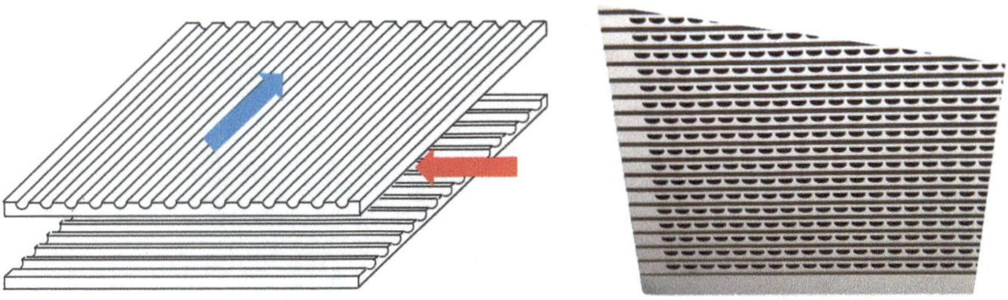

Fig. 6.75 Printed circuit heat exchanger. (Courtesy Heatric)

barrier, isolating flows, as shown in Fig. 6.75. Fluids are distributed and recovered by headers that are welded on the block. This very sturdy structure allows operating in conditions that are often forbidden with other kinds of heat exchangers, especially when a very cold fluid, gas or liquid, is sent into a warmer heat exchanger.

Characteristics of Printed Circuit heat exchangers

- Pressures: HP and LP, more than 100 bar
- Size: a few m^3
- Maximum acceptable temperature difference: no limitation

6.4.2 The Aluminium Alloy Plate and Fin Heat Exchanger

Almost all modern refrigerators and liquefiers are constructed with aluminium alloy plate and fin heat exchangers. The original idea came from an Italian mechanic, Paolo Fruncillo. A plate-fin heat exchanger is made of layers of corrugated sheets that are separated by flat parting sheets, both made out of aluminium alloy, to create a series of finned channels that are closed at the edges by so-called side bars. Warm and cold fluids flow countercurrent through alternating layers of the heat exchanger. Heat is transferred from one fluid through the fins (acting as a secondary heat exchange surface) of its channel and then through the parting sheet then to the fins of the other channel and finally into the other fluid. The fins also serve to withstand the pressure efforts of the heat exchanger.

6.4.2.1 Basic Elements

There are three basic components.

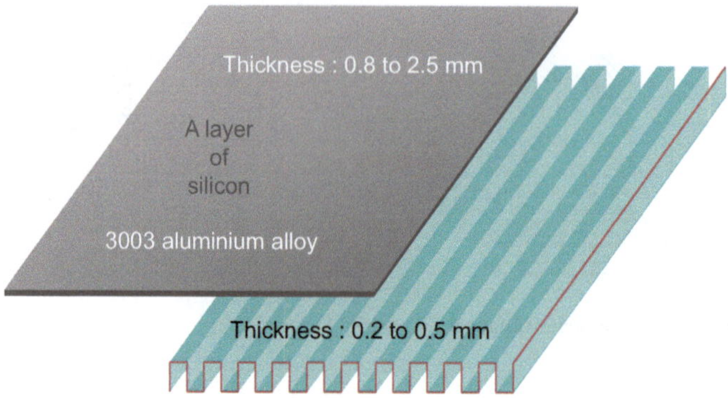

Fig. 6.76 Parting sheet and plain fins

6.4.2.1.1 The Parting Sheet

The parting sheet is made out of 3003 aluminium alloy (see Fig. 6.76 left). The thickness is between 0.8 and 2.5 mm, depending on the heat exchanger operating pressure.

Each side of the parting sheet is covered with a thin layer of silicon in anticipation of the brazing process.

6.4.2.1.2 The Fins

Heat is transferred from one stream through the fin interface to the parting sheet and through the next set of fins into the adjacent fluid. The fins also serve to increase the structural integrity of the heat exchanger and allow it to withstand high pressures while providing an extended surface area for heat transfer (see Fig. 6.76 right).

The main types of fins are (see Fig. 6.77):

- Plain, which refers to simple straight-finned rectangular designs (see Fig. 6.76).
- Perforated which refers to cuts and perforations in plain fins.
- Serrated (see Fig. 6.78) are made by cutting the fins every 3.2 mm and displacing the second fin to a point half way between the preceding fin. This shape provides a dramatic increase in the heat transfer coefficient (see Fig. 6.78 bottom, right).
- And herringbone, where the fins provide a zigzag path.

In helium refrigeration, plain and perforated fins are used for distribution zones, and only serrated is used for heat exchange zones.

Fins are generated by folding a thin aluminium alloy plate in special (and very noisy) machines (see Fig. 6.79).

A typical ratio of area versus volume is 850–1500 m^2/m^3.

The maximum ratio is now 2000 m^2/m^3 with serrated fins, for helium.

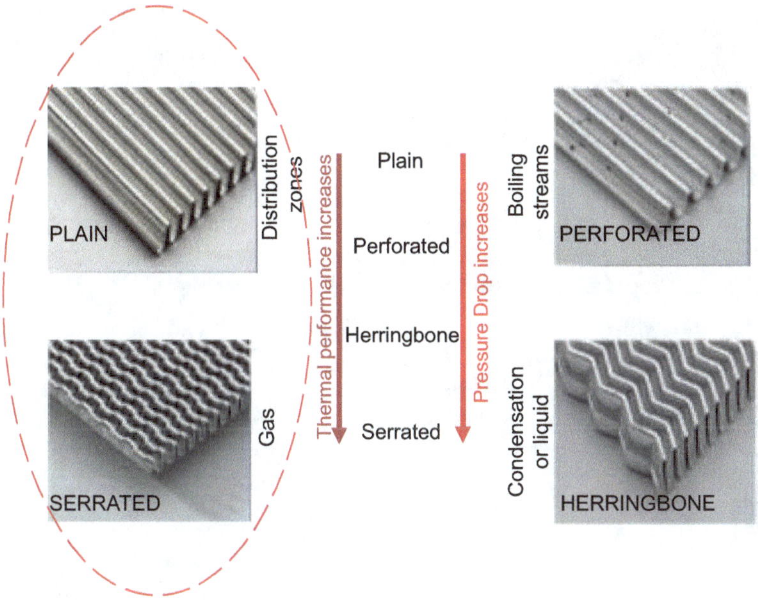

Fig. 6.77 The various types of fins

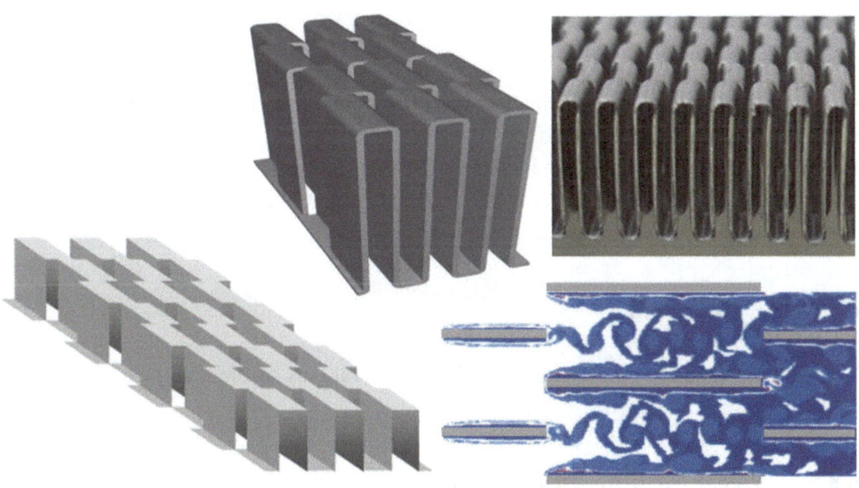

Fig. 6.78 Detail of the serrated fin structure. Turbulences that improve the heat exchange in serrated fins (bottom, right)

Fig. 6.79 Fin machining. (Credit photo Fives)

For a given fin pattern:

- Increasing the number of Fins Per Inch (FPI) increases the thermal performance.
- Increasing the fin height reduces the pressure drop.
- Increasing the fin thickness increases the pressure drop but enhances the performance. It also allows operating the heat exchanger at higher pressure.

6.4.2.1.3 The Side Bar

The side bar (see Fig. 6.80, right) closes the lateral part of the channels.

6.4.2.2 Structure of the Heat Exchanger

The heat exchanger is arranged by manually piling up the various components, a parting sheet, side bars, a second parting sheet and so on, to get the basic structure (see Fig. 6.80). Here, one can see that this is a pure countercurrent heat exchanger.

The heights of corrugations and bars may vary between plates, as shown in Fig. 6.81 right. Aluminium plate and fin heat exchangers can be made up to

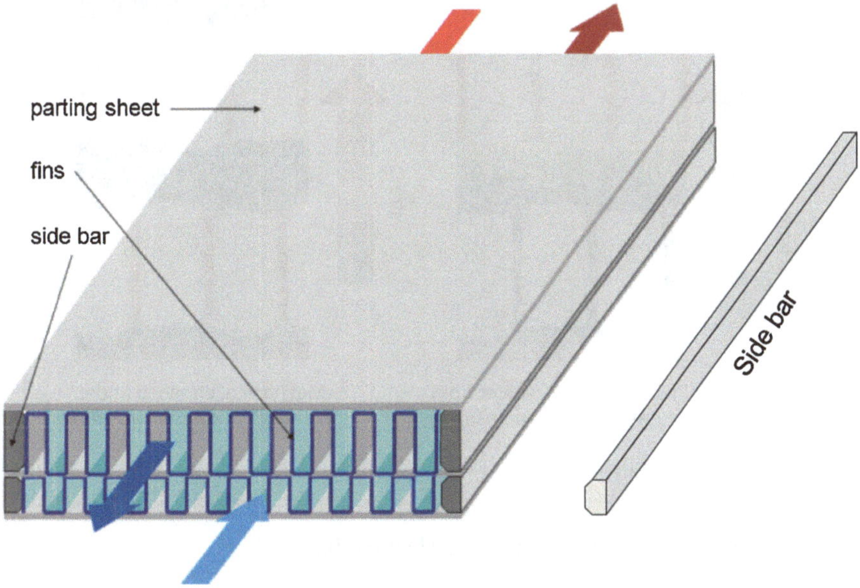

Fig. 6.80 Basic structure of the plate and fin heat exchanger

Fig. 6.81 Basic elements of a plate and fin heat exchanger

1.2 m × 1.2 m cross section and around 7.0 m long (the limitation comes from the vacuum furnace; see Sect. 6.4.2.2.2).

The fins have more than one duty: they fight against pressure and transfer heat from the parting sheet to the gas and conversely (see Fig. 6.82).

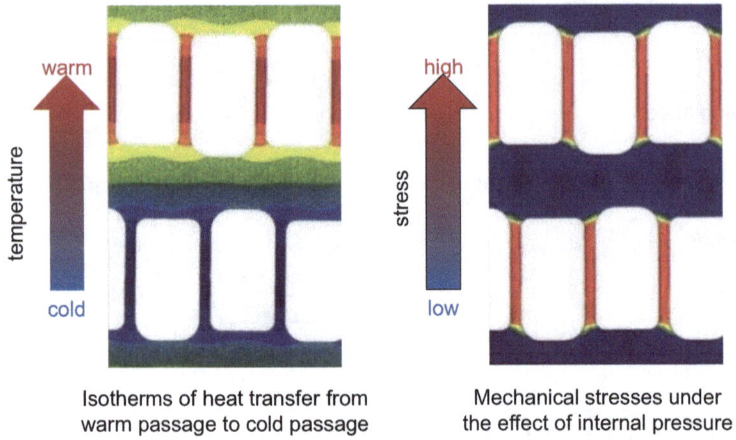

Fig. 6.82 Duties of the fins

6.4.2.2.1 Balancing Fluid Through the Heat Exchange Fins

Now comes the question: how to evenly distribute the fluids in the heat exchange channels? This issue must be considered in two folds: "lateral" balance inside one channel and balance between the channels. Lateral balance inside one channel is reached by using other kind of fins as "distributors". On Fig. 6.83, one can see that plain fins are used at the ends of the heat exchange zone that is made with serrated fins: any path in the system distributor-exchange zone-distributor has same length, therefore almost same hydraulic conductance.

However, as in any distribution issue, it is important that the pressure drops in the distribution zones be negligible compared to the pressure drop in the heat exchange zone. On Fig. 6.84, top, where the heat exchange zone is represented as a series of identical pipes in parallel, and a distribution zone has an insufficient cross section, the pressure drops in the distributor unbalance the flow in each elementary pipe.

On Fig. 6.84, bottom, where the distributor cross section is larger, the flow in each elementary pipe is equal, even if there is a default decreasing slightly the cross section of the distributor.

In Fig. 6.85, one can see how three channels are arranged and piled up to make a heat exchanger.

A real example of another distribution pattern is shown in Fig. 6.86 with the use of perforated fins for the distribution zone.

Figure 6.87 shows a very common structure: how two heat exchangers can be combined into one block. This arrangement that allows sparing connection piping and relevant pressure drops is typical for the warmest heat exchangers of Claude cycle where some helium is withdrawn from the high-pressure circuit to feed a turbine (remember heat exchangers HX1 and HX2 in Fig. 3.51).

Balance between channels of one fluid is done by means of the headers that allow collecting all the flows issued from the channels of a given fluid (see Fig. 6.91).

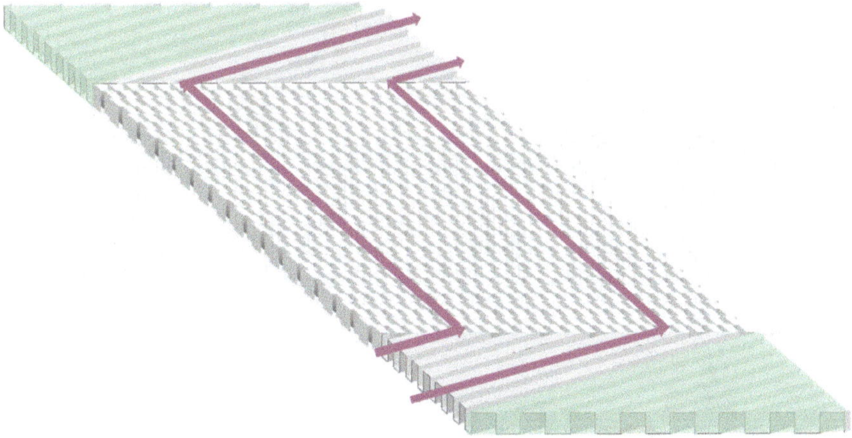

Fig. 6.83 Balancing of fluids among the heat exchange fins

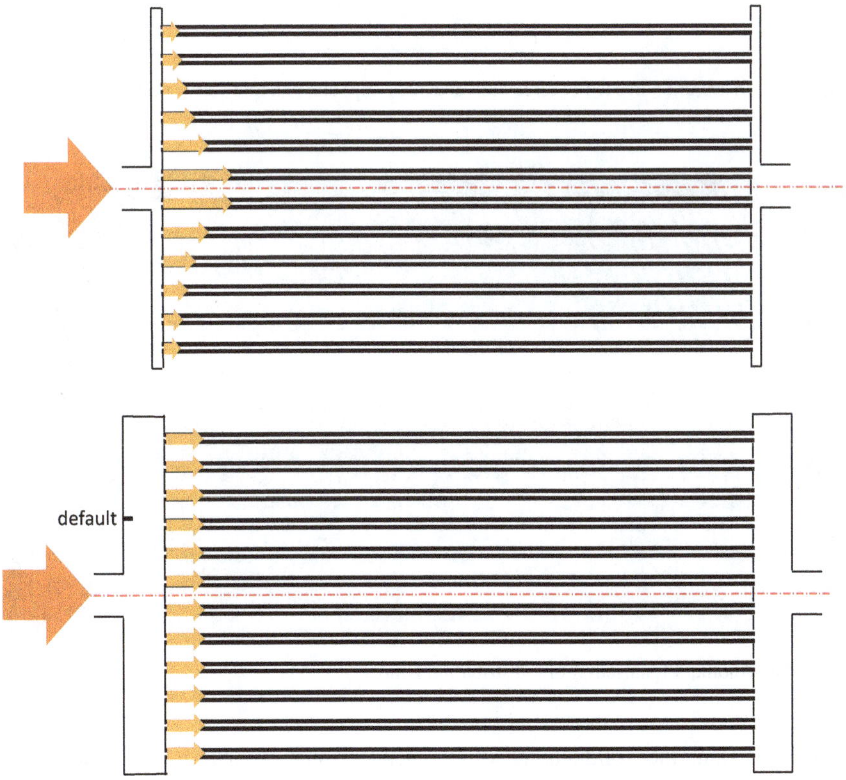

Fig. 6.84 Incidence of the header in balancing fluid flows in the heat exchange fins

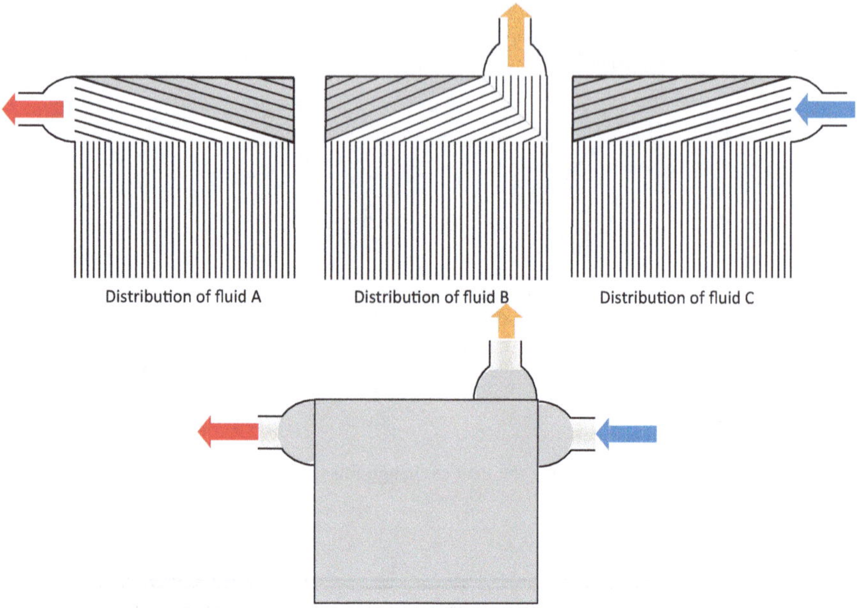

Fig. 6.85 Distribution in each of three fluid channels (top) and external view of one end of the heat exchanger (bottom)

Fig. 6.86 Another organisation of a distribution zone

Here, again, the conductance of the header must be lower than that of the channels. When several fluids are used in one unit, the channels of each stream are to be evenly arranged across the stack height. It is a skill of the heat exchanger maker.

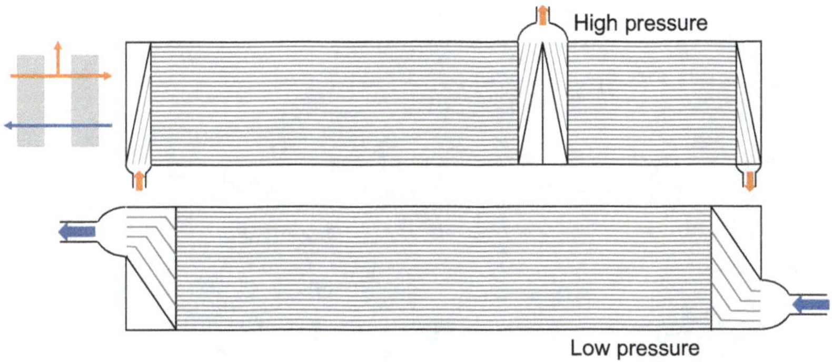

Fig. 6.87 A combination of two heat exchangers in one block

A Very Important Remark For heat exchangers that operate at rather low temperature (less than 20 K) and rather high pressure (around 15 bar), one must be very careful about a possible incidence of the fluid densities that can create an important maldistribution (see Sect. 2.4.4) in a horizontal heat exchanger. The vertical orientation is heavily recommended under 20 K.

6.4.2.2.2 Assembling the Heat Exchanger Components

The parting sheets have been covered with a brazing alloy (silicon). Fins, parting sheets and side bars are perfectly cleaned.

The fins are manually piled up in combination with the parting sheets (see Fig. 6.88). Side bars are manually arranged on the edges and ends of the block.

Fins, side bars and parting sheets are pressed together into a special assembly and then inserted into a vacuum furnace heated up to 620 °C (see Fig. 6.89). The aluminium alloy/silicon melts at a temperature that is only 20 °C lower than the aluminium alloy!

All contact points braze together (see Fig. 6.90) to form a compact block. The brazing process is monitored with thermocouples that are located inside the core to ensure an even temperature distribution (Fig. 6.91).

After brazing, the headers are welded on the block, and then a pressure test is performed before a helium leak test.

During the helium leak test (see Fig. 6.92), various checks are performed:

- Leak tightness from channel to channel
- Leak tightness from any channel to vacuum

Heat exchanger makers are used to deliver blocks that are already connected between each other. This is called a heat exchanger battery. The connections are made with aluminium alloy pipes; the blocks are supported by a frame (Fig. 6.93).

Fig. 6.88 Piling up the fins. (Credit photo Fives, M. Laurent)

Fig. 6.89 The brazing vacuum furnace. (Credit photo Fives)

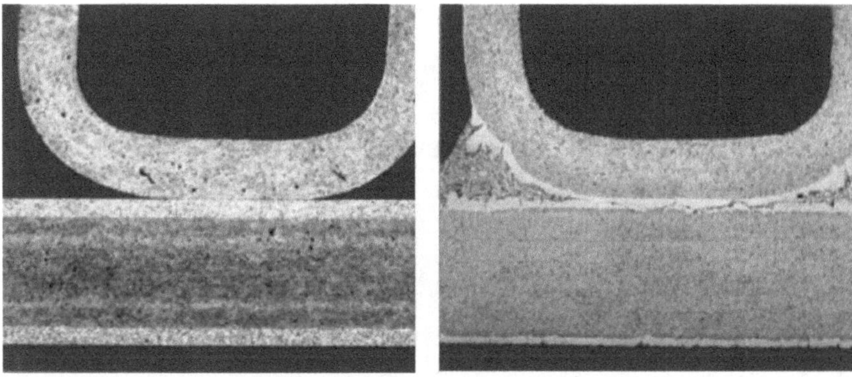

Fig. 6.90 Before and after brazing

Fig. 6.91 How a header collects the flow issued from various channels of the same fluid. (Credit photo Fives)

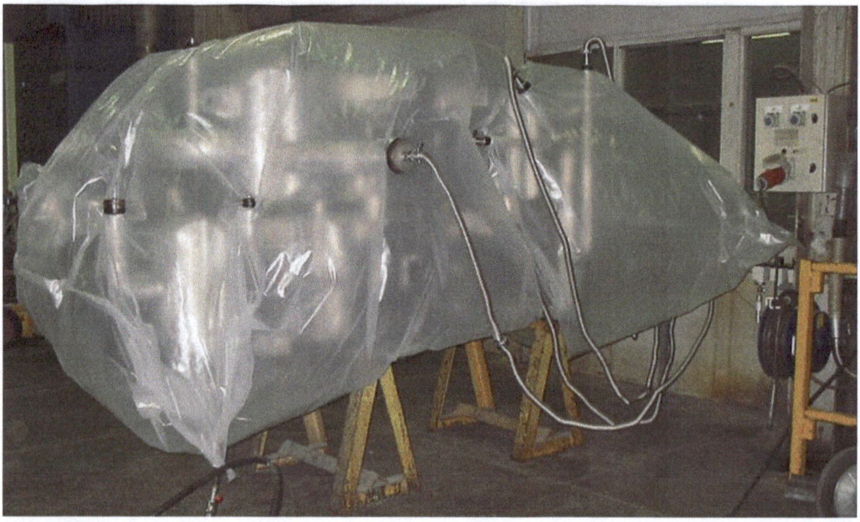

Fig. 6.92 The helium leak test

Bimetal Transitions

Because these heat exchangers are manufactured out of aluminium alloy, while all the other elements of the cold box are made out of stainless steel, the interfaces of the exchanger block (see Fig. 6.94 left) are fitted with stainless steel-aluminium alloy bimetallic transitions (Fig. 6.94 right). Some of these bimetallic transitions are made by just pushing (incrustation) the stainless steel part into the aluminium part at high temperature. Other transitions are made by friction welding. Thus, the welds connecting the block to the other cold box components that are made at the cold box assembly workshop are "conventional" stainless steel to stainless steel welds.

The kind of "label" that is stuck on the transition is a memory thermometer that allows making sure that the transition has not been submitted to too high a temperature during the welding process.

6.4.2.3 A Remark on Aluminium Alloy Plate and Fin Heat Exchangers

Heat exchanger makers have a general rule: the temperature difference of two fluids must not be higher than 50 K. What is the reason for this rule? Let us suppose that some liquid nitrogen is sent into a room temperature brazed plate and fin heat exchanger as shown in Fig. 6.95.

It is obvious that the header, which has a low mass compared to the heat exchanger block, cools faster, and therefore, its length reduces faster, creating stresses at the ends of it that can destroy the heat exchanger structure and "tear" the brazed parts. However, it is also obvious that the stress is lower when the header is shorter. Consequently, it is possible to send liquid nitrogen into a room

Fig. 6.93 Heat exchanger blocks for a large helium liquefier. (Credit photo Fives)

temperature heat exchanger, provided the stack height is small enough. Printed circuit heat exchangers (remember Sect. 6.4.1.5) are the good choice for such situation.

6.4.2.4 Characteristics

Pressures: HP and LP, 80 bar maximum
Size: up to very large
Maximum acceptable temperature difference: 50 K

Fig. 6.94 Aluminium alloy to stainless steel transitions (Thevenet Clerjounie)

Fig. 6.95 Risk of deterioration of a heat exchanger by a large temperature difference

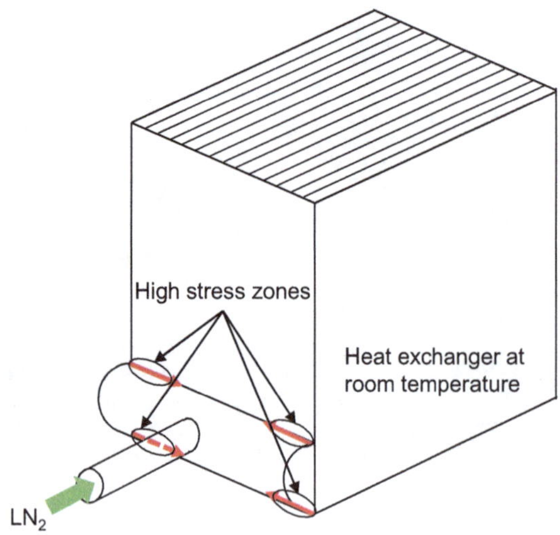

6.5 The Bearing Systems for Cryogenic Rotating Machines

The general structure of a bearing system is a shaft that is supported by two radial bearings and positioned longitudinally by a thrust bearing. The journal bearing acts against forces of gravity, if the axis is horizontal, unbalance of the supported system (shaft and both turbine and brake compressor wheels). The thrust bearing acts against the pressure difference between both ends of the shaft, gravity if the axis is vertical, and the dynamic forces that results of the operation of wheels.

6.5.1 Bearing System Technologies

There are various kinds of cryogenic rotating machines: expansion turbines, centrifugal compressors, gas circulators and pumps. All these machines need peculiar bearing systems:

- The shaft rotational speed might be rather high, up to 6000 Hz (360,000 rpm) for small turbines; therefore not any contact is allowed between shaft and cushion.
- No pollution is allowed into the cycle gas.
- Stiffness must be sufficient in order to control tiny gaps.
- There is a high temperature gradient between the wheel and the bearing system.
- The bearing system must accept various temperature map situations from room temperature down to operating temperatures. The axial position of the shaft is a very important parameter: it governs the clearance between the wheel wings and the fixed part that is in front of them.
- The bearing system must be perfectly leak tight to the atmosphere.
- The reliability must be high.

Table 6.6 shows the various requirements that are linked to each machine.

There are not many bearing candidates to fulfil such high-level duties. This is the reason why most of the technologies related to cryogenic rotating machine bearing have been developed in-house, generally by companies dealing with cryogenic refrigeration.

The most usual bearing systems are:

- Special high rotating speed rolling element bearings.
- Static oil bearings, historically the first ones but almost no more used presently, for helium. This topic is not discussed here.
- Static gas bearings.
- Dynamic gas bearings.
- Active magnetic bearings.

Table 6.7 shows the peculiar characteristics of each type of bearing system. Presently, most of cryogenic expansion turbines operate on static or dynamic gas bearings, cryogenic centrifugal compressors and circulators on active magnetic bearings (a few on dynamic bearings). Ceramic ball bearings are also used for cryogenic compressors or circulators for which the rotational speed is lower than for expansion turbines.

Table 6.6 Specificities of cryorotating machines

Small-size wheel	Medium-size wheel	Large-size wheel	High rotational speed (operation)	Medium rotational speed (operation)	Low rotational speed (cool down)
Turbines			Turbines		
		Compressors		Compressors	Compressors
	Circulators and pumps				Circulators and pumps

Table 6.7 Bearing technology choices

	Supporting strength	Stiffness	Rotational speed high	Rotational speed low			
Ball bearings	High $^{+}$	High $^{+}$	Yes $^{-}$	Yes		*Compressors*	*Circulators and pumps*
Oil bearings							
Static	High	High	Yes $^{-}$	Yes	*Turbines (large)*		
Gas bearings							
Static	High	High	Yes	Yes	*Turbines*		
Dynamic	High $^{-}$	High $^{-}$	No	No	*Turbines*		
Magnetic bearings							
Active	High	High	Yes $^{-}$	Yes	*Turbines (large)*	*Compressors*	*Circulators and pumps*

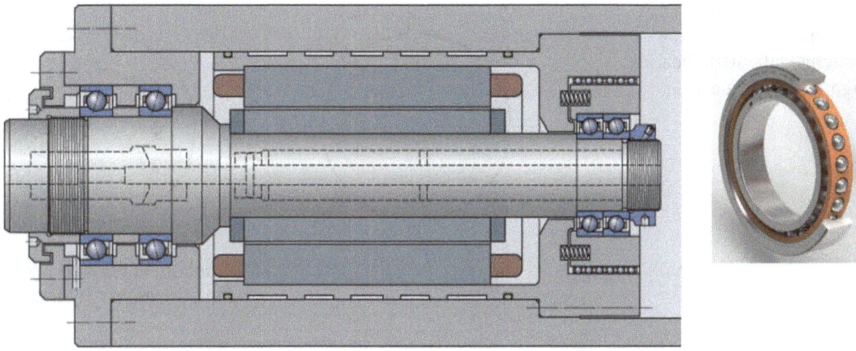

Fig. 6.96 A high-speed machining spindle

6.5.1.1 Ball Bearings

Such systems are generally based on high-speed machining spindle bearings as shown in Fig. 6.96, left, that are fitted with ceramic ball bearings as shown in Fig. 6.96, right. Such spindles are adapted to an operation in helium atmosphere with no lubrication. The casing must be cooled with water.

It is the simplest of any bearing systems used for cryogenic rotating machines but its rotational speed and Mean Time Between Maintenance (MTBM) are limited.

6.5.1.2 Gas Bearing Systems

Gas bearing systems are mainly used for cryogenic expansion turbines. There are two types of gas bearing systems.

6.5.1.2.1 Gas Static Bearing System

Gas static bearing systems are fed with a small flow rate of helium (2 to 3% of the cycle mass flow) taken at or near the high pressure of the cycle and returned to the low pressure (see Fig. 6.97). The layer of gas that is located between shaft and bushing behaves like a lubricant pad, avoiding any contact between the two parts.

A pressure profile is established as shown in Fig. 6.98, left. When the shaft leaves the geometrical axis of the bushing (decentration), the cross section through which the gas is fed into the bushing decreases on one side. Consequently, as the gas flow increases, the pressure field is changed, and the supporting strength is reduced. Conversely on the diametrically opposite side, the cross section decreases, the pressure field is changed, and the supporting strength is increased. The system is auto-stable. Same principle is used for the thrust bearing. The whole bearing system is composed of a shaft that is supported by two radial bearings and a thrust bearing

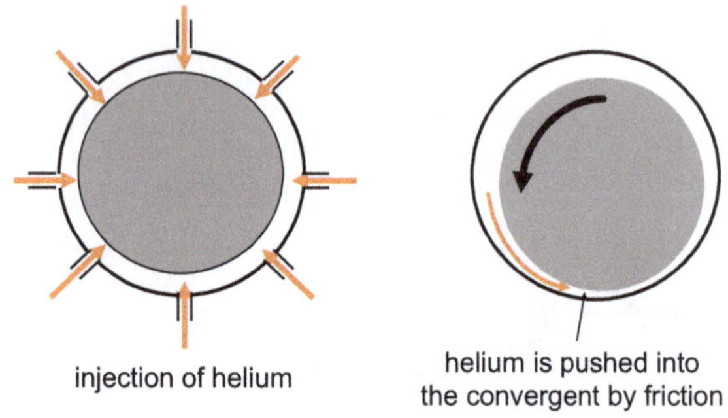

Fig. 6.97 Operating principle of gas bearing systems, left, static bearing, right, dynamic bearing

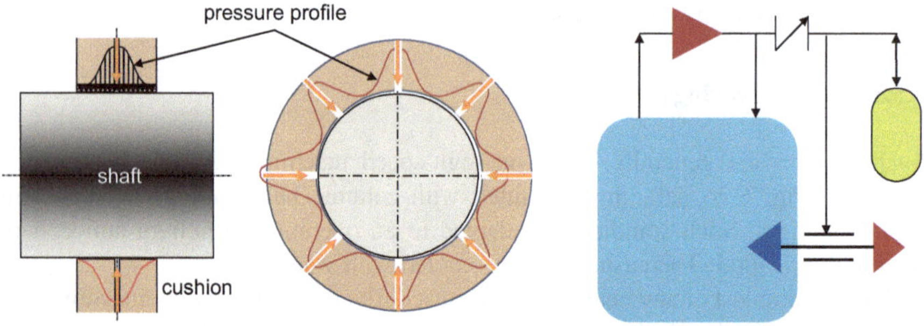

Fig. 6.98 Pressure profiles in a static gas radial bearing system, left; the bearing capacity connection, right

that positions the wheels relatively to the fixed parts. The stiffness of the bearing system can be varied by changing the supply pressure.

When the turbine is not operated, the shaft is "floating" above the gas layer. There is no contact between the shaft and the bushing. The shaft is generally horizontally oriented (history?) even if a static bearing system can indifferently operate either horizontally or vertically.

Such a gas static bearing system needs to be fitted with a dedicated capacity that supplies the bearing system in case the cycle compressor stops, as shown in Fig. 6.98, right.

In Fig. 6.98, left, one can see the operating principle and a cross section of a static bearing assembly and, in Figs. 6.99 and 6.100, a shaft and a bushing. The bearing system operates at high and low pressures that are different from the cycle and the brake compressor; therefore, labyrinths are fitted at both ends of the shaft. See Sect. 6.6.3.2.1 (Fig. 6.118) about the setting of the labyrinth.

The power that is dissipated by gas friction in the bearing system is naturally dissipated into the gas that is circulated in the system.

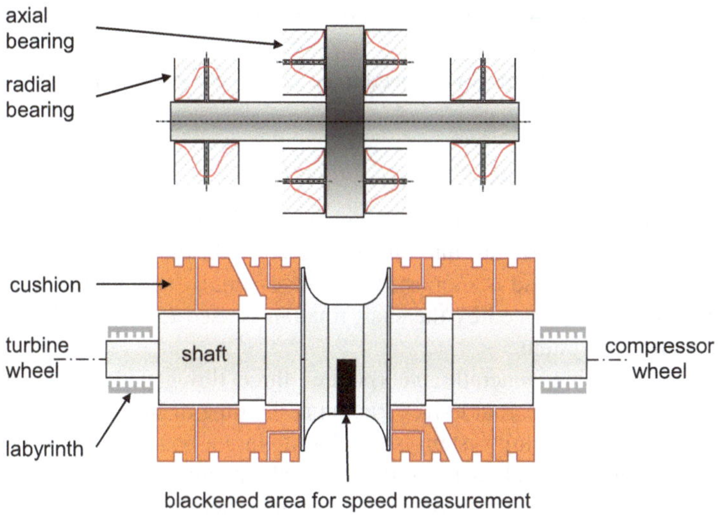

Fig. 6.99 A static gas bearing system. (Courtesy Air Liquide)

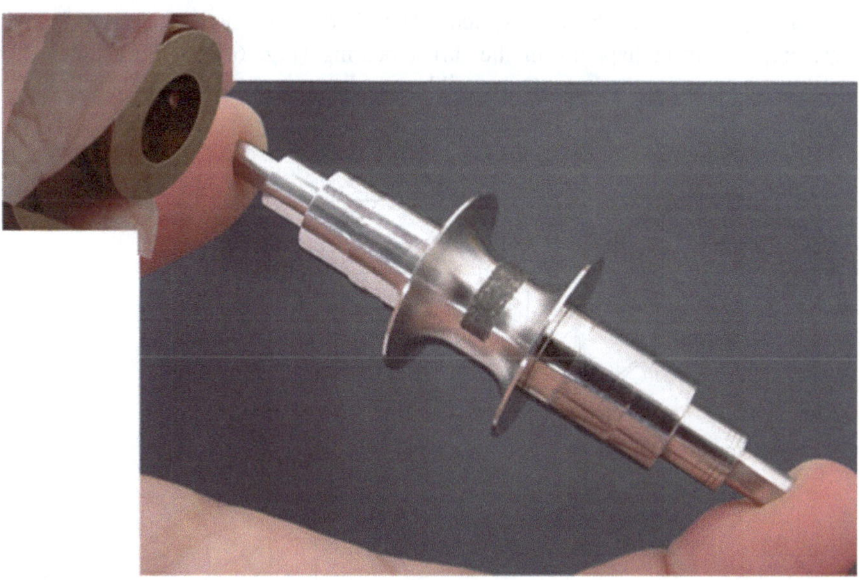

Fig. 6.100 A shaft and one bushing. (copyright Air Liquide)

6.5.1.2.2 Gas (or Self-Acting) Dynamic Bearing Systems

Gas dynamic bearing systems do not need to be fed by any gas for their sustentation.

6.5.1.2.2.1 Gas Dynamic Bearing with Tilting Pads

The tangential velocity of the shaft combined with the gas viscosity and a tricky shape of the pad pushes gas into the gap and produces a pressure field (see Fig. 6.101). The tilting pad is supported into the casing by another tricky solution: the pressure that is generated by the shaft rotation is used to support the pad with kind of a static gas system.

Each radial bearing generally incorporates three tilting pads made out of a material that allows some trouble-free friction at the time of start and stop.

The gas dynamic bearing system is "enclosed" into a box (see Fig. 6.124). The operating pressure is equal to the turbine wheel tip pressure, therefore variable according to the operating conditions of the turbine. There is no pressure difference between the various areas of the system, and, therefore, labyrinths are not needed.

The power that is dissipated by gas friction in the bearing system is to be rejected by means of a cooling water circulation in the bearing casing.

When the turbine is not operated, the shaft that is submitted to gravity is in contact with fixed parts of the bearing system. The shaft is always vertically oriented; therefore, a contact happens in the thrust bearing (Fig. 6.103). Sometimes, for small turbines, a magnetic system equilibrates the weight of the shaft in order to

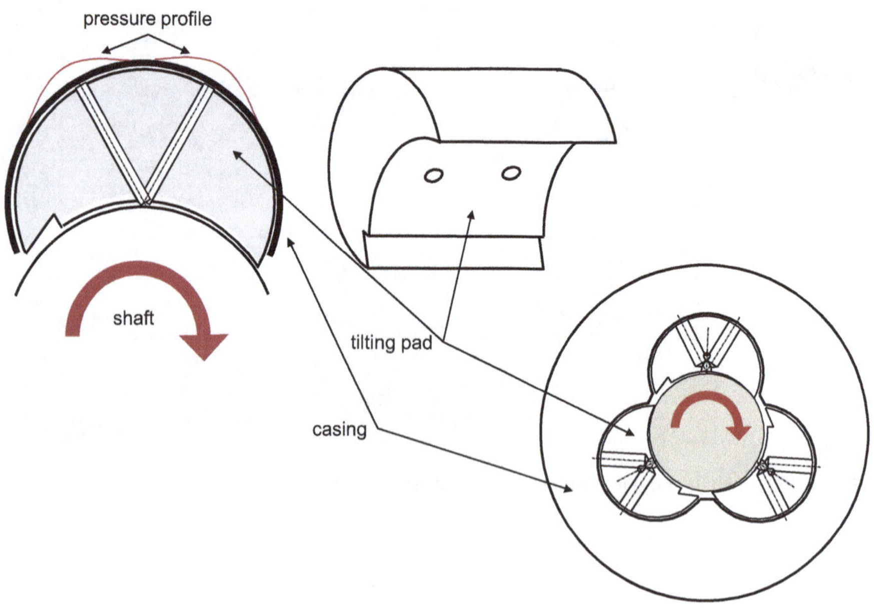

Fig. 6.101 A gas dynamic radial bearing system with tilting pads

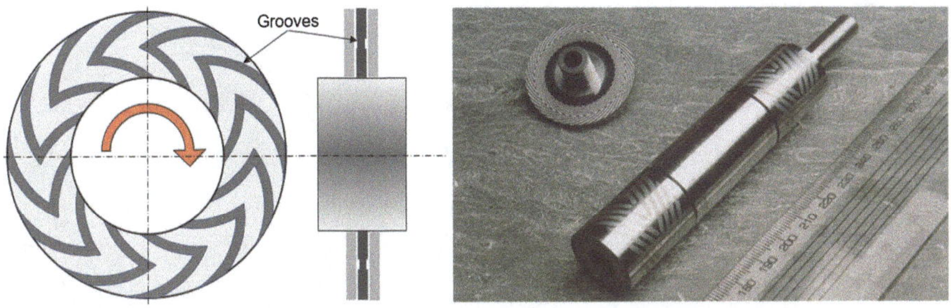

Fig. 6.102 A gas dynamic bearing system with grooves. (copyright Air Liquide)

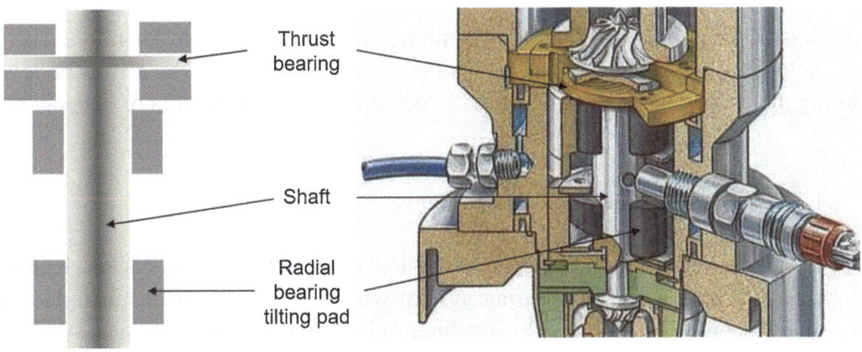

Fig. 6.103 Structure of a gas dynamic bearing system. (© Linde Kryotechnik AG)

avoid any contact on the thrust bearing during start-up. For big turbines, injection of helium into the bearing system reaches same target.

6.5.1.2.2.2 Gas Dynamic Bearings with Grooves

The thrust bearing operates similarly as the radial bearings, but the gas "compression" is performed by means of grooves that are engraved on the shaft, both for radial and thrust bearings, as shown on Fig. 6.102, left.

In Fig. 6.102, right, one can see a shaft with grooved radial and thrust bearings (Air Liquide). Another example can be seen in Fig. 16.28.

6.5.1.2.2.3 Gas Dynamic Bearing with Foils

There is another type of gas dynamic bearing system: the foil bearing system, the principle of which is shown on Fig. 6.104.

As any of the gas dynamic bearing system, it does not need to be supplied by an external source of gas, but it does not allow an important rotational speed reduction. For this reason, the cool down process of a cryogenic compressor or circulator, which necessitates a gentle rotation for temperature homogeneity, is very delicate. Its

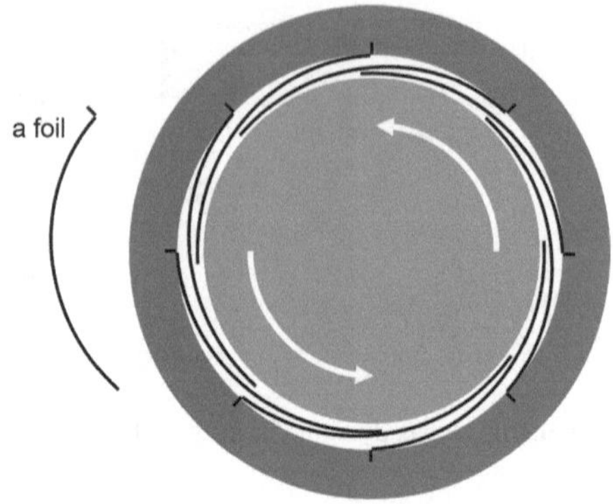

Fig. 6.104 A gas foil bearing system (gaps are exaggerated)

stiffness is lower than the dynamic gas tilting pad bearing system. Therefore, such technology is not advised for machines the rotational speed of which is to be low during the cool-down phase and varied in operation like compressors or circulators.

6.5.1.3 Active Magnetic Bearing Systems

A very interesting solution comes with the active magnetic bearing system. Unfortunately, building a simpler bearing system with only permanent magnets is not possible (theorem of Earnshaw); therefore, only active magnetic bearing systems are used.

The radial position of the shaft is measured with four sensors (see Fig. 6.105). The position information is processed, and current is injected into the supporting coils in order to keep the shaft axis in a fixed position related to the bearing system axis.

The very first active magnetic bearings (Tore Supra, see Sect. 12.8.1) were operating at liquid nitrogen temperature. Presently all of them operate at room temperature; they are cooled by circulation of water in the casing.

When the rotational speed increases, both efforts and frequency of correcting actions increase. However, as efforts increase with the second power of frequency, the control of the shaft position becomes more and more difficult. Here, a very interesting trick is to be noticed: instead of trying to keep the shaft geometrical axis fixed in relation with the geometrical axis of the bearing system, the control system allows the shaft to rotate around its inertia axis. As a consequence, there are no more big efforts to control the position of the shaft. Another consequence is that the vibration level is almost reduced to zero.

It is obvious that, in the case of the occurrence of an electrical breakdown, the bearings have to be fed until the rotation speed is zero. Such duty can be provided by a buffer battery, or in a modern and smarter way, the electrical motor operates as a generator to feed the bearings during the time it rotates with the remaining kinetic energy. However, in case the shaft is no longer supported, roll bearings that are fitted with a gap between the shaft and the internal raceway, which means that they do not rotate at normal operating conditions, can support the shaft during the time the

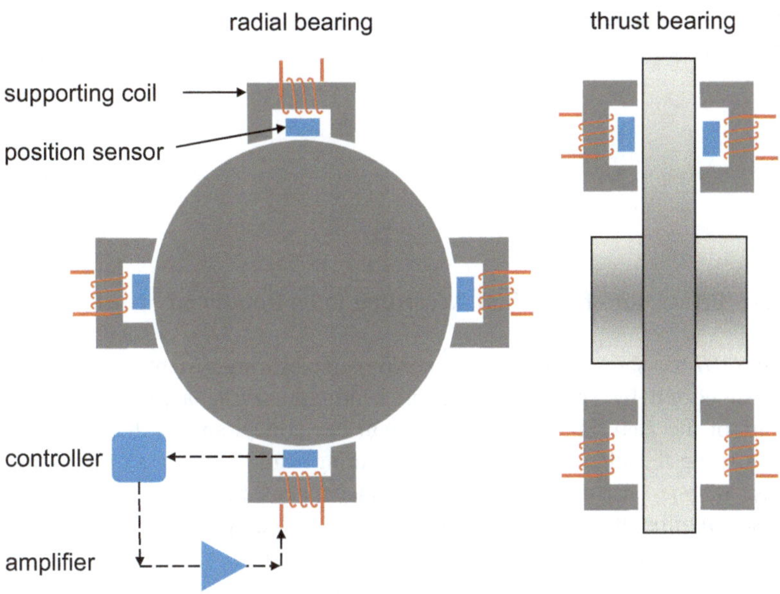

Fig. 6.105 Active magnetic bearing system principle

rotational speed comes to zero. It is obvious that the number of such "landings" is limited. After a few landings, the roll bearings must be replaced.

The operating principle of active magnetic bearings brings big advantages: in addition to accurately monitoring the position of the shaft, both radially and axially, it allows an accurate and controlled axial position of the shaft that makes possible controlling the clearance between the wheel and the fixed part that is in front of it. Such a clearance is generally an important parameter in the operation of a cryogenic centrifugal compressor.

6.5.1.4 A heat Intercept on the Shaft

When the heat leak along the shaft might deteriorate the machine efficiency, a heat intercept is implemented, as shown in Fig. 6.106. A ring that is located around the shaft, without contact with it, is circulated by a cold fluid (liquid nitrogen or gaseous helium). Heat is transferred from the shaft to the ring mainly by gas convection and conduction.

Such an arrangement is mainly used for compressors or circulators. An example is shown in Fig. 6.146.

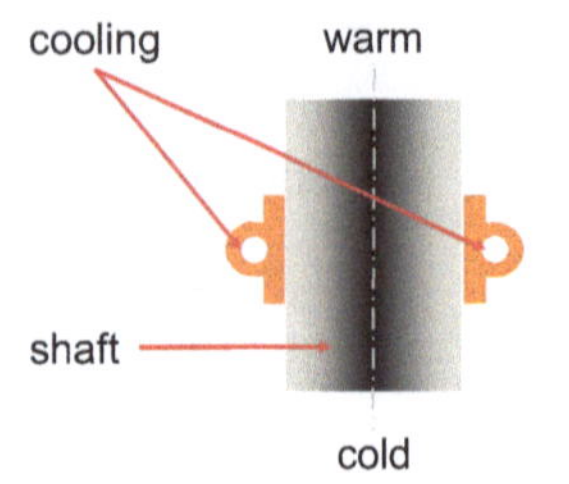

Fig. 6.106 A heat intercept on a shaft

6.5.1.5 Behaviour of the Shaft According to its Rotational Speed

Figure 6.107 top shows a mechanical analogy of a bearing system: a rigid beam (the shaft) supported by springs (the radial bearings) having a given stiffness (the pressure of the gas). Such a bearing system has critical speeds that are related to $\sqrt{k/m}$, k being the stiffness of the spring (bearing) and m the mass of the shaft plus wheels. At one bearing critical speed (that happens first when rotational speed increases) regime, the shaft axis moves according to a cylinder (Fig. 6.107 bottom left), the axis of which is same as the bearing system. At the other bearing critical speed, the shaft axis moves according to a cone, the axis of which is same as the bearing system one (Fig. 6.107 right).

Figure 6.108 shows the variation of decentration of a shaft in a gas static system, according to the rotational speed. One can easily see the two critical speeds where the decentration is maximum. Operators of such turbines are used to hear the "crossing" of critical speeds when the rotational speed increases during start-up.

Even if it is obvious that a high rotational speed shaft is to be correctly balanced, one understands that it is at the time the critical speeds are "crossed" that the risk of a contact between shaft and bearing is maximum.

The behaviour of the bearings differs according to their technology. In a gas static system, as k is constant, $\sqrt{k/m}$ is also constant: the critical speeds are located at fixed rotational speeds (see Fig. 6.109, blue arrows). However, it is different for a dynamic bearing system where k is related to the rotational speed. Therefore, the critical speeds vary with the rotational speed. This is shown in Fig. 6.109 where critical speeds that vary according to the shaft rotational speed are indicated (red arrow). One can see that such critical speeds are not "crossed" because they are always lower than the rotational speed of the shaft.

It is rather usual to "help" the dynamic bearing system at low rotational speeds in order to increase the bearing supporting strength by injecting helium into it, turning it in kind of a static one. But helium is injected at room temperature and can only be evacuated into the cycle, where it can bring perturbations. Therefore, such a procedure can be only used without consequences on the cycle operation when starting or stopping the turbine.

In a dynamic system, as all parts are at same pressure (the wheel tip pressure), there is no gas circulation between the various areas of the system (see Fig. 6.124). However, in static bearing system, pressures along the shaft are different. As the rotational speed of the shaft is high, there is no way to perform positive leak tightness

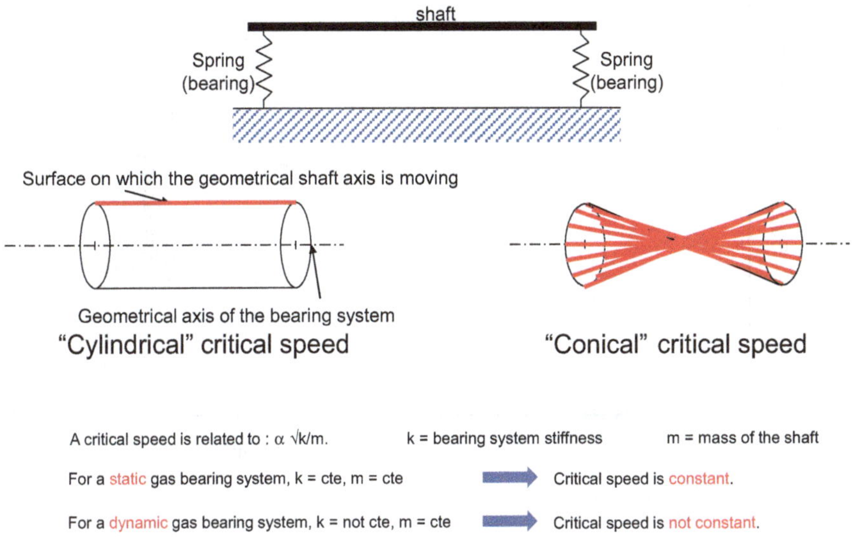

Fig. 6.107 Analogy of a bearing system

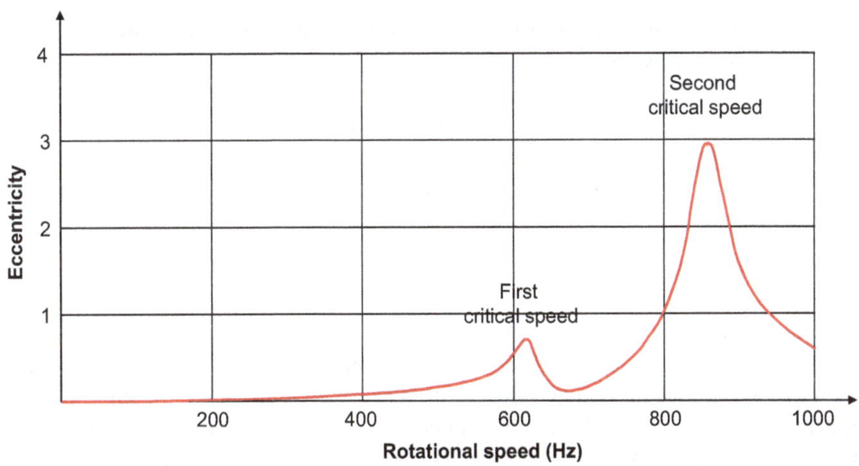

Fig. 6.108 An example of gas static bearing critical speeds and their consequence on decentration

between the various areas of the system. Only labyrinth systems can be used. That means that a permanent gas leak flows from high-pressure to low-pressure areas, along the shaft. Such gas flow must be accurately monitored (see Sect. 6.6.3.2.1 (Fig. 6.118)).

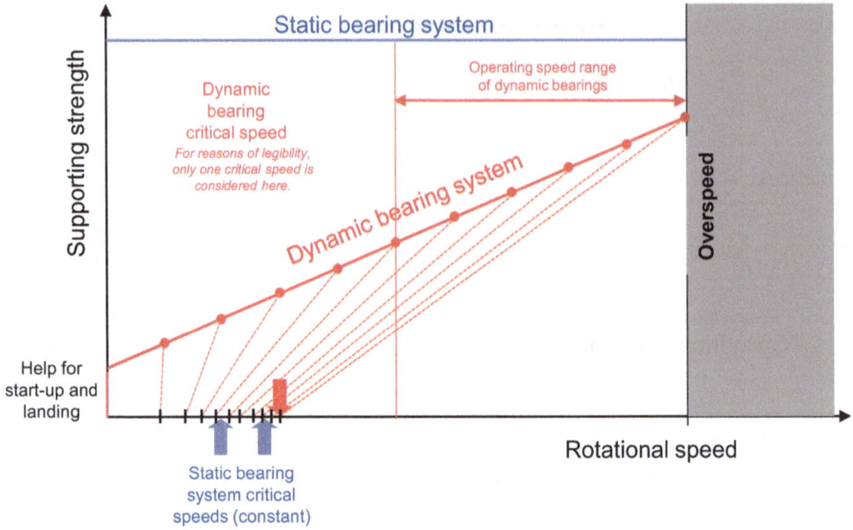

Fig. 6.109 Comparing bearing systems and their critical speeds

6.5.1.6 Comparison Between Gas Static and Dynamic (Tilting Pad) Bearing Systems

It is of interest to compare the two systems (see Table 6.8).

In an operational point of view, experience shows that each of these two technologies having some plus and minus happens to be both very reliable with a very high MTBF.

Rotational Speed Measurement

For gas bearing technology, as no contact is allowed with the shaft, the speed is picked up either by an optic or magnetic process. In the optical way, the shaft is blackened on half the circumference (see Fig. 6.100). A light beam is brought by half the cross section of an optical fibre on it. When the beam falls on a shiny part, it reflects and returns along the second half cross section of the fibre to a light detector. Conversely, when it hits a blackened part, there is no beam reflection. The light detector generates a square signal the frequency of which being that of the shaft rotation. In the magnetic way, a small magnet is embedded into the shaft. A pick-up coil is located near the shaft and generates two signals per revolution.

For magnetic bearings the frequency of the converter gives the rotational speed.

Table 6.8 Comparison of bearing technologies

	Static	Dynamic
Feed gas	Yes	No
Axis orientation	Any	Vertical
Bearing stiffness	High	Lower
Bearing stiffness vs rotational speed	Independent	Related to
Bearing critical speeds	Yes	Never crossed
Pollution of the cycle gas	No	No
Contact when starting up or stopping	No	Possible
Compliance with temperature change	Yes	Mind
Compliance with fast cycle pressure change	Yes	Mind
Power during cool-down	Nominal	1,5 x nominal
Bearing safety capacity	Yes	Not needed

6.6 The Cryogenic Expansion Turbine

The cryogenic expansion turbine is a machine that has to cope with a very difficult challenge: it must be able to operate within a very large range of conditions, between the start-up at room temperature and the nominal operating conditions. Furthermore, such off-design operation during cool-down of heavy loads may last for long periods of time.

6.6.1 Thermodynamic Aspect of a Cryogenic Expansion Machine

On a T-s diagram, the cryogenic expansion machine operation can be shown as in Fig. 6.110 left, at point 1, gas at pressure P_1 and temperature T_1 is expanded down to pressure P_2. On the T-s diagram, point 1 is located on the P_1 isobar, at T_1 temperature. An isentropic expansion would follow a vertical line until it crosses the P_2 isobar at point h_{is}. The *isentropic* enthalpy drop is h_1–h_{is}. Such an ideal expansion machine would reach the T_{is} discharge temperature.

As the actual expansion machine is not ideal, the amount of enthalpy that it can extract is lower than the isentropic one; consequently, its discharge enthalpy and temperature are, respectively, higher than h_{is} and T_{is}. The actual discharge point is located on the P_2 isobar, at point 2, but its enthalpy h_2 is obviously higher than h_{is}.

The isentropic efficiency of the expansion machine is the ratio of the actual enthalpy drop versus the available enthalpy drop (isentropic):

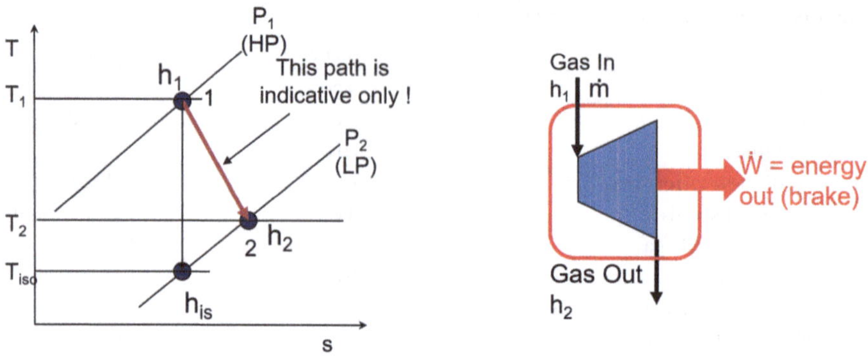

P (bar)	T (K)	h (J/g)	ṁ (g/s)	Power (W)	Δh real (J/g)	Δh isent. (J/g)
14.00	25.00	130.49	50.00			
0.75						
1.20	12.93	70.85		-2981.63	-59.63	-79.51

Fig. 6.110 Turbine on the T-s diagram

$$\eta_{\mathrm{is}} = \frac{(h_1 - h_2)}{(h_1 - h_{\mathrm{is}})} \tag{3.12}$$

The actual power of an expansion machine is:

$$\dot{W} = \dot{m} \times \eta_{is} \times (h_1 - h_{\mathrm{is}}) \tag{6.8}$$

$$\dot{W} = \dot{m} \times (h_1 - h_2) \tag{6.9}$$

$$\dot{W} = \dot{m} \times \Delta h \tag{6.10}$$

Δh is usually called the *enthalpy drop* of the turbine.

Such power is to be removed by the brake system (see Fig. 6.110, right).

One of the tools of the helium tool box (see Sect. 15.8.3.2) is the turbine spread sheet shown in Fig. 6.110, bottom. It can be used to predict the discharge temperature (here: 12.97 K), but also, by using the "Goal seek" in the "What if analysis", one can evaluate the isentropic efficiency of a cryogenic expander the operating conditions of which are known. Obviously, such figure accuracy is linked to the measurement accuracy, especially for temperatures. With such a tool, the operator can compare various expander operations provided that, within a small range, the inaccuracy of the sensors is not changed.

6.6.2 Various Cryogenic Expansion Machines

An expansion machine removes energy from the gas that it expands. A reciprocating expander or a turbine is a gas engine. As the energy (enthalpy) of a gas is mainly related to its temperature (see Sect. 1.4.1.2), expansion with work extraction cools the gas. The energy that is extracted from the gas is to be dissipated at room temperature.

The very first expansion machines were reciprocal. Obviously, they were easy to build and maintain (see Fig. 16.13). However, as for any machine, when moving towards higher flows, the size and the frequency increase, but the frequency is rather limited. The solution was to circulate at higher velocities, therefore decreasing the size. That was possible with turbines (Fig. 16.15). Since the 1990s, almost all helium refrigerators or liquefiers have been fitted with expansion turbines. Therefore, we will only concentrate on this type of equipment.

6.6.3 Cryogenic Expansion Turbines

As for any machine, there is a very detailed literature about turbines, but it mainly refers to power turbines, which are generally large or very large machines, generally operating at far above room temperature.

This chapter is not a course on turbines. It is only dealing with the most important and/or specific issues related to cryogenic expansion turbines that interest someone dealing with helium refrigeration. Some simplifications or assumptions are made that allow easier understanding. For example, most of the information in this chapter relates to turbines fitted with *radial inlet* wheels and *sonic distributors*.

There are some specificities about cryogenic expansion turbines:

- Expansion of helium requires high tangential speeds (a few hundred m/s) with rather small volume flows (a few L/s) and rather high expansion ratios. To reach high tip speeds, a small turbine wheel must rotate at very high speeds that need special bearing systems, making use of special very sophisticated technologies (see Sect. 6.5.1.2).
- Cryogenic expansion turbines must operate, sometimes for a rather long time during the cool-down phase of a system, at any temperature between room temperature and their nominal operating temperature. One has to take special care of the turbine during the cool-down phase.
- Cryogenic expansion turbines are sensitive to impurities that are diluted into the processed gas. During the cool-down phase, the coldest location of the system is one of the turbines. As any impurity in helium is turned into a solid before helium reaches its boiling temperature, solid can appear into the turbine. One has to take special measures in order to avoid formation of any solid in the turbine. In static bearing systems, impurities can also deposit into the bearing at a location where the shaft is at a "cool" temperature (remember Sect. 6.5.1.2.1).

To deal with such specificities, one has to take care of the mechanical behaviour of the turbine but should not forget about the quality (purity) of helium that is expanded.

These various aspects are taken into consideration in this chapter.

6.6.3.1 Operating Principle of an Expansion Turbine

A turbine operates in three steps:

1. The energy of pressure is converted into kinetic energy.
2. Most of the kinetic energy is converted into mechanical energy,
3. The low remaining kinetic energy is converted into pressure. (Of course, the remaining gas velocity is related to the limitation of the diameter and, therefore, the cross section of the discharge side of the wheel.)

Consequently, a turbine is made out of three components, each of them fulfilling the specific duty of one of the steps indicated here above:

1. The **distributor** converts the gas pressure energy into gas velocity (kinetic energy).
2. The **wheel** converts the gas velocity into mechanical energy (torque).
3. The **diffuser** converts the low remaining gas velocity at the discharge of the wheel into pressure.

6.6.3.1.1 The Distributor

The distributor, which is a *fix nozzle* (see Figs. 6.111 right and 6.112), fulfils a *triple duty*:

First, it accelerates helium: as energy is conserved, enthalpy is converted into kinetic energy (see $v = \sqrt{2\Delta H}$ (Refers to equation in Chapter 17).

$$\Delta h_{\text{dist}} = ½ \left(C_{\text{disch}}{}^2 - C_{\text{in}}{}^2\right) \tag{6.11}$$

C_{disch} is the gas velocity at the discharge side of the distributor
C_{in} is the gas velocity at the inlet of the distributor
Δh_{dist} is the enthalpy drop of the distributor

As a consequence, the temperature drops. (Remember what happens at the discharge of the valve seat as described in Sect. 1.4.2.2).

Second, it gives a correct direction to the gas velocity in order that gas enters the wheel at a correct angle (tangential to the wheel fins).

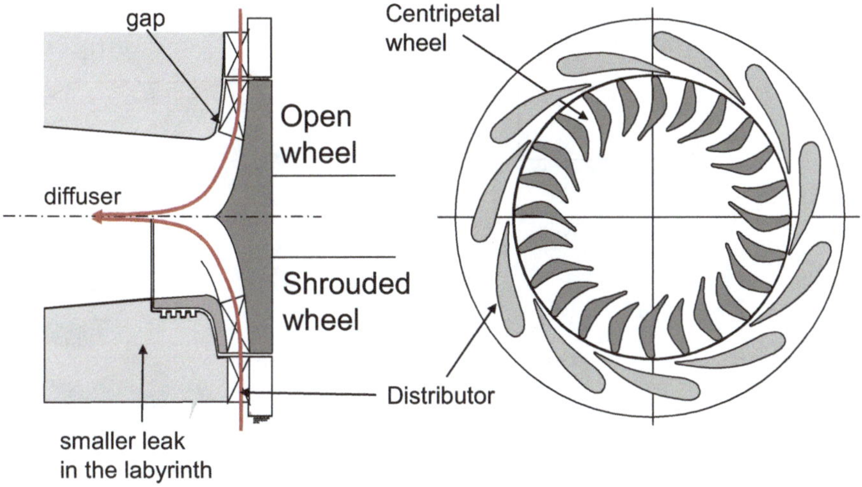

Fig. 6.111 Open and shrouded wheel, distributor and wheel

Fig. 6.112 A distributor and a wheel. (copyright L. Lelong Air Liquide)

Third, it calibrates the helium flow according to its cross section. If the distributor operates in a sonic regime, the mass flow is proportional to the inlet pressure and inversely proportional to the square root of the inlet temperature (see Sect. 6.6.3.7).

The relatively small size of helium cryogenic expansion turbines does not allow fitting them with a directional wing distributor. However, even with a fixed distributor, it is possible to efficiently control the power that is extracted by a turbine (see Sect. 8.7).

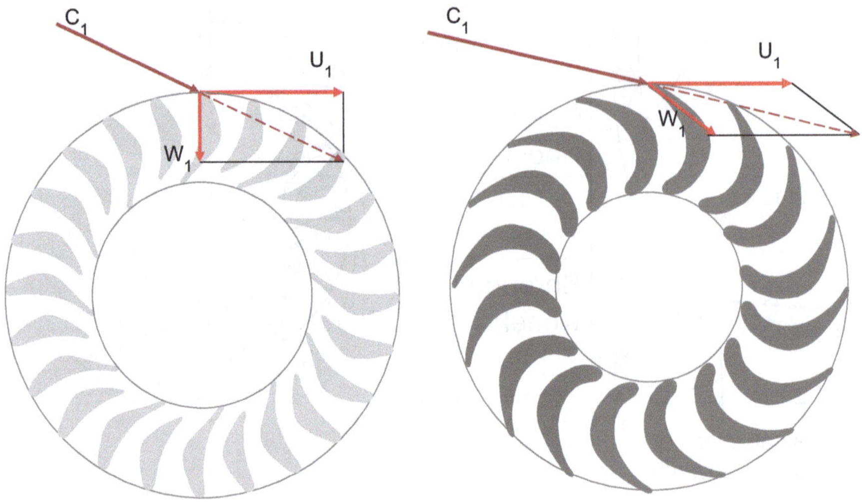

Fig. 6.113 Velocity composition at the inlet of the wing

6.6.3.1.2 The Wheel

The wheel is a *rotating nozzle*. Gas circulates from the periphery to the centre of the wheel: it is a *centripetal* wheel in which the gas fights against centrifugal forces; therefore, it loses more energy.

In Fig. 6.113, C_1 is the helium velocity at the discharge side of the distributor. It depends on the distributor helium inlet temperature and pressure. U_1 is the tip speed of the wheel. It is related to the rotational speed and the wheel diameter. W_1, which is the inlet velocity into the wheel, is the composition of C_1 and U_1. W_1 must be *tangential* to the inlet of the wing (for a radial inlet wheel as in Fig. 6.113 left, the W_1 velocity is radial).

To avoid any resonance problem, the number of wings of the distributor and the wheel must be different and not multiple.

The enthalpy drop varies as $U_1{}^2$, so a high enthalpy drop means a high tip speed. The material of the wheel, which is submitted to centrifugal forces, imposes a limit of U_1, usually around 450 m/s. In the case of a radial inlet wing wheel, such tip speed corresponds to an enthalpy drop of 100 J/g.

A way to process higher enthalpy drops is to modify the inlet orientation of the wings and use tilted wings as shown in Fig. 6.113 right. With such a trick, the acceptable enthalpy drop can rise up to about 400 J/g.

Remarks

1. It is sometimes of interest to have access to high enthalpy drops, but one should not forget that fulfilling a more difficult duty leads to a lower turbine efficiency. At the time of designing a cycle, some trade-off is necessary (as very often in helium refrigeration design) between enthalpy drop and efficiency.
2. Another way to process high enthalpy drops is to operate turbines in a pure series arrangement (see Sect. 4.2.1.3).

Various Types of Wheels

There are two types of wheel: open and shrouded (see Fig. 6.111 left and Fig. 6.114). In the open wheel, there is a gap between the top of the wings and the fixed part (top of the figure). The small quantity of gas that flows through this gap (g) is not processed into the wheel, and, therefore, the turbine efficiency is reduced. The gap in front of the wheel depends on the bearing system technology. For a bearing system size, it is almost constant whatever the wheel dimensions are.

One way to reduce the parasitic leak is to use a shrouded wheel (bottom of Fig. 6.111) where the leak is reduced thanks to a labyrinth.

A wheel can also be purely radial (Fig. 6.115 top) or radial-axial (Fig. 6.115 bottom). In a purely radial wheel, gas fights against centrifugal forces; this is good to lose its energy. Therefore, why go to radial-axial wheels? One of the sizing parameters of a wheel is related to the discharge conditions: the cross section, and so the diameter, is calculated according to an acceptable velocity at the discharge side of the wheel for a given volume flow. However, the diameter of the discharge part must be so that the wings have a reasonable length. This leads to a possible too large outer diameter. To keep a reasonable wing length, one answer is the radial-axial structure that allows to "keep" a rather small diameter with a correct wing length.

The smaller diameter of a radial-axial wheel has two interesting consequences.

1. Heat leaks

When the wheel diameter is smaller, the overall size of the cold part of the turbine assembly is smaller, and, as heat leaks come mainly through the surface, they are smaller. Less heat leaks means a higher efficiency.

2. Wing height

Gas gets into the wheel through the lateral cross section: $A = \pi \times d \times h$, with A area, d wheel diameter and h height of wings (see Fig. 6.114). When the wheel diameter is smaller, in order to keep the gas inlet at same cross section A, the wing height is inversely proportional to the diameter: $h_2 = d_1 / d_2$. The wheel wings must

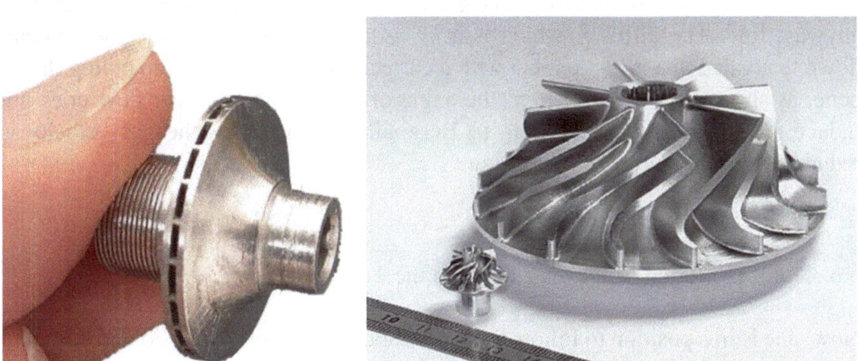

Fig. 6.114 Turbine wheels: left, shrouded radial; right, open radial-axial. (copyright Air Liquide)

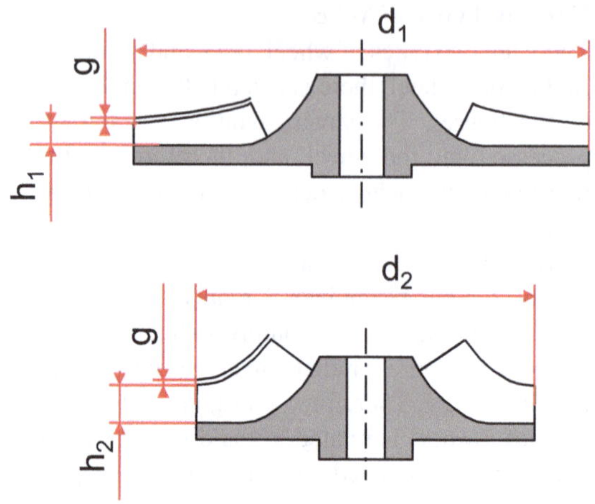

Fig. 6.115 Comparison between radial and radial-axial wheels

not have any contact with the fixed part that is in front of it. There is a gap: g. Such gap is related to the bearing technology. In other words, for similar bearing system dimensions, it is constant, whatever the wheel diameter is. When the wings are higher, the ratio g/h is smaller, which means that the flow that circulates in this gap is not processed into the wheel. Therefore, again, the turbine efficiency is increased.

Furthermore, it makes possible to fit the wheel on a smaller-size bearing system, which is a positive move towards both smaller heat leaks coming by solid conduction and cost.

Machining of such wheels is a very nice job that is, fortunately, made easier with numerical controlled machines, as shown in Fig. 6.116. Have a look at spectacular videos of machining on the Internet!

6.6.3.1.3 The Diffuser

At the discharge side of the wheel, the gas velocity is obviously not equal to zero but, anyway, low. The diffuser, which is a *fix nozzle*, reduces efficiently the velocity down to a value that is compatible with a correct flow in the discharge piping. It is a cone, as shown in Fig. 6.111 *left*, the angle of which is so that the flow does not induce any turbulence. According to Bernoulli's theorem, as the gas velocity is reduced, both pressure and temperature rise.

6.6.3.1.4 Evolution of the Gas Through the Turbine

Now, one is in a position to follow more accurately the whole expansion process into the turbine.

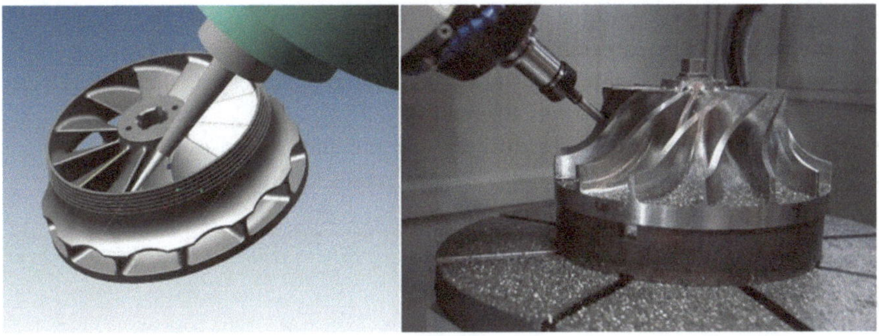

Fig. 6.116 Programming the machining sequence of a turbine wheel, left. Machining on a five-axis numerically controlled machine, right

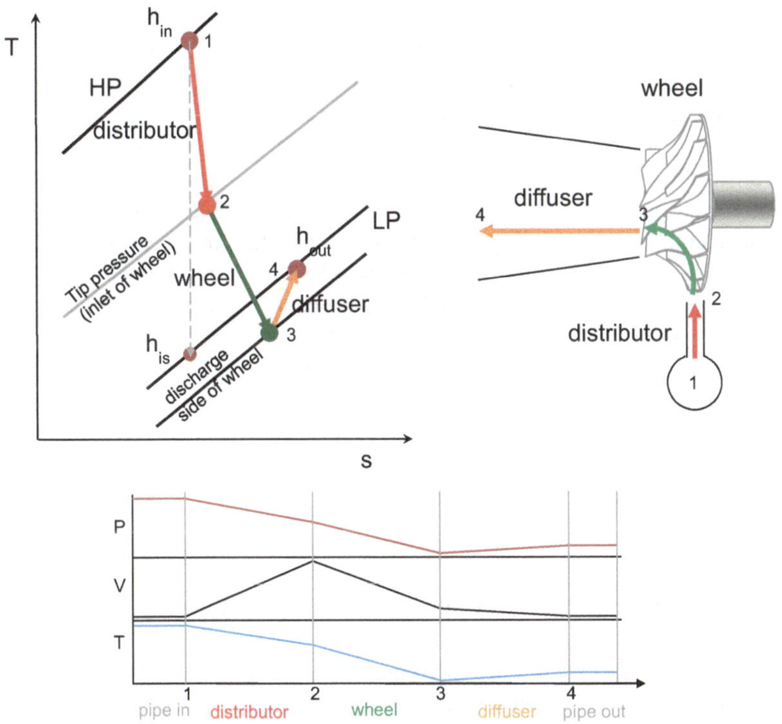

Fig. 6.117 Evolution of the gas through the turbine

In Fig. 6.117, one can follow the evolution of gas through the turbine, simultaneously in the T-s diagram (left), the turbine sketch (right) and another diagram showing the evolution of pressure, gas velocity and temperature (bottom):

- From point 1 to 2, into the distributor, the pressure decreases, the gas velocity increases, and the temperature decreases.
- From point 2 to 3, into the wheel, the pressure decreases, the gas velocity decreases, and the temperature decreases.
- From point 3 to 4, into the diffuser, the pressure increases slightly, the gas velocity decreases, and the temperature increases slightly.

6.6.3.1.5 The Turbine Brake

The turbine extracts energy from the gas that is expanded. This energy is to be taken out of the system. This is the duty of the brake that is submitted to the torque generated by the turbine wheel, transmitted by the shaft.

The brake system dissipates the power that is extracted by the wheel, but it must fulfil this duty at the *same rotational speed* (this is discussed in Sect. 6.6.3.3)!

Recovering the energy that is extracted by the expansion turbine wheel is difficult, especially for small machines. In a refrigerator, it would allow to increase the efficiency by 1.4% and even more for a liquefier: up to 18.4% (remember Sect. 3.2.4). The difficulty sits in the high rotational speed. It is difficult to pick up the torque with a simple, reliable solution.

Therefore, the most usual technologies that are available for the brake are the centrifugal compressor and eddy current system, the latter limited to around 2 kW. In the case of a centrifugal brake compressor, the wheel is located at the opposite end of the shaft, compared to the turbine wheel. The centrifugal brake compressor of a helium expansion cryogenic turbine generally dissipates energy into a water cooler; it does not allow recovering the energy that is extracted. Only in the case of rather high power, it is possible to connect the brake compressor in series with the cycle compressor, as it is commonly done in air separation plants or some special systems as a turbo Brayton cycle (see Sect. 4.6). Here, we will only consider the brake compressor solution with no energy recovery.

6.6.3.2 The Structure of a Cryogenic Expansion Turbine

All cryogenic expansion turbines have the same structure: a shaft supported by a bearing system with the turbine wheel at one end and the compressor wheel at the other end. The equipment that can be most often found on the market differs by the bearing system.

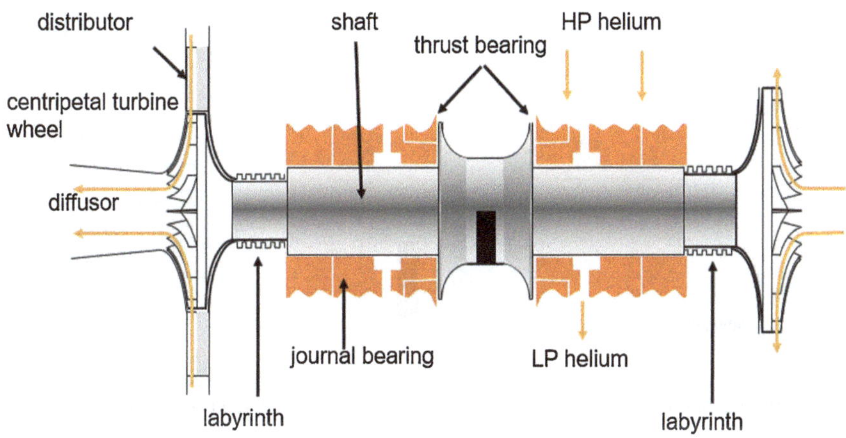

Fig. 6.118 Structure of a static gas bearing turbine. (copyright Air Liquide)

6.6.3.2.1 Static Gas Bearing System (Fig. 6.118)

There is a special feature on static gas bearing turbines that is to be discussed. In Fig. 6.119, one can see that the helium leak of the labyrinth, which is never zero, is related to a pressure difference, that is, wheel tip pressure – bearing discharge pressure. A small helium flow through the labyrinth is interesting because it absorbs part of the heat leak that is brought by conduction from the warm part of the shaft, in the bearing, to the wheel. This is a similar process as the one that has been explained in the operation of current leads (remember Sect. 3.2.3). However, if the pressure difference is high and/or the operating temperature is rather low, such helium flow could be too large, and there is a possibility that such cold gas cools the bushing and reduced its diameter and sizing of the shaft in the bushing might happen. It is necessary to control the labyrinth leak. For that, a room is located in the labyrinth, where room temperature helium is injected. This has two interests:

- The pressure difference between the wheel tip and the labyrinth room can be brought to zero; therefore the helium leak is zero.
- In the labyrinth part that is located between the wheel tip and the labyrinth room, helium is very cold, but in the part between the room and the labyrinth end, the helium is warmed by addition of room temperature gas.

There are two ways to control the labyrinth:

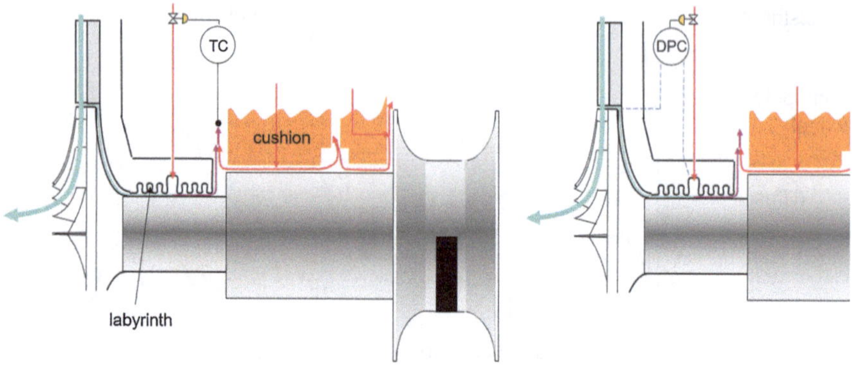

Fig. 6.119 Tuning the labyrinth of a static gas bearing turbine (copyright Air Liquide)

- Controlling the labyrinth discharge temperature: a control loop keeps the discharge temperature of the mixture of both the labyrinth and part of the bearing system flows above a given level (Fig. 6.119, left).
- Controlling the pressure difference between the wheel tip and the labyrinth room (Fig. 6.119, right).

Figure 6.120 shows the main components of an Air Liquide assembly. The turbine enclosure is welded on the cold box enclosure, therefore using the same vacuum.

In Fig. 6.120, one can see a usual scope of supply for an Air Liquide turbine. The turbine cartridge can be exchanged without breaking the insulation vacuum.

The most well-known maker of gas static bearing turbines is Air Liquide Advanced Technology, France.

6.6.3.2.2 Dynamic Gas Bearing System (Fig. 6.121)

The most well-known maker of gas dynamic bearing turbines is Linde Kryotechnik AG, Switzerland.

The Czech company PBS (Prvni Brnenska Strojirna) makes small turbines fitted with gas dynamic bearings and an eddy current brake (Fig. 6.122).

6.6.3.3 Setting the Rotational Speed of a Turbine

The power that is absorbed by a centrifugal compressor is proportional to:

- The processed mass flow rate: $\dot{m}$
- The gas density: ρ
- The third power of the rotational speed: N^3

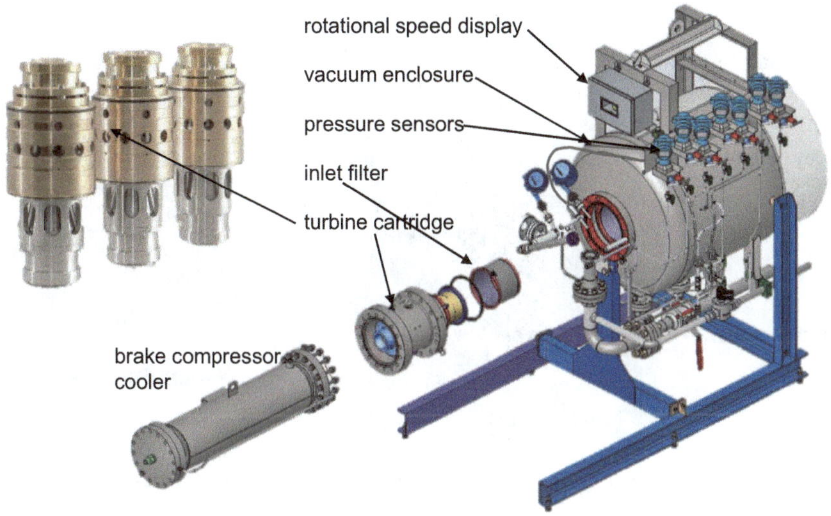

Fig. 6.120 An Air Liquide turbine assembly. (copyright Air Liquide)

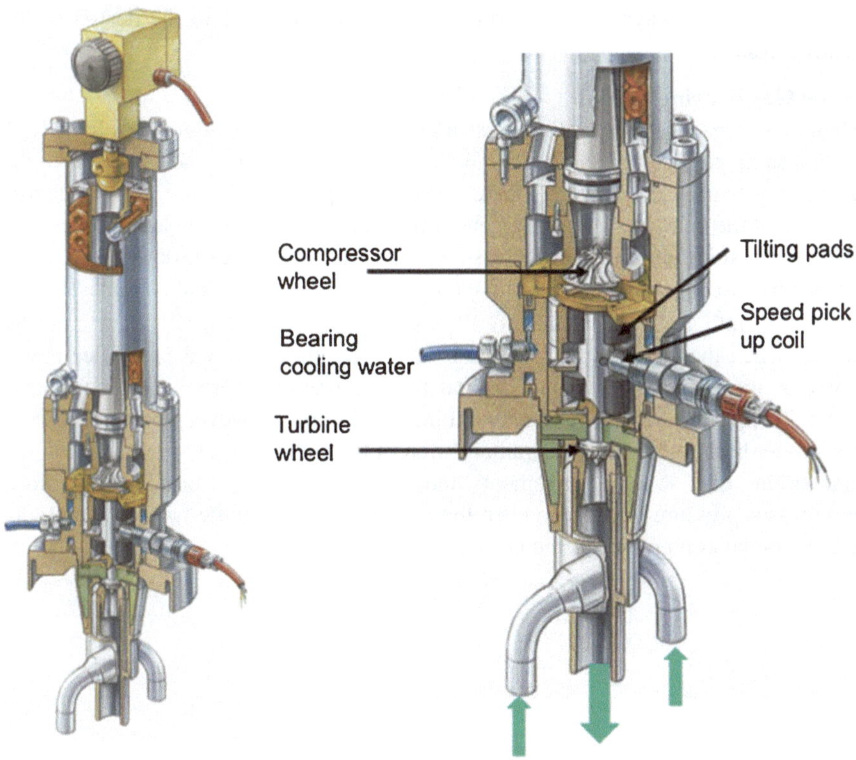

Fig. 6.121 A gas dynamic bearing turbine. (© Linde Kryotechnik AG)

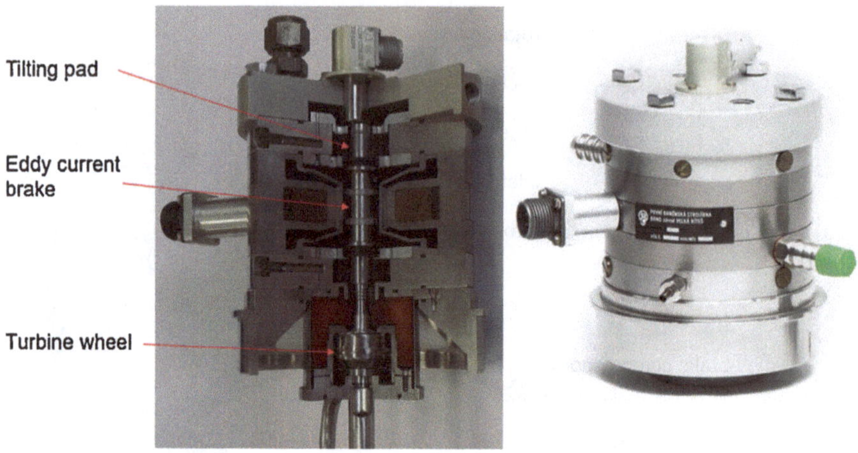

Fig. 6.122 A gas dynamic bearing turbine with an eddy current brake. (Courtesy PBS)

$$\dot{W} = \propto \dot{m} \times \rho \times N^3 \quad (6.12)$$

There are two ways to set the rotational speed, according to the design of the bearing system.

Static Gas Bearing System

Method: control of the operating pressure of the brake compressor (see Fig. 6.123)

In a static gas bearing system, the brake compressor circuit is separated from the bearing system by a labyrinth (brake compressor pressure on one side, discharge pressure of the bearing system; i.e. generally, the cycle compressor suction pressure, on the other side). The brake compressor pressure and therefore the density of the gas that is processed by the compressor are controlled with valve A that compensates the small labyrinth leak. According to $\dot{W} =\propto \dot{m} \times \rho \times N^3$ Eq. 6.12, for a constant power, when the compressor pressure is increased, the rotational speed decreases. Valve A is generally set manually, and the compressor operating pressure stays generally constant for any turbine operating conditions. However, in some systems where the discharge pressure is rather different from atmospheric pressure (e.g. the first turbine of a series arrangement), the process controller can adjust the brake compressor pressure in order to keep the difference between the wheel tip pressure and the brake compressor as a constant.

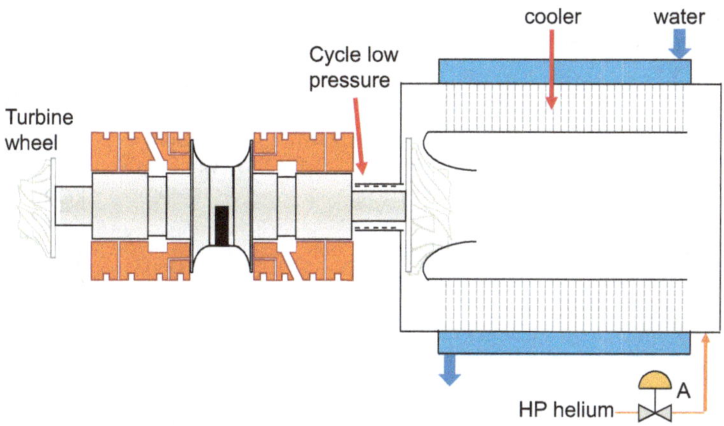

Fig. 6.123 Setting the rotational speed of a static gas bearing turbine by varying the brake compressor pressure

Dynamic Gas Bearing System

Method: control of the mass flow rate that is processed by the brake compressor (see Fig. 6.124)

In a dynamic gas bearing system, one can see that all components of the bearing system and the brake compressor are operating at a pressure that is around the wheel tip pressure. The wheel tip pressure depends on the cycle operating pressures. Therefore, there is no way to control the gas density of the brake gas. The control is performed by varying the compressor processed mass flow rate with valve B. According to the equation, by increasing the opening of the control valve, the processed mass flow rate increases, and the rotational speed decreases for same power. Valve B is permanently controlled by the process controller to be in a position to react correctly when the brake compressor operating pressure varies. An integrated by-pass ensures that a minimum mass flow always circulates in the brake circuit.

Remark In a static gas bearing system, the brake compressor operating pressure is controlled. However, it could also be possible to control the compressor flow, bringing the advantage of getting more power during the cool-down phase (see Fig. 6.137).

6.6.3.4 Tuning the Rotational Speed of a Turbine

When the rotational speed of a turbine is correct, the inlet velocity is tangent to the wing inlet (see Fig. 6.125 top centre).

How to set the optimum rotational speed of a turbine? At the optimum rotational speed, the turbine efficiency is at maximum, or, in other words, the temperature difference that is generated by the turbine is at maximum. Therefore, when the

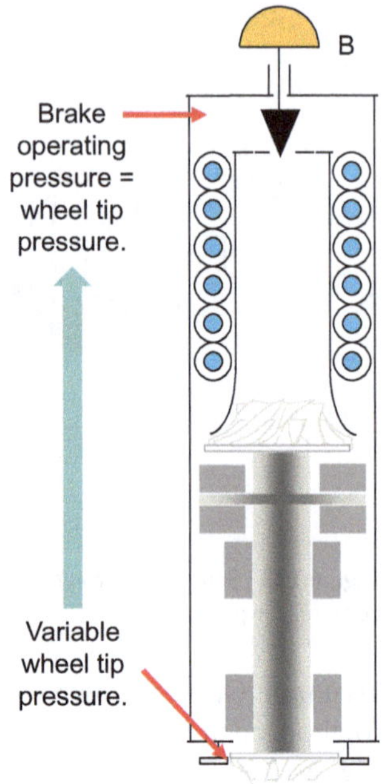

Fig. 6.124 Setting the speed by varying the brake compressor flow

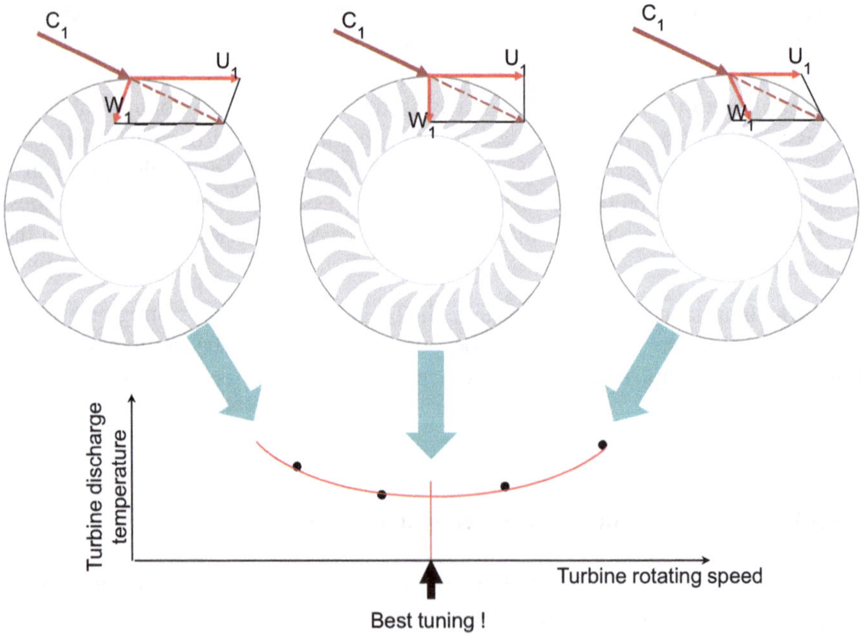

Fig. 6.125 Rotational speed tuning

refrigerator is operating in a *steady state*, one simply looks at the turbine discharge temperature indication. Various discharge temperature measurements corresponding to rotational speed settings (see Fig. 6.125 bottom) allow identifying the optimum rotational speed of the turbine *for the operating conditions at the time of the test*. As one compares the indications of the same sensor, a possible wrong indication of it has no incidence on the result accuracy.

6.6.3.5 Efficiency of Helium Expansion Turbines

An expansion turbine is generally custom designed for the system operating conditions in order to reach the highest possible efficiency. Obviously, such calculations are rather complicated, and it is not easy to get an estimation of the efficiency without running them. However, it can be said that the efficiency of an expansion turbine is related to various parameters, among them:

- The enthalpy drop that is performed by the turbine. The higher the enthalpy drop, the lower the efficiency. Remember that the enthalpy drop has a maximum limit (see Sect. 6.6.3.1.2).
- The operating temperature. The lower the temperature, the higher the heat leaks by conduction and radiation.
- The size
  - The smaller the size, the higher the ratio of area versus volume; therefore, the heat leak compared to the mass flow rate is higher.
  - The gap between the wheel and the fix part that is opposite depends on the bearing technology and therefore is constant for a given bearing system size. As explained in 6.6.3.1.2, the gap incidence is larger when the wheel dimension is smaller.

The relative weight of a few parameters is shown in Fig. 6.126.

In a few words, the higher the temperature, the lower the enthalpy drop, the larger the size, and the higher the efficiency is. Cryogenic expansion turbines can reach isentropic efficiency's as high as 85%.

Remark A possible recovery of the energy that is extracted by the turbine would not change the turbine efficiency but would improve the refrigerator efficiency.

6.6.3.6 Power of Cryogenic Expansion Turbines

Turbines that are fitted with gas static bearing systems can extract up to 200 kW (Fig. 6.127). Formerly such a high power was only possible with oil static bearings (a technology that is presently almost not anymore used in helium refrigeration).

Magnetic bearings are, presently (2020), generally not yet used for helium expansion. When the request for more powerful turbines comes, there are chances that this technology will be used.

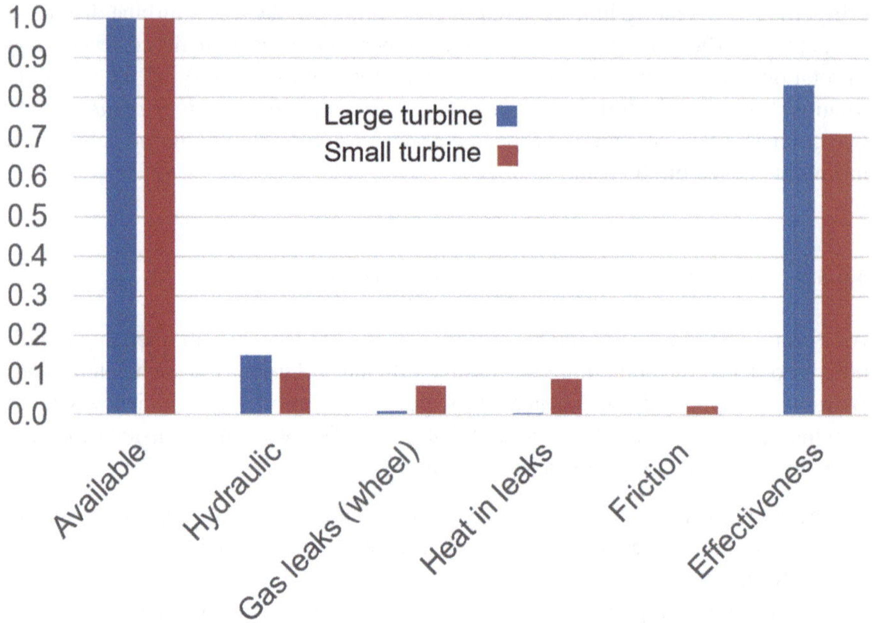

Fig. 6.126 Various losses affecting the turbine efficiency

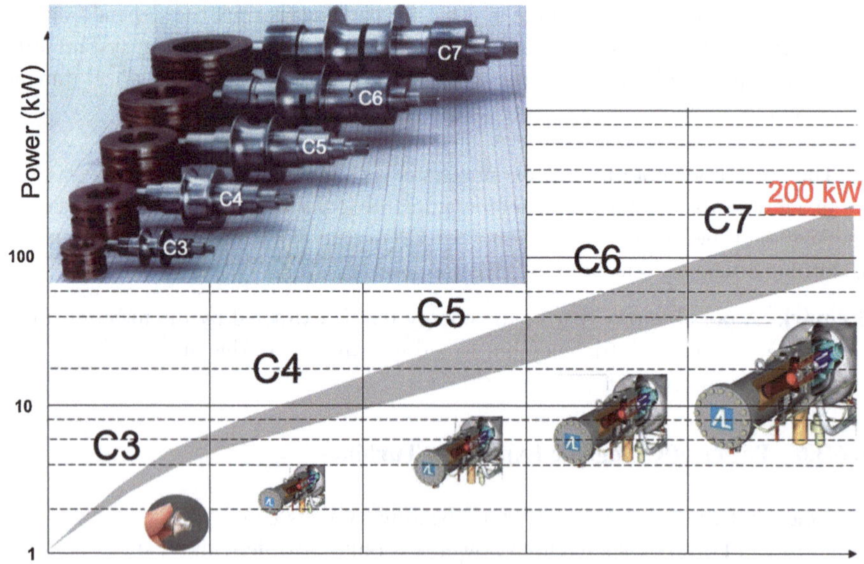

Fig. 6.127 The range of cryogenic expansion turbines from Air Liquide (copyright Air Liquide)

6.6.3.7 The Cool-Down Procedure of an Expansion Turbine

A turbine is obviously designed for its nominal operation, but it has also to operate in very different conditions, especially at the time of the system cool-down. One should not forget that measures have to be taken against possible generation of solid impurities in the turbine during cool-down!

In order to ease the explanations, the distributor is assumed to operate in a sonic regime.

6.6.3.7.1 How Does the Enthalpy Drop of a Turbine Vary According to the Inlet Temperature?

Let us perform ideal turbine (efficiency =1) calculation with various inlet temperatures, keeping same inlet and discharge pressures in Table 6.9.

One can see that the *enthalpy drop varies proportionally according to the temperature*, except at low temperatures where helium properties are affected by the proximity of liquid (Fig. 6.128). This is a thumb rule to keep in mind.

6.6.3.7.2 How Does the Processed Flow of a Turbine Vary According to the Inlet Temperature?

In paragraph Sect. 6.6.3.1.1, it has been shown that the cross section of the distributor calibrates the turbine mass flow rate.

When the distributor operates in a sonic regime, its operation can be simulated with its flow coefficient[1]:

$$fc = \dot{m} \times \frac{\sqrt{T_{\text{in}}}}{P_{\text{in}}} \tag{6.13}$$

Calculation in Table 6.10 shows that, for a given distributor cross section (which has same flow coefficient) and constant inlet pressure, the processed mass flow rate decreases from its nominal value, when the temperature increases.

6.6.3.7.3 How Does the Power of a Turbine Vary with Temperature?

Remember that the power of a turbine is the product of the processed mass flow rate times the enthalpy drop (Sect. 6.6.1).

[1]The flow coefficient is strictly $fc = \dot{m} \times \frac{\sqrt{Z \times T_{\text{in}}}}{P_{\text{in}}}$ Z being the compressibility factor of the gas. In the book, the incidence of the compressibility factor is neglected for simplification. Z influences the flow coefficient at temperatures that are near the liquid region.

Table 6.9 Calculation of the enthalpy drop of a turbine (HEPAK)

P	T	h	ΔT	Δh_{real}	P	T	h	ΔT	Δh_{real}	P	T	h	ΔT	Δh_{real}
(bar)	(K)	(J/g)	(K)	(J/g)	(bar)	(K)	(J/g)	(K)	(J/g)	(bar)	(K)	(J/g)	(K)	(J/g)
5.00	***300.00***	1574.81	−130.47	−678.77	***10.00***	300.00	1576.42	−171.50	−893.48	***20.00***	300.00	1579.65	−202.59	−1058.17
1.00					1.00					1.00				
1.20	169.53	896.04			1.20	128.50	682.94			1.20	97.41	521.48		
5.00	200.00	1055.52	−86.97	−452.93	10.00	200.00	1057.15	−114.32	−596.64	20.00	200.00	1060.40	−135.04	−707.55
1.00					1.00					1.00				
1.20	113.03	602.58			1.20	85.68	460.51			1.20	64.96	352.86		
5.00	100.00	536.00	−43.48	−227.03	10.00	100.00	537.42	−57.15	−299.58	20.00	100.00	540.25	−67.51	−356.37
1.00					1.00					1.00				
1.20	56.52	308.97			1.20	42.85	237.84			1.20	32.49	183.88		
5.00	20.00	116.53	−8.75	−44.52	10.00	20.00	114.18	−11.52	−57.42	20.00	20.00	110.52	−13.61	−65.83
1.00					1.00					1.00				
1.20	11.25	72.01			1.20	8.48	56.75			1.20	6.39	44.69		
5.00	***10.00***	58.47	−4.44	−18.98	10.00	10.00	50.42	−5.59	−20.63	20.00	10.00	44.88	−5.59	−22.66
1.00					1.00					1.00				
1.20	5.56	39.49			1.20	4.41	29.79			1.20	4.41	22.22		

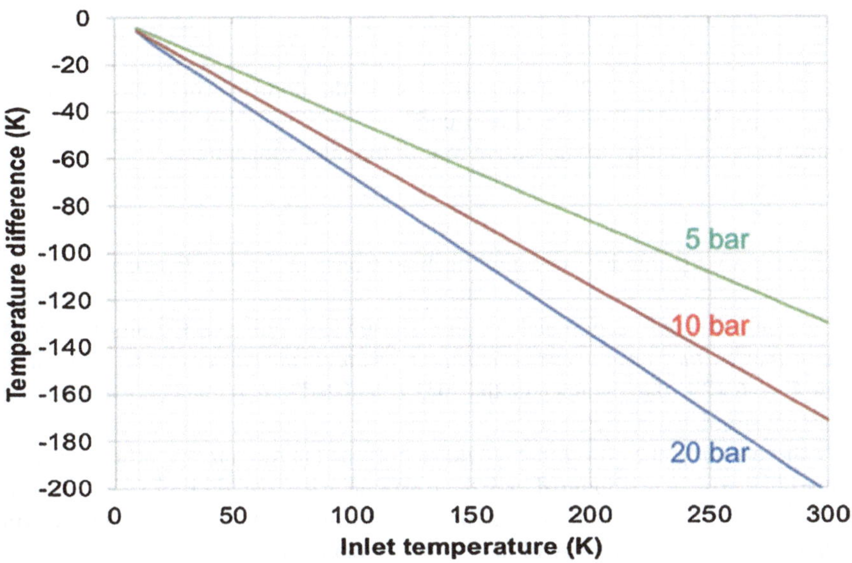

Fig. 6.128 Ideal turbine temperature difference versus inlet temperature

Table 6.10 Variation of mass flow rate versus temperature for same flow coefficient

P (bar)	*T* (K)	*h* (J/g)	$\dot{m}$ (g/s)	Flow coefficient (–)
5.00	***300.00***	1574.81	***25.82***	89.44
0.75				
1.20	202.21	1065.73		
5.00	***200.00***	1055.52	***31.62***	89.44
0.75				
1.20	134.83	715.82		
5.00	***100.00***	536.00	***44.72***	89.44
0.75				
1.20	67.44	365.73		
5.00	***60.00***	327.75	***57.74***	89.44
0.75				
1.20	40.46	225.39		
Design conditions				
5.00	***20.00***	*116.53*	***100.00***	*89.44*
0.75				
1.20	*13.32*	*83.14*		
5.00	***10.00***	58.47	***141.42***	89.44
0.75				
1.20	6.32	44.24		

$$\dot{W} = \dot{m} \times \Delta h$$

Calculation in Fig. 6.129 shows also that, in the domain where the thermodynamic properties of helium are almost constant, the power that is extracted by the turbine varies according to the square root of the inlet temperature.

6.6.3.7.4 An Assumption for Simplification: Constant Turbine Efficiency

For reasons of simplification and for an easier understanding of the process, let us assume that the *turbine efficiency is constant, whatever the operating conditions are*.

In Fig. 6.129, the parameters that are calculated are plotted along with the rotational speed.

In this example, the power that would be extracted at room temperature would be 4 times higher than the nominal one, the rotational speed would be 1.6 time higher, and the enthalpy drop ratio would be 16 times: obviously, the turbine cannot perform such a heavy duty so far away of its nominal operating conditions!

An easy way to stay into comfortable operating conditions is to cool the turbine at a *constant power*, which means at constant rotational speed (remember that the power that is extracted by the brake compressor varies as the third power of the speed; see Sect. 6.6.3.3).

Consequently, the turbine power must be controlled at its maximum. For a static bearing system where the brake compressor pressure is constant, such condition can be achieved by reducing the inlet pressure by throttling the inlet valve. Operating conditions are shown in Table 6.11 and Fig. 6.130.

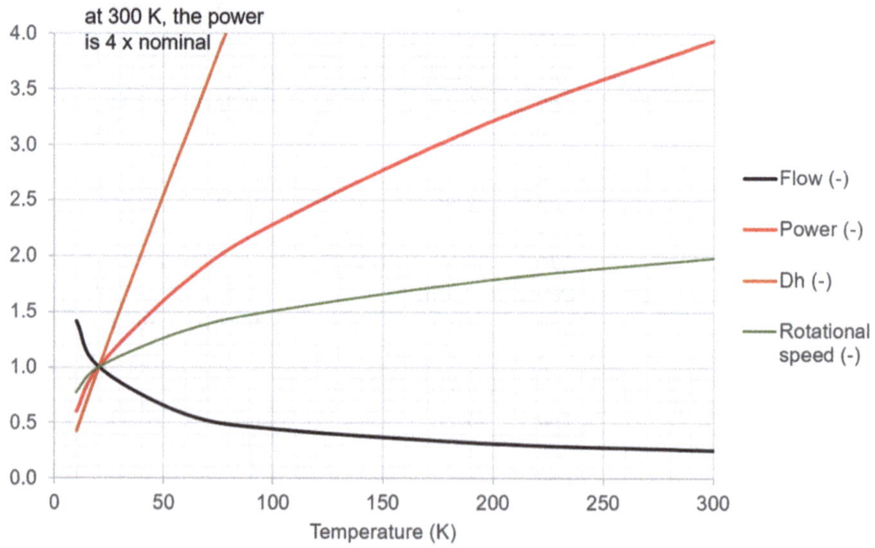

Fig. 6.129 Changing conditions during cool-down

Table 6.11 Turbine cool-down at constant power, brake compressor pressure and efficiency

P	*T*	*h*	$\dot{m}$	Flow coefficient	Δh real	Power
(bar)	(K)	(J/g)	(g/s)	(–)	(J/g)	(W)
2.35	***300.00***	1573.95	***12.13***	89.44		
0.75						
1.20	247.04	1298.55			–275.40	–3339
2.57	***200.00***	1054.73	***16.28***	89.44		
0.75						
1.20	160.58	849.55			–205.17	–3339
Design conditions						
5.00	***20.00***	*116.53*	***100.00***	*89.44*		
0.75						
1.20	*13.32*	*83/14*			*–33.39*	*–3339*
5.00	***10.00***	58.47	***141.42***	89.44		
0.75						
1.20	6.32	44.24			–14.23	–2013

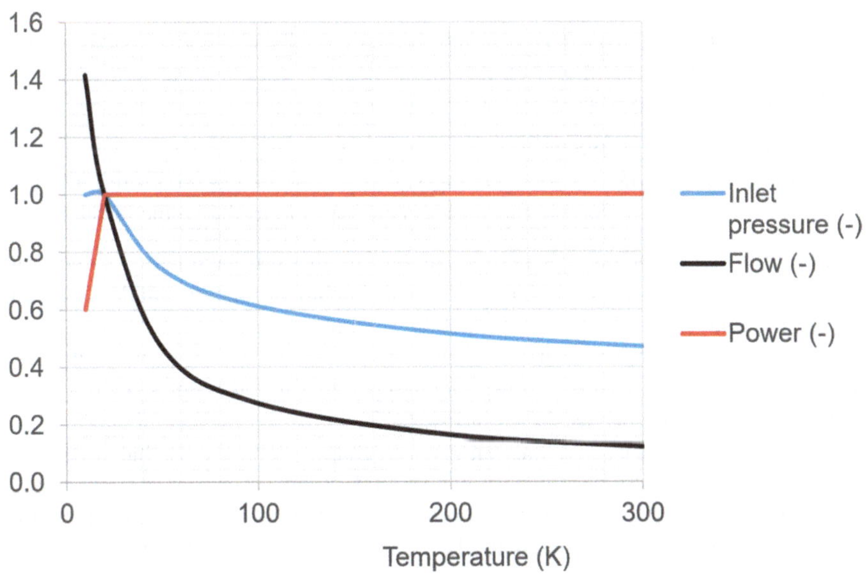

Fig. 6.130 Cool-down at constant power, brake compressor pressure and efficiency

6.6.3.7.5 A Realistic Turbine Cool-Down: The Efficiency Is Not Constant

Up to now, we have assumed a simple, but unrealistic, behaviour of the turbine: a constant efficiency whatever the operating conditions are. Obviously, this cannot be right: the efficiency changes with the turbine operating conditions because there are possibilities of re-circulation, velocity mismatch at the inlet of the wheel and others.

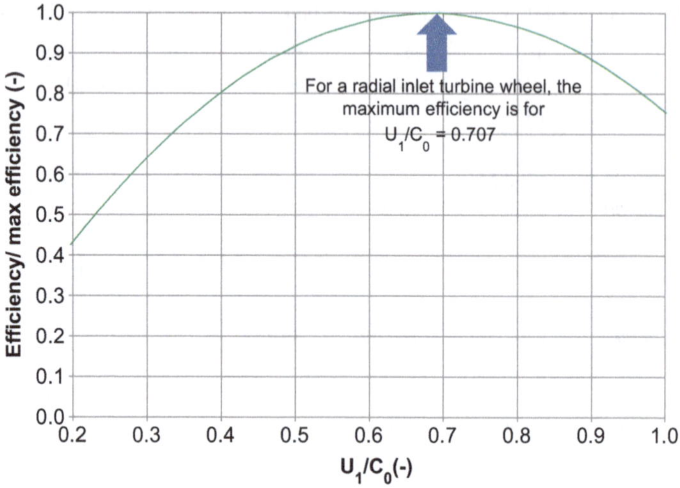

Fig. 6.131 Typical relative efficiency of a radial inlet expansion turbine

Turbo machinery specialists are used to indicate the efficiency change of an expansion turbine as shown in Fig. 6.131.

Turbo machinery specialists use the ratio: U_1/C_0.

- U_1 is the tip speed of the wheel.
- C_0 is the *theoretical velocity* (spouting velocity) that the gas would reach at the outlet of a distributor having the same expansion ratio as the whole turbine.

For a non-turbo machine specialist (as most of us are), this parameter is somewhat mysterious! Let us try to understand the physical behaviour that is "hidden" behind such a formula.

Consider what happens at the discharge side of the (sonic) distributor: the gas velocity is related to the square root of the temperature, but it keeps same direction (given by the distributor), as shown in Fig. 6.132.

At 300 K, the velocity combination gives an entering velocity W_1 that is very far from being tangential to the wheel wings. At 200 K, such a discrepancy is not as bad as for 300 K. At 100 K the situation improves, and at the nominal inlet temperature 20 K, W_1 is perfectly tangential to the wing. For all temperatures higher than nominal, the turbine wheel does not rotate fast enough.

However, if the inlet temperature is lower than nominal, we can see that W_1 is tilted in the other direction. In this case, the rotational speed of the turbine is too high.

Now, it is easier to understand the meaning of the U_1/C_0 diagram (see Fig. 6.133)!

Remark This is another way to understand how the fine-tuning of the efficiency of a turbine is done, as described in Sect. 6.6.3.4.

The cool-down procedure of a turbine can be performed in various ways according to the way the brake compressor is controlled.

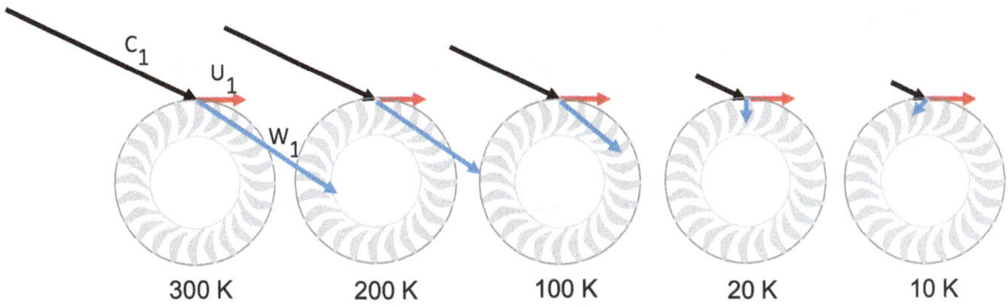

Fig. 6.132 How helium enters the wheel according to the inlet temperature

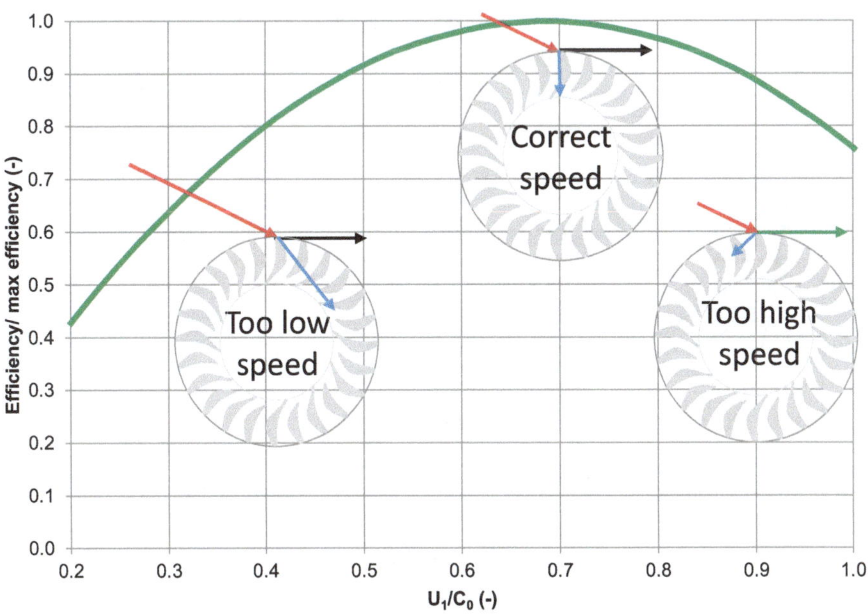

Fig. 6.133 An easier understanding of the U_1/C_0 curve

6.6.3.7.6 Cool-Down Procedure at a Constant Brake Pressure, Non-constant Efficiency

As above, the constant power operation is performed by controlling the rotational speed (see Sect. 6.6.3.1.5) by throttling the inlet valve.

The calculation in Table 6.12 shows that the inlet pressure ranges from 3.63 bar at room temperature to the 5-bar nominal inlet pressure and, simultaneously, the mass flow rate increases from 9.37 g/s at 300 K to 50 g/s at nominal (Fig. 6.134).

Procedure (see Fig. 6.135)

Assumption: the brake pressure is constant.

Table 6.12 Turbine cool-down at constant power, brake compressor pressure. Efficiency according to temperature

P	T	h	$\dot{m}$	Δh real	Power
(bar)	(K)	(J/g)	(g/s)	(J/g)	(W)
3.63	***300.00***	1574.36	***9.37***		
0.32					
1.30	267.27	1403.66		170.70	1600
3.61	***200.00***	1055.06	***11.41***		
0.40					
1.30	173.14	914.84		140.22	1600
Design					
5.00	*20.00*	*116.53*	***50.00***		
0.75					
1.30	*13.61*	*84.53*		*32.00*	*1599.9*
5.00	15.00	88.67	***57.74***		
0.71					
1.30	10.29	66.62		22.05	1272.9

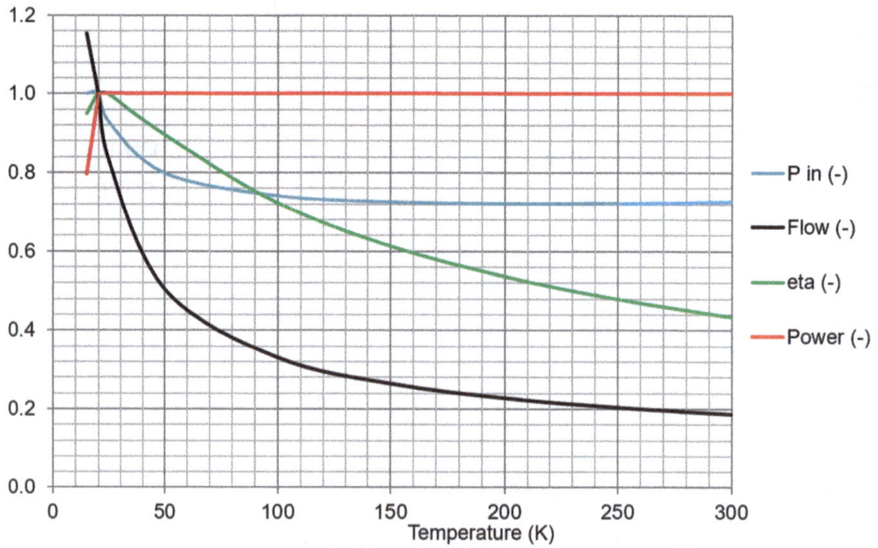

Fig. 6.134 Change during cool-down at constant power and brake compressor pressure. Efficiency according to temperature

The “active” component is the inlet valve.

In phase 0, the turbine is stopped, but due to gas viscosity, helium flowing into the bearings runs the shaft at low speed (a few tens hertz).

In phase 1, the starting procedure begins by opening the inlet valve according to a ramp until the rotational speed reaches a given threshold (red dot).

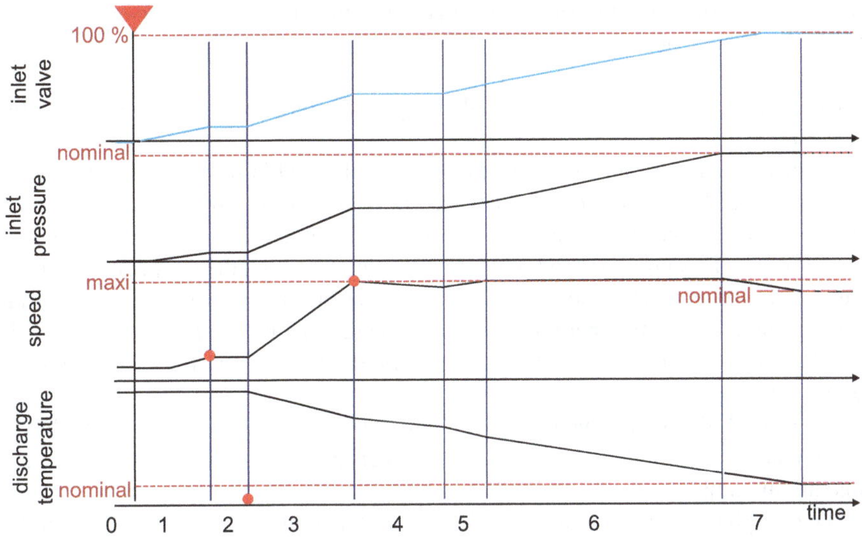

Fig. 6.135 Cool-down procedure at constant speed and constant brake compressor pressure

In phase 2, the position of the inlet valve is kept constant. If the turbine were starting from room temperature, this "waiting" phase is of no use, but if the cold box was already cold or partially cold, this time period allows the inlet temperature to reach a stable value (piping and valves warm up generally faster than the heat exchanger blocks).

In phase 3, when the time period is elapsed, the inlet valve opens again according to a ramp until the rotational speed reaches the maximum.

In phase 4, as the position of the inlet valve is constant, the turbine inlet pressure is also constant. But, as the turbine is now "working", its temperatures decreases. Consequently, its enthalpy drop decreases and similarly its power decreases. As the rotational speed is related to the power (see Sect. 6.6.3.1.5), it decreases too.

In phase 5, it is therefore possible to open the inlet valve in order to reach again the turbine maximum rotational speed.

By the way, this is easily performed by the turbine speed control loop that controls the rotational speed of the turbine at its maximum all along the cool-down procedure in phase 6.

In phase 7, when the inlet valve is almost fully open, its pressure drop becomes negligible, and the turbine inlet pressure equals that of the cycle high pressure. Consequently, as the turbine operating temperature still decreases, its rotational speed decreases too. When the inlet valve is fully opened, the inlet pressure does not change anymore. Both the turbine inlet temperature and rotational speed reach their nominal values.

6.6.3.7.7 Cool-Down Procedure at a Variable Brake Pressure, Non-constant Efficiency

Assumption: the brake pressure is floating, according to the turbine wheel tip pressure.

In phase 0, the turbine is stopped; its rotational speed is zero (Fig. 6.136).

In phase 1, the inlet valve is opened according to a ramp until it reaches 80% of its maximum rotational speed. At that time, the brake valve that controls the rotational speed opens. This phase lasts about 5 minutes during which the temperatures of the turbine decrease slowly and stabilise in case the turbine is being re-started at an intermediate temperature between room and operational temperatures.

The "active" components are, both and *simultaneously*, the inlet valve and the brake valve.

In phase 3, the inlet valve is opened up to 100% according to a ramp. The brake valve reacts accordingly (it opens more) in order to keep the rotational speed at the maximum value.

During phase 4, the systems cool down. Consequently, for same inlet and outlet pressures and because the temperature decreases, the enthalpy drop decreases; therefore, simultaneously, the processed mass flow rate increases. To keep a constant rotational speed, the brake compressor valve closes. Same situation occurs until the turbine reaches its operating conditions in phase 5.

As it can be seen in Fig. 6.137, the extracted power during cool-down is higher when the brake compressor is controlled by its processed flow.

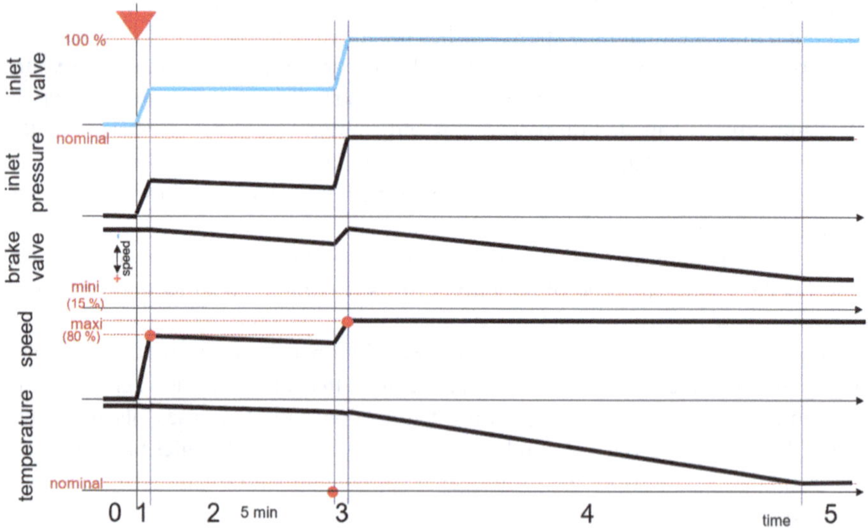

Fig. 6.136 Cool-down procedure at constant speed and variable brake compressor pressure. (© Linde Kryotechnik AG)

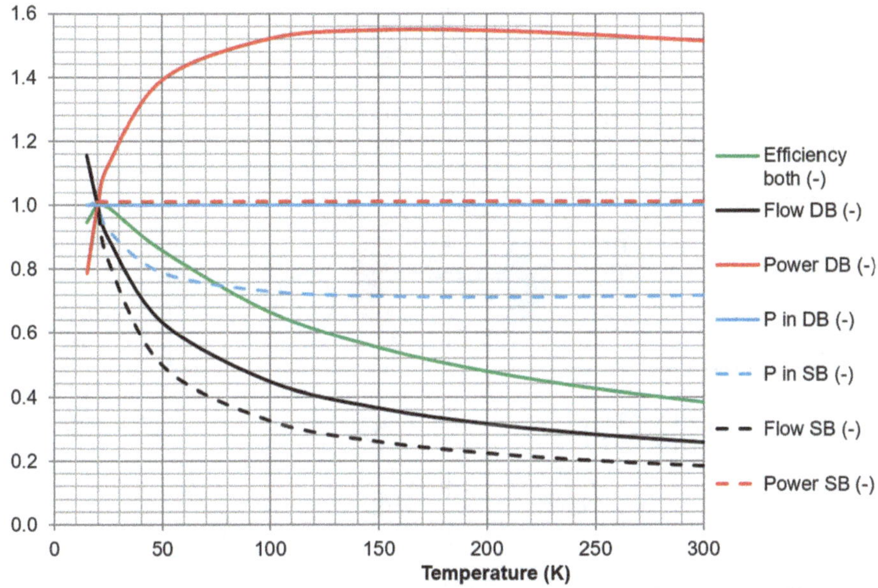

Fig. 6.137 Comparing behaviours during cool-down, according to bearing technology. (*SB* static bearings, *DB* dynamic bearings)

6.6.3.8 Special Operation of Some Turbines

In Sect. 4.2.1.1, turbines expanding directly in the two-phase region have been used.

Such an expansion can be performed by either a reciprocating expander or a turbine. The limit of size of a reciprocating expander brings a limitation towards high mass flow rates. Furthermore, such machines do not have the high reliability that is required for large refrigerators. Such a solution has been used formerly because, for not very clear reasons, it was believed that a cryogenic expansion turbine could not discharge in two-phase. By the way, it is possible to operate a cryogenic expansion turbine in conditions where liquid appears at the discharge side.

An example of such an arrangement is shown in the flow diagram of Fig. 12.18 for turbine T8.

6.6.4 Monitoring of a Turbine

These are two possible monitoring ways for turbines.

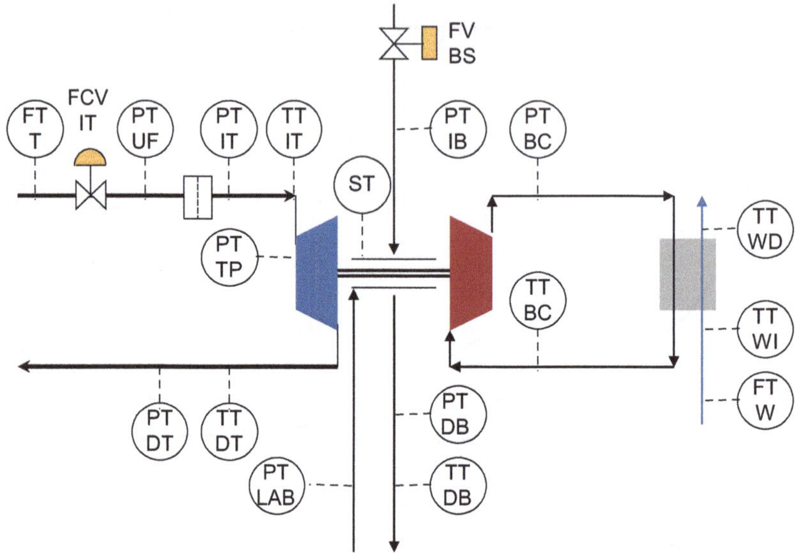

Fig. 6.138 Monitoring of a static bearing turbine

6.6.4.1 A Gas Static Bearing Turbine

In Fig. 6.138, and Table 6.13, an example of the parameters to be monitored is shown. The low-switch and high-switch values should be arranged in two levels: first an alarm and then a stop action.

6.6.4.2 A Gas Dynamic Bearing Turbine

In Fig. 6.139, and Table 6.14, an example of the parameters to be monitored is shown. The low-switch and high-switch values should be arranged in two levels: first an alarm, then a stop action.

6.6.5 *Turbine Digest*

- The processed mass flow rate of a cryogenic expansion turbine is set by the cross section of the distributor, the inlet pressure and temperature when the gas flows in a sonic regime in the distributor (pressure ratio through the distributor >2). If the pressure ratio is <2 (not a sonic regime), the distributor discharge pressure is also to be considered. There is no direct incidence of the rotational speed on the processed mass flow rate.

Table 6.13 Parameters to be monitored for a static bearing turbine

Sensor	Low switch	High switch	Duty
FT IT		X	Helium inlet pressure
TT IT	X		Helium inlet temperature
PT DT		X	Helium discharge pressure
TT DT		X	Helium discharge temperature
FT			Helium mass flow rate (range)
FCV IT	X		Inlet valve position
PT UF			Helium pressure inlet filter
PT UF - PT IT		X	Filter pressure difference
ST		X	Turbine rotational speed
PT TP	X		Wheel tip pressure
PT IB	X		Bearing feed pressure
PT DB		X	Bearing discharge pressure
TT DB		X	Bearing discharge temperature
PT LAB	X		Helium pressure in labyrinth
TT BC		X	Helium temperature in brake circuit
PT BC			Helium pressure in brake circuit
PT TP - PT BC	X	X	Trust bearing load
TTWI		X	Water temperature inlet of after cooler
TT WD		X	Water temperature discharge of after cooler
FT W	X		Water flow rate brake cooler

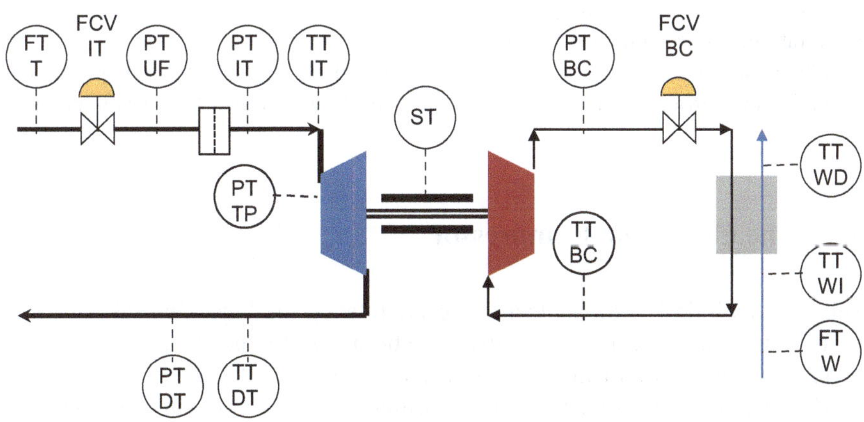

Fig. 6.139 Monitoring of a dynamic bearing turbine

- The power that is extracted by a cryogenic expansion turbine is the product of the processed mass flow rate times the actual enthalpy drop (or the isentropic enthalpy drop times the efficiency).
- For constant inlet and discharge pressures,

Table 6.14 Parameters to be monitored for a dynamic bearing turbine

Sensor	Low switch	High switch	Duty
PT IT		X	Helium inlet pressure
TT IT	X		Helium inlet temperature
PT DT		X	Helium discharge pressure
TT DT		X	Helium discharge temperature
FT			Helium mass flow rate (range)
FCV IT	X		Inlet valve position
PT UF			Helium pressure inlet filter
PT UF - PT IT		X	Filter pressure difference
ST		X	Turbine rotational speed
FCV BC			Brake compressor flow valve
PT TP	X		Wheel tip pressure
TT BC		X	Helium temperature in brake circuit
PT BC			Helium pressure in brake circuit
TT WI		X	Water temperature inlet of after cooler
TT WD		X	Water temperature discharge of after cooler
FT W	X		Water flow rate brake cooler

  - The flow varies with the reverse of the square root of the inlet temperature.
  - The enthalpy drop varies approximately according to the inlet temperature.

- Control of the rotational speed is made by varying the pressure or the mass flow rate of the brake compressor.
- Rotational speed must be optimised.
- When turbines of same efficiency are arranged in series, the efficiency of the equivalent turbine is higher than that of each of the turbine of the arrangement.

6.7 The Cryogenic Compressor

This paragraph deals with another cryogenic rotating machine. General principles are obviously the same as the ones that have been used for the cryogenic expansion turbine; therefore, the structure of this chapter is rather similar.

The operating range of a cryogenic centrifugal compressor or circulator is not so large as that of a cryogenic expansion turbine. Prior to operation, it is cooled at almost its nominal operating temperature.

6.7.1 Thermodynamic Aspects of a Cryogenic Compressor

On a T-s diagram, the cryogenic compressor operation can be shown as in Fig. 6.140, left: at point 1, gas at (low) pressure P_1 and temperature T_1 is compressed up to (high) pressure P_2. On the T-s diagram, point 1 is located on the P_1 isobar, at T_1 temperature. An isentropic compression would follow a vertical line until it crosses the P_2 isobar at point h_{is}. The *isentropic* enthalpy gain is h_{is}–h_1. Such an ideal compressor would reach the T_{is} discharge temperature.

As the actual compressor is not ideal, the enthalpy that it transfers into the gas is higher than the isentropic one; consequently, its discharge enthalpy and temperature point that is located on the P_2 isobar, at point 2, is at higher both enthalpy and temperature than h_{is} and T_{is}.

The isentropic efficiency is the ratio of the theoretical enthalpy gain (isentropic) versus the actual enthalpy gain:

$$\eta_{is} = (h_{is} - h_1)/(h_2 - h_1) \tag{6.14}$$

The actual power absorbed by a compressor is:

$$\dot{W} = \dot{m} \times \eta_{is} \times (h_1 - h_{is}) \tag{6.15}$$

$$\dot{W} = \dot{m} \times (h_1 - h_2) \tag{6.16}$$

$$\dot{W} = \dot{m} \times \Delta h \tag{6.17}$$

Such power is provided by the motor (see Fig. 6.140, right).

One of the tools of the Cryo toolbox (see Sect. 15.8.5) is the cryogenic compressor spread sheet shown in Fig. 6.140, bottom.

As for the turbine, this spread sheet can be used to predict the discharge temperature (here, 4.96 K), but also, by using the "Excel goal seek", one can evaluate the isentropic efficiency of a compressor the operating conditions of which are known. Obviously, such figure accuracy is linked to the measurement accuracy, especially temperatures. With such a tool, the operator can compare various compressor operations provided that, within a small range, the inaccuracy of the sensors is not changed.

6.7.2 Various Kinds of Cryogenic Compressors

The first attempts of cryogenic compressors were reciprocating machines. Only a few machines were built, and the reciprocating volumetric cryogenic compressor was soon abandoned in favour of the centrifugal compressor.

As for the cryogenic expansion turbines, one must have a bearing unit capable of running at rather high speeds. The technologies that are currently the most often used

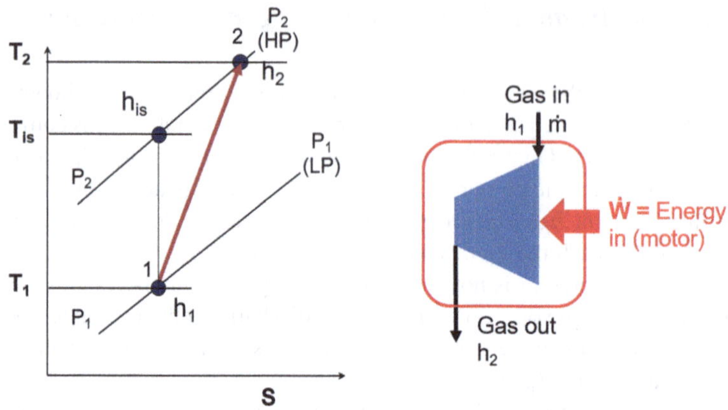

P	T	h	ṁ	Power	Δh real	Δh isent.
(bar)	(K)	(J/g)	(g/s)	(W)	(J/g)	(J/g)
0.016	4.60	28.87	150.00			
0.70						
0.056	8.87	51.00		3320	22.14	31.62

Fig. 6.140 A compressor on the T-s diagram

are active magnetic bearings, ball bearings and foil bearings (see paragraph 6.5.1), operating at room temperature. Foil bearing, in addition to their low stiffness, cannot operate at reduced speed, a necessary feature to allow smooth and safe cool-down of the machine. In former times, some magnetic bearings have been operated at about 100 K. The author is really in favour of active magnetic bearing systems; even their cost, which is slowly decreasing, might be higher.

The compression ratio generally obtained with one stage is around three, so it is necessary to have several stages connected in series to achieve the high compression ratios that are required (see Sect. 4.3.1.1) to reach temperatures of 2.0 or even 1.8 K.

6.7.3 Centrifugal Compressors

Cryogenic centrifugal compressors must, such as compressors operating at room temperature, be kept in their stable operation domain. Many very good books deal with centrifugal compressors, the majority of them operating at temperatures near or above room temperature. The following paragraphs are intended to "visit" the centrifugal compressors with the minimum material that is necessary to understand the specific operation of cryogenic machines.

After having gone through the reminders on conventional centrifugal compressors, specificities of cryogenic centrifugal compressors are discussed.

6.7.3.1 Operating Principle of a Centrifugal Compressor

The general operating principle of a centrifugal compressor is:

- Imparting *kinetic* energy to the fluid in an impeller
- Converting this energy into *pressure* energy by controlled decreasing velocity

Therefore, a centrifugal compressor operates in three steps:

- Some energy of pressure is converted into kinetic energy.
- Mechanical energy is transferred to the gas as kinetic energy.
- The kinetic energy is converted into pressure.

Consequently, a centrifugal compressor is made out of three components, each of them fulfilling the specific duty of one of the steps indicated here above:

- The **convergent** converts some gas pressure energy into gas velocity (kinetic energy).
- The **wheel** transfers mechanical energy to the gas in order to increase its velocity (kinetic energy).
- The **diffuser** converts the high gas velocity at the discharge of the wheel into almost static pressure.

6.7.3.1.1 The Convergent

The convergent is a fixed nozzle. It gives the right velocity to the gas in order that it enters correctly into the wheel, tangentially to the wing (or blade), as shown in Fig. 6.141. As velocity is increased, according to Bernoulli's theorem, the pressure and temperature decrease.

6.7.3.1.2 The Wheel

The wheel (or impeller) imparts velocity to the gas with wings that are attached to a rotating disk (see Fig. 6.143, top). It is a rotating nozzle that converts mechanical energy into kinetic energy. The gas, which is submitted to centrifugal forces, circulates from the centre to the periphery (Fig. 6.142).

The enthalpy gain:

$$\Delta h = 0.5 \times \left(U_2^2 - U_1^2\right)$$

(with U_1 inlet gas velocity at the inlet of the wheel) varies with the square of the tip speed U_2, so a high enthalpy gain means a high U_2. Due to material properties, the higher limit of U_2 is, as for expansion turbines, around 450 m/s; therefore there is a limitation in the compression ratio for one wheel (or one stage).

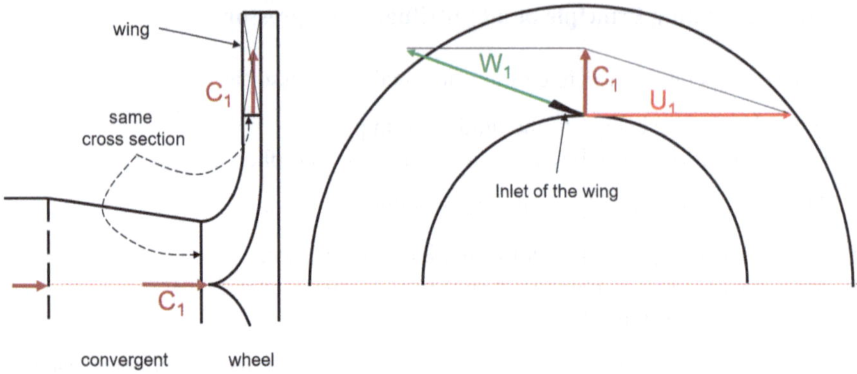

Fig. 6.141 Velocity composition at inlet of the wheel, for design conditions, showing the duty of the convergent

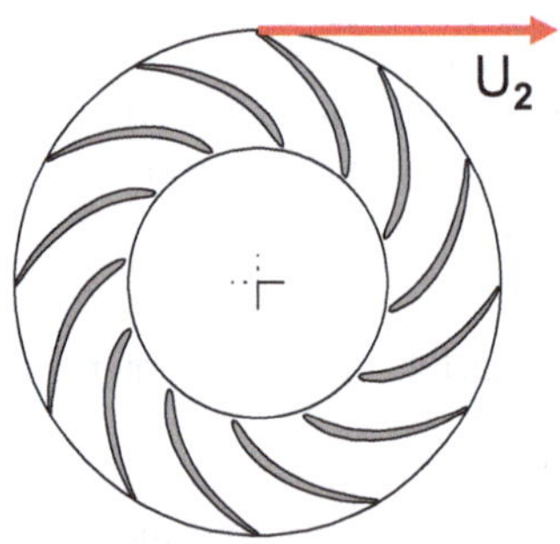

Fig. 6.142 Tip velocity of the wheel

Various Types of Wheels

As for turbines, the suction conditions govern the inlet diameter of the wheel. It is of interest to arrange the wheel in an axial-radial configuration (see Fig. 6.143 right) in order to have sufficiently long wings for a limited outer diameter.

6.7.3.1.3 The Diffuser and the Scroll

The diffuser is a fixed nozzle that converts kinetic energy in static pressure by gradually decreasing the gas velocity. It can be either vaneless or vanned (see Fig. 6.144 left). A vanned diffuser has a higher efficiency, but its operational range is reduced, while a vaneless diffuser has a lower efficiency but over a larger operational range (Fig. 6.144 right). The scroll is a volute casing of increasing cross section that collects the gas to the discharge pipe.

Fig. 6.143 Wheels: top, a pure radial wheel and its diffuser Tore Supra, (copyright Air Liquide), bottom, an axial-radial wheel (LHC, courtesy PBS)

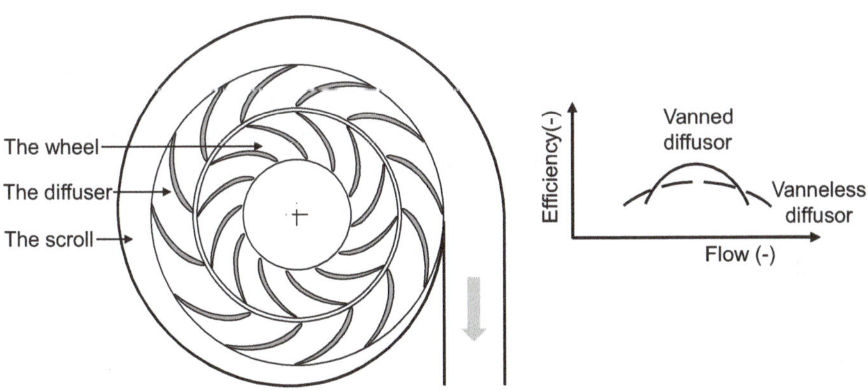

Fig. 6.144 The wheel, the diffuser and the scroll

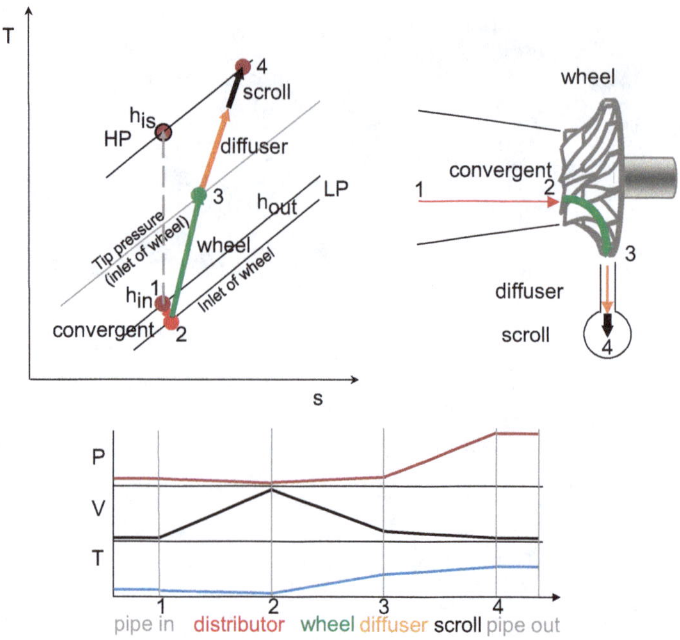

Fig. 6.145 Evolution of the gas

6.7.3.1.4 Evolution of the Gas Through the Centrifugal Compressor

In Fig. 6.145, one can follow the evolution of gas through the compressor, simultaneously in the T-s diagram (left), the compressor sketch (right) and another diagram showing the evolution of pressure, gas velocity and temperature (bottom):

- From point 1 to 2, into the convergent, the pressure decreases slightly, the gas velocity increases slightly, and the temperature decreases slightly too.
- From point 2 to 3, into the wheel, the pressure increases, the gas velocity increases, and the temperature decreases.
- From point 3 to 4, into the diffuser and the scroll, the gas velocity decreases, and the pressure and the temperature increase.

6.7.3.2 The Structure of a Cryogenic Centrifugal Compressor

The structure of a cryogenic centrifugal compressor is similar to that of an expansion turbine: a bearing system supports a shaft on which a wheel is attached at one end. The electric motor is located in between the journal bearings. The bearing system is, now, always operating at room temperature. The separated assembly that is shown in Fig. 6.146, right, incorporates the bearing system and the impeller. The convergent (inlet) and diffuser (discharge) if any are located in the fixed part. Such a structure allows an easy assembly and disassembly without breaking the insulation vacuum.

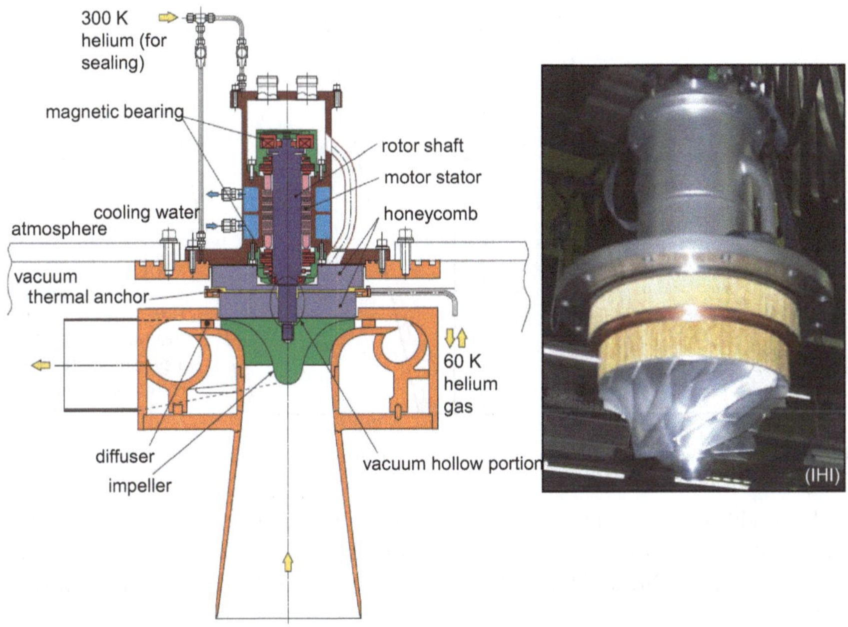

Fig. 6.146 A cryogenic centrifugal compressor by IHI. (Quoted from IHI Engineering review)

6.7.3.3 The Operational Limits of a Centrifugal Compressor

The centrifugal compressor has to be operated within very strict borders, inside its operating domain. Let us try to understand why.

6.7.3.3.1 Low Flow Rates: The Surge Situation

The centrifugal compressor is a *dynamic* compressor. Energy is transferred to the gas in the form of *velocity*.

The surge situation is not trivial to explain. Let us try an explanation on Fig. 6.147. The sketch at left represents a centrifugal compressor that feeds a capacity that is emptied through the valve. The pressure into the volume is the consequence of the equilibrium between the compressor flow and the valve flow.

Let us make an analogy: a ball (representing some gas) in Fig. 6.147, 1, located at the bottom of a slope, is launched along the slope with an initial velocity V (it is the duty of the wheel that imparts a velocity to the gas). According to its velocity, the ball can reach a maximum height (or pressure difference). In conditions 1, at centre, the velocity of the ball allows it to reach the maximum height and makes possible for it to cross the threshold. Point 1 is plotted on the compressor map (see Sect. 6.7.3.3.2), at right of Fig. 6.147.

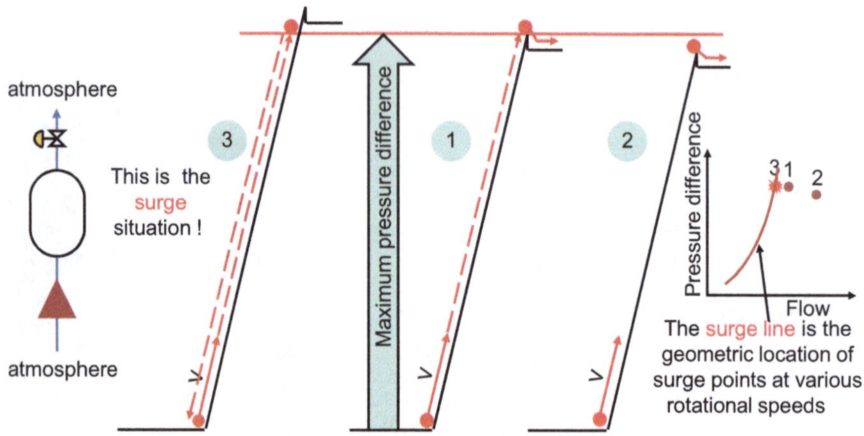

Fig. 6.147 An analogy for the surge situation of a centrifugal compressor

In condition 2, the valve is slightly opened: the compressor discharge pressure equilibrates at a lower value. It is like is the threshold would be lower. The ball crosses easily the threshold and for same amount of energy, a higher mass flow rate is processed.

Now, let us slightly close the valve: for a same flow processed by the compressor, the new equilibrium pressure would increase. In the compressor analogy, it is as if the slope would have a higher threshold, as shown in condition 3. Obviously, the ball, with the same initial velocity, cannot cross the threshold; it comes back to the bottom of the slope; therefore, the processed flow comes to zero. This is the *surge* situation. In the compressor map, at right of Fig. 6.147, this point is on the surge line. But as flow continues leaving the downstream capacity through the valve, the pressure in the capacity decreases until the compressor can, again, process gas. A cyclic situation establishes, called *pumping*, the frequency of which depends on the processed flow and the volume of the downstream capacity. Prolonged operation in this unstable mode can cause serious mechanical damage, especially to the bearing system. When operating in a surge condition, the compressor discharge temperature increases significantly, and the compressor experiences erratic and severe vibrations. A compressor can be brought out of surge by increasing either the flow (by-pass opening) or the rotational speed.

There is a flow limitation by low values.

6.7.3.3.2 High Flow Rates: Choke Situation

When the valve, downstream the capacity in Fig. 6.147, left, is more open, the flow increases until a sonic neck establishes somewhere in the compressor (impeller or diffuser, the first that is encountered) that forbids a higher flow, whatever the discharge pressure is. This is said to be the *choke* point (or *stonewall*).

There is also a flow limitation for high values!

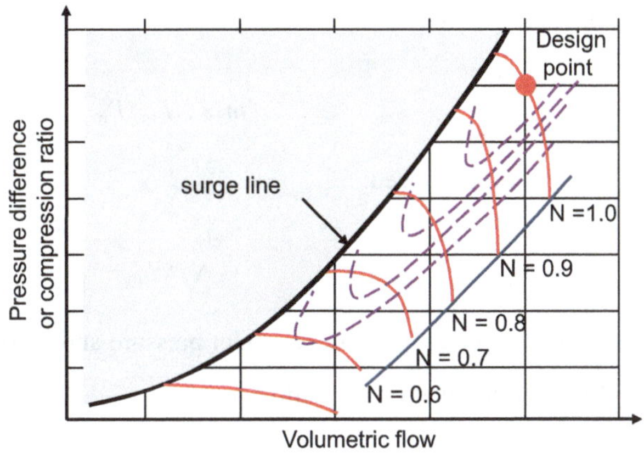

Fig. 6.148 A basic centrifugal compressor map

6.7.3.3.3 The Map of a Centrifugal Compressor

Centrifugal compressor characteristics for various operating conditions are generally displayed on a so-called compressor map, as shown in Fig. 6.148, where the pressure difference (or compression ratio) is plotted versus the compressor volume flow.

Iso-speed curves show the rise in pressure (or compression ratio) versus the volumetric suction flow. As the rotational speed increases, the pressure difference (or compression ratio) increases. Of course, the operating conditions are not limited to that shown by the few curves plotted but are continuously adjustable to any intermediate value. Each iso-speed curve has a negative slope. It ends at the point where the slope is near zero. It is not possible to operate the compressor at the left of this point (remember Sect. 6.7.3.3.1). The curve that links the point of zero slope of each iso-speed curve is called the *surge line*. Such a curve is, roughly, a parabola. The determination of the compressor's actual surge limit is a very important aspect of the manufacturer's shop testing program.

At the choke flow, the pressure versus volume iso-speed curves become vertical: and it is not possible to develop pressure at any higher flow. However, when the rotational speed decreases, the processed flow decreases, the gas velocity decreases too, and the sonic neck disappears: the iso-speed curve is less and steep towards low compression ratios.

It is of interest to plot the iso-efficiency curves (in dotted lines) that make the "efficiency hill" because it looks like level curves on an Earth map.

6.7.3.3.4 Reduced Parameters

It is of highest interest to be able to easily compare the operating conditions of various centrifugal compressors. For such a purpose, the turbo machine specialist uses laws of similitude dealing with reduced parameters:

- Dimensionless flow:

$$\dot{m} \times \sqrt{T_{\text{in}}}/P_{\text{in}} \tag{6.18}$$

- Dimensionless speed:

$$N/\sqrt{T_{\text{in}}} \tag{6.19}$$

with T_{in} inlet temperature, P_{in} inlet pressure and N rotational speed and

- Normalised dimensionless flow:

$$X = \left(\dot{m} \times \sqrt{T_{\text{in}}}/P_{\text{in}}\right)/\left(\dot{m}_{\text{dis}} \times \sqrt{T_{\text{indis}}}/P_{\text{indis}}\right) \tag{6.20}$$

- Normalised dimensionless speed:

$$Nu = \left(N/\sqrt{T_{\text{in}}}\right)/\left(N_{\text{dis}}/\sqrt{T_{\text{indis}}}\right) \tag{6.21}$$

with T_{in0} design inlet temperature and P_{in0} inlet design pressure.

The normalised dimensionless compressor map (later, in the book, it is referred as the compressor map) is drawn on compression ratio versus normalised dimensionless flow X for various normalised dimensionless speeds Nu as shown in Fig. 6.149.

Centrifugal compressors are often arranged in series in order to reach a high compression ratio. The normalised dimensionless compressor map is of interest when discussing the operation of such a train of compressors.

In Fig. 6.150, where three compressors are arranged in a pure series (no heat exchanger in between them), one can see that the design volume flow at the inlet of

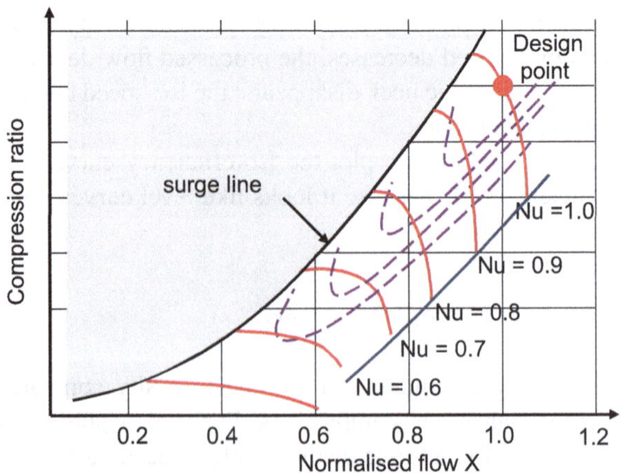

Fig. 6.149 A normalised reduced centrifugal compressor map

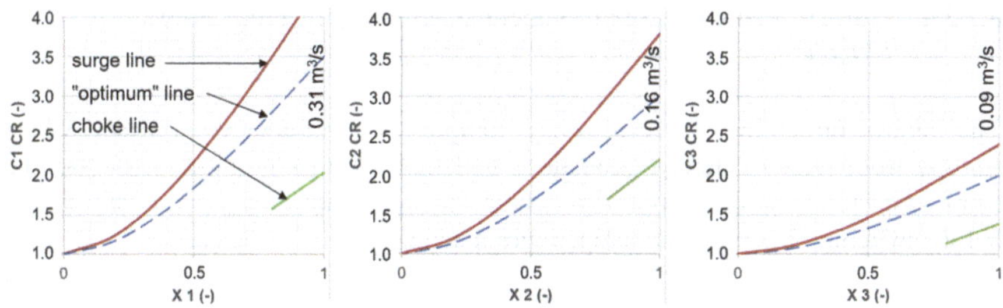

Fig. 6.150 Compressor normalised reduced maps to compare various machines

Fig. 6.151 Efficiency of an equivalent compressor

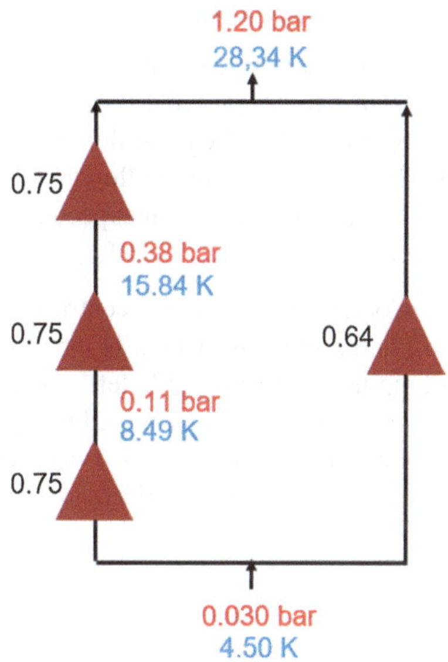

each compressor is very different (0.31 m^3/s for C_1 which is at the lowest pressure, then 0.16 and finally 0.09 m^3/s for C_2 and C_3), but using the normalised dimensionless maps allows an easy comparison between them.

A Peculiarity of Compressors Arranged in Pure Series

In Sect. 4.2.1.3, pure series arrangement has been discussed for expansion turbines. Let us compare the situation with cryogenic compressors.

In Fig. 6.151, one can see that the equivalent compressor, at right, that performs the same duty (same inlet and outlet conditions) as the three machines having each one an efficiency of 0.75, arranged in pure series at left, has an equivalent efficiency of 0.64. Contrary to turbine series arrangement (remember Sect. 4.2.1.3), the efficiency of the equivalent compressor is lower than that of the elementary one.

6.7.3.4 Control Principles for Cryogenic Centrifugal Compressors (CCC)

The control principles of a centrifugal compressor are rather complicated when starting from scratch. Here, explanations will follow a gentle progression, starting from room temperature compressor where the temperature excursions are small, before switching to cryogenic temperature machines.

6.7.3.4.1 Introduction

The control procedure of a CCC has to deal with various situations:

- Cool-down, where the suction pressure decreases (or the compression ratio rises from 1 to design value)
- Steady state, where the flow variations are small
- Turn-down, where the flow reductions are large
- Re-starting after an unexpected stop, where the inlet conditions can be heavily changed

Contrary to most of the centrifugal compressors, CCC operate either at a constant discharge pressure, which is the atmospheric pressure in case of a pure cryogenic pumping (see Fig. 4.17, left), or the suction pressure of the room temperature compressor in case of a mixed pumping (see Fig. 4.17, centre), which is conversely proportional to the processed mass flow rate. During the pump-down process, it is the suction pressure that decreases. For this reason, CCC are sometimes called “cold pumps”.

6.7.3.4.2 Pump-Down With a Room Temperature Centrifugal Compressor

Just before starting a centrifugal compressor, the flow is zero and the compression ratio is one. The target of the pump-down procedure is to reach the design point by following a path that avoids hitting either the surge line or the choke area, as shown, for example, in blue-dotted line in Fig. 6.150.

6.7.3.4.2.1 Pumping from Atmosphere

Consider an air compressor with its suction side that is connected to atmosphere (that means that, whatever the flow, the suction pressure stays constant).

It is fitted with a calibrated orifice that can process the compressor flow at design conditions (see Fig. 6.152, top right). The calibrated orifice discharges at atmospheric pressure. Rotational speed, discharge valve position, compression ratio and flow are plotted against time in Fig. 6.152, left. The corresponding positions of the operating point are shown on the compressor map, at bottom right.

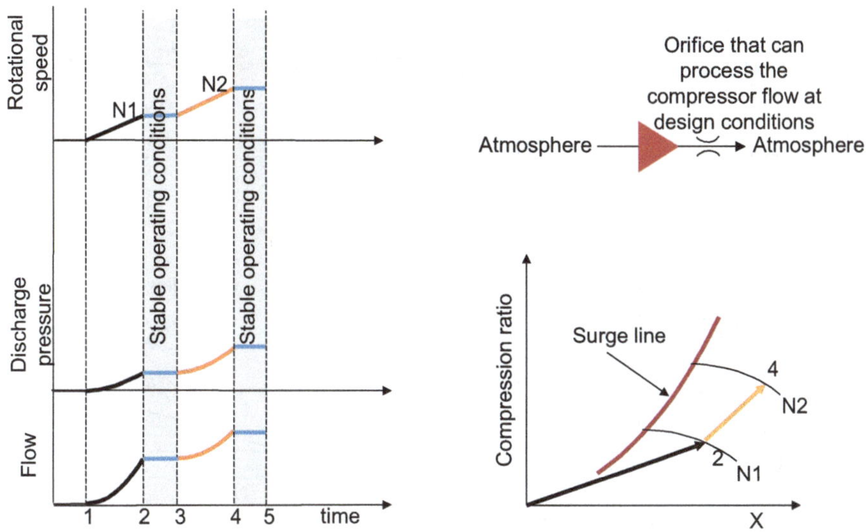

Fig. 6.152 Pumping from atmosphere through a calibrated orifice

At time 1, the rotational speed is progressively increased to N1. Consequently, the discharge pressure and the flow increase. The situation is shown at point 2, on the compressor map. It is possible to *stay indefinitely* at time 2 conditions (constant rotational speed and valve opening), because atmosphere is an infinite air reservoir that provides a constant suction pressure to the compressor, whatever the flow it processes. At time 3, the rotational speed is increased up to N2, corresponding to point 4 on the compressor map, where conditions are stable again and can last forever as in point 2. This procedure can be repeated until the operating conditions reach design.

Let us now a little bit complicate the system, as shown in Fig. 6.153, top right: the calibrated orifice is replaced by a control valve that is at a fixed opening at the beginning. As in the former example, rotational speed, discharge valve position, compression ratio and flow are plotted against time on Fig. 6.153, left. The corresponding positions of the operating point are shown on the compressor map, at right.

At time 1, the rotational speed is progressively increased to N1. Consequently, the compression ratio and the flow increase. The situation is shown at point 2, on the compressor map. It is possible to *stay indefinitely* at time 2 conditions (constant rotational speed and valve opening), because atmosphere is an infinite air reservoir that provides a constant suction pressure to the compressor, whatever the flow it processes. At time 3, the discharge valve is slightly closed. The compression ratio increases slightly and the flow decreases. The operating point moves on the N1 iso-speed curve to the left, up to point 4, where the operating conditions are stable. (Notice that, if the valve would be too much closed, the operating point would hit the surge line!) At time 5, the rotational speed is increased up to N2, corresponding to

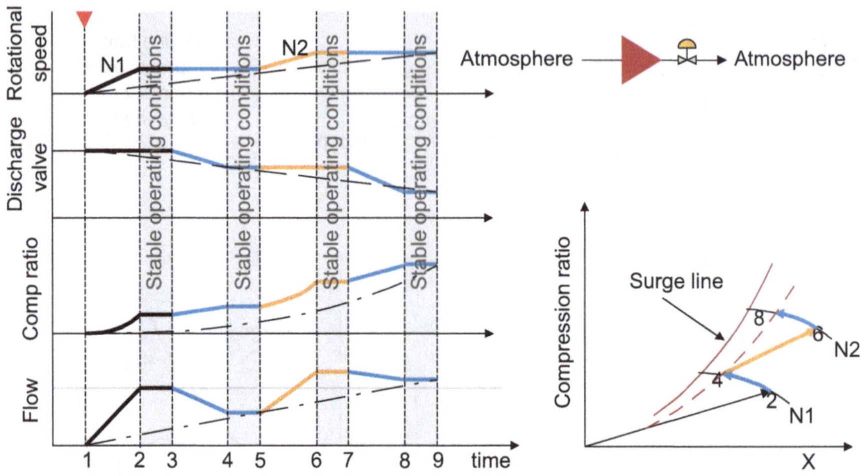

Fig. 6.153 Pumping from atmosphere through a control valve

point 6 on the compressor map, where conditions are stable and can last forever as in points 2 and 4. This procedure can be repeated until the operating conditions reach design.

Instead of varying the rotational speed and the valve opening by steps, it is possible to make such increases simultaneously and in a continuous way, provided that both variations are compatible between each other. If such variations are made in a way to follow the "optimal" curve in dotted line on the compressor map, the pump-down process is very comfortable.

6.7.3.4.2.2 Pumping into a Volume

Let us make a similarity between the operation of a CCC that is pumping on a liquid bath and that of a room temperature compressor.

The flow of helium that is processed by the CCC is provided by the liquid helium bath (see Fig. 6.154, left). It can be sorted into two parts:

- The constant static heat load that is the consequence of a non-ideal thermal insulation
- The dynamic heat load that is related to the operation of the object to be cooled

One can propose a similar system operating at room temperature (see Fig. 6.154, right) where the compressor is fed by a volume and discharges at atmospheric pressure. The volume gets atmospheric air through:

- A calibrated orifice, therefore allowing a constant air flow, simulating the static heat load
- A control valve, allowing a variable air flow, simulating the dynamic heat load

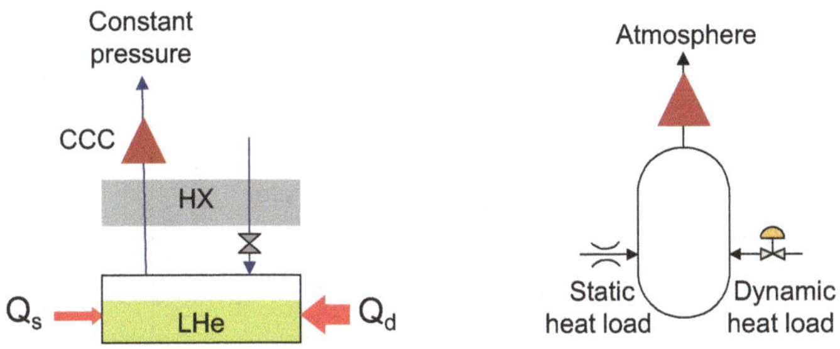

Fig. 6.154 Origin of the flow that is processed by a CCC

In Fig. 6.156, left, rotational speed, suction pressure and flow are considered in the related plot.

At time 1, the rotational speed is increased to N1 within a "short time". Simultaneously, the suction pressure decreases and the flow reaches a value. If, during the same time period, the rotational speed is increased up to N2, higher than N1, the suction pressure decreases, and the flow increases faster. These changes are even faster at speed N3. This small example shows that when pumping on a finite capacity, the processed flow depends on the *derivative* of the rotational speed.

If the same compressor is connected to a larger capacity, one can see, on the right part of Fig. 6.155, that the variation of the suction pressure and of the flow is slower.

Therefore, setting the flow is performed by changing the derivative of the rotational speed. For a given compressor, the pumping duration is inversely proportional to the volume of the capacity (or the volume of liquid).

Let us now study the situation where the compressor is connected to a closed capacity that is, initially, at atmospheric pressure.

Rotational speed, suction pressure and flow are plotted against time in Fig. 6.156, left. The positions of the operating points are shown on the compressor map, at right.

At time 1, the rotational speed is progressively increased up to N1 until time 2 and then kept constant. Simultaneously, the suction pressure decreases, and the flow increases, but, as air is taken out, the pressure in the capacity decreases, and, consequently, the flow decreases too. The operating point moves along the N1 iso-speed curve to the left, up to point 3a where it reaches the surge line.

Therefore, contrary to the former situation, the compressor *cannot* be in a stable and indefinitely long-lasting situation.

In order to avoid such a situation, the rotational speed must be increased from N1 to N2 (point 3 to point 4) in order to increase the processed flow and bring the operating point up on the N2 iso-speed and to the right. Such steps have to be repeated until the design pressure is reached. Of course, a permanent flow must be provided by any means into the capacity in order to stay at design steady state.

Instead of varying the rotational speed by steps, it is possible to make such speed increase in a continuous way, as shown in dotted lines, provided that the variation is compatible with a correct operation (not too fast, but also not too slow!).

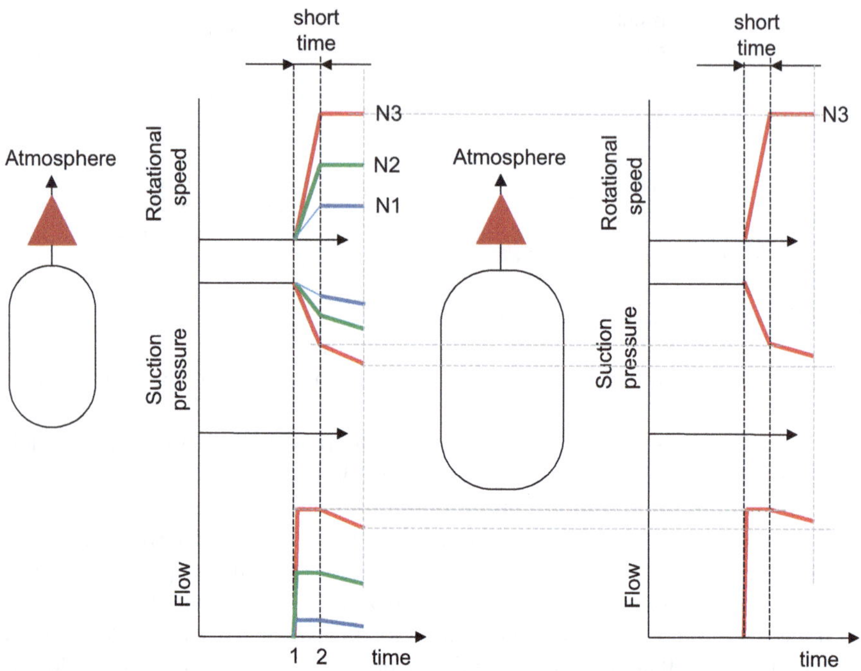

Fig. 6.155 Creating the flow

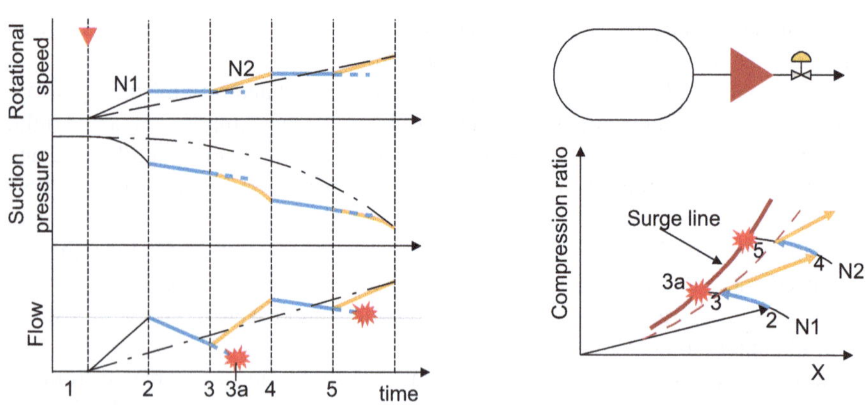

Fig. 6.156 Pumping into a volume: non-stable conditions

A Partial Digest on Control

Here, one sees that such a process is a *dynamic* one: the flow to correctly "feed" the compressor is provided by the way the rotational speed of the compressor is varied. It is a "self-provided" flow. The driving parameter is the *derivative* of the rotational speed!

Note that in these two examples dealing with air, the suction temperature is constant; therefore, the normalised reduced flow X depends only on mass flow and pressure.

6.7.3.4.3 Pump-Down with a Centrifugal Compressor Operating with Helium at Cryogenic Temperatures

Compared to a room temperature centrifugal compressor, the control procedures of a cryogenic centrifugal compressor are more complicated because the temperatures of the system may vary strongly (remember that, at low temperatures, the temperature ratios might be important even if the temperature differences are small). The author has filed a few patents about such procedure (US Patent 5,499,505 of 19/03/1996).

By the way, pumping out of a volume is a similar situation as what happens when pumping on a liquid helium bath:

- Natural heat leaks on the system provide a constant and rather small helium flow.
- Decreasing the pressure makes liquid helium be turned into vapour in order to cool the remaining liquid. This is the big amount of flow.
- The liquid that has been vaporised is replaced by new liquid that is expanded and provides some extra vapour due to its expansion flash.
- The compressor suction temperature is almost constant even if it is depending on the operating conditions of the JT heat exchanger, especially when pumping down, where the HP and LP flows can be unbalanced, therefore changing the first compressor suction temperature.

In Sect. 4.3.1.2, it has been shown that there are two possible arrangements for cryogenic centrifugal compressors: mixed pumping (Fig. 6.157, left) with a volumetric room temperature compressor, the suction pressure of it being floating according to the processed flow, or pure cryogenic pumping (Fig. 6.157, right) where the cryogenic centrifugal compressor train discharge pressure is constant, slightly above atmospheric pressure.

In order to compare both arrangements and various operating conditions in the following part, let us consider two arrangements each processing 100 g/s from 0.030 bar (or 1.98 K). The pure cryogenic system is a four-stage pumping system that compresses up to 1.15 bar. The mixed system is a three-stage pumping system that compresses up to the suction pressure of the room temperature compressor (RTC).

The following assumptions for simplification are made:

- No pressure drops
- Ideal room temperature compressor (constant volumetric efficiency)
- Constant C1 inlet temperature: 4.5 K
- Efficiency of each compressor: 0.75
- Constant cryogenic compressor efficiencies

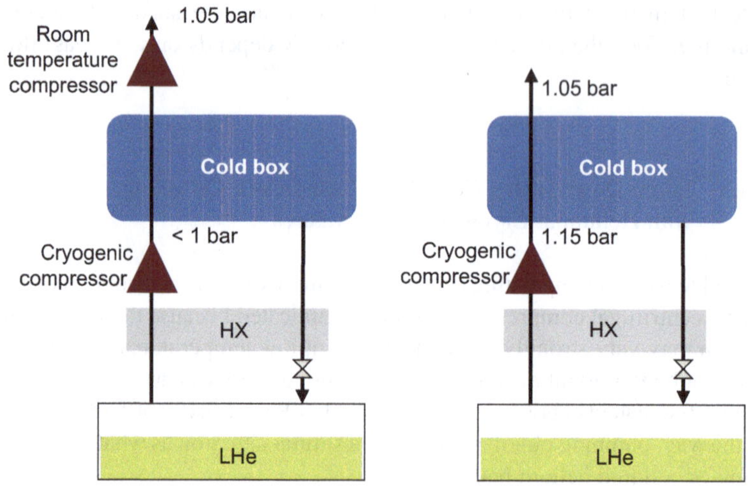

Fig. 6.157 Arrangements of cryogenic compressors, mixed pumping, left, pure cryogenic pumping, right

Remark When pumping on a liquid bath, the time to reach the design pressure is proportional to the volume of liquid. However, nobody knows the quantity of liquid that is sitting in the bath: it means that the pumping procedure must be able to run without this information.

6.7.3.4.3.1 Pure Cryogenic Pumping

As understood in Sect. 6.7.3.4.2.2, the processed flow is "self-provided" by the derivative of compressor rotational speeds. It must be such that the operating point on the map of each compressor is correctly located apart from the surge line and, if possible, at a *constant relative distance* expressed in X. It is a curve that is a homologous of the surge line and goes through the design operating point. For a same compression ratio, the operating point is, for example, located at +20%, expressed in X, from the surge line (can be checked on the chart). Let us call this curve the *optimum* operating curve. Figure 6.158 displays the compressor maps of four machines arranged in a pure series (no heat exchanger between machines). One can see that the operating points that have been calculated (dots on the dotted curve) are located at a constant relative distance (expressed as 20% of X flow) of the surge line.

As an important consequence, the process flow that is necessary to operate in such conditions is *substantially* higher than the design flow during the pump-down procedure, here 1.6 times (see Fig. 6.159 that shows evolution of the various compression ratio and the processed mass flow rate versus suction pressure). It is important that the refrigerator is able to take care of such an important return flow without too many perturbations. In Fig. 6.159, one can also notice that it is C4 that

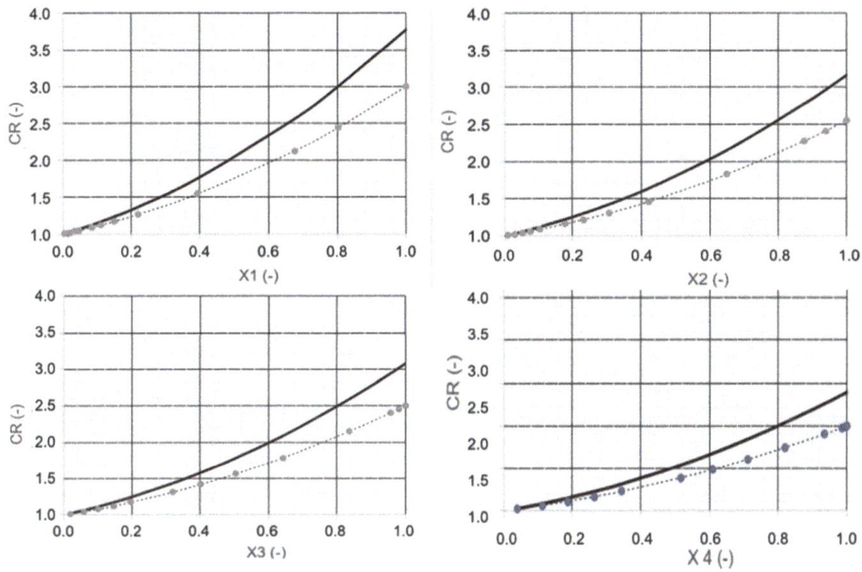

Fig. 6.158 Compressor maps for a 4-CCC-train with no RTC, showing the operating points and the surge lines

Fig. 6.159 Evolution of CR and relative flow for each compressor

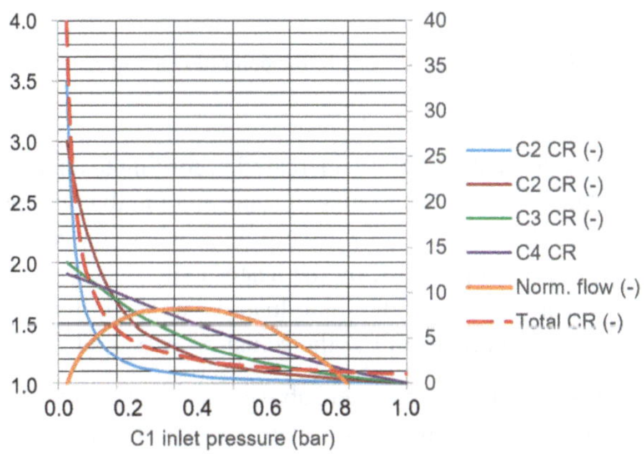

starts the compression job earlier than others because it is the one that has the highest X (the nearest from the design conditions) and, therefore, that can generate the highest compression ratio.

When reaching the steady state at design conditions, the design processed mass flow is to be provided, either by normal operation (difficult) or by an electrical heater (located in the saturated helium bath) load as shown in Fig. 6.160, top. It would be difficult to set accurately the power that is to be instantaneously delivered by the heater; therefore, the procedure that is shown in Fig. 6.160, bottom, is smoother. At

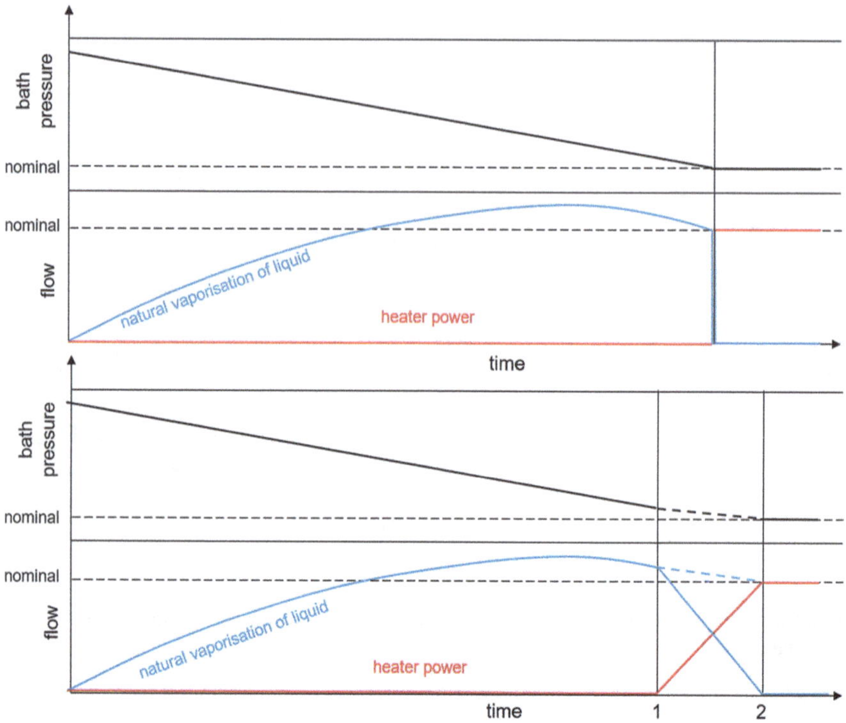

Fig. 6.160 Smoothing the end of the pump-down sequence

time 1, the heater power is increased according to a ramp: the saturated vapour that is produced by the heater replaces same amount that was provided by liquid expansion until the latter comes to zero at time 2, where the nominal pressure is reached. All vapour is now provided by the heater load.

When the useful thermal load is applied, the heater load is progressively decreased in order to fit to the actual process load.

Remarks The example here above runs on the assumption that there is a 20% safety margin expressed in X flow apart from the surge line. Such safety margin could be lower, and, as a consequence, the necessary maximum processed flow would be lower, but the system would be more sensitive to perturbations. To avoid such a large oversizing of the flow, it is possible to increase the reduced flow X by one of the ways that are shown in Fig. 6.164.

6.7.3.4.3.2 Mixed Pumping

To run the room temperature compressor at its nominal power, its suction pressure must be kept at nominal value, here 0.630 bar, either by recycling or controlling its rotational speed. Therefore, the cool-down procedure is performed at constant nominal mass flow rate (here 100 g/s).

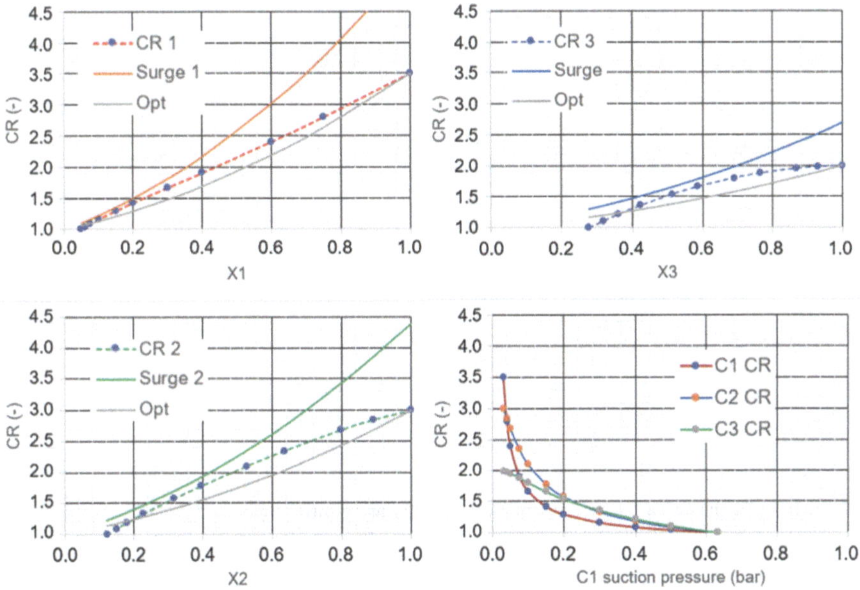

Fig. 6.161 Operating points in the compressor maps, compression ratios versus C1 suction pressure, bottom right

During the first phase of the pump-down, the room temperature compressor runs alone until it reaches its design suction pressure: the bath pressure reaches 0.630 bar; therefore, the cryogenic centrifugal compressors can start up.

Figure 6.161 shows the evolution of the compression ratio of each stage versus the reduced flow X. In Fig. 6.161, bottom right, one can see how the compression ratios vary during the pump-down process. Again, when the nominal conditions are reached, the nominal flow must be provided to the compressor train.

Remarks Remember the assumptions that have been made and, among them, the constant efficiency during the pump-down sequence, which is certainly wrong. As a consequence, when the efficiency is lower, the discharge temperature of compressor Cn, which is the inlet temperature of compressor Cn + 1, is higher, and the Cn + 1 reduced flow is higher, allowing a higher compression ratio. Therefore, such situation helps during the pump-down procedure.

6.7.3.4.4 Turn-Down Operation (Reduced Mass Flow at Constant Suction Pressure)

Turn-down means that the processed flow that is provided by the final user thermal load decreases but the liquid helium bath pressure *must stay constant*. The two possible arrangements, mixed or pure cryogenic, behave in very different ways during such turn-down situations.

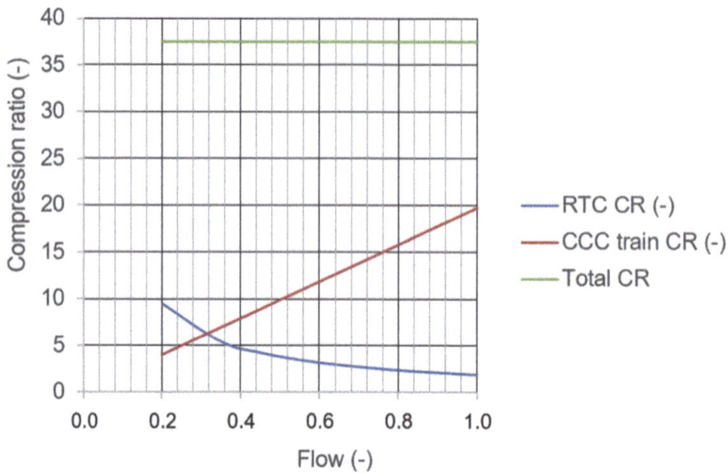

Fig. 6.162 Evolution of CRs versus mass flow rate during turn-down

6.7.3.4.4.1 Mixed Compression Arrangement

In a mixed arrangement, as the room temperature compressor is a volumetric machine, its suction pressure is proportional to the processed flow (theoretical machine with a constant volumetric efficiency). Therefore, when the processed mass flow rate decreases, the compression ratio that is to be generated by the CCC train decreases too (see Fig. 6.162).

In Fig. 6.163, one can see that all operating points are naturally located at the right of the surge line.

Remark In Fig. 6.163, the safety margin expressed in X is 30%. If it were lower, the safety would be very small even zero at X values around 0.3 for compressor C2 and around 0.6 for compressor C3.

In a mixed arrangement with a train of cryogenic centrifugal compressors backed by a volumetric room temperature compressor, the processed flow can be significantly and easily reduced.

Remark One should not forget the assumption about the ideal room temperature compressor: the constant volumetric efficiency. With an actual compressor, its suction pressure would be higher, and there might be a limit towards low suction pressures.

Of course, with an actual machine, the volumetric efficiency decreases with the pressure difference; therefore, the curves are slightly different towards low flow.

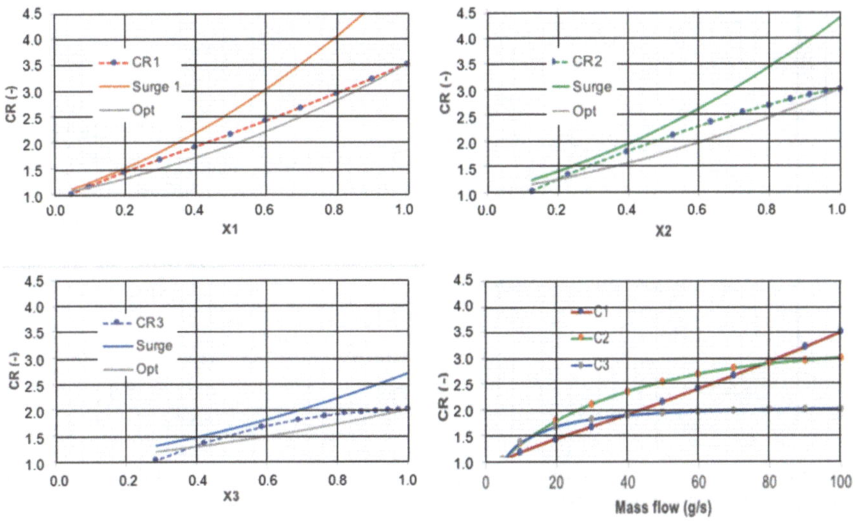

Fig. 6.163 Compressor maps during turn-down, evolution of compression ratios versus processed flow, bottom right

6.7.3.4.4.2 Pure Cryogenic Compression Arrangement

In this arrangement, the situation is rather different: when the flow decreases, the CCC train has to keep the same compression ratio. It is obvious that the operating point moves to the left, along the iso-speed curves, and hits rather rapidly the surge line. The possible flow reduction is about 20% only, except if the compressor has been designed with a higher margin which, in any case, cannot be much larger. It is therefore necessary to increase the reduced flow $X = \dot{m} \times \sqrt{T_{\text{in}}}/P_{\text{in}}$ of each compressor in order not to hit the surge line.

There are various possibilities: increase $\dot{m}$ or/and T_{in} or decrease P_{in}.

Of course, if any of these inlet parameters is modified, all compressors of the train will be affected.

Increase $\dot{m}$: (see Fig. 6.164, left) a by-pass around the compressor train can be incorporated: both flow and temperature are increased at the suction side of C1. It is also possible to inject warmer gas at the suction side of C1. Due to the large temperature difference, it is mainly the inlet temperature that increases much more than the mass flow (be careful; it is difficult to make, in short distance, a correct mixture of gases at so different temperatures!).

Increase T_{in}: a heater can be used.

Decrease P_{in}: this is not possible because it would increase the compression ratio.

All these solutions lead to processing either a higher flow or at a higher temperature; consequently the power that is dissipated by the compressor train is increased, and the refrigerator has a heavier duty to fulfil, which is not really the target for a turn-down operation.

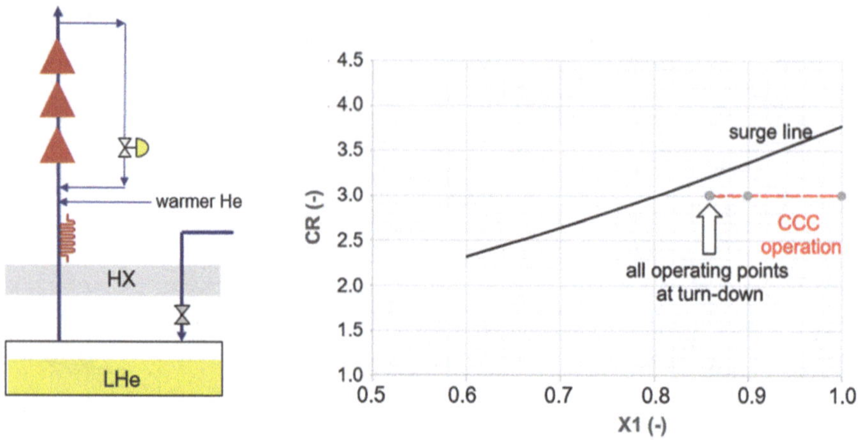

Fig. 6.164 Various ways to increase the compressor train reduced flow X in turn-down conditions, left; operating points at turn-down

Remark Remember the assumptions that have been made and, among them, the constant efficiency which is certainly wrong. As a consequence, the discharge temperatures are higher than calculated, and consequently the reduced flows are higher too, allowing a reduced mass flow rate.

Let us try various methods to keep the X flow at an acceptable value.

Recycling Helium From Discharge of C4 into C1 Suction Side
Keeping the X flow at constant value needs recycling up to almost 6 g/s, as shown in Fig. 6.165. The consequence is that the inlet temperature of C1 increases, and so does the discharge temperature of C4. Obviously, such temperature that would reach about 170 K at 30 g/s is too high to be taken care by the refrigerator. Furthermore, the power that is absorbed by the compressor train is almost doubled. Such an arrangement is not realistic for a large turn-down ratio but might be acceptable at lower values

Injecting Room Temperature Helium at C1 Suction Side
In Fig. 6.166, room temperature helium is injected at the inlet of C1 (mind about the correct mixing of gases with such a large temperature difference!). The room temperature flow to be injected is lower than in the former case, but, again, the discharge temperature of C4 is high: about 190 K! Here, again, the power of the compressor train is doubled.

Such an arrangement is not realistic.

Injecting 10 K Helium at the Suction Side of C1
This time, operating conditions are more acceptable for the refrigerator: the compressor train discharge temperature reaches 45 K for 30 g/s, and the compressor train power stays almost constant (Fig. 6.167).

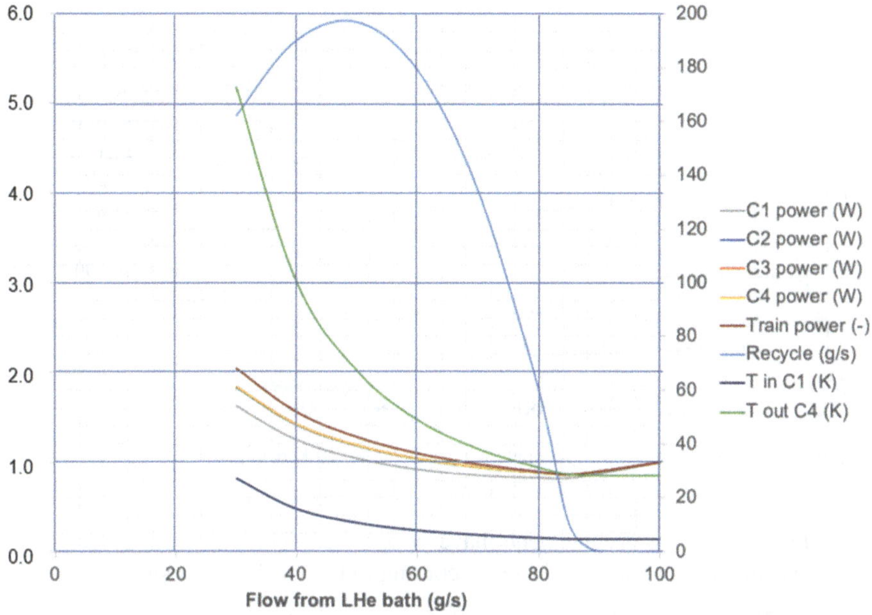

Fig. 6.165 Recycling helium

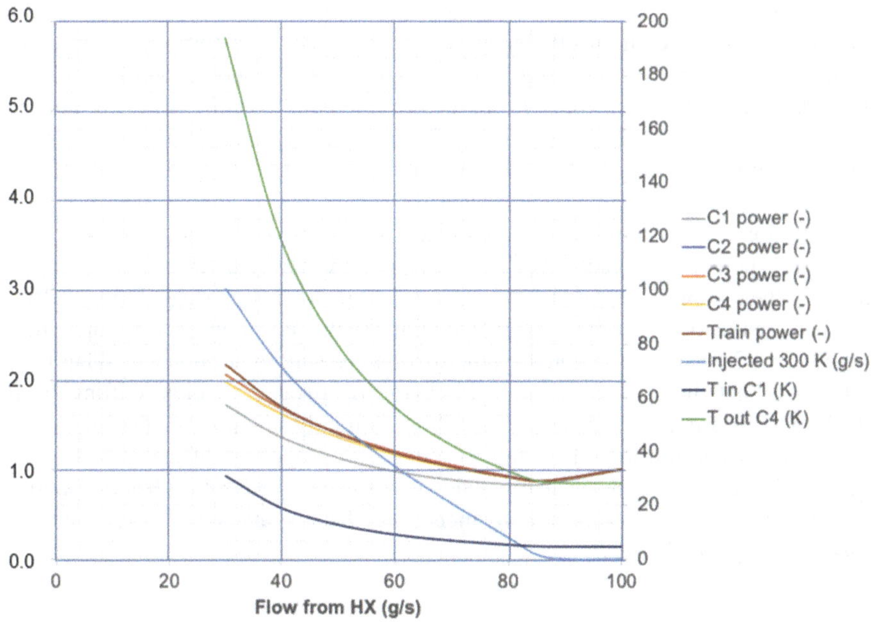

Fig. 6.166 Injecting room temperature helium at C1 suction side

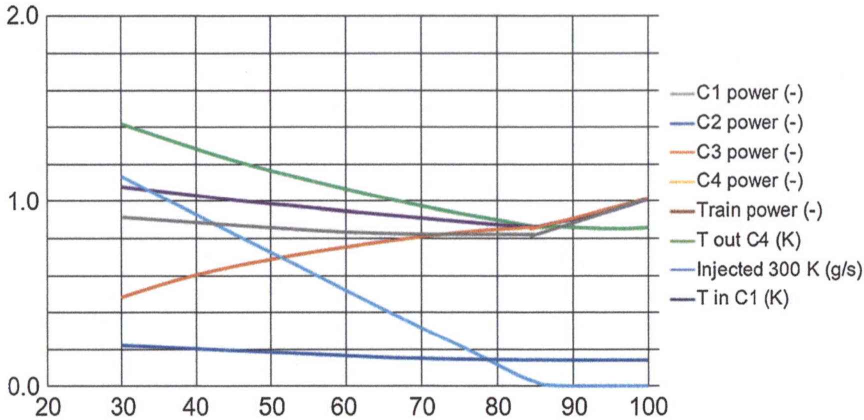

Fig. 6.167 Evolution of various parameters when injecting 10 K helium

Here above, some arrangements have been tested. There are, of course, other arrangements that should be selected according to the local conditions.

Another Way to Cope Efficiently With Low Flows at Turn-Down Conditions The above arrangements have various disadvantages, especially a higher compressor train discharge temperature that disturbs the operation of the refrigerator. They can however be used in the case of moderate turn-down (low flow reduction).

In 1999, the author proposed to operate efficiently, that is to say keeping almost the same compressor train efficiency at reduced flow, by incorporating an extra compression stage at the discharge of the compressor train, in other words, adding a fifth stage to a four-stage train. In the design operating conditions, as the duty is fulfilled by the four stages, the extra stage is not operating; it has a compression ratio of 1. Conversely, at reduced flow, for example, at 50%, the fifth stage is at its design operating conditions, and the first stage of which the *X* flow is half brings a compression ratio that is sufficient to reach the expected discharge pressure say 1.15 bar. Another way to describe the system is to say that from design to turn-down, a part of the compression duty is shifted from the first to the fifth stage. In Fig. 6.168, one can see the location of the design operating points (red circles) and the operating points for a 50% mass flow rate at same suction pressure (blue squares). What is of interest is that the compressor train discharge temperature is only shifting from 28.34 K for design to 28.54 K at 50% flow. Similarly, the power injected by the compressor train into the refrigerator is reduced in proportion of the flow. Consequently, it is possible to reach an efficient whole system (4.5 K and sub atmospheric) turn-down operation. This arrangement has been patented by the author: 5941095 August 24, 1999.

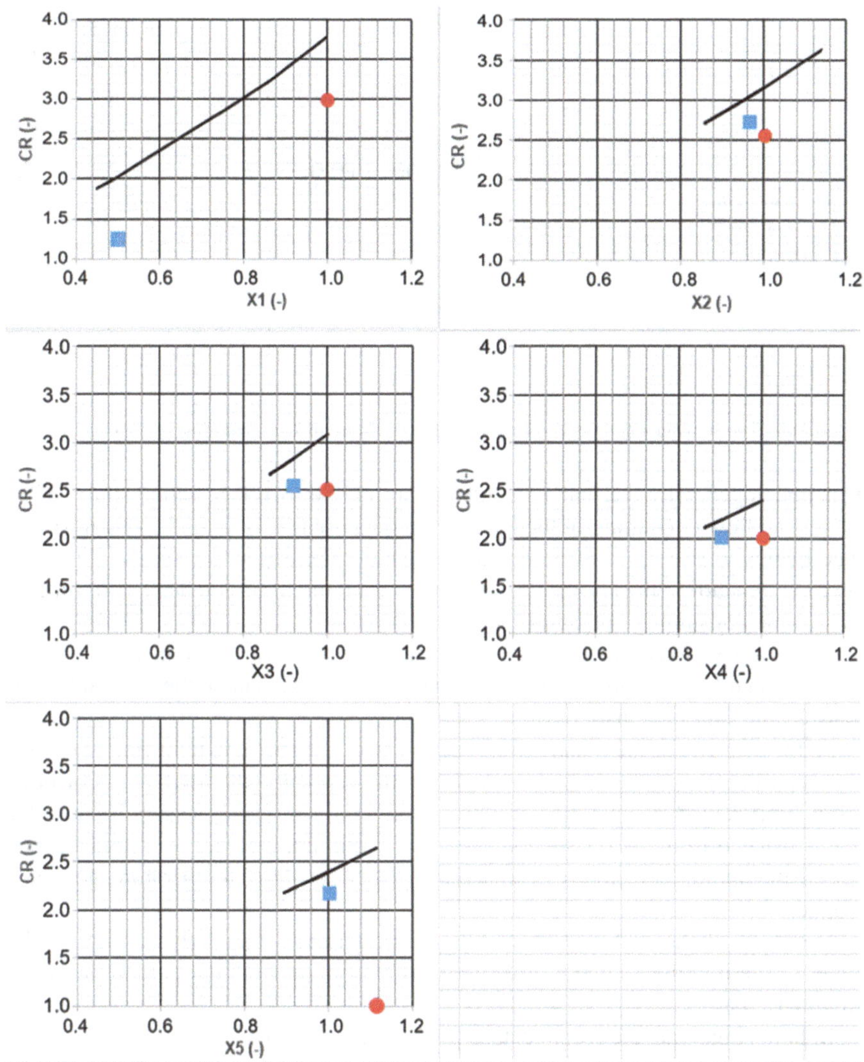

Fig. 6.168 Turn-down with an extra stage. Design, red circles; turn-down: blue squares

6.7.3.4.5 Reconnection to the System

When a compressor train shutdown occurs, the usual procedure is to close the inlet valve of compressor C1 in order to avoid a reverse flow towards the cryostat. How to re-start up the compressor train and re-connect it to the helium bath?

In Fig. 6.169 it is shown that, between the JT heat exchanger(s), located in the cryostat, and the pumping system, there is a transfer line that might be rather long.

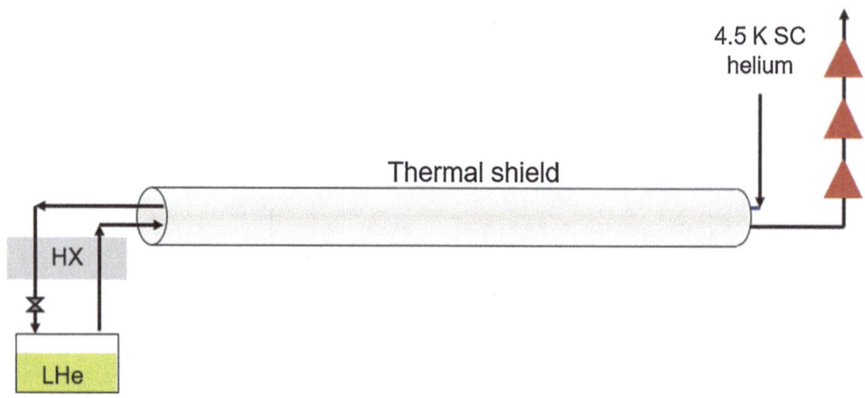

Fig. 6.169 The transfer line between the liquid helium bath and the pumping system

This line is generally thermally shielded at around 80 to 100 K temperature. When there is no flow circulating in the line, its temperature that is around 5.0 K in steady state begins to rise up to 80 K maximum. Therefore, depending on the shutdown duration, re-starting the cryogenic compressor train might be a delicate and, furthermore, time-consuming operation.

There are various situations to be considered, as shown in Fig. 6.170, where pressures and temperatures, respectively, at the inlet of C1 and at the near end of the transfer line (line) are displayed.

At the near end of the transfer line, suction side, pressure and temperature change slowly, according to the static heat leaks on the cryostat, whether there is or not a room temperature compressor downstream the cryogenic compressors because the line is isolated from the C1 suction side. The maximum pressure is around 1.2 bar, but the maximum temperature is related to the thermal shield temperature, generally around 80 K.

However, at the suction side of C1, the pressure rises quickly and drastically either to the minimum suction pressure of the room temperature compressor or to around 1.2 bar.

In both situations, the reconnecting procedure is similar:

- Decreasing the pressure at the suction side of the first stage compressor C1 and both controlling the inlet temperature of C1 in order that each compressor stays in its operating domain and providing an adequate helium flow. Such flow is to be provided by a special phase separator that is located in the cold box and connected as shown in Fig. 6.171. Heater HTR LIQ allows controlling the flow; heater HTR VAP allows controlling the C1 suction side temperature.
- When the pressures are equalised on each side of valve V1, connection is done, by slightly opening first V2 that is a smaller valve than V1 (lower heat to be extracted from the valve body material in order to bring the smallest possible temperature variation at inlet of the first stage compressor) and then V1.

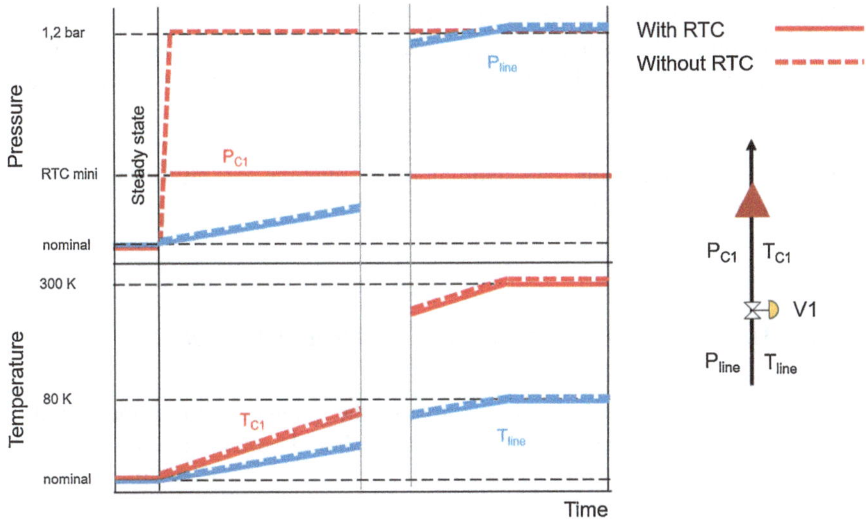

Fig. 6.170 Re-start situations

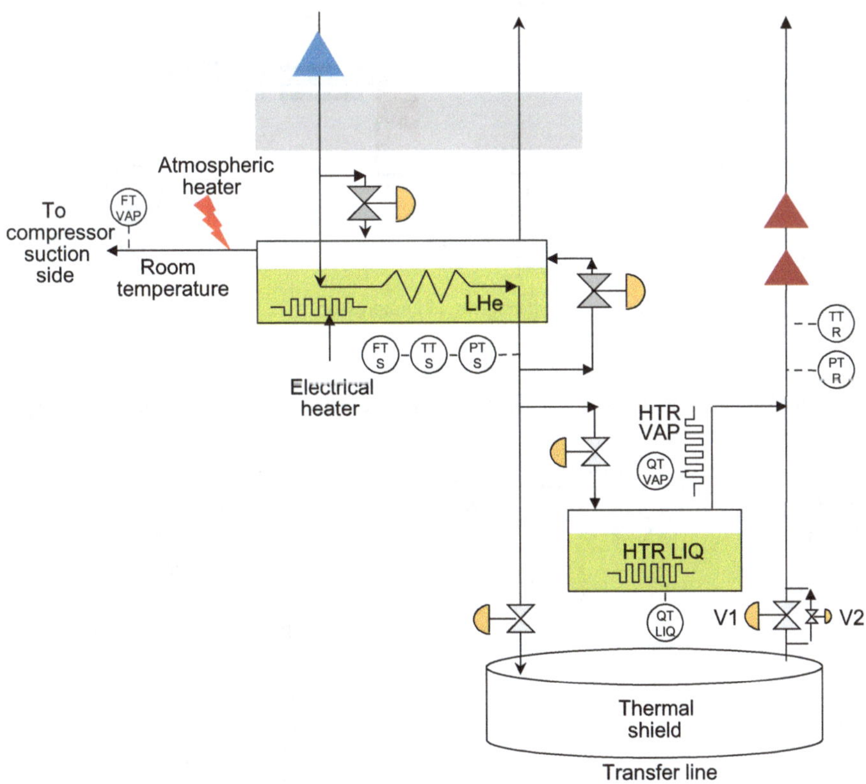

Fig. 6.171 A possible arrangement with a phase separator for re-starting the compressor train

- When the connection is done, decrease the liquid helium bath pressure, and, simultaneously, reduce to zero the flow that was provided by the phase separator. This phase is similar to the end of the pump-down procedure already described.

Remark In order to be able to re-start without losing time, it is advised to keep this equipment permanently cold.

6.7.4 Monitoring of a Cryogenic Centrifugal Compressor

In Fig. 6.172, and Table 6.15, an example of the parameters to be monitored is shown. The low-switch and high-switch values should be arranged in two levels: first an alarm and then a stop action.

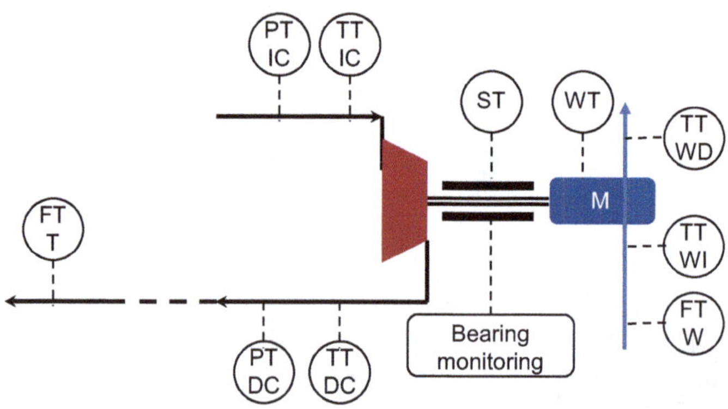

Fig. 6.172 Monitoring of a cryogenic centrifugal compressor

Table 6.15 Parameters to be monitored for a magnetic dynamic bearing cryogenic centrifugal compressor

Sensor	Low switch	High switch	Duty
PT IC		X	Helium inlet pressure
CTT I	X		Helium inlet temperature
PT DC		X	Helium discharge pressure
PT DC		X	Helium discharge temperature
FT			Helium mass flow rate (range)
ST		X	Turbine rotational speed
WT		X	Motor power
TT WI		X	Water temperature inlet
TT WD		X	Water temperature discharge
FT W	X		Water flow rate
Bearing system		X	Operating conditions of bearings

6.7.5 A Cryogenic Compressor Digest

The operating domain of a CCC is limited, both towards low mass flow rates and, sometimes, also towards high mass flow rates. The compressor diagram shows the operating domain.

When compressors of same efficiency are arranged in pure series, the efficiency of the equivalent compressor is lower than that of each compressor of the arrangement.

6.8 The Cryogenic Circulator

The duty of such a component has been introduced in Sect. 4.2.4.2.

A cryogenic circulator is a cryogenic compressor that compresses a low compressibility fluid (supercritical helium), with a small compression ratio.

The structure of a cryogenic circulator is very similar to that of a cryogenic compressor; however, as the compression ratio is lower, the rotational speed is also lower, allowing to incorporate not so high rotational speed bearings as roll bearings.

In a hydraulic point of view, it looks like a water pump, except that it operates at low temperature.

The low compression ratio gives the circulator map a special shape: the iso-speed curves are rather flat and do not "fall down" vertically towards high flows, as shown in Fig. 6.173.

There is a big difference in controlling the cryogenic circulator because it generally circulates super-critical helium into a closed circuit of almost constant impedance. Consequently, the circuit pressure drop is related to the square of the

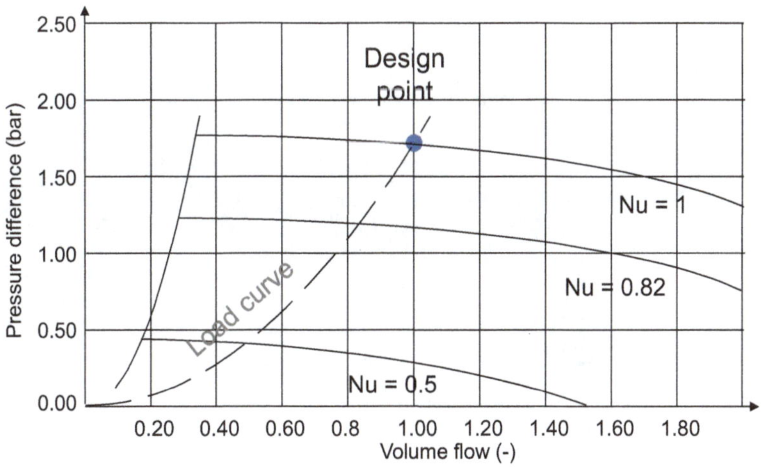

Fig. 6.173 Map of a cryogenic circulator

mass flow rate, and the operating point of the circulator stays naturally in a safe area as shown in Fig. 6.173.

Do not forget that the power that is dissipated by the circulator is to be absorbed by the refrigerator. In some large systems, such load is around 25% of the total power!

6.9 General Structure of a Refrigerator or a Liquefier

6.9.1 The Compression Station

The compressor station consists of one or more cycle compressors, each usually driven by its electric motor, the helium oil-removal system, oil and helium cooling and, finally, cycle pressure control system (see Fig. 6.174) which includes cycle buffer capacity.

Compressors and control valves are very noisy objects, so they are either enclosed in sound-absorbing chambers or housed in a special room. The cumbersome cycle buffer capacity is generally installed outdoors.

Fig. 6.174 A CERN LHC refrigerator compression station. In the foreground, to the left, part of the oil management system with pumps and horizontal bulk oil separator (courtesy Air Liquide CERN)

The compression station is connected to the cold box by high pressure, low pressure and, sometimes, middle pressure if it is a three-pressure cycle.

6.9.2 *A Room Temperature Full Flow Dryer*

The duty of the room temperature full flow dryer is:

- During the cool-down phase, purifying the initial helium inventory and the helium that is circulated into the objects to be cooled that might not be perfectly clean, in order to avoid that any moisture is trapped into the cold parts
- During long-run operation, keeping the cycle gas perfectly dry

The room temperature full flow dryer is located between the compression station and the cold box. It is generally located in the compressor room.

6.9.3 *The Cold Box*

The cold box houses in its vacuum enclosure all the components that operate at low temperatures: heat exchangers, expansion machines, valves, adsorbers, filters, connecting pipes and, in some models, the internal phase separator. Thermal insulation is achieved by combining an insulation vacuum at very low pressure, provided by permanent pumping, and a multi-layer insulation also known as "super insulation" (Fig. 6.175).

The vacuum enclosure is cylindrical. It can be vertical axis (see Fig. 12.3) or horizontal (Fig. 6.175). Horizontal layout sometimes allows better integration into buildings; it is widespread for very large machines, but attention must be paid to the possibility of defective operation of heat exchangers operating in a horizontal position due to parasitic re-circulations (see Sect. 2.4.4). Cold components are suspended or supported in such a way as to reduce heat input by conduction. The part of the cryogenic valves operating at room temperature (stem and actuator) protrudes the top plate in the case of a vertical box or is near the top generator in the case of a horizontal box, as shown in Fig. 6.176.

A cold box can be split in parts, either for reasons of transportation, local environmental conditions or others.

6.9.3.1 The Cycle Cold Adsorbers

The duties of the cold box adsorbers are:

- During the cool-down phase, purifying the initial helium inventory in order to avoid that any impurity is trapped into the cold parts
- During long run operation, keeping the cycle gas perfectly pure

Fig. 6.175 A cold box. One can see four expansion turbines with horizontal axis. (courtesy Air Liquide)

Fig. 6.176 Cryogenic valve actuators out of the top part of an LHC cold box (courtesy Air Liquide)

There are generally two stages of cold box adsorbers:

- The "warm" adsorber that takes care of air impurities (oxygen, nitrogen, argon)
- The "cold" adsorber that takes care of hydrogen and neon.

Remark The cold box adsorbers are not intended to purify helium that is to be liquefied. Such a duty is fulfilled by a special external or internal purifier that processes only the gas to be liquefied (see Sect. 9.4.1).

In Sect. 10.3, the cool-down operation is discussed, emphasising the necessary purity of helium. Such purity is reached with various adsorbers; the operating principle is discussed in Sect. 9.4.

6.9.3.2 A Cryogenic Valve

The volume that exists between the bellows and the sliding O-ring on the stem used an air lock when the valve is connected to a sub-atmospheric pressure circuit (see Sect. 6.9.3.4) (Fig. 6.177).

6.9.3.3 A Phase Separator

It is of interest to incorporate a phase separator into the cold box of a refrigerator. It might fulfil various functions:

- Separate liquid and vapour.
- Store a quantity of liquid, the level of which is measured. The behaviour of the level gives information on the power balance of the system (see Sect. 8.7.1). The volume of stored liquid can also allow for some extra time of operation, for example, during the time a superconducting coil is safely unloaded in case the refrigerator is down. This is generally the criterium to size the volume of the phase separator.
- House a subcooling heat exchanger in case the refrigerator has to deliver supercritical helium (remember Sect. 5.3.2.3),
- Make a refrigerator performance test, at the time of commissioning but also at any time in the life of the plant to check if the initial performance has been kept (see Chap. 14).

A heater, located in the liquid phase, allows checking the performance of the cold box alone, independently of the system to be cooled (the final user). See various examples in Chap. 5. It makes possible a refrigerator performance check at any time.

In another structure shown in Fig. 5.6, bottom, the subcooling heat exchanger is out of the phase separator and cooled by natural thermosyphon circulation.

Remark The design of the phase separator must take into account the liquid-vapour separation efficiency.

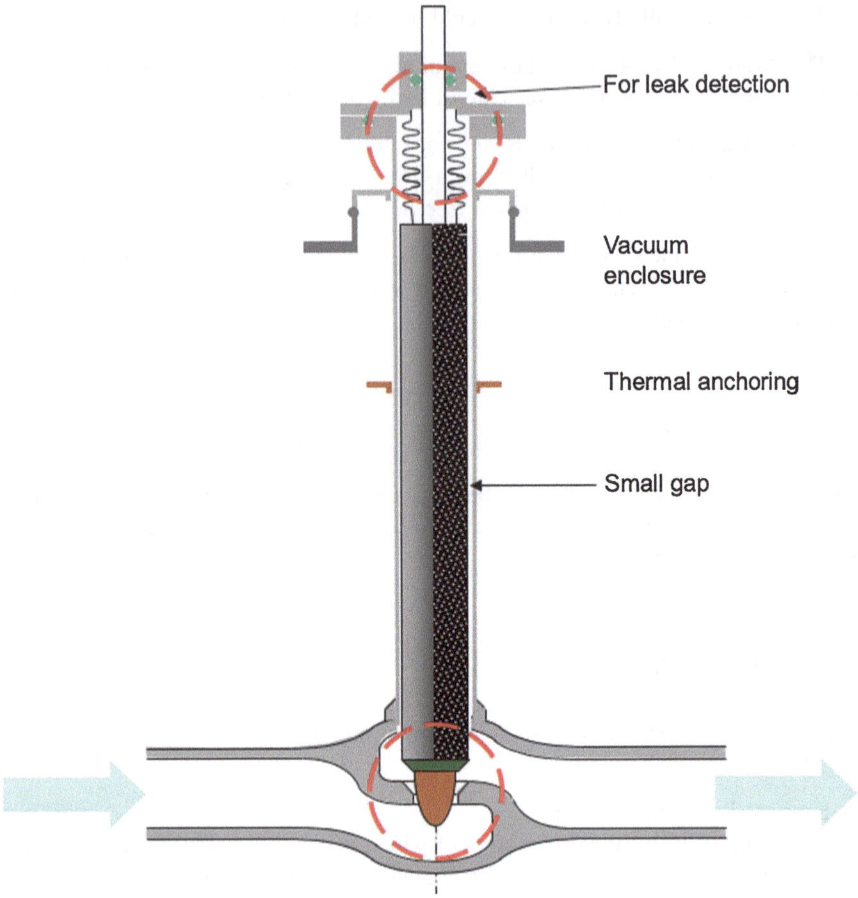

Fig. 6.177 A cryogenic valve

6.9.3.4 Specific Arrangement for Operation at Sub-atmospheric Pressure

When a circuit is at sub-atmospheric pressure, air might get in if there is a leak. If the circuit is a cryogenic temperature, air components (water, nitrogen and oxygen) are turned into solid as soon as they reach their solidifying temperature. If it is a pressure sensor, the measurement becomes wrong. If it is a safety device, it is dangerous because it cannot protect the circuit anymore.

The conventional arrangement for a pressure sensor and a safety device is shown in Fig. 6.178, left and centre. The sensor or the safety device is sitting in a volume, at

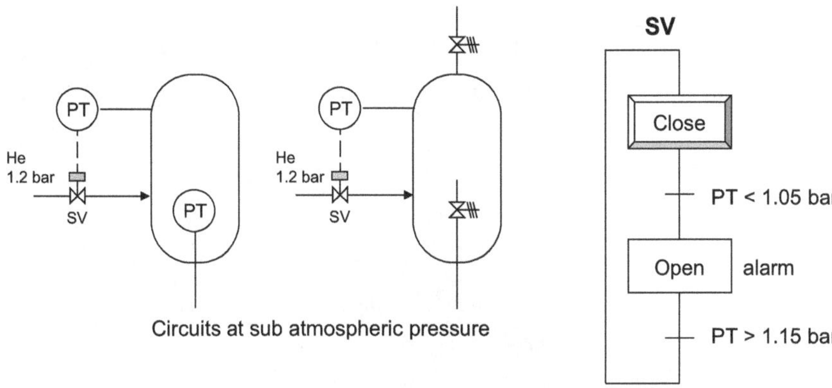

Fig. 6.178 The air lock for sub-atmospheric circuits

the pressure of around 1.15 bar, which is measured by PT. When PT is lower than 1.05 bar, it means that a leak has developed, either from the volume to air or, more likely, from the capacity to the sub-atmospheric circuit. As far as helium from the volume flows into the sub-atmospheric circuit, there is no risk of plugging the pipe, only a possible heat input.

To avoid that the pressure in the volume becomes lower than atmospheric pressure, the solenoid valve SV opens, and an alarm is issued until PT reaches 1.15 bar again. The sequence, described in GRAFCET (see Sect. 8.9.2), is shown in Fig. 6.178, right.

In the case of a safety device, a safety valve is connected from the volume to air. Make sure that the total of the cracking pressures of the two safety valves that are arranged in series equals the requested protection pressure.

6.9.3.5 The Insulation Vacuum Set

The vacuum set is generally composed of a primary vane pump in series with an oil diffusion pump as shown in Fig. 6.179 or a turbo pump. A remote-controlled or manual valve between the vacuum set and the vacuum enclosure allows maintenance of the vacuum set.

Fig. 6.179 A vacuum set with a vane primary pump and an oil diffusion secondary pump

6.9.3.6 The Test Equipment

The test equipment is mandatory in most of the cases (see Sect. 14.1). It allows operating the plant at the time the client components are not yet connected. It is obviously used for the commissioning tests and for checking the performances of the refrigerator at any time, during the lifetime of the system.

Chapter 7
Off-Design Operation

Abstract It is both surprising but logical that, except for a pure liquefier, a plant is almost never operated at its design conditions due to the safety margin that the client takes. Operation in turn-down regimes, when the final user load decreases, must cope with keeping the highest possible efficiency. Calculations performed in the former chapters were related to one-off machines operating at their design conditions. Here, the behaviour of actual systems in various operating conditions is discussed.

7.1 Introduction

Let us take a simple example: the calculations of a cryogenic system result in a theoretical cryogenic heat load that must be removed by the refrigeration plant. However and especially when the system is complicated or based on new concepts, some safety margin must be taken by the system designer. If, for example, the safety margin is 1.5, it means that, if the assumptions and calculations were correct, the plant will spend all its lifetime operating at maximum $1/1.5 = 0.66$ of the designed power. Furthermore, there are situations where the object to be cooled does not dissipate the usual amount of power because it is in a standby mode.

All calculations that have been performed in Chaps. 3 and 4 are related to machines operating in a steady state. Each calculation refers to a one-off machine operating in a one-off given set of operating conditions.

How does such a real refrigerator or liquefier, that is to say with same compressor (s), heat exchanger(s) and expander(s), behaves if the operating conditions shift from design operating conditions to any other operating conditions?

For example, as there is no fundamental difference between the structures of a liquefier and a refrigerator, the operating regime can be moved towards more refrigeration or more liquefaction compared with the design conditions.

G. Gistau Baguer, *Cryogenic Helium Refrigeration for Middle and Large Powers*, International Cryogenics Monograph Series,
https://doi.org/10.1007/978-3-030-51677-2_7

Every component of a machine behaves according to its characteristics and interacts with all others, which can make complex changes to the whole system operation.

For example, it has already been learned that:

- If the intake pressure of a compressor decreases, it processes a lower mass flow rate (remember Sect. 6.2.1.1)
- If, for a constant inlet temperature, the pressure at the inlet of a turbine decreases, the turbine processes a lower mass flow rate and extracts a lower power (remember Sect. 6.6.3.7.2). And its efficiency varies.

One can expect that if a heat exchanger operates in off-design conditions, changes will appear in temperature differences.

Such calculations, which require multiple iterations, are generally performed using specialised software for thermodynamic processes like HYSIS from AspenTech USA. The examples discussed in this chapter are simple and indicative and include a lot of simplifications (steady state, no pressure drops, constant volumetric efficiency of the compressor, approximation of the variation of the overall heat transfer coefficient of heat exchangers, sonic turbine distributor, constant turbine isentropic efficiencies).

The interest of this chapter is to get information on how refrigeration plants behave (again, the idea of "feeling" the plant behaviour!). It is mainly a pedagogic chapter. It is not intended to reach accurate results, a target that is out of capability when using simple tools as Excel and REFPROP or HEPAK. It does not take into consideration practical safety margins on components generally considered during the design phase.

An Important Remark In this chapter, where the target is to discuss the behaviour of systems that are operating in off-design conditions, it is assumed that the control system is not operating, except the low-pressure control loop. It is also assumed that the components (compressor, heat exchanger, turbine) can operate in the new and possibly unrealistic conditions (e.g. a higher pressure or power). The main target is to understand how a refrigerator behaves according to changes in the operating regimes.

7.2 The Cycle Compressor

The mass flow rate that is processed by a compressor:

$$\dot{m} = \eta_{\text{vol}} \times \Delta V \times P_{\text{suc}}/T_{\text{suc}} \tag{6.1}$$

7.3 The Heat Exchanger

7.3.1 Behaviour of a Heat Exchanger

The main assumption that is considered here is related to the heat exchange coefficient U: it varies with the 0.7 power of the mass flow rate of the low pressure (because the heat exchange coefficient is generally lower on the low-pressure side of a heat exchanger and, therefore, calculations results are on the safe side).

Let us consider some off-design operating conditions (see Fig. 7.1).

- On the left, the design conditions are displayed.
- In the top middle (off-design 1), the LP mass flow rate: m_3, is reduced. As a consequence, T_2, T_4 and U are changing.
- Similarly, at the top right (off-design 2), the LP flow is increased, and one can expect modifications on the same operating parameters.
- On the bottom middle (off-design 3), both HP and LP flows are increased, the heat exchanger stays balanced.
- On the bottom right (off-design 4), it is the temperature difference along the balanced heat exchanger (difference between warm to cold ends or higher temperature gradient) that is increased.

Let us perform some calculations on a balanced heat exchanger that is shown on Fig. 7.2, operating in its design conditions.

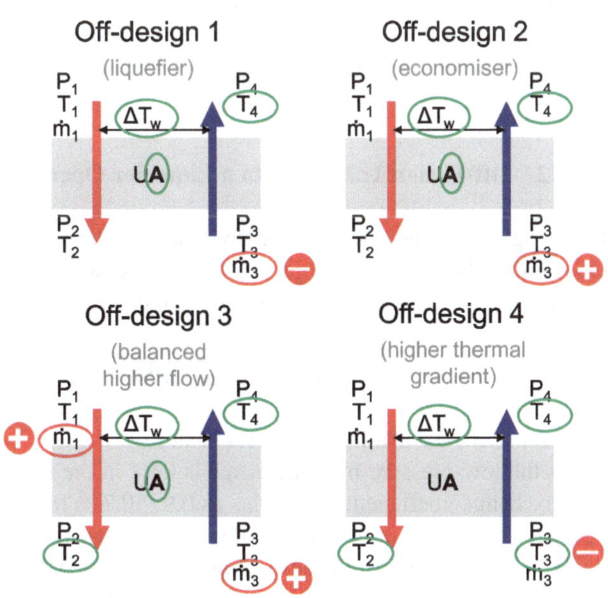

Fig. 7.1 A few off-design operations of a heat exchanger

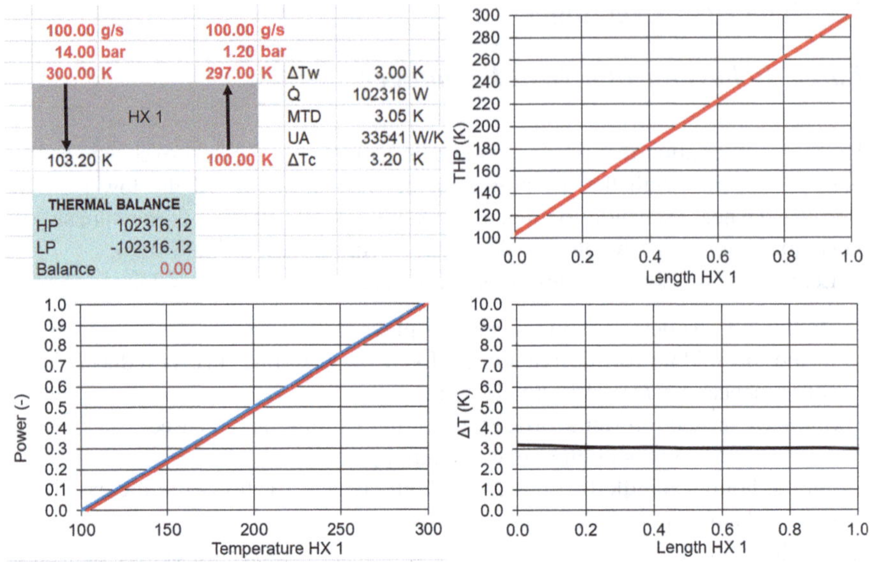

Fig. 7.2 Design of a balanced heat exchanger

7.3.1.1 Design Operating Conditions: Refrigerator Operation

Such heat exchanger characteristic is its *UA* value: 33541 W/K (see Fig. 7.2). As it is balanced and operating in a domain where the helium characteristics are almost constant, its warm and cold end temperatures differences are almost equal (not far from 3.0 K), and the temperature profile along it is linear (remember Sect. 2.4.1.2).

7.3.1.2 Off-Design 1: Moving to a Liquefier Operation

In this operation, the low-pressure flow rate is reduced by 3 g/s or 3% (see Fig. 7.3). Obviously, the heat exchange area (*A*) does not change, but the heat transfer coefficient is affected by the mass flow rate change. Here, the selected law is:

$$U_{\text{new}} = U_{\text{design}} \times \left(\dot{m}_{\text{new}}/\dot{m}_{\text{design}}\right)^{0.7} \tag{7.1}$$

(The exponent 0.7 is a little bit pessimistic)

As the low-pressure mass flow rate is 0.97 in the new operating conditions, the heat exchange coefficient (*U*) varies as 0.97^0.7 = 0.9789. Calculation of the new operation is made using a solver that varies the warm low-pressure temperature until the UA equals the one of the design operation, times 0.9789 (here, 32,833 W/K). The new heat exchange diagram shows a larger temperature difference at the cold end and a smaller temperature difference at the warm end.

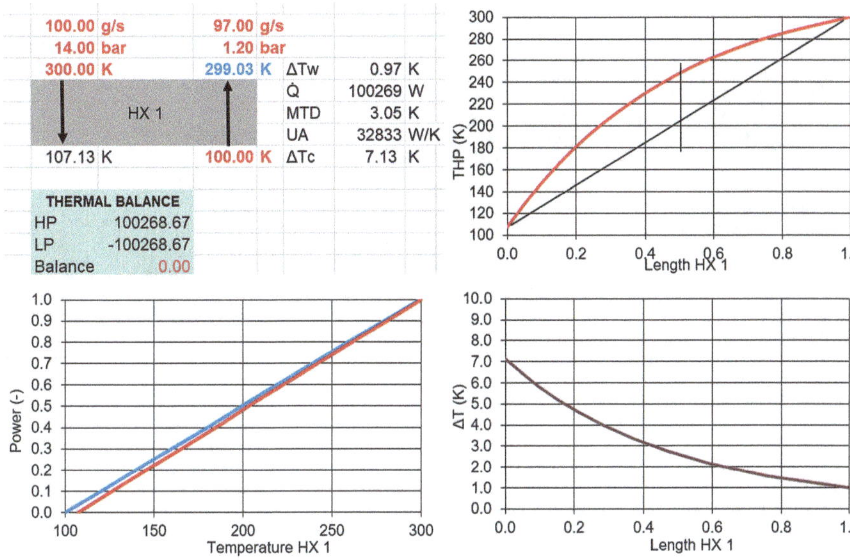

Fig. 7.3 Same heat exchanger with a 3% flow unbalance, in a "liquefier" mode

The important point to be noticed is the way the temperature varies along the heat exchanger length: the temperature variation is no longer linear, and at the middle of the heat exchanger, the temperature is increased by 34.8 K, compared to the balanced one, although the LP flow has been changed by 3% only!

7.3.1.3 Off-Design 2: Moving to an Economiser Operation

In this operation, the heat exchanger is unbalanced in the reverse way: the high-pressure flow rate is reduced by 3 g/s (see Fig. 7.4).

The UA value is unchanged because the LP flow is unchanged. Using the same way of calculation, one can notice that the temperature variation at the middle of the heat exchanger is 47.2 K! Again, do not forget that the mass flow rate has been unbalanced by only 3%!

7.3.1.4 The Dynamic Behaviour of a Heat Exchanger that Is Suddenly Unbalanced

To have an idea of the response time of a heat exchanger that is unbalanced, let us consider the aluminium alloy balanced heat exchanger that is operating in conditions stated in Fig. 7.2, the weight of which is 700 kg.

At time 0 in Fig. 7.5, the operating conditions are switched from conditions of Fig. 7.2 (balanced heat exchanger) to that of Fig. 7.3: a reduction of the low-pressure

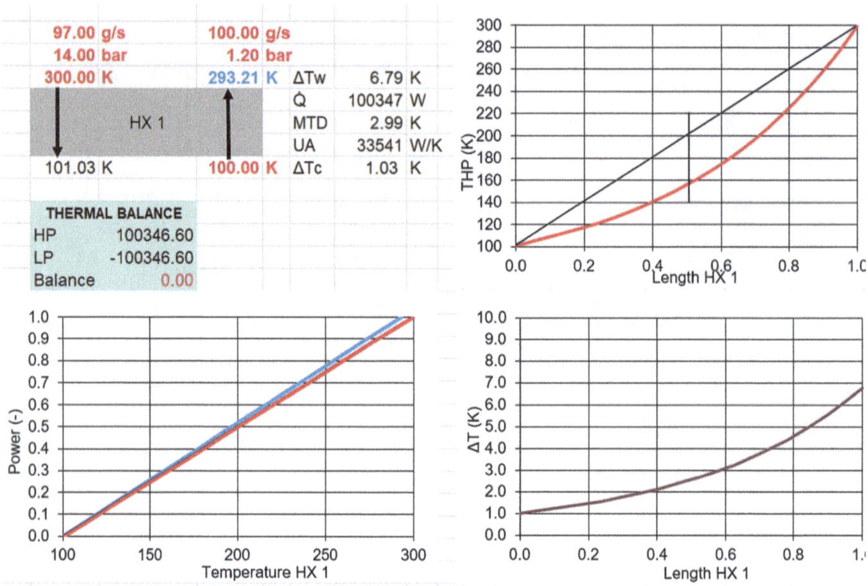

Fig. 7.4 Same heat exchanger with a 3% unbalance in an "economiser" mode

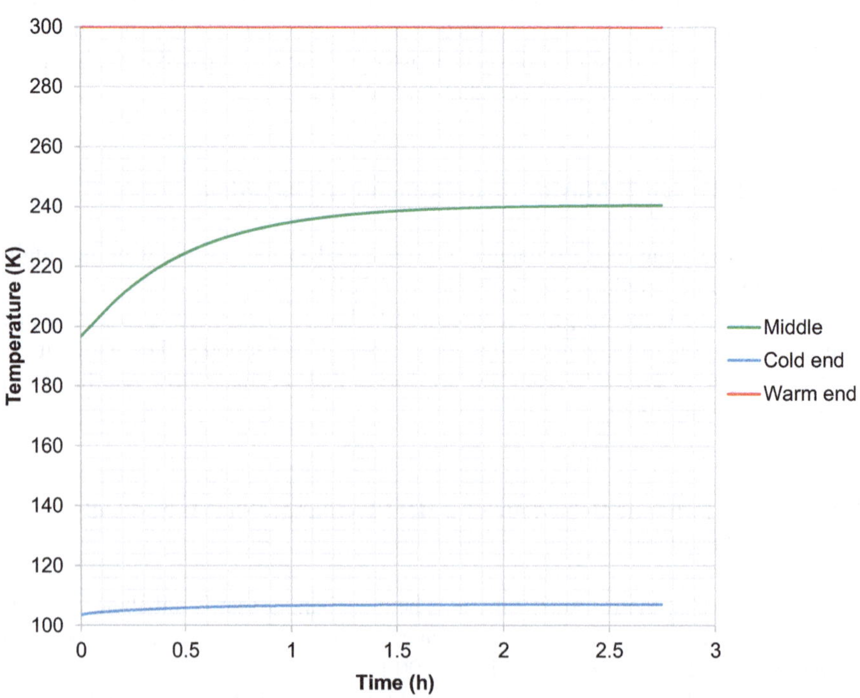

Fig. 7.5 Dynamic behaviour of a heat exchanger that is unbalanced by 3%

flow by 3 g/s (3%). These calculations have been performed by Benjamin BRADU from CERN. Thanks to him!

One can see that the temperature map is not yet stable after 1 hour! The heat exchanger needs 2 hours to reach a new steady state.

When a heat exchanger goes through such an unbalanced operation, as the middle temperature of the heat exchanger changes, some heat is either stored into or released from both the heat exchanger material and the gas that is stored inside it. One should not forget such a phenomenon that brings some hysteresis (delay) when dealing with transient regimes!

7.3.1.5 Off-Design 3: Changing the Mass Flow Rate

In this balanced operation, the helium mass flow rate is increased by 20% (Fig. 7.6, left). U increases as 1.2^0.7 = 1.136. The LMTD is increased from 3.00 to 3.17 K. The power is increased in the proportion of the mass flow rate.

Remark If the heat exchange coefficient would not vary, one could expect that the temperature difference would increase in the same ratio as the mass flow rate: 3.6 K. Therefore, be careful when performing thumb rule approximations!

7.3.1.6 Off-Design 4: Changing the Temperature Gradient

In this operation, the temperature of the cold LP flow is decreased to 50 K (see Fig. 7.6, right). As the flows stay constant, U is not affected, but because the temperature gradient along the heat exchanger is increased and, therefore, the exchanger power, the temperature difference increases from 3.0 to 3.71 K.

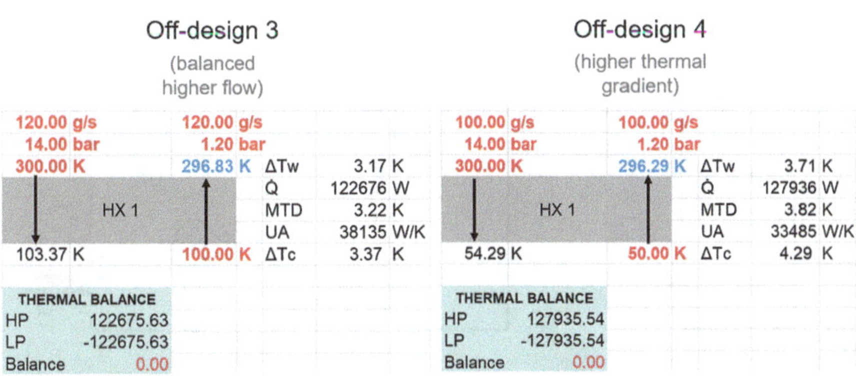

Fig. 7.6 Same heat exchanger with a mass flow rate increased by 20% (left) and a higher temperature gradient (right)

7.4 The Expansion Turbine

The mass flow rate that is processed by a turbine is such that the flow coefficient:

$$f_c = \dot{m} \times \frac{\sqrt{T_{in}}}{P_{in}} \tag{6.13}$$

remains constant.

7.5 The Brayton Cycle

Contrary to what has been done for the basic cycle study in Sect. 3, here the Brayton cycle is considered first.

Let us consider some off-design operating conditions (see Fig. 7.7).

- On the left, the design conditions are displayed.
- On the middle, the thermal load is higher.
- On the right, the cycle LP pressure is lower.

7.5.1 *Off-Design 1: Changing the Thermal Load*

To run off-design calculations for a Brayton cycle (Fig. 7.8), one has to consider not only the heat exchanger behaviour as in Sect. 7.3.1 but also the cycle compressor and the expansion turbine. The compressor is supposed to have a constant volumetric efficiency, which means that the ratio of the processed mass flow rate versus the suction pressure stays constant. The turbine is characterised by its flow coefficient

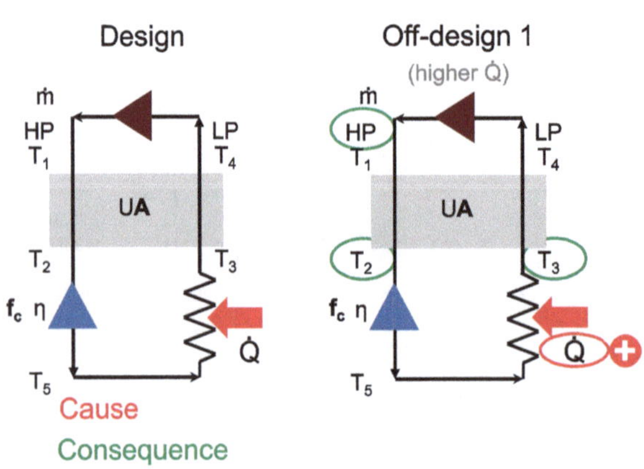

Fig. 7.7 A few off-design operations of a Brayton cycle

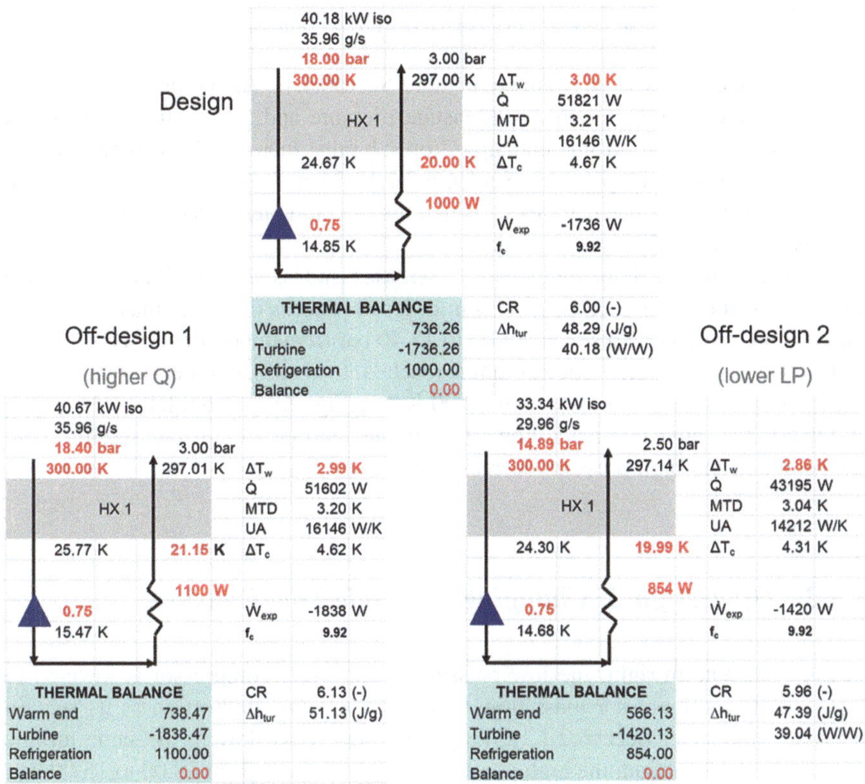

Fig. 7.8 Calculations of a Brayton cycle (reference, top) operating at a higher power (bottom, left) and lower LP (bottom, right) than design (top)

(remember Sect. 6.6.3.7.2): $f_c = \dot{m} \times \frac{\sqrt{T_{in}}}{P_{in}}$(here: 9.92); its value is constant whatever the operating conditions are. Running the calculation consists in fulfilling all such conditions simultaneously. This is made possible by using a solver.

What happens when the heat load of a Brayton cycle increases (see Fig. 7.8, bottom, left)?

In equation:

$$\dot{Q}_u + \dot{m}_1 \times (h_4 - h_1) = \dot{W}_{exp} \tag{3.15}$$

in order the equation stays balanced, when $\dot{Q}_u$ increases, it is necessary that the power $\dot{W}_{exp}$, extracted by the turbine increases by the same amount.

$$\dot{q} + \dot{Q}_u + \dot{m}_1 \times (h_4 - h_1) = \dot{W}_{\text{exp}} + \dot{q}$$

The turbine power is $\dot{m} \times \Delta h$ (remember Sect. 6.6.1). As the processed helium mass flow rate $\dot{m}$ does not change (constant pressure and temperature at the cycle compressor inlet), the turbine enthalpy drop Δh must increase. Remember that Δh varies with the inlet temperature (Sect. 6.6.3.7.1). Therefore, the only possibility in order the thermal balance stays exact is that the operating temperature of the turbine increases (see 6.6.3.7).

Since the turbine inlet temperature increases and as the distributor acts as a calibrated orifice (see Sect. 6.6.3.1.1) in order to process the same mass flow rate, the high pressure increases from 18.00 to 18.40 bar (remember: the HP control loop is deactivated), and a new equilibrium is reached, to absorb a higher power, but at a higher temperature 21.15 instead of 20.00 K (see Fig. 7.8, bottom left).

Of course, same rationale shows that if the thermal load decreases, the cold temperatures and high pressure decrease too.

7.5.2 *Off-Design 2: Changing the Cycle Low Pressure*

In Fig. 7.8, bottom right, the low pressure of the cycle compressor is lowered to 2.50 bar. The compressor mass flow is reduced in same proportion as its suction pressure (remember Sect. 6.2.1.1). To process less mass flow in the same turbine distributor cross section, the high pressure <u>should</u> decrease from 18.00 to 15.00 bar.

One could expect that as a result of processing a fraction of the design mass flow rate, the available new cryogenic power is in the same ratio (here 2.5/3.0 = 0.833, leading to 833 W). By the way the power is slightly higher (854 W) because the power that is getting into the system through the warm end is lower due to a lower cycle mass flow rate, therefore, a lower warm end temperature difference: 556 instead of 738 W. The turbine inlet temperature is slightly lower (24.30 K instead of 24.67 K), and that explains why the HP is not exactly 15 bar as expected: the turbine distributor needs a lower pressure to process the compressor flow.

The expansion ratio is almost unchanged (CR = 6.00 and 5.96) so as the enthalpy drop (Δh = 48.29 J/g and 47.39 J/g).

The important point is that, by only changing the low-pressure set point, the system has adapted automatically to the new thermal load. Such an adaptation is made at almost same efficiency: comparing the thermal load to the compressor isothermal input mower gives almost same figures: 40.18 W/W and 39.04 W/W.

This is a very important principle that is used to run refrigeration cycles at turn-down conditions (see Sect. 8.7).

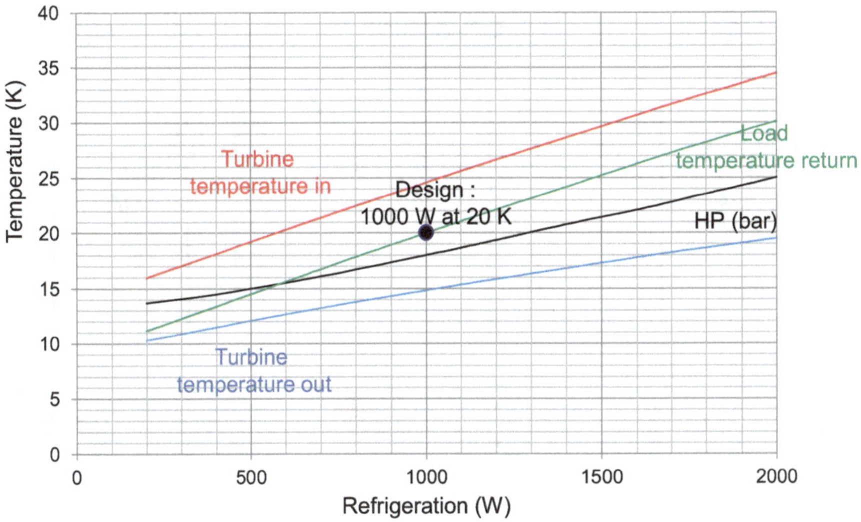

Fig. 7.9 Evolution of various parameters of a Brayton cycle versus cryogenic refrigeration power

7.5.3 Evolution of Parameters

In Fig. 7.9, one can see how various parameters vary according to the cryogenic power when varying the cycle low pressure.

7.6 The Joule Thomson Cycle

The Joule Thomson cycle is used for refrigeration or liquefaction or a combination of the modes. Let us consider some off-design operating conditions (see Fig. 7.10).

- On the left, the design conditions are displayed,
- On the middle, the refrigeration rate is increased.
- On the right, the high pressure is increased.

Remark The increase of the thermal load on a Brayton cycle has for consequence an increase of the operating temperatures (remember Sect. 7.5.1); but in case of a JT cycle, a higher heat load would only vaporise the existing liquid: there is no way to get more power for the same HP.

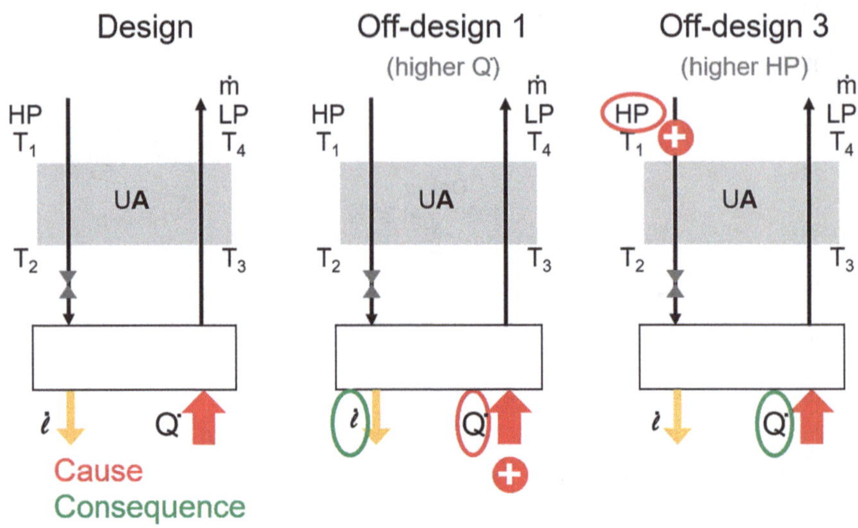

Fig. 7.10 A few off-design conditions of a Joule Thomson cycle

7.6.1 *Off-Design 1: Changing the Thermal Load Repartition – Moving Towards More Refrigeration*

The consequence of increasing the refrigeration rate is, not surprisingly, a decrease of the liquefaction duty. As a consequence, the heat exchanger becomes less unbalanced, the warm end temperature decreases, and the cold end temperature decreases. The temperature upstream the JT valve decreases from 5.66 K to 5.35 K, and the liquid concentration after expansion increases from 0.51 to 0.57.

Conversely, increasing the liquefaction rate decreases the warm end temperature difference 0.17 K instead of 0.20 K), increases the temperature upstream the JT valve (5.95 K instead of 5.66 K) and gives lower liquid concentration (0.45 instead of 0.51) (Fig. 7.11).

7.6.2 *Off-Design 2: Changing the Cycle High Pressure*

Here, it is not surprising that a higher HP, compared to design, gives a higher refrigeration power for same liquefaction rate. In Fig. 7.12, the lower temperature upstream the JT valve decreases from 5.66 to 5.11 K, but, as the high pressure is different, one cannot strictly compare the liquid concentration after expansion.

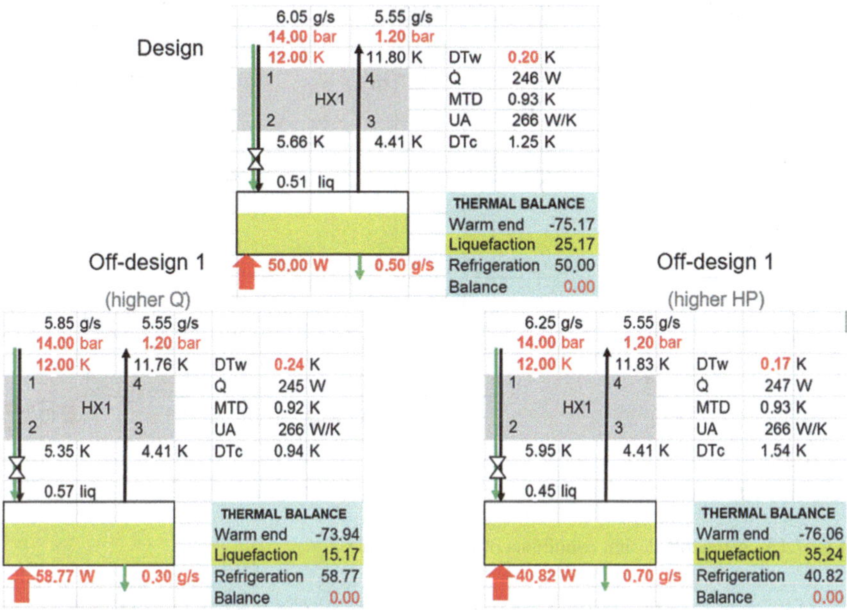

Fig. 7.11 Calculations of a Joule Thomson cycle operating at a higher thermal load (left) and higher liquefaction rate (right) than design (top)

Fig. 7.12 Calculations of a Joule Thomson cycle operating at a higher high pressure

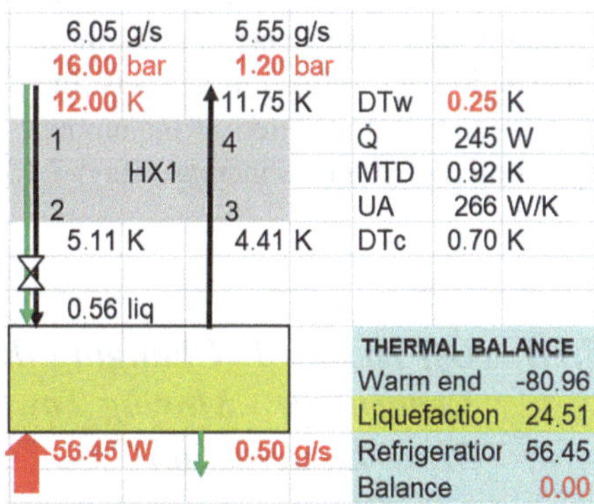

7.7 The Claude Cycle

In the case of a Claude cycle, the basic mechanisms are obviously the same, but the situation is more complicated, because of the higher number of components. Here, too, the heat load is split into refrigeration and liquefaction. These simple exercises are performed on cycles with turbines in a parallel arrangement. Results would be similar with series-arranged turbines, but calculations would be heavier considering the simple calculation tools that are used here.

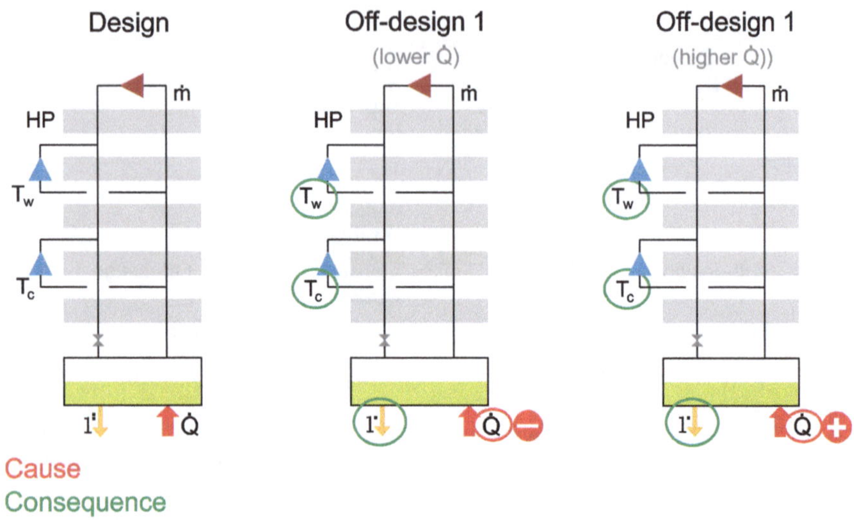

Fig. 7.13 A few off-design conditions of a Claude cycle

One understands intuitively that, as it has been noticed in the JT cycle behaviour, for an unchanged total "equivalent heat load", if the refrigeration load is reduced, the liquefaction rate increases and vice versa (remember Sect. 3.2.4).

However, as in the case of the Brayton cycle, if the total equivalent load is different, the high pressure of the cycle varies: it increases as the liquefaction load increases and vice versa. Furthermore, according to the relative values of the liquefaction and refrigeration loads, the power extracted by each turbine (and thus its temperatures) is varied: if the amount of liquefaction load increases, the turbine operating temperatures increase (Fig. 7.13).

7.7.1 *Off-Design 1: Changing the Thermal Load Repartition – Moving Towards More Refrigeration or More Liquefaction*

Consider the system (liquefier/refrigerator), during steady state mixed operation, roughly half and half in liquefier and refrigerator mode, as defined in Fig. 7.14.

To run off-design calculations for a Claude cycle (Fig. 7.15), as for a Brayton cycle, one has to consider simultaneously the behaviours of each heat exchanger, each cycle compressor and each expansion turbine. It is the same procedure as the one used for the Brayton cycle but, of course, much more complicated due to the higher number of components.

When the thermal load is reduced to 30 W, how does the liquefaction load change? The reduction of the thermal load at the temperature of liquid helium bath

Fig. 7.14 Claude cycle at design conditions

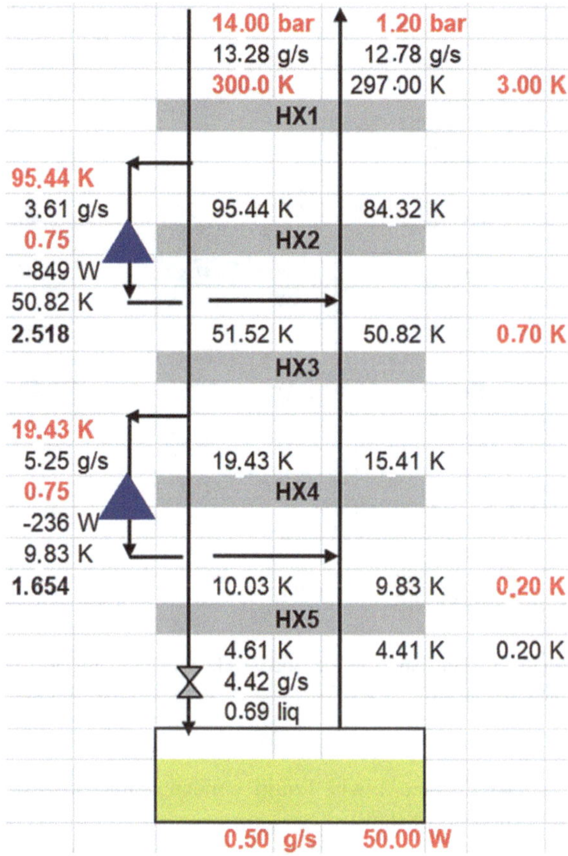

(refrigeration) results in less liquid vaporisation. The additional liquid product is withdrawn such that the level remains constant to maintain thermodynamic equilibrium.

As there is more liquid that is produced, consequences are (see Fig. 7.15, left):

- Turbines must extract more energy, consequently, their operating temperatures increase (remember Sect. 7.5.1). The inlet of the warm turbine increases from 95.44 K to 112.35 K; the inlet of the cold turbine increases from 19.43 K to 21.01 K.
- As the inlet temperatures of the turbines increase, the cycle high pressure rises from 14.00 bar to 17.35 bar, in order to process the same mass flow rate. Remember that the HP control loop is deactivated.
- All the heat exchangers are more unbalanced (the HP flow increases), consequently:
  - In HX 5 heat exchanger (JT) the cooling of the high-pressure flow is lower: the temperature upstream the JT valve is higher (4.62 K instead of 4.61 K),

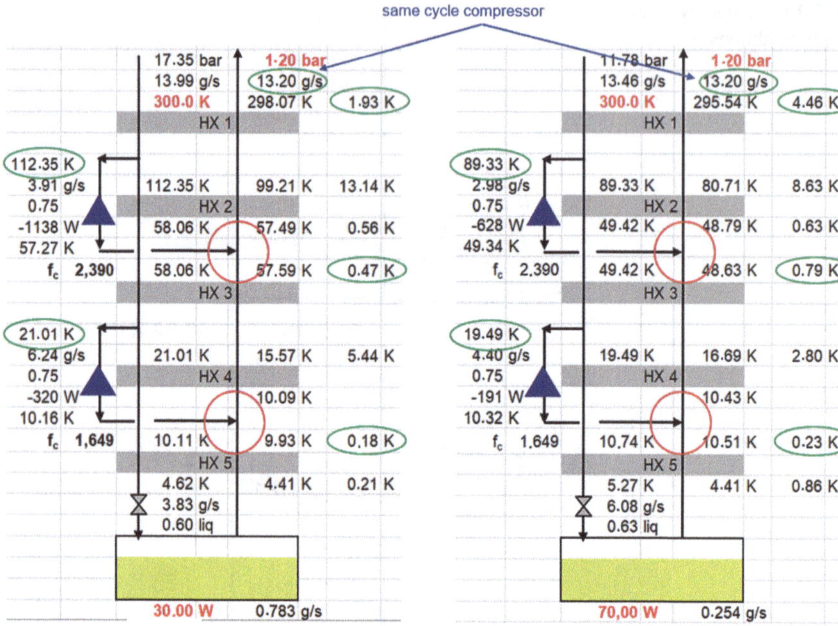

Fig. 7.15 Calculation of a Claude cycle operating "towards liquefaction", left, or refrigeration, right

resulting a lower liquid concentration downstream of the JT valve (0.60 instead of 0.69),

- The warm end temperature difference of HX1 decreases from 2.99 K to 1.93 K.

Remark that the discharge temperature of the turbines is no more equal to the temperature of helium getting out of the low pressure of the heat exchanger; for example, the flow that gets out of T1 is at 57.27 K; it mixes with LP flow from HX 3 that is at 57.59 K.

The system finds a new equilibrium with a modified temperature map.

The resulting liquefaction rate is higher than the one for design: 0.78 g/s.

Conversely, if the thermal load is higher:

- The turbine operating temperatures decrease.
- The cycle high pressure decreases.

In Fig. 7.16, one can see the evolution of a few parameters versus power when the refrigeration power increases:

- The temperatures decrease.
- The turbine powers decrease.
- The high pressure decreases.

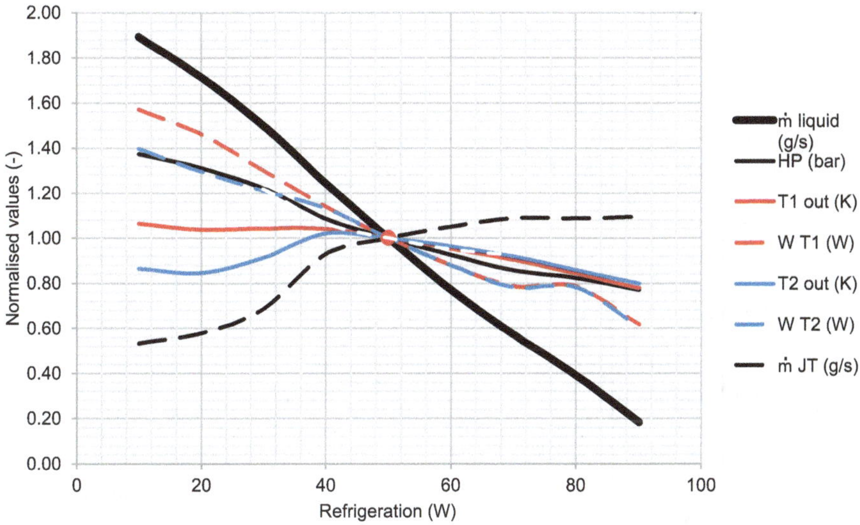

Fig. 7.16 Evolution of various parameters of a Claude cycle versus power

- The JT mass flow rate increases.
- The liquefaction flow decreases almost linearly.

The variation of some of these parameters is not so "clean" (not almost linear) due to the changes of helium properties in this operating range.

An Important Remark The above calculations are pure theory, actual machines could not operate on such a wide range of conditions; the results are more and more unrealistic as far as the refrigerator operates away from the design point. The interest of this study is to show the way in which the cycle parameters vary.

7.7.2 Off-Design 2: A Pure Liquefier Operating at Constant or Rising Level

Cycle process calculations that have been performed in Sect. 3.5.4 consider that the liquid level is constant. It is interesting to see how a liquefier behaves when switching from constant to rising level regime; in Fig. 7.17, one can see that:

- The apparent liquefaction rate increases by 17%.
- All temperatures decrease due to the extra pre-cooling that is brought by vapour that is "pushed out" of the dewar by liquid.
- As the temperature upstream the JT valve is lower, the liquid concentration after expansion is higher: 0.47 instead of 0.30.
- The total power of turbines is almost constant.

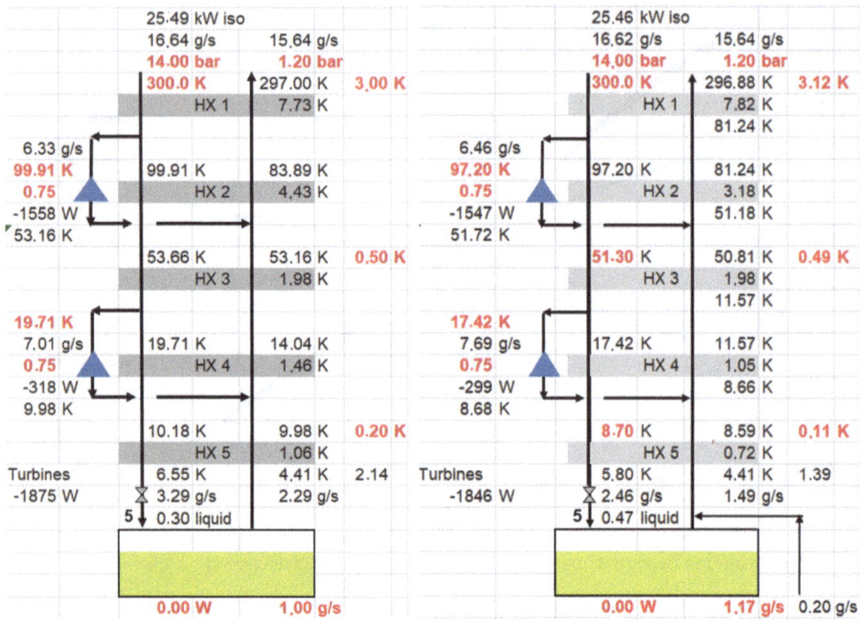

Fig. 7.17 Pure liquefier operating at constant or rising level

- At the place where a turbine flow mixes with an LP flow, there are small temperature differences.

7.7.3 A Special Situation Happening During a Pure Liquefier Operation: Liquid Withdrawal

During night time, when no liquid is withdrawn from the stationary dewar, a liquefier operates at rising level, where liquid is displacing the saturated vapour phase out of the dewar, through the low-pressure circuits, helping pre-cooling of helium prior to the JT expansion. Such operation has been considered for the JT cycle in Sect. 3.3.2.4.3.

But, during day-time, when liquid helium is withdrawn from the stationary dewar, each volume of liquid that is leaving is replaced by the same volume of vapour. The rate at which liquid helium is withdrawn is generally higher than the liquefaction rate (some hundreds of litres per hour can be withdrawn from a dewar that is fed by a 100 L/h liquefier). If one considers an increasing withdrawn rate, for low values, the liquefier can provide saturated vapour and liquid. When the withdrawn flow rate increases, the proportion of liquid versus saturated vapour decreases until only saturated vapour is produced: there is no more liquid production. For even higher withdrawal rates, the liquefier will only be able to provide helium that is no more saturated vapour but only cold gas, the latter being warmer and warmer as far as the withdrawn flow increases.

Therefore, in order to roughly estimate the actual liquid helium production during a 24-hour period of time, it is better to use the constant level calculation result.

One must be careful with measurement of liquid production that is made by observing the level increase in the stationary storage: the figures must be compared with the corresponding rising level calculation result provider by the maker.

7.7.4 *Output Increase by Liquid Nitrogen Pre-Cooling*

7.7.4.1 Brayton Cycle

In Fig. 7.18, left, the refrigerator operates without LN2 pre-cooling. Here the UA of the warmest heat exchanger is the sum of HX 1 and HX 0, which makes a larger than usual heat exchanger. This is the reason why the warm end temperature difference at room temperature is smaller than usually: 1.00 K instead of 3.00 K. Note that HX00 and HX VAP do not bring any extra UA.

In, Fig. 7.18 one can see that, when the temperature gradient along heat exchanger HX 1 is reduced form 300–23 = 277 K to 80–22 = 58 K, the warm end temperature of the heat exchanger decreases from 3.00 K to 0.36 K and the heat load at the warm end is reduced from 295 W down to 167 W. In order to balance the cycle, the turbine is less powerful; its temperature is lower: to circulate the same helium mass flow rate in it, the inlet pressure is lower: 16.95 bar instead of 18.00 bar.

One should also notice that the discharge temperature of the turbine is lower: 13.28 K instead of 13.70 K. Be careful such a temperature change could bring trouble in some situations related to solidification of hydrogen, for example.

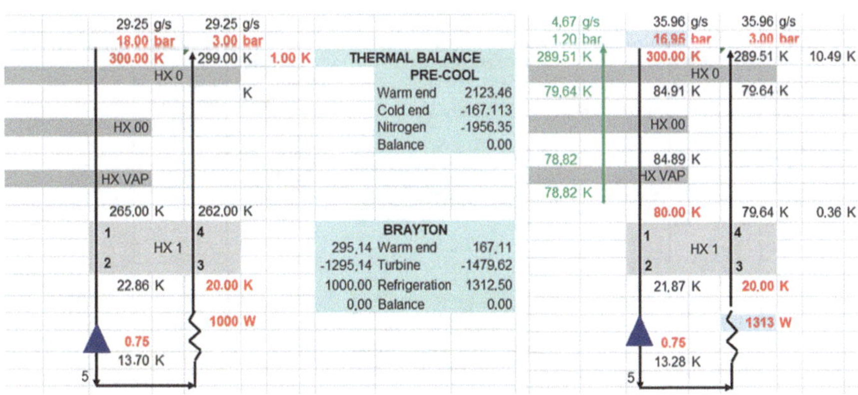

Fig. 7.18 Liquid nitrogen pre-cooled Brayton cycle

7.7.4.2 Claude Cycle

7.7.4.2.1 Pure Liquefier

The operation of a liquefier without or with liquid nitrogen pre-cooling is a usual situation in small size standard liquefiers.

When the liquefier is pre-cooled, the energy of helium to be liquefied is removed by heat exchange with nitrogen, down to around 80 K. Therefore, in a thermal way, turbines "take care" of the helium to be liquefied from a lower temperature (80 K instead of 300 K). Furthermore, the warm end temperature difference of heat exchanger HX 1 is reduced because its thermal gradient is lower, so the heat load at the warm end is decreased (remember Sect. 7.3.1.5).

In Fig. 7.19, left, the liquefier operates without LN2 pre-cooling. Here the UA of the warmest heat exchanger is the sum of HX 1 and HX 0, which makes a larger than usual heat exchanger. This is the reason why the warm end temperature difference at room temperature is smaller than usually: 1.50 K instead of 3.00 K. Note that HX 00 and HX VAP do not bring any extra UA.

When the calculation is performed with liquid nitrogen pre-cooling, in Fig. 7.19, right, the warm end temperature of the turbine-cooled cycle is the warm end of the heat exchanger HX 1 which is now at about 80 K.

As a result:

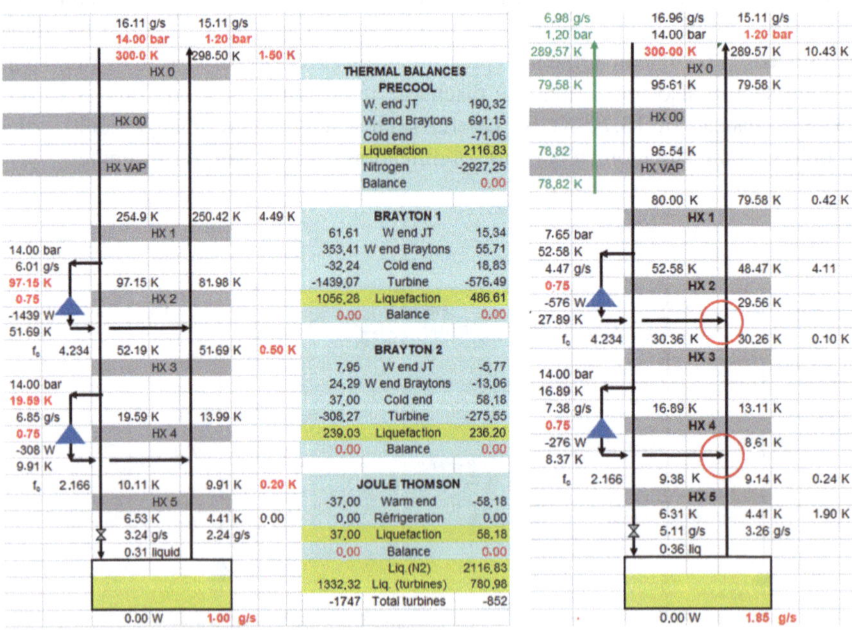

Fig. 7.19 A pure liquefier pre-cooled with liquid nitrogen

- The heat exchanger HX 1, which processes the same mass flow rate, sees its temperature gradient (temperature difference between the hot and cold ends) reduced. The result is a reduction of its temperature difference at the warm end (remember Sect. 7.3.1.6). As seen in Sect. 3.2.4, the result is that the heat input to the warm end of the Brayton cycle is reduced.
- The energy required to cool the helium to be liquefied by the turbines is reduced, since it is cooled down by nitrogen from 300 K to 80 K.
- Both effects contribute to a reduction of the thermal load of turbines; consequently, their operating temperatures decrease (see Fig. 7.19).

More flow is necessary for the JT cycle, but the compressor cannot provide more; therefore, it is necessary to throttle upstream the warm turbine the inlet pressure of which is now 7.65 bar instead of 14.00 bar.

A new equilibrium is reached.

Therefore, with same cycle compressor mass flow rate, the production of a liquefier designed with no liquid nitrogen pre-cooling is approximately doubled when pre-cooled with liquid nitrogen.

Remark If the cycle compressor that operates at 50 Hz is speeded up to 60 Hz, the liquid output rises up to 2.07 g/s: it is interesting to add a frequency converter on the system.

7.7.4.2.2 Pure Refrigerator

As the cycle is balanced and there is no heat load due to the cooling of the helium to be liquefied, the reduction of the turbine thermal load is only due to the smaller warm end temperature difference of the exchanger HX 1B (Fig. 7.20).

The increase in capacity of the refrigerator is of the order of 10 to 20%.

7.8 Behaviour of an Almost Actual Brayton Refrigerator

Remember that some simplification assumptions have been made in Sect. 2.4. They applied for all former calculations in this chapter:

- No pressure drops, heat leaks or longitudinal conduction in the heat exchangers
- No change in turbine efficiency, compressor volumetric efficiency or material properties

Actually, all these parameters change.

As an example, let us look at a simple Brayton cycle operating in design conditions (Fig. 7.21, left). In order to simulate an "almost actual" machine, pressure drops and heat leaks have been introduced:

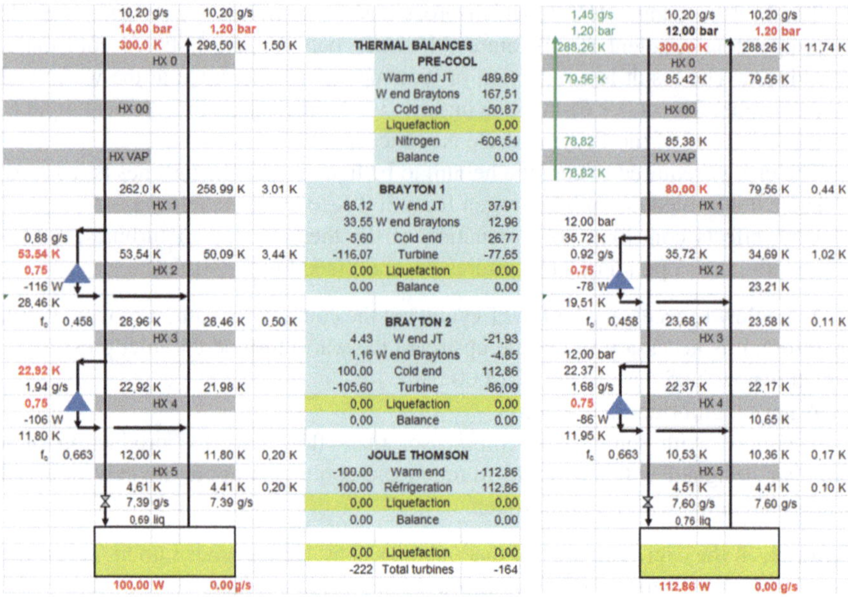

Fig. 7.20 A pure refrigerator pre-cooled with liquid nitrogen

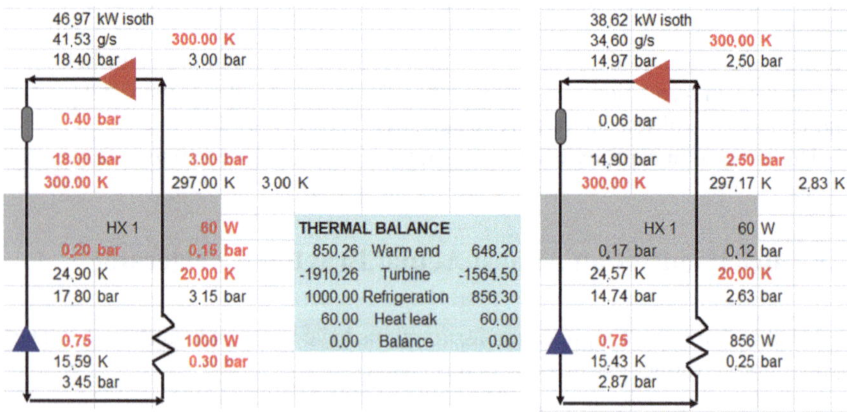

Fig. 7.21 An almost actual Brayton refrigerator with pressure drops and heat leak

- A pressure drop of:
  - 0.40 bar in the oil removal system
  - 0.20 bar in the HP circuit of the heat exchanger
  - 0.30 bar in the user heat load
  - 0.15 bar in the LP circuit of the heat exchanger

- A heat leak of 60 W on the heat exchanger

One can make a comparison with the cycle that is shown in Sect. 3.4.3 for a refrigerator that fulfils same duty: 1000 W at less than 20.00 K, which has neither pressure drops nor heat leaks, where the cycle mass flow rate is lower 35.96 g/s and, consequently, needs 40.18 kW isothermal instead of 46.97 kW.

Now, let us decrease the compressor suction pressure to 2.50 bar, as it is done in Sect. 7.5.2. The heat leak is, obviously, unchanged, but the pressure drops vary according to mass flow rate and pressure. The new operation is shown in Fig. 7.21, right.

Some remarks on the new operating conditions:

- The cycle mass flow rate is lower.
- The warm end temperature difference decreases from 3.00 K to 2.83 K.
- As the thermal load return temperature (20 K) is the same for both configurations and as the cycle mass flow rate is lower, the discharge temperature of the turbine is lower: 15.43 K instead of 15.59 K.

Compared with same operation, without pressure drops and heat leak (remember Fig. 7.8, right), the cryogenic power is very similar: (856 W and 854 W), but the isothermal compressor power is 38.62 kW instead of 33.34 kW.

7.9 A Digest on Off-Design Operation

This chapter is the most important in terms of "feeling" the behaviour of a cryogenic refrigeration system.

7.9.1 The Cycle Compressor

The mass flow rate that is processed by a compressor is proportional to the suction pressure.

7.9.2 The Heat Exchanger

The heat exchange coefficient varies roughly as the 0.7 power of the low-pressure mass flow rate.

Change of Mass Flow Balance When unbalancing a heat exchanger that has been designed for a balanced operation (*pure refrigerator*), for same LP mass flow rate (Fig. 7.22, left):

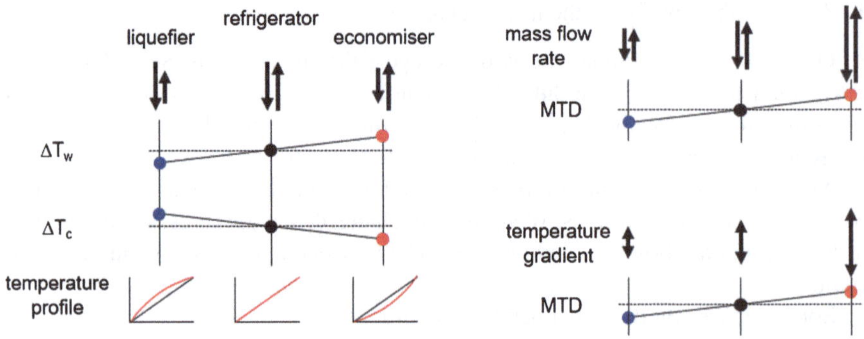

Fig. 7.22 Evolution of some parameters of a heat exchanger according to the operating regime

- The warm end temperature difference decreases when going towards liquefier operation or increases when going towards economiser operation.
- The cold end temperature difference increases when going towards liquefier operation or decreases when going towards economiser operation.
- The temperature profile changes, as shown in Fig. 7.22, left, bottom. When going towards liquefier operation, the temperature at the middle of the heat exchanger increases or decreases when going towards economiser operation.

Changing the flow balance of a heat exchanger makes a rather large temperature change inside the heat exchanger.

As a consequence, even for a few per cent flow unbalance, a not negligible transient time due to thermal inertia of the heat exchanger material plus gas can occur.

Change of Mass Flow Rate When increasing the mass flow rate of a flow balanced heat exchanger, the mean temperature difference increases as shown in Fig. 7.22, top, right, and, conversely, the MTD decreases when the mass flow rate is lower.

Change of Temperature Gradient When increasing the temperature gradient of a flow balanced heat exchanger, the mean temperature difference changes as shown in Fig. 7.22, bottom, right, and, conversely, the MTD decreases when the temperature gradient is lower.

On top of the changes that have been identified here above, one should not forget that, in real heat exchangers, other parameters change too: pressure drops, change of thermal conductivity of the heat exchanger material and effect of the longitudinal conduction.

7.9.3 *The Expansion Turbine*

The flow coefficient governs the flow that is processed in the turbine.

When, for same HP and LP pressures, the thermal load on a turbine increases, the operating temperatures increase too, and the mass flow rate decreases (see Fig. 7.23).

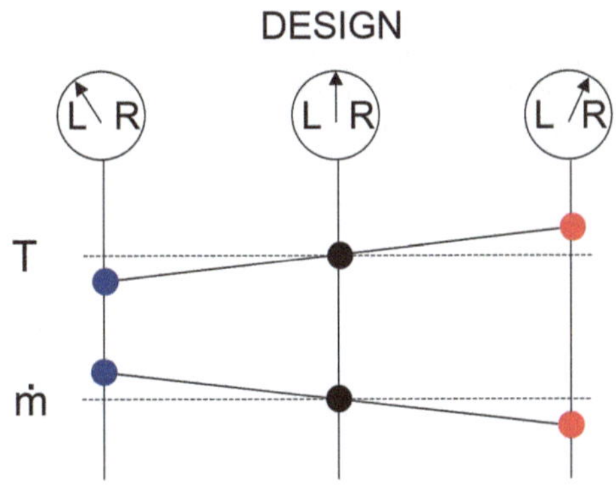

Fig. 7.23 Evolution of some parameters of a turbine according to the thermal load

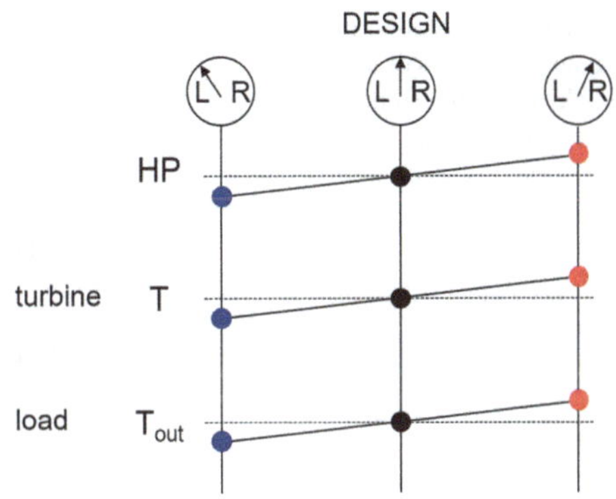

Fig. 7.24 Evolution of some parameters of a Brayton cycle according to the operating regime

7.9.4 The Brayton Cycle

When increasing the heat load, the cycle high pressure, the operating temperature of the turbine and the return temperature from the load increase (Fig. 7.24).

When the compressor low pressure decreases, the power that is absorbed decreases in almost the same ratio.

When the cycle compressor suction pressure is decreased, the cryogenic power of the cycle decreases almost proportionally.

When pre-cooled with liquid nitrogen, a Brayton cycle can absorb up to 30% more power at same temperature.

7.9.5 The Joule-Thomson Cycle

For same HP, when the operating regime moves towards more refrigeration, the heat exchanger warm end temperature difference and the liquid concentration at the discharge of the JT valve increase, the temperature upstream the JT valve decreases (Fig. 7.25)

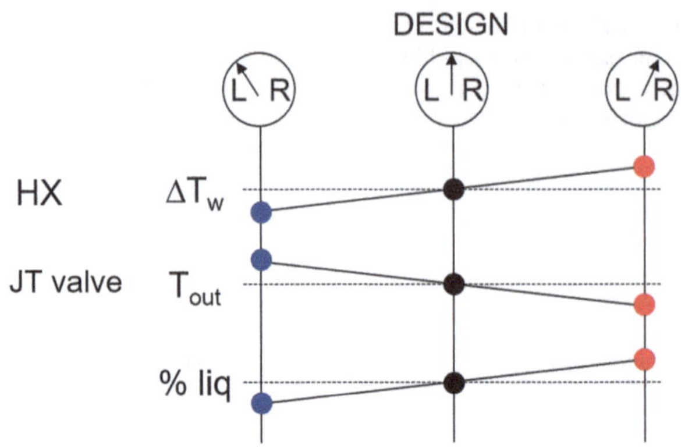

Fig. 7.25 Evolution of some parameters of a Joule-Thomson cycle according to the operating regime

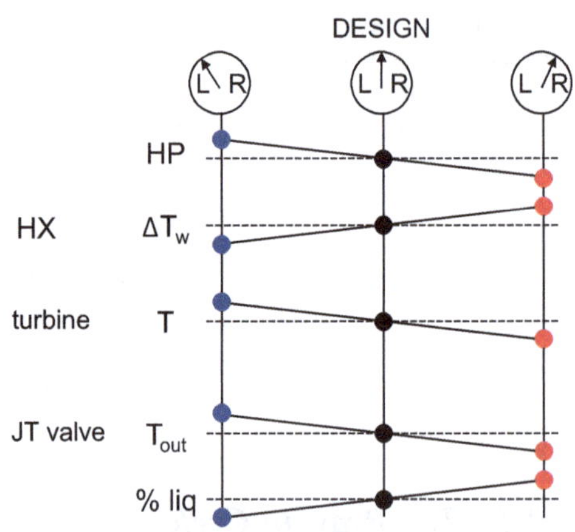

Fig. 7.26 Evolution of some parameters of a Claude cycle according to the operating regime

7.9.6 The Claude Cycle

When the operating regime moves towards liquefaction, the cycle HP increases, the warm end temperature difference of heat exchanger decreases, the temperature of the turbines and upstream of the JT valve increase, and the liquid concentration of the discharge of the JT valve decreases and conversely (see Fig. 7.26).

This sentence alone is a digest of the behaviour of the components of any refrigeration cycle.

When pre-cooled with liquid nitrogen, a liquefier can roughly double its production; however, a refrigerator would only absorb, at maximum, around 20% more cryogenic power.

Chapter 8
System Control

Abstract Modern refrigeration or liquefaction systems are automatically controlled, irrespective of their size. The modern control system is not only in charge of the control of the machine but also of many ancillary tasks dealing with the operating parameters. The quality of the control system procedures allows reliable operation and high availability of the system, making them part of the important features of a modern cryogenic system.

8.1 Introduction

Up to this place, the book described calculation of cycles and technology of the main components. Let's suppose the refrigerator has been built: it is time to run it. In former times, the systems to be controlled were rather simple, but instrumentation was expensive, therefore, scarce and processing of information with electrical relays was a difficult task. See the aspect of a control panel at the end of the 1970s in Fig. 16.33! Since the 1980s, the evolution of computer technology moved so fast that it became possible, with a reasonable investment, to get more information on the process and access to systems that could perform controlling (Boolean combinations, loop control, calculations, etc.) actions.

The power and the complexity of the refrigeration systems have very much increased (from 1965 to 2013, the author has seen liquefier output rising from 7 to 8000 L/h: this is three orders of magnitude!); therefore, a manual control became soon really no longer possible: for example, a large refrigerator as one that is operating at LHC, CERN, is operated with some 40 control loops! When dealing with so high powers taken from the grid, it is mandatory to operate the plant in an efficient way: efficient power control is a must.

On top of that, plants have to operate at least all day long and up to all year long with minimum human actions to be done. On very large systems as the former CERN LEP cryogenic system, after a very short interruption (an electrical power outage of 1 minute), the plant needed 1 hour in order to recover the former steady state of the system. If the shut-down time was higher, the time to recover increased

G. Gistau Baguer, *Cryogenic Helium Refrigeration for Middle and Large Powers*, International Cryogenics Monograph Series,
https://doi.org/10.1007/978-3-030-51677-2_8

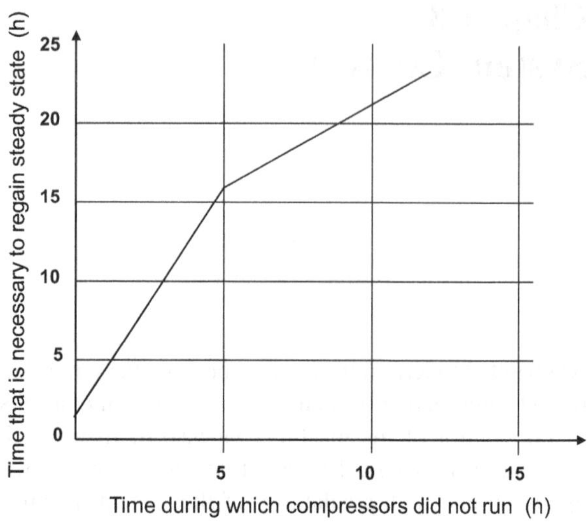

Fig. 8.1 Time that is necessary to recover a steady state after a power breakdown in a big cryogenic system

strongly (see Fig. 8.1). This means that a cryogenic system should be able to re-start automatically. The author knows that operators generally do not like to allow a system to re-start until a human check of the system has been performed. However, today, there are a number of new sensors, including image processors, that can be combined with smart procedures and would allow to reach the same level of confidence as that of a human check. Furthermore, this process is faster and applies whatever the time it is, including night, making re-starting the system in the smallest possible time period.

A control system allows performing *accurately*, and *at the correct time*, process control actions. Today, small- or middle-size modern refrigerators or liquefiers are always controlled by a process control system that is generally integrated into the cold box module.

In order to emphasise the importance of the control system, let's assume that a person who is in charge of a project has got two offers for a refrigeration plant. Both of them fulfil all the requirements and have the same capital expenditure (CAPEX) and operational expenditure (OPEX). According to this information, one has no means to select the best suited plant. However, among the two of them, the one that is fitted with the most "clever" set of control procedures is the best, because it would react to perturbations by actions that would preferably reduce the cryogenic power instead of stopping the plant, giving this plant the highest reliability and availability, which are the most important characteristics of a cooling plant, especially when dealing with large and complex system.

The control system makes also possible to witness, store into a data bank and plot parameter evolutions for the operator. See in Fig. 16.32 the antic ways to follow a cool-down operation!

Even if this does not exist yet, to the author's knowledge, it is possible to think about autodiagnosis procedures to help the operator. A lot of room exists to improve the process control in helium refrigeration plants!

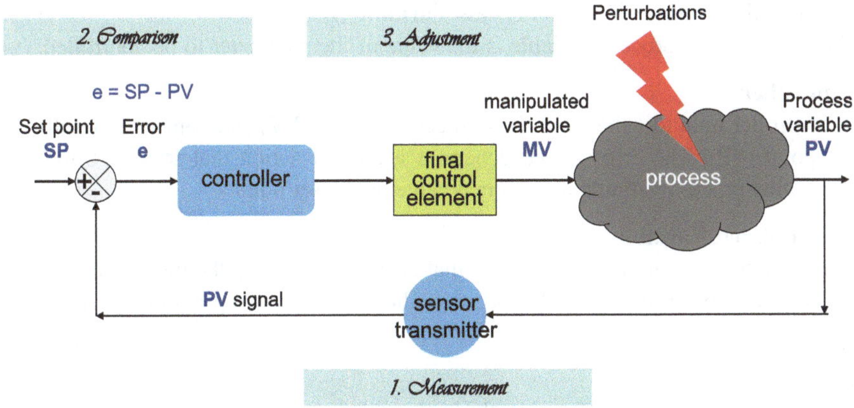

Fig. 8.2 Block diagram of a PID control loop

Let's have a glance on some specific aspects of conventional process control for helium refrigeration plants. Special procedures as full state feedback or predictive systems that are used for systems having very long response time are not considered here.

Remark In Sect. 6.6.3.7 turbine and Sect. 6.7.3.4 cryogenic compressors, various very special procedures to control these machines have already been described. In the present chapter, the system general control procedures integrate these specific procedures.

8.2 A Reminder About Process Control

8.2.1 Terminology

See Fig. 8.2.

Process Variable (PV)
A process variable is a condition of the process fluid (a liquid or a gas) that can change the manufacturing process in some way. A pressure or a temperature can be a process variable.

Sensor or Transmitter
A sensor is a component that measures the process variable and translates it into a signal (PV signal), generally electric, either analogical or digital, that is the image of the process variable, for example, a pressure transmitter.

Set Point (SP)
The set point is the value at which a process variable is to be maintained, for example, 12.5 bar for a pressure.

Measured Variable (MV) or Process Variable (PV)
The measured or process variable is the value of the parameter to be controlled.

Controller
A controller monitors the controlled process variable (PV) and compares it with the set point (SP). The error signal generates a control action that tries to bring the controlled process variable at the same value as the set point.

Manipulated Variable
The manipulated variable is the factor that is changed to keep the measured variable at set point. For example, a valve position allows to control a pressure (or any other parameter).

Error (e)
The error is the difference between the measured variable and the set point. It can be either positive or negative.

Correcting Element/Final Control Element
The correcting or final control element is the part of the control system that acts to physically change the manipulated variable, for example, a valve.

Actuator
An actuator is the part of a final control device that causes a physical change in the final control element when signalled to do so, for example, a valve actuator.

8.2.2 A Conventional PID Control Loop

A proportional-integral-derivative controller (PID controller) is a generic control loop feedback mechanism widely used in industrial control systems. The controller attempts to minimise the error by adjusting the process control inputs (see Fig. 8.2). The theory of PID control was developed by Nicolas Minorsky (1885–1970), an American engineer, to improve the automatic steering of boats in 1922.

The PID controller calculation algorithm involves three separate parameters: the proportional (P), the integral (I) and the derivative (D) values. The weighted sum of these three actions is used to adjust the process via a control element such as the position of a control valve or the power supplied with a heater.

The sensor transmits the process value signal to the controller in which a comparison of the process variable and the set point is done, then processed according to the P and I values and sent to the final control element (Fig. 8.3).

The target of a control system is to control a parameter in order that it stays stable, but, if some action values are not correctly set, there are some risks like generating process instabilities that could lead to mechanical damages in the system.

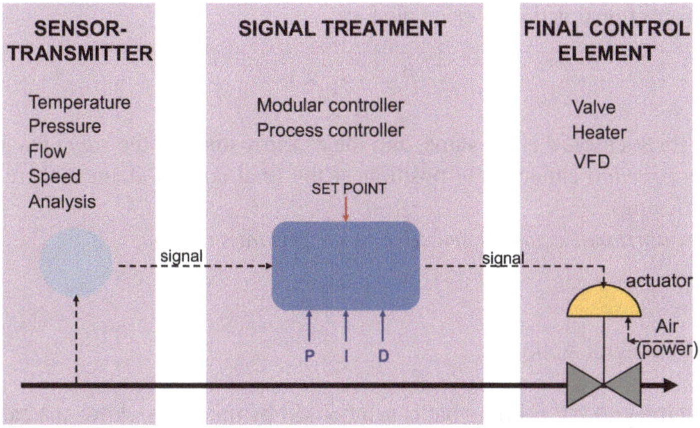

Fig. 8.3 The physical components of a conventional PID control loop

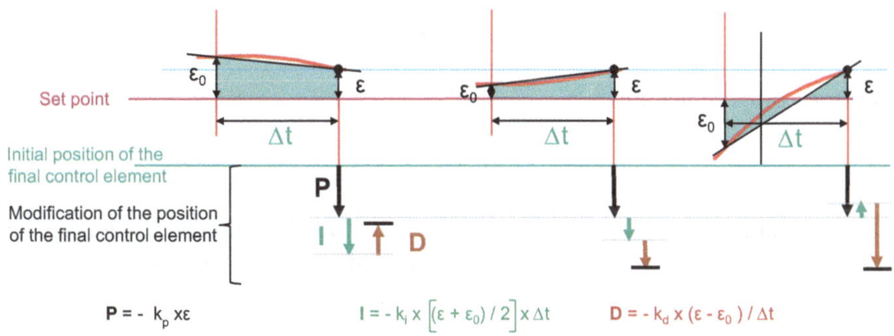

Fig. 8.4 The basic settings of a PID control loop

8.2.2.1 A Light Theory of the PID Control

In order to get an idea about the behaviour of a control loop, let us look at the three situations in Fig. 8.4 where, at the time that is considered, the value of the process variable is indicated by the black dot. The set point level is given by the red solid line. The initial position of the final control element is shown by the green solid line. In all three examples a, b and c, the error ε has the same value and is positive. Let us look at the way each action is acting.

8.2.2.1.1 Proportional Action

To get the proportional ("*P*") action that is symbolised by the black arrow (direction and intensity) in Fig. 8.4, the error is multiplied by a negative constant k_p and then algebraically added to the position of the control element.

The proportional action is calculated as:

$$P = -k_p \times \varepsilon \tag{8.1}$$

In the three cases, ε is the same, and the P arrow that has the *same intensity* and the *same direction* changes the position of the final control element (here, it goes towards closing).

The proportional action is sensitive to the instant situation.

8.2.2.1.2 Integral Action

To get the integral ("I") action that is symbolised by the green arrow, the calculation considers the value of the area located between the process variable curve and the set point line during a time interval Δt. Such an area is evaluated as:

$$\Delta t \times (\varepsilon + \varepsilon_0)/2,$$

therefore, proportional to:

$$(\varepsilon + \varepsilon_0),$$

making it an *average over the time period.* This value is multiplied by a negative constant k_p and then algebraically added to the position of the final control element.

The integral action is calculated as:

$$I = -k_i \times (\varepsilon + \varepsilon_0) \times \Delta T \tag{8.2}$$

In case (a), integral action is shown by the green arrow that is proportional to a_1. In case (b), the area a_2 has a value that is half that of $a_{1.}$ Consequently, the integral action in case (b) is half that of case (a). In case (c), the area a_3 is negative (algebraic sum of areas).

The integral action is sensitive to the past situation.

It is important to know that a system that is only proportionally controlled oscillates more and less about the set point due to the fact that upon reaching it, as the error is zero, the correction is zero. However, as the error change, the tendency has the same value for a while, and the system oscillates or stabilises at too low or too high a position. The addition to the input of the integral action allows reducing the average difference between the measurement and the set point output. So, a well-regulated PI loop slowly reaches the value of the set point.

8.2.2.1.3 Derivative Action

To get the derivative ("*D*") action that is symbolised by the brown arrow, the calculation considers the value of the *average* slope during the time interval, Δt, as $(\varepsilon - \varepsilon_0)/\Delta T$.

The derivative action is calculated as:

$$D = -k_d \times (\varepsilon - \varepsilon_0)/\Delta t \tag{8.3}$$

In cases (a) and (b), α_1 and α_2 have the same value but opposite directions: the derivative action has the *same intensity* but *opposite directions*. In case (c), a larger slope has for consequence a stronger action of the derivative.

The derivative action tries to take care of the future situation.

In other words, these values can be interpreted in terms of time:

- P depends on the present error.
- I depends on the accumulation of past errors.
- D is a prediction of future errors, based on current rate of change.

Remark In most conventional cryogenic systems, the derivative action is not mandatory. It makes tuning more delicate for a not very much better result.

8.2.2.1.4 Tuning a Control Loop

Figure 8.5 shows various behaviours of a control loop.

There are a lot of possibilities based on very sophisticated theories, but this is generally a matter for a control specialist. A simple but easy-to-use tuning method is known as the Ziegler-Nichols method. The k_i and k_d are first set to zero. The proportional k_p gain is increased until its critical value k_{pc} when the output of the control loop starts to oscillate with a critical oscillation time period P_c. For a PI control loop, the correct settings are not far from:

- $k_p = 0.45\ k_{pc}$
- $k_i = 1.2\ P_c$

This "recipe" is generally good for a one-off control loop operating alone. Such a situation is rather unusual because, in most of the cases, there are several control loops that inter-react in the system, so one should not expect a miraculous straight-forward method.

8.2.2.2 Other Possibilities in Using Control Loops

One can do a lot of nice things with a simple PID loop and some additional logic:

- Set the PID parameters according to the process conditions (see Sect. 8.3.6).

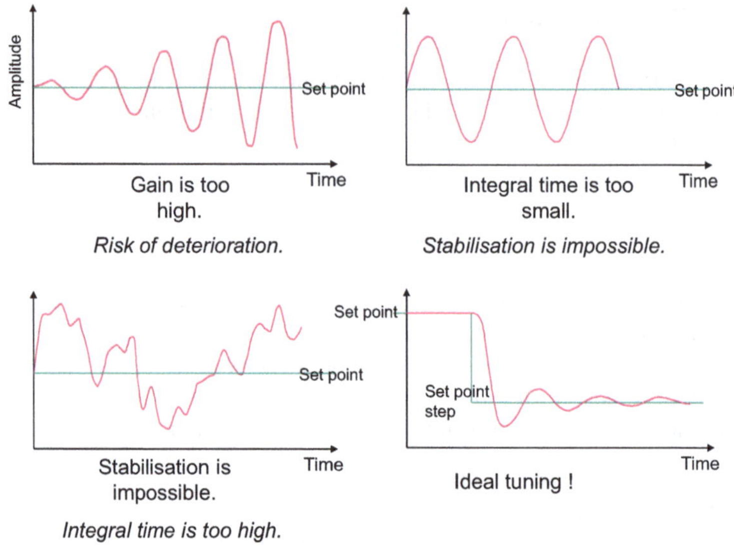

Fig. 8.5 Bad and good control loop tunings

- Adjust a set point according to a ramp in order that the process is not disturbed.
- Adjust a set point dynamically, for example, according to a calculation.
- Use an "attenuator" (see Sect. 8.2.3).

8.2.3 A Simple and Interesting Tool: The Attenuator

When more than one process variable is to be controlled by one final control element, a conventional solution is to arrange one loop per process variable and operate them in a cascade arrangement.

There is a simplest way to behave, which is often implemented in helium refrigeration: the attenuator.

Consider a control loop (CL) that actuates a final control element FV, as shown in Fig. 8.6 (top). The CL set point is PV1 SP. If the final control element brings also changes in other process variables as PV2 and PV3, it is possible to take care of all three process variables at once. Suppose one wants PV2 to be always lower than a maximum value $PV2_{max}$. The attenuator multiplies the PV1 set point by a coefficient "A PV2", the value of which is between 1 and 0. Such coefficient is calculated according to the function that is shown on the bottom left: as far as PV2 is lower than $PV2_{max}$, the coefficient equals 1; therefore, the PV1 set point is transmitted as it is to the control loop. When PV1 reaches $PV2_{max}$, the coefficient starts to decrease, reducing the set point that is sent to CL and, consequently, avoiding PV2 to be higher than $PV2_{max}$.

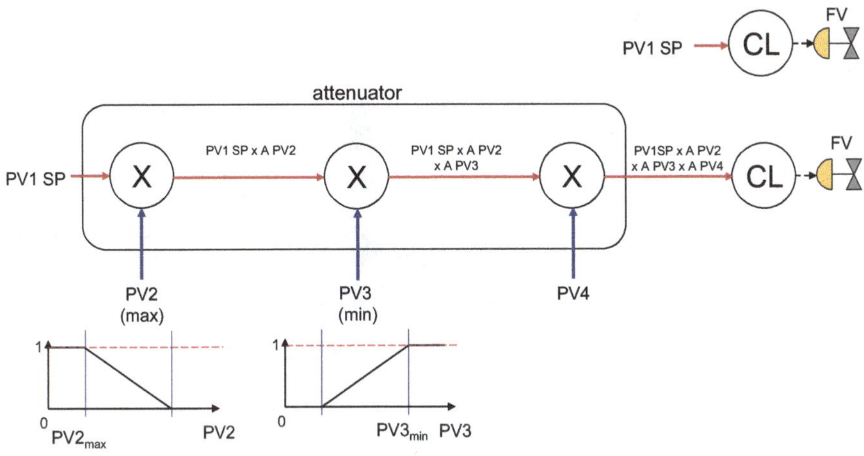

Fig. 8.6 The attenuator

Similarly, if one wants, simultaneously, another process variable PV3 to stay always above $PV3_{min}$, the same solution applies but with a different function as shown in the bottom right.

And it is possible to add any other process variables to be maintained over or lower than a threshold in this way.

An example of using such attenuator is shown in Sect. 8.4.3

8.2.4 A Control Valve

A control valve is a very common component. It is interesting to see how it works. The valve stem is generally driven by a pneumatic actuator. The air pressure fights against a spring that brings the valve into a position if the instrument air pressure comes to zero. This position is the "fail" position: either fail closed (FC) or fail open (FO). The choice is made at the time of the functional analysis of the system.

When moving, the stem fights against pressure difference and also friction (that can be rather erratic); therefore, there is not a two-way relationship between the valve position and the air pressure acting on the actuator.

The solution sits in a device, the positioner (see Fig. 8.7), that measures the position of the valve stem, compares it to the signal that is delivered by the process controller (error) and changes the air pressure on the actuator in order to get a zero error. Therefore, a control loop is "hidden" into any control valve!

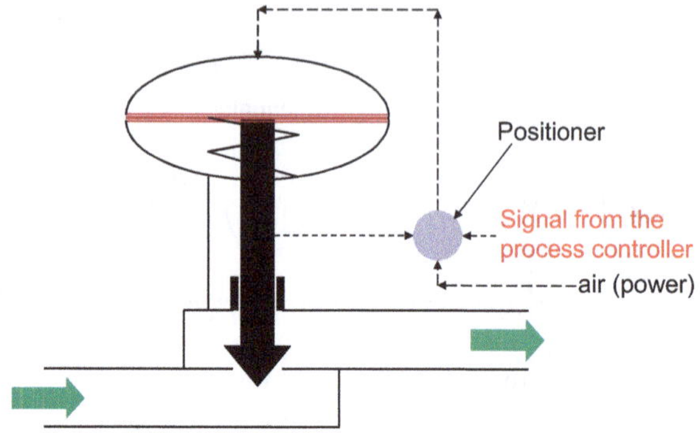

Fig. 8.7 A control loop is "hidden" into a control valve

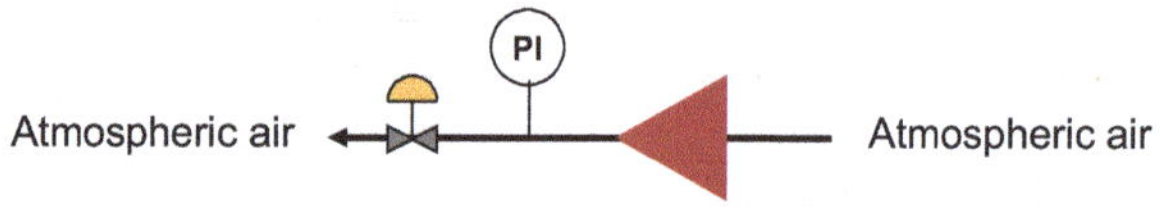

Fig. 8.8 A very simple system

8.3 Refrigerator or Liquefier Cycle Pressure Control

8.3.1 General

The control of the cycle pressures is the first one to be implemented in a refrigeration or liquefaction system. It must keep the cycle pressures at the design values, if possible, and control the gas inventory, whatever the situation is.

First, let's have a look at a very simple system: a compressor processing atmospheric air and discharging into a pipe on which a valve is fitted (Fig. 8.8). Such a compressor can reach up to 200 bars.

What is the compressor discharge pressure when the valve is full open? If the valve has a very high conductance, the answer is, obviously, atmospheric pressure (or the same pressure as the suction pressure). However, as soon as the valve begins to be closed, the compressor discharge pressure rises. Therefore, the discharge pressure of a compressor is related to the equipment that is fed by the compressor, not to the compressor itself.

Keep in mind that:

- The discharge pressure is set by the valve (the final user of the compressor flow).
- The flow is set by the compressor.

Let's now consider another simple circuit in Fig. 8.9. It is a closed circuit, initially filled at 12.00 bars. When the compressor runs with the valve full open, the pressure does not change as in the above situation. If the valve is closed until the high pressure reaches 14.00 bars, some helium is stored into the high-pressure circuit. Such helium comes from the low-pressure circuit where the pressure decreases down to 8.00 bars. In steps 2 and 3, the valve is again closed until the low pressure is 1.00 bar. This kind

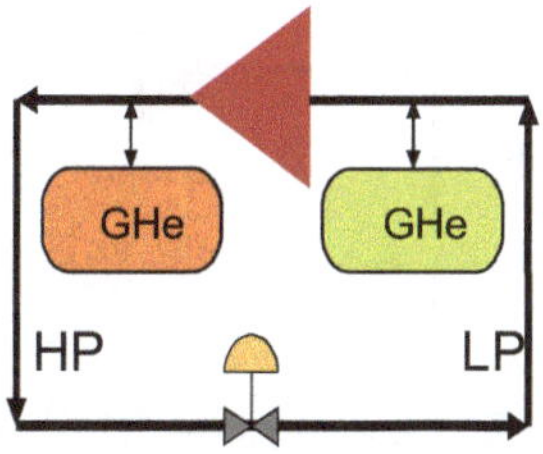

	Initial	Step 1	Step 2	Step 3
HP Volume	4.0			
HP	12,00	14,00	16,00	17,50
BP Volume	2,0			
BP	12,00	8,00	4,00	1,00

Fig. 8.9 A simple closed circuit

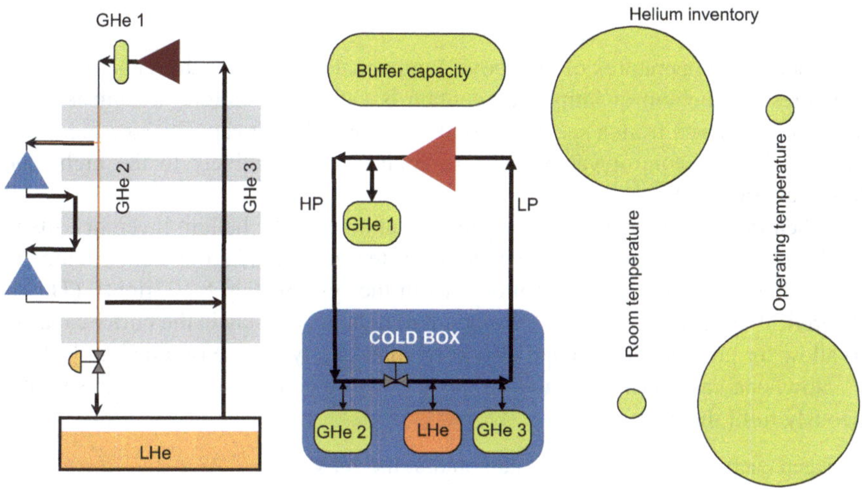

Fig. 8.10 Where is helium stored into the system

of arrangement is not acceptable for a refrigerator because the low pressure reaches high values that would, among others, make necessary to design the low-pressure circuits to a high pressure.

In a refrigeration or liquefaction system, the cycle helium sits in various places that can be at variable temperatures (see Fig. 8.10, left):

- At room temperature: the room temperature piping and the capacities of the helium coolers and the oil removal system (GHe 1)
- At variable temperature between room and operational: the high-pressure piping, heat exchangers, adsorbers (GHe 2)
- At constant temperature: the phase separator (LHe)
- At constant temperature: the low-pressure piping and heat exchangers (GHe 3)

(In Fig. 8.10, right, the valve in the cold box simulates both the turbines and the JT valve). The discharge pressure of the compressor is *set by the valve*, i.e. by the cold box components, not by the compressor alone.

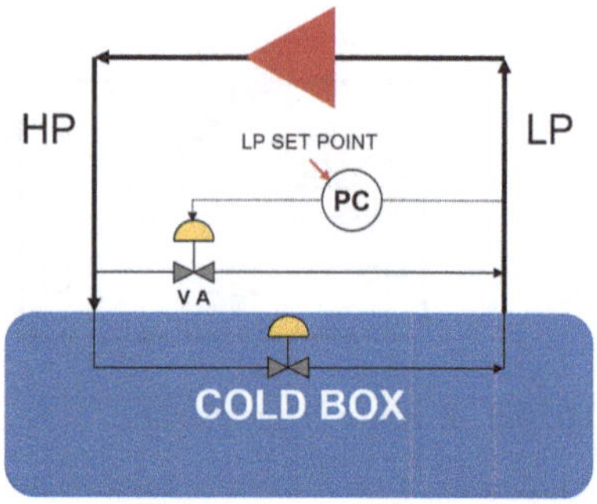

Fig. 8.11 Low-pressure control

When the temperatures of the above components change for the same operating pressures, the helium quantity they contain is variable; therefore, helium is to be stored or de-stored from a special capacity, the cycle buffer capacity (see Fig. 8.10, centre), through control valves that are connected respectively to the high- and low-pressure piping.

When the system is at room temperature, almost all the helium inventory sits in the buffer capacity. Conversely, when the system is at nominal operating temperatures, almost all the helium inventory sits in the system (Fig. 8.10, right). (This is theoretical; practically, a minimum amount of helium is to stay in the buffer capacity to allow for pressure control and also keep some safety in case of a helium leak.)

Now, one can understand that the cycle pressure control system has to simultaneously fulfil three functions:

- Keep each cycle pressure constant and at the set value.
- Supply the cold box at the required mass flow rate.
- Manage the helium inventory.

In the following paragraphs, it is assumed that the volumetric flow of compressors is constant (they are supposed to run at constant volumetric efficiency and rotational speed).

8.3.2 The Cycle Low-Pressure Control

In Fig. 8.11, valve VA is connected from downstream of the high-pressure oil removal system to the low pressure. It is controlled by a control loop that compares the actual low pressure to the low-pressure set point.

Remark It is a good practice to control the low pressure, i.e. the suction pressure of the compressor, at a value that is slightly higher than the local atmospheric pressure (30–50 mbar) in order to avoid any air leakage into helium by pressure difference.

The low-pressure control loop fulfils part of the functions: it keeps the low pressure constant at the set value and supplies the cold box with the required mass flow rate.

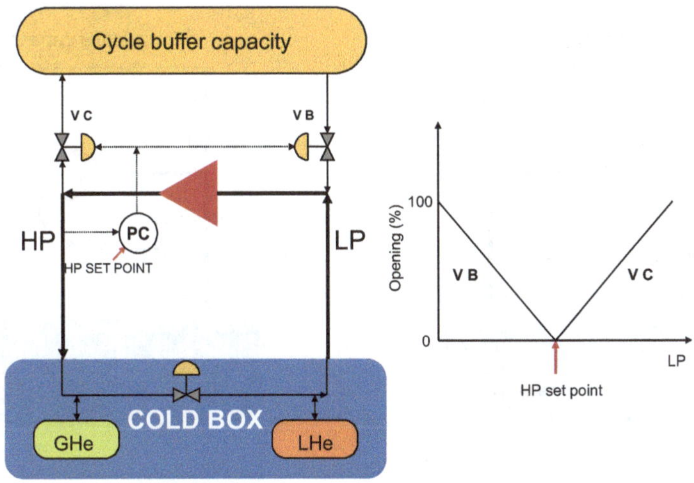

Fig. 8.12 High-pressure control

Remark Helium flowing through the HP by-pass has been compressed: it costs energy. Therefore, such flow must be *minimised* (see Sect. 8.7.2.1).

8.3.3 The Cycle High-Pressure Control

As the low pressure is kept constant by the by-pass, the quantity of helium that is stored into the LP circuits is constant in a short time interval.

Valve VB connects the buffer capacity to the low-pressure piping; valve VC connects the high-pressure piping (downstream of the oil removal system) to the buffer capacity. Both valves, which operate in opposition, are controlled by a common control loop that compares the actual high pressure to the high-pressure set point.

During cool-down, as helium is stored into the cold circuits, the high pressure tends to decrease: the control loop opens VB as indicated in the right part of Fig. 8.12. Conversely, during warm-up, when the cold circuits release helium, the high-pressure increases, and the control loop opens valve VC to dump helium into the buffer capacity. Valves VB and VC are arranged in a so-called complementary split range.

The high-pressure control loop fulfils part of the functions: it keeps the high pressure constant at the set value and manages the helium inventory.

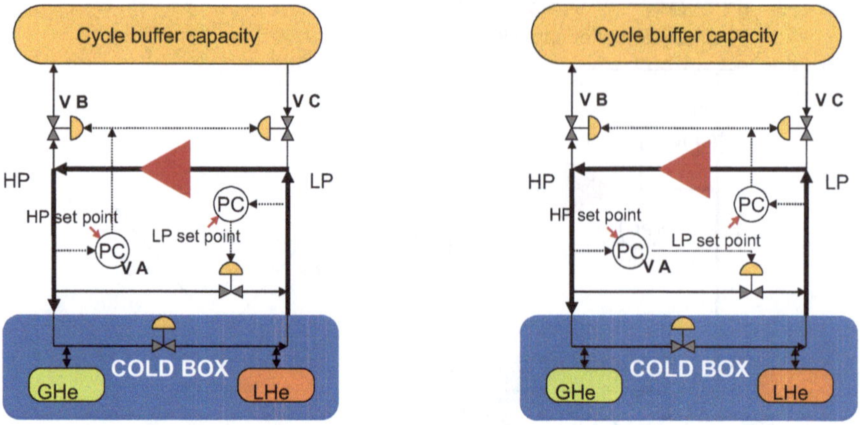

Fig. 8.13 High- and low-pressure controls

8.3.4 *Simultaneous Control of Both High and Low Pressures*

The whole cycle pressure control system is shown in Fig. 8.13 (left), where both systems are integrated.

Why not play with this system and think about its behaviour when, from a steady state (VB and VC closed, VA at a constant opening), the operator increases slightly the high pressure set point?

Let's consider two cases:

- The cold box is not circulated by the compressor flow.
- The cold box is circulated by the compressor flow.

CB is not circulated.
As a consequence, actions are:

- Larger opening of valve VB.
- LP tendency to increase.
- Smaller opening of VA (keeps the LP circuit helium quantity constant, therefore any extra helium coming from VB goes to HP circuits).
- When the new HP set point is reached, VB closes, and VA has reached a smaller opening compared to initial situation.

CB is circulated.
To make things simple, let's consider that the cold box is considered to be a sonic orifice. Therefore, the flow that is processed into the cold box (but also into VA for a same opening) is proportional to the HP value.

As a consequence, actions are:

- Larger opening of valve VB.
- LP tendency to increase.

- Smaller opening of VA (keeps the LP circuit helium quantity constant, therefore any extra helium coming from VB goes to HP circuits).
- When the new HP set point is reached, VB closes, and VA has reached a smaller opening than when the cold box is not circulated.

Remark One can guess that if the HP set point is too much increased, VA will reach a closed position!

There is another possible arrangement
It is also possible to control the high- to low-pressure by-pass valve according to the high pressure and the control valves VB and VC according to the low pressure, as shown in Fig. 8.13 (right).

Why choose one or the other arrangement?
It is interesting to consider the behaviour of these two systems in case the HP set point is too high or if the cold box needs more gas than the compressor can process (e.g. if temperatures are lower than design). In both cases, the system reaches a situation where the HP-LP by-pass comes to a closed position. At this time, only the other control loop that takes care of VB and VC can operate. If the loop controls the LP, LP stays constant but HP decreases. If the loop controls the HP, HP stays constant, but LP increases.

In some systems, it is thought that the cycle low pressure is more stable (can react faster to perturbations) when it is controlled by the compressor by-pass. However, for a liquefier, it is better to use the compressor by-pass to control the high pressure and avoid the low-pressure rising when the compressor flow is not sufficient to keep the high pressure at the set point (see Sect. 8.6). Therefore, the system behaviour in the expected situations is an argument to make the choice between the two arrangements.

It is very fruitful for an operator to perform such reasoning in order to get a good "instinctive" feeling of the behaviour of the components of any system.

8.3.5 *Control of a Three-Pressure Cycle*

The MP by-pass valve deals with the flow that is processed into the Brayton turbines, and the HP by-pass valve deals with the flow that is processed into the JT cycle, either a valve or turbines (Fig. 8.14).

8.3.6 *Adapting the P and I Settings According to the Configuration of the Circuits*

When valves FV 2 and FV 3 in Fig. 8.15 are closed, the volumes of the operating circuits are low: for a given perturbation, pressure changes can be important. The proportional gain must be rather high in order to react correctly. When the

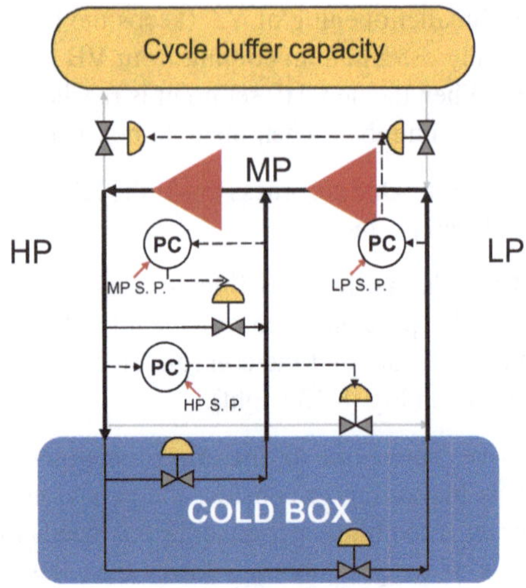

Fig. 8.14 A three-pressure control arrangement

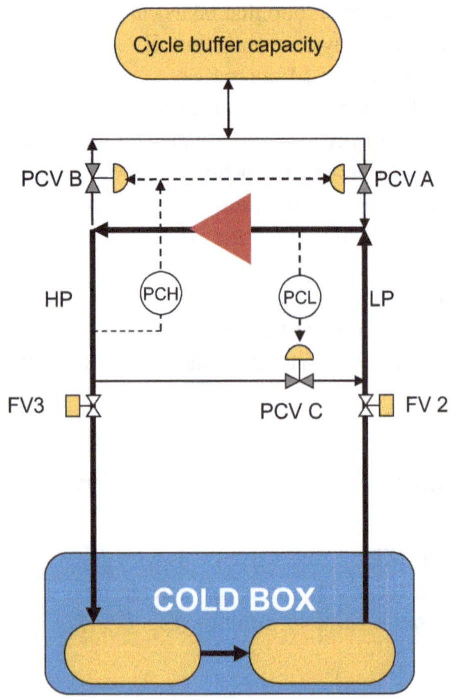

Fig. 8.15 Adapting the P and I pressure control settings

valves are open, the operating circuit has a larger equivalent volume, especially if the cold box is at design operating temperatures because for a same geometric volume, the quantity of helium that is stored is larger, acting as a larger equivalent volume at room temperature: the proportional gain is to be reduced. Obviously, such settings can be automatised.

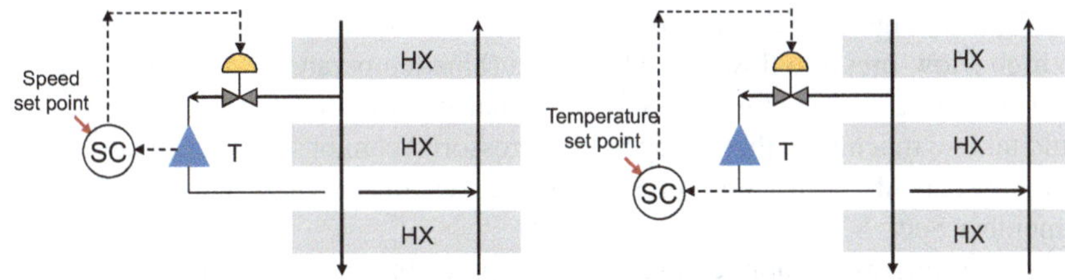

Fig. 8.16 Control of the rotational speed of a turbine, left; avoiding too cold a turbine discharge temperature, right

8.4 Few Controls Around Turbines

In Sect. 6.6.3.7, the control of expansion turbines for starting up has already been explained.

8.4.1 Rotational Speed Control

When the enthalpy drop to which a turbine is subject is higher than its nominal value (when operating at a higher than nominal temperature), the extracted power exceeds the turbine rated power. The speed of rotation, related to the power that is extracted (see Sect. 6.4.1), exceeds its nominal value, and the turbine may hit an overspeed. This situation happens typically during the cooling phase. Indeed, we have seen that for the same high and low pressures, the enthalpy drop is roughly proportional to the temperature (see Sect. 3.4.1). A simple way to keep the power extracted at a constant value is to control the turbine rotational speed (remember Sect. 6.6.3.7). This control is performed by a loop which, depending on the rotational speed, acts on the position of the inlet valve of the turbine (see Fig. 8.16 left), if the speed tends to exceed the set point value, action on the valve goes in the closing direction and vice versa.

Remark Such an arrangement is used as a "limiting" procedure. The most usual situation is the control of the rotational speed of the turbine at the time of the cool-down phase (see Sect. 8.4.2.1). In steady-state operation, the turbine inlet valve *must be 100% open* in order not to generate any pressure drop.

8.4.2 Discharge Temperature Control

8.4.2.1 Avoiding Too Cold a Turbine Discharge Temperature

As for the rotational speed, this is not exactly a temperature control but merely a system that avoids the discharge temperature of a turbine to reach too low a value (see Fig. 8.16 right). Such a control can be used when the cold box is unbalanced by too high a low-pressure flow and when the turbine temperature drops lower than the usual operating temperature. This is particularly sensitive with cold turbines that could take so much flow that the cycle compressor(s) cannot supply sufficient helium and, consequently, the high pressure decreases or the low pressure increases (remember Sect. 8.3.4).

Another interesting application of such a procedure happens during the cooling down phase of the cold box; one does not want water ice to be generated in the turbine (see Sect. 10.3.1.1).

8.4.2.2 Tuning the Optimum Discharge Temperature of a Turbine

In paragraph "Calculation of the Thermodynamic Balance", it has been noticed that when calculating the process, the designer had to determine the optimal exhaust temperature of each turbine in order to get the smallest cycle mass flow rate and, therefore, the smallest equipment sizes and power consumption.

When operating a refrigerator, the first idea that comes in mind is to run the actual machine at the optimum calculated design turbine discharge temperature, but there are two reasons not to do this way. Firstly, the accuracy of the turbine discharge temperature measurement might be not good (which is not harmful for a correct operation of a control loop as soon as the procedure that is described here under is implemented). Secondly, when the actual machine is to be operated, the operator faces a rather similar situation as has been found during design: run the real machine at optimum conditions. The real machine is fitted with components that have been specified according to the process calculations, but which might not be exactly as expected (the compressor might process a slightly higher or lower flow, the heat exchangers and expanders could have a slightly higher or lower efficiency, etc.). Therefore, the actual optimised turbine discharge temperature might be slightly different from the calculated value. How to find this temperature?

It has been shown that the operating temperature of a turbine varies with the thermal load (see Sect. 7.5.1). At the cold end of the cycle, the thermal load on the turbine is due to the temperature difference at the warm end of the exchanger located immediately upstream the turbine (thermally speaking) and the cooling and condensation of the helium that is liquefied (remember Sect. 3.5.3.1.1, where one can see in Fig. 3.50, the incidence of the warm end heat load and the liquefaction load on the cold Brayton cycle). Control of the exhaust temperature of the cold turbine is performed by a loop which, depending on the exhaust temperature, changes the position of the Joule Thomson valve, thereby varying the JT flow rate and, consequently, the thermal load of the turbine. If the temperature value tends to exceed the set point, the action of the valve is towards closing, and conversely towards more opening if the temperature is lower than the set point (see Fig. 8.17, left).

When starting a new refrigerator or having changed the temperature sensor, one has to make sure it operates at its optimum turbine discharge temperature by

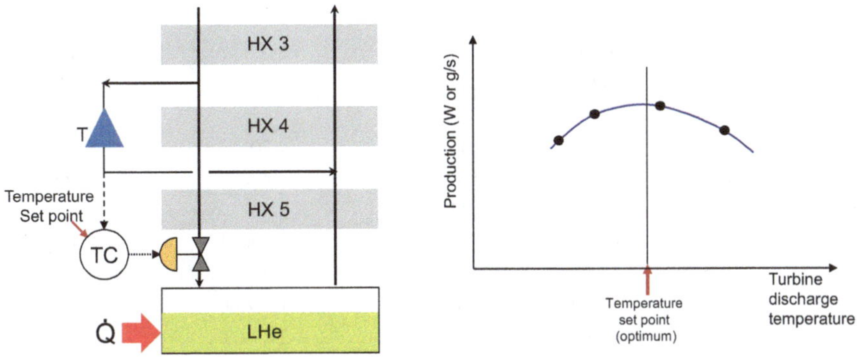

Fig. 8.17 Control of the optimum cold turbine operating temperature

measuring the production according to various temperatures and selecting the temperature that gave the maximum production as the set point (see Fig. 8.17, right).

8.4.3 Cool-Down of Two Turbines in a Series Arrangement

Before considering the cool-down of a cold box, let's have a glance at the special situation that happens when two turbines that are in a series arrangement, at room temperature, begin cooling down (this is a good opportunity to rehearse what has been discussed on turbines!). These turbines have been designed in order that they process the same mass flow rate at design condition; therefore, their flow coefficients (Fc) (see Sect. 6.6.3.7.2) are known. They are calculated in the column "design" of Table 8.1.

When starting up at room temperature, the inlet temperature of each turbine is at 300 K, much higher than design. Therefore, in the operating conditions shown in column 2 of Table 8.1, it appears that at room temperature, the processed flow is lower, due to a higher temperature (remember Sect. 6.6.3.7.2), but it is also shown that in order to circulate the same flow in both turbines, the inlet pressure of the cold turbine would raise up to 12.00 bars, resulting in a theoretical power of 8503 W, which is not possible: the rotational speed would be much too high!

Remark These calculations are not strictly accurate because the turbine distributors do not operate in a sonic regime. However, it allows to understand the situation.

A way to behave is to run the cold turbine at its nominal power, as it has been done in the cool-down procedure that is discussed in Sect. 6.6.3.7.6, by controlling its rotational speed. The result is shown in column 3 where the cold turbine power at cool-down is equal to that at design conditions.

As the turbines are in a series arrangement, such control can be performed by the valve that is upstream the warm turbine. This is a situation where the attenuator

Table 8.1 Cool-down of two turbines in a series arrangement

		Design	2		3		4
$\dot{m}$	(g/s)	*50.00*	25.82		8.66		22.95
T1							
P_{in}	(bar)	*14.00*	14.00		4.70		12.44
T_{in}	(K)	*80.00*	300.00		300.00		300.00
$\dot{m}$	(g/s)	*50.00*	25.82		8.66		22.95
h_{in}	(J/g)	*434.20*	1577.71		1574.71		1577.21
eta	(–)	*0.75*	0.35		0.35		0.35
h_{out}	(J/g)	*342.89*	1544.87		1542.00		1378.24
Fc	(–)	*31.94*	31.94		31.94		31.94
W	(W)	*−4566*	−848		−283		−4566
T2							
P_{in}	(bar)	*6.00*	12.00		4.03		4.03
T_{in}	(K)	*20.00*	300.00		300.00		300.00
$\dot{m}$	(g/s)	*50.00*	25.82		8.66		
h_{in}	(J/g)	*116.04*	1577.07		1574.49		1574.49
eta	(-)	*0.75*	0.35		0.35		0.35
P_{out}	(bar)	*1.20*	1.20		1.20		1.20
h_{out}	(J/g)	*79.73*	1247.72		1364.94		1364.94
Fc	(-)	*37.27*	37.27		37.27		37.27
W	(W)	*−1815*	***−8503***		−1815		−1815
bypass	(g/s)	*0.00*	0.00		0.00		14.29
W total		*−6381*	−9351	0	−2098	0	−6381

(remember Sect. 8.2.3) can be used, as shown in Fig. 8.18, by injecting the value of ST2 and using a "maxi" function as shown in the bottom left.

When an attenuator is implemented, it is possible to use it for other purposes. For example, if one wants the discharge temperature of T2: T_d2, to stay over a minimum value T_d2_{min}, it is possible to implement another "mini"-function as shown in the bottom right: if the temperature gets lower than T_{min}, the attenuator multiplies the ST1 set point by a value lower than 1.

In Sects. 10.3.4 and 10.3.5, there are other examples of using an attenuator.

Remark The above procedure has an inconvenience: as the control limits the cold turbine rotational speed, the warm turbine is operated at a lower power (here, 283 W instead of 4566 W), increasing the cool-down time of the cold box. By connecting a valve between discharge of T1 and the low-pressure circuit, as shown in Fig. 8.19, the helium that cannot be processed by T2 is by-passed, allowing T1 to rotate at its maximum speed and therefore at its maximum power: the cool-down phase is performed at nominal power.

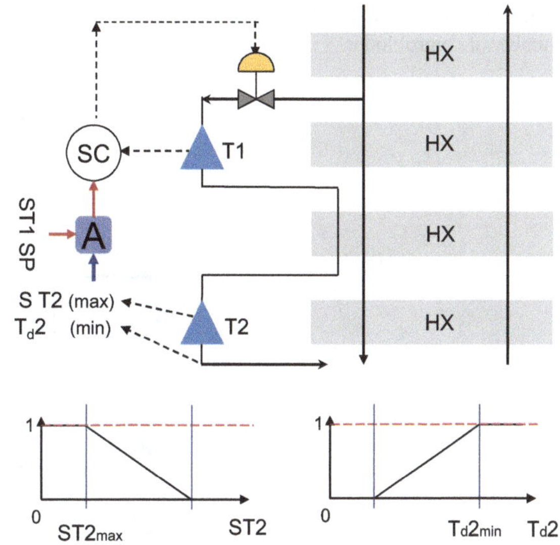

Fig. 8.18 Cool-down of two turbines in a series arrangement

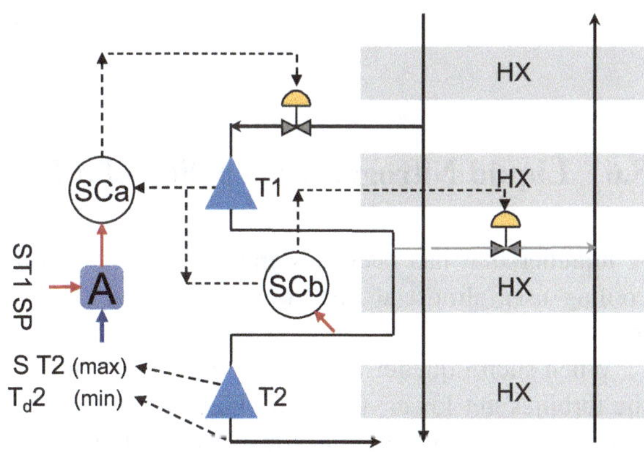

Fig. 8.19 Cooling with the warm turbine at full power

8.5 The Minimum Number of Control Loops for a Claude Cycle

According to what has been seen, one understands that there is a minimum number of control loops to be implemented in order to run automatically a refrigerator in a steady state, provided there is no major perturbations. It is only necessary to control the cycle pressures and the discharge temperature of the coldest turbine (Fig. 8.20).

By the way, it is how former refrigerators were controlled.

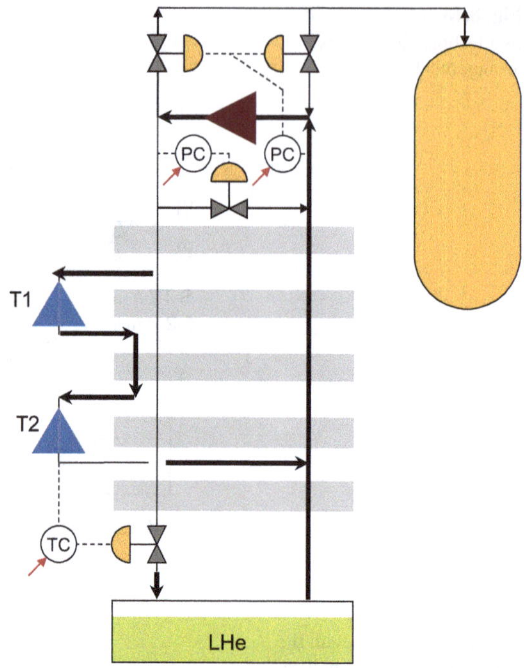

Fig. 8.20 Minimum number of control loops

8.6 Liquid Nitrogen Pre-cooling of a Liquefier

A liquefier that has been designed for an operation without liquid nitrogen pre-cooling uses almost all the compressor flow. The compressor by-pass is near to totally closed.

When such liquefier is operated with pre-cooling, the operating temperatures of the turbines get lower; therefore, they can process a higher mass flow rate for the same high pressure. Simultaneously, the JT flow is higher (remember Sect. 7.7.4.3). As a consequence, the high pressure decreases (provided that the compressor by-pass controls the high pressure as shown in Fig. 8.13) until it reaches a steady state.

When the turbines are fitted with gas static bearings, it is necessary to keep the supply of the bearings over a minimum pressure (see Sect. 6.5.1.2.1). For this reason, the valve that is upstream the warm turbine avoids the HP to get too low, as shown in Fig. 8.21.

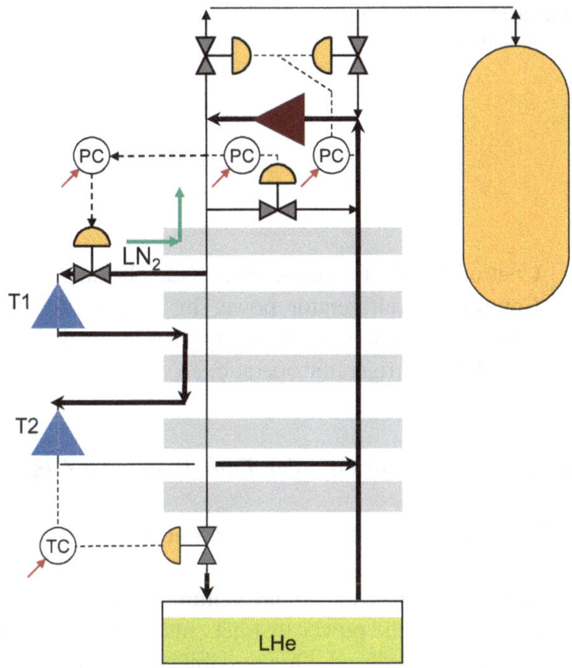

Fig. 8.21 Liquid nitrogen pre-cooling of a liquefier

8.7 Efficient Cryogenic Power Control

8.7.1 General

Remember the assumption that has been made at the beginning of Chap. 7: a refrigerator might spend all its life operating at 66% of its nominal power. If such plant absorbs a few megawatts, one understands that such an off-design operation must be as efficient as possible. This situation happens only with refrigerators, generally operating in mixed regime, that are connected to loads as magnets or resonant cavities, the power they dissipate being able to vary in a very large range.

This is not the case for a liquefier that is always operated at full capacity, i.e. at best efficiency.

How to know if the refrigerator extracts more or less power than needed? When it is operating in an isothermal mode as shown in Fig. 5.6, or in a non-isothermal mode as shown in Figs. 5.7 or 5.9, the liquid helium level variation in the phase separator is a significant parameter: if it rises, it means that the thermal load is lower than the power that is absorbed by the refrigerator conversely.

Controlling directly the power by the liquid level might have some disadvantage as the refrigerator response time is rather long.

A better way is to perform the two steps (see Fig. 8.22):

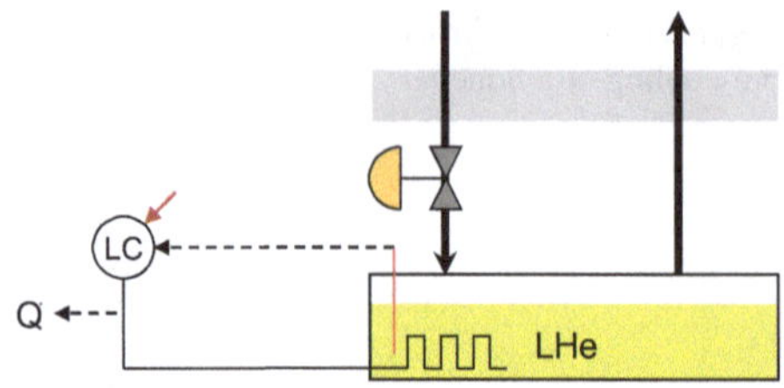

Fig. 8.22 Control of the liquid helium level

- Control quickly the liquid level with a heater.
- Control the refrigerator power by keeping the heater power amount Q at a constant and low value (a safety margin for the operator) at a rate that does not disturb the refrigerator operation.

8.7.2 Turn-Down of a Two-Pressure Claude Cycle

8.7.2.1 Using the Turbine Inlet Valve

In Fig. 8.23, when the liquid helium level rises, the refrigerator power must be reduced. As all the power that gets into the cycle is extracted by the turbines, this power must be reduced. One possibility is to decrease the turbine high pressure by closing its inlet valve.

Consequently, the turbine processed mass flow rate is decreased (remember Fig. 1.19), and, simultaneously, the turbine expansion ratio is decreased too: the turbines extract less power. If an attenuator (remember Sect. 8.2.3) is implemented, the maximum power value that is allowed in steady state Q is injected into the attenuator, with a law that is shown in Fig. 8.23 (right top).

The mass flow rate that is not processed into the cold box should not be compressed (keeping the HP-LP by-pass at the same almost closed position); therefore, the compressor slide valve or, better, the rotational speed of the compressor should be tuned in order that the by-pass stays at the same minimum tolerated opening. The high pressure is not changed.

In case of a liquefier, the same procedure is applied by considering the liquid helium dewar level with the attenuator shown in Fig. 8.23 (right bottom).

Remark Suppose it is 4 PM on a Friday. The amount of helium to be liquefied will allow the liquefier to run until Saturday noon, or the dewar should be almost full Saturday noon. It would be stupid to run the liquefier in a standby mode, which means absorbing some electrical power for nothing, until Monday 8 AM when liquid helium will be withdrawn. The good solution is to stop the liquefier and start it up on Monday 7 AM, knowing that 1 hour is sufficient to reach operating conditions. Obviously, if no gas or full dewar situation happens on Monday 6 AM, the liquefier should not be stopped.

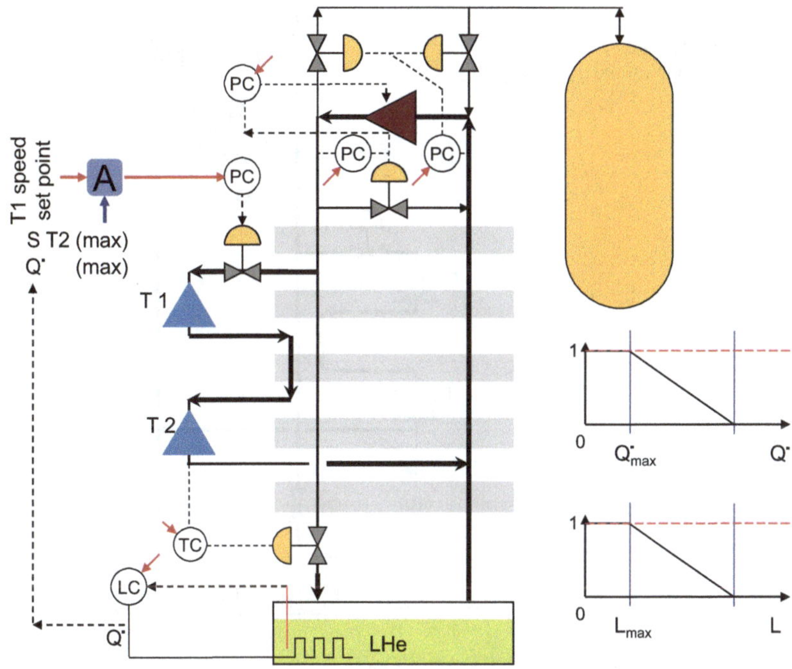

Fig. 8.23 Simple turn-down of a two-pressure Claude cycle, acting on the turbine inlet valve

8.7.2.2 Changing the High Pressure

The cycle power can also be controlled by varying the cycle high pressure: when the electrical power that is injected into liquid reaches the set value (maximum power that is allowed in liquid), the control loop decreases the compressor flow, either with the slide valve or, better, by the rotational speed. Of course, using the by-pass would not be efficient! (Fig. 8.24).

Remark This procedure can be implemented as soon as there is not a necessity to keep the high pressure over a minimum value as it is the case with a gas static bearing system that must be permanently fed at a minimum pressure (remember Sect. 6.5.1.2.1). In such a case, the cycle high pressure is kept at a minimum value. If the electrical power injected into the liquid is still too high, the turbine inlet valve can be used as in Sect. 8.7.2.1, but the compressor absorbed power will not decrease anymore.

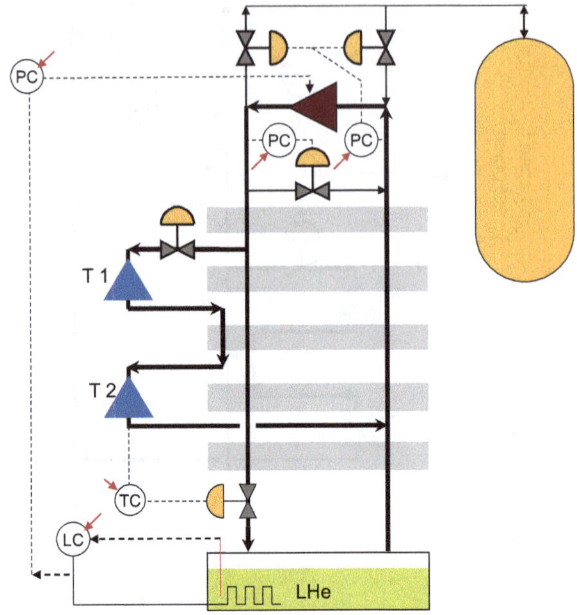

Fig. 8.24 Simple turn-down of a two-pressure Claude cycle, changing the high pressure

	NOMINAL	TURN-DOWN
MP	4	2
HP	20	10
Turbine flow :	1	0.5
Turbine expansion ratio :	1	1
Turbine enthalpy drop :	1	1
Turbine power :	1	0.5

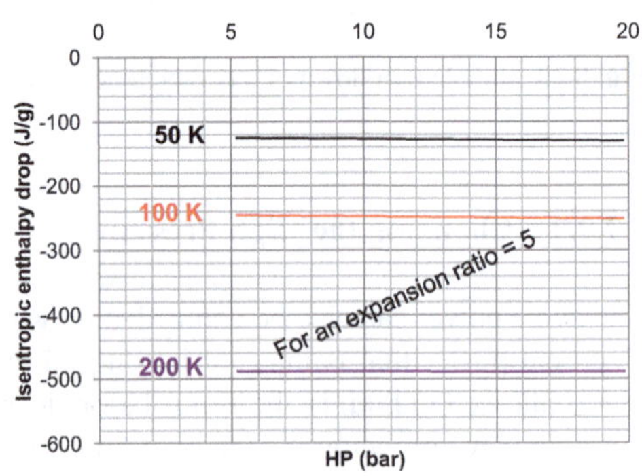

Fig. 8.25 Efficient turn-down of a Brayton cycle (left). Enthalpy drop of the turbine is related to the expansion ratio (here: 5/1), not to the high pressure (right)

8.7.3 *Efficient Turn-Down of a Brayton Cycle*

Consider a simple Brayton cycle with one cycle compressor and one expansion turbine. The respective operating pressures are low pressure = 4 bar and high pressure = 20 bar.

Let us lower the set point of the control loop of the low pressure to 2 bars (see Fig. 8.25). As the suction pressure of the compressor is divided by 2, the compressor processed mass flow rate is also divided by 2 (see Sect. 6.2.1.1).

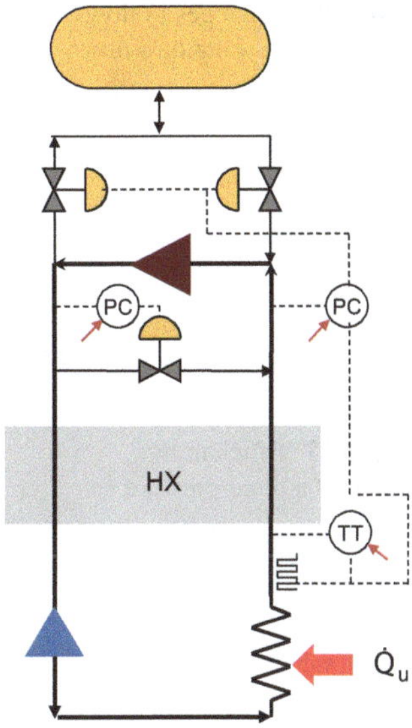

Fig. 8.26 Efficient control of a Brayton cycle

When the flow that is processed by the turbine is divided by two, as we can consider that a turbine behaves as a sonic orifice (see Sect. 6.6.3.7.2), high pressure establishes at half of its original value, that is to say 10 bar: the expansion ratio to which the Brayton cycle turbine is submitted is unchanged (20/4 = 10/2).

The enthalpy drop of a turbine operating at various high pressures, under the *same expansion ratio,* stays almost constant (see Fig. 8.25, right). When dealing with a half mass flow rate and the same enthalpy drop, the turbine power is divided by two (see Sect. 6.6.3.6). It should be noticed here that the compressor operates not only at constant compression ratio but also at a constant volume flow rate: thus, its efficiency is unchanged and remains at maximum. Remember that such off-design regime has been described in Sect. 7.5.2.

The efficient control of a Brayton cycle is performed by setting the low pressure according to the power input: controlling the load return temperature (see Fig. 8.26, left). Operators like to have some reasonable extra power amount that is "burned" by a heater (see Fig. 8.26, right). Here, the control loop keeps the heater power at the "reasonable" set value.

8.7.4 *Efficient Turn-Down of a Three-Pressure Claude Cycle*

A three-pressure Claude cycle is the combination of at least one Brayton and a Joule Thomson cycle, as shown in Sect. 3.5.1. As most of the power is extracted by the turbines of the Brayton cycles, the efficient turn-down is performed, as for the simple Brayton cycle discussed in Sect. 8.7.3, by floating the middle and high pressures, keeping the same ratio between them (see Table 8.2). Of course, it is not

Table 8.2 Changes in flows and powers during turn-down

	Nominal	Turn-down	
LP	1	1	
MP	4	2	
HP	20	10	
		Brayton	JT
Turbine flow	1	½	½
Turbine expansion ratio	1	1	½
Turbine power	1	½	< ½

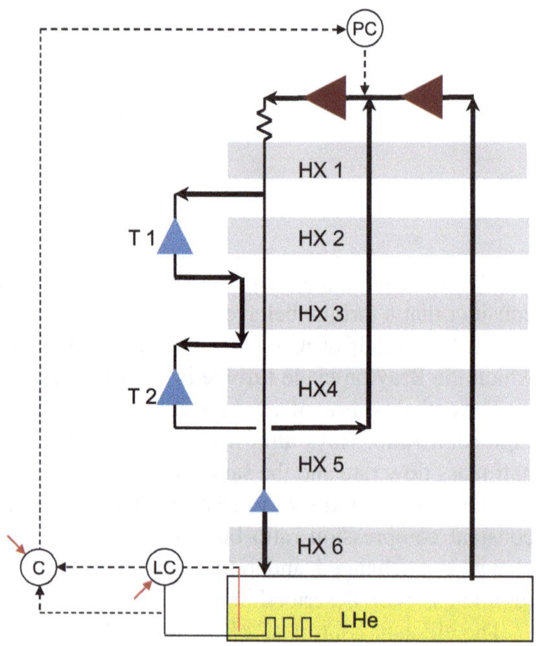

Fig. 8.27 Efficient turn-down of a three-pressure cycle

allowed to change the low pressure that is minimum in order to get the coldest possible liquid.

For turbines operating between the cycle high and low pressures, the mass flow rate is also halved, but the expansion ratio is also halved. Thus, the power of these turbines decreases faster than the expansion ratio. This floating pressure control was successfully implemented by the author in the CERN LEP refrigerators in 1990.

The middle pressure control loop in Fig. 8.27 keeps the heater power at a reasonable level by slow changes that do not affect the system stability.

Figure 8.28 shows the results of measurements that were performed on the LEP refrigerators at the time of commissioning. The LP flow was tuned by operating the compressor slide valves. At the present time, a better result could be reached when using a frequency converter that would not decrease the compressor efficiency.

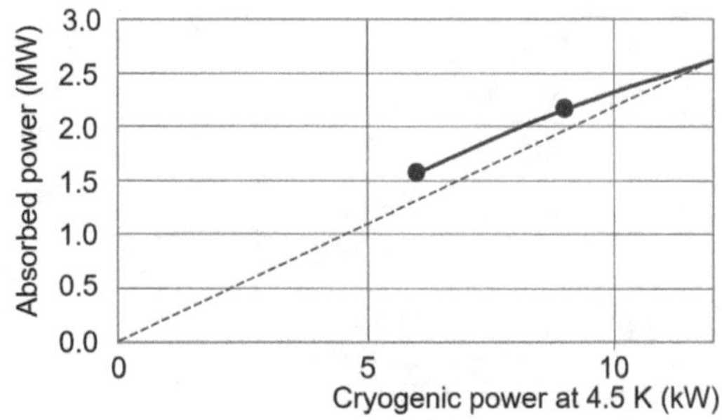

Fig. 8.28 Measured absorbed electrical power versus cryogenic power (LEP refrigerators at CERN)

8.8 Displays

8.8.1 Mimic Displays

It is important to have comfortable displays to check easily the operation of the plant.

A general display gives, at a glance, most of the main parameter values (Fig. 8.29). Such general display should not be too much crowded.

Then, subsystem displays allow to get into details (see Fig. 8.30) for various components or parts of the system.

In case of a centrifugal cryogenic compressor, the compressor diagram should be displayed with the position of the current operating point and the possibility to display previous points during a time period that can be set according to the operating situation (on a long stable regime, display can include points since a few hours; however in a transient regime, the time period should be reduced).

With time and operating experience, such displays are to be slightly modified to add parameters that are of interest or delete parameters that are not frequently used and impede the perception of useful information, according to the operating team.

8.8.2 Trends

Up-to-date systems allow to easily plot any parameter or set of parameters. This is very interesting at the time an abnormal behaviour occurs (or has occurred!). However, as for mimic displays, it is interesting to have a selection of plots that provide the important information on either the general situation of details on some part of component. As for mimic displays, the "quality" of the selected parameters for these plots will be improved with operating experience.

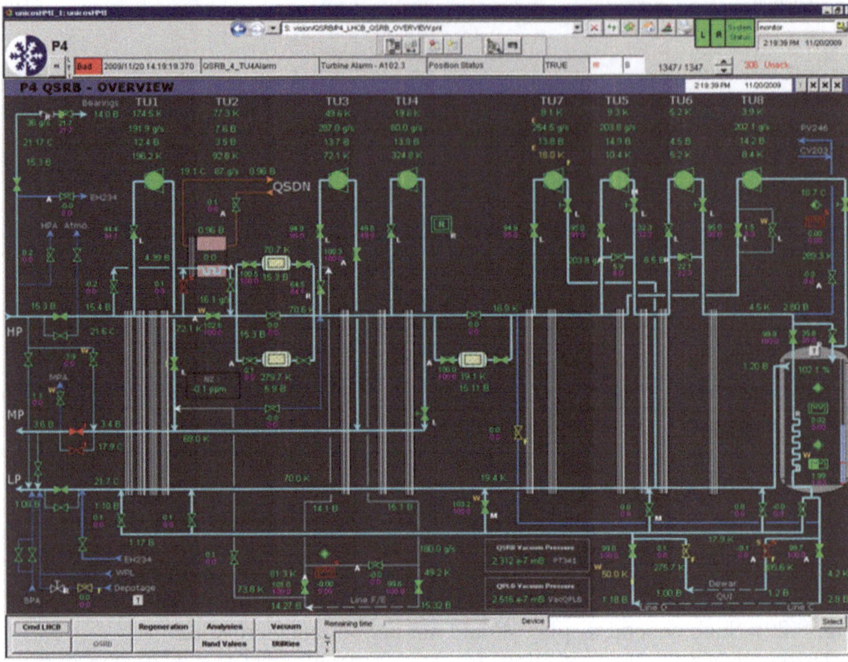

Fig. 8.29 General display operating conditions (LHC, CERN)

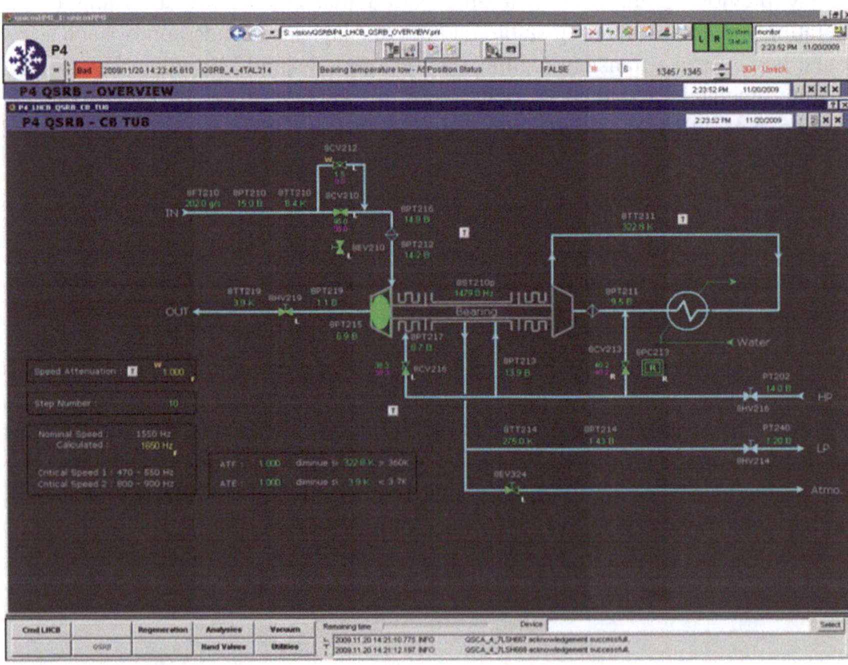

Fig. 8.30 Turbine display operating conditions (LHC, CERN)

8.9 An Example of a Simple Control Procedure

8.9.1 The Operating Procedure Written in a Human Language

Suppose one wants to start a Brayton refrigerator (Brayton is chosen for simplicity).

What are the conditions to be fulfilled prior to starting the first machine: the compressor? One needs:

- Electrical power
- Cooling water
- Instrument air
- Insulating vacuum
- Helium

This information is either digital (on-off) or analogical (see Table 8.3). For example, electrical power is a digital data coming from the electrical gear system; air could be either a digital data provided by a pressure switch set at the minimum acceptable pressure or a threshold on air pressure that is an analogical data.

There are a lot of not remotely actuated components, the position of which must be checked prior to starting the system. For example, if the circuit conditioning procedures are not performed automatically, one must be sure that all hand valves that are used for such purpose are closed. A good method is to have all these not remotely actuated components fitted with a position sensor that is checked by the control system. Only when all such components are known to be in the correct position the starting procedure is allowed.

The combination of these information is made through a Boolean operation:

Table 8.3 Information that is needed prior to starting-up

	Digital	Analog	Actions
Pre-starting checks			
Utilities:	Power		Boolean
	Water		
	Air		
Vacuum:		Vac. pressure	
Helium:		He pressure	
Circuit conditioning			
	Valve positions	He pressures	Sequential
		He analysis	
"START" order			
Start compressor			
Reach cycle pressure stability			Sequential + analog
Connect cold box			
Start turbine			

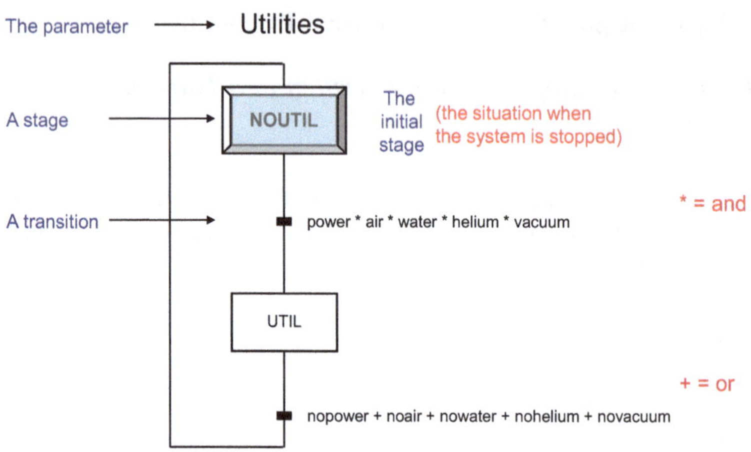

Fig. 8.31 The elementary graph

$$\text{Utilities} = \text{power and water and air and vacuum and helium.}$$

that is generally written:

$$\text{Utilities} = \text{power} * \text{water} * \text{air} * \text{vacuum} * \text{helium} \; (* = \text{and}).$$

But one must be sure that the helium circuits are filled with sufficiently pure helium and so on.

By the way, such action is called the functional analysis of the system, a very important step in the process of automation.

8.9.2 The Operating Procedure Translated into a Machine Language: GRAFCET as an Example

GRAFCET (*Graphe Fonctionnel de Commande des Étapes et Transitions*) is a sequential function chart operating as a stage and transition language. It is intended to describe operating procedures.

It is a graphic language showing the operation of a system based on a set of:

- Stages to which actions are associated
- Transitions between stages
- Connections that link the various stages and transitions of a graph (see Fig. 8.31)

Let's, simultaneously, describe the elementary graph and refer to the checks that are to be made prior to start the compression station.

Each elementary graph deals with one parameter, the name of which is displayed on top of the graph.

When the system is at rest, the initial stage is *active*. To discriminate the initial stage, it is enclosed into a double line (here, it is underlined in light blue). In the initial stage, no action is launched, all the system is stopped, and all components are in their fail mode (e.g. a fail-closed valve is closed).

Each stage has a tag that could be anything like a number or a mnemonic tag (here, NOUTIL is the mnemonic of NO UTILities).

The transition, which is under (always under) a stage, can be open or closed. As long as the transition is closed, the stage that is upstream it stays active; it does not change state. An open transition is underlined in light blue. When the transition opens, the stage that is upstream the transition is deactivated, and the stage that is downstream is activated.

A transition opens as soon as the set of conditions that is written at its right is fulfilled. Here the condition is, as already stated above:

Power <u>and</u> water <u>and</u> air <u>and</u> vacuum <u>and</u> helium.

Power, water, air, vacuum and helium are parameter names that are naturally stating their availability.

The usual way to write it is:

$$\text{Power} * \text{water} * \text{air} * \text{vacuum} * \text{helium (the} * \text{means: and).}$$

As soon as these conditions are fulfilled, the transition opens, the NOUTIL stage is deactivated, and the UTIL (UTILities) is activated. The UTIL stage does not launch any action, but its activity will be used as a condition in another graph. The system stays in the UTIL stage as long as the downstream transition is closed; however, as soon as one of the conditions is no more fulfilled, the transition opens. This is written:

$$\text{Nopower} + \text{noair} + \text{nowater} + \text{nohelium} + \text{novacuum (the} + \text{means: or).}$$

As soon as the transition opens, the UTIL stage is deactivated and the NOUTIL stage is activated. This very simple graph allows to know if the compression station can be started.

8.9.3 Example of the Compression Station

Let's now start the compression station, the sketch of which is shown in Fig. 8.32. Remember that the operation of this subsystem has been described in Sect. 8.3.4.

A number of new graphs are necessary. They are shown in Fig. 8.33. The first graph is the general command graph. It is the "boss" who will order all subsystems to run. The first transition is START order * UTIL, which means that if both conditions are fulfilled (the start order has been sent *and* all utilities are available), the OFF stage is deactivated, and the COOLING stage is activated. The second transition

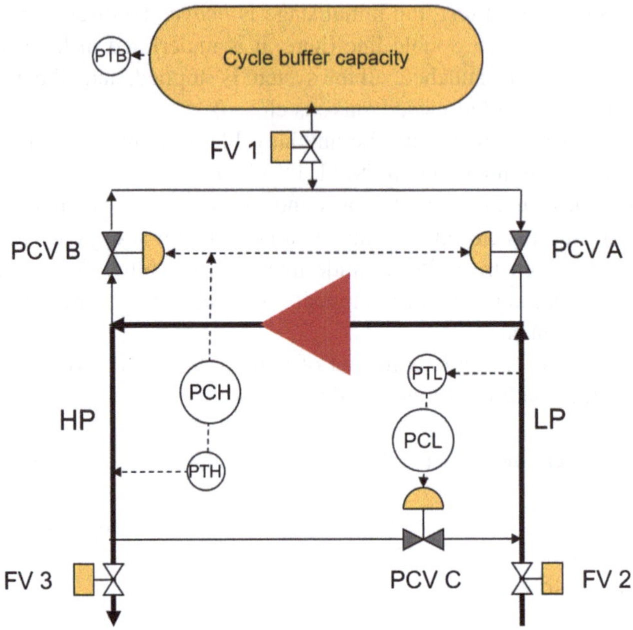

Fig. 8.32 Compressor station sketch

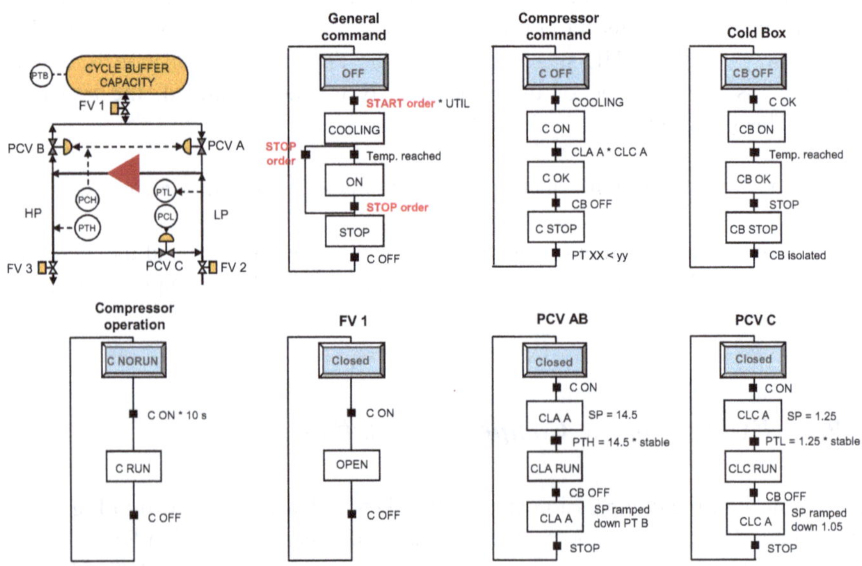

Fig. 8.33 The initial situation

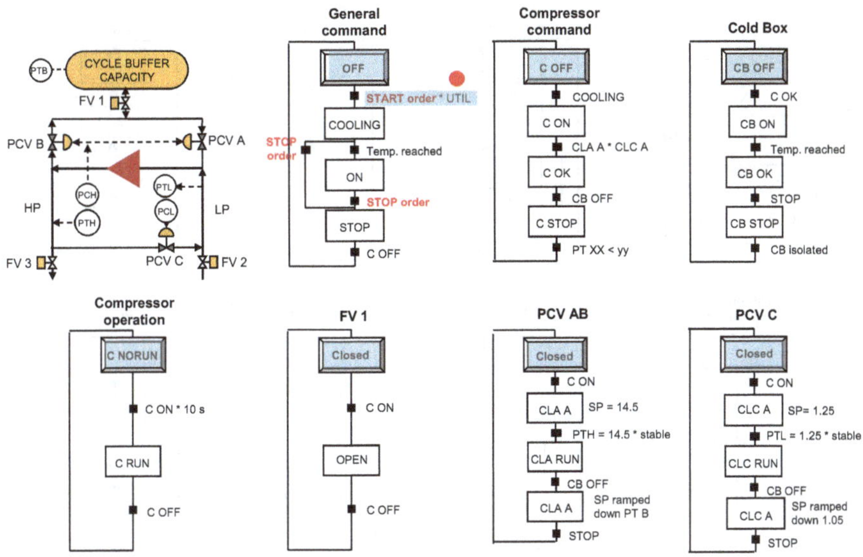

Fig. 8.34 Step 1

means that the system has reached the operating temperature (a significant temperature is lower than a set level). The ON stage is where the system is sitting during steady state at cold conditions. The STOP stage is activated when the STOP order is issued. It can also be activated when the COOLING stage is active (see Figs. 8.34 and 8.35). The STOP stage launches action to stop the system. Among the actions to stop the system, the compressor stop is the last one. The C OFF (compressor off) stage in the compressor GRAFCET is activated, the C OFF transition opens, and the initial stage is again activated.

Now let's look at the other graphs in Fig. 8.33:

- The "compressor command" graph takes care of the compressor electrical breaker.
- The "cold box" graph gives the status of the cod box.
- The "compressor operation" graph gives the status of the compressor.
- The "FV 1" graph operates valve FV 1.
- The "PCV AB" graph operates both PCVA and PCV B valves (HP control).
- The "PCV C" graph operates PCVC valve (LP control). CLC A means that CLC is operating in automatic mode. The set point is indicated at the right.

All these graphs are in their initial stage: the system is at stop.

Step 1 (Fig. 8.34). When "START order" is issued and Utilities are available, the transition: "START order * UTIL" Opens.

Any change is identified by a red dot.

Step 2 (Fig. 8.35). When the "COOLING" stage is active, transition "COOLING" opens.

Step 3 (Fig. 8.36). The compressor command graph moves from "C OFF" to "C ON".

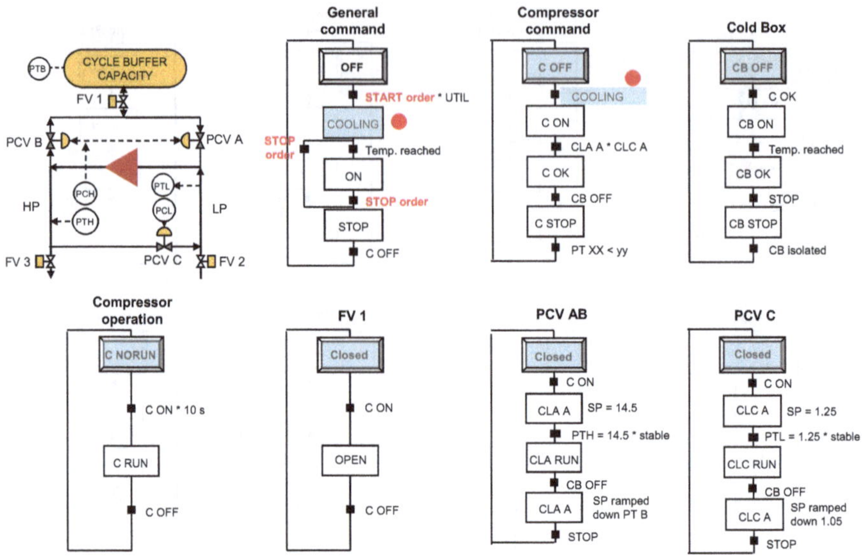

Fig. 8.35 Step 2

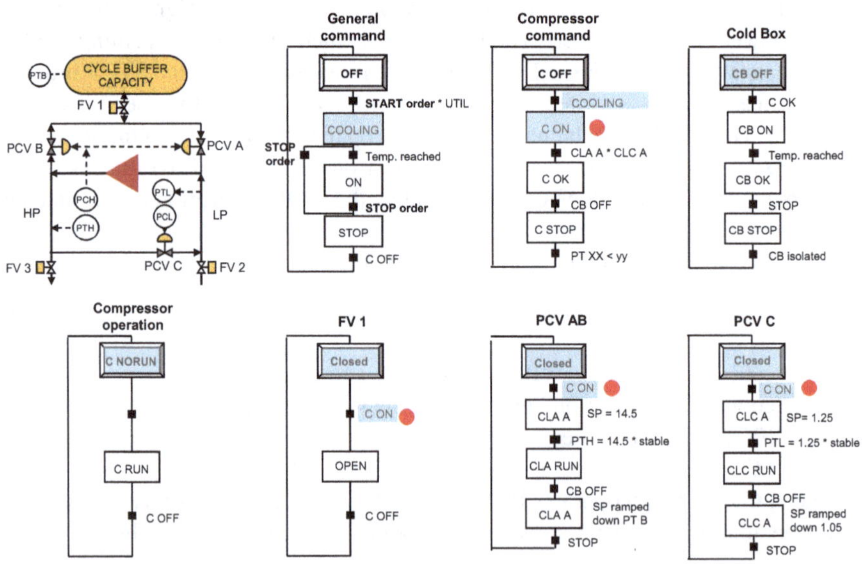

Fig. 8.36 Step 3

Step 4 (Fig. 8.37). Stage “C ON” is active, transitions “C ON” open

Step 5 (Fig. 8.38). Stages “OPEN”, “CLA A”, “CLC A” are active (control loops switch to automatic). After 10 s, stage “C RUN” is active: the compressor runs. The control loops try to keep the pressures at the set point, the system is not yet stable.

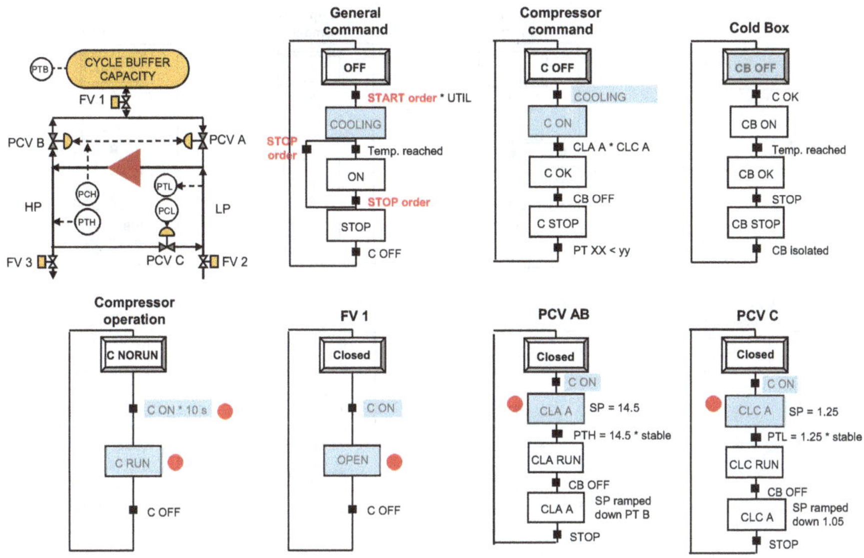

Fig. 8.37 Step 4

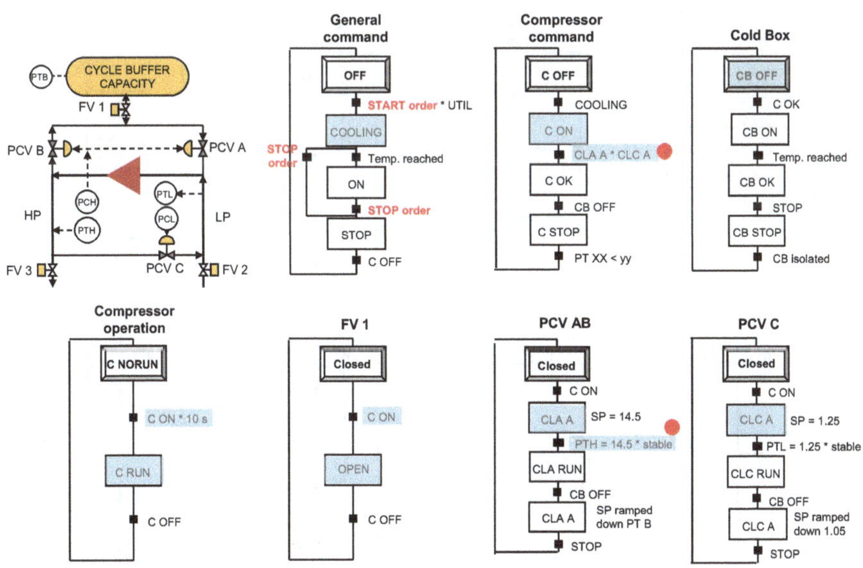

Fig. 8.38 Step 5

Step 6 (Fig. 8.39). As soon as one pressure is stable (stable means that the process variable is within ± 5 % from the set point during a period of time. This condition is checked by a small dedicated graph, not shown here.), transition "PTH =14.5 * stable" opens. Stage "CLA RUN" is activated.

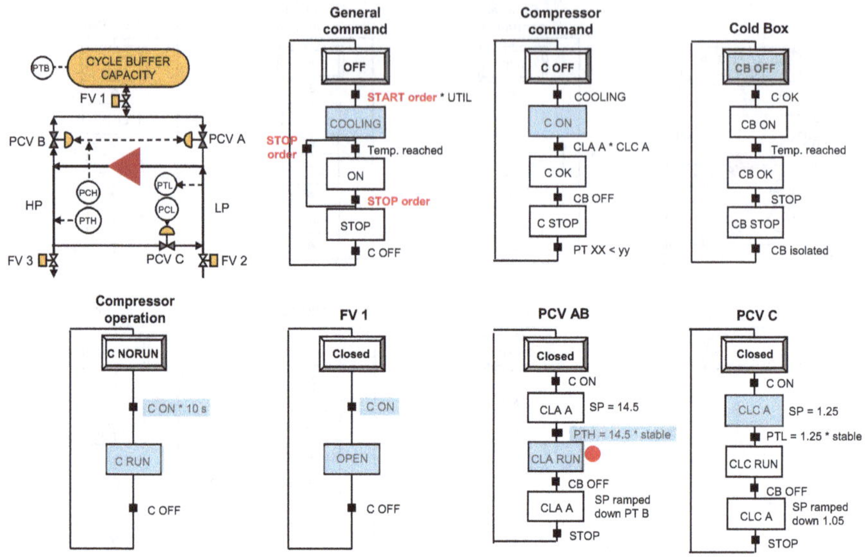

Fig. 8.39 Step 6

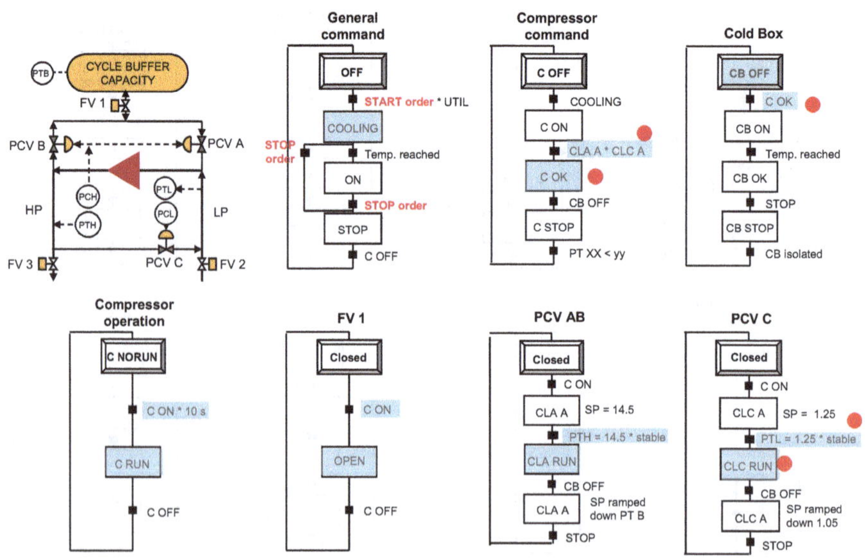

Fig. 8.40 Step 7

Step 7 (Fig. 8.40). When LP is stable, stage "CLC RUN" is activated. Transition "CLA A* CLCA" opens. Stage "C OK" is activated. Transition "C OK" opens.

Step 8 (Fig. 8.41). Stage "CB ON" is active. Nothing will move until the system reaches a given temperature: the system cools down.

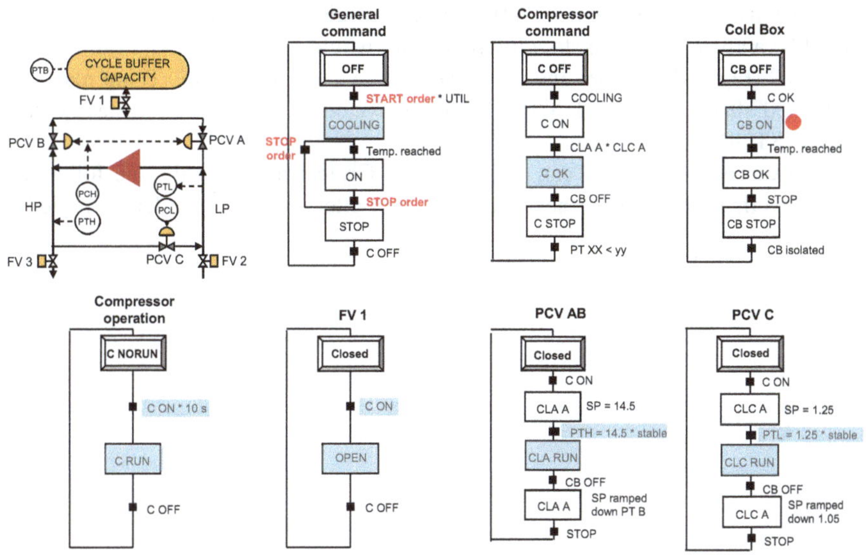

Fig. 8.41 Step 8

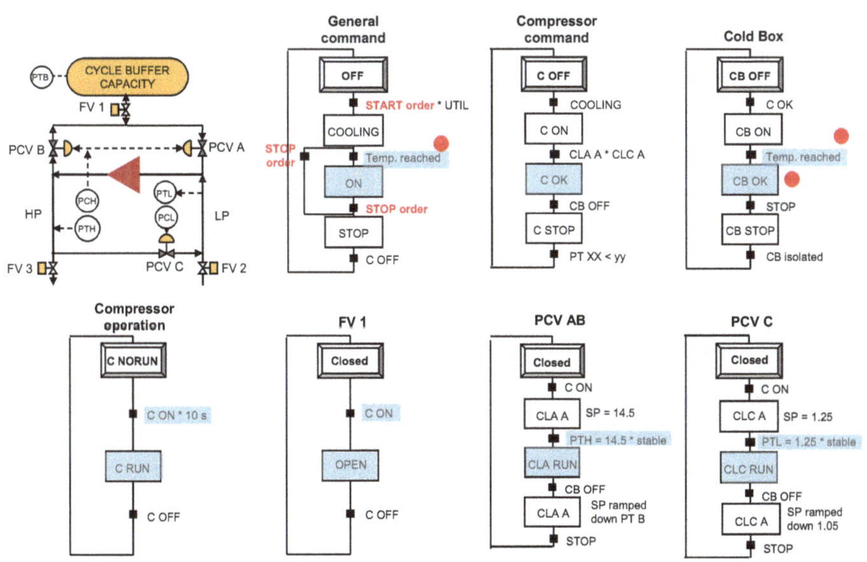

Fig. 8.42 Step 9

Remark: If, at any time, a "STOP order" would be issued, the "STOP" stage would be activated and the stopping procedure would be launched.

Step 9 (Fig. 8.42). When the temperature is reached, transition "Temp reached" opens and stages "ON" and "CB OK" are activated. The system is in steady state until a "STOP order" issued.

An exercise: the reader can play with the Grafcet and describe the following sequences that would bring the system to a stop after the "STOP order" is issued.

Chapter 9
Helium Management

Abstract When helium reaches very low temperatures, any impurity becomes a solid and may plug the cold circuits. The cycle helium pollution is discussed. Helium purification technologies (adsorption and cryo-trapping) and analysis are considered.

9.1 Introduction

Helium is the cycle gas for liquefiers and refrigerators. As it reaches very low temperatures, any impurity becomes a solid, may plug the cold circuits and is responsible of deterioration in case of refrigerators incorporating expansion turbines. Therefore, helium purity is a paramount issue in the operation of helium refrigeration or liquefaction systems. Helium purification and analysis are discussed here.

9.2 Helium as a Fluid

9.2.1 A Very Short History of Helium

Although it is rare on Earth, helium is the second most abundant element in the universe, especially in the sun and stars. It is substantially a quarter of the mass of the universe. However, helium was not discovered until 1868 in the chromosphere of the sun, simultaneously and independently by Joseph Norman Lockyer an American and Jules Janssen, a French. The helium name comes from the Greek word "helios" meaning "sun".

Some 25 years later, helium was also found, on Earth, by W. Ramsay into cleveite, an ore which contains uranium, yttrium, erbium, cerium and thorium. In 1905 H. Cady showed that natural gas produced from a well in Kansas contained 1.84% helium, thus paving the way for industrial production.

G. Gistau Baguer, *Cryogenic Helium Refrigeration for Middle and Large Powers*, International Cryogenics Monograph Series,
https://doi.org/10.1007/978-3-030-51677-2_9

Helium was, after hydrogen, the last of the “permanent” gas to be liquefied. The first liquefaction was successfully performed by Kamerlingh ONNES on May 10, 1908, in the laboratory of Leiden, Netherlands.

The thermodynamic characteristics of helium are displayed in Sect. 1.4.2.1.

Terrestrial helium production is mainly due to the decay of uranium and thorium which produces alpha particles. Each particle, capturing two electrons, is transformed into a helium atom. The source of helium is almost unlimited! The concentration of helium in the atmosphere is around 5 ppm volume (5×10^{-6}). This concentration results from the balance between production and dissipation of the terrestrial helium atmosphere in the space. Although the atmosphere contains about 0.2×10^{14} Nm^3 helium; its low concentration makes extraction from air an expensive process. Therefore, the only source of helium is currently being processed from natural gas, the concentration of which, depending on the origin location, varies from a few hundredths of a percent to a few percent by volume.

The amount that is stored in the ground is currently estimated at about 1.0^{14} Nm^3. Natural helium contains two isotopes ^3He and ^4He. The $^3He/^4He$ ratio is of the order of a millionth. In conventional helium refrigeration, only ^4He is used.

9.2.2 Helium Production and Consumption in the World

It is rare that natural gas is only processed for helium extraction: the cost involved would be too high. Since natural gas contains impurities (heavier gases, various hydrocarbons, carbon monoxide and dioxide, hydrogen sulphide, water; lighter gases, nitrogen, hydrogen, helium…), they must be separated before distribution in natural gas networks in order not to take unnecessary volume and consume energy compression in the gas pipelines. When nitrogen, which is the heavier of the light impurities, is separated from natural gas, it brings with it all other light gasses including helium. The concentration of helium in waste nitrogen increases in the inverse proportion of nitrogen versus methane; it is therefore multiplied by 10–5. A first separation between nitrogen and helium is made by low-temperature distillation, followed by purification by adsorption on activated carbon or molecular sieve maintained at the temperature of liquid nitrogen to remove the traces of nitrogen. Other methods operating at room temperature, such as membranes or PSA (Pressure Swing Adsorption), can also be employed. It is sometimes necessary to make a special treatment with respect to the neon whose concentration can reach a few ppM Volume. Helium is then compressed at high pressure (200 bar) or liquefied for distribution (see Sect. 12.5).

In the liquid state, helium is almost pure. However, transfers to which it is subjected between the plant and the location where it is used may generate pollution. The impurities are then in a solid form and accumulate at the bottom of the various storage tanks.

The recognised global resources and their locations, estimated in 2017, are 59 billions of m^3 ($MNm^3 = 10^6$ Nm^3) and are shown in Table 9.1.

Table 9.1 Helium production, reserves and resources in million cubic meters (MNm^3)

	Production 2016	Reserves	Resources
Algeria	10	1800	8200
Australia	4	NA	NA
Canada	<1	NA	2000
China	NA	NA	1100
Poland	2	25	NA
Qatar	50	Qatar	10100
Russia	3	1700	6800
USA	85	3900	31300
Others	NA	NA	NA
World total	**>154**	**>7425**	**>59500**

US Geological Survey, Mineral Commodity Summaries, January 2017

In 1960, the Congress of the United States of America decided to launch a "helium conservation programme" to form a reserve of 1700 MNm^3: impure helium was stored in a natural reservoir in Cliffside Field, near Amarillo, Texas. This programme was stopped in 1973. Since then, the storage is used as a buffer, sometimes storing the product and sometimes yielding to complete the application.

The evolution of the global helium production is shown in Table 9.1. Cryogenics uses around 35% of these quantities. Note, however, that the amount of helium that is used for cryogenics is actually much more important because, whenever the recovery is possible, the helium is recycled after purification or re-liquefaction.

The availability of helium depends on the operation of the large helium supplier separation plants, themselves depending on the operation of the natural gas plants. Helium availability is variable and sometimes there are almost helium shortages.

In 2016, the Air Liquide company started a helium underground storage in Gronau-Epe, Germany. A quantity of pure helium, corresponding to more than 1 year of Air Liquide helium sourcing, is stored into a salt reservoir located 1300 m underground, using natural brine to adjust the storage volume. This system allows smoothing the variable availability of helium.

9.3 Why Is Helium Polluted?

Helium is delivered to a refrigerating system, either in a gaseous or in a liquid form. In the liquid form, it is obviously an almost pure material. In the gaseous form, it is generally rather pure, but there is always some risk that such gas is not perfectly pure; therefore it must be either analysed or, better, purified prior to be injected into the system. However, even if the initial load of gas were pure, the cycle gas becomes polluted after some time of operation of the system.

Fig. 9.1 The effusion phenomenon

9.3.1 The Effusion (or Back-Diffusion) Phenomenon

The effusion phenomenon is not obvious. It is a usual statement that if a leak exists in a wall separating two volumes at different pressures, the gas flows from the higher-pressure side to the lower one. It is what happens in case of a "big" leak, say, for example, a 0.2 mm diameter hole.

However, when the leak is very small, a so-called micro-leak that could be the result of, for example, a not perfect weld, the above phenomenon is to be considered according to the *partial* pressure of each gas.

If helium at 18 bar is sitting in a pipe where a micro-leak exists (see Fig. 9.1), each gas flows according to the partial pressure difference between both sides of the micro-leak:

- Partial pressure of helium in the pipe is 18 bar, almost zero outside: helium flows from inside to outside. This is obvious.
- Partial pressure of nitrogen in the pipe is almost zero, but in the air, partial pressure of nitrogen is 0.8 bar: nitrogen flows into the pipe.

Same behaviour applies to other gases as oxygen and water.

Such an effusion phenomenon occurs also through a porous material like rubber. The leak flow is proportional to $\frac{1}{\sqrt{M}}$ (M is the molar weight of the gas).

Both the effects of partial pressures and diffusion make possible ingress of impurities, mainly issued from ambient air, into the helium circuit. Such a process happens only with a micro-leak, not with a large leak where the exiting gas has such a high velocity that any other gas cannot get in.

Figure 9.2 shows how pure gas that is circulated through various pipes made out of different materials is polluted.

It is obvious that *only metal pipes* must be used to keep the purity of a gas.

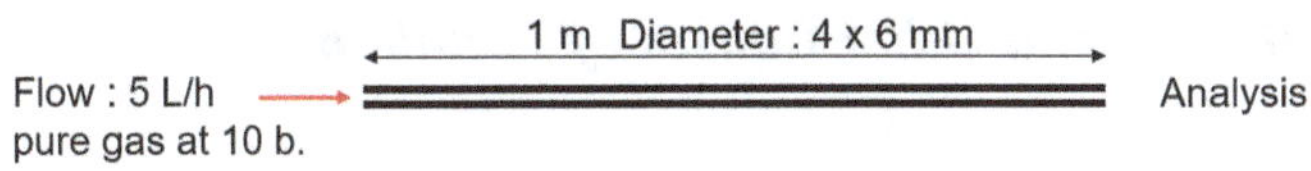

Tube material	Oxygen pollution (ppM V)
Stainless steel	0
Copper	0
Micro-leak 20 µm	20
Kel-F	0.6
Neoprene	7
Polyethylene	11
Teflon (PTFE)	13
Polyvinyl	27
Natural rubber	40

Fig. 9.2 Pollution through various materials

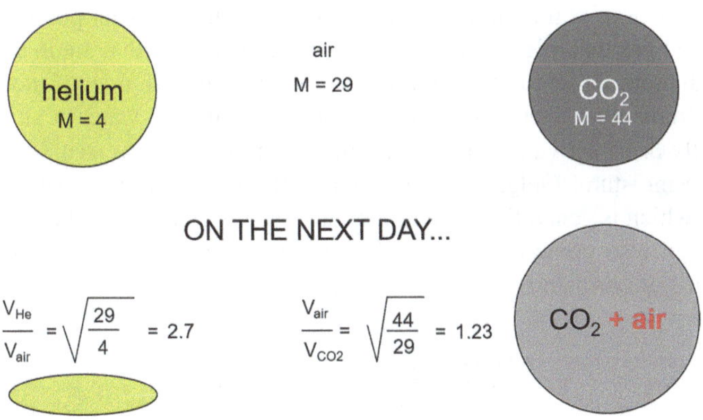

Fig. 9.3 A simple and surprising experiment

There is a simple experiment to illustrate the effusion phenomenon. Let us fill, up to the same diameter, two rubber balloons for kids, one with helium and the other with carbon dioxide, and wait (see Fig. 9.3, top). On the next day, the helium balloon is slightly depleted because the diffusion process, which is proportional to the inverse of the square root of the molar weight, allows more helium ($M = 4$) to get out than nitrogen ($M = 28$) or oxygen ($M = 32$) to get in $(1/\sqrt{4} > 1/\sqrt{28})$. However, the situation is different for the carbon dioxide-filled balloon where the high molar weight of carbon dioxide ($M = 44$) allows nitrogen and oxygen to get in faster $(1/\sqrt{28} > 1/\sqrt{44})$. As a consequence, the carbon dioxide-filled balloon's diameter increases!

9.3.2 *Effusion and Other Causes of Pollution*

Let us consider a re-liquefaction plant in Fig. 9.4 as an example. Some parts of circuits as the LHe syphon are disconnected and reconnected at each filling operation of a mobile dewar. The consequence is that some helium is lost in atmosphere and, simultaneously, that some air and moisture get into the syphon. The leak tightness of gas couplings is, or should be, generally good when a system is new, but during operation, vibrations, sometimes helped by temperature changes as in compressors, can initiate small leaks through which helium is lost and air gets in. Porous materials (that should be avoided) like seals or membranes can also bring impurities into the circuits. One must be aware of a special situation concerning the gas holder. This component is made out of a tissue that is impregnated with rubber of equivalent material that is not perfectly leak tight, through which effusion occurs. The effusion flow is proportional to the area of the gas holder that might be rather large. Therefore, whatever the helium quantity of helium that is stored in the gas holder, the mass flow rate of impurities coming from air is constant. The gas holder is, generally, the most important source of pollution for a re-liquefaction centre. Another possibility comes from the feed gas that might not be perfectly pure. Materials that are in contact with helium can contribute to its pollution as compressor oil that, if not correctly processed (remember Sect. 6.3.9), can release hydrocarbons, water or air. Similarly, not correctly dried activated charcoal in the oil removal system (see Sect. 14.4.8.3) can release moisture. Outgassing of metal walls might release small amounts of hydrogen, which is generally not an important issue. An accidental helium pollution by water can also come from a non-leak-tight helium or oil cooler, generally a consequence of corrosion (see Sect. 11.6.5.1).

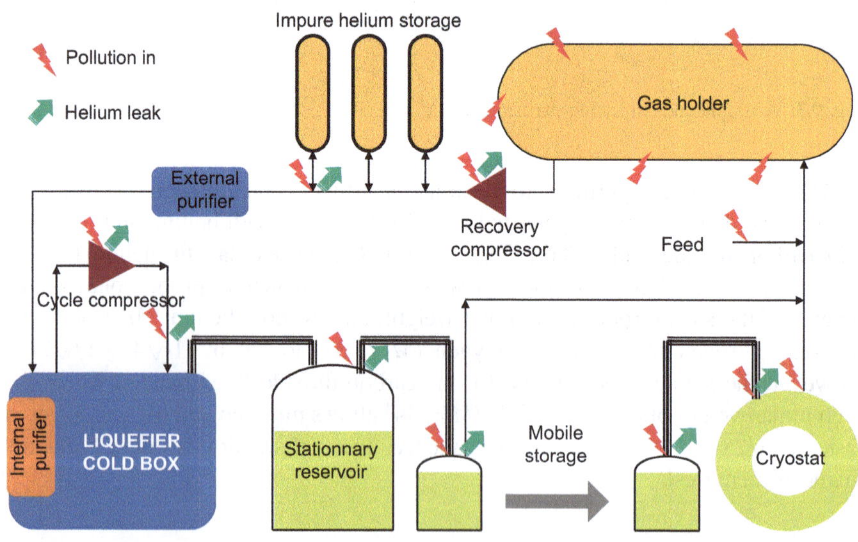

Fig. 9.4 A helium re-liquefaction centre: helium leaks and air in-leaks

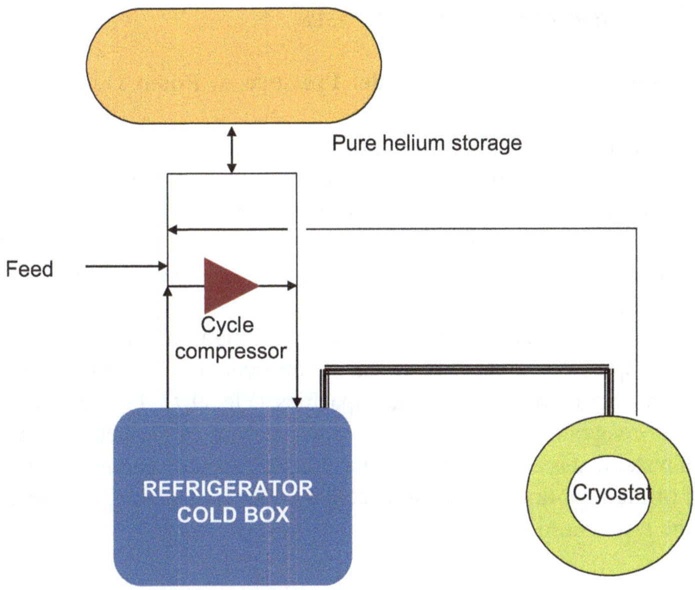

Fig. 9.5 A helium refrigerator closed circuit

If one considers a closed-circuit system as shown in Fig. 9.5, same pollution possibilities exist except those coming from disconnections of LHe transfer lines and gas holder.

9.4 Helium Purification

According to what has been discussed here above, one understands that, as soon as a system is operating, there are chances that leaks develop and impurities get into the system. A first consequence is that leak hunting is an imperative, even if bothering, everyday task. This is recalled into operating and maintenance actions (see Chap. 11).

If one wants a system to be operated on long time periods with a high availability, such, even if minor, pollutions, are to be permanently taken into consideration. The method to be used depends on the impurity to be dealt with. In this part, purification processes are described and then implemented to deal with the cycle helium and the helium to be liquefied.

9.4.1 Helium Purification Processes

9.4.1.1 Condensation of Water Under Pressure, at Room Temperature

This process is naturally applied during the compression of recovered helium that is generally almost saturated with moisture at room temperature, due to the permeability of the gas holder material as said in Sect. 9.3.1. A recovery compressor is generally a multi-stage reciprocating machine where helium is cooled after each compression stage. In each cooler, water may condense, according to Fig. 9.6, and is then separated as a liquid.

Similarly, when the recovery helium is stored in high-pressure vessels, it is good to run the compressor recovery at its maximum operating pressure by installing a back-pressure regulator upstream the capacities (Fig. 9.7). Figure 9.6 shows the residual water vapour in the temperature and pressure of the helium content. The amount of water that is drained in liquid form in the recovery compressor reduces the workload of the final drying system. If cooler water is available, it is interesting to use it in the last helium cooler.

9.4.1.2 Adsorption of Gases on Solid

9.4.1.2.1 Definition

Adsorption is a physical process involving an "adsorbent" material, generally in the form of spheres or pellets of a few millimetres. Adsorption is the adhesion of gas

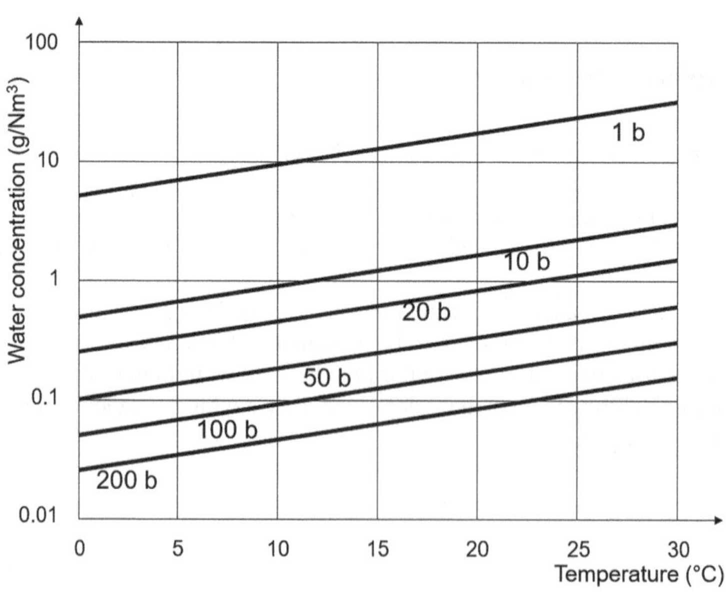

Fig. 9.6 Residual water concentration in a gas, according to temperature and pressure

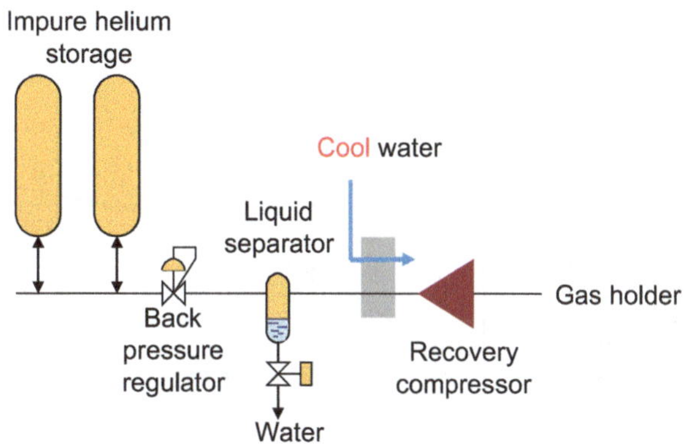

Fig. 9.7 Partial drying by condensation

molecules on a surface. This process creates a layer of the "adsorbate" on the surface of the "adsorbent". An adsorbent is an almost "empty" material with the area of the trapping surface that can reach more than 1000 m^3/g (difficult to figure out)! Physical adsorption differs from chemical absorption, in which the absorbate permeates the absorbent. Adsorption is a surface-based process, while absorption involves the whole volume of the material. The word "adsorption" was proposed in 1881 by the German physicist Heinrich Kayser (1853–1940). Desorption is the reverse process of adsorption. The optimum adsorption operating temperature is the temperature at which the impurity is liquefied at atmospheric pressure (room temperature for water and oil vapour, liquid nitrogen temperature for air components, 20 K for neon and hydrogen).

The adsorption process is exothermic (it liberates energy), but as the impurity concentration in helium is generally low, the temperature rise is negligible.

The adsorption phase has, obviously, to be followed by a regeneration phase prior to the next adsorption phase.

9.4.1.2.2 Adsorption Isotherms

Remark The adsorption isotherm plots that are displayed in this document are given as *examples*. For actual process calculations, one must use accurate information.

An adsorbent characteristic is usually quantified by means of an isotherm that shows the maximum amount of adsorbate that can be trapped, at equilibrium, by the adsorbent as a function of its partial pressure at a constant temperature (see Fig. 9.8). For low pressure, the quantity of adsorbate that is trapped is proportional to its partial pressure: it is the situation where adsorbate molecules sit nearby one to the other until they constitute a monolayer on the free surface of the adsorbent. When the first layer is fully occupied, a second layer takes place above the first one, but as the distance to the surface is larger, attraction forces are weaker. The more the number of

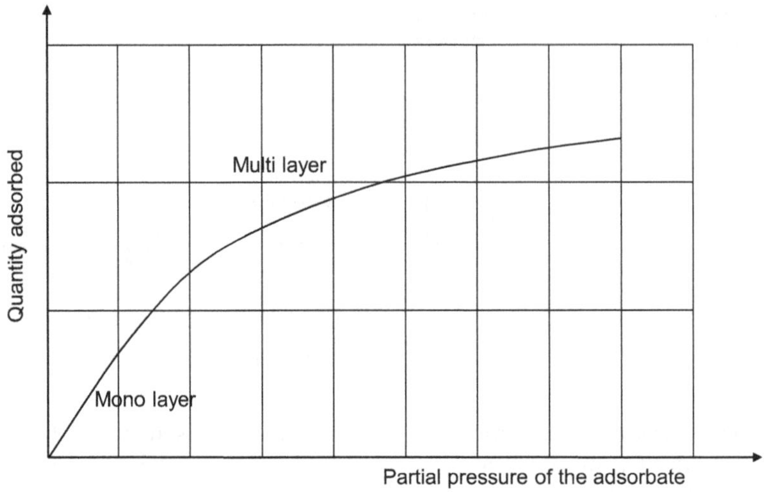

Fig. 9.8 Example of an adsorption isotherm

layers, the less the attraction forces. Consequently, the adsorption capacity increases slower, as what can be seen by looking at the slope of the isotherm in Fig. 9.8.

One can see that the quantity of impurity that can be trapped decreases strongly with its partial pressure. Isotherms are experimentally plotted. Each adsorbent material has its specific adsorption isotherm.

Remark Keep in mind that the adsorption capacity depends on the *partial pressure* of the impurity, not on its concentration!

A typical adsorption isotherm for water on molecular sieve 5A is shown in Fig. 9.9.

On Fig. 9.10, one can see that, according to the adsorption capacity of various materials, it is possible to select the best suitable adsorbent for the foreseen operating conditions.

Figure 9.11 shows the incidence of the temperature on the isotherms of molecular sieve 13×.

Figure 9.12 gives same information as Fig. 9.11, but the capacity scale is logarithmic. This is only to drive the attention of the reader about various usual ways to display adsorption isotherms.

Important When dealing with an adsorption system, keep in mind that the operating conditions are defined by a *triplet of parameters*: adsorbent, adsorbate and temperature.

On Fig. 9.13, one can see how the adsorption capacity of activated charcoal for nitrogen is strongly affected by temperature.

Important The adsorption capacity of an adsorbent increases when the pressure increases and when temperature decreases.

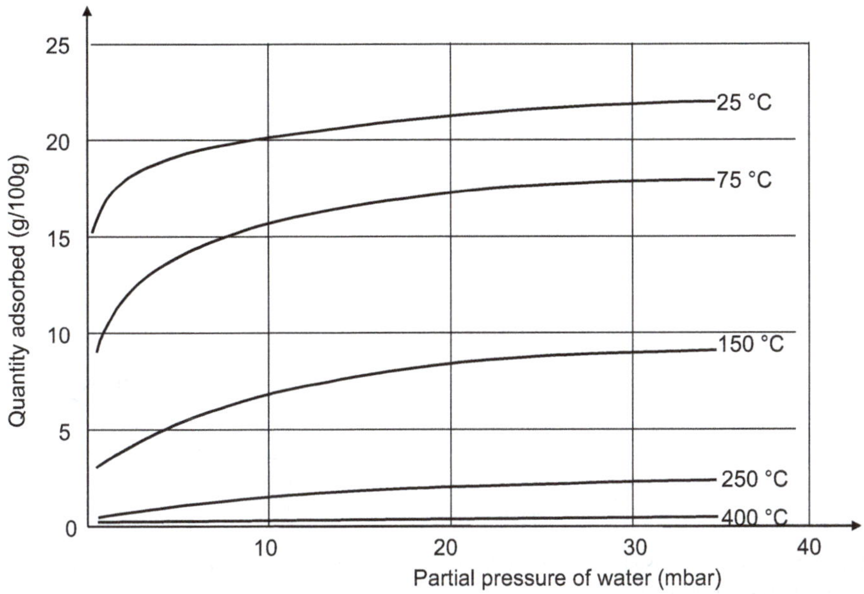

Fig. 9.9 Adsorption isotherms of water on molecular sieve 5A, at room temperature

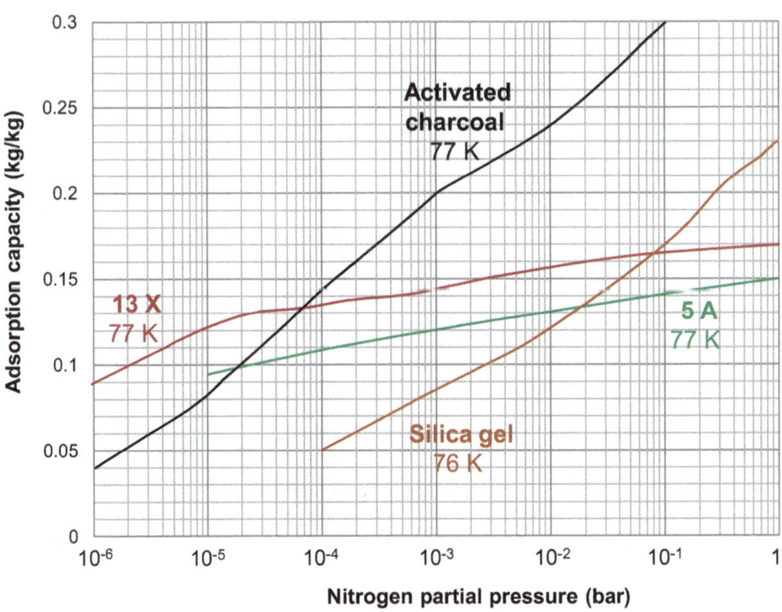

Fig. 9.10 Adsorption isotherms for nitrogen on various adsorbents, at liquid nitrogen temperature

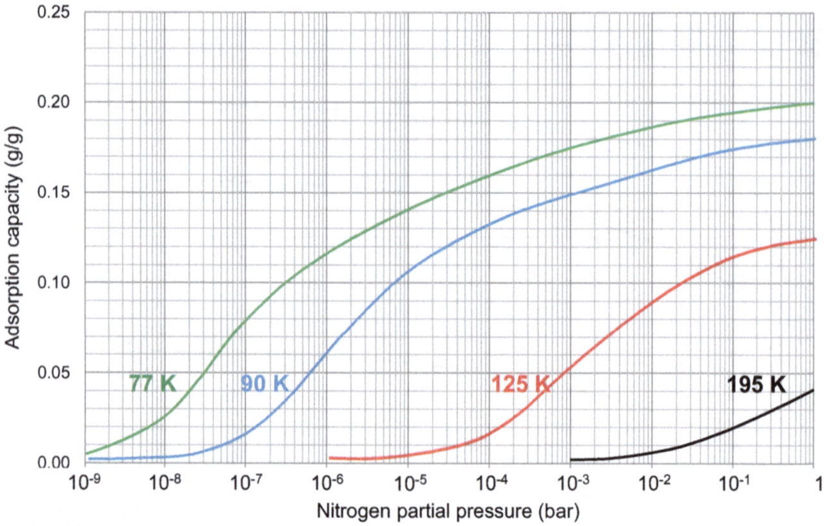

Fig. 9.11 Adsorption isotherms for nitrogen on molecular sieve 13× at various temperatures. Capacity on a linear scale

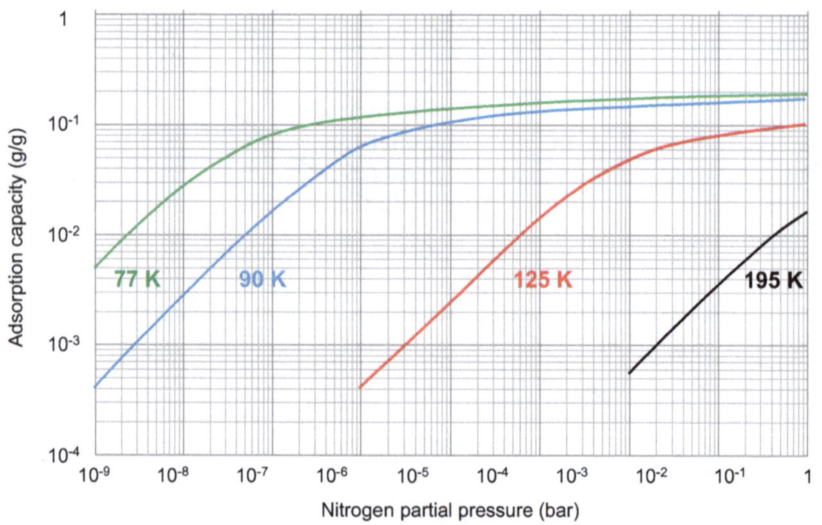

Fig. 9.12 Adsorption isotherms for nitrogen on molecular sieve 13× at various temperatures. Capacity on a logarithmic scale

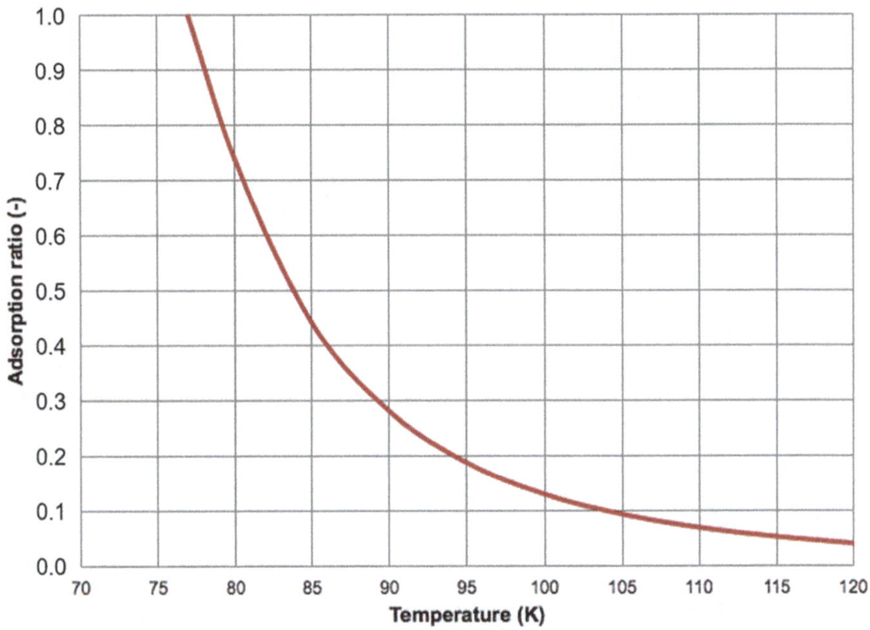

Fig. 9.13 Variation of activated charcoal adsorption capacity for nitrogen according to the temperature

9.4.1.2.3 Operating Principle

Let us consider a bed of perfectly "clean" adsorbent, which means that not any impurity is trapped in it and helium circulates through it, at a temperature that is generally near the condensation temperature of the impurity that is expected to be trapped (see Fig. 9.14).

Mind An adsorber bed (or column) must be installed with a vertical axis in order to avoid any gas to by-pass the bed near the top generatrix (see bottom of Fig. 9.14)! The horizontal position is only to make an easier drawing.

After some time of operation, one can differentiate three zones along the bed:

- The saturated zone, where the adsorbent is loaded with impurity until it reaches its saturation state corresponding to the operating conditions (pressure, temperature): all trapping sites are occupied by impurities. If one refers to the adsorption isotherm (Fig. 9.14, bottom right), the saturated zone corresponds to the isotherm value at the partial pressure of the impurity. Helium that comes out of the saturated part of the adsorbent has, obviously, the same concentration of impurity as at the entrance of the bed.

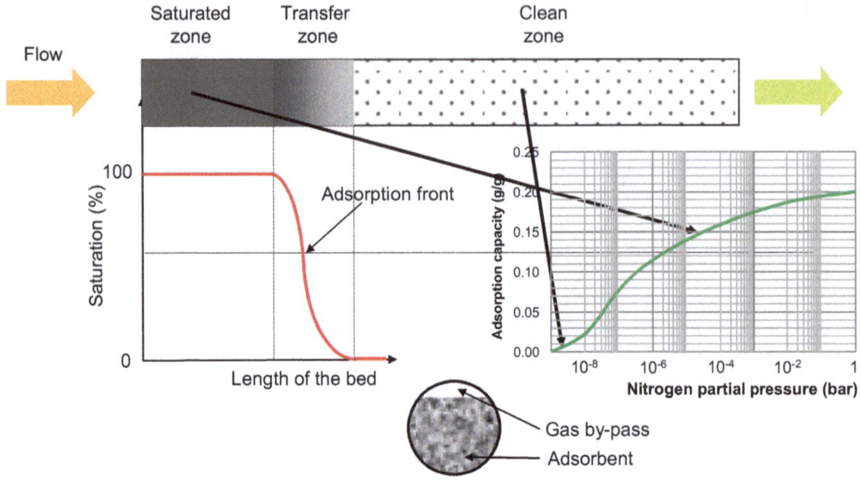

Fig. 9.14 The adsorption process

- The transfer zone, where adsorption takes place. Helium getting out of the transfer zone is "pure". On the adsorption isotherm, it corresponds to an unknown part of the curve, for very low partial pressure.
- The clean zone through which pure helium circulates.

At the bottom, left of the figure, the "saturation" of the adsorbent is plotted. Saturation is, obviously, 100% in the saturated zone and then decreases to zero along the transfer zone and stays at zero later.

It is obvious that, when the transfer zone reaches the end of the bed, impure helium gets out of it. In order to avoid such a situation, a helium sample is taken nearby the end of the bed (e.g. 4/5) and analysed. When the analysis value reaches the switching threshold, helium is no more allowed to circulate (see Fig. 9.15).

One must be careful with the meaning of the word "saturated"; it means that, for a given temperature and pressure duet, the adsorbent cannot adsorb anymore impurity. But it means also that if pressure and/or temperature changes, the adsorbent can trap more impurity or release part of the impurities that have been formerly trapped!

Let us consider the operating conditions at the top of Fig. 9.16. The adsorbent is at P_1 and T_1. The saturated zone has a given length. On top of the saturation curve, let us plot the impurity concentration along the bed. The plot has a similar shape as for saturation. Suppose, now (Fig. 9.16, bottom), that the operating conditions change, either a lower pressure or a higher temperature or both, simultaneously a lower pressure and a higher temperature. According to the adsorption isotherms, one knows that the trapping capacity of the adsorbent is lower; therefore, some impurity that had been trapped is released. Such released impurity flows along the saturated zone (at new conditions) and is trapped into the new transfer zone that has extended and moved towards the end of the bed. As the quantity of impurity in the new saturated zone is unchanged, the grey area between the concentration plot and the

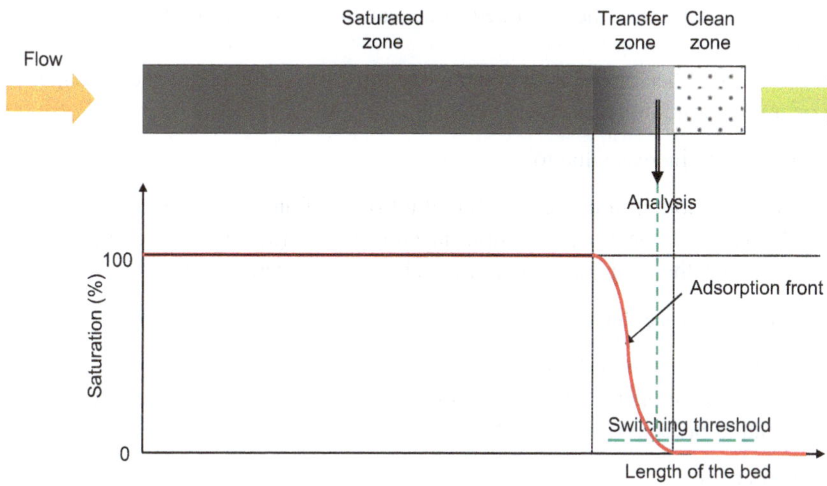

Fig. 9.15 End of the adsorption phase: analysis to detect the end of the adsorption phase

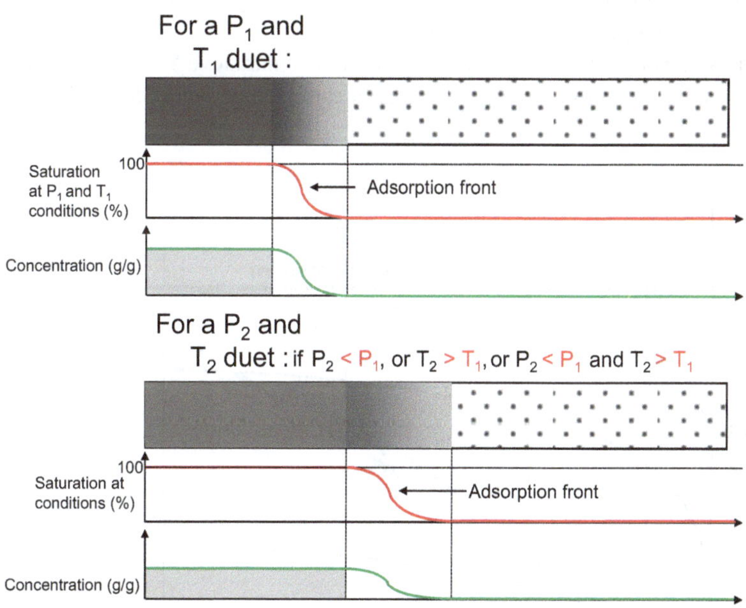

Fig. 9.16 Change of operating conditions

bed is unchanged too. As a consequence, the adsorption front has moved towards the end of the bed, and it is now farther from the inlet. If the transfer zone had been nearer the end, it could have happened that impurity gets out of the bed. This phenomenon is also valid in the reverse situation, when pressure increases or temperature decreases: the saturated zone can trap more impurity – it is not a dangerous situation.

Be careful! When operating a system, one must be careful about such changes in the operating conditions, as:

- Temperature increase due to:
  - An interruption in the feed of liquid nitrogen of an external purifier
  - A change of operating conditions in a refrigerator (larger liquefaction load that increases the temperature map in the heat exchangers; see Sect. 7.7.1)
- Pressure decrease due to:
  - A lack of helium to be liquefied
  - A change of operating conditions in a refrigerator (lower cryogenic power by floating pressures; see Sect. 8.7.2)

Regeneration

When the impurity threshold is reached (see Fig. 9.15), the bed must be regenerated, which means that impurities that are trapped into the adsorbent must be discarded. According to the isotherms, one knows that impurity is released when pressure decreases or temperature increases or both.

Let us consider the isotherm of water on molecular sieve 5A in Fig. 9.17. If one decreases the pressure from 33 to 3 mbar at same temperature 25 °C, the adsorbed

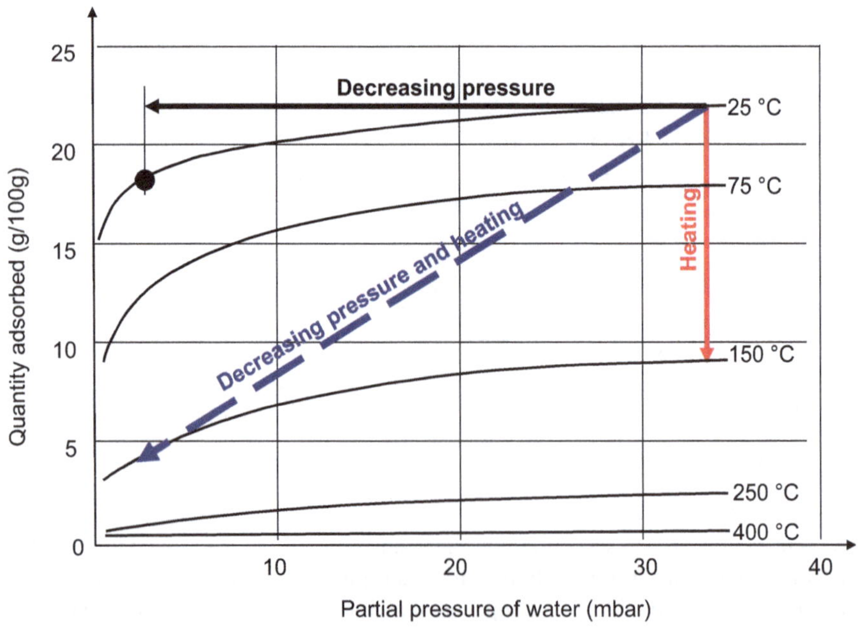

Fig. 9.17 Regeneration of water on molecular sieve 5A

quantity decreases from 22 to 18 g/100 g. If the temperature is increased from 25 to 150 °C, the adsorbed quantity decreases from 22 to 9 g/100 g.

When changing simultaneously both parameters, pressure and temperature, the adsorbed quantity decreases from 22 to 4 g/100 g. It is obviously the good procedure. In this example, one can see that the regeneration could be better by heating the bed at a higher temperature. As usually, a trade-off must be performed between the resulting capacity of the adsorbent and the energy to be paid for heating. Remember that adsorption is an exothermic phenomenon. Therefore, it is not surprising that regeneration is endothermic (it requires energy).

Regeneration by heating usually lasts several hours, because the thermal inertia of the vessel plus the adsorbent is important. Regeneration by depressurisation at constant temperature is faster but as the retention capacity of the adsorbent is far from being used to its maximum level (around 20 % in case of Fig. 9.17), it is generally not used for helium, except in some dryers that are sometimes used to feed cryogenic purifiers.

The operating procedures for an adsorption system are described in Sect. 9.4.3.4.

9.4.1.2.4 Sizing Parameters of an Adsorber Bed

Inputs

- Processed gas flow and pressure (maxima and minima)
- Nature and concentration of impurities at inlet
- Expected concentration of impurities at discharge

Choice of the adsorbent

- According to impurities
- Process conditions (pressure, temperature)

Choice of the regeneration procedure
Duration of one cycle

- Quantity of adsorbent
- Number of beds

Finally, sizing of the adsorbent capacity diameter is to be done in order to avoid attrition (wearing or grinding down by mutual friction between the pellets that are moved by the dynamic forces resulting of gas circulation). In the case of a gas circulation from top to bottom, the apparent front velocity can be twice the velocity from bottom to top. This allows to use a smaller capacity diameter. It is a good practice to have a ratio length/diameter >2 in order to avoid maldistribution of the flow.

9.4.1.2.5 Adsorbent Materials

Adsorbent materials are numerous. They are mainly used in chemical processes. Here, only conventional adsorbents used for helium purification are considered.

Activated charcoal

In 1908, when Kamerlingh Onnes liquefied helium for the first time (see Sect. 16.2), he was already using activated charcoal to purify helium! Activated charcoal is certainly the oldest adsorbent material. It is a highly porous, amorphous solid consisting of microcrystallites with a graphite lattice that is manufactured out of coconut shell or peat, usually prepared in small grains (see Fig. 9.18 left) or pellets.

The manufacturing process includes drying and then heating to separate by-products, including tars and other hydrocarbons from the raw material, as well as to drive off any gases generated. Then this material is heated over 400 °C in an oxygen-free atmosphere that avoids combustion. The carbonised particles are then "activated" by exposing them to an oxidising agent, usually steam, at high temperature. This agent burns off the pore blocking structures created during the carbonisation phase and so develops a porous graphite lattice structure, resulting in a high surface area.

Activated charcoal, which is a cheap material, is used for trapping impurities in helium as oil vapour at room temperature or nitrogen, oxygen, argon, neon and hydrogen at cryogenic temperatures.

Safety The main drawback of activated charcoal is that it reacts with oxygen at moderate temperatures (over 300 °C). Such a situation might happen (and unfortunately happened) during the regeneration phase (see Sect. 9.4.1.2.3), when the bed that releases oxygen is warmed up. For such reason, and for applying the precautionary principle, some organisations or companies forbid the use of activated charcoal to purify helium from air gases. However, if one analyses the incidents that happened, it is generally because the temperature, at the time activated charcoal was in presence of desorbed gaseous oxygen, was too high, sometimes a consequence of an electrical heating element that was not protected against high

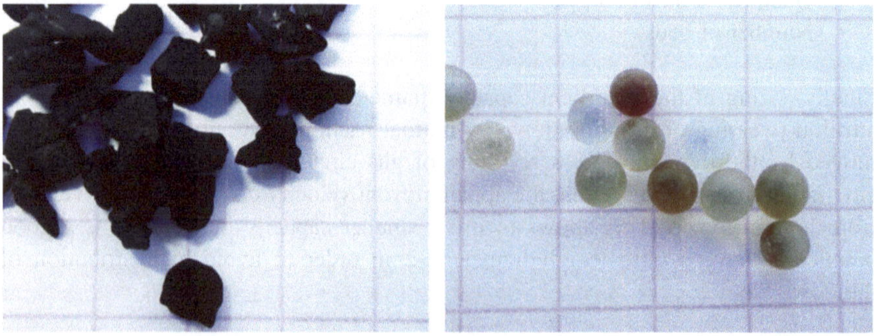

Fig. 9.18 Activated charcoal and silica gel (grid pattern is 5 mm)

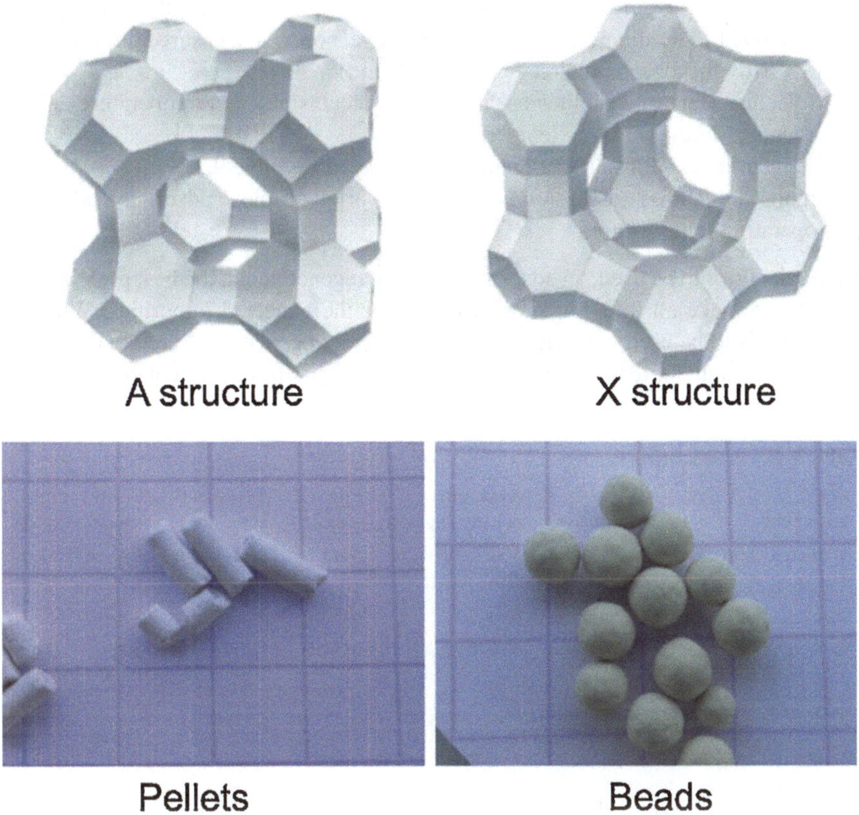

Fig. 9.19 Molecular sieve structures on top, pellets and beads bottom (grid pattern is 5 mm)

temperature or which protecting device had failed. However, a correctly designed regeneration system using a coil inserted into the adsorbent and circulated by warm gas or by flushing the bed with nitrogen is, to the author's sense, perfectly safe.

Silica gel
Silica gel is a chemically inert, amorphous form of SiO_2 (see Fig. 9.18 right). It is prepared by the reaction between sodium silicate and acetic acid, which is followed by a series of after-treatment processes resulting in various pore size distributions (wide or narrow pores).

Narrow-pore silica gel is used for drying and adsorption of air components.

Molecular sieves or zeolites are synthetic crystalline highly porous materials which have a repeating pore (very small hole) of uniform size network. They are manufactured by hydrothermal synthesis of sodium alumina silicate or another silica source followed by ion exchange with certain cations that is followed by drying of the crystals, which are pelletised with a binder to form macro-porous pellets or beads (see Fig. 9.19). There are two structures A and X that depend on the ratio Si/Al (see

Fig. 9.19, top). The pore diameters are similar in size to molecules that are expected to be trapped. The diameter of molecular sieve pores is measured in angstroms (1 Å = 10^{-10} m) or nanometres (1 nm = 10^{-9} m).

Molecular sieves or zeolites are used as desiccants at room temperature and air adsorbents at liquid nitrogen temperature.

Characteristics of usual adsorbents are displayed in Table 9.2.

Figure 9.20 shows adsorption isotherms of various triplets: adsorbent, adsorbate and operating temperature.

Remark Prior to be used, an adsorbent must be regenerated in the best possible way in order to make all trapping sites available for the expected impurities. When an adsorbent is to be used at cryogenic temperature, it is generally "saturated" with air components and, more bothering, water at room temperature. Consequently, in order

Table 9.2 Characteristics of most commonly used adsorbents

Adsorbent	Impurity	Apparent density (kg/m^3)	Pore size (Å)	Specific area (m^2/g)	Regeneration temperature (K)
Activated charcoal	Air, oil	500	5–40	900–1200	200–300
Narrow-pore silica gel	Water, air	700–750	10–40	600–800	200–300
Molecular sieve 4A	Water	650–700	4	700–800	500
Molecular sieve 13×	Air	600–650	8.5–10	500	500

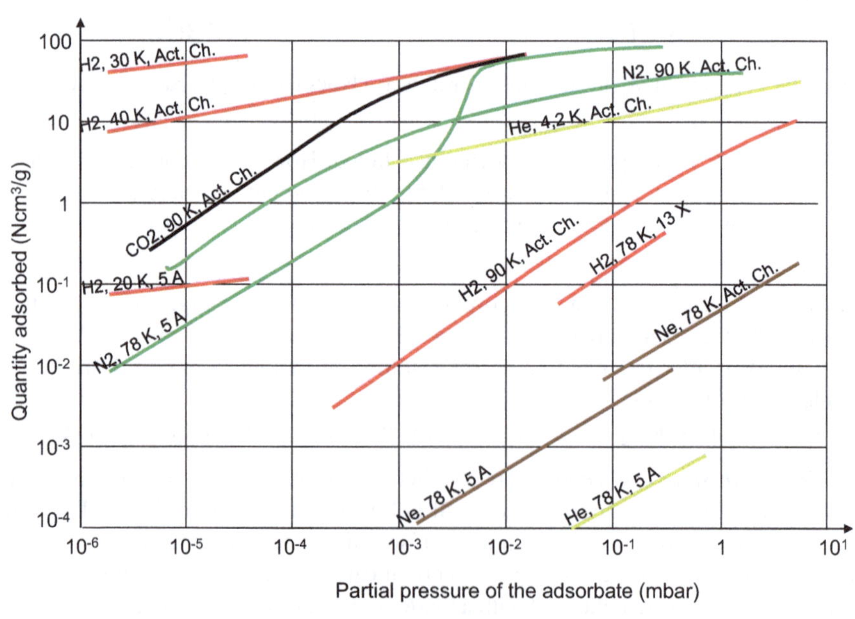

Fig. 9.20 Adsorption isotherms of various adsorbent operating at various conditions

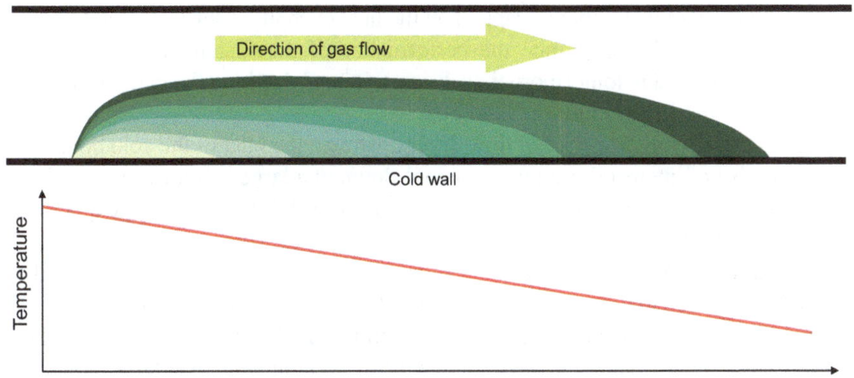

Fig. 9.21 Principle of cryo-trapping

to be rid of water, a "water" regeneration must be performed. For activated charcoal, such regeneration temperature is about 100 °C, but for a molecular sieve, it is about 200 °C. That means that the system must be able to reach a higher temperature than for usual regeneration and that the insulating material and temperature sensors must withstand the temperature without any deterioration.

9.4.1.3 The Cryo-trapping Process Operating Principle

The impure helium flows along a cold surface. Impurities are deposited on the surface, first in liquid and then in solid form, depending on the temperature of the surface (see Fig. 9.21). When the deposit reaches a certain thickness, the cross section of the channel in which the gas circulates decreases. Therefore, the pressure drop increases: this is the signal to start the regeneration process which involves heating the surfaces until they reach the melting temperature of the deposits. Liquefied impurities flow by gravity and are then discarded.

9.4.2 Cleaning and Keeping the Cycle Helium Pure

The cycle helium, which must be perfectly pure, is cleaned (from low impurity values) and kept clean by means of the cold box adsorbers (see paragraph Sect. 9.4.2.2).

9.4.2.1 Moisture

If moisture enters the cold box, it is trapped into the high-pressure channels of the heat exchangers the temperature of which is between 300 and 200 K. If the usual run time between two machine stops is lower than a few months (3–6), experience shows that the quantity of water that is trapped on the heat exchanger surfaces is not so high that it could heavily deteriorate the performance of the system. However, if the running time period is long (more than 1 year), it is advised to take care of moisture prior to allowing helium into the cold box.

This is done by inserting a full flow dryer upstream the inlet of the cold box. Such a dryer is also of interest during the first cool-down of a large system where the large area of walls that are in contact with helium, as in coils, might release large quantities of adsorbed water.

The structure of such a dryer is similar to the equipment that is described in Fig. 9.26, provided that, for large cycle flows, the regeneration procedure is using the right-side solution: flushing with warm dry nitrogen. Except in situations where helium is to be permanently dried (e.g. during cooling down of large not clean systems), one-off bed is sufficient. A by-pass allows operation without dryer because the regeneration time period is short compared to the time during which the system is operating.

9.4.2.2 Air Gases, Neon and Hydrogen

Air gases (nitrogen, oxygen and argon) that condense at about liquid nitrogen temperature are trapped into an adsorber that is operating at about 80 K. In Fig. 9.22, left, for a two-turbine conventional cycle, the warm adsorber is located upstream the warm turbine. Consequently, it protects the cold turbine (the warm turbine operating temperatures are generally higher than the temperature at which air impurities turn into solid).

The cold adsorber is to trap neon and hydrogen that are generally at rather low concentrations. Therefore, its operating temperature is about 20 K. As one can see in Fig. 9.22, left, the cold adsorber protects the lowest-temperature circuits and components. This is of interest on more complicated cycles where some turbines operate at temperatures as low as 5 K (see Chap. 4).

Let us now consider a Brayton cycle operating at about 20 K. In a commonly used structure that is shown in Fig. 9.22, top right, the adsorber is located upstream the expansion turbine that it protects during cool-down. During steady state, the temperature is around 25 K which is a correct temperature for adsorption of neon and hydrogen. By the way, one has little risk to be bothered by such impurities because their concentration is very low and none of them will turn into solid at these temperatures. However, air gases are trapped into the heat exchanger instead. Therefore, it seems that the location of the adsorber is not correct.

To trap air gases, a better location of the adsorber is shown in Fig. 9.22, bottom right. The reason why the top right location is often selected sits in the simplification of the cold box: inserting the adsorber between two heat exchangers is a little bit more complicated. However, the top right location is acceptable in the case of a refrigerator for a Cold Neutron Source because the usual duration of a run is about 6 weeks, a time during which the system can operate without any trouble due to impurities.

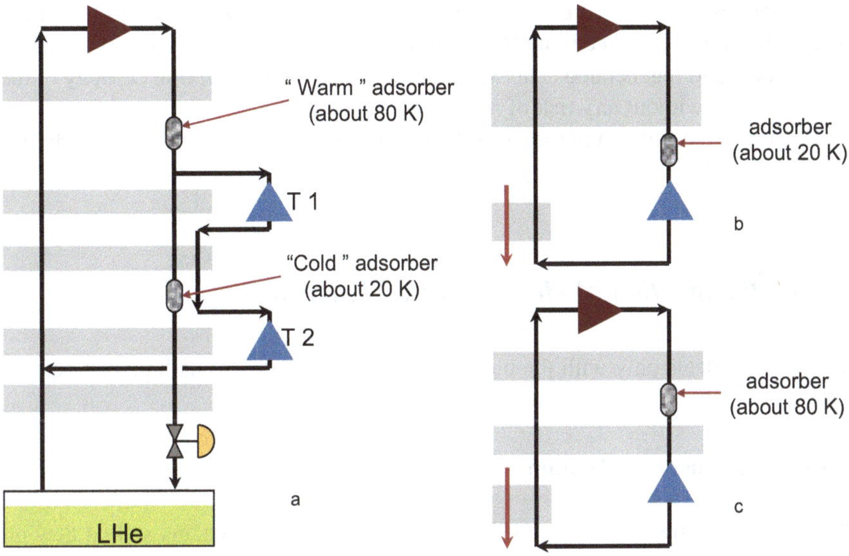

Fig. 9.22 Location of cryogenic adsorbers for protection of the cycle

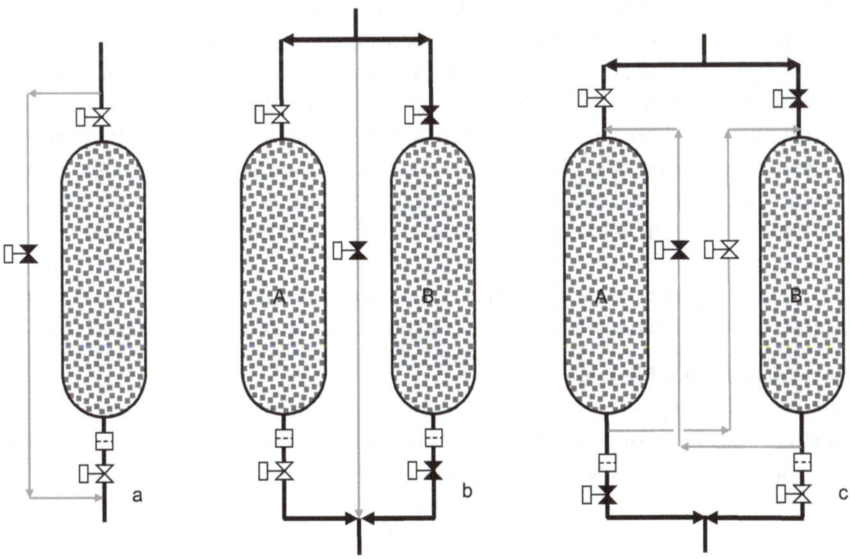

Fig. 9.23 One or two adsorbers

Decision about using one-off (Fig. 9.23, left) or two switchable adsorbers (Fig. 9.23, centre) is the same as for the full flow dryer (remember Sect. 9.4.2.1). A hint that has sometimes be used is shown in Fig. 9.23, right. When the impurity concentration threshold is reached in adsorber A, instead of switching to B, helium issuing from A is circulated into B. When impurity is detected at the discharge of A, it is isolated and regenerated. This allows using totally the adsorption capacity of the adsorber beds without any risk of releasing impurity.

During the cool-down phase, operation of the adsorbers is described in Sects. 10.3.2.1 and 10.3.2.2.

9.4.3 Purification of Helium to Be Liquefied

This chapter deals only with the purification of helium prior to its liquefaction.

9.4.3.1 Impurities in Helium

Prior to liquefaction, helium must be freed of impurities it contains so that they do not clog the system during cooling down and, later, steady-state operation. Main impurities contained in helium depend on its origin.

When helium is extracted from natural gas, the impurities are essentially water, nitrogen, hydrocarbons, carbon dioxide and monoxide, hydrogen and neon.

When liquefying helium that has been recovered from laboratories, the impurities are mainly water and air gases, especially when users do not correctly connect their systems or, even worse, if they pump on liquid helium bathes through leaky systems.

If helium has been re-compressed to a high pressure with recovery lubricated reciprocating compressors, it may also contain hydrocarbons, carbon dioxide and carbon monoxide, the latter two being generated by a combination of oil with oxygen due to the high temperature reached during the compression phase.

Helium purification can take place into an external adsorption purifier that is cooled with liquid nitrogen or with a cryo-trapping purifier that is housed into the liquefier cold box and cooled with some cycle high-pressure helium (see Fig. 9.24).

Necessity of a double train

As a regeneration phase is necessary when the adsorber bed is almost saturated, two identical systems must be implemented if a continuous flow is to be processed, one operating and the other being regenerated, as shown in Fig. 9.25.

9.4.3.2 Moisture

9.4.3.2.1 The Structure of an Adsorption Dryer Regenerated by Heating

Drying is the first purification operation that is performed on recovered helium. The compression process, which is the first one, is utilised to separate the greater part of the water, as discussed in Sect. 9.4.1.1. The final drying is generally carried out by

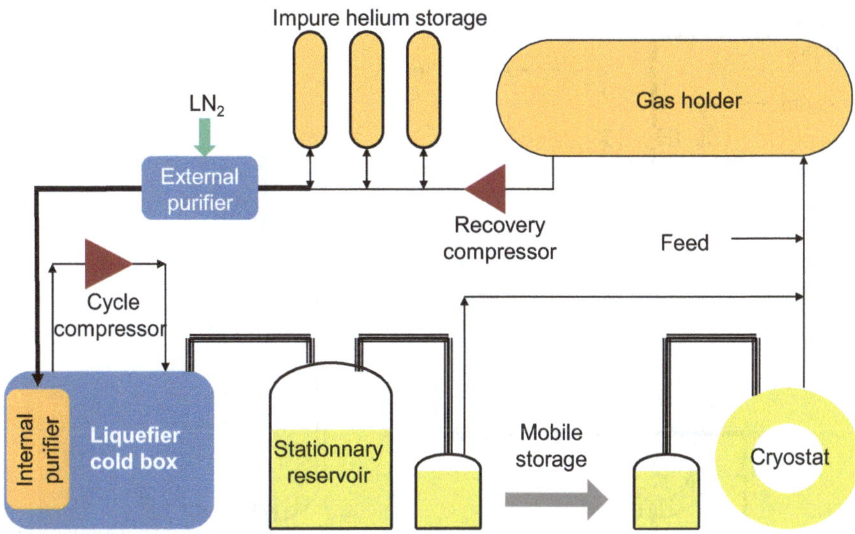

Fig. 9.24 Location of external and internal purifiers

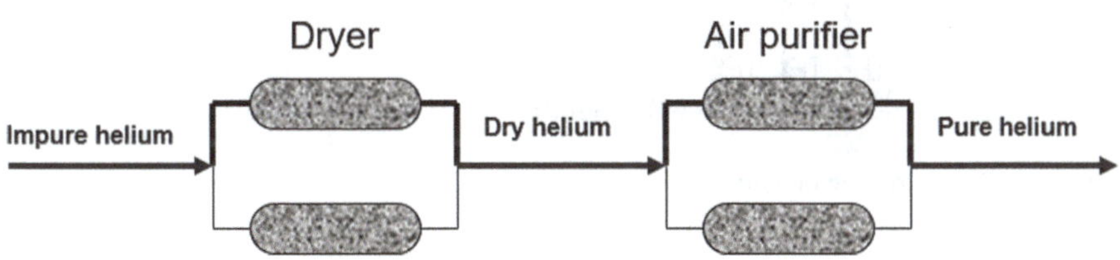

Fig. 9.25 Double train for continuous flow

adsorption at room temperature on a suitable adsorbent. It may also, more rarely, take place by condensation and solidification at moderate low temperature.

The basic structure of an adsorption dryer is shown in Fig. 9.26, left.

The adsorbent material is contained into a volume that is isolated by valves to fulfil the operating sequences. The adsorbent volume is always in a vertical position to avoid a gas by-pass if the cross section of the volume is not totally filled by adsorbent. A filter is fitted at the discharge side of the volume. Valves allow purging to atmosphere and pumping.

VA1 recovery of helium	VA6 elution
VA2 pumping of the bed	VA7 sampling
VA3 inlet of humid helium	VA8 nitrogen
VA4 discharge of dry helium	VA9 evacuation of impurities
VA5 filling of A	

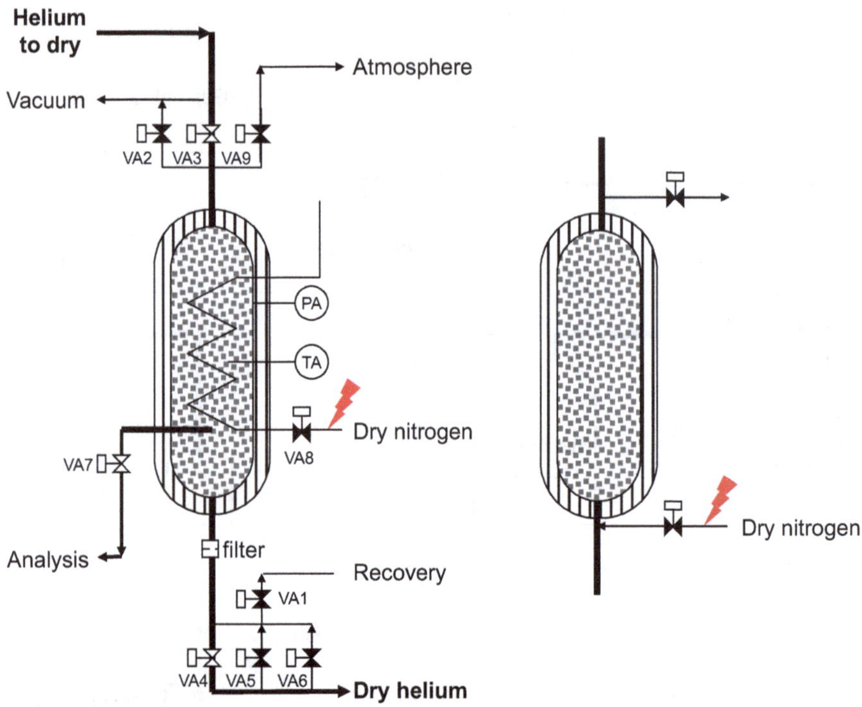

Fig. 9.26 Basic structure of a dryer

Pressure and temperature are measured. Heating for regeneration can be performed in two ways. For small equipment (diameter < 300 mm), heated nitrogen or air can be circulated into a coil that is immersed into the adsorbent (Fig. 9.26, left). For large equipment, nitrogen must be directly circulated through the adsorbent (Fig. 9.26, right). To reduce heat loss during the regeneration phase, the volume is thermally insulated. It is, of course, protected against overpressure due to a possible action of fire with a safety device, not shown in the diagram. A sampling connection, located before the end of the bed, brings helium to the analyser (remember Fig. 9.15).

9.4.3.2.2 Operation of an Adsorption Dryer at Room Temperature

The different phases of the operation are shown in Figs. 9.27, 9.28, 9.29, 9.30, 9.31, 9.32 and 9.33. Here, only the system with circulation of nitrogen into a coil is discussed.

In Fig. 9.27, helium flows through the A volume where it is dried. The capacity B is maintained at the operating pressure through valve VB4. A helium sample is collected continuously at the 4/5 of the bed A by VA7 and analysed. When the

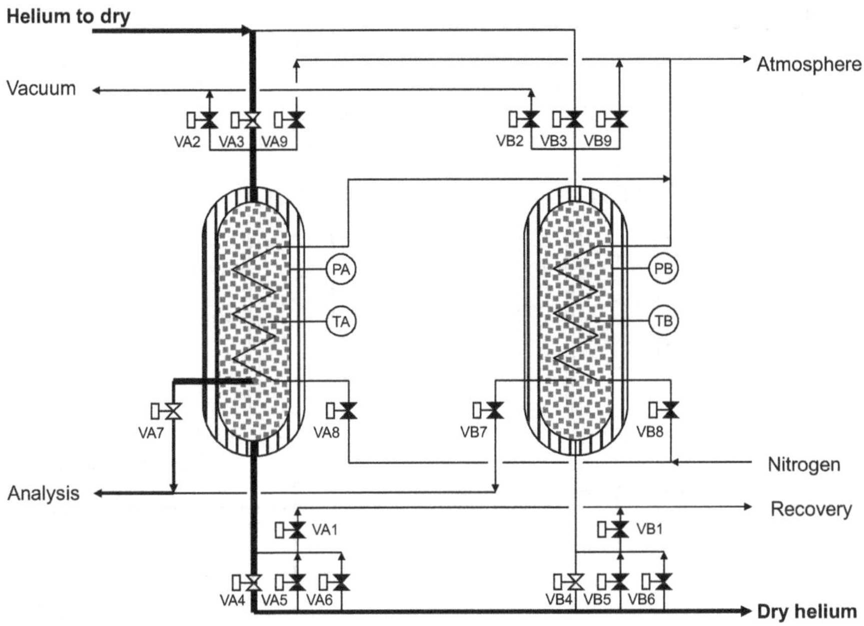

Fig. 9.27 Phase 2: A operating, B waiting

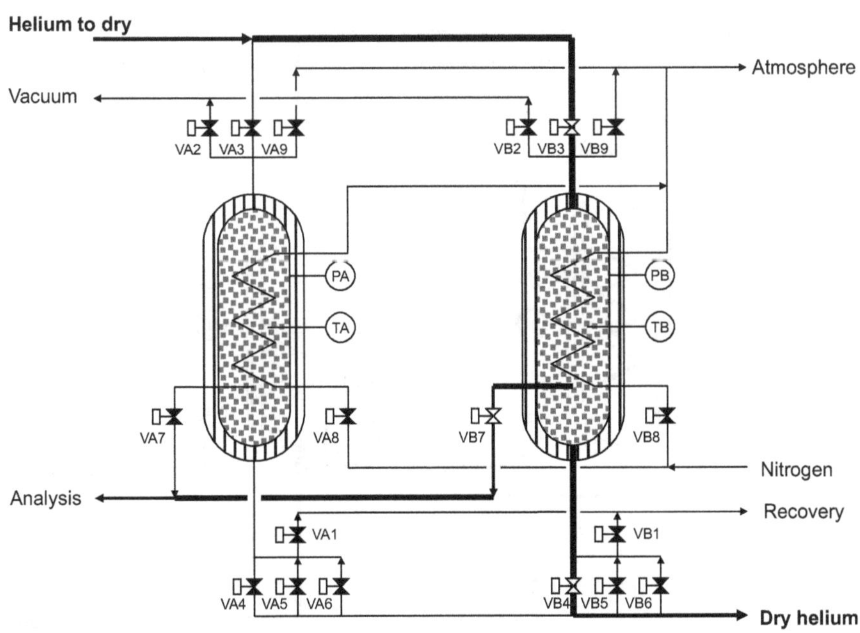

Fig. 9.28 B operating, isolation of A

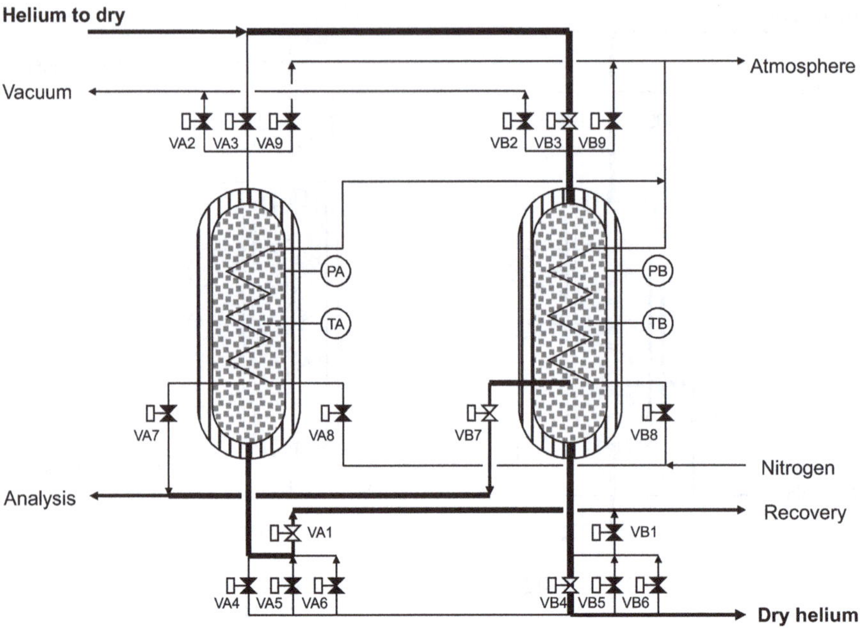

Fig. 9.29 B operating, depressurisation of A

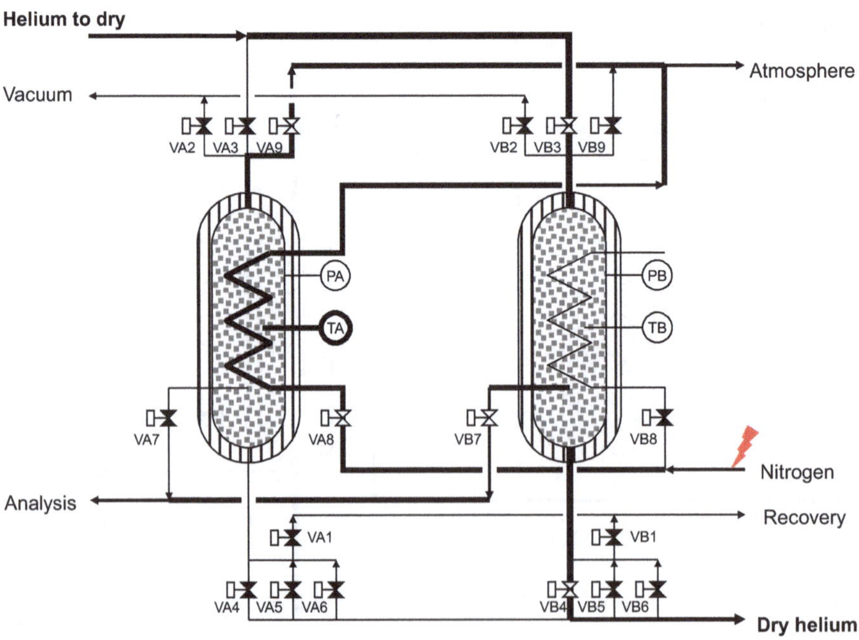

Fig. 9.30 B operating, heating A

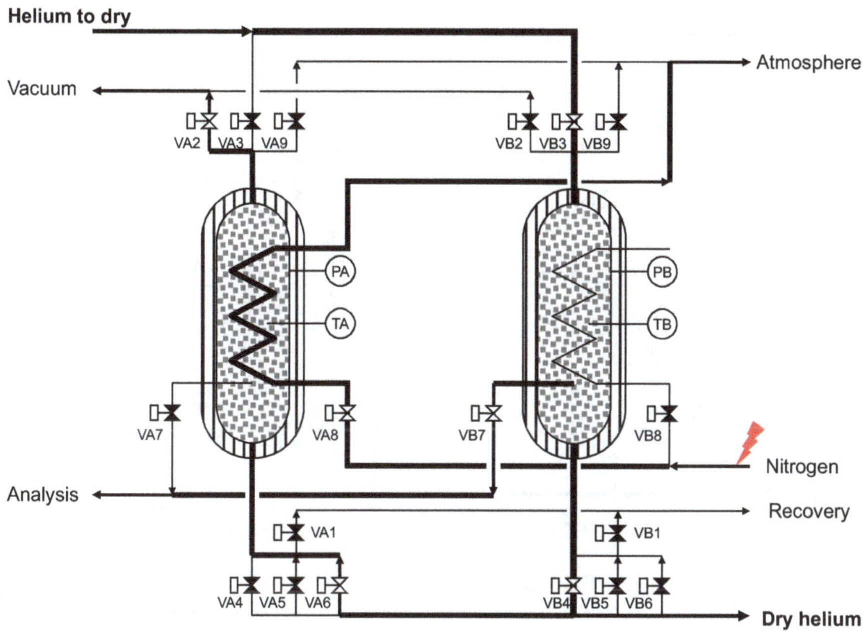

Fig. 9.31 B operating, pumping on A

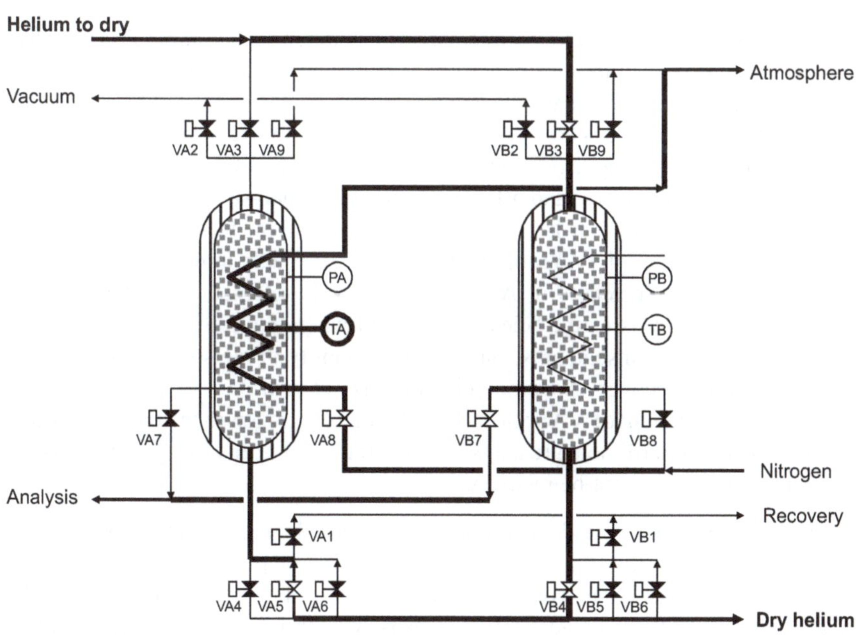

Fig. 9.32 B operating, cooling A

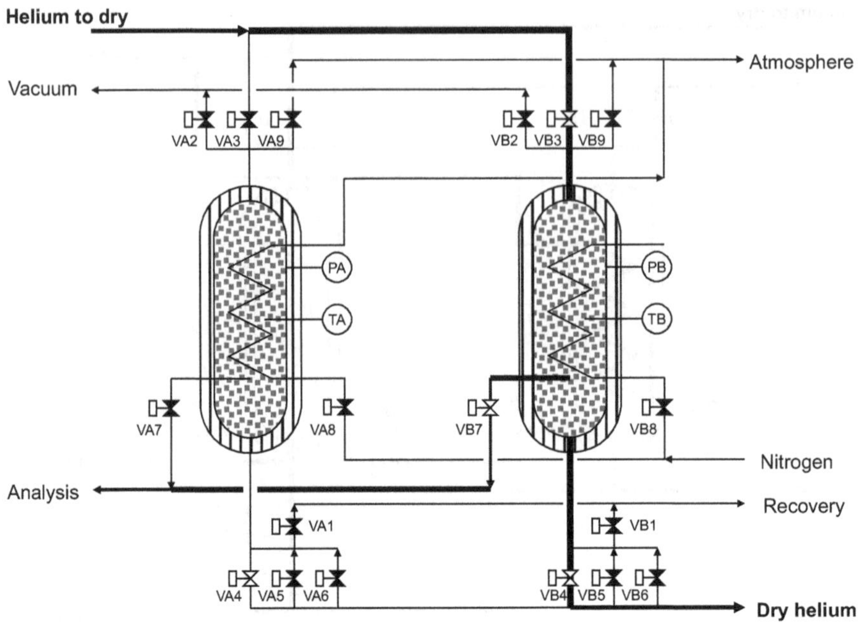

Fig. 9.33 B operating, A ready to operate

moisture threshold is reached, the procedure for switching the volume B is initiated. Since the sample of helium is taken before the end of the bed, there is a run time on A that is sufficient to allow for a safe switch over B.

In Fig. 9.28, volume B is connected by opening of VB3. Helium circulates now in parallel through both capacities, and then, A is isolated by VA3 and VA4. The sampling valve VA7 closes, and VB7 opens to allow analysis on B.

In Fig. 9.29, helium that is still in A is to be released. It is discharged to the recovery system through VA1 until pressure in A equals atmospheric pressure. When the pressure decreases in A, some moisture is released from the saturated zone. Such helium crosses the "clean" adsorbent where part or all of this moisture quantity is trapped. Consequently, only a small quantity of moisture might possibly be returned to the recovery system, and helium is not lost.

In Fig. 9.30, opening of VA8 allows circulation of nitrogen in the coil. The heater is activated. To discard moisture and some helium, VA9 opens to atmosphere (such very humid helium cannot be recovered).

Remark It would be a mistake to pump down *during* the heating phase, because the adsorbent has a low thermal conductivity and grains or pellets have a very small contact area between themselves. Therefore, heat cannot be transferred by solid conduction. Gas in between adsorbent grains is necessary to allow a correct temperature homogeneity.

In Fig. 9.31, when the temperature reaches the TA threshold, VA9 is closed, and VA2 is open in order to pump on A. VA6 is used to circulate a *very low flow* of pure helium that allows, as a gas piston, to push out the water vapour that has been released from the adsorbent to the pump: it is the *elution* process. Heating is hold on.

In Fig. 9.32, after a pre-determined pumping time, VA2 and VA6 are closed, and VA5, with a small diameter, is opened to fill in A without disturbing the system pressure and avoid too high gas velocity into the adsorbent bed (attrition). The flow of nitrogen gas is maintained but the heating is stopped, in order to cool B.

Remark If the B volume is not to be used within a short time, one can spare gaseous nitrogen for cooling by letting the volume return naturally to room temperature.

In Fig. 9.33, when A is back to room temperature, VA8 and VA5 are closed, and VA4 is opened. A is now ready for operation.

Remark For reasons of comprehension, the two functions, providing a small flow of gas for elution (VA 6) and filling up the volume (VA 5) are fulfilled by two separate valves. In a real machine, they can be fulfilled by the same valve.

Another possibility
During regeneration, instead of transferring heat by means of an exchanger immersed in the adsorbent bed, it is possible to circulate warm dry nitrogen directly through the bed of adsorbent (remember Sect. 9.4.3.2.1). Of course, after this phase, nitrogen must be carefully removed by flushing helium or pumping before introducing helium.

Remark The adsorbent, even if it has been selected for a drying purpose, adsorbs also other gases, particularly nitrogen that is flushed through the adsorbent bed. Consequently, if the bed is cooled down by nitrogen circulation, some nitrogen stays trapped into the adsorbent. When the system is circulated by nitrogen-free helium, the adsorbent is kind of "regenerated" and releases nitrogen that is sent into the cold box. Fortunately, nitrogen is trapped into the "warm" adsorber. To avoid such inconvenience, it is advised, if possible, *to cool down the adsorbent with helium.*

9.4.3.3 Air Gases

The selected process is adsorption at cryogenic temperature.

9.4.3.3.1 The Structure of a Cryogenic Adsorption Purifier

Principle
The helium to be purified is circulated through a bed of adsorbent, at low temperature (Fig. 9.36). The optimum operating temperature is close to the vaporisation temperature of the impurity trapping: 80 K for air. The adsorbent is loaded with impurities

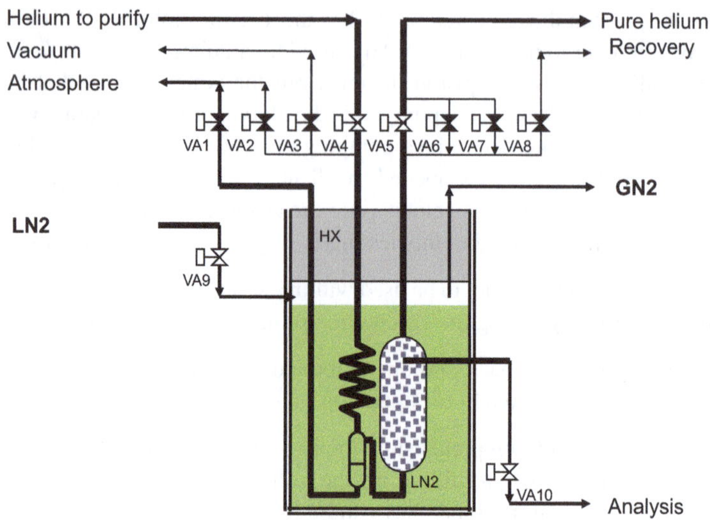

Fig. 9.34 The structure of a cryogenic adsorber

until it reaches its saturation state corresponding to the operating conditions (pressure, temperature).

A volume containing the adsorbent and circulated by helium to be purified is maintained at low temperature using cold gas or immersed in liquid nitrogen (see Fig. 9.34). To recover the helium enthalpy and minimise the consumption of liquid nitrogen, a full flow heat exchanger is used. Upstream the adsorbent volume, a phase separator is inserted. It separates liquid air, if any, that is discarded through valve VA1.

9.4.3.3.2 Operation of an Adsorption Purifier

Interest of the phase separator

Let us assume that the helium pressure is 20 bar and, for simplicity, that the only impurity is nitrogen. The phase separator operates at liquid nitrogen temperature: 77.4 K. When the nitrogen concentration increases at the inlet of the purifier and the partial pressure of nitrogen in helium reaches 1 bar, nitrogen liquefies and is separated and discarded. Such a situation happens for a 5% nitrogen concentration (see Fig. 9.35).

If the nitrogen concentration increases, all nitrogen over 5% liquefies and is discarded. As a consequence, the nitrogen concentration into helium that is processed into the adsorbent rises from 0% to 5% and stays at this value whatever the concentration at the inlet of the purifier is! In other words, the time that is

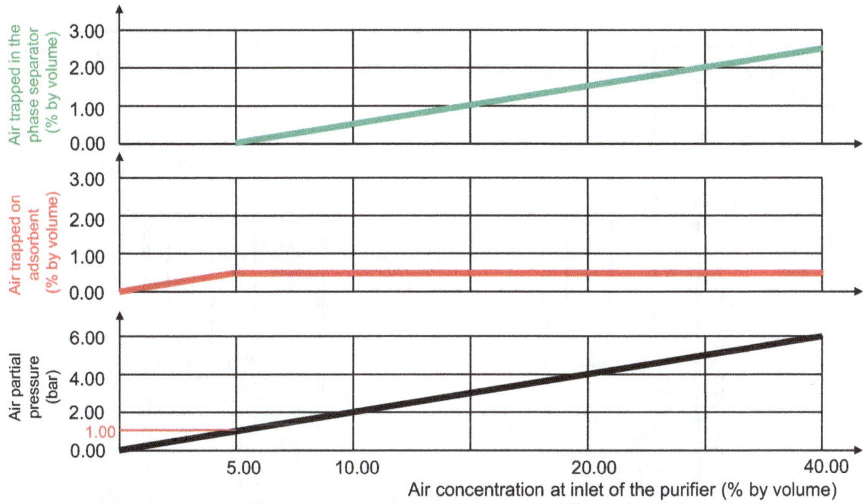

Fig. 9.35 Interest of a phase separator operating at 20 bar, 77.4 K

necessary to saturate the purifier decreases with the nitrogen concentration up to 5% and then stays constant for whatever any higher impurity concentration.

Various valves, usually operated by a process controller, allow scrolling through the different sequences of operation.

VA1 evacuation of air	VA6 filling
VA2 evacuation of impurities	VA7 elution
VA3 pumping of the bed	VA8 recovery of helium
VA4 inlet of impure helium	VA9 LN_2 feed
VA5 discharge of pure helium	VA10 sampling

9.4.3.4 Operating Procedure of a Cryogenic Adsorption Purifier

The different phases of the operation are shown in Figs. 9.36, 9.37, 9.38, 9.39, 9.40, 9.41, 9.42 and 9.43. In Fig. 9.36 the adsorber A is in operation. VA10 allows the sampling of helium in the last part of the volume of A. VA9 controls the level of liquid nitrogen. Adsorber B is waiting; it is cold and kept under pressure by VB4. VB9 controls the level of liquid nitrogen.

When analysis of A reaches the threshold, B is circulated by opening VB5, as shown in Fig. 9.37. A and B operate in parallel.

In Fig. 9.38, A is isolated by closing VA4, VA5 and VA10. Liquid nitrogen feed is stopped by closing VA9. Sampling on B is started by opening VB10.

In Fig. 9.39, helium under pressure that is stored in A is sent to recovery through VA8 (same remark as in Sect. 9.4.3.2.2 about purity of helium that is recovered);

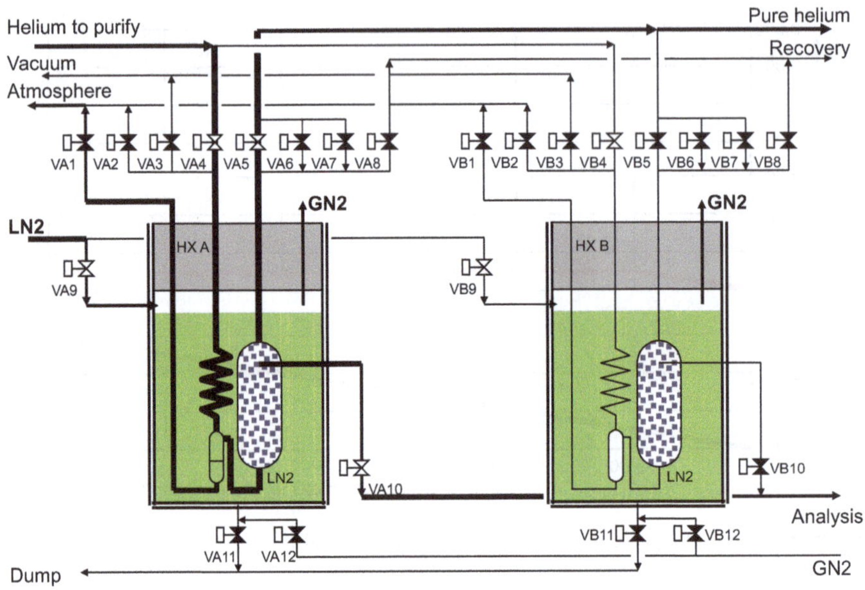

Fig. 9.36 Structure of an adsorption purifier

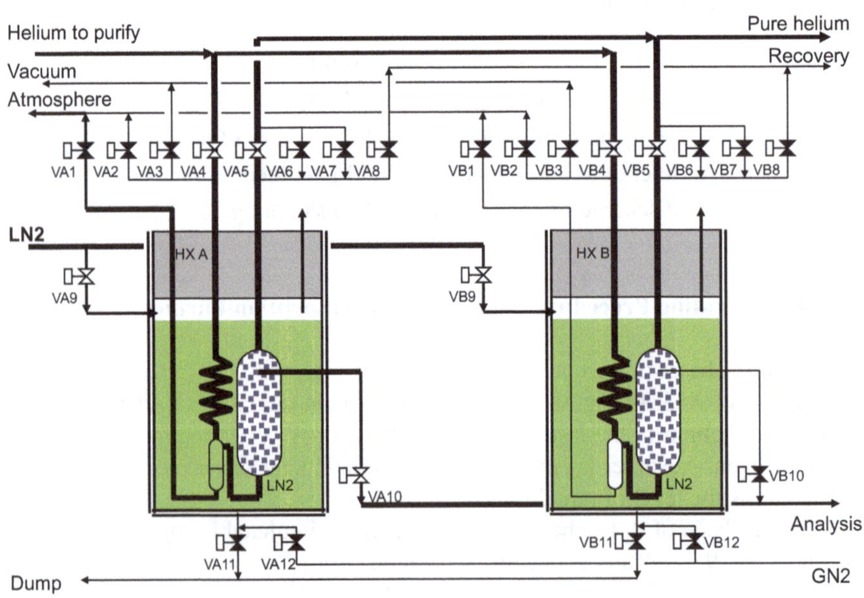

Fig. 9.37 Mise en service de B

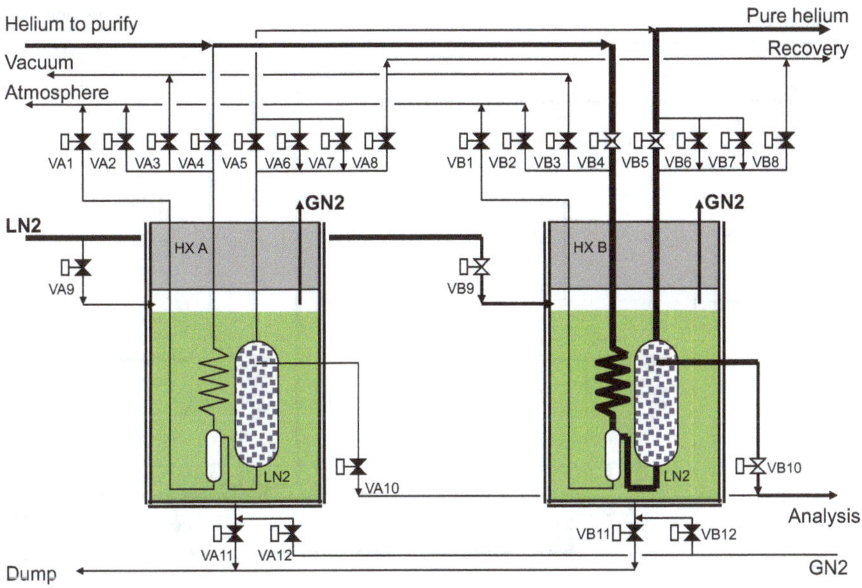

Fig. 9.38 B operating, isolation of A

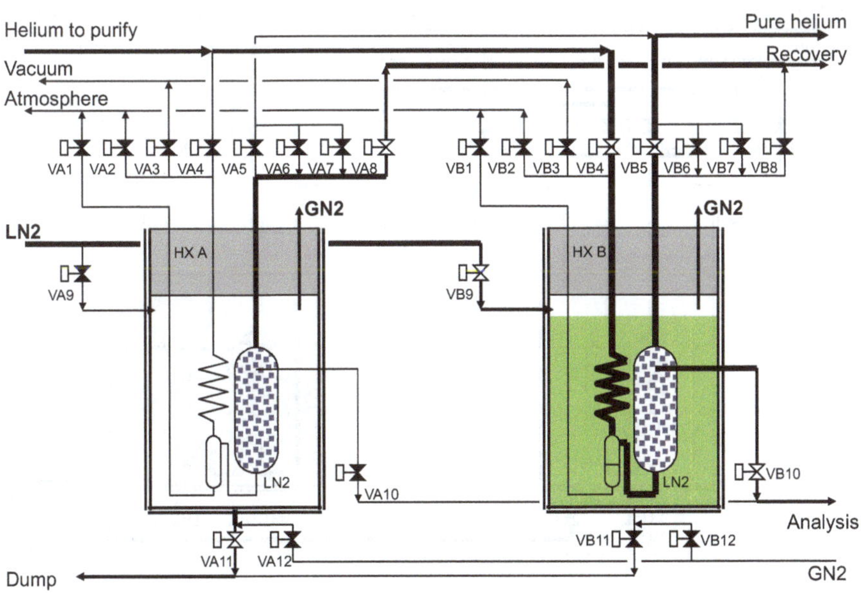

Fig. 9.39 B operating, emptying A

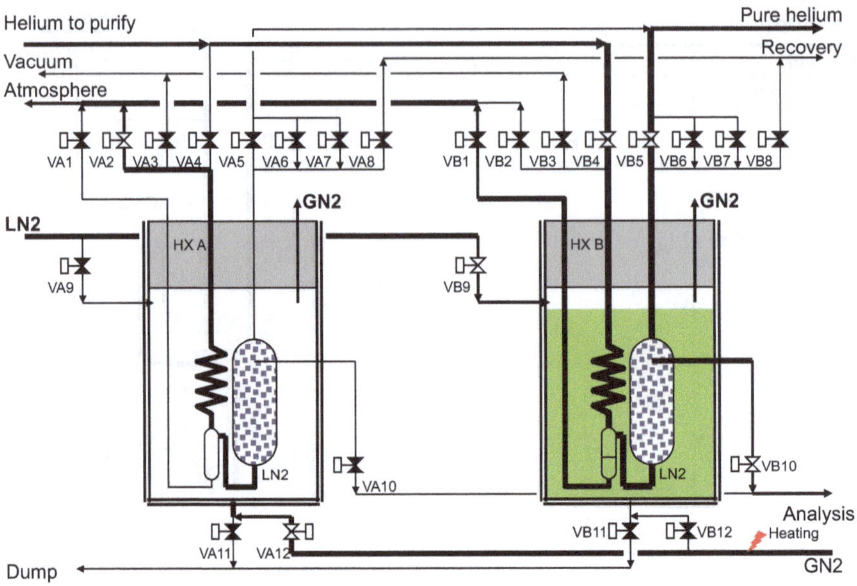

Fig. 9.40 B operating, heating A

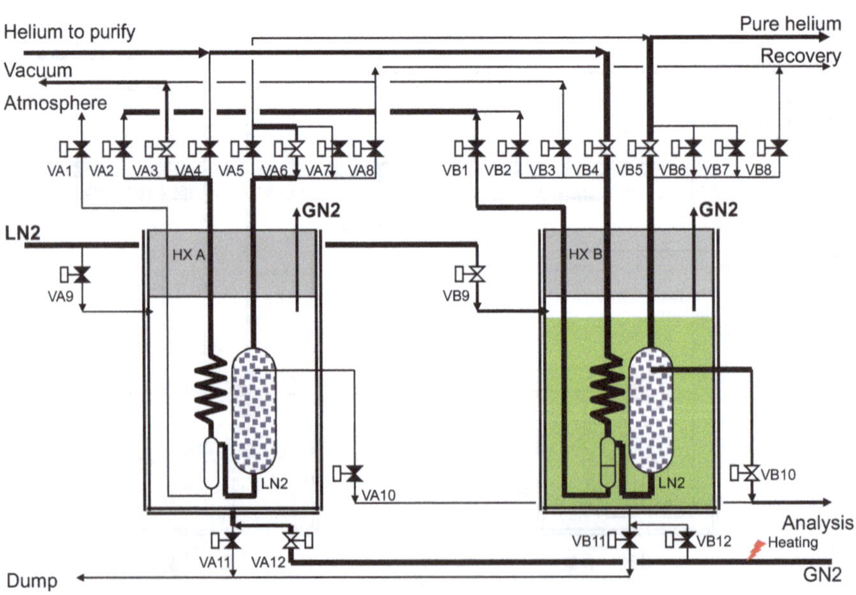

Fig. 9.41 B operating, pumping on B

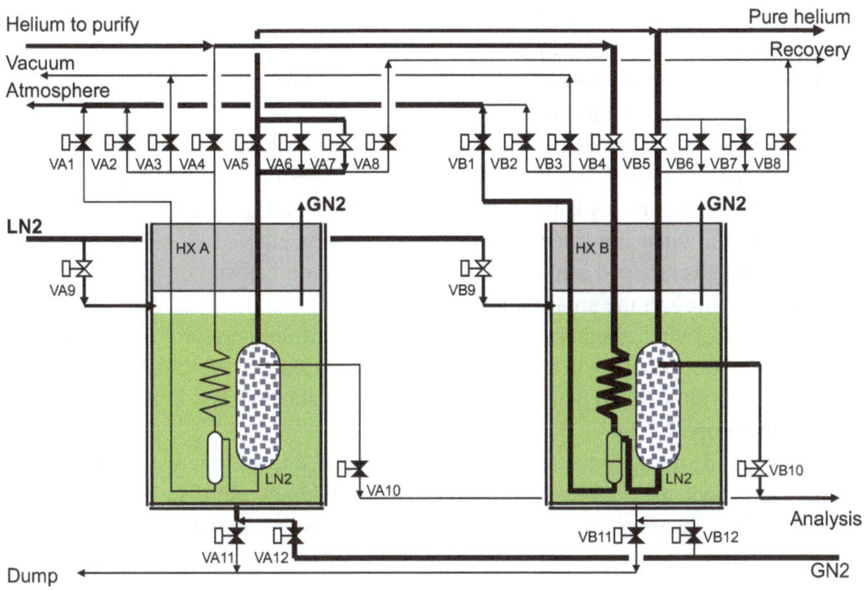

Fig. 9.42 B operating, cooling and LN2 filling A

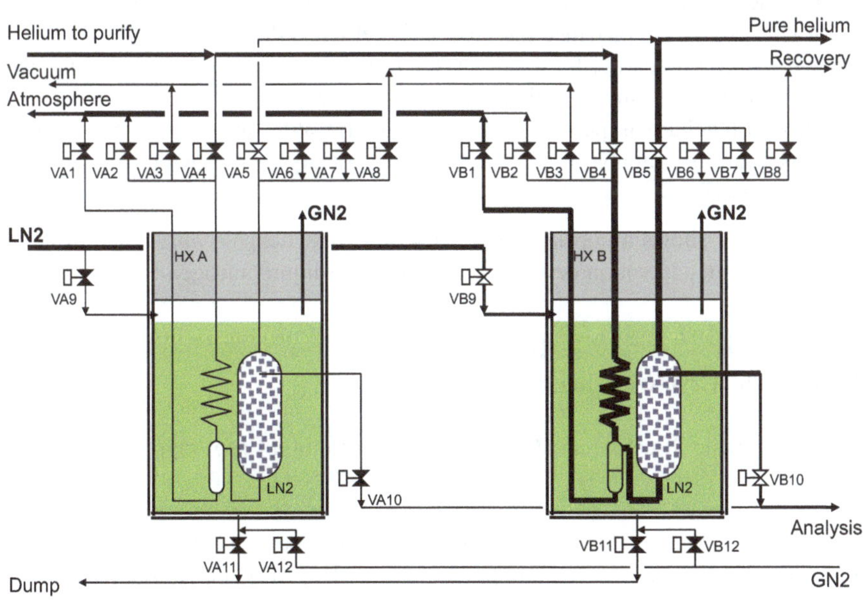

Fig. 9.43 B operating, A waiting

liquid nitrogen in A is discarded by opening VA11. Liquid nitrogen is generally lost, but it is also possible to recover it.

In Fig. 9.40, warm dry air or nitrogen is circulated through VA12. VA2 is opened to discard impure helium to atmosphere.

In Fig. 9.41, when A reaches the requested temperature, it is isolated by closing VA2 and then pumped down through VA3. As for the dryer, elution allowing the displacement of impurities is performed through VA6. Heating is kept "on".

In Fig. 9.42, when the selected time of pumping is elapsed, VA3 and VA6 are closed. Heating is stopped but air or nitrogen circulation is resumed in order to cool A. A is filled through the small section valve VA7.

In Fig. 9.43, when temperature of A is near room temperature, air or nitrogen circulation is stopped by closing VA12. VA7 is closed and VA5 is opened. A is cooled down with VA9. When the liquid nitrogen level is reached in A, the adsorber is ready to be operated.

Of course, if the operator (or the control system) knows that B is only needed within a few days, the cool-down procedure by room temperature air or nitrogen can be spared. The adsorber cools naturally. Similarly, cooling down to liquid nitrogen temperature must be made only at the time the adsorber operation is needed.

Remarks

1. The system that is described here above consists in two identical sets, each of them incorporating a heat exchanger. It is also possible to use one heat exchanger for both adsorbers, provided the processed helium is dry.
2. As has been said earlier for the dryer, the functions of valves VA 6 (elution) and VA 7 (filling up) can be fulfilled by one-off valve.
3. Drying of helium is performed by room temperature adsorption, but, as low temperature is available in the system, why not dry helium with same process that is described in Sect. 9.4.3.5.2?

Figure 9.44 shows a purifier operating with two switchable lines, each of them composed of an adsorption dryer and an adsorber at liquid nitrogen temperature.

9.4.3.5 The Cryo-trapping Purifier

9.4.3.5.1 Basic Structure and Operational Principle of a Cryo-trapping Purifier

In order to implement the cryo-trapping principle, one must look at the vapour pressures of the impurities to be trapped in Fig. 9.45, left scale. On the right scale, corresponding concentrations are displayed.

It shows that air is practically dry under 200 K and air gases are free of any impurity except hydrogen under 30 K (concentration is less than 10^{-8} or 10^{-2} ppM V).

In order to implement the operating principle that is described in Sect. 9.4.1.3, let us take a pipe, the temperature distribution along which is going from 300 to 30 K

Fig. 9.44 A liquid nitrogen cooled adsorption helium purifier (copyright P. Aravian Air Liquide)

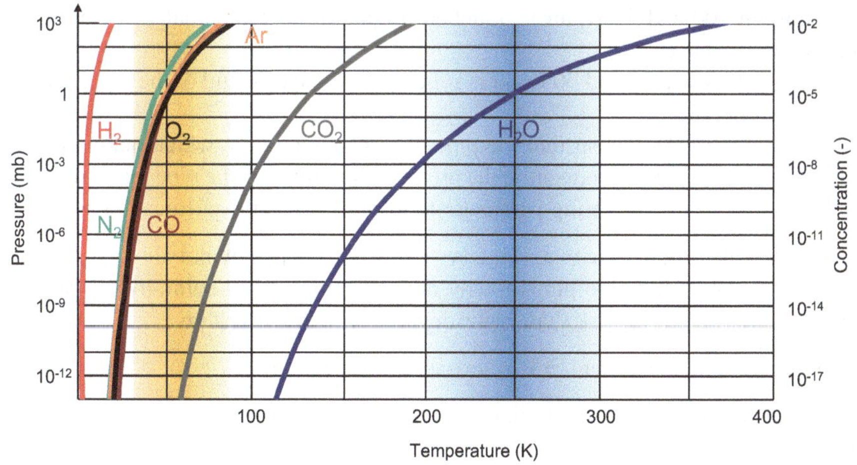

Fig. 9.45 Partial pressures (left scale) and concentrations of impurities for a helium pressure of 20 bar (right)

(Fig. 9.46 left). Helium cooling is performed by using some cycle helium that is taken in the high-pressure circuit, at about 20 K.

In the first part, between 300 and 200 K, water first condenses (turns into liquid) and then solidifies (turns into water ice). Liquid water that is separated from helium is permanently discarded by pressure difference. At 200 K, helium is dry.

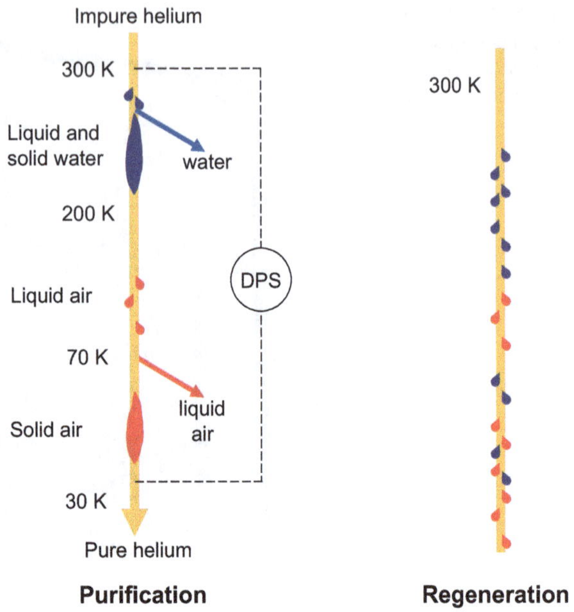

Fig. 9.46 Tentative scheme for building a cryo-trapping purifier

Between 200 and 70 K, air is liquefied. Liquid air is permanently discarded by pressure difference.

Finally, between 70 and 30 K, air is solidified. At the cold end of the pipe, helium is practically pure.

When impurities deposit into the pipe, the cross section in which helium circulates is reduced, and, consequently, the pressure drop through the pipe increases. The pressure drop is the triggering parameter to decide to stop the purification phase.

Remark One should note that, if the system is not stopped by not allowing helium to get through it, the pipe plugs totally and, therefore, stops naturally the helium circulation. There is no risk that impurities get through the system.

To regenerate the system, it is necessary to (see Fig. 9.46, right):

- Turn solid water and air into liquid
- Discard such liquids

This is done by warming up the pipe. One sees that, due to the vertical position of the pipe, all liquids go down, towards the coldest part, resulting in plugging the pipe. Therefore, in order to liquefy all solids, it is necessary to warm up all the device up to room temperature.

In order to avoid the inconvenience that has been witnessed during regeneration of the "simple pipe purifier", one should avoid that liquids that are flowing down go to colder parts. The solution sits in the orientation of the temperature gradient, as shown in Fig. 9.47. During regeneration, all liquids, water and air, flow down,

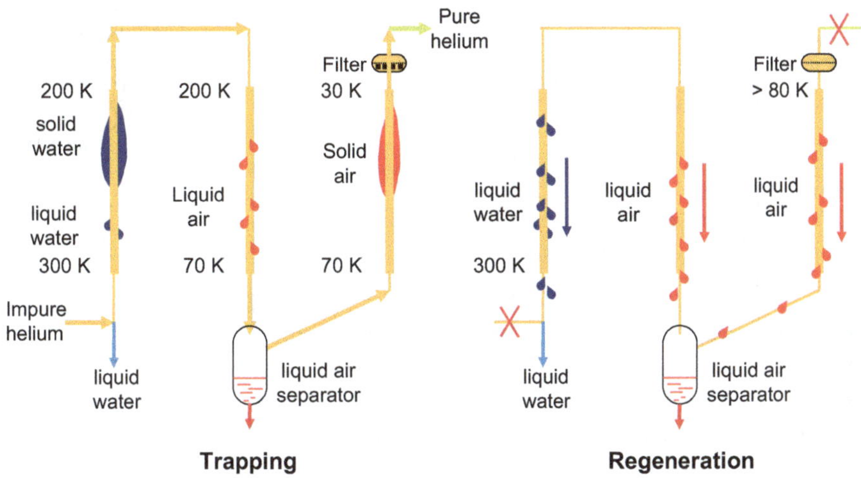

Fig. 9.47 Correct structure for a cryo-trapping purifier

towards higher-temperature parts. Another advantage is that it is not necessary to warm all the system up to room temperature: only 80 K is sufficient. Remember Sect. 1.5.1, where it is shown that thermal capacity of materials decreases very much at low temperature; therefore, the energy consumption for regeneration is minimised.

9.4.3.5.2 Actual Structure of a Cryo-trapping Purifier

The following procedure is basic. Actual procedures might be more complicated.

A cryo-trapping purifier consists of three heat exchangers in a series arrangement (see Fig. 9.48). Staggered temperatures allow insuring the different steps of purification. The heat exchanger HX1 operates between room temperature and about 200 K. In this temperature range, water condenses first and then solidifies. Because the exchanger HX1 is installed with its warm end down, water, as a liquid, flows down by gravity to the warm end and is discarded continuously by pressure difference, during the purification process. In the coldest part of the exchanger (top), the water is retained in solid form.

Heat exchanger HX2 operates between 20 and 70 K. In this temperature range, the air condenses. It is constantly discarded downstream of the exchanger, during the purification process, via a liquid phase separator. In order to recover its enthalpy, liquid air is vaporised and warmed up along HX2 and HX1.

As in the case of the purifier by adsorption, when the air content in helium to be purified is so high that liquid appears, the action of the liquid separator allows a saturation time (or quantity of air that is trapped) of exchanger HX3 that remains constant whatever the air concentration is (see Fig. 9.35).

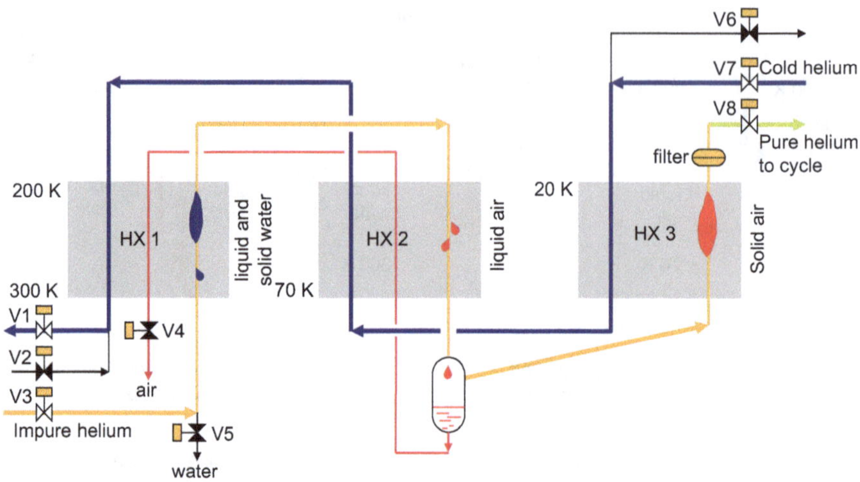

Fig. 9.48 Structure of a cryo-trapping purifier. Purification phase

Heat exchanger HX3 operates between about 70 and 30 K. In this temperature range, air solidifies. A very small volume (not shown on the sketch) filled with adsorbent is circulated for trapping hydrogen or neon by adsorption.

These three heat exchangers have a special structure that offers the impure helium a large area for the condensation of impurities but also a large section of passage for storing a large amount of impurities. Figure 9.49 shows one type of finned tube that can be used. Helium for cooling circulates inside the pipe; impurities that were deposited by impure helium are condensed on the external fins.

9.4.3.5.3 Operation of a Cryo-trapping Purifier

In Fig. 9.48, the impure helium flows through valves V3 and V8. Pure helium is injected into the cycle high-pressure circuit. Possible solid air particles are stopped by the filter. Water in liquid form is discarded through the valve V5, and air as liquid is discharged through V4. Valve V1 controls the flow of helium coolant to control the temperature of the cold end of the heat exchanger HX3. When the pressure difference between the inlet and the outlet of the impure helium circuit exceeds a pre-determined value, the regeneration is decided.

In Fig. 9.50, V3 and V8 are closed to isolate the helium impure circuit. V4 and V5 are open for discarding impurities. Helium at room temperature is injected to the warm end of exchanger HX1 and flows to the cold end of heat exchanger HX3. Regeneration is considered as finished when the cold end temperature of heat exchanger HX3 is 80 K.

The impurities collect in liquid and gaseous form and are removed through V4 and V5.

Fig. 9.49 Example of a finned tube (Wieland)

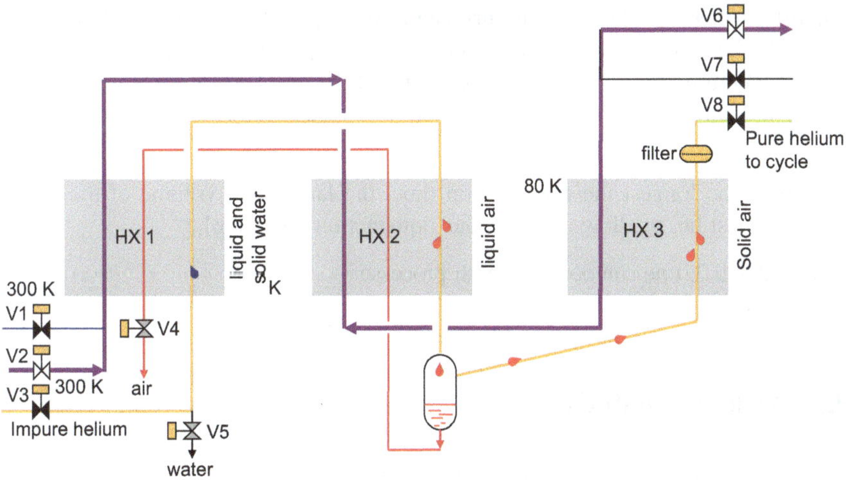

Fig. 9.50 Regeneration phase

In Fig. 9.51, the purge valves V4 and V5 are closed. The circulation of helium at room temperature is stopped by closing V2 and V6. The helium-impure circuit is pressurised by V8, and cooling the heat exchangers is done by opening V1 and V7. When the operating temperature at the cold end of the heat exchanger HX3 is reached, the purifier is returned to service by opening the valve V3.

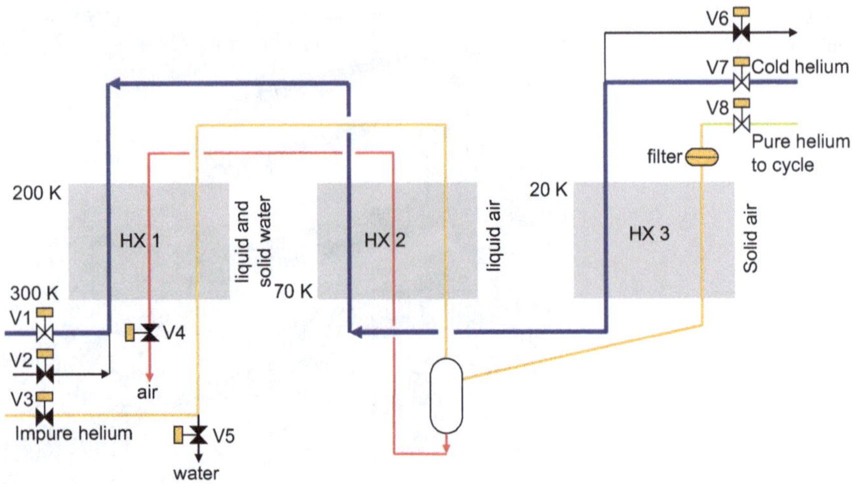

Fig. 9.51 Cool-down phase

Due to the low weight of material and the low temperature levels at which it is cycled (lower than 80 K where the thermal capacity of materials is reduced, remember that the specific heat of stainless steel is reduced by a ratio of 6 from 300 to 20 K), the time required to perform a heating and cooling cycle is about 15–20 min. This feature allows to perform a continuous operation with only a one-off purifier line. During the purification phase, the flow of purified helium is set at a higher value than the flow of helium that is liquefied (see Fig. 9.52, left). The excess helium is stored in the buffer capacity of the cycle. During the regeneration phase, helium that has been stored in the buffer capacity is liquefied (Fig. 9.52, right).

Remark The longest the regeneration time, the largest the volume of the buffer capacity must be, to allow a continuous liquefaction operation!

In Table 9.3, one can compare both processes.

9.5 Helium Analysis

This chapter deals with "ordinary" or "everyday" helium analysis

It does not deal with special situations where deep investigations to find unusual impurities are necessary (better call a gas company who can offer special mobile analysis services).

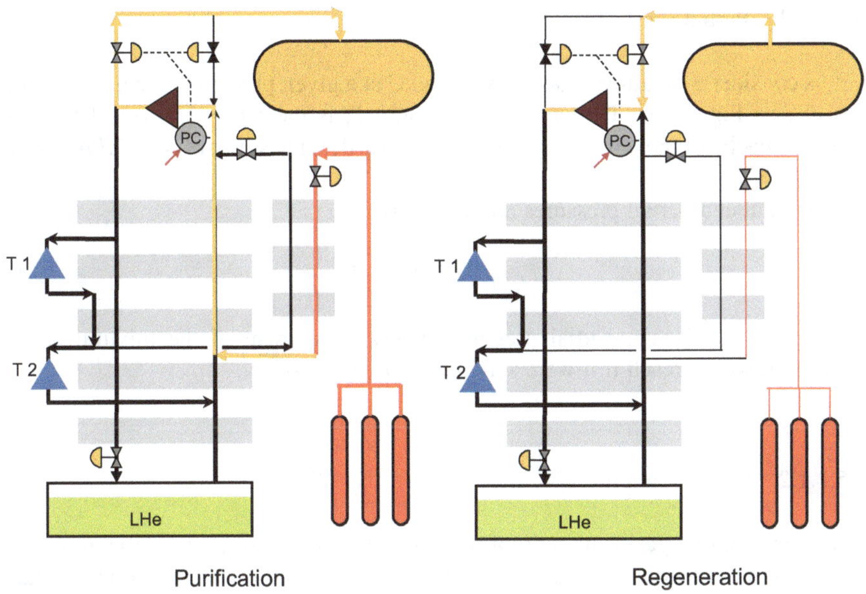

Fig. 9.52 Helium management during purification (left) and regeneration (right) phases

Table 9.3 Comparison between adsorption and cryo-trapping processes

	Adsorption	Cryo-trapping
Impurities	Air	Air *and water*
Temperature	80 K	30 K
Detection	Analysis	*Pressure drop*
Possible release of impurities	Yes	*No*
Regeneration	Heating	Heating
Temperature	From 80 to >200 K	From 30 to 90 K
Duration	A few hours	*15–20 min*
Number of lines	2	*1*

9.5.1 Impurity Levels in Helium

9.5.1.1 Expression of Impurity Levels in Helium

It is of importance to know how impurity levels are expressed, even if helium that is delivered to a refrigerator is generally pure. The main impurities that are found into helium are:

- Moisture (water)
- Air gases (nitrogen, oxygen, argon)
- Various hydrocarbons

9.5.1.1.1 Partial Pressure

Let us consider a mixture of gases: A, B and C at a given pressure P_t enclosed into a vessel (see Fig. 9.53). The *measured* pressure P_t is the *total pressure* of the gas mixture. Each component of the gas mixture (A, B, C) has its own *partial pressure* (P_A, P_B, P_C).

The total and partial pressures are linked by:

$$P_t = P_A + P_B + P_C \tag{9.1}$$

In other words, the partial pressure of any component of the mixture is the pressure it would reach if it were staying *alone* in the vessel.

9.5.1.1.2 Concentration

Concentration of a gas is the ratio of one gas quantity of this gas (e.g. the impurity) versus the total amount of gas. As concentration is a ratio, it is important to state which quantity is dealt with: volume or weight. For gases, it is generally volume.

Concentrations are expressed as:

- Percent: $1\% = 0.01 = 10^{-2}$
- Part per Million: $1\text{ ppM} = 10^{-6} = 10^{-4}\%$ ($1\% = 10{,}000$ ppM)

(As one might not be familiar with such units, remember that 1 ppM of 1 km is only 1 mm!)

Concentrations are to be written as:

- ppM V (**p**arts **p**er **M**illion by **V**olume)
- ppM W (**p**arts **p**er **M**illion by **W**eight)

For gases, ppM V are used. However, for oil, ppM W are used because ppM V would not be significant. In the helium refrigeration activity, the impurity levels range from a part of ppM V to a few tens of ppM V in the cycle helium.

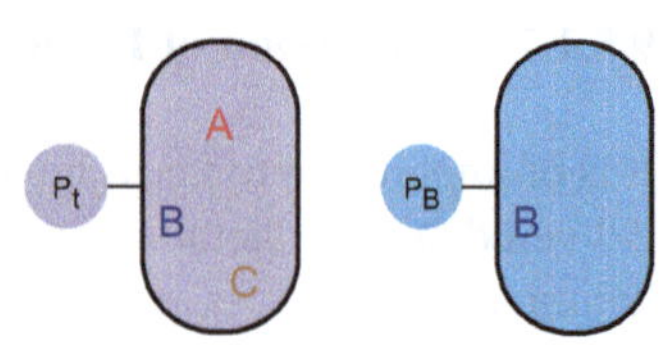

Fig. 9.53 Total and partial pressures

9.5.1.1.3 Relation Between Partial Pressures and Concentrations

To understand the relationship between partial pressures and concentrations, let us consider the system that is represented in Fig. 9.54, 1. It is a volume V_1, filled with helium at 1 bar, containing 10 ppM V of nitrogen. The nitrogen partial pressure is:

$$1 \times 10 \times 10^{-6} = 0.01 \text{ mbar.}$$

In Fig. 9.54, 2, the volume is reduced by means of a piston. $V_2 = 0.5 \times V_1$; consequently, the total pressure rises to 2 bar, the nitrogen partial pressure rises to 0.02 bar, but the nitrogen concentration is unchanged: 10 ppM V.

Same reasoning in the case of Fig. 9.54, 3, gives, respectively, 4 bar, 0.04 bar, but still 10 ppM V. In Fig. 9.54, 4, starting from 1, a quantity of pure helium is added, in order to increase the total pressure to 2 bar: the nitrogen partial pressure is unchanged at 0.10 mbar, but its concentration is divided by two at 5 ppM V.

If the concentration of component A c_a is indicated as a volume ratio, the partial pressure of the component is:

$$P_A = P_t \times c_A \tag{9.2}$$

Example: There are 10 ppM V of nitrogen in helium at 15 bar. The partial pressure of nitrogen is $10 \times 10^{-6} \times 15 = 15 \times 10^{-5} = 0.000015$ bar

9.5.1.1.4 A Special Impurity: Water

Things are a little bit more complicated when dealing with water (moisture) because water may condense at room temperature. As far as the water partial pressure is lower than its saturation pressure, water behaves as a gaseous impurity like nitrogen that was discussed in Sect. 9.5.1.1.3, but as soon as the saturation pressure is

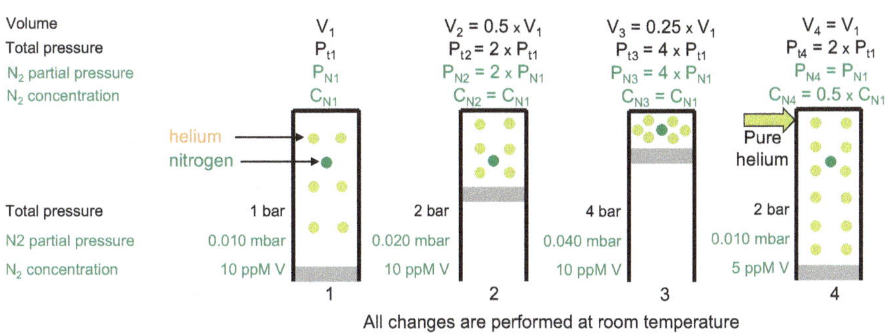

Fig. 9.54 Changes in total and partial pressure and concentration for a permanently gaseous impurity

reached, water partial pressure does not increase anymore because condensation happens.

The temperature at which water appears as a liquid is called the *dew point*. For a given pressure, there is a correspondence between dew point and water partial pressure that is given in Fig. 9.55, on the right axis. On the left axis, one can read the water concentration. Therefore, to know the water concentration in a gas, one can measure the dew point with a hygrometer (see Sect. 9.5.2.1).

As we did earlier with nitrogen, let us consider now the system that is represented in Fig. 9.56, 1. It is a volume V_1, filled with helium at 1 bar, containing 10 ppM V of water. The water partial pressure is:

$$1 \times 10 \times 10^{-6} = 0.01 \text{ mbar}.$$

If one refers to the diagram shown in Fig. 9.55, for such conditions, the dew point is −60.6 °C; therefore, water is exactly at saturation.

In Fig. 9.56, 2, the volume is reduced by means of a piston. $V_2 = 0.5 \times V_1$. The total pressure rises to 2 bar, but, as the conditions were exactly at saturation in case 1, the water partial pressure stays constant *because a part (half) of the vapour is condensed*. Consequently, in the gas phase, the water concentration decreases to 5 ppM V. If a moisture analysis were performed at atmospheric pressure, the dew point would be – 65.6 °C.

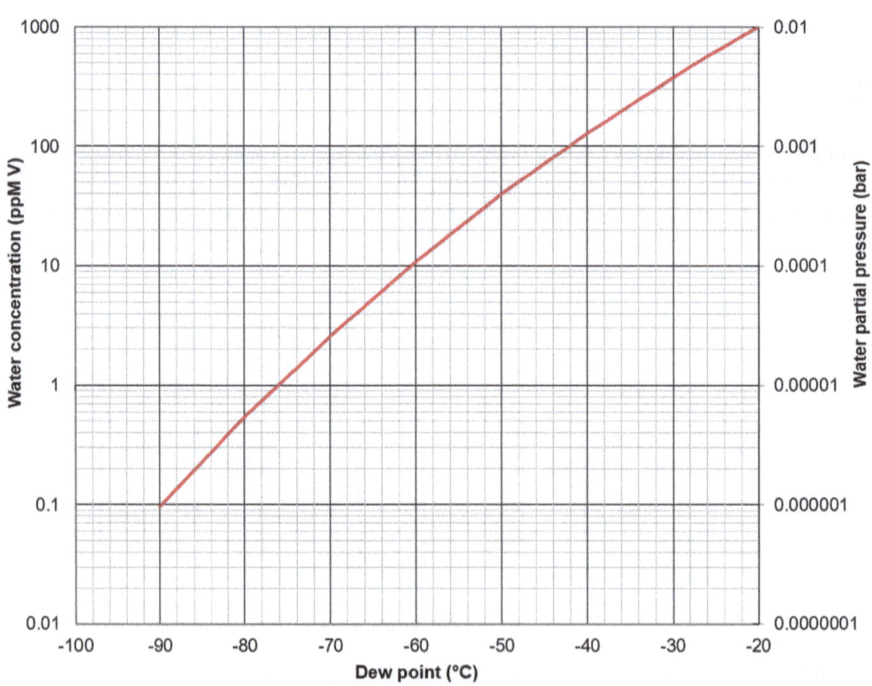

Fig. 9.55 Correspondence between water partial pressure, concentration and dew point at atmospheric pressure

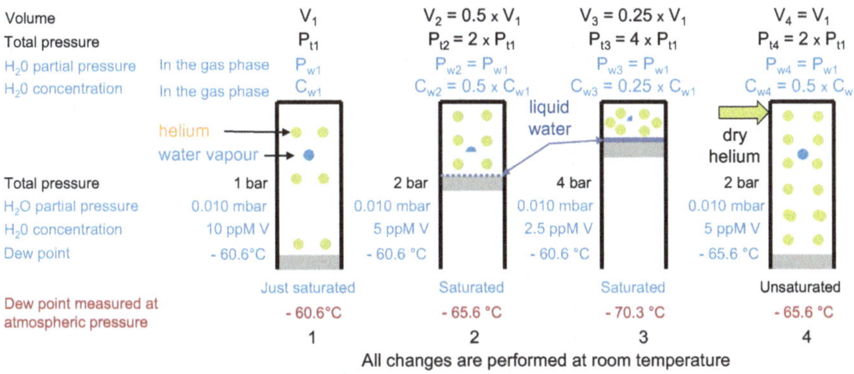

Fig. 9.56 Changes in total, partial pressure and concentration for moisture

In Fig. 9.56, 3, the volume is again decreased by the piston: $V_3 = 0.25 \times V_1$; the total pressure rises to 4 bar, but as the conditions were already at saturation in case 2, the water partial pressure stays constant *because a part (half) of the remaining vapour is condensed.* Consequently, in the gas phase, the water concentration decreases to 2.5 ppM V. If a moisture analysis were performed at atmospheric pressure, the dew point would be – 70.3 °C.

In Fig. 9.56, 4, same quantity of dry helium is added: the total pressure doubles, the water partial pressure stays constant, and, obviously, the water concentration is half that of case 1. Such a behaviour is same as that of non-condensable nitrogen.

Remarks

1. About measurement of the dew point, see Sect. 9.5.2.1.
2. If the dew point of helium in conditions 1, 2, 3 and 4 were measured at atmospheric pressure, the result would be different from the measurement performed under pressure. By the way, as most of hygrometers operate at atmospheric pressure, it is usual to deal with dew point at atmospheric pressure.

It is interesting to compare, in Fig. 9.57, the behaviour of an impurity that stays permanently in the gaseous phase, as nitrogen and water.

9.5.1.2 Impurity Concentrations in Helium Plants

One must discriminate the cycle helium and the recovered helium.

9.5.1.2.1 Cycle Helium

According to CERN cryogenic operating team experience, Table 9.4 gives the acceptable levels of impurities during operation.

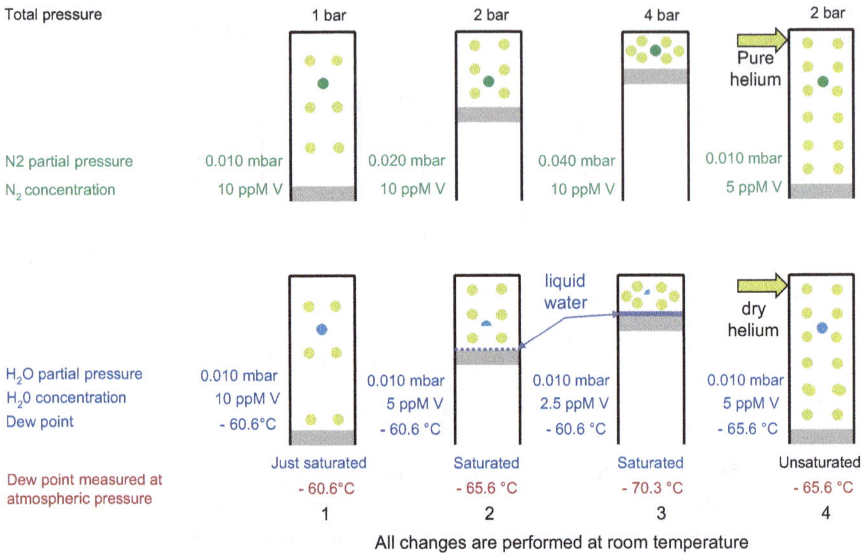

Fig. 9.57 Comparison of behaviours between a gaseous impurity and moisture

Table 9.4 Acceptable impurity concentrations (CERN)

	During a *steady* state (*long* duration) (ppM V)	During a *transient* state (*short* duration) (ppM V)
Air	1	<20
Water	0.1	<10

9.5.1.2.2 Recovered Helium

Helium that is recovered is generally polluted. In addition to air impurities, water, nitrogen, oxygen and argon, it may contain impurities that have been released by the users, for example, when some of them pump on liquid helium bath through circuits that are not leak tight. Rather large amounts of air can be sucked in to which some oil vapour that is released by the vacuum pumps can add. And finally, impurities that are generated by the oil lubricated high-pressure recovery compressor are hydrocarbons and carbon monoxide and dioxide that are the consequence of the combination of some oxygen from air with oil at high temperature in the compressor.

Figure 9.58 shows the range of analysers that are of interest for a helium plant operation.

For analysing the cycle helium, the usual range is 0–100 ppM V.

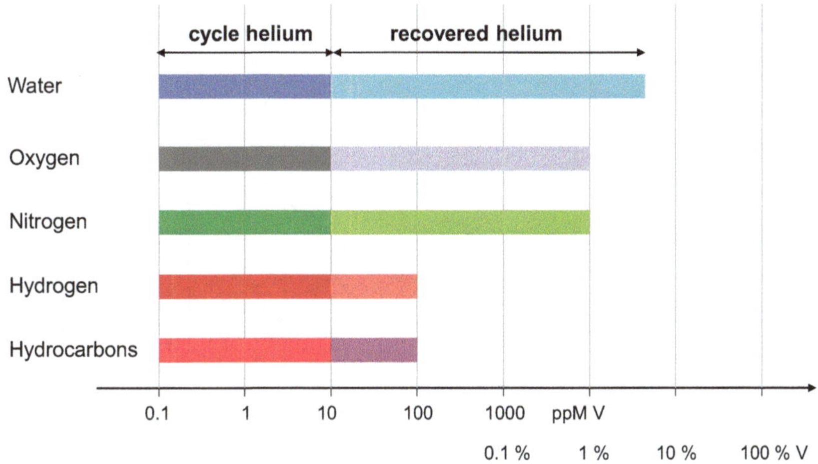

Fig. 9.58 Range of analysers for helium plants

9.5.2 Dedicated Mono-component Analysers

9.5.2.1 Water Analysis

The dew point is the temperature at which the water vapour that is into the gas condensates (is turned into dew). It means that, at the dew point temperature, the water vapour is saturated. The dew point is expressed as a temperature (°C or K). For a given pressure, there is a relationship between dew point and water concentration (see Fig. 9.59 where this relation is given at atmospheric pressure).

What happens if the pressure changes?

Such relation is generally given at atmospheric pressure.

Let us consider a dew point 220 K (−53.15 °C) at atmospheric pressure. According to Fig. 9.59, the moisture content is 26 ppM V. This means that the partial pressure of water is:

$$26 \times 10^{-6} \times 1000 = 26 \times 10^{-3}\,(\text{mbar}).$$

Remember:

$$P_{\mathrm{t}} = P_{\mathrm{gas}} + P_{\mathrm{water}}$$

Therefore:

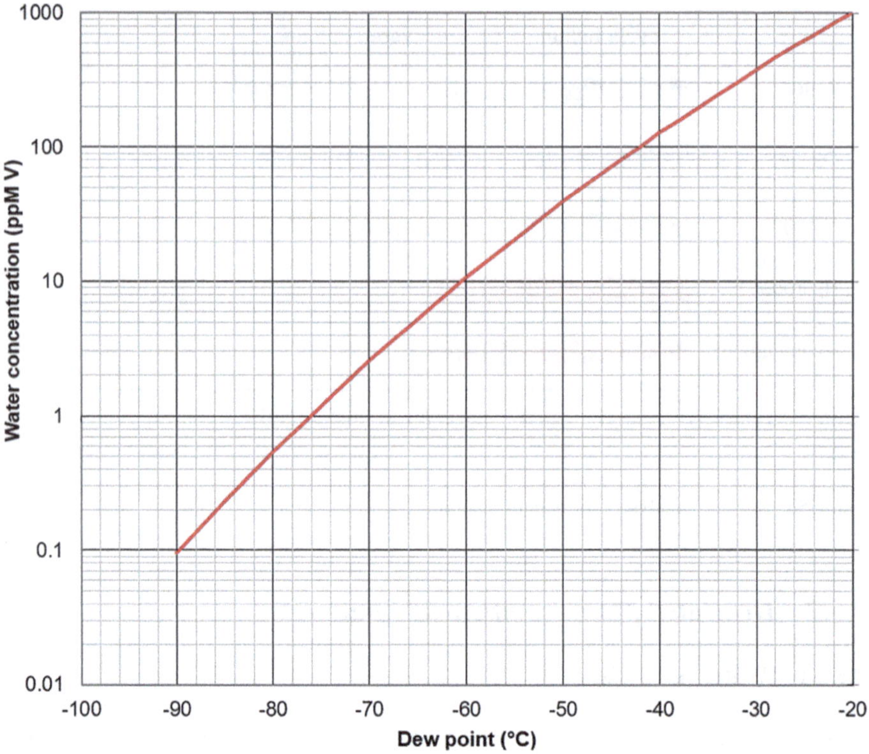

Fig. 9.59 Reminder: dew points at atmospheric pressure

$$1000 = 999.974 + 0.026\,(\text{mbar})$$

Water concentration = 0,000001 * 0.128/999.872 = 128 ppM V

If the gas pressure changes, the partial pressure of water remains the same: 0.026 mbar

If the gas pressure doubles:

$$P_t \times 2 = (1999.974 + 0.026) = 2 \times P_{\text{gas}} + 0.128$$

Water concentration is:

$$0,000001 * 0.128/2 \times 1999.974 = 64 \text{ ppM V}$$

For a given dew point, the water concentration varies as the *reciprocal of the total pressure*.

9.5.2.1.1 The Capacitive Hygrometer

9.5.2.1.1.1 Operating Principle

The gas sample is circulated around the sensor that is made out of a small piece of oxidised aluminium (alumina), generally a wire (see Fig. 9.60). Alumina is covered by a very thin layer of gold. Aluminium + alumina + gold make a capacitor. Alumina is a water adsorbent; it traps water that is brought by the gas. The water content in the alumina modifies the characteristic of the capacitor.

The capacitive hygrometer is not sensitive to the flow rate of the sample gas.

Rating of the device
Range: <1–1000 ppM V
Sample flow: ~20 L/h (advised)

Such device is perfectly suited for helium plant operation.

9.5.2.1.1.2 The Bell Portable Capacitive Hygrometer

This device is almost a must for a helium plant operator, especially at the time of starting a plant. Its most interesting characteristic, on top of being really portable, is that when it is not used, the sensor is housed into a perfectly dry gas capacity; therefore, prior to being used, its indication must be zero, as a good indication of the device health. If the gas to be measured has a low moisture concentration, there is no need to wait a long time until the sensor reaches equilibrium (refer to Sect. 9.5.5).

On Fig. 9.61 left top, the analyser is not operating. One can see that the sensor is not exposed to the sample gas but housed into a capacity (the bell) partially filled with an adsorbent: the sensor is in contact with a very dry gas.

When closing the discharge side of the sample gas, the gas pressure exercises on the bell that rises (see Fig. 9.61 right top), putting the sensor in contact with the sample gas. Photos in Fig. 9.61, bottom, show the device bell in the two positions and the device dial that gives an easy correspondence between dew points and moisture concentration at atmospheric pressure.

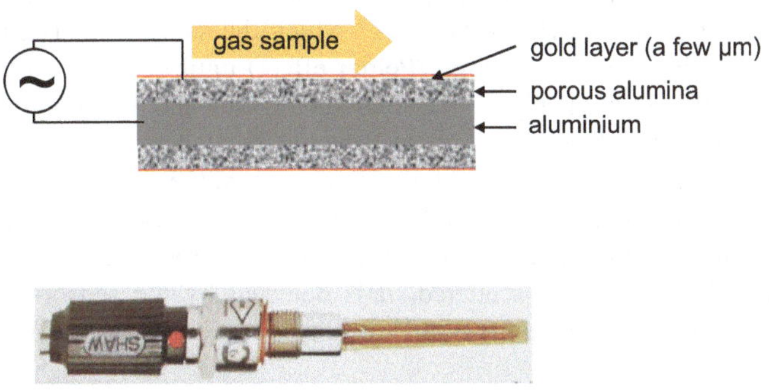

Fig. 9.60 Operating principle of the capacitive hygrometer capacitive cells (Shaw, Panametrics)

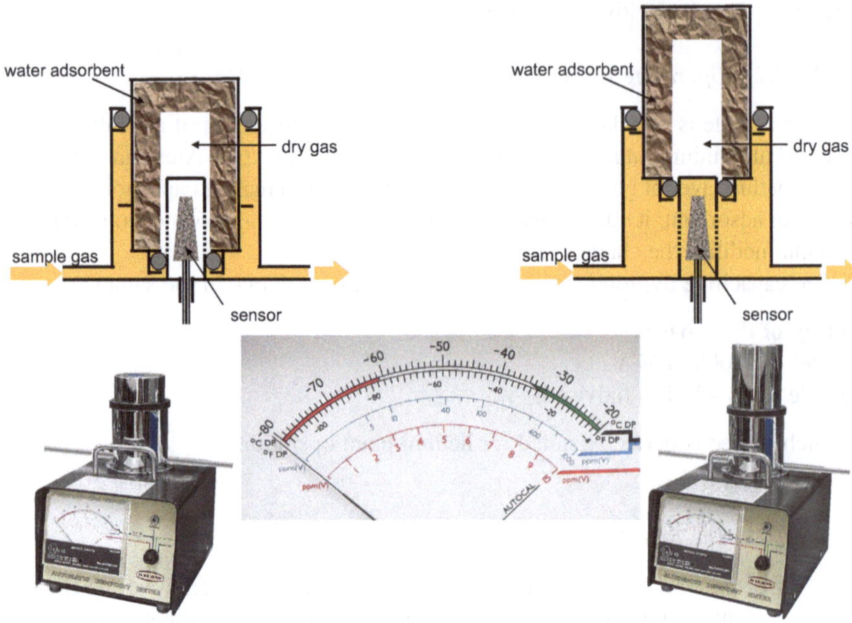

Fig. 9.61 Operation of the bell capacitive hygrometer

A modern version of the bell capacitive hygrometer exists (see Fig. 9.62). It is smaller and can be connected to a personal computer for easy logging.

9.5.2.1.1.3 The Chilled Mirror Absolute Hygrometer

The chilled mirror hygrometer is an absolute device because it measures the temperature at which water condenses (or frosts when temperature is lower than 0 °C) on a gold mirror. The reader has few chances to see one of them, but it is interesting to know how it works. The mirror is constructed out of a good thermal conductivity material such as silver or copper, plated with gold to prevent tarnishing and oxidation (see Fig. 9.63). The mirror, over which the sample gas is circulated, is chilled using a thermoelectric cooler (Peltier effect) until dew (or frost) begins to form. A beam of light, typically from a solid-state broadband light emitting diode, is aimed at the mirror surface, and a photo detector, located around the light source, monitors the reflected light. When the mirror temperature is higher than the dew point, it stays clean and reflects the light perpendicularly, back into the light emitter, but when the temperature is slightly lower, dew droplets form on the mirror surface, and as the reflected light is scattered, it is detected by the sensor. This allows controlling the mirror temperature that is displayed as dew or frost point.

Fig. 9.62 A more modern version, waiting, left; operating right. (Alpha Moisture Systems)

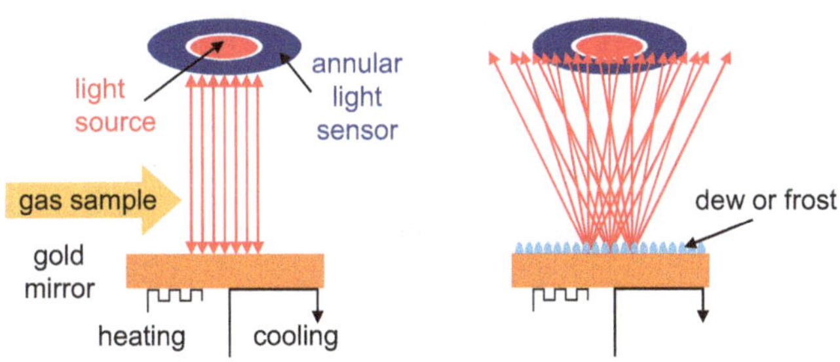

Fig. 9.63 The chilled mirror absolute hygrometer

9.5.2.2 Oxygen Analysis

Air is one of the major impurities that can be found in helium. As measuring the oxygen concentration is simpler that nitrogen concentration, it has been a habit to use an oxygen analyser and consider that when one measures a given amount of oxygen, the equivalent air amount is five times the oxygen concentration.

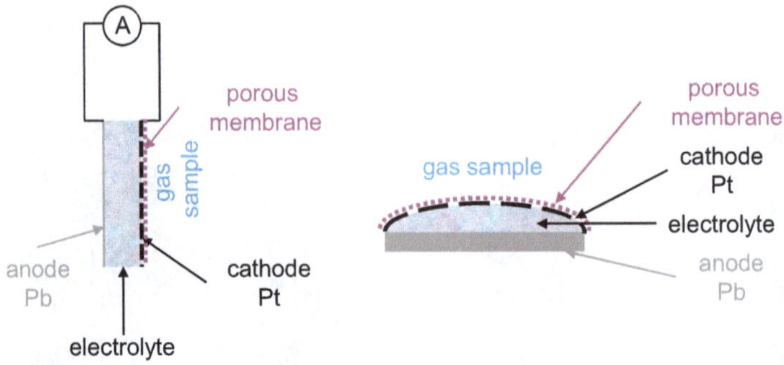

Fig. 9.64 Electrochemical oxygen analyser, principle, left, structure, right

However, whatever the analyser that is used, one should never forget that nitrogen may come from another source than atmospheric air, for example, from the liquid nitrogen pre-cooling circuits, and, therefore, this "five times" rule might be wrong.

In some analysers like the zircon cell analyser that operate at high temperature, oxygen may combine with any combustible impurity (hydrogen, hydrocarbons…), due to high operating temperature, making the result wrong by a lower value than in reality. Such situation might be dangerous.

The most often presently used device is fitted with an electrochemical sensor that incorporates a lead anode and a platinum or silver cathode, both of them in contact with some electrolyte (see Fig. 9.64). The sample gas diffuses into the sensor through the Teflon porous membrane. The electrochemical reaction (oxidisation of the lead anode) results in an electric current that is proportional to the oxygen concentration. The analyser incorporates a temperature compensation module. The cell is discardable and not sensitive to the flow of sample gas.

Rating of the device
Range: 1–1000 ppM V
Sample flow: ~20 L/h (advised)

Such device is perfectly suited for helium plant operation.

9.5.2.3 Nitrogen Analysis

The plasma emission detector
When a gas is energised by a plasma, it produces an intense emission spectrum which relates directly to specific molecular bonds within the sample gas. The emission spectrum of the target gas is isolated with an optical filter.

The plasma sensor consists of a plasma generator, which creates a controlled electromagnetic field, that excites the plasma gas in order it emits spectral emission lines and a plasma detector cell, tuned to recognise the emission line of a specific gas,

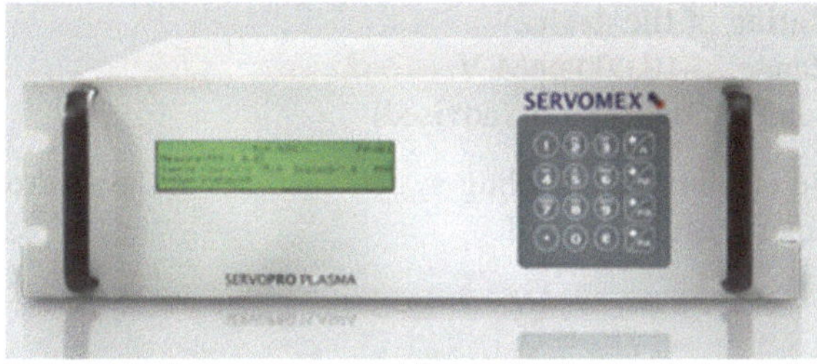

Fig. 9.65 Nitrogen analyser. (Courtesy Servomex)

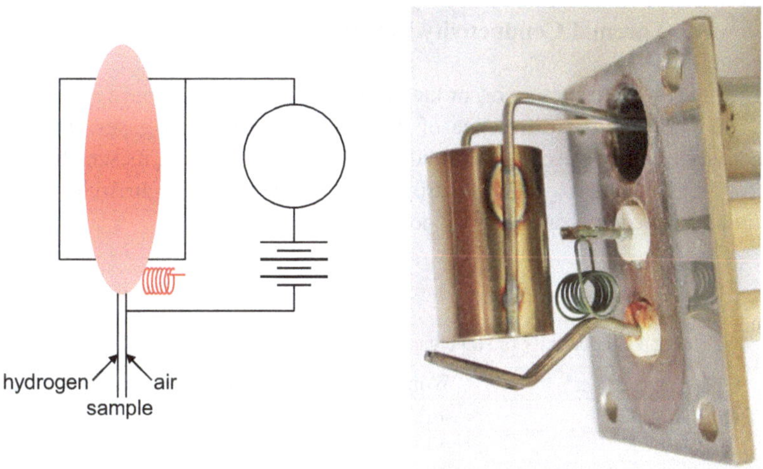

Fig. 9.66 The flame ionisation hydrocarbon detector

here nitrogen. The detector cell is made out of quartz, which is transparent to emitted light spectra (Fig. 9.65).

Rating of the device
Range: 1–1000 ppM V
Sample flow: ~5 L/h (advised)

9.5.2.4 Hydrocarbon Analysis

The Flame Ionisation Detector (FID)
The operating principle of the FID is based on the detection of ions that are generated during combustion of organic compounds in a hydrogen flame (see Fig. 9.66). The generation of these ions is proportional to the concentration of organic species in the sample gas stream. FID measurements are usually reported as equivalent methane, meaning the quantity of methane which would produce the same response of the detector.

Rating of the device
Range: 1–10,000 ppM V
Sample flow: ~5 L/h (advised)

Be careful: the sampling system surfaces act as hydrocarbon traps – the sampling pipes must be heated!

9.5.3 *Multi-component Analysers*

9.5.3.1 The Thermal Conductivity Detector

The thermal conductivity sensor, or katharometer, is a global impurity measurement device. It works on the difference of thermal conductivity of gases. The thermal conductivity of a few gases is indicated in Table 9.5. Therefore, its sensitivity is not same for all gases, but except for hydrogen that has a similar conductivity as helium, other impurities can be rather easily detected (Fig. 9.67).

Table 9.5 Thermal conductivity of some gases

	W/m.K (at room temperature and atmospheric pressure)
Hydrogen	0.18
Helium	*0.15*
Nitrogen	0.026
Oxygen	0.026
Argon	0.017
Carbon dioxide	0.017
Methane	0.036

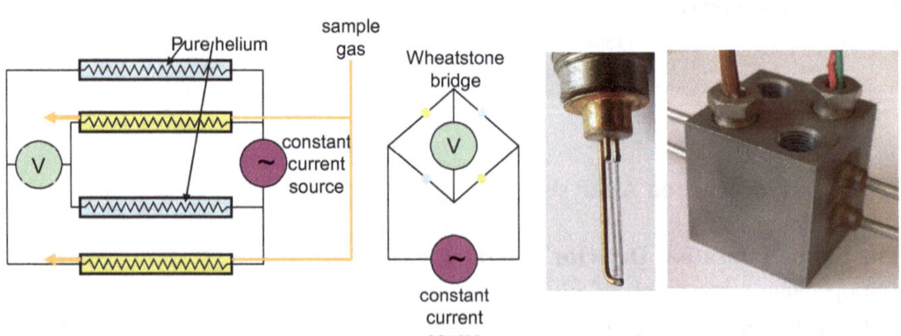

Fig. 9.67 Operating principle of the thermal conductivity detector, right; filament, centre, cell block, right

Rating of the device
Range: 0.1% V to a few % V, according to impurity
Interest: cheap
Sample flow: 20 L/h
Drawbacks: not very sensitive, sensitivity is not constant for all gases, sensitive to temperature changes

This kind of analyser is generally used to get rough and global information on impurity levels, mainly for recovery helium. It is also sufficient to give an OK for the cool-down of a cold box that is fitted with cold adsorbers.

9.5.3.2 The Multi-component High-Frequency Discharge Detector

This device has been specially developed for operation of helium plants by the operating team of the cryogenic system of Fermilab, near Chicago, USA, in the 1970s.

The key element of the multi-component detector is a measurement cell where the sample gas is set to an excited state of luminescence by a discharge of alternating current (see Fig. 9.68). The light emission that is significant for the component is selectively assigned a photoelectric current in the detector, which contains sensors that consist of an interference filter located in front of a photodiode. By the way, it is the same operating principle as the nitrogen analyser in Sect. 9.5.2.3, but here the light emission is analysed by three detectors, each of them tuned for a different signal (Fig. 9.68).

An extra module, shown at the bottom of Fig. 9.69, is dedicated to measurement of aerosols (see Sect. 9.5.4.1).

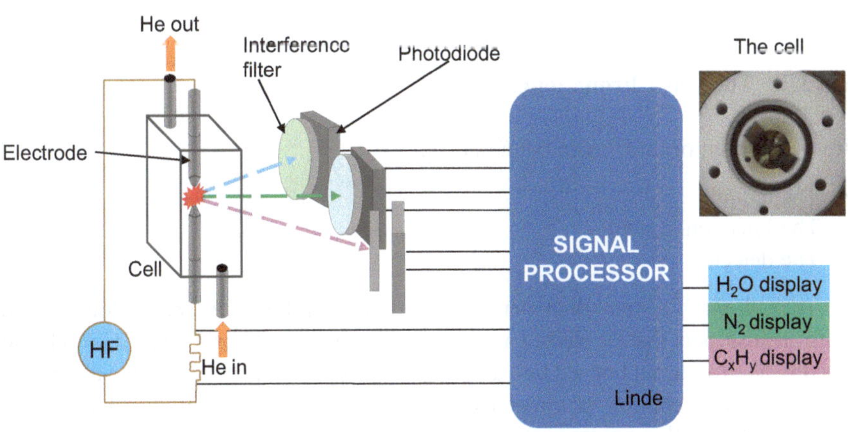

Fig. 9.68 Operating principle of the Linde multi-component analyser, cell at top right

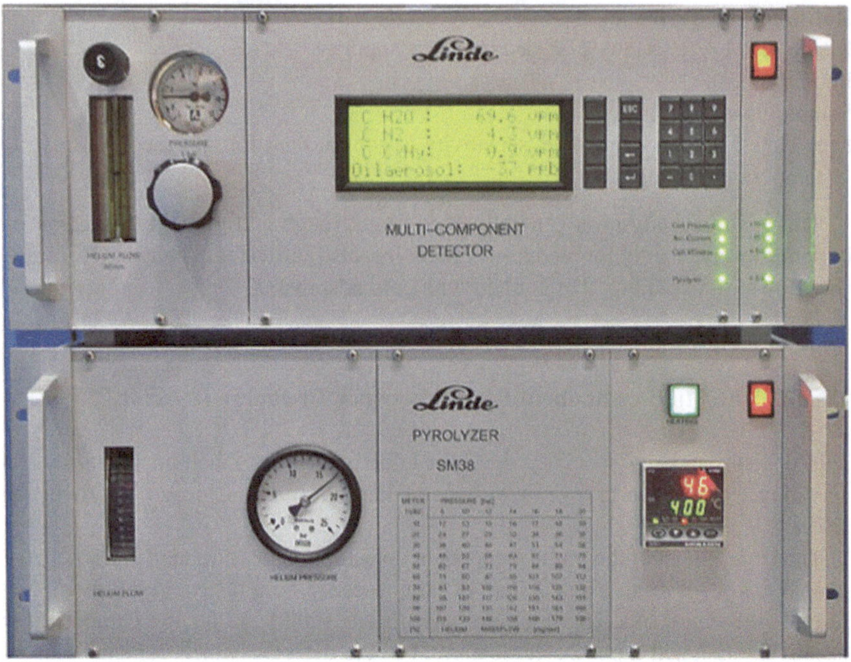

Fig. 9.69 The Linde multi-component analyser

Rating of the device
Range: 1–100 ppM V for N2 and H2O, 1–30 ppM V for C_nH_n
Interests: multi impurity
Sample flow: 20 L/h
Drawbacks: sometimes, mutual incidence of components

9.5.3.3 The Gas Chromatograph

It is composed of three main components:

- The column
- The injection system
- The detector

The column is a small diameter pipe (around 2 mm ID, a few meters long), filled with an adsorbent that is selected according to the operating conditions. It is coiled to occupy less room and housed into a temperature-controlled enclosure (the oven); see Fig. 9.70. The column is circulated by a small very pure helium flow (a few 10 cc/min) so-called bearer gas.

Fig. 9.70 The column (left) and the oven (right)

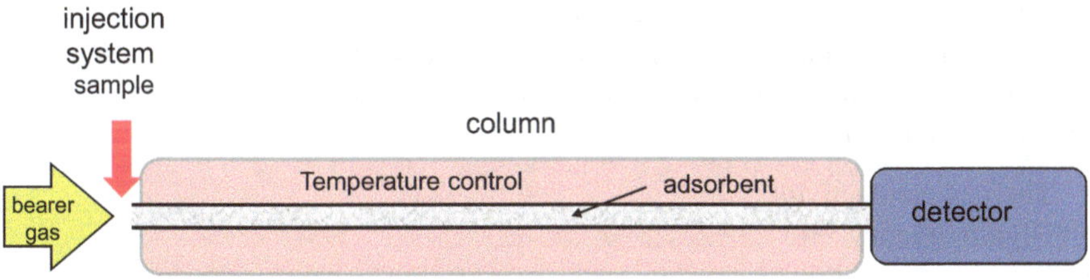

Fig. 9.71 The structure of a chromatograph

At the inlet side of the column (see Fig. 9.71), an injection device introduces a known amount of sample gas. At the discharge side of the column, a detector is connected.

The bearer gas has a known velocity that is directly related to the operating conditions (pressure, temperature, nature and size of the adsorbent, flow).

The A impurity sample that is introduced into the bearer gas is successively adsorbed and desorbed. Such process reduces the transit velocity of the impurity sample along the column: the impurity velocity is *lower* than the one of the bearer gas.

At the end of the column, the detector indicates a *change in a parameter of the* gas it processes (see Fig. 9.72 top).

If the sample gas contains two impurities A and B, each one transits at its specific velocity for a given column material, for given operating conditions (temperature, pressure, bearer gas flow), as shown in Fig. 9.72 bottom. Therefore, the detector "sees" the impurities at various times that are specific of the impurity. On the detector signal record, each impurity gives a peak. A gas is identified by the *time*

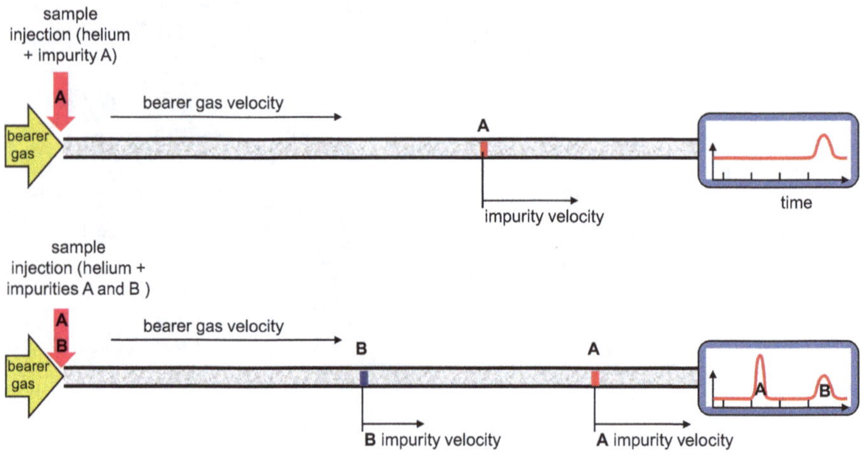

Fig. 9.72 Operating principle of the gas chromatograph

it needs to circulate along the column. The quantity of gas (or concentration) is indicated by the *area* of the signal peak.

The gas chromatograph must be calibrated: helium containing a known concentration of known impurities is "analysed". The specific transit time of each impurity is measured. The detector is selected according to the sensitivity for the impurities that it has to detect:

- Thermal conductivity
- High frequency
- Flame ionisation
- And so on

The injection system is a rotating valve that allows introducing a known volume of sample gas into the bearer gas. On the left side of Fig. 9.73, the sample is circulated into the sampling loop. The chromatograph is ready to operate. At the time of injection (Fig. 9.73, right), the valve is rotated: the sample volume that was sitting into the sampling loop is now injected into the bearer gas.

When using loops of various volumes, it is possible to vary the analysis range of a chromatograph.

The result of an analysis is shown on the chromatogram (see Fig. 9.74) that is the record of the detector signal versus time. When reading a chromatogram, one needs to know all operating conditions:

- Nature of the column: size, nature of the adsorbent
- Temperature
- Bearer gas: nature and flow
- Detector: type

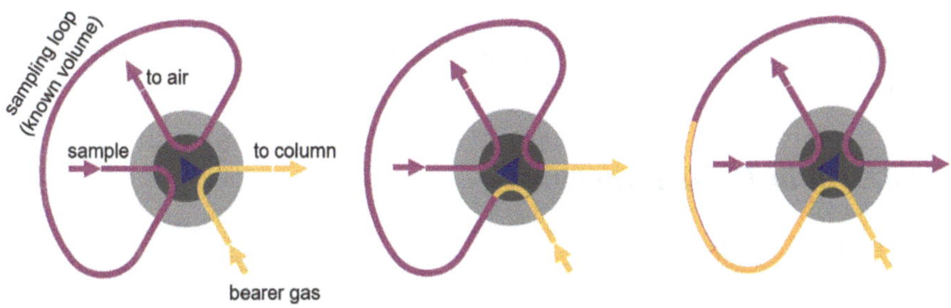

Fig. 9.73 The injection system

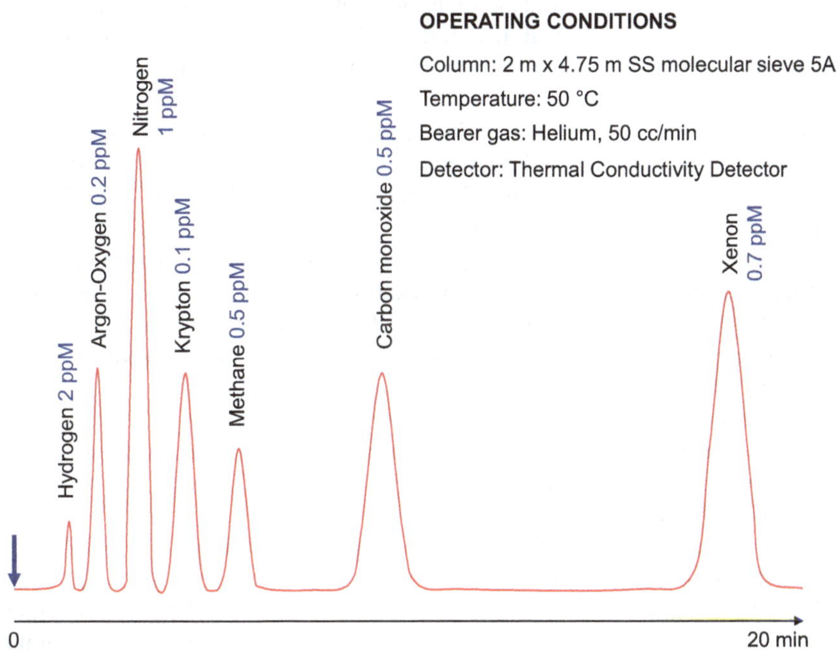

Fig. 9.74 A chromatogram

It might be surprising to see that the sensitivity is not same for each impurity. This is related to the nature of the detector; here, it is a katharometer that is less sensitive to hydrogen (high thermal conductivity) than to xenon (low thermal conductivity).

Remark With other operating conditions (adsorbent, temperature, detector, etc.), the chromatogram would have a very different aspect.

Rating of the device
Range: <1 to >1000 ppM V
Interests: multi impurity

Sample flow: 20 L/h
Drawbacks: expensive, must be often calibrated, only two or three measurements per hour

9.5.4 *Oil in Helium Analysis*

9.5.4.1 Oil Aerosols

Oil aerosol measurement is very difficult to perform without experience and special equipment. Both sampling and measurement are difficult. To give an idea of the difficulty in sampling, the author heard about a bad experience where helium aerosol was sampled and sent into a flame ionisation detector. The result was perfect: no hydrocarbons were detected. But, unfortunately, when dismantling the sampling circuit, oil was found dripping form the pipes: the cold sampling pipes acted as a good aerosol separator!

The only accurate method is an in situ measurement, where the aerosol measuring device is fitted on the process pipe. A laser beam interacts with the aerosols; light is diffused by aerosols and measured. For small aerosol concentration, the signal is proportional to the aerosol concentration.

Offline laser counter can be used, provided the sampling procedure is correct.

A membrane filter can be circulated by a known amount of helium and then rinsed with CCl4 (check for safety rules!) that is analysed by infrared absorption.

The Linde multicomponent analyser (see Sect. 9.5.3.2) can be operated with its pyrolyser that cracks hydrocarbons prior to analysis, as explained in Fig. 9.75.

In the first phase (see Fig. 9.75 top), the sample is circulated at room temperature through the sintered metal filter that is supposed to trap the oil aerosols, for a given time period, therefore, a certain quantity of sample gas. The helium is sent to the helium recovery system.

In the second phase (middle), the sintered metal filter is heated up to 150 °C in order that water is desorbed if any. The helium is sent to the helium recovery system.

In the third phase (bottom), the filter is heated up to 400 °C. The oil desorbs, but also the hydrocarbons are cracked into methane, while the sample is, now, sent into the detector.

This equipment makes correct measurements as far as the sample is *correctly brought to the cell.*

Be careful about sampling:

- The velocity in the sampling pipe should be about the same as in the process pipe.
- The sampling system surfaces act as aerosol traps. It is advised to keep them at around 100 °C to reduce aerosol trapping
- And, of course, the sampling circuit must be the shortest possible with minimum number of devices (no pressure reducer).

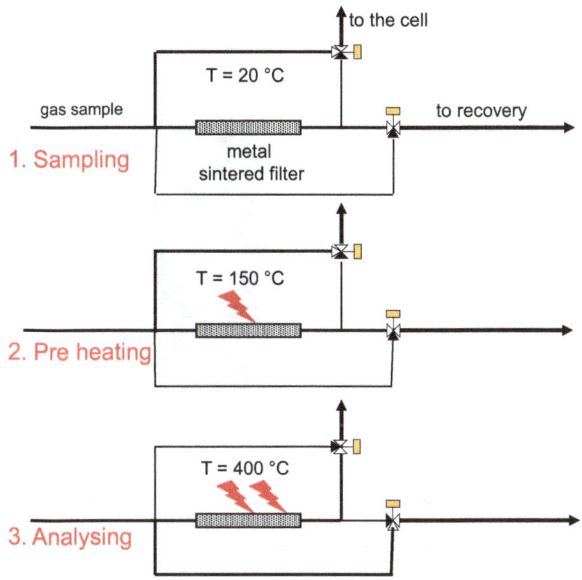

Fig. 9.75 Operating principle of the Linde pyrolyser

9.5.4.2 Oil Vapour

The flame ionisation detector is the most sensitive detector for hydrocarbons.

Be careful about sampling: the sampling pipe surfaces act as aerosols traps – the sampling circuit must be kept at around 100 °C.

Mind! An analysis of oil vapour is very difficult to perform without experience. . .

9.5.5 Helium Sampling

9.5.5.1 Getting a Relevant Sample

The analyser must process a *representative* sample, i.e. totally identical to the gas that is circulating in the process pipe.

That means that the sample must not be "polluted":

- "Positively" by environment (air addition)
- "Positively" by the surfaces of materials of the sampling circuit (desorption)
- "Negatively" by the surfaces of materials of the sampling circuit (*adsorption*)

Consequently:

- The sampling circuit must be perfectly leak tight; only metallic fittings should be used, for example, as shown in Fig. 9.76,

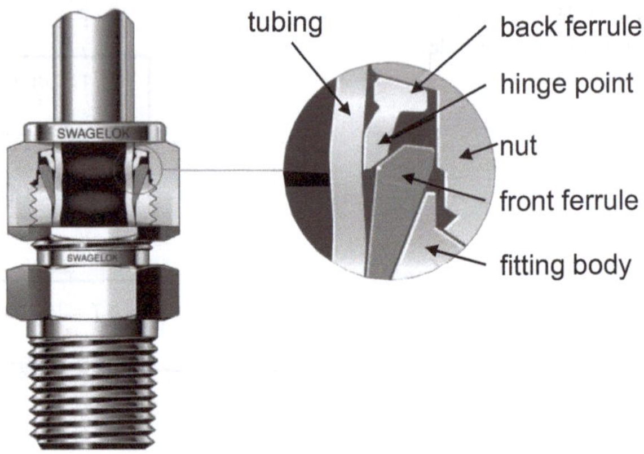

Fig. 9.76 Compression metallic fitting. (Image courtesy of Swagelok Company)

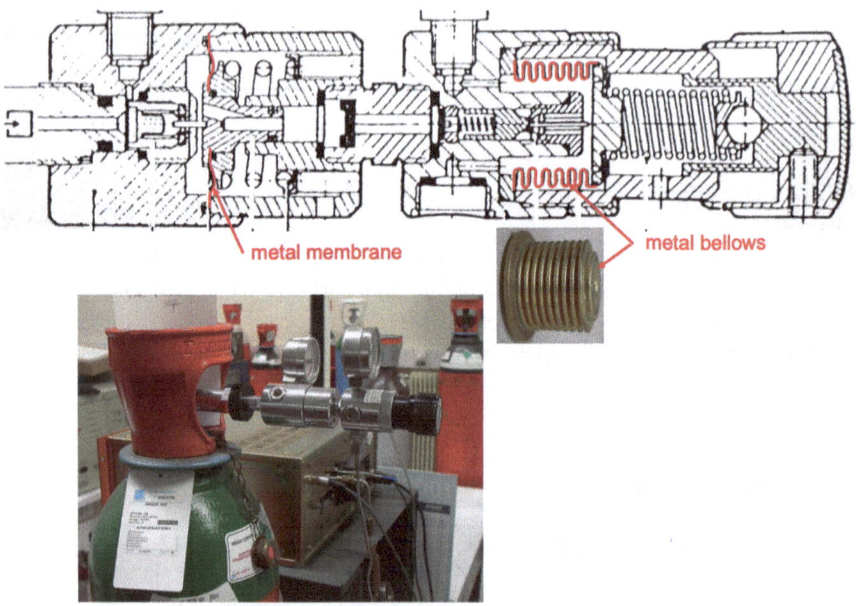

Fig. 9.77 A metallic bellow pressure controller (courtesy Air Liquide)

- The internal pipe surfaces must be perfectly clean, absolutely oil free and, if possible, electropolished
- The gas handling equipment as valve and pressure regulator must be fitted with metallic membrane or bellows (see Fig. 9.77).

9.5.5.2 The Response Time of the Analysis System

The analysis system response time is the total of the sampling time plus the analyser response time. One must be cautious about it because it could be rather long and, therefore, not acceptable.

9.5.5.2.1 Response Time of the Sampling Circuit

We want the detector to have a fast answer.

The sample must not "travel" too long, prior to reach the detector.

So, it means:

- The sampling circuit must have the smallest possible volume and avoid any capacity.
- One must avoid dead end that needs long time to reach same concentration as the sample gas.
- If necessary, a circuit flushing to shorten the response time must be added, as shown in Fig. 9.78, where the travel time is indicated according to the pipe diameter and the sampling pressure.

9.5.5.2.2 Response Time of the Analyser

Consider a hygrometer: the active material of the cell is hygroscopic (alumina).

The response time of the hygrometer is short when the cell is flown by a gas that is more humid because alumina traps moisture at high rate. Consequently, water is rapidly adsorbed, and the cell is rapidly at same water concentration equilibrium as

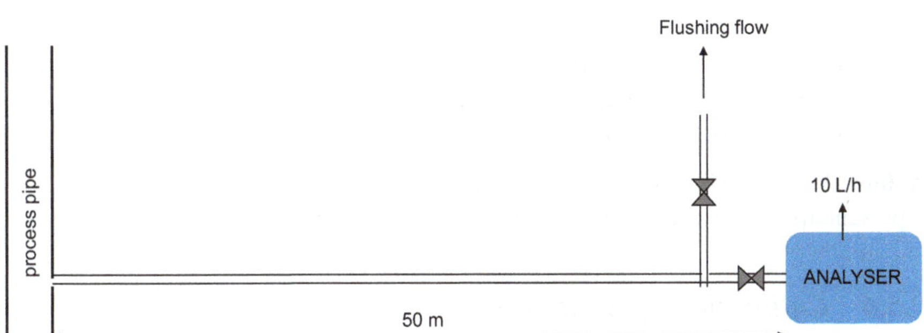

P = 1 bar	Flush flow	L/h	0	20	50	0	50
	I D	mm	4	4	4	2	2
	Time	sec	226	75	38	57	9
P = 15 bar	Flush flow	L/h	0	20	50	0	50
	I D	mm	4	4	4	2	2
	Time	sec	3393	1131	565	848	141

Fig. 9.78 Incidence of flushing flow on the response time

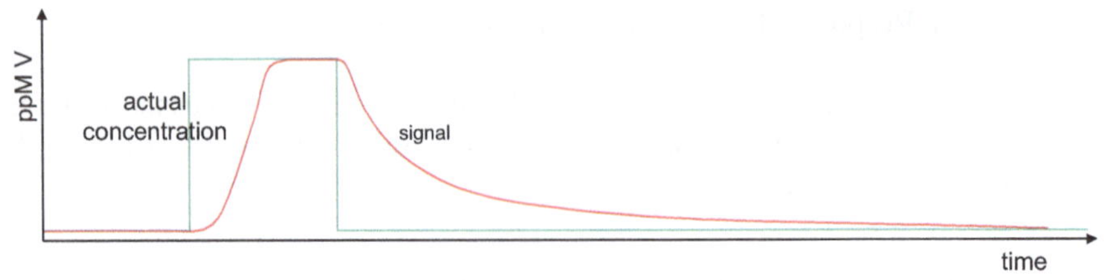

Fig. 9.79 Evolution of the signal compared to sample actual concentration

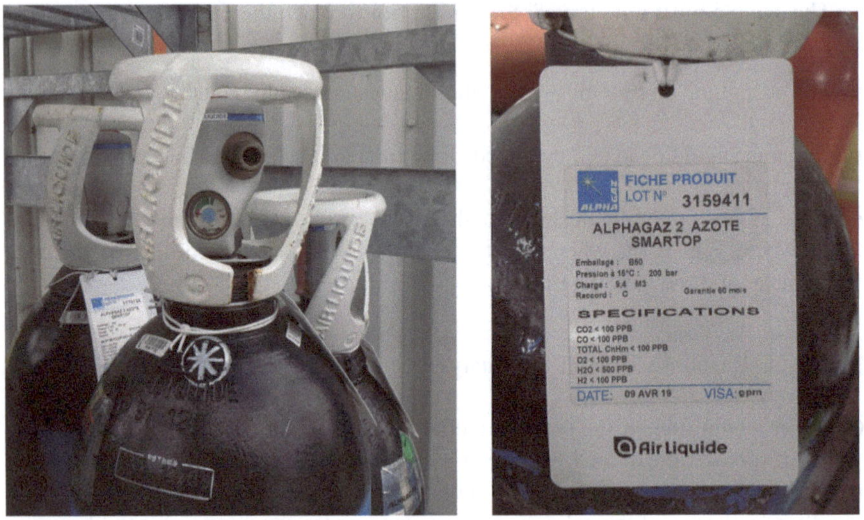

Fig. 9.80 Calibrated gases

the sample gas (see Fig. 9.79). However, the response time of the hygrometer is long when the cell is flown by a gas that is drier, because desorption is a slower process than adsorption and there is not yet heating to help fast desorption.

A hint When dealing with very low concentrations, consider the reading to be representative if the measured value has not changed for at least 1 hour.

9.5.6 Calibration of Analysers

An analyser is to be calibrated with a given gas that contains a known level of known impurity. Calibrated gas mixtures are available in special cylinders (see Fig. 9.80).

Remark There is no moisture calibration gas stored in cylinders because moisture would adsorb on the cylinder walls and desorb according to pressure and temperature, therefore making impossible to get a constant concentration in the gas that is withdrawn.

Chapter 10
Operation of a Helium Refrigeration Plant

Abstract In this chapter, one follows the operations that lead to operating conditions from room temperature state: circuit conditioning, cool-down, steady state and warm-up. Purity of the cycle helium in operation and everyday monitoring are discussed.

10.1 Introduction

The operation of a helium plant is the everyday job of the operator(s) whom actions can be sorted between:

- Launching commands (start, stop, changing operational regime, etc.) according to the requirements of the final user
- Changing set points
- "Helping" the control system in situations where it is not able to react correctly
- Monitoring the plant through the control system and, also, by human checks around equipment: looking at and listening to
- Performing some human actions (looking for helium leaks)
- Performing some periodical tests that necessitate measurements (oil leak at the compressor shaft seal, efficiency of oil separation, etc.)
- Tuning the operation of some components (optimum operating temperature of a turbine, tuning of the labyrinth flow, etc.)

The three later actions are difficult to sort between operation and maintenance; therefore, they have been considered as maintenance actions. They are all described in Chap. 11, with other pure maintenance actions.

According to info gathered during operation, some maintenance actions can be planned for the next maintenance period (see Chap. 11) or urgently performed (the plant is to be stopped); therefore, operation and maintenance actions are very tightly connected.

G. Gistau Baguer, *Cryogenic Helium Refrigeration for Middle and Large Powers*, International Cryogenics Monograph Series,
https://doi.org/10.1007/978-3-030-51677-2_10

10.2 For Comfort in Operation: All the System "at a Glance"

The refrigerator supplier generally provides flow diagrams that incorporate ALL components (main components, but also pipe tags and diameter changes). It is an important and necessary tool, but as it is "crowded", it is not convenient to see comfortably how the main components are connected together and "feel" how they interact. For example, one does not need to see the safety valves to think about the operating procedures. Therefore, the author suggests the operator to draw himself his (her) own "simplified flow diagram" as an example shown in Fig. 10.1 and local flow diagrams as compression station. Such diagrams should be permanently displayed on the main spots of the system as compression station, cold box, etc. Furthermore, when introduced to a new system, self-drawing it is a good way to "learn" it.

Obviously, these sketches should be part of the mimic displays on the HMI.

10.3 Cool-Down

As said is Sect. 8.9.1, one must check that all utilities are available.

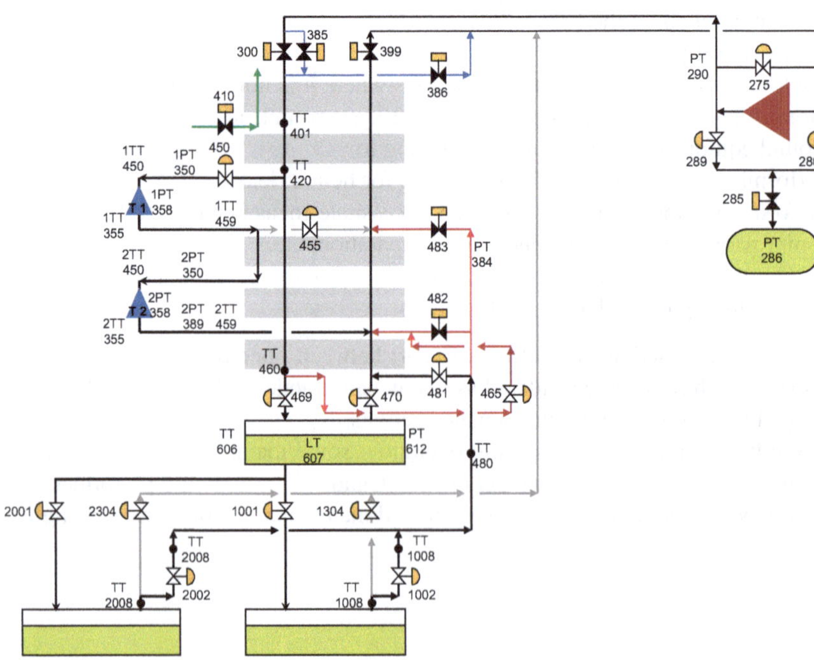

Fig. 10.1 A simplified and easy-to-use flow diagram

The cool-down sequence is the most dangerous in the operation of a helium refrigeration system because if the cycle helium is not perfectly pure, gaseous impurities will be turned into solid as soon as the temperatures decrease and such solid particles can plug circuits, reducing the performance of the system, or, even worse, induce incidents, for example, on cryogenic rotating expanders.

One solution is to start the plant cool-down procedure with perfectly pure helium. This is a condition that is very difficult, if not impossible, to fulfil, especially in large and complicated systems. The circuit conditioning procedure described in Sect. 11. 4.1 cannot ensure pure helium in the whole system. One could also purify all the helium inventory by circulating it through an external liquid nitrogen-cooled purifier that is an extra investment. It is, anyway, an energy and time-consuming process. On top of that, one must be sure to circulate through any circuit portion (e.g. helium circulation is not possible into the part of the turbine circuit that is upstream and downstream the inlet valve).

A practical solution consists in being able to cool down the plant with helium that is not perfectly pure.

Common sense shows that, as soon as the temperature is not lower than the impurity dew point, there is not any risk of operation. Starting from this statement, there are two ways to behave: either avoid that any part of the system reaches a temperature that is lower than the dew point or purify helium in order that its impurity dew point is higher than the temperature of any part of the system. In the first way, used for moisture, one controls the *cool-down speed* in order not to hit the dew point, and in the second way, used for air impurities, one *purifies* helium in order no condensation occurs in the expanders.

Let us consider what happens during a cool-down operation regarding the possible impurities that can be present in the cycle helium.

10.3.1 Purity of the Cycle Helium

10.3.1.1 Water

If helium that is expanded is not perfectly dry, water ice will be created at the coldest part of the system: in the turbine wheel. Ice deposition can partially plug the wheel that might be unbalanced by the uneven ice deposition.

In case of static bearings, when the tip pressure increases due to a partial plugging of the wheel, the pressure difference across the labyrinth (see Sect. 6.6.3.2.1 (Fig. 6.118)) increases and consequently the helium flow that is leaking through it. As a consequence, the bushing that is on the side of the wheel can be so cooled that its diameter reduces, and seizure of the shaft in the bushing might happen.

The shaft length between the wheel and the journal bearing is small. As a consequence, there is an important temperature gradient on the part of the shaft between the wheel and the bushing. In case of a static bearing system, if the bearing feed gas brings some impurity the dew point of which is higher than the temperature on a part of the shaft, such impurity can condense or even, solidify. The cross section where the gas circulates between the shaft and the bushing is reduced or totally

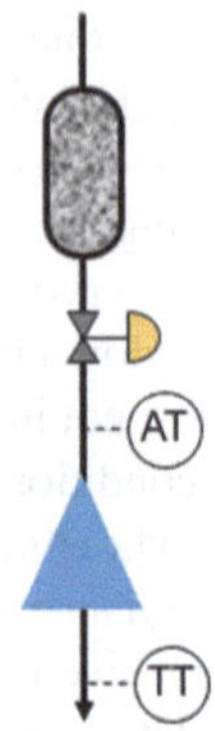

Fig. 10.2 Taking care of the helium dew point to protect a turbine during the cool-down phase

plugged, the suspending strength of the bearing system decreases, and seizure of the shaft in the bushing might happen.

To take care of moisture during the cool-down process, a cycle helium dew point measurement is performed upstream the turbine inlet valve: AT (see Fig. 10.2). The control system does not allow the turbine discharge temperature: TT to be lower than the cycle helium dew point by closing the turbine inlet valve if necessary, reducing the turbine power and, consequently the cool-down speed (see Sect. 8.4.2). When that happens, as helium is still circulated into the cold box and the adsorber(s), it is dried, its dew point decreases progressively, and, consequently, the turbine temperature is allowed to decrease further. When the dew point reaches around 210 K, this procedure is disabled because the water concentration in helium is so low that it is no longer dangerous for the turbine operation.

Such a procedure is acceptable when cooling the cold box alone because the water quantity that is "stored" into the cold box circuits is generally limited and can be easily trapped into the cycle adsorber. However, when cooling very large amounts of materials as magnets that can release a big quantity of impurities, one must make possible to trap all water into a dedicated system, for example, a full flow dryer that is located at the HP inlet of the cold box (see Sect. 9.4.2.1).

According to the water quantity that could be released during a system cool-down, such a dryer might be a switchable two-bed device in order to be able to operate continuously during all the cool-down period. (See also in Sect. 9.4.2.1 the interest of this solution for long runs.)

In case there is no full flow dryer, one must keep in mind that, as soon as the turbine temperature is lower than around 200 K, the remaining water is trapped on the heat exchangers surfaces, somewhere upstream the turbine, a possible source of reduction of the efficiency of the plant.

10.3.1.2 Air

An adsorber is located upstream the turbine as shown in Fig. 10.2. It is operating at the inlet temperature of the turbine.

Remember the triple points:

- Nitrogen: 63,2 K, 0,13 bar
- Oxygen: 54,4 K, 0,0015 bar

When the turbine discharge temperature reaches about 63 K, i.e. the possibility to turn impurities into solid, the upstream adsorber is at about 80 K, a temperature at which helium is already perfectly purified except, obviously, the very small amount of neon and hydrogen that cannot be totally adsorbed at 80 K (a colder adsorber will do the job). Therefore, the turbine is protected against solid plugging during the cool-down sequence.

A personal experience: At the time the author was developing the helium liquefier HELIAL, he wanted to prove the efficiency of such process. He cooled down the liquefier with *2.0% by volume of air* into the cycle helium, including buffer capacity, and reached liquid helium production without any damage! Of course, after such a cool-down, it is advised to make a full regeneration in order to discard all impurities that have been trapped (see Sect. 10.4.1.2)!

10.3.2 Examples of Cool-Down Procedures

10.3.2.1 A Liquid Nitrogen Pre-cooled Cycle

Two cool-down procedures are described here under as examples. They deal with a Brayton cycle for reasons of simplicity, but issues are same as for any other more complicated cycle.

An LN2 pre-cooled Brayton cycle is shown in Fig. 10.3, left.

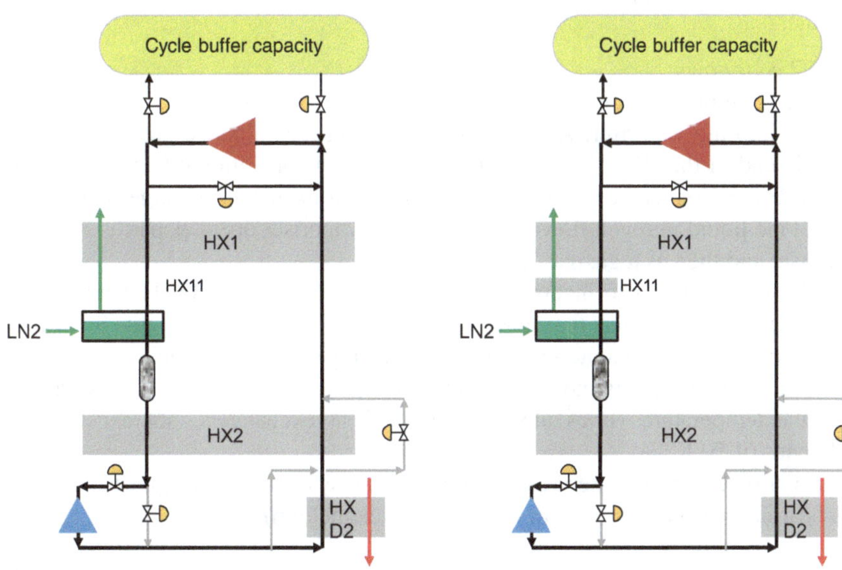

Fig. 10.3 A liquid nitrogen pre-cooled Brayton cycle

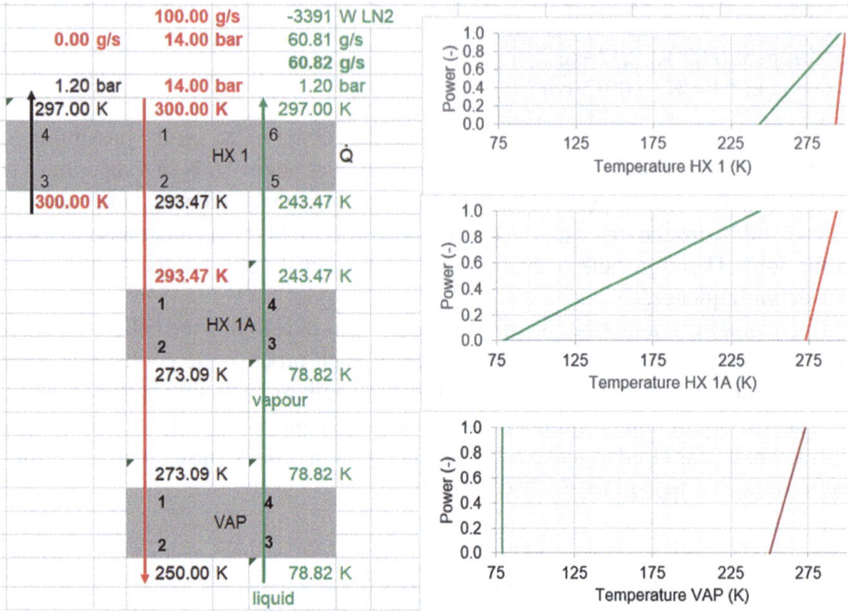

Fig. 10.4 Controlling the HX 1 temperature difference

Initial situation:

- All circuits are at room temperature,
- The helium level impurity is “reasonable” (<100 ppM V),
- The cycle buffer capacity holds a helium quantity that is enough to reach a steady state.

LN2 cool-down:

Cycle helium is circulated through the turbine by-pass valve, while LN_2 is allowed into the vaporiser. At this time, one might face a dangerous situation for the cold end of the HX 1 heat exchanger where the temperature difference reaches around 200 K (remember Sect. 6.4.2.3) if this heat exchanger is large. One could control the liquid nitrogen flow in order that the vaporiser operates partly as a pure vaporiser and then as a gas-gas heat exchanger.

A better solution is to incorporate another heat exchanger: HX 11 between the main heat exchanger and the vaporiser, as shown in Fig. 10.3, right. Such heat exchanger must be able to accept safely a large temperature difference; it can be, for example, a printed circuit one as introduced in Sect. 6.4.1.5. In Fig. 10.4, one can see how the temperature varies along the three heat exchangers, not exceeding a difference of 50 K.

The turbine does not operate even if its bearing system is active (in case of static gas bearings, the bearing system is fed with cycle helium).

The adsorber is cooled at LN2 temperature and acts as an almost full flow purifier. Only the cycle compressor by-pass and the turbine bearing flows are not processed through the adsorber; however, by a mixing process, the impurity concentration decreases in all the circuit helium inventory.

Such operating conditions can be kept until the cycle helium analysis shows correct values (typically <10 ppM V nitrogen and <5 ppM V moisture).

At this time, as a consequence of temperatures decreasing in the cold box, some helium has been transferred from the cycle buffer capacity to the cold box and eventually purified.

Turbine cool-down:

When the temperature of the adsorber is around 80 K and the cycle impurity level is acceptable, the turbine is started up. More helium is taken from the cycle buffer capacity to the cold box. As explained formerly, this gas is purified into the cycle adsorber.

Remarks At the end of the LN2 cool-down procedure, the HX 2 heat exchanger is not yet cooled (one end is at 80 K; the other end is still at about 300 K; such a situation is also met in Sect. 10.3.4). It is later cooled by the turbine.

1. It is possible to simultaneously cool HX 2 down to 80 K by using a by-pass around it.
2. Another simpler way is to run the turbine since the beginning of the procedure. Therefore, both HX2 by-pass and turbine by-pass can be saved.

10.3.2.2 A Non-liquid Nitrogen Pre-cooled Cycle

Differences with the flow diagram in Fig. 10.5: the nitrogen circuit and the turbine by-pass do not exist; the adsorber is generally located immediately upstream the turbine.

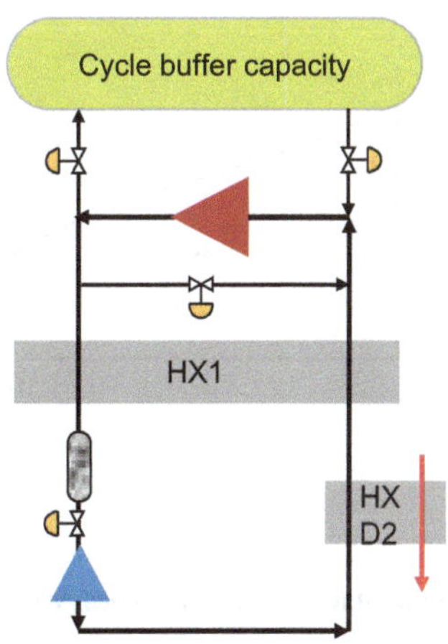

Fig. 10.5 A Brayton cycle that is not liquid nitrogen pre-cooled

Cooling down is totally performed by the turbine that is protected against water ice and solid air by the arrangements that are described in Sect. 10.3.1.1.

Initial situation:

- All circuits are at room temperature,
- The helium level impurity is “reasonable” (< 100 ppM V),
- The cycle buffer capacity holds a quantity of helium that is sufficient to reach a steady state

Turbine cool-down:

When the turbine is started up, its operating temperature decreases. The turbine is protected against solid impurity generation according to the above procedures. During the cool-down process, helium taken from the buffer capacity is allowed into the low pressure (see Sect. 8.3.2.), at the suction side of the cycle compressor and therefore purified when flowing through the adsorber. There is no need for a turbine by-pass.

10.3.3 Cooling Down Heavy Loads with LN2

Objects to be cooled down, as big magnets can weight up to several tons. A very high amount of energy is to be extracted from theses loads. This duty could be performed by the turbines, but their power is limited, and, consequently, the cool-down time period might be too long. It is possible to “help” the system with liquid nitrogen.

A magnet cooling procedure generally must fulfil two constraints: the helium temperature difference between in and out of the coil must be lower than 50 K, for thermal constraint limitation, and the cooling down rate is generally limited to 0.5 K/h. An arrangement as shown in Fig. 10.6, right, can be used.

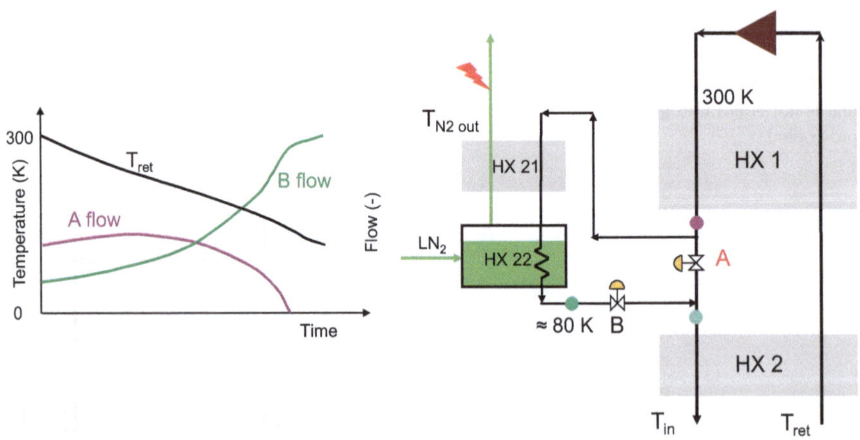

Fig. 10.6 Cooling down heavy loads with LN2

At the beginning of the cool-down procedure, helium returning from the coil is at $T_{ret} = 300$ K. In the left part of Fig. 10.6, one can see how the mass flows from valves A and B vary, respectively, according to the temperature when circulating through an object to be cooled down that has a constant conductance.

At the beginning of the cool-down phase, as helium getting into HX 21 is at room temperature, gaseous nitrogen is warmed up to almost temperature before being vented: its enthalpy is almost totally recovered.

Remark There is no way to recover the enthalpy of saturated vapour (sensible heat) at temperatures higher than the temperature of helium returning from the load; therefore, in order to avoid ice or water on the discharge pipe, a heater, if possible atmospheric, must be used to warm gaseous nitrogen up to room temperature before venting out.

10.3.4 Cool-Down of a Liquefier Without Liquid Nitrogen Pre-cooling

In the following figures, the circuits that are circulated with helium are thicker, warm heat exchangers are red, and cold heat exchangers are blue.

10.3.4.1 The Dewar Is Cold

Initial situation: the dewar is cold and contains some liquid helium. Only the cold box is to be cooled down and then connected to the dewar. This is the most frequent situation.

Cool-down phase 1. At the very beginning of the cooling-down procedure, helium is circulated through the Brayton cycles (see Fig. 10.7, left). The JT valve is kept closed. The warm turbine rotational speed is controlled by the inlet valve control loop, the set point of which is amended by the attenuator (remember Sect. 8.4.3) that, at this time, is only taking care of the rotational speed of the cold turbine: S T2.

Cool-down phase 2. Later (see Fig. 10.7, right), as temperatures start decreasing, the attenuator checks also the dew points at the inlet of both turbines: DP T1 and DP T2.

But why the JT heat exchanger is not circulated by opening the JT valve in order to cool it down? Because, if the JT valve would be open, HP helium that is, for example, at 100 K would be warmed up to room temperature in the HP circuit, expanded (here we neglect the temperature change) and circulated in the LP circuit where it would be cooled down to around 100 K by the HP helium! By the way the JT heat exchanger would be performing its duty that is to separate cold and warm temperatures and not allow its end that is warm at this time to cool down!

Cool-down phase 3. In order to avoid such a situation where the JT heat exchanger cannot be cooled down, one has to "neutralise" the heat exchanger

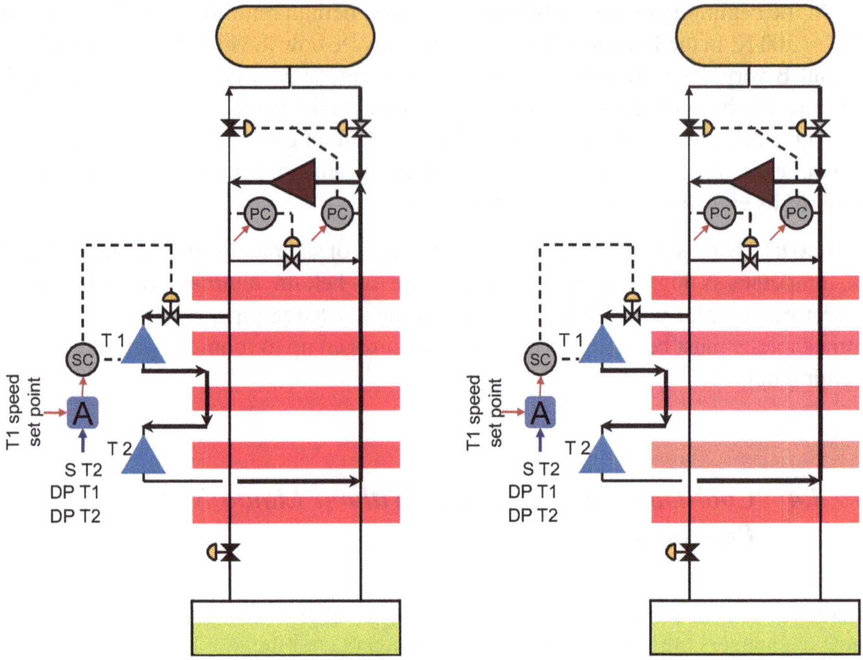

Fig. 10.7 Liquefier cool-down phases 1 and 2

function by circulating in one circuit only, transforming it into a simple piece of metal. In Fig. 10.8, left, a by-pass circuit is installed, allowing to by-pass the LP circuit of the JT heat exchanger: now, the temperature of the JT heat exchanger can be decreased around that of discharge of the cold turbine. The pipe that feeds the by-pass is thermally linked with the pipes to and from the dewar in order to cool them down.

Remark Such by-pass can also be implemented as shown in Fig. 10.8, centre , top.

Cool-down phase 4. The system cools down until the temperature upstream the JT valve reaches a value where the expansion cools helium, for example, around 15 K. When the cold turbine discharge temperature nears its set point, the control loop is set in automatic operation (see Fig. 10.8, right).

10.3.4.2 The Dewar Is at Room Temperature

Initial situation: the dewar and the cold box are at room temperature. Not frequent: only when the plant is brand new or after a maintenance action.

To send cold helium into the dewar, the JT valve must be slightly open; the flow that has circulated the dewar is directly returned to the low-pressure circuit, at room

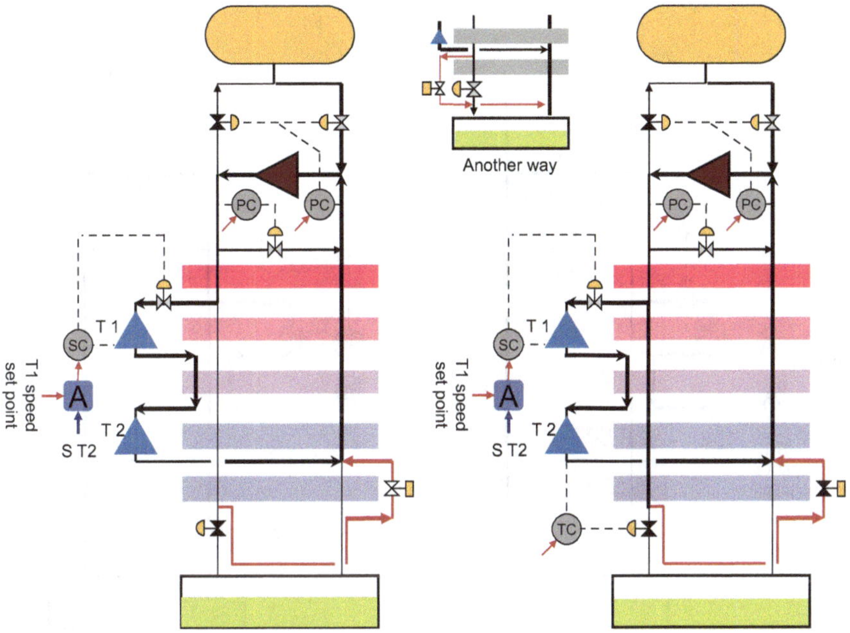

Fig. 10.8 Liquefier cool-down phases 3 and 4

temperature by an external circuit in order to unbalance the heat exchangers. When the temperature upstream the JT valve is around 15 K, the JT control loop is activated, and the heat exchanger by-bass is closed (Fig. 10.9).

10.3.5 Cool-Down of a Refrigerator Without Liquid Nitrogen Pre-cooling

Initial situation: the dewar and the object to be cooled are at room temperature. In this situation, the refrigerator has to deal with a heavy load (e.g. a magnet). To cope with efficiency, it is necessary to cool the object with helium as warm as possible. Furthermore, coil builders have very stringent rules to be fulfilled during cool-down: the maximum temperature difference between helium getting into the coil and out of the coil must be 50 K at maximum, and the cool-down speed is limited to 0.5 K/h.

Consequently, it is necessary to circulate through the load from the very beginning of the cool-down procedure, therefore, circulating into the JT heat exchanger.

Cool-down phase 1. In order to "neutralise" the heat exchanger stack, a by-pass circuit is added as shown in Fig. 10.10, left. The Brayton cycle operates under the control of the warm turbine control loop amended by the attenuator, as for the liquefier cool-down explained above. However, the attenuator can also take care of the helium temperature difference between inlet and outlet of the coil and the

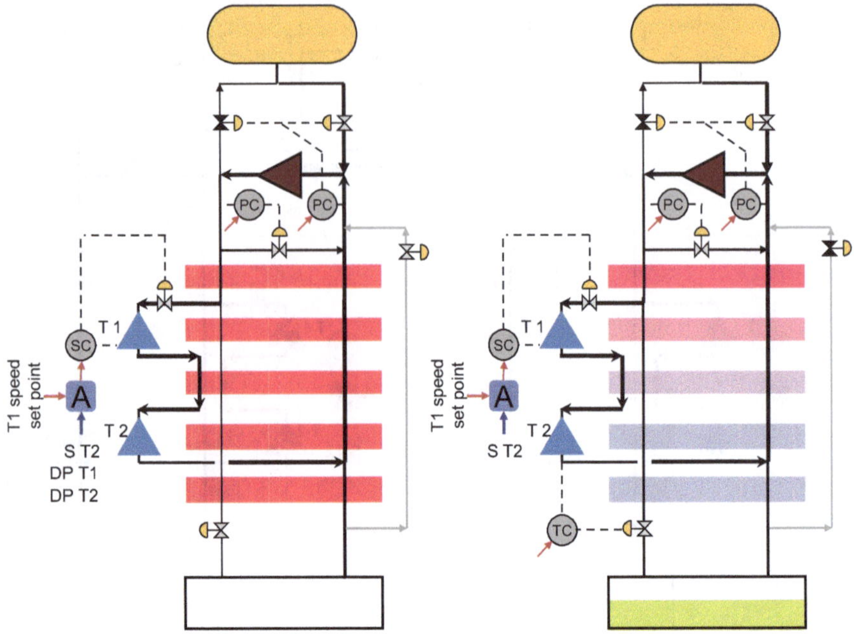

Fig. 10.9 Simultaneous cool-down of cold box and dewar

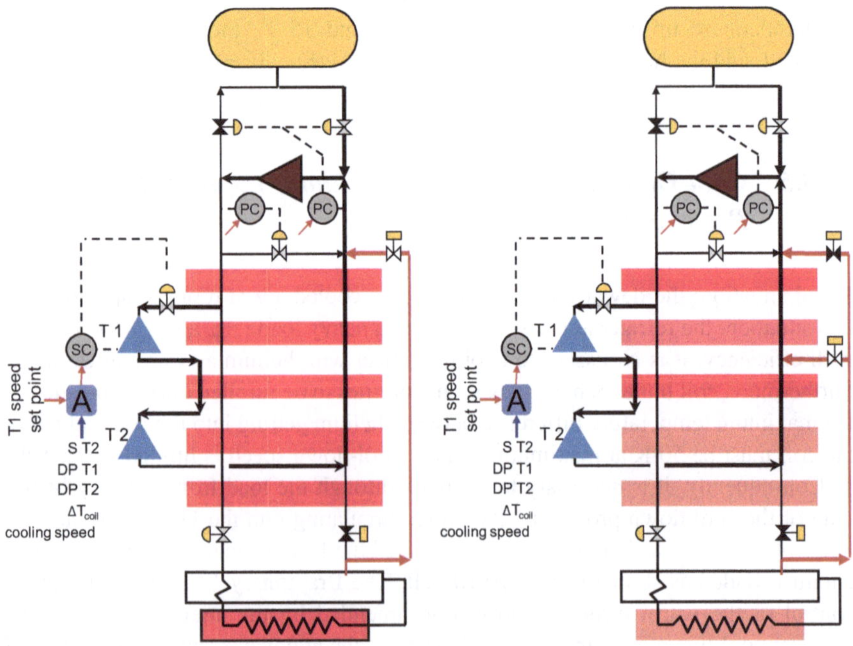

Fig. 10.10 Heavy load cool-down phase 1 and 2

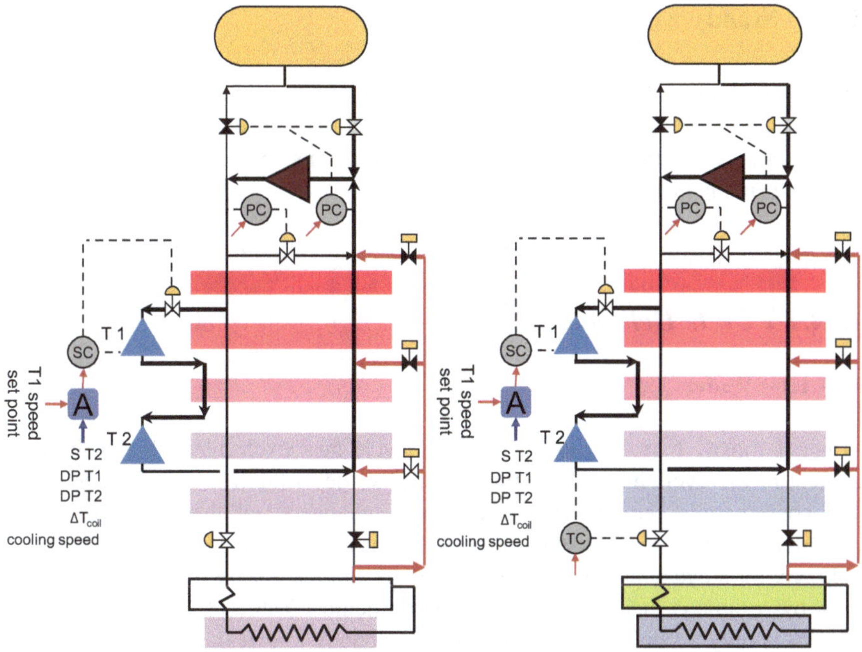

Fig. 10.11 Heavy load cool-down phase 3 and 4

cooling-down speed. The return temperature from the load is obviously room temperature: this is why the flow is directly sent to the suction side of the cycle compressor.

Cool-down phase 2. After some time, as the return temperature from the load is lower, it would not be efficient to return helium to the suction side of the compressor: the helium enthalpy would be lost, and the thermal load on the refrigerator would increase until it reaches a steady state, or, in other words, the temperature would not decrease anymore. Recovery of helium enthalpy is done by recovering cold helium at an intermediate level of the heat exchangers as shown in Fig. 10.10, right.

Cool-down phase 3. The return temperature being colder and colder, another by-pass is operated as shown in Fig. 10.11, left.

Cool-down phase 4. The system cools down until the temperature upstream the JT valve reaches a value where the expansion cools helium, for example, around 15 K. When the cold turbine discharge temperature nears its set point, the control loop is set in automatic operation (see Fig. 10.11, right).

10.4 Steady State Operation

In this paragraph, only the checks to be performed during operation are discussed. In a large plant, most of these checks, sometimes including special calculations, should be performed automatically and issue-related alarms. Settings are explained in Chap. 11, maintenance.

10.4.1 Cycle Helium Purity

10.4.1.1 Water

Water is mostly trapped into the warmest cycle heat exchangers. If the moisture content of the cycle helium is rather high, this affects the heat transfer coefficient and, consequently, the efficiency of the heat exchangers and, therefore, reduces the performance of the refrigerator. Remember that the water quantity that gets into a cold box for a cycle mass flow rate of 2000 g/s and a moisture concentration of 0.5 ppM by volume is 115 kg for a 1-year run (remember Table 11.2).

This is only a calculation, but, obviously, nobody knows the real water concentration of the cycle helium upon such a long period of time. There could be accidental situations where the water concentration rises suddenly, for example, if a water leak occurs in an oil or helium cooler. The full flow dryer, that has been introduced in Sect. 9.4.2.1, will save the run, provided that a moisture measurement is performed upstream to give a warning and allow the operator to react accordingly. For a long run protection, only one dryer can be used because its regeneration time is short compared to its operating time.

In other words, a full flow dryer is advised if one wants to reliably perform long runs (more than half a year) or very long runs.

There is another advantage in case of cooling large components as coils in which some rather high quantity of water can stay adsorbed. Such water is desorbed during the beginning of the cool-down process, where temperatures are higher than 200 K and released into the cycle helium that is dried when circulated into the full flow dryer.

This can, sometimes, be witnessed by checking the heat exchanger warm end temperature difference at steady state. Remember that in order to be in a position to compare measurements made on a heat exchanger, the operating conditions must be identical (see Sect. 7.3.1); therefore, as it is difficult to reach such identical situation, this test is not very accurate. However, when the heat exchanger is really fouled by water ice, it is very easy to notice.

10.4.1.2 Air

At the end of the cool-down process, the concentration of impurity (mainly air and water) in the helium that is still stored into the cycle buffer capacity has not changed because no pure helium from the cycle has been allowed to get into it.

The design of the cycle adsorber allows purifying a given amount of helium at an initial given impurity concentration (this is indicated into the plant specification). *The cycle helium impurity concentration prior to start up must be less than such design quantity.*

As a consequence of the cool-down process, most of the cycle helium impurities are trapped into the high-pressure circuit, between the warm end of the heat exchanger and the adsorber. Running a regeneration allows to be rid of all impurities and start again with almost all pure cycle helium. It is wise to perform this procedure when:

- A plant has been cooled down for the first time.
- A maintenance operation on the cold box has been performed.
- The oil vapour adsorber has been dried with nitrogen gas.
- The plant has been stopped for a long period of time.
- Once a year

The regeneration procedure must, obviously, not spread the impurities in the whole cycle (refer to Sect. 10.6.1).

If the source of impurity is permanent (as a water leak into helium or oil cracking into the compressor), be aware that *the turbine bearing flow is not purified*. This could be a dangerous situation for the turbine static gas bearing system (see Sect. 6.5.1.2.1).

10.4.2 Helium Inventory

It is a wise behaviour to perform a periodic calculation of the whole helium inventory. The cycle pressure control system compensates automatically any leak of the system; therefore, by monitoring the buffer capacity pressure, a general leak rate estimation can be easily done, assuming the refrigerator is operating in a steady state during the time of the measurement. A good trick is to measure the buffer capacity pressure at same hour of the day, preferably in the very morning, before sun hits the buffer capacity. The calculation of the helium inventory might not be very accurate; therefore, a trend is to be plotted that will give a more reliable correct information. Another way to increase temporarily the sensitivity is to operate with the minimum volume of buffer capacity when the latter is composed of multiple units.

However, it is a sometimes-tedious task, even if simple, for liquefaction centres because the helium inventory is located in many various places. Such action can be automatically done by a special part of the control system.

Topping up with pure, or at least "clean gas", can be performed in the buffer capacity or by allowing a flow of helium into the low-pressure circuit. Do not trust compressed helium purity unless an analysis has been performed or purify systematically any helium prior to topping up. Analysis during top-up is a possibility.

10.4.3 Helium Leaks

Any gas processing system that is an assembly of various components might be leak tight at a time, but operation generally creates leaks by various phenomena as vibrations, pressure and temperature changes.

Leaks are the biggest and usual trouble an operator has to face permanently. Leaks not only allow helium to get out of the circuits, but also, some air can get into the circuits by partial pressure difference (see Sect. 9.3.1).

Consequently, a permanent and periodic leak hunting action must be performed, especially if the helium inventory calculation shows a leak. Such leak hunting should not be performed with bubbling liquid because it is a possibility to introduce water into the circuits by partial pressure difference (see Sect. 9.3.1). It is better to use a portable helium leak detector as the one that is shown in Fig. 10.12, right.

Helium leak hunting is not an exciting activity; therefore, a wise way is to sort the system in various parts (compressor, oil removal system, external components of the cold box, recovery system, etc.) and, for example, every week, check one of the parts. Use tricks as shown on Fig. 10.12 to ease detection on a flange.

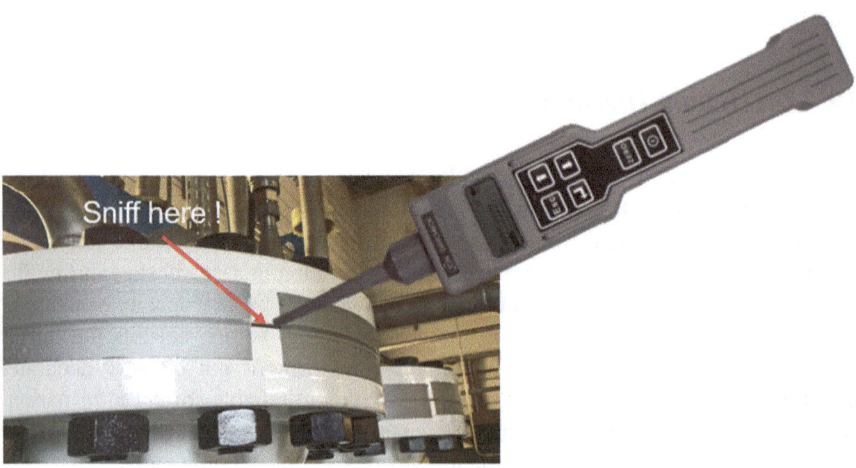

Fig. 10.12 How to check a flange with a portable helium leak detector (courtesy Ion Science Ltd)

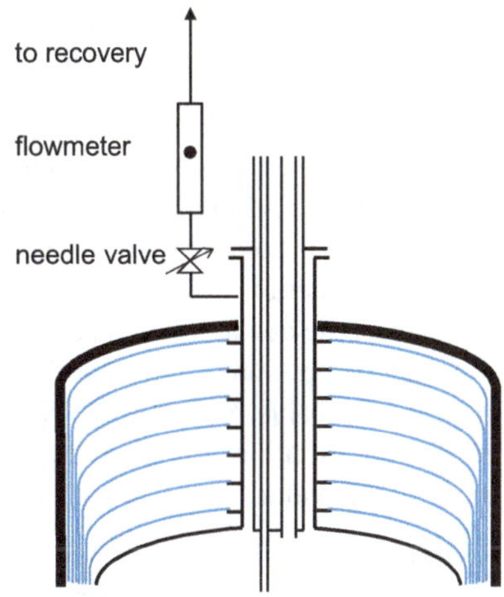

Fig. 10.13 Equipment to control the neck flow rate of a stationary helium dewar

In systems that are operating at temperatures lower than 4.2 K, some parts of the circuits are at sub-atmospheric pressure: here, the trouble is air in leaks due to pressure difference. Special arrangements must be made against such risk (remember Sect. 6.9.3.4).

10.4.4 The Stationary Liquid Helium Storage

A stationery dewar is generally permanently connected to a liquefier or a refrigerator cold box. Most of such dewars are fitted with thermal shields that are thermally connected to the neck (see Fig. 10.13, left). Heat leaks falling onto the liquid generate a flow of helium vapour. There is no way to reduce the heat load of an existing dewar, and therefore, that means that the helium vapour flow cannot be reduced. (Obviously, if it were reduced, the dewar pressure would increase dangerously.) The helium vapour gets out of the dewar by flowing along the neck, absorbing the heat that is brought by conduction along the neck. (This process is similar as the one that is used to cool current leads for a solenoid!) In a standalone configuration (helium is allowed to atmosphere), the neck flow stabilises at the equilibrium amount. If the system is correctly designed, helium that gets out is at about room temperature. It is the way the dewar has been tested at the makers (if it has ever been tested).

But, when the dewar is connected to a cold box, various piping has been introduced into the neck. This, obviously, increases the total heat leak of the dewar and, consequently the neck helium flow. When the cold box is operating, any heat leak is taken by the cold box, and, consequently, nobody notices if the flow has been incorrectly set. However, when the refrigerator is not running, the dewar performance can be rather affected. Let us consider two possibilities:

- The flow is too high: helium that gets out is cold; ice appears.
- The flow is too low: helium that gets out is at room temperature, but the heat leak along the neck is larger.

In both cases, the cold box has to compensate for the bad tuning.

The usual question is: How to correctly set the helium flow when such a dewar is connected to a cold box? The proposed procedure is to be performed when the cryogenic system is not in operation.

Connect the dewar to the recovery system. Open the valve at 100% (provided the atmospheric pressure is constant during the test). Log the helium flow until it comes to a constant value. Such a value corresponds to the natural helium loss of the dewar in this configuration When the liquefier is again in operation, the helium flow must be set at this value.

Remark Depending on the liquefier operation, the dewar pressure can be different, and, consequently, when the control is made by a simple needle valve, the neck flow changes! The operator has to check that the flow is correctly set in each operating condition or, better, use a flow controller instead of the simple needle valve.

10.4.5 An Example of a Conventional Periodic Human Check of a Refrigeration System

During the tour of a plant, it is possible to perform very simple checks as:

Compressor

- Noises
- Shaft oil leak
- Other leaks
- Vibrations, machines, pipes (see below)
- Oil temperature at injection (with the palm of the hand)
- Motor bearing temperature (with palm of the hand)

Oil removal system

- Check the oil drop frequency in coalescer oil sight glass if any
- Oil pump shaft leak
- Other leaks
- Stability of control valves (see here under)

Cold box

- Temperature of the low-pressure pipe out of the cold box (with the palm of the hand)
- Vacuum enclosure wall temperature

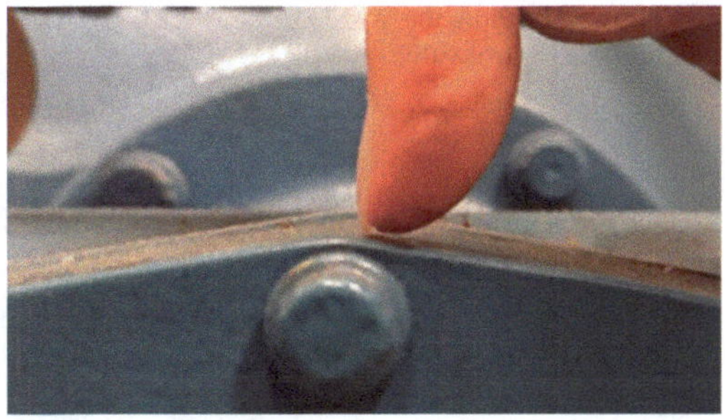

Fig. 10.14 An easy check of vibration level

- Parts of vacuum enclosures with water condensation or ice
- Turbine noise (not at critical speed)
- Turbine bearing LP temperature (on static bearing machines, with the palm of the hand)
- Stability of control valves (see below)
- Listening to instrument air leaks

Dewar

- No ice on the neck
- No pressure oscillations

A few tricks

Getting an idea of the vibration level of a machine by touching, with the end of the nail (not the finger), as shown in Fig 10.14 in the three directions. It is rather easy to acquire a good vibration-feeling sense!

Listening to the positioners of control valves, or if there is too much noise, putting the finger on the place where the stem gets into the stuffing box, as shown in Fig. 10.15, in order to feel if the valve is stable or oscillating permanently, which is the indication of an incorrect setting of PI values of the control loop.

10.5 A Special Situation for Liquefiers: Back Flushing the Cold Box Adsorbers

When a liquefier has been stopped, for example, in the situation that is described in Sect. 8.7.2.1 (automatic stop and re-start), the internal adsorber temperatures rise. Depending on the time duration and operating conditions prior to the stop, the quantity of impurities that have been stored into the adsorbers is variable; therefore, during the time the liquefier is stopped, some impurities can be released by the adsorbent and diffuse in the adsorber volume and the neighbour components. At the time of re-starting the turbines, these impurities may condense or solidify into the turbine, leading to a possible damage.

Fig. 10.15 Checking the stability of a control valve

An interesting solution is to circulate the circuits where impurities might be trapped in a *reverse* direction, with pure helium. When impurities are released, they are brought out of the circuits by the small helium flow, as shown in Fig. 10.16. In operating conditions, valve A is closed; three-way valve B sends the small helium flow that has cooled the thermal shields of the dewar to the recovery. When the liquefier is stopped, after a period of time, high-pressure circuit is vented to the recovery through valve A that sends helium to the recovery (right location in Fig. 10.16). Valve B changes position to send pure helium from the dewar into the high-pressure circuit. The high-pressure circuit is back flushed with pure helium. Any impurity that is released is taken out of the circuits.

It is, therefore, possible, at any time, to re-start the liquefier without any risk of processing impure gas into the turbines.

Such a procedure can also be implemented in other situations than a liquefier.

10.6 System Warm-Up

The goal is to bring (sometimes rather quickly) all the circuits back to room temperature, prior to maintenance or to any intervention, in order to prevent any condensation when air gets in, either on the gas or the vacuum side.

One has to bring energy into the cold box and into the objects to be warmed, in almost all cases, by circulation of warm gas.

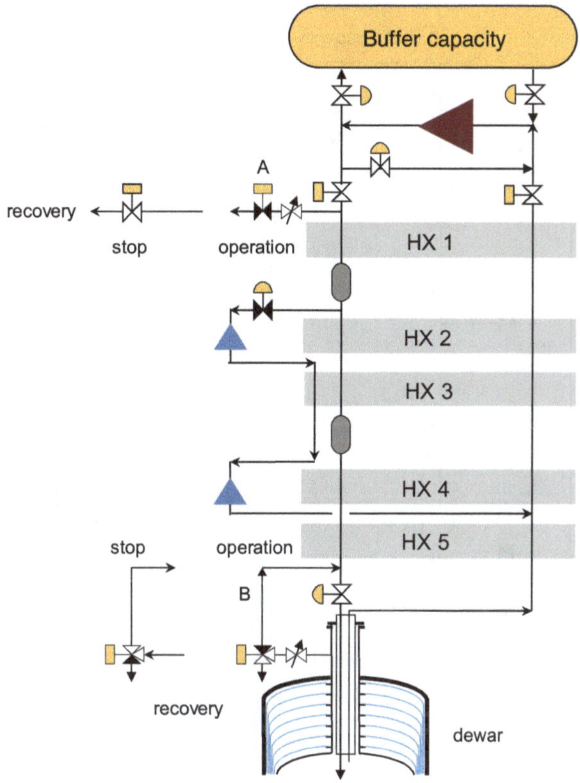

Fig. 10.16 Back flushing adsorbers

There are very few situations where warming up can be performed by heating components. For example, one could think to warm up a coil by circulating some electrical current; however the author has never experienced such a situation.

Warming up by breaking the insulating vacuum, even in a controlled way with dry gas, is not advised for large systems. The risks are condensation of air or water on internal cold surfaces (at the next pump down, it will be difficult to desorb water from the multi-layer insulation), condensation of water on the external part of the vacuum enclosure, a possible deterioration of the multi-layer insulation by dynamic pressure differences and, finally the risk of anoxia (oxygen deficiency) if a human intervention takes place in the enclosure when a neutral gas as nitrogen has been used.

Let us consider various warm-up procedures and compare them.

10.6.1 Circulating in All Circuits

If helium is only circulated into the cold box through high- and low-pressure circuits (see Fig. 10.17, left), in the normal circulating direction, the heat load to warm up the

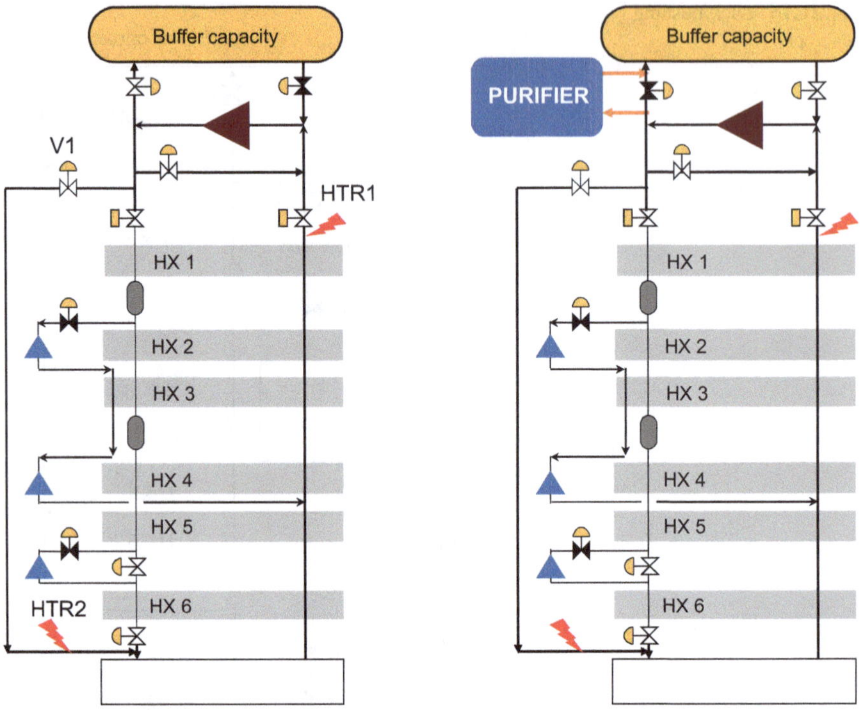

Fig. 10.17 Warming up a cold box

system is brought to the cold box by the warm end temperature difference of the warmer heat exchanger and the heat leaks due to not perfect insulation. (Such heat leaks are generally low and, therefore, will not be considered here.)

Such a heat load is small, too small to be in a position to warm up the internal components (heat exchanger pack, adsorbers, piping, valves) at a reasonable speed. Therefore, it is necessary to bring heat by injecting room temperature helium into the cold part of the system through V1 (see Fig. 10.17, left). Such action unbalances the heat exchangers, and consequently, helium that gets out of the LP circuits is colder than room temperature. It is to be warmed up to room temperature (heater HTR1) in order to avoid any trouble due to too low temperature on the piping to the compressor. When the temperature difference between heat exchangers and room temperature helium becomes too small, the heater HTR2 on injected helium is activated.

Such a process does not generate a thermal shock on the heat exchangers (remember Sect. 6.4.2.3) but, unfortunately, spreads out the impurities that were trapped and confined in the high-pressure circuits in all other circuits (including the cycle buffer capacity that should be permanently kept clean!). Therefore, such a procedure is not advisable.

10.6.2 Circulating in all Circuits and Using an External Purifier

When, by chance, an external liquid nitrogen cooled purifier is available, it is interesting to use it in the following way; the discharge valve that connects HP to the cycle buffer capacity is closed (see Fig. 10.17, right). Helium is circulated into the cold box, and heat is brought in by room temperature helium as described in the former procedure. Helium that is released by the cold box is circulated into the purifier prior to be stored, as pure, into the buffer capacity.

Such a procedure does not generate a thermal shock on the heat exchangers and traps all impurities into the purifier. Therefore, it is a good one, provided a purifier is available.

10.6.3 Circulating into the LP Circuits

During operation, impurities are trapped and "stored" into the HP circuits (in the 80 K and 20 K adsorbers and on heat exchanger surfaces). Therefore, these circuits are isolated by the cold box inlet valve, the turbine inlet valves and the JT valve (see Fig. 10.18, left). Room temperature helium is injected into the cold part of the cold box as explained in the former procedure. During warming up, impurities are released through valve V2 to atmosphere and do not pollute the cycle helium. Of course, pump down is advised prior to re-starting the plant. Such a procedure generates a thermal shock on the coldest heat exchangers but rejects correctly the impurities that were trapped in the circuits. Therefore, such a procedure is acceptable only with *small machines* that have small size heat exchangers.

10.6.4 Circulating Back-Way into the LP Circuits

HP circuits are isolated as explained in former procedures. In order to avoid a thermal shock on the coldest part of the cold box, room temperature helium is injected into the LP circuits in a reverse direction through valve V3 (see Fig. 10.18, right). Cold helium is withdrawn at the cold end of the LP circuits through valve V4 and warmed up prior to be sent to the compressor. Such a procedure reduces the thermal shock on heat exchangers. It might be necessary to finish the warm-up procedure by injecting heated helium at the cold end through V1 and HTR1.

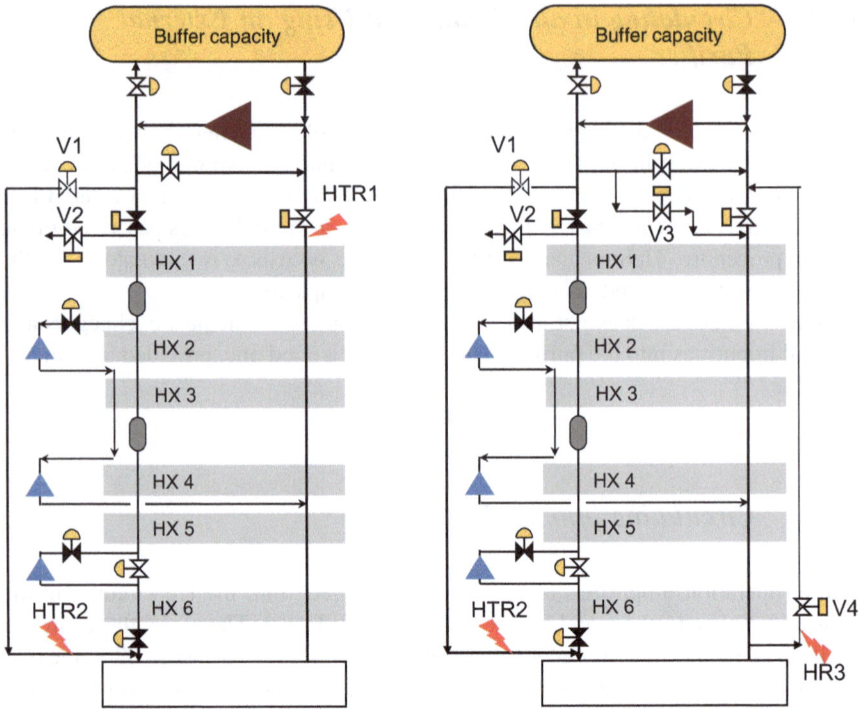

Fig. 10.18 Better ways for warming up a cold box

10.6.5 Circulating Back-Way into the MP Circuits

The HP circuits, where impurities are "stored", are isolated. The MP cold box valve is closed. Room temperature helium is injected into the MP circuits in a reverse direction through V 5 (see Fig. 10.19) and then to the LP cold end (V6). Helium circulates in the LP circuits in normal direction and gets out at the warm end of the LP circuits. Unbalancing of the heat exchangers and bringing heat into the cold box are performed by injecting room temperature helium at the cold end of the cold box (V1). Helium getting out of the LP circuits is to be warmed up (HTR1) prior to be returned to the compressor. Such a procedure avoids any thermal shock on the heat exchangers.

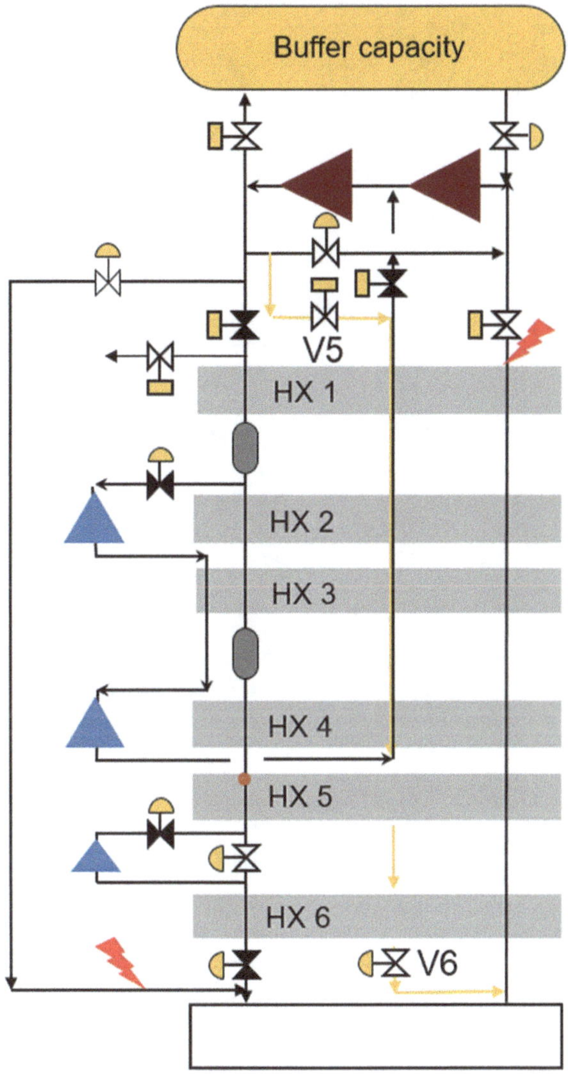

Fig. 10.19 On a three-pressure cycle

Chapter 11
Maintenance of a Helium Refrigeration Plant

Abstract The target of this chapter is to point out general practice and also common-sense behaviour in order to reach the highest reliability of a cryogenic plant and, therefore, to prevent troubles during the operation. Emphasis is put on the specific points that are related to a cryogenic refrigeration system.

11.1 Introduction

The maintenance target is to reach the highest availability of the system at the minimum cost. There are a number of methods and software that help this process. In this chapter, one will look at points that are specific of cryogenics or related equipment. As said in Sect. 10.1, some actions that could be part of the operation:

- Performing some human actions (looking for helium leaks)
- Performing some periodical tests that necessitate measurements (oil leak at the compressor shaft seal, efficiency of oil separation, etc.)
- Tuning the operation of some components (optimum operating temperature of a turbine, tuning of the labyrinth flow, etc.)

are gathered in this chapter.

One should not forget that preventive maintenance actions are indicated into each machine operating manual (e.g. the cycle compressor). It is obvious that the operator MUST carefully read these documents and follow the instructions.

11.2 A Reminder About Maintenance

This is not a course on maintenance; however, let us make some reminders about the various types of maintenance.

Maintenance is the care and servicing by personnel for the purpose of maintaining equipment in satisfactory operating condition by providing for

G. Gistau Baguer, *Cryogenic Helium Refrigeration for Middle and Large Powers*, International Cryogenics Monograph Series,
https://doi.org/10.1007/978-3-030-51677-2_11

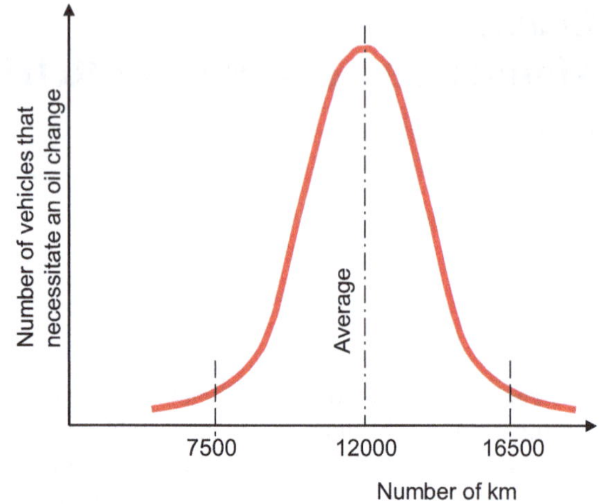

Fig. 11.1 Taking care of a vehicle fleet

inspection, detection and correction of incipient failures either before they occur or before they develop into major defects.

Corrective maintenance, called “repair” in everyday language, is conducted after a failure, to get equipment working again. Failure can occur at any time: a correct operation is not granted during the whole run. Cost of repair is generally maximal, both by spare parts, possible work during night hours and non-operation.

Preventive maintenance is conducted to keep equipment working and/or extend the life of the equipment. Maintenance actions are planned according to time duration or several elapsed hours. To stay on the safe side, maintenance is generally performed according to the statistical minimum lifetime of a component. Therefore, it is obviously not optimised and can be costly, even less than corrective maintenance.

Condition-based maintenance is based on using *real-time data* to maintain the correct equipment, at the right time.

Observing the state of the system is known as condition monitoring. Such a system determines the equipment’s health and acts only when maintenance is actually necessary.

An example of condition-based maintenance.

Let us consider the issue of changing motor oil in a vehicle fleet. On Fig. 11.1, one can see the number of vehicles that necessitate an oil change versus distance.

Only 2.5% of vehicles need a change of the oil at 7500 km

- Most of them could reach 12,000 km.
- And some of them could even reach 16,500 km!

Changing oil at 7500 km for all vehicles, as it is generally the case in preventive maintenance, is expensive and obviously not very smart!

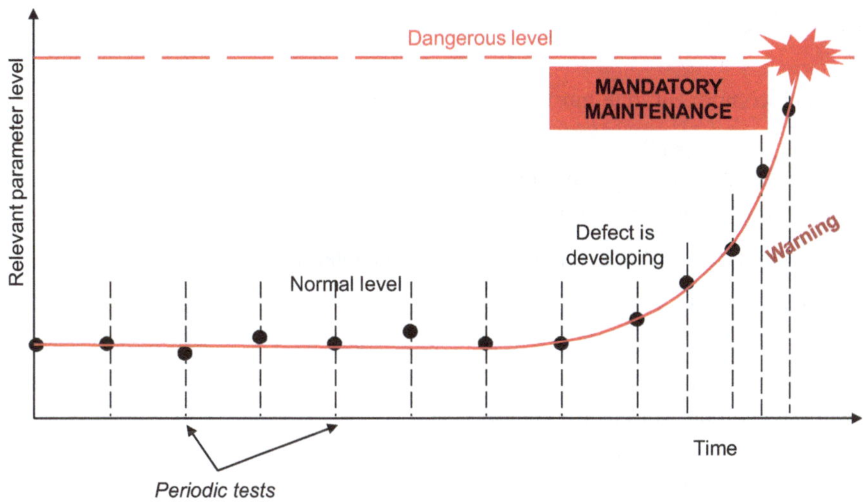

Fig. 11.2 Evolution of a relevant parameter

It is of interest to find a *relevant parameter* that can be monitored (for the given example: oil analysis results). The variation of such relevant parameter can be used to trigger an optimised maintenance operation.

On Fig. 11.2, the relevant parameter level is roughly stable for some time (7 tests), and then it begins to rise, here for the eighth test. A ninth test confirms the level change. At this time, it is of interest to reduce the test period because the slope change increases and one reaches rather fast a warning situation and the time of a mandatory maintenance happens.

Monitoring a set of relevant parameters induces, obviously, some cost.

Therefore, common sense suggests implementing such a condition-based maintenance only for the main or expensive components of a helium refrigeration plant, provided conditions exist:

- Cycle compressor (oil analysis, vibrations)
- Compressor electric motor (vibrations)
- Cycle heat exchangers (temperature measurements)

But, unfortunately, a cryogenic expansion turbine has *no relevant parameter* that can be monitored.

All other "ordinary" components for which getting relevant parameters could be too much expensive compared to the benefit it could bring should follow the preventive maintenance method, for example, for vacuum sets.

11.3 Method

For each component, one must ask for *the most frequent situations*:

- What can happen?
- What to check during operation?
- Which maintenance action?

The components that are addressed in this chapter are:

- The cycle helium
- The compression station:
  - The cycle compressor
  - The compressor electric motor
  - The oil management system
  - The oil separation system
  - The full flow dryer
- The cold box:
  - The cycle heat exchanger
  - The cryogenic expansion turbine
  - The vacuum set
  - The liquid helium storage
- The instrumentation equipment
- The electrical equipment

At the end of an important maintenance action, it is advised to make a general performance test of the system.

11.4 The Cycle Helium

What Can happen?
Impurities circulating with the cycle helium are trapped into the cold box.
Leaks can release helium into atmosphere, and, also, let air in.

What to Check During Operation?
The purity of the cycle helium
The leak tightness of the circuits

Which Maintenance Action?
Make circuits leak tight
Periodic calibration of analysers
Top-up with pure helium

11.4.1 Circuit Conditioning

A paramount condition to operate a helium refrigeration system is that the cycle helium must be as pure as possible. The cycle helium purity is one of the most important parameters influencing the plant reliability.

At the time of construction of the equipment, all circuits are full of air. This could also be the case for part of the circuits when a maintenance operation happened on the system. Helium can also be polluted when a system has not been operated since a long period of time (months).

Therefore, in all such situations, a total or partial circuit conditioning operation is to be performed, prior to a system cool down.

The goal is to condition the circuits in order that they contain only pure helium, prior to the cooling down operation.

The risks are of not remove all the impurities (air, water, etc.) that are present into the circuits and let some impurities be trapped on surfaces that are still cold.

Do not forget that such a conditioning procedure does not grant that all parts of all circuits are filled with pure helium.

A good practice is to aim at concentrations indicated in Table 11.1.

Speaking about impurity concentration can be misleading. What is of importance is the *accumulated quantity* of impurities into the cold box according to the operating time. Table 11.2 gives an idea of these values for various contexts.

The first action is conditioning all the circuits of the system.

The most common procedure is to perform a series of successive purges and back-fillings. For an expansion ratio of 100/1, the *theoretical result* that is shown on Table 11.3 shows that three cycles should be enough.

However, such a calculation does not represent reality: vacuum is not perfect, especially at the end of small diameter pipes (e.g. connection to pressure sensor); therefore, a small amount of air stays in the system.

Let us look at what happens (see Fig. 11.3):

Step 1, *initial situation*: all circuits are full of air.

Step 2, *purging*: vacuum is made. However, some air stays at the far dead end of circuits, for example, near a pressure sensor that is connected by a small internal diameter pipe.

Table 11.1 Acceptable impurity concentration levels according to CERN's experience

	During a *steady* state (*long* duration)	During a *transient* state (*short* duration)
	(ppM V)	(ppM V)
Air	1	<20
Water	0.1	<10

Table 11.2 Integration of impurities into the cold box over time

Cycle gas mass flow rate	g/s	*40*	*100*	*1000*	*1800*
Air concentration	*ppM V*	*1.0*	*1.0*	*1.0*	*1.0*
Water concentration	*ppM V*	*1.0*	*0.1*	*0.1*	*0.1*
Operating time	h	*700*	*700*	*2100*	*7000*
Cycle gas volume flow rate	Nm3/h	816	2040	20.400	36.720
Air volume flow rate	*Nm3/h*	*0.001*	*0.002*	*0.020*	*0.037*
Air mass flow rate	*g/h*	*1.06*	*2.64*	*26.41*	*47.54*
Air trapped volume	*kg*	*0.74*	*1.85*	*55.46*	*332.78*
Liquid air trapped quantity	*L*	*0.64*	*1.60*	*48.14*	*288.85*
Water vapour volume flow rate	*Nm3/h*	*0.001*	*0.000*	*0.002*	*0.004*
Water mass flow rate	*g/h*	*0.66*	*0.16*	*1.64*	*2.95*
Water trapped quantity	*kg*	*0.46*	*0.11*	*3.44*	*20.66*
Liquid water trapped quantity	*L*	*0.46*	*0.11*	*3.44*	*20.66*

Table 11.3 Theoretical calculation of evolution of the concentration of impurities

Cycle number	1	2	3
Conc. at beginn. cycle	1	10^{-2}	10^{-4}
Pressure ratio	100	100	100
Conc. at end of cycle	10^{-2}	10^{-4}	10^{-6}

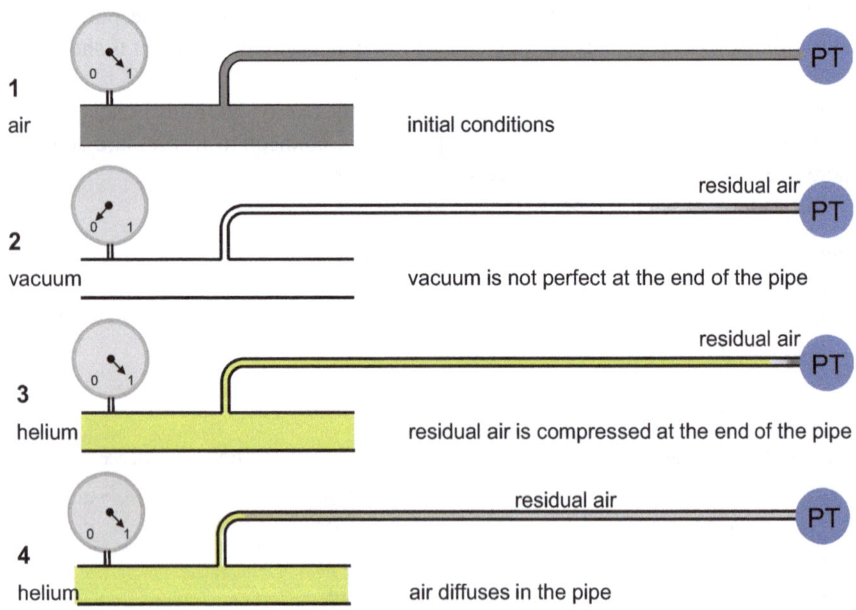

Fig. 11.3 Importance of the gas diffusion process in the conditioning procedure

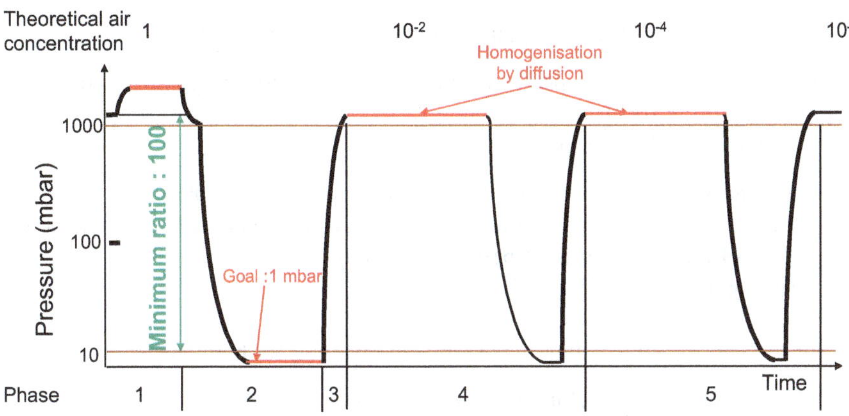

Fig. 11.4 Pressure changes during a circuit conditioning sequence

Step 3, *back-filling*: pure helium at atmospheric pressure is allowed into the circuits. The residual air is now *pushed* to the far dead ends of the circuits by helium. If, right now, vacuum were immediately pulled, residual air would expand, and the circuit status would be very similar as that found in step 2; *such a procedure would be of no interest, only discarding pure helium and loosing time*!

Step 4, *waiting*: Instead of pumping at once, one has to *wait* that allows gaseous diffusion to spread the residual air all over the circuits. Therefore, at the next pumping phase, a part of the remaining air is pulled out of the circuits.

Keeping this in mind, the correct procedure to condition circuits is as follows (see Fig. 11.4).

Initial situation: the circuits are at a pressure slightly higher than atmospheric.

Phase 1, *looking for big leaks*: *roughly checking* the circuit leak tightness at atmospheric pressure:

- Fill circuits with pure helium up at around 1.10 bar (0.10 bar gauge)
- Wait and check the pressure:
  - If there is no pressure change within minutes, go to next phase
  - If pressure changes within minutes, that means that there is a big leak (e.g. a valve that is full open, a fitting that is not connected, etc.). Find the large leak and fix it before continuing.
  - Test again and, if necessary, repeat phase 1 until no more leak is detected.

It is obvious that if such a rough test is not performed and if there is a big leak, pumping will allow air into the circuits!

Remark One can decide to perform same test with a higher pressure (1 or 2 bar) in order to increase the sensitivity.

Phase 2, purging: *checking* again the circuit leak tightness at vacuum:

- Pump the circuits to the pressure limit of the conditioning pump. A minimum pressure ratio is 100/1.
- Wait and check the pressure:
  - If there is no pressure change within minutes, go to next phase.
  - If pressure changes within minutes, that means that there is leak. Look for the place of the leak, and fix it before continuing.
  - Test again.

Phase 3, back-filling: filling the circuits and waiting for homogenisation:

- Fill circuits with pure helium up to atmospheric pressure,
- Wait until natural gas diffusion spreads impurities in the circuits

Phase 4, purging back-filling: pumping, then filling the circuits and waiting for homogenisation:

- Fill circuits with pure helium up to atmospheric pressure,
- Wait until natural gas diffusion spreads impurities in the circuits.

Phase 5, purging back-filling: repeat phase 4.

Remarks

1. The usual question is: How long does the operator has to wait for a correct homogenisation? Never less than 30 min for a small plant and never less than 1 hour for large plants. Such a homogenisation procedure is, sometimes, somewhat neglected (especially, waiting time is often shortened or even forgotten), except if it is automatically carried out by the control system.
2. Mind: water is a special impurity!

In the gaseous phase: water has a polar molecule that is *easily adsorbed* on the circuit surfaces but is *hardly desorbed* if no energy (heating) is brought (remember that desorption needs energy).

In the liquid phase: one has to bring energy to vaporise water. Remember that the cold box circuits are difficult to dry by pumping because they are thermally insulated; therefore, heat cannot come naturally from ambient. If some liquid water quantity is accumulated in a lower part of the circuit, for example, a pipe bend, vaporisation can turn it into ice, which is even more difficult to pump out by sublimation.

11.4.1.1 Leak Hunting

There are two ways to look for leaks; either by a global test that gives information on the total amount of the leak or by looking for individual leaks with a leak detector all around the plant.

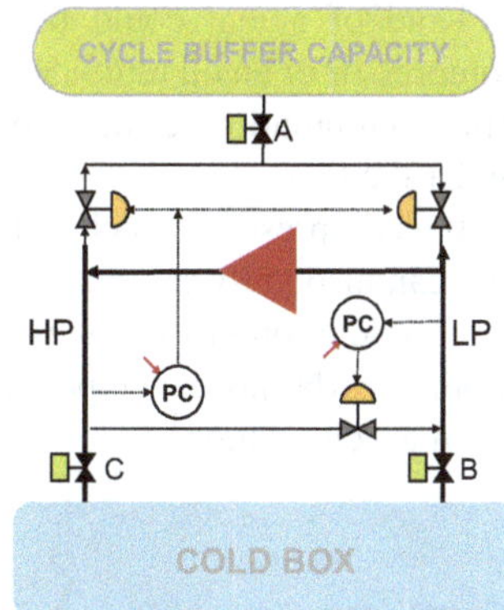

Fig. 11.5 Checking the leak tightness of the compression station

11.4.1.1.1 Global Testing

Example of a leak tightness test of the compression station (see Fig. 11.5).

Such test, that is more sensitive than the one described in Sect. 10.4.2, can only be performed when the cryogenic system is not in operation.

Isolate the cold box, by closing valves B and C.

Run the compression station at nominal operating cycle pressures.

When the system is in a thermal steady state, isolate the buffer capacity by closing valve A.

Keep the compressor(s) running at nominal pressures.

Monitor the cycle pressures: as the cycle buffer capacity is isolated, there is no leak compensation, and consequently, the cycle pressure decreases if there is a leak.

As the operating volume is small, a leak can be more easily detected and evaluated.

11.4.1.1.2 Individual Leak Hunting

Refer to Sect. 10.4.3 for leak hunting during operation.

Helium leak hunting is not an exciting activity; therefore, a wise way is to sort the system in various parts (compressor, oil removal system, external components of the cold box, recovery system, etc.), and, for example, every week, check one of the parts.

11.4.1.1.3 Some Practical Tricks

It is very useful and common to stick a tape around a flange to recover the possible helium leak at one point, making the test easier and more sensitive (remember Fig. 10.12, right).

Kind of a tent could be also built that encloses a part of the plant, like a compressor or an oil removal system. Helium would raise to the top and concentrate. This accumulation-testing procedure allows making a general test of the part that is under the tent.

If a compressor is enclosed into a sound proof enclosure (like a Kaeser machine), one can also sniff cooling air at the discharge port (possibly, stop the air fan for a short time in order to increase the sensitivity of the test). In a large size compressor room, CERN has developed a helium detection at the discharge of the compressor room air circulation[1].

11.5 The Compression Station

What Can Happen?

A compressor is submitted to temperature changes and vibrations that may cause helium or oil leaks and mechanical damages.

Misalignment of the coupling.

Oil leaks.

Oil pollution or deterioration.

Oil or helium cooler leak.

Incorrect oil separation.

What to Check During Operation?

Oil leak tightness of compressor shaft seals

Oil leak tightness of oil pump shaft seals

Oil quality

Vibrations

Efficiency of oil and helium coolers

Efficiency of oil separation

Quality of cooling water

Which Maintenance Action?

Electric motor bearing lubrication

Oil top-up or change only with processed oil if necessary, as requested by the manufacturer

Oil or/and helium cooler descaling

Periodic activated charcoal change

[1]L. Delprat and al., in LHC cryogenic availability and helium management from run 1 to run 2, 7th international workshop on cryogenics operation, 2015, Fermi National Accelerator Laboratory

11.5.1 The Compressor

The screw compressor comprises quite few moving parts. It is a rather simple machine that can run for extended periods of time. But it does not mean that it is a maintenance-free machine!

11.5.1.1 The Shaft Seal

The shaft seal is one of the delicate parts of a screw compressor and needs a special attention. A shaft seal usually leaks (reasonably): see the compressor operating manual to know which leak rate is considered as acceptable by the maker.

The shaft seal oil leak must be collected into a recipient (see Fig. 11.6). Once a week, the oil quantity is measured. When the shaft seal degrades, the oil leak flow rate might increase quickly.

On large plants, it is of interest to have an automatic data logging and processing in order to calculate the oil leak rate and automatically issue a warning.

Fig. 11.6 Collection of the shaft seal oil leak

Remarks

1. When the shaft seal is located towards the suction side of the screws, leakage can lead to an air in leak. A reaction of oxygen brought by air with oil might generate aldehydes that are *not trapped* into the ORS activated charcoal and can deposit on the cold end part of a static bearing turbine shaft, leading to trouble.
2. When a compressor has not been running for a rather long time (a few months), it is possible that the vibrations induced by neighbour running machines have destroyed the oil film that prevents any contact between the sliding parts of the seal. When the compressor is started, chances are big that the seal is destroyed during the very first revolutions. It is therefore important, as generally advised by the manufacturer, to rotate the compressor by hand every few weeks and prior to starting in order the seal is again lubricated.

11.5.1.2 Vibrations

Checking the vibrations that are generated by a machine is a very interesting way to get information about its "health". There are various ways to measure vibrations.

A vibration sensor must be operated always in same conditions. The best way is to have it screwed on the machine, at the same location, permanently or not (see Fig. 11.7).

See in Fig. 11.7 the incidence of the way the sensor is either hold or screwed. Another very important point is that there must not be any damping material between the sensor and the machine casing, for example, paint. And finally, the sensor is to be installed in a way that it is correctly oriented in relation with the vibration direction that is to be measured.

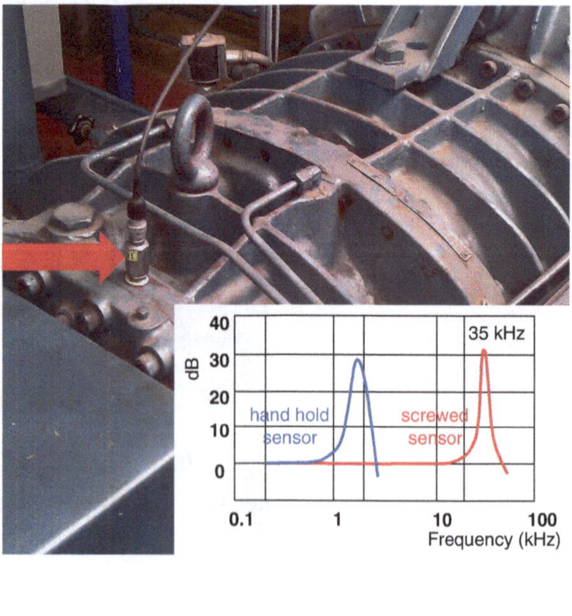

Fig. 11.7 Permanent fixation of a vibration sensor

Vibrations must be considered in various points of view:

- Vibrations that are induced by the machine itself
- Vibrations that are induced by neighbour machines

11.5.1.2.1 Vibrations that Are Induced by the Machine Itself

Spectral Analysis

Let us look at the machine assembly that is shown in Fig. 11.8.

One can easily understand that the vibration peaks can be related to some specific aspect of the machine. For example, the peak at about 25 Hz is due to a slight unbalance of the electric motor rotor. The peak at around 2000 Hz is related to the operation of the gear system (the combination of teeth number and rotational speeds). Table 11.4 shows the mains causes of vibration.

Example: A centrifugal pump is driven, through a coupling, by a two-phase electric motor running at 2960 rpm.

A spectral analysis is performed on the system: see Fig. 11.9.

There is high amplitude at two times the rotational speed. According to the table in Fig. 11.10, it is a coupling defect.

Result of coupling alignment can be seen on Fig. 11.10.

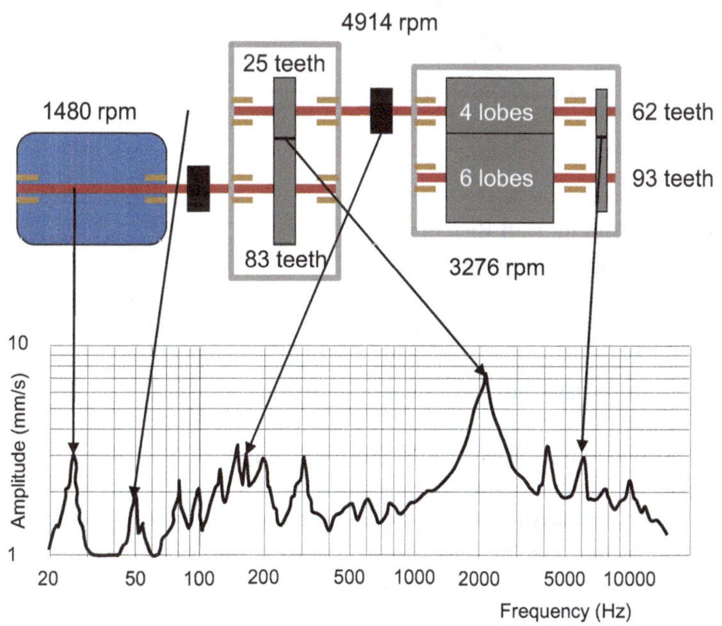

Fig. 11.8 Example of a system

Table 11.4 Origin of vibrations

Origin	Frequency	Direction
Unbalance	1 × Fr	Radial
Alignment	2 × Fr	Axial and radial
Electrical excitation	1, 2, 3, 4 × Fr	Axial and radial
Hydrodynamic	Wing frequency	Axial and radial
Roll bearing	High frequency	Axial and radial

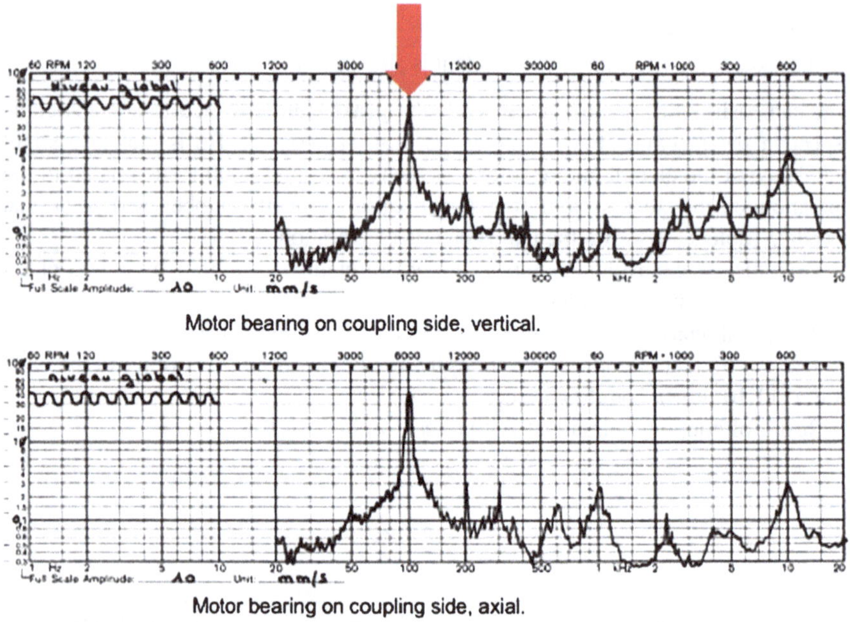

Fig. 11.9 Diagnosis

Monitoring of Ball or Roll Bearings

Principle of the shock pulse method (SPM).

A bearing race track damage cannot be detected by elevated levels of low-frequency vibration until the damage is quite severe. The reason for this is that when a rolling element passes over a damaged area of the race, a shock pulse is created that can be detected only in the high-frequency range.

The transitory shock wave is quickly amortised. When it reaches the sensor, it creates a signal as shown in Fig. 11.11. The peak amplitude is related to the velocity of the impact (v).

In Fig. 11.12, it is easy to discriminate the unbalancing of the rotor from the shock that is generated by some bad operating rolling element bearing.

Figure 11.13 shows the various situations that are related to the operation of a rolling element bearing.

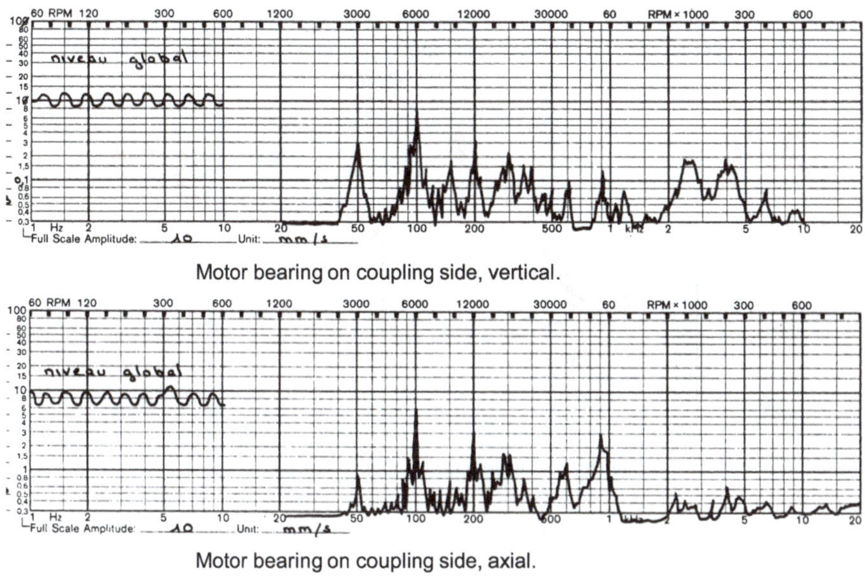

Fig. 11.10 After maintenance

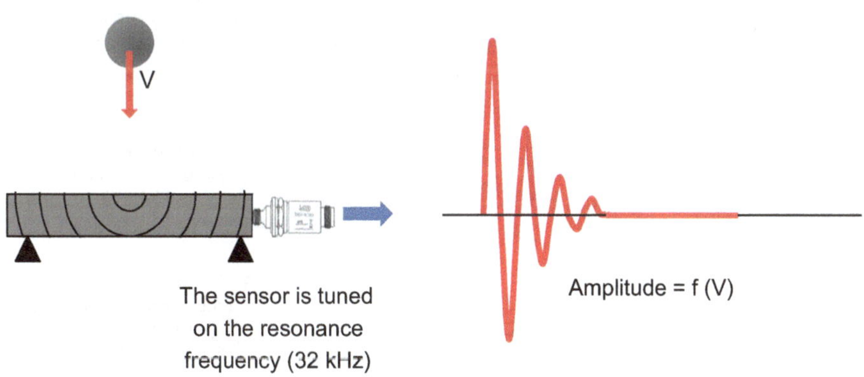

Fig. 11.11 Principle of the shock pulse method

- In the top of the figure, the lubricant layer is correct. There is almost no contact between the rolling element and the race track, only from time to time. The signal that is collected by the vibration sensor is rather low (dB_c carpet) with a small difference between dB_m (maximum) and dB_c. It indicates a correct operation of the rolling element bearing (very few direct contacts between the rolling element and the track) with a correct lubricant layer.
- In the middle part of the figure, the signal that is collected by the vibration sensor is rather high (dB_c) with a small difference between dB_m and dB_c. It indicates

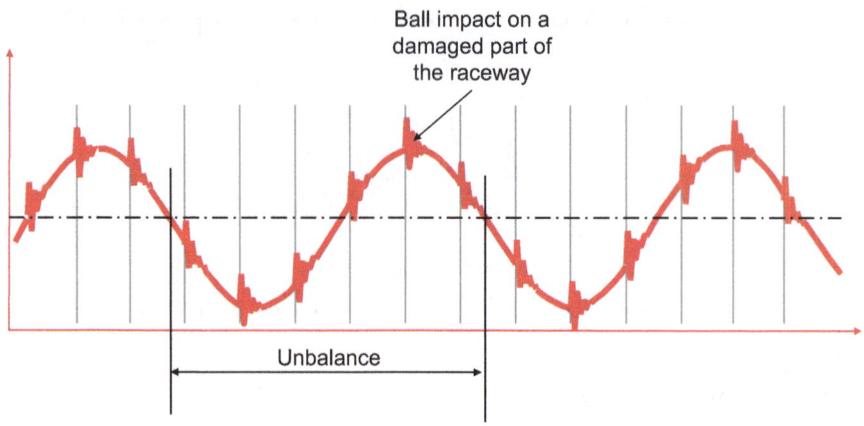

Fig. 11.12 An easy discrimination of unbalance and ball bearing vibrations

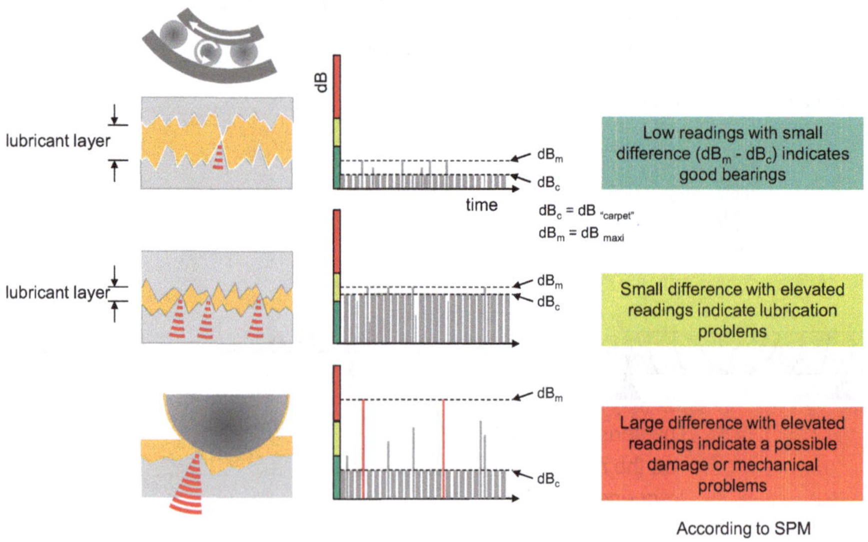

Fig. 11.13 The various levels of investigation

more frequent direct contacts between the rolling element and the track, which is an indication of a too thin lubricant layer. Lubrication must be urgently performed.

- In the middle part of the figure, the signal that is collected by the vibration sensor is rather low (dB_c) with a very large difference between dB_m and dB_c, which is an indication of high shocks between the rolling element and the track: possible damage.

Remember the "nail feeling" method for checking vibrations that is explained at the end of Sect. 10.4.5!

11.5.1.2.2 Vibrations that Are Induced by Neighbour Operating Machines

In a normal operation, a rolling-element bearing has the rollers and races tracks separated by a thin layer of lubricant such as grease or oil. Although these lubricants normally appear liquid (not solid), under high pressure, they act as solids and avoid any contact between the rolling element and the rings.

When a machine fitted with rolling-element bearings is stopped and that, simultaneously, another is running not far away, "brinelling" (or false brinelling) can happen: between the rolling-element and the rings that are submitted to vibrations originated from neighbour machines in operation, the lubricant is not replaced by movement and is progressively pushed out of the loaded region. Without lubricant, wear increases; balls or rolls make imprints into the rings (Fig. 11.14). It is even possible for the resulting wear debris to oxidise and form an abrasive compound which further accelerates wear. When the bearing is running again, balls roll on a surface that is not anymore flat, and the bearing can quickly deteriorate.

False brinelling is said to have been discovered in the 1930s, when brand-new cars were transported by train on long distances like Detroit to Los Angeles. When they were unloaded, they would soon develop wheel bearing troubles. It turned out that many wheel bearings were damaged as a consequence of the regular impact every time a railroad car wheel passed a track joint.

As a consequence of possible false brinelling, during a long period of time where a machine is not running (a main machine or a spare), there are some precautions to be taken: regularly switch on the machine before the oil film is degraded, for example, monthly. Damping studs can also be incorporated at the time of the machine erection.

In case of compressors that are fitted with an integrated oil pump (remember Sect. 6.3.6.2), the hand pump must be used to feed the bearings prior to starting after a very long stop.

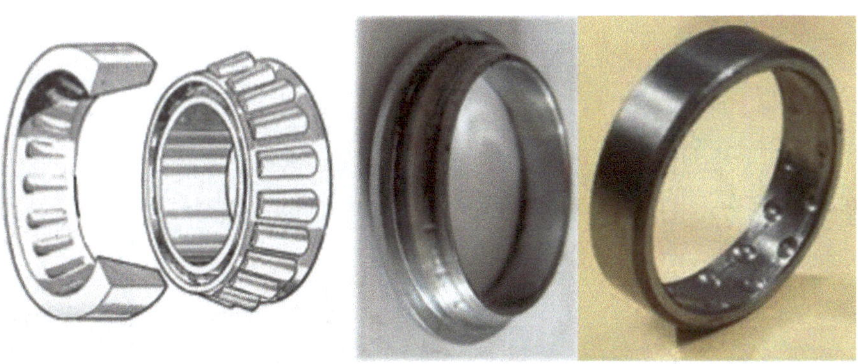

Fig. 11.14 Imprints made by false brinelling

11.5.1.3 Oil Injection Control

The more oil is injected, the lowest the discharge temperature is. However, one should avoid a situation where oil (a *non*compressible fluid) might be “squeezed” between the screws and the casing at the end of the compression phase! This would create a very high pressure that would excessively load both the axial and thrust bearings leading to a possible destruction. Vibrations are generated by too much oil injection.

11.5.2 The Electric Motor

Roll or ball bearing lubrication must be made:

- At the correct time period
- With the correct grease
- With the correct amount, that is to say a number of pump strokes in order to be sure to replace all “old” grease by new one (see Fig. 11.15)

Such information is given in the maintenance manual of the motor. However, the number of pump strokes is to be selected by the operator, according to the grease pump that he uses.

As grease is slowly injected into a bearing, the change in sound or lack of sound change informs the operator when enough grease is present.

Automatic greasing devices are of interest (see Fig. 11.15, right).

Coupling

A bad alignment of the coupling might also lead to a deterioration of the shaft seal (remember Sect. 11.5.1.2.1).

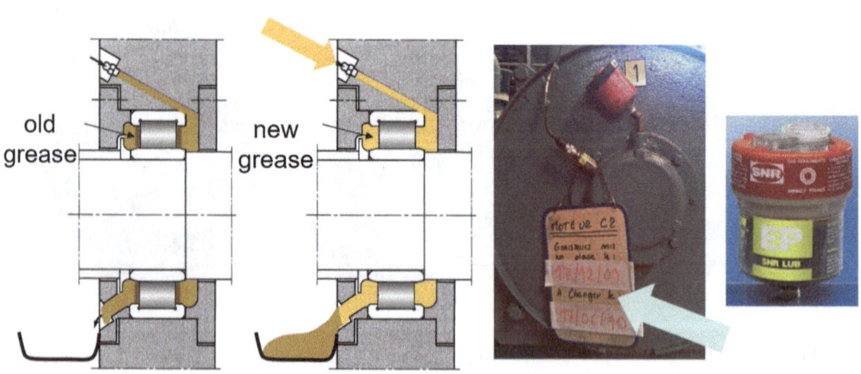

Fig. 11.15 Correct amount of grease injected into a roll bearing and an automatic greasing device (note the label telling when the device is to be replaced)

11.5.3 Compressor Overhauling

Even if the compressor is a reliable machine, it is advised to perform an overhauling at the manufacturer workshop at rather large time period intervals (order of magnitude is 50,000 operating hours). After such a heavy maintenance action, it is wise to perform a general test (see Sect. 14.4.4) in order to check its results with a same test that should have been performed just prior to the overhauling action.

11.6 The Oil Management System

The oil management system can be split into two parts: the management of oil as a machine lubricant (circulation, filtering, cooling, actuation of the slide valve) and the oil removal system (ORS) where separation of oil takes place in order to get pure helium.

An ORS is an almost static system and is generally operating smoothly. However, one must know that if such ORS does not operate perfectly, chances are that oil gets into the cold box, and this is a very big trouble because cleaning the internal parts of a cold box that has been polluted by oil means a few weeks of plant non-operation on top of big expenses[2].

11.6.1 Oil

Remember that the oil that is used for helium compression is a special one (see Sect. 6.2.2.2.7)! Avoid any contact with skin or mucous membranes.

Do not forget that such oil deteriorates paint; use only seals made out of Nitrile or Viton.

The operating conditions of such an oil are very comfortable: neutral gas, no oxygen, moderate temperature and no high mechanical load between moving parts. Every time oil is injected into the compressor, it undergoes a treatment that is similar to the one that has been carried out at the suppliers. Consequently, the more such oil has been operating, the better it is!

Therefore, as soon as the mechanical characteristics of the oil are acceptable, a very long lifetime (>100,000 h) can be expected. Topping up must be made only with correctly processed oil ONLY (see Fig. 11.16)!

Oil analysis
A periodic oil analysis must be performed to check:

- "Mechanical" properties, related to the compressor operation

[2]CMS cleaning and consolidation, C. Fabre et al., 2016 CryoOps workshop

Fig. 11.16 The label on a barrel of processed oil

- Viscosity
- Acidity
- Particles and their nature
- Particular properties related to the helium compression, especially the water content (by the direct injection Karl Fischer method, not the oven method)

Oil sampling necessitates precautions in order not to pollute the oil sample, especially with atmospheric moisture. The sample holder is to be perfectly dried (heated up to 100 °C and flushed with dry nitrogen), and the operator must make sure that the pipe from which it collects the sample is also clean and dry. This part of the pipe downstream the sampling valve must be permanently isolated from atmosphere by a tape that is removed just prior to oil sampling.

Be careful with oil plus oxygen combination: when the processed helium contains some oxygen (around 1% volume), a chemical reaction could happen with Breox oil.

The result is an aldehyde production that can be detected by smelling the cycle helium: it smells like alcohol. Aldehyde is not adsorbed into the oil vapour adsorber. It can condense on the cold part of the turbine shaft (turbine wheel side) and bring trouble to the bearing system.

A trick When a mechanical part has been wetted with some Breox oil, it reacts by shining when illuminated by UV light!

11.6.2 The Bulk Oil Separator (BOS)

If the BOS has no coalescing cartridges incorporated, its separation efficiency cannot vary with time. Such a check has normally been performed at the time of the commissioning tests. However, if the BOS incorporates cartridges, one must check that they operate correctly. Such measurement is explained in Sect. 6.3.8.5.2 (Fig. 11.17).

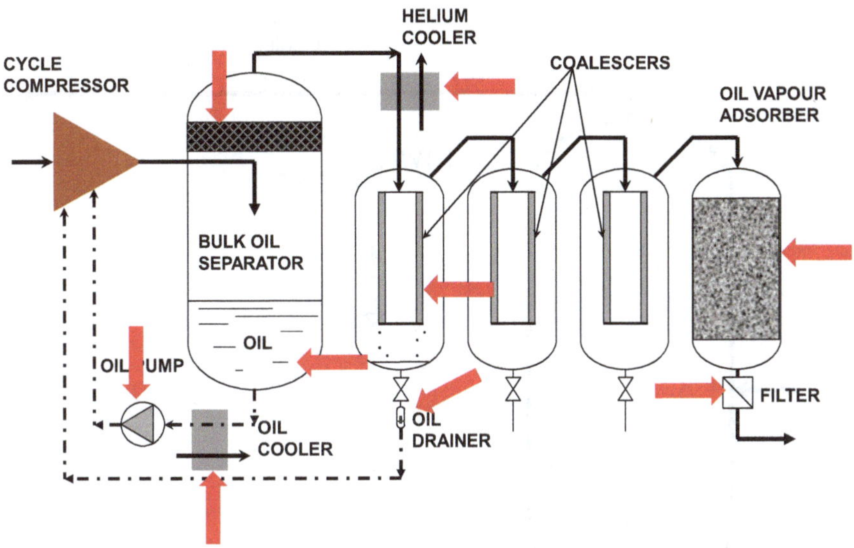

Fig. 11.17 The oil removal system. Arrows show components to be checked

11.6.3 The Oil Pump

As for the compressor, the shaft seal is the delicate part of the machine.

When an oil pump is stopped for a long time, it is possible that the vibrations induced by neighbour machines have destroyed the oil film that prevents any contact between the sliding parts of the seal. If the pump is started (that could happen automatically), chances are big that the seal is destroyed during the very first rotations. It is therefore important to rotate the pump for 1 hour every few weeks. (This can be implemented into the process control system!)

11.6.4 The Coolers

11.6.4.1 Balancing Water Flows in the Cooling Water Circuit

When the cooling water circulates in series into the coolers, there is obviously no balance to be performed.

However, in a parallel arrangement, as shown in Fig. 11.18, balancing is of high importance because most of the water has to circulate into the oil cooler.

The cooling water flow balancing procedure is:

- Have both valves V_h and V_o full open.

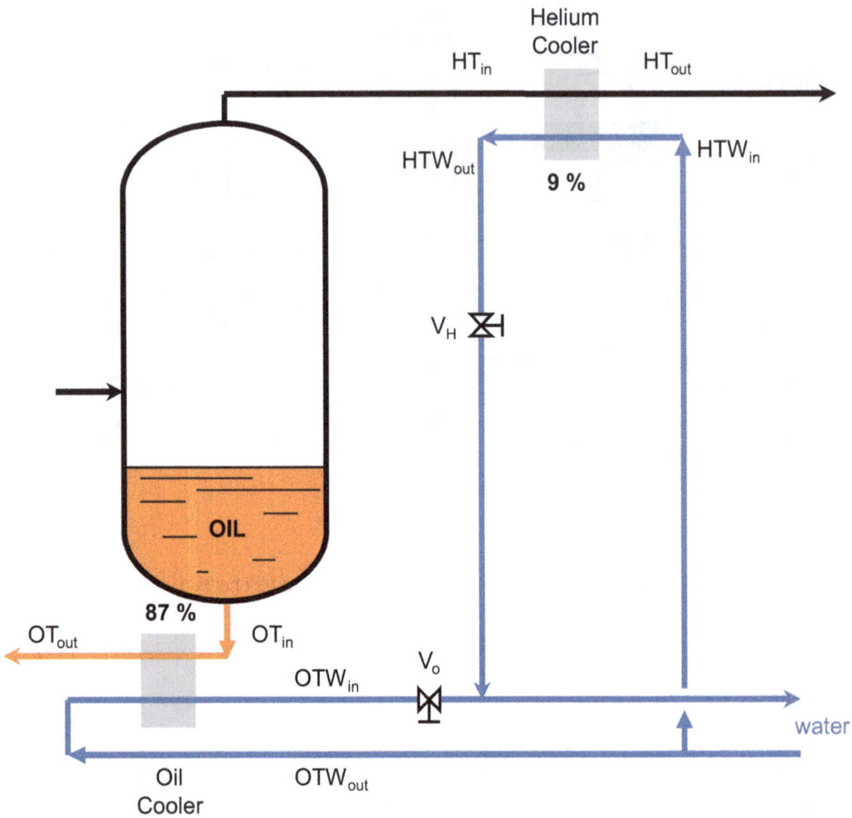

Fig. 11.18 Cooling water circuits in parallel

- Check the oil injection temperature. There are chances that it is too high because too much water circulates into the helium cooler.
- Control and set the oil injection temperature at the value that is indicated by the compressor supplier (around 45 °C) by closing the water valve V_h to the helium cooler.

11.6.5 Cooler Performance Check

It is a good advice to monitor the oil and helium cooler operation by filling periodically a table as in Fig. 11.19.

A periodical calculation (e.g. once a month) of the heat exchanger LMTD (remember Sect. 2.4.1.3) is suggested. The LMTD plot gives information to detect such a damage as shown in Fig. 11.20 or in Fig. 11.21, right. Of course, such a check can also be automatically performed by the control system.

Date		01/01/2011	01/02/2011	01/03/2011	01/04/2011	01/05/2011	01/06/2011	01/07/2011	01/08/2011
OIL COOLER									
Water flow	m^3/h	27	27	27	27	27	27	27	27
Water flow	kg/s	7.50	7.50	7.50	7.50	7.50	7.50	7.50	7.50
Water inlet temp	(°C)	25.00	25.25	25.50	25.75	26.00	26.25	26.50	26.75
Water outlet temp	(°C)	31.00	31.00	31.00	31.00	31.00	31.00	31.00	31.00
Oil inlet temp	(°C)	66.80	66.80	66.80	66.80	66.80	66.80	66.80	66.80
Oil outlet temp	(°C)	45.00	45.50	46.00	46.50	47.00	47.50	48.00	48.50
Warm DT	(K)	35.80	35.80	35.80	35.80	35.80	35.80	35.80	35.80
Cold DT	(K)	20.00	20.25	20.50	20.75	21.00	21.25	21.50	21.75
LMDT	(K)	27.14	27.29	27.44	27.59	27.75	27.90	28.04	28.19
HELIUM COOLER									
Water flow	m^3/h	2.00	2.00	2.00	2.00	2.00	2.00	2.00	2.00
Water flow	kg/s	0.56	0.56	0.56	0.56	0.56	0.56	0.56	0.56
Water inlet temp	(°C)	25.00	25.25	25.50	25.75	26.00	26.25	26.50	26.75
Water outlet temp	(°C)	33.00	33.00	33.00	33.00	33.00	33.00	33.00	33.00
He inlet temp	(°C)	66.80	66.80	66.80	66.80	66.80	66.80	66.80	66.80
He outlet temp	(°C)	27.00	27.50	28.00	28.50	29.00	29.50	30.00	30.50
Warm DT	(K)	33.80	33.80	33.80	33.80	33.80	33.80	33.80	33.80
Cold DT	(K)	2.00	2.25	2.50	2.75	3.00	3.25	3.50	3.75
LMDT	(K)	11.25	11.64	12.02	12.38	12.72	13.05	13.36	13.67

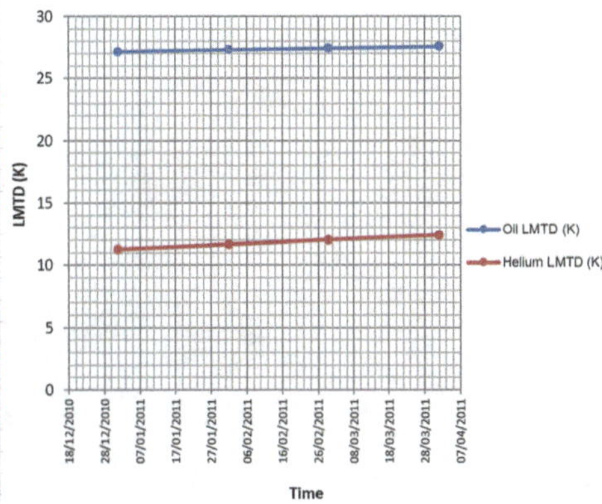

Fig. 11.19 Monitoring the oil and helium cooler operation

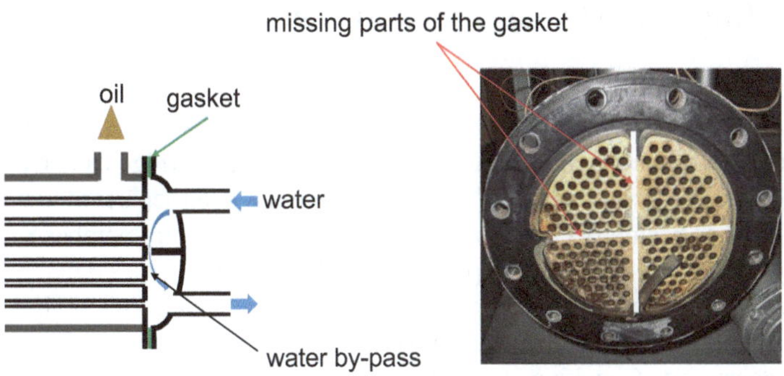

Fig. 11.20 Deteriorated of a parting gasket (white parts have disappeared!)

11.6.5.1 Cooling Water Quality

Dealing with cooling water is often considered as trivial and, therefore, is somewhat neglected. If the water is corrosive, consequences are generally heavy because water gets into the oil or helium circuits and, later, into the cold box (see Fig. 11.21, right). Remember that such leakages are independent of the circuit pressures (Sect. 9.3.1). If water is depositing scale, the thermal performance of the cooler degrades, and the cooled oil or helium temperature increases (see Fig. 11.21, left).

A correct water treatment is of *top importance to insure a reliable operation.*

A good cooling water treatment results in the trade-off between:

- A corrosive water
- A scale depositing water

On top of that, cooling water must be algae free.

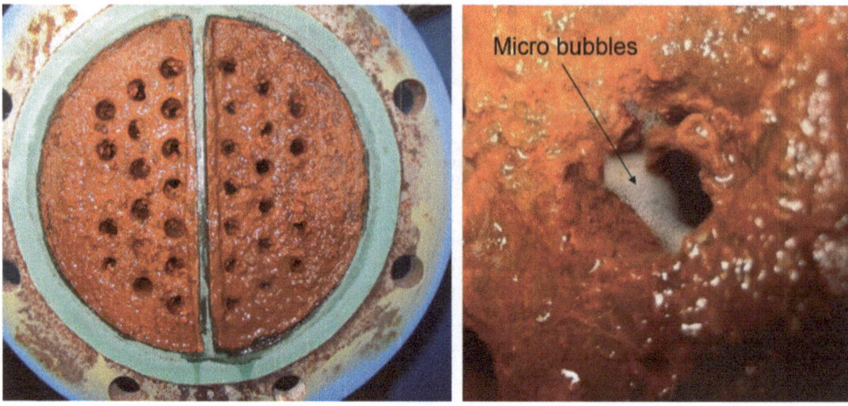

Fig. 11.21 A consequence of bad cooling water quality!

11.6.6 The Final Oil Separation System

11.6.6.1 The Coalescers

The calculation of the oil concentration at the inlet of the cartridge can be performed for any coalescer by comparing the oil quantity that has been separated to the helium quantity that has been processed during the same time period (remember Sect. 6.3.8.5.2). Such calculation can also be performed automatically!

A leak in the seal of a coalescer cartridge connection (see Fig. 11.22) is an aerosol by-pass. Therefore, some aerosols are not collected. The efficiency of the coalescer is partially destroyed. Checking carefully the correct connection of each coalescer cartridge is very important.

Question

Is it necessary to replace periodically the coalescer cartridges?

As there are no solid particles carried by helium, there is no risk of plugging the cartridge medium, therefore a long lifetime (>100,000 h) can be expected and has been experienced.

However, there are two risks related to such cartridges:

- Loss of leak tightness at the connection seal.
- Some wrong operation, generally manual, might induce a very high instantaneous reverse flow that can collapse one or more cartridges (see Fig. 11.23).

The wise behaviour that is advised is:

- Check periodically the efficiency of the coalescers.
- ONLY if an abnormal situation happens (either the oil quantity collected decreases, meaning that the collection efficiency of the cartridge decreases, leak

Fig. 11.22 Be careful with the O-ring! (image supplied by Parker Hannifin)

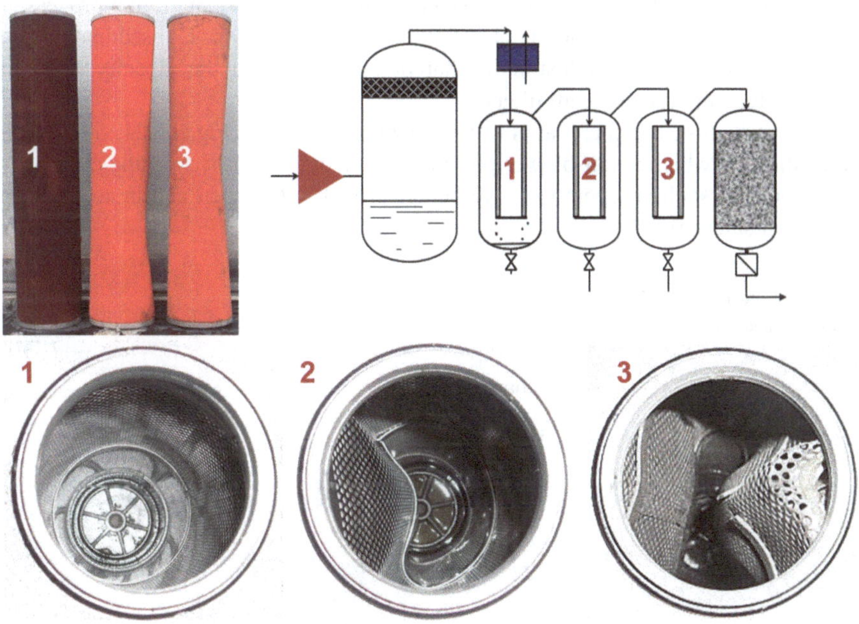

Fig. 11.23 Coalescer cartridges that have been collapsed by a high reverse flow

at the seal or collapsed cartridge, or increases, the upstream cartridge collection efficiency has decreased), check the status of the cartridges and the correct location of the seals. Sometimes, one can consider that it is an opportunity to replace the cartridge when it is a cheap component. Be sure to replace a cartridge with same type (compactness)!

11.6.6.2 The Oil Vapour Absorber

The lifetime of the adsorbent depends on the quantity of impurities that have been released by the oil load (or the quality of oil treatment if it was new oil). As it is difficult to know which impurities have been trapped, a conventional regeneration process that is supposed to release all trapped impurities cannot be performed because it might happen that not all impurities are released.

Therefore, it is advised to periodically replace the adsorbent:

- After the first 8000 operating hours with a *new oil load*
- Then, every 20,000 operating hours

11.6.6.2.1 Replacement of the Activated Charcoal

Dust Removal

Even if the activated charcoal has been clean at the end of the elaborating process, it has been subjected to vibrations at the time of transportation. Consequently, one must check that there is no dust in the charcoal, prior to filling the capacity (if not, it quickly plugs the downstream filter).

If there is dust, the charcoal is to be blown as shown on Fig. 11.24: on a mesh plate, through which an air blower provides a soft swift, pour slowly the charcoal.

Obviously, such an operation is to be performed in the open and in accordance with local safety rules.

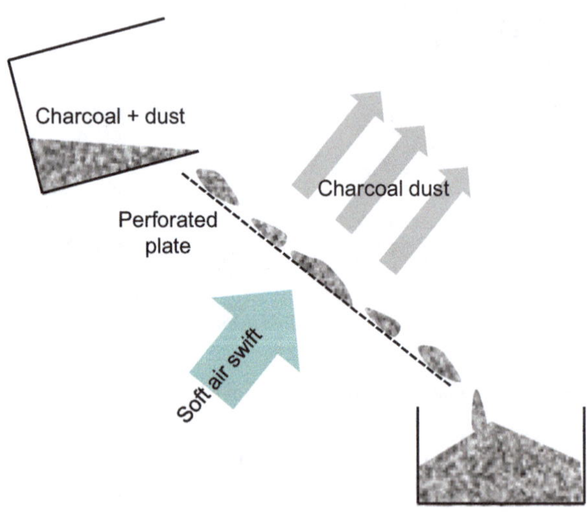

Fig. 11.24 Removing dust from activated charcoal

Filling the Capacity

Charcoal that is dropped into the capacity might again generate dust. Pouring must be done gently. A slight vibrating phase would ensure that the charcoal load is at maximum.

Water Removal

A common-sense remark: activated charcoal that is generally stored in contact with atmospheric air (or in a container that is not leak tight) does its job; it adsorbs atmospheric moisture. Therefore, it is *saturated* with moisture at room temperature conditions (can trap up to 0.005 g/g at atmospheric pressure and room temperature!). When such saturated activated charcoal is flown by dry helium during the cold box operation, it releases the water it had formerly adsorbed. Obviously, such water is sent *into the cold box*. Consequently, activated charcoal must be *water regenerated* prior to re-starting the plant!

The procedure is flushing with *dry and hot* air or nitrogen at >100 °C. See Fig. 11.25.

Attention! Regeneration is correctly finished when *both*:

- The discharge temperature is >100 °C.
- And the moisture level is <5 ppM V, provided that the measurement is made when only a small flow of nitrogen is being flushed!

Obviously, re-condition the circuit by pumping and filling with pure helium (see Sect. 11.4.1) before allowing helium in!

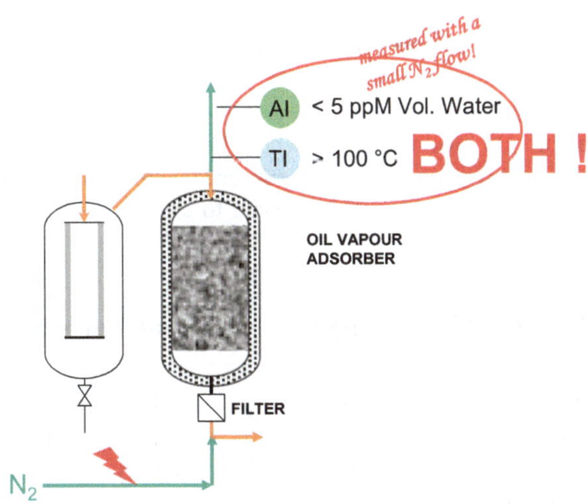

Fig. 11.25 In situ regeneration of the activated charcoal

The Filter Located Downstream the Adsorber

The cartridge of the filter that is located downstream the activated charcoal capacity has to be replaced when the pressure drop in operation, at nominal conditions (mass flow rate and pressure), increases.

11.7 The Cold Box

What Can Happen?

Impurities (water, air) can be trapped on the heat exchangers.
Helium for shield cooling of the dewar might get out at too low a temperature.
In liquid nitrogen pre-cooled systems, a leak might occur between nitrogen and helium circuits.

What to Check During Operation?

Temperature of helium that is released at the neck of the stationary dewar that is connected to the cold box
Turbine helium coolers

Which Maintenance Action?

Set correctly the neck flow
Periodic replacement of the adsorbent
Periodic maintenance of the insulation vacuum system

11.7.1 The Cycle Heat Exchangers

Deposition of impurities brought by cycle helium on the exchange surfaces: loss of performances. In most of cases, impurities come from air (nitrogen, oxygen, water). The usual regeneration procedure takes care of such trouble. But impurity could also be oil. Oil cleaning of the whole HP circuits is mandatory: it is a very heavy job, obviously not part of a "conventional" maintenance. Of course, one must identify the reason for bad oil separation prior to re-starting!

11.7.2 The Cryogenic Expansion Turbine

During normal operation of a gas bearing turbine, there is no contact (friction) between rotating and static parts. Therefore, in such conditions, the lifetime of a turbine has no limitation. Trouble generally comes from impurities sitting in the cycle gas.

- Replacing the cycle and bearing system filter cartridge as indicated in the operating manual

- Checking the actual setting of the trip levels according to the manufacturer's documentation
- Tuning of the labyrinth leakage for static gas bearing turbines (see Sect. 6.6.3.2.1) (Fig. 6.118))
- Efficiency optimisation (rotational speed tuning) (see Sect. 6.6.3.4)

11.7.3 The Cold Adsorbers

Replacing the adsorbent is a rare action.

11.7.4 The Insulation Vacuum System

Pump maintenance
Periodic oil change for vane primary pumps and oil diffusion secondary pumps
Periodic gauge check (primary and secondary)
Performance check (lowest vacuum value that can be reached at room temperature)

11.8 The Non-specific Components

11.8.1 Pressure Vessels

According to each country regulations, periodic operations such as pressure tests are to be performed on pressure vessels. Such a pressure test is generally performed with water, but one knows that all helium circuits MUST be perfectly clean, and especially DRY. Therefore, special procedures have to be implemented in order to reach this "ALL DRY" target. Most of them deal with drying by heating and pumping after a water test pressure has been performed. However, do not forget that heating a large component as a buffer capacity might be difficult, especially in winter. Liquids other than water can be used, but one has always to figure out how to behave in order to be totally rid of them before cooling down again!

A special remark for plants operating with static gas bearing expansion turbines: be careful with the safety bearing capacities; do not forget to dry them!

All pressure vessels must be recorded into a list that is updated after each maintenance operation.

11.8.2 Instruments

11.8.2.1 Pressure Sensor

Each sensor must be periodically calibrated against a reference pressure indicator.

In order to avoid disconnecting pressure sensors for each calibration operation (time consuming and leak generating), it is of interest to have pressure sensors connected as indicated in Fig. 11.26. Such an arrangement that is standard at CERN, Switzerland, allows an easy calibration without any sensor disconnection.

Calibrating valves are generally part of a special manifold that is supplied by the pressure sensor maker (see Fig. 11.26, right).

Such a procedure can also be implemented for differential pressure sensors as for level or flow measurements.

11.8.2.2 Temperature Sensor

Non-accessible cryogenic temperature sensors are generally doubled.

11.8.3 Control Valves

11.8.3.1 Cryogenic Valve

Remember the structure of a cryogenic valve in Fig. 11.27.

A cryogenic valve has, generally, a double barrier: a bellows and a sliding seal, very often an O-ring (see Fig. 11.27, right). The volume in between the bellows and the seal is connected to atmosphere by a threaded orifice, plugged by a screw. Such an orifice is used to check the leak tightness of the stem. If helium is detected (or if there is some pressure), it means that the bellows has been deteriorated. To check if

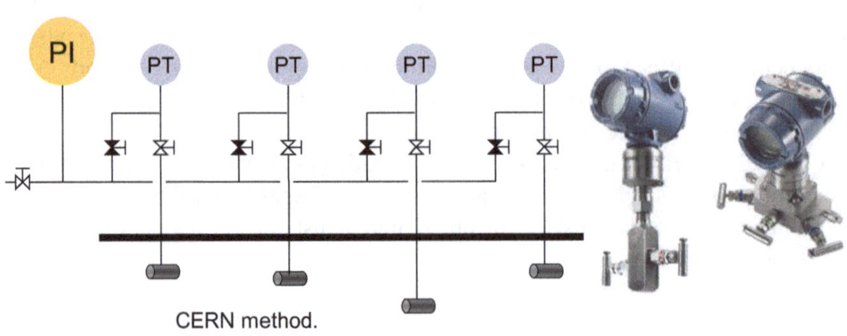

Fig. 11.26 On-site calibration principle, left, pressure sensors with calibration manifold, right

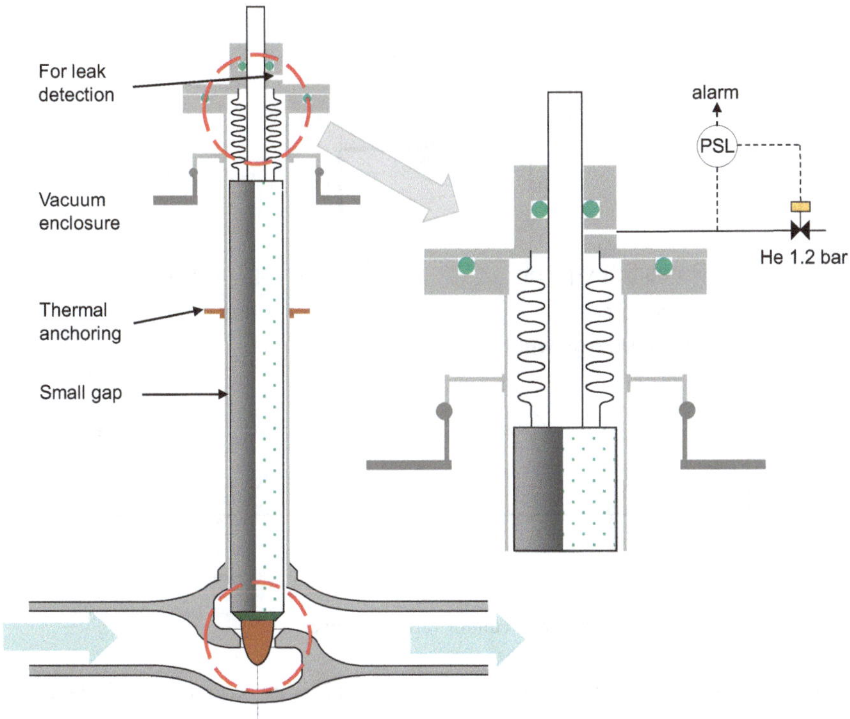

Fig. 11.27 A cryogenic valve and its stem leak tightness system

the sliding seal is leak tight, helium is to be blown into the volume in between the bellows and the seal. Helium detection is to be made on the atmosphere side.

In case the circuit is operating at sub-atmospheric pressure, this double barrier is used differently: helium at a slightly over atmospheric pressure is blown into the detection volume. If there is a leak through the bellows, the pressure decreases; the PSL issues a signal that, both, opens the solenoid valve to re-set the pressure and sends an alarm to the operator (remember Sect. 6.9.3.4). Such an arrangement is also used for pressure sensors and safety valves that are housed in a special enclosure.

11.8.3.2 The Valve Positioner

It is of high importance that the positioner of a control valve is correctly calibrated. Incorrect calibration can lead to auto-perturbations that are the consequence of a too late actual valve action compared to the signal issued by the control system. In some case, the accuracy of the controlled parameter can be very bad.

Let us have a glance at the two valves that control either the low or high pressure of a cycle shown in (see Sect. 8.3.).

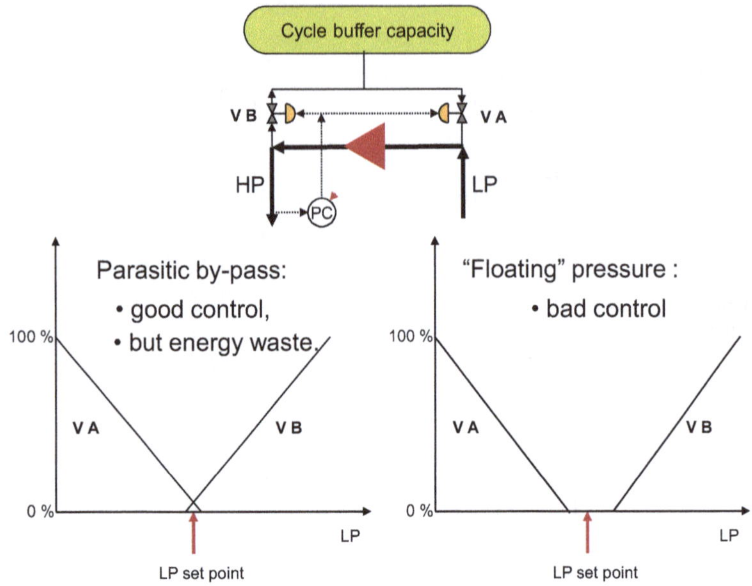

Fig. 11.28 Possible bad calibration of control valve positioner

If load valve A and/or unload valve B positioner is not correctly calibrated (see Fig. 11.28):

- Left: there is a permanent helium by-pass (energy waste) but the pressure control is correct.
- Right: the pressure floats between the two pressures where the valve actually starts to open.

Therefore, each control valve positioner must be periodically calibrated.
Some suppliers show calibrating procedures on the Internet.

11.8.4 Safety Valves

A periodic safety valve cracking pressure check is to be performed according to the local and applicable safety regulation.

Such periodic check is facilitated when there are two safety valves connected with a three-way hand valve: one valve can be disconnected during the time the other is calibrated (see Fig. 11.29, centre).

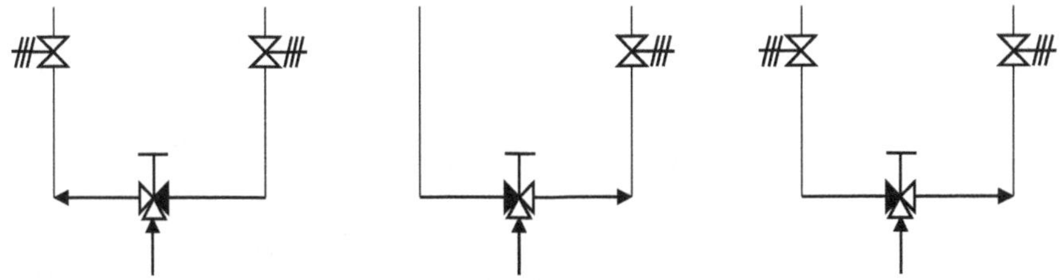

Fig. 11.29 Easy way for managing safety valves

11.8.5 The Oil and Helium Filters

They must be maintained as any filter: periodic change of the cartridge.

The helium filters can be plugged by activated charcoal dust, especially if the activated charcoal that has been newly loaded was not dust free.

11.8.6 Analysers

An analyser can be a very dangerous equipment if its measurement is a wrong one.

Examples:

- Pessimistic: a "high" indication for a "zero" value is just worrying.
- Optimistic: a "zero" indication for a "non-zero" value due to a loss of sensitivity, happens sometimes. This is very dangerous.

The analysers must be periodically returned to the manufacturer's, but they also have to be on-site re-calibrated against calibration gas mixtures (see Fig. 11.30).

Remember that there is *no calibrated mixture for moisture*!

11.9 The Refrigerator Performance Check-Up After Maintenance

It is of interest to perform a periodic refrigerator performance test after a heavy maintenance operation to check if the system performance has been modified in time. The results of such a test are to be compared with the values that have been recorded during the commissioning tests.

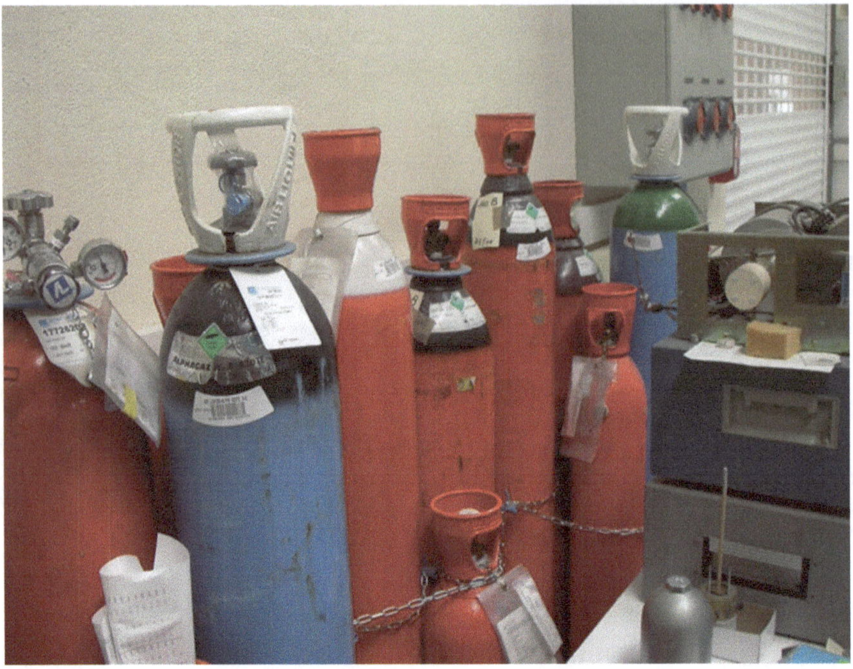

Fig. 11.30 Calibrated gas cylinders

11.9.1 Cycle Compressor(s) Performance

Operate the compressor at *nominal conditions*.

Remark It is possible to meet all operating conditions but the suction temperature that is generally related to the cooling water temperature (or air). A correction must be applied to the processed mass flow rate (see Sect. 6.2.1).

11.9.2 Whole System Performance Test

Operate the system at *nominal conditions*.

The above remark about the compressor suction temperature applies too.

- The heat exchanger warm end temperature difference is to be checked.
- The turbine discharge temperature is to be optimised (see Sect. 8.4.2.2).
- The turbine labyrinth leak is to be tuned (see Sect. 6.6.3.2.1).
- Compare the test results to the initial values (see Sect. 14.1).

Chapter 12
Examples of Various Plants

Abstract In Chaps. 3 and 4, the theory of thermodynamic cycles has been developed. In Chap. 5, the connection of these cycles to the object to be cooled was described. In this chapter, a few plants that are representative of their class are shown. The thermodynamic cycles of some of them are commented.

12.1 Introduction

This chapter gives information on some refrigeration systems that operate or have been operated in the world. A sample selection has been made in order to make possible comparisons with a wide range of both cryogenic powers and operating temperature wise.

12.2 An Industrial Turbo Brayton Refrigerator

Re-liquefaction of methane boil-off on large tankers is a solution that has become more and more popular. Similarly, applications involving high-temperature superconductors as power transportation lines seem promising.

Another application could be a pre-cooling stage for a refrigerator operating at colder temperatures as suggested in Fig. 4.31.

The turbo Brayton refrigerator that has been introduced in Sect. 4.6.1.1 that can fulfil these duties is shown in Fig. 12.1. Its performances are displayed in Fig. 12.1, bottom, right.

G. Gistau Baguer, *Cryogenic Helium Refrigeration for Middle and Large Powers*,
International Cryogenics Monograph Series,
https://doi.org/10.1007/978-3-030-51677-2_12

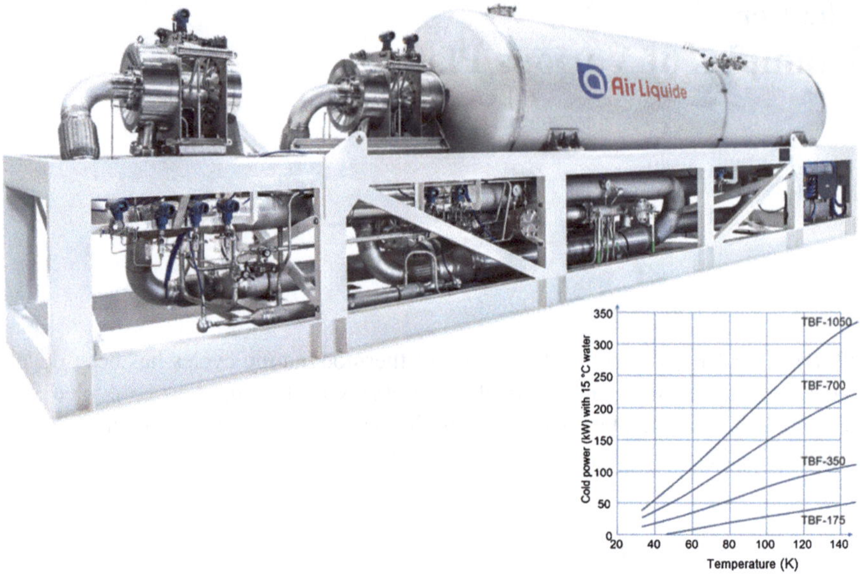

Fig. 12.1 Turbo Brayton refrigerator TBF 350, 50 kW at 77 K and its performance range (copyright L. Lelong, Air Liquide)

12.3 A 20 K Refrigerator

The main application of 20 K refrigerators is the cooling of cold neutron sources (remember Sect. 4.5.1) (Fig. 12.2).

12.4 A Middle-Size Refrigerator/Liquefier: HELIAL

It is a standard machine, one range being optimised for helium liquefaction from 15 to 300 L/h and the other range being optimised for helium refrigeration from 110 to 1000 W at 4.5 K (see Fig. 12.3). The cycle is a conventional Claude cycle with two turbines arranged in series and a JT valve (remember Sect. 3.5.3.3). These machines can be pre-cooled with liquid nitrogen according to the needs of the customer (remember Sect. 3.6.4).

Linde Kryotechnik AG (LKT), Switzerland, is building a similar machine.

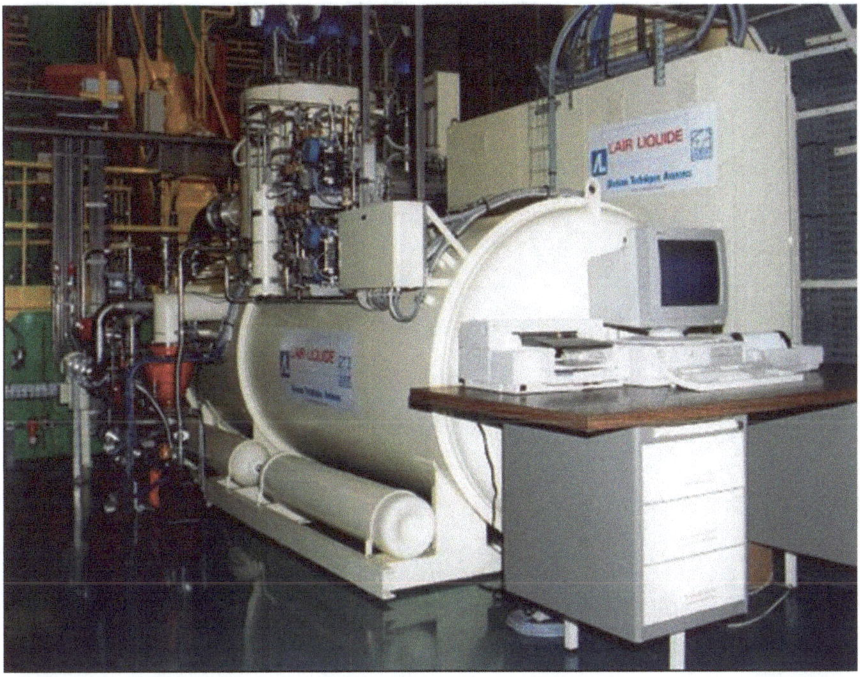

Fig. 12.2 The cold box of the ORPHÉE refrigerator, Saclay, France, 1900 W at 20 K (Courtesy ORPHÉE Air Liquide)

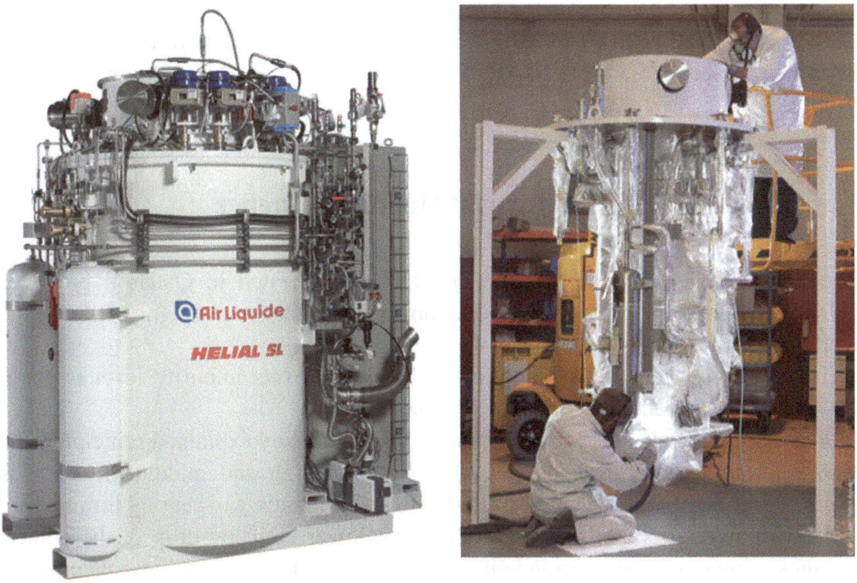

Fig. 12.3 A HELIAL 4.5 K liquefier/refrigerator. Cold box, left, and cold components, right (courtesy P. Avanian Air Liquide)

Fig. 12.4 A 1500 L/h hydrogen liquefier (courtesy Air Liquide)

12.5 A Middle-Size Hydrogen Liquefier

This hydrogen liquefier incorporates a two-turbine helium Brayton cycle as described in Sect. 4.5.3 (Fig. 12.4).

12.6 A Very Large Industrial Helium Liquefier

The largest industrial helium liquefier (2020) is operated in Qatar; it produces 7000 L/h. At the time it began production in 2013, that was roughly 20% of the world production.

Helium to be liquefied is supplied under pressure; its impurity level is about 100 ppM V of which about 70 ppM V of neon.

The selected cycle (see Fig. 12.5) makes use of already explained principles: the warmest turbines are arranged in series, between high and middle pressure, as explained in Sect. 4.2.1.4. One can notice that these static gas-bearing turbines are the most powerful: slightly above 200 kW. A lower enthalpy drop of the coldest turbines allows a parallel arrangement. The cold end is one of the conventional arrangements shown in Sect. 4.2.2.2).

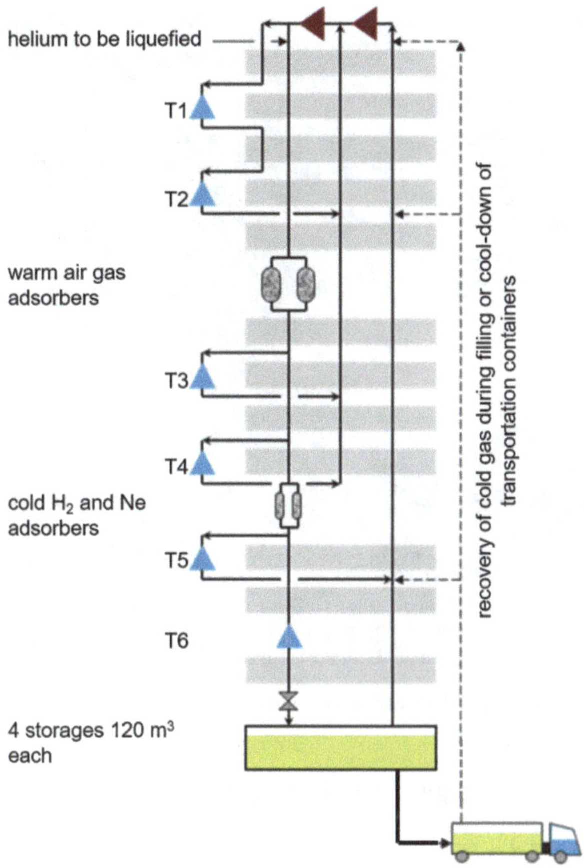

Fig. 12.5 A very large industrial liquefier (courtesy Air Liquide)

The final purification of helium is performed in two switchable "warm" adsorbers for air gases and two switchable "cold" adsorbers for neon and hydrogen.

The huge stationary liquid helium storage of 480 m^3 consists of four tanks that are at a higher level than the transportation containers in order to fill them by simple gravity, allowing a natural recovery of both molecules and enthalpy of the helium vapour that is displaced by liquid when getting into the container as shown in Fig. 5.3. The cold vapour is returned to the cold box at adequate temperature levels: one can see the two flexible thermally insulated pipes in Fig. 12.6. This arrangement allows also cooling containers in a similar way as heavy loads (remember Sect. 10. 3.5), with cool-down by-passes shown in Fig. 12.5.

Fig. 12.6 A filling station for containers

12.7 Large 4,5 K Refrigerators

12.7.1 The RHIC Refrigerator

The RHIC (Relativistic Heavy Ion Collider) is cooled by the refrigerator that had been built in the 1970s for ISABELLE, a project that was later modified in CBA (Colliding Beam Accelerator) and finally into RHIC, Brookhaven National Laboratory (BNL), USA [5]. It was initially planned to be operated at 25 kW at 2.5 K, but today, it is operated at almost half the power and at 4,5 K. At the time, it has been the largest helium refrigerator. It incorporated a train of oil lubricated screw compressors, plate fin brazed heat exchangers and oil-bearing expansion turbines (Fig. 12.7). The cold parts were housed into three separated horizontal cold boxes.

It is a conventional arrangement with four Brayton cycles grouped into two sets with turbines arranged in series, followed by a JT expansion. There is no liquid nitrogen pre-cooling (Fig. 12.7).

One could notice the way heat exchangers are arranged on the turbine flow between the turbines of the Brayton cycles. Is it because the designer was concerned by a three-fluid heat exchanger? Or because he was fearing oil breakthrough from the turbine bearings: a small two-stream heat exchanger is easier to clean than a large three-stream one?.

Fig. 12.7 The expanders of the RHIC refrigerator. In the foreground, the oil management system of the expansion turbines is painted red. (Courtesy Brookhaven National Laboratory)

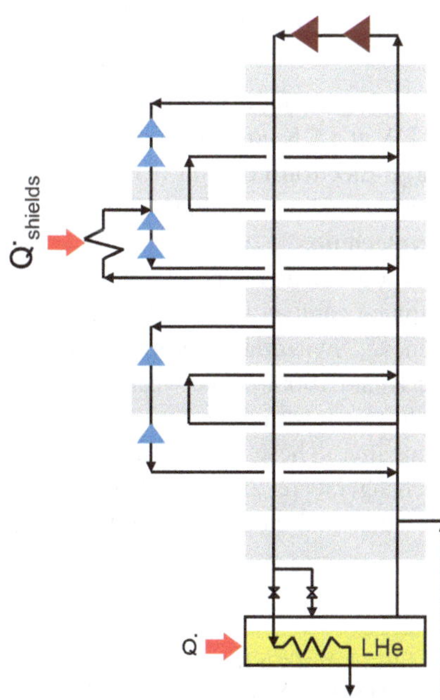

Fig. 12.8 The RHIC refrigerator

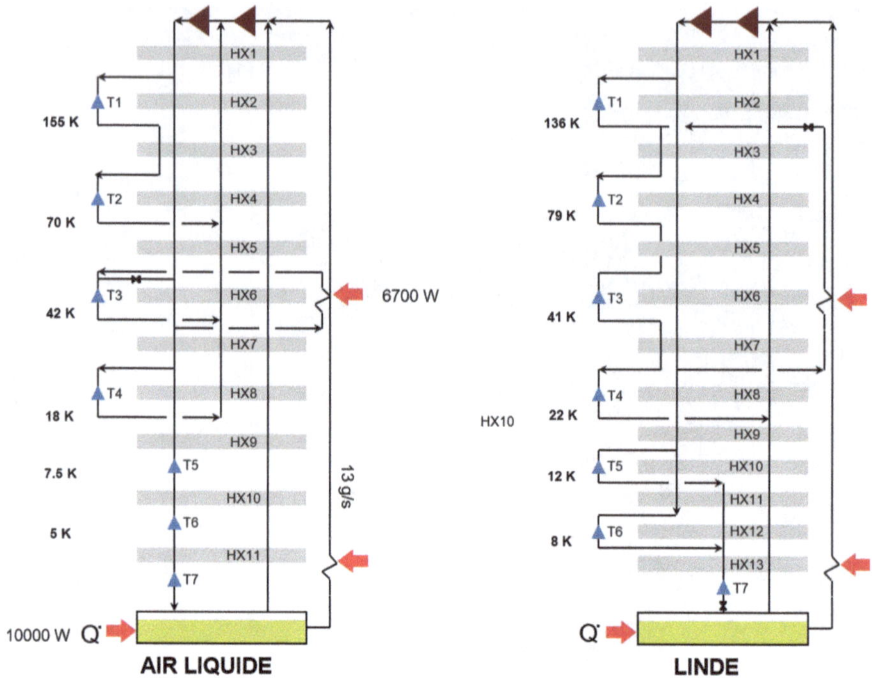

Fig. 12.9 The LEP refrigerator

12.7.2 *The LEP Refrigerators*

In the beginning of the 1990s, CERN issued a request for quotation for four large 12 kW at 4.5 K equivalent power (see Sect. 4.2.3) refrigerators to be used to cool the Large Electron Positron (LEP) superconducting resonant cavities.[1] Two solutions, which are shown in Fig. 12.9, were implemented both by Air Liquide and Linde Kryotechnik.

These refrigerators were used to cool down and keep at 4.5 K superconducting resonant cavities for the colliding ring of LEP (Large Electron Positron) at CERN, Genève, Switzerland. In the ultimate version, each of them absorbed 15,000 W at 4,5 K and 10000 W between 55 and 75 K and liquefied 20 g/s.

It is interesting to compare the structure of the cycles that were proposed by each company. These machines are almost pure refrigerators; therefore, the heat exchangers are almost balanced. The Air Liquide machine operates with three pressures and the Linde with two.

[1]A 12/18 kW at 4.5 K helium refrigerator for CERN's LEP superconducting accelerating cavities, G. M. Gistau, J. Veaux, Cryogenics1994 Vol 34 ICEC 15 proceedings supplement.

Starting from room temperature, the Air Liquide cycle incorporates two turbines arranged in series into a Brayton cycle operating from high to middle pressure (T1 and T2). The discharge temperature of T2 is related to the operation of a cold adsorber that is on the high-pressure circuit, between heat exchangers HX 4 and HX 5. A third Brayton cycle takes care of the thermal shield heat load as discussed in Fig. 3.69. One sees that turbine T3 is exactly operating at temperatures of the thermal shield, which is the optimum solution. The T3 discharge temperature is a constraint coming from the customer and related to the thermal shield operating temperatures. The fourth Brayton cycle operates down to 18 K which is another constraint given by CERN because all equipment at temperatures higher than 20 K is located in a cold box that sits at ground level. An adsorber is located on the high-pressure circuit, between heat exchangers HX 5 and HX 6. It protects the colder part of the cycle. The warm turbines, T1 and T2, are not sensitive to air impurities because they operate at temperatures where air does not solidify.

For temperatures lower than 20 K, the equipment sits into another cold box that is located at tunnel level, between 50 and 150 m lower, depending on the location. The two cold boxes are connected by cryogenic transfer lines located in shafts. Into the second cold box, three turbines are arranged in series with a heat exchanger in between each of them (remember Sect. 4.2.2.1). The coldest turbine T7 expands directly in two phases.

The Linde cycle is both similar and different. Down to 22 K, there are four Brayton cycles with four turbines arranged in series. The shields are cooled by high-pressure helium that is withdrawn between heat exchangers HX 7 and HX 8, circulated into the shield circuits and injected at the discharge side of T1. The adsorber that is located between heat exchangers HX 5 and HX 6 protects the components that are at lower temperature except turbines T3 and T4 that may have trouble with solid air. The cold end incorporates two Brayton cycles, and finally, a supercritical turbine processes the flow, in series with T5 and T6. A valve that controls T7 discharge pressure avoids operation of the turbine in two phases.

These refrigerators have been later modified in order to help cooling the LHC machine. One cold box is shown in Fig. 6.175.

12.7.3 A Special Distributed System: The Fermilab Tevatron

There has been a historical cryogenic large system partially operating in an economiser way for the Tevatron (Fermilab, Batavia, USA).[2] The Tevatron operation ended on September 2011. However, it is interesting to have a glance at this special structure. A central helium liquefier (CHL) with two cold boxes, the first one 3775 L/h and then a second one, later, 5220 L/h, that were pre-cooled with liquid

[2]Cryogenic system for the Tevatron, Geynisman M. and al. NIFS Symposium 96, Toki City, Japan, May 27–28, 1996.

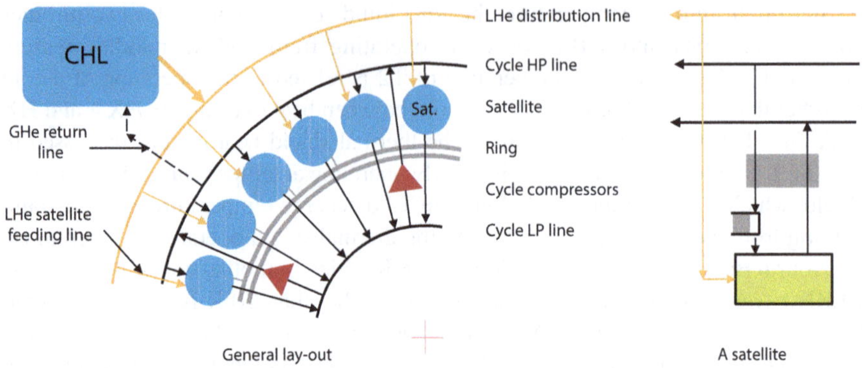

Fig. 12.10 The Fermilab Tevatron system layout, left; a satellite, right

nitrogen provided liquid helium to 24 one-kW "satellite" refrigerators located along the circular accelerator, as shown in Fig. 12.10, left. These liquid nitrogen pre-cooled refrigerators were fed by a high-pressure pipe and discharged helium into the low-pressure pipe running around the accelerator. Twenty-four decentralised located cycle compressors were processing helium from the LP to the HP line.

Each satellite refrigerator incorporated a two-phase reciprocating expander (Fig. 12.10, right). Liquid helium was fed at the cold end, making the refrigerator operating in an economiser regime (Remember Sect. 3.3.2.4.5). In a second step, cryogenic centrifugal compressors were added in order to operate at 3.56 K (not shown here).

12.7.4 ITER

In the beginning of the 1920s, ITER is the largest and may be the most complicated cryogenic system [6].

The ITER cryoplant provides a cooling power at 80 K, 50 K and 4.5 K for the thermal shields, the current leads, the superconducting magnets and the cryopumps.

It incorporates three liquid nitrogen pre-cooled horizontal cold boxes, each one able to absorb an equivalent power of 25 kW at 4.5 K and 600 kW at 80 K (see Fig. 12.11). They are fed by a nitrogen re-condenser (remember Sect. 3.6.7) of 1100 kW.

The thermal shields are cooled by two helium loops that are circulated by two room temperature centrifugal compressors of 2 kg/s each (remember Sect. 3.7.2.3).

The three cold boxes are connected to a horizontal distribution box where the cooling powers of the two 80 K loops and the three LHe cold boxes are combined and distributed to the various circuits towards the tokamak.

Fig. 12.11 The three ITER cold boxes and the distribution box at right (© ITER Organization)

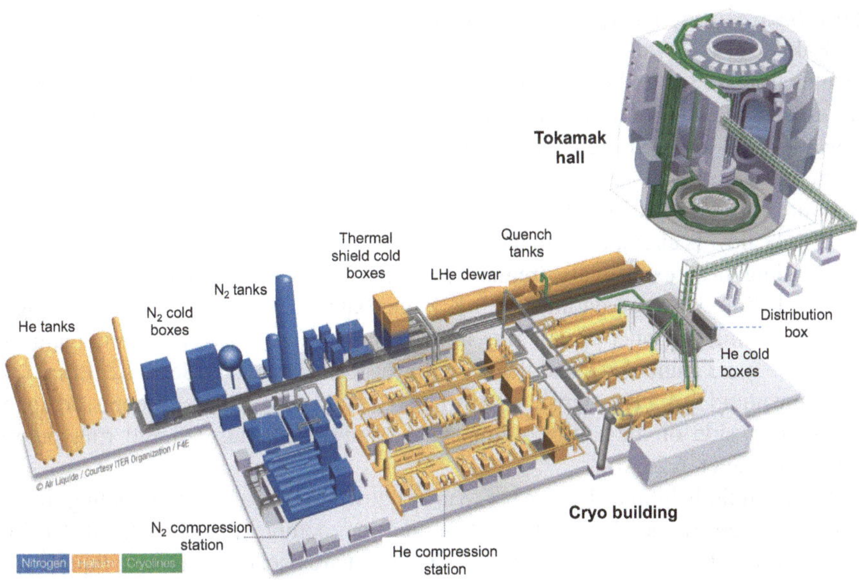

Fig. 12.12 General layout of the ITER cryogenic system (copyright Publicis Air Liquide, courtesy ITER Organisation)

Figure 12.12 shows the layout of the plant. It occupies a 120 × 46 m area. At full power, the whole cryogenic plant absorbs 35 MW.

Part of the heat rejected by the compressor coolers is recovered for building heating.

12.8 Large Refrigerators Operating at Less than 4.5 K

12.8.1 Tore Supra (WEST since 2013): The Very First Refrigerator with Cryogenic Compression

In the early 1980s, Euratom decided to build the Tore Supra fusion experiment [2]. The necessary cryogenic power was 300 W at 1.8 K. At this time, the existing refrigerators operating at 1.8 K were built in the "all room temperature" pumping system with root machines. Air Liquide, who was one of the potential suppliers, thought that if fusion was really a future solution, larger powers would soon be necessary. As explained in Sect. 4.3.1.2, the size of the root machines for the sub-atmospheric stream pumping station would then be unrealistic. It was necessary to go towards cryogenic compression (see Sect. 6.7). After development on own funding by Air Liquide, two prototypes were built and incorporated into the Tore Supra refrigerator. To reduce the number of cryogenic pumping stages (and therefore, the difficulty for this first trial), it was decided to go to a mixed pumping structure (see Fig. 12.13). At this time, the use of screw compressors operating at sub-atmospheric suction pressure was not common; therefore, liquid ring pumps (see Sect. 6.2.6.5) were incorporated for the first time into a 1.8 K refrigerator.

Baking of the torus vacuum chamber, which increases the load on the thermal shields from 11 to 30 kW, was to be performed once a week, for half a day. In order not to size the warm turbine and the corresponding components for the maximum power, liquid nitrogen pre-cooling was incorporated to punctually deliver such an extra refrigeration load. Two Brayton cycles operating between high and low pressure cool-down to 10.5 K.

The JT flow is circulated into the 18 thick casings that house the coils, using a set of dividing heat exchangers (remember Sect. 2.4.3.4). After circulating in one casing, helium is re-cooled into a 4.5 K liquid bath and, finally, expanded through a reciprocating expander discharging in two phases. Note that such an arrangement allows to spare a circulator and the related heat load (remember Sect. 4.2.4.2).

Some liquid at 4.5 K is expanded down to 0.8 bar, and vapour is compressed up to the suction side of the low-pressure cycle compressor with a liquid ring pump (remember Sect. 6.2.6.5).

Some liquid helium at 1.2 bar is expanded down to 0.016 bar after cooling into a heat exchanger (remember Sect. 4.3.1.3) and then compressed up to 0.090 bar by two centrifugal cryogenic compressors (CCC) that are backed by another liquid ring pump. The CCC are located in a box extension, on the side of the cold box as can be seen in Fig. 12.14, left.

The connection to the torus cryostat is explained in Sect. 5.3.4.2.1.

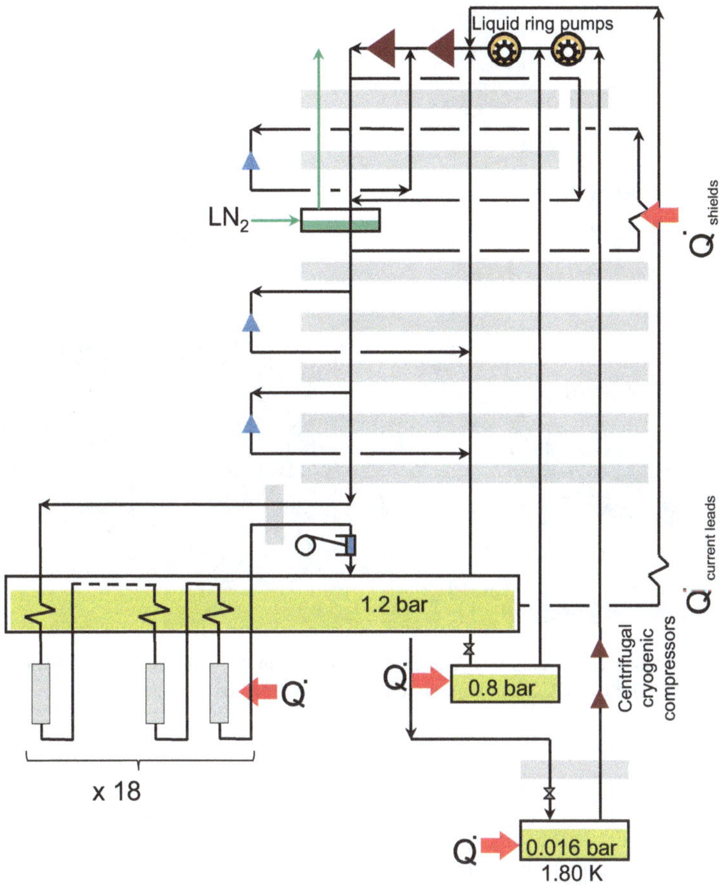

Fig. 12.13 The Tore Supra (now WEST) refrigerator, Cadarache, France (Air Liquide)

12.8.2 CEBAF (Continuous Electron Beam Accelerator Facility)

After Tore Supra showed that centrifugal cryogenic compression was a viable and good solution, when CEBAF, JLAB, Newport News, VA, USA, thought about the refrigerator they needed, it was decided to go for a full cryogenic compression [7]. Figure 12.15 shows the structure of this refrigerator capable of absorbing 5000 W at 2.00 K. Some supercritical helium at 4.50 K is expanded down to 0.032 bar. The full cryogenic compression is performed with four centrifugal compressors arranged in pure series, which process 250 g/s. The full cryogenic compression allows to have no part of helium circuits at subatmospheric pressure in air.

Fig. 12.14 The Tore Supra cold box. The CCC are housed in the white box extension bottom right of the cold box. The two CCC can be seen at right, where the box is open

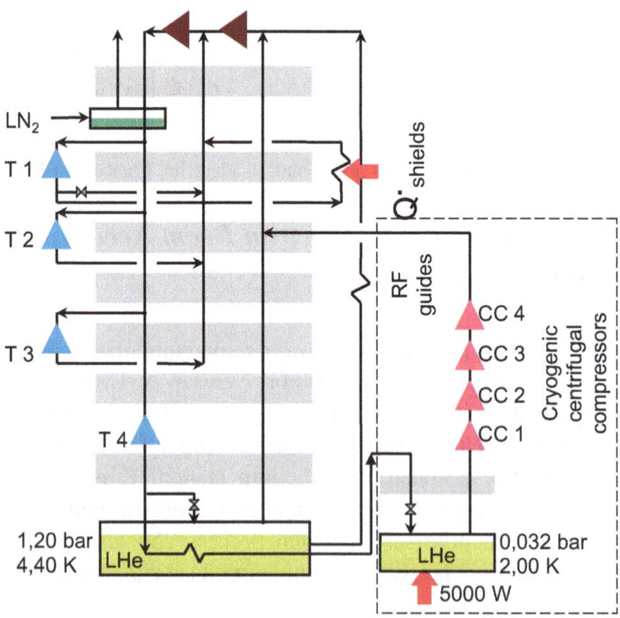

Fig. 12.15 The CEBAF refrigerator, Newport News, USA (CVI, Air Liquide)

Fig. 12.16 CEBAF refrigerator. In the foreground, the 2 K cold box where the CCC are housed, farther, part of the 4.5 K cold box

However, the control of the train of compressors happened to be delicate, particularly during the phase of decreasing pressure on the bath (see Sect. 6.7.3.4).

The reduction of the processed flow, at constant suction pressure, is limited by the phenomenon of surge, which is specific of centrifugal compressors (see Sect. 6.7.3.3.1). The addition of a fifth compression stage allows to reduce the flow in a ratio of two, at constant compression ratio, while staying comfortably far from the surge conditions (remember Sect. 6.7.3.4.3).

The "2.0 K" parts of the cycle (cryogenic compressors and JT heat exchanger) are located into a separate cold box shown in dotted lines (see Fig. 12.16). Note that this almost pure refrigerator, which has been designed in the USA, is liquid nitrogen pre-cooled!

12.8.3 The CERN LHC

A few years after the LEP refrigerators had been built, CERN issued another request for quotation for the Large Hadron Collider (LHC) refrigerators. Again, two solutions were proposed.

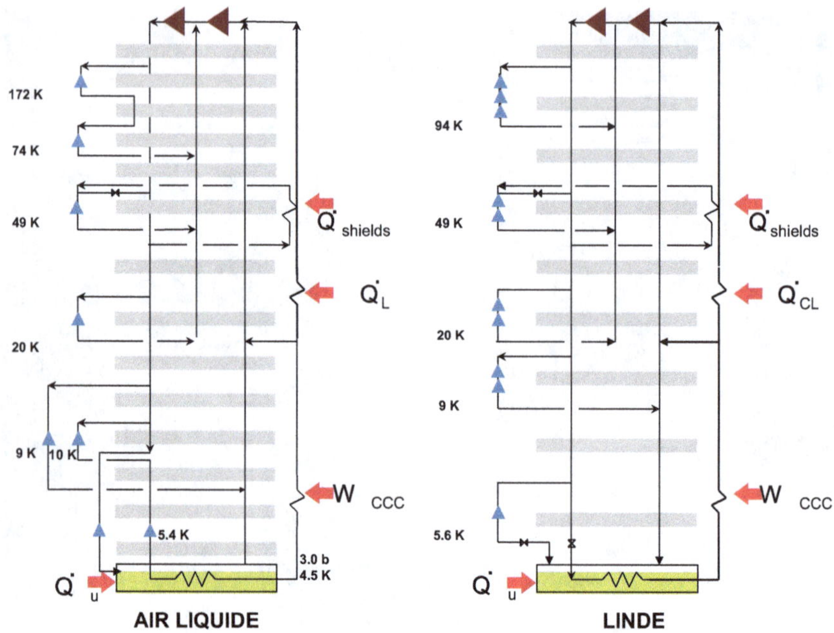

Fig. 12.17 Comparison of two cycles fulfilling the same duty for the CERN LHC

These refrigerators are used to cool and keep at 1.8 K one eighth of the coils of the colliding ring of the LHC (Large Hadron Collider) operating at CERN, Geneva, Switzerland. Each of them absorbs 2400 W at 1,8 K and 33000 W between 50 and 75 K and liquefies 41 g/s. See Sect. 12.8.3. Each 1.8 K refrigerator is a combination of a 4.5 K refrigerator and a 1.8 K cold box supplied with supercritical helium.

The 4.5 K Refrigerator

The cool-down of the very high coil mass (5000 tons per refrigerator), down to 100 K, is performed with liquid nitrogen (the special heat exchanger is not shown in Fig. 12.17, but it can be seen in Fig. 10.6), and then the cool-down phase is finalised with turbines.[3]

In the Air Liquide's designed refrigerator, turbine 8 expands directly into two phases.

The 4.5 K cold boxes are located at the ground level.

The 1.8 K Refrigerator

The mixed pumping solution is, today, the one which is the most used (see Sect. 4.3. 1.3.1) for the big machines. The room temperature compression stage, which is

[3]TECHNICAL NOTE LHC Cryogenic System: Architecture and process functional analysis LHC-CRYO-OP-N-109.

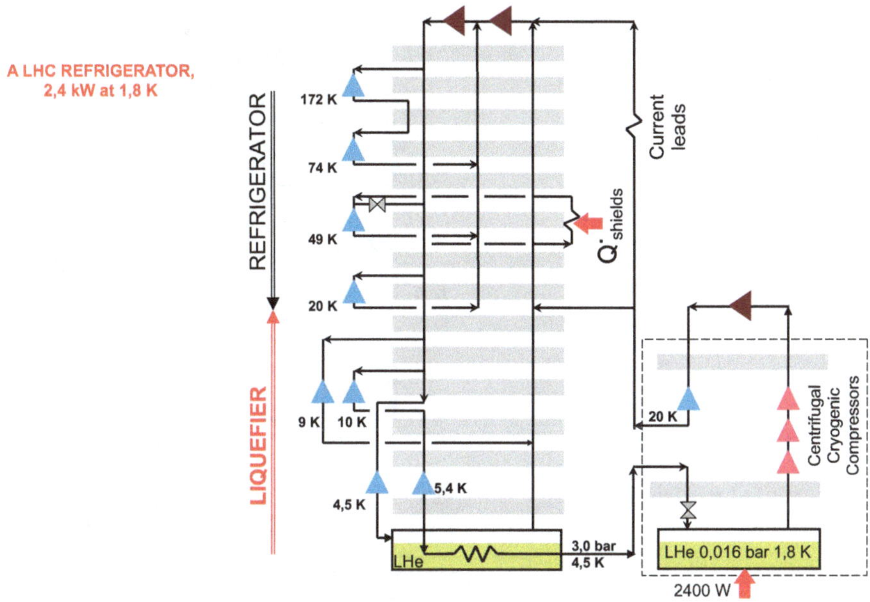

Fig. 12.18 A CERN LHC, Geneva, Switzerland, refrigerator (Air Liquide)

volumetric (screw compressors), brings a large flexibility of control (see Sect. 6.7.3. 4.3.2) and allows working safely on a larger flow range.

The compression station is shown in Fig. 6.174 and the cold box in Fig. 6.175.

Supercritical helium (3 bar, 4,5 K) is sent to the 1.8 K cold box where it is expanded down to 0,016 bar after having been cooled in the JT heat exchanger. The saturated vapour is compressed by three cryogenic centrifugal compressors, then, after having been warmed up to room temperature, by 2 oil room temperature oil lubricated screw compressors in parallel. Such flow is then cooled into the main heat exchanger and expanded into a turbine in order for it to return at 1.2 bar, 20 K to 4,5 K refrigerator cold box.

Remark The cold part of the cycle is operating in a quasi-pure liquefier, because most of the helium is returned at 20 K, although the warm part is operating almost as a pure refrigerator.

For reasons of layout, the sub-atmospheric system is housed in a separate cold box (in dotted lines in Fig. 12.18).

A photograph of the 4.5 K cold box is shown in Fig. 6.175.

The Linde solution incorporates four cryogenic centrifugal compressors, one compound two-stage oil lubricated screw compressors and one expansion turbine.

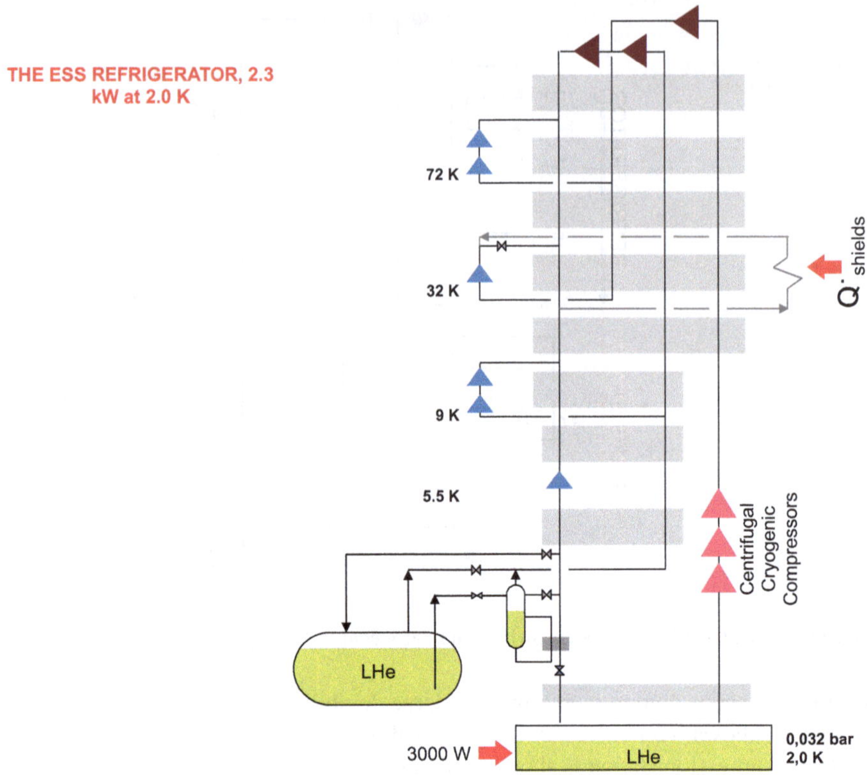

Fig. 12.19 The European Spallation Source (ESS) refrigerator, Lund, Sweden (Linde Kryotechnik AG)

12.8.4 The European Spallation Source

The European Spallation Source (ESS) generates neutrons by striking a target made out of tungsten with high-energy protons (2.5 GeV).[4] Neutrons are then cooled (slowed down) through liquid hydrogen as a moderator. Such a system needs cryogenic refrigeration for cooling the accelerator cavities at 2.0 K and at 20 K to re-condense hydrogen. Here, the refrigerator that provides 3.0 kW at 2.0 K plus 11.4 kW between 40 and 50 K for shields and 9.0 g/s LHe is described.

It is a very simple arrangement of three Brayton cycles, the second one taking care of the thermal shields, followed by a conventional supercritical turbine and a JT valve. There is no liquid nitrogen pre-cooling (Fig. 12.19).

[4]ESS accelerator cryoplant process design, X. L. Wang et al 2015 IOP Conf. Ser.: Mater. Sci. Eng. 101 012012.

Fig. 12.20 The ESS compression station (Linde Kryotechnik AG)

Fig. 12.21 The ESS cold box (Linde Kryotechnik AG)

The cryogenic compressors, which are housed in the cold box, pump on the liquid bath through the JT heat exchangers that are located in the cryostats and are backed by a room temperature oil lubricated screw compressor discharging into the middle pressure of the compression station (Figs. 12.20 and 12.21).

Part of the heat rejected by the compressor coolers is recovered for building heating.

Chapter 13
A Helium Plant (System) Technical Specification

Abstract This is not a recipe to write a correct full technical specification but, mainly, warnings on what not to forget. Only *technical issues* are considered.

13.1 The Structure of a Technical Specification

Here, the wording "possible supplier" means the company that makes an offer according to the technical specification.

The main points are:

- A presentation of the context in which the system will operate
- The scope of delivery
- The technical requirements
- The documentation
- The available utilities and local operating conditions
- The commissioning tests to be performed (see Chap. 14 where the most usual ones are described)
- Even it is not a technical issue, it is interesting to set a rule stating how the supplier selection will be made according to the relative incidences of CAPEX (capital expenditure) and OPEX (operational expenditure), provided the operational functions are fulfilled

13.2 Examples of Possible Important Specific Requirements

Such requirements should be adapted to the size and operating conditions of the system.

The expected performance at the system interfaces, measured as suggested in Chap. 14:

G. Gistau Baguer, *Cryogenic Helium Refrigeration for Middle and Large Powers*, International Cryogenics Monograph Series,
https://doi.org/10.1007/978-3-030-51677-2_13

- The cryogenic powers: heat load, outlet and return pressures and temperatures, for *all* requested operating regimes
- The power that is absorbed by machinery (at least compressors), for *all* requested operating regimes

Remark: When specifying the heat loads, the final user should not forget to include a safety margin (at same pressures and temperatures)!

The operating situations that the control system must be able to handle:

- Stability conditions during steady states (pressure and/or temperature stability criteria)
- Transient situations: in some situations where a parameter changes with time the rate of change should be accurately described (e.g. ± x % of heat load change per minute during a time period)
- The adsorber regeneration (warm and cold)
- The reconnection to the user system after any short or long disconnection time in case CCC are part of the supply
- Any other identified perturbations

The expected features at some locations in the system (maximum total helium leak, maximum oil leak).

The minimum time period during which the system is to be operated without any stop (utility breaks down are not considered).

Minimum MTBM and MTBF values.

The rules, regulations and standards to be fulfilled.

13.2.1 Oil Separation

Measurement of oil concentration at the discharge of the bulk oil separator and coalescers, preferably automatically. Oil concentration at the discharge of the bulk oil separator should be around 100 ppM W.

13.2.2 Instrumentation

13.2.2.1 Instrumentation for Measurements at the Interfaces of the System

At the limits of supply, implement the most accurate sensors or instrument measuring chain (pressure, temperature, flow) that will be used for the commissioning tests. The cryogenic temperature sensors should be interchangeable without breaking the insulation vacuum.

13.2.2.2 Instrumentation for Performance Measurements of Special Components

The correct operation of components as expansion turbines, cryogenic compressors and circulators is of utmost importance. Their performances, even if not guaranteed, must be measured in the most accurate way possible. In case of cryogenic compressors, it is advised to have two sensors located at same point, one (A) being operated from the control system and the other (B) connected up to the feed-through. The later (B) can be either read with an external equipment or switched if the one in operation (A) shows that it gives wrong readings.

Such instrumentation must comply with same rules as for interface measurements.

13.2.2.3 Standard Cryogenic Temperature Measurement

Each cryogenic temperature sensor must be duplicated with a second sensor located at same place and wired until the electrical connector located at room temperature.

13.2.2.4 Environment

For example:

- Maximum noise levels
- Oil retention systems in case of a leak
- Possibility to recover the heat that is released by the compression system
- Incitation to be environment compatible in the various phase: design, construction, erection, commissioning, operation, maintenance, material recycling possibility

13.3 Examples of Important Expected Information from the Possible Supplier

Such expected information should be adapted to the size and operating conditions of the system.

13.3.1 Cycle Compressors

For each compressor: expected efficiency measured in the conditions stated in the technical specification.

13.3.2 Expansion Turbines

For each turbine:

- Expected and guaranteed efficiency for each operating regime
- U_1/C_0 curve
- MTBM, MTBF

13.3.3 Cryogenic Compressors and/or Circulators

For each cryogenic compressor or circulator:

- MTBM, MTBF
- A confirmation that the bearing system complies with the operating conditions of the machine, especially MTBM, MTBF, cool-down (remember Sect. 6.5.1. 2.2.3)
- The expected efficiency for each of the requested regimes
- A compressor field (compression ratio versus reduced flow) with location of the operating points for the requested regimes
- Measurement of at least 5 operating points in order to identify the surge line on the compressor field

13.3.4 Test Equipment

A description of the commissioning device(s) (test equipment: cryostat or heater) that is (are) to be supplied and their location (inside the cold box or not). See Sects. 14.5.5 and 14.5.6 for examples of test equipment.

13.3.5 Liquid Helium Storage

Guaranteed helium mass flow that is released by the dewar according to the volume of liquid (see Sect. 14.6.2.2).

13.3.6 Margins Taken by the Possible Supplier on the Absorbed and Cryogenic Powers

The possible supplier of the cryogenic system has to give the margins on the absorbed power and the mass flow rate of each cycle compressor. Similarly, he has to give the heat load margins at same pressure and temperatures that have been used for the process calculations.

13.4 Interfaces Requirements and Expected Information

This is a possibility to display the mains requirements and expected information, as shown in Fig. 13.1.

Shields

Requirements: the power, the conductance of the circuit in order that the pressure drop can be calculated according to the mass flow rate (an equivalent circuit can be defined as the inner diameter, the length and the rugosity of a straight pipe) and the maximum return temperature and inlet pressure if any

Expected information: pressures and temperatures at interfaces and mass flow rate

Liquefaction

Requirements: the mass flow rate and the maximum pressure

Expected information: pressure

Isothermal-Duty Refrigeration

Requirements: the power, the temperature, the maximum return temperature and the non-isothermal power if any (transfer line)

Expected information: the mass flow rate and the outlet temperature

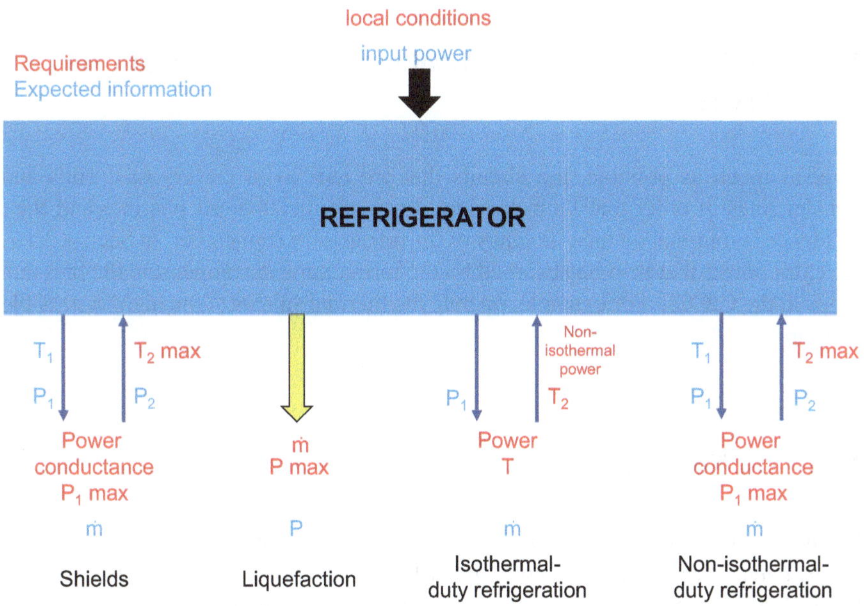

Fig. 13.1 A simple way to present general requirements (in blue) and expectations (in red)

Non-isothermal-Duty Refrigeration

Requirements: the power, the conductance of the circuit in order that the pressure drop can be calculated according to the mass flow rate (or the return pressure) and the maximum return temperature and inlet pressure if any

Expected information: pressures and temperatures at interfaces and mass flow rate

Requirements and expected information are to be given for *each operating regime*.

Decision-Making Related to CAPEX OPEX

A very simple but anyway very effective rule is to consider the minimum of the sum CAPEX + OPEX. One has to set:

- For the CAPEX, an accurate limit of supply.
- For the OPEX, the time duration (h), the actual cost of the electrical power (money unit/kWh). The actual cost of electrical power should take in consideration not only the cost of electrical power, but also the distribution losses and the amortisation of the electrical distribution equipment. Same consideration has to be made regarding the cooling water and the liquid nitrogen if any.

More sophisticated (and complicated!) rules can be set, according to the specificities of the system.

13.5 Do Not Specify a Plant Looking Only at the Final User Needs

Size, as much as possible, the circuits that are located in the cryostat, coils and transfer lines in order that their pressure drops are the smallest possible and their incidence is minimal on the efficiency of the refrigerator (remember, in Sect. 4.2.4.2, the extra power that is to be absorbed by the refrigerator to compensate the pressure drops in the CICC!). For example, operate the thermal shields at the highest possible pressure.

13.6 Do Not Over-Specify

For helium leaks in room temperature systems as a compressor station, specify a *realistic leak rate* in Sm^3/day, not in mbar.L/s. Evaluation of leaks in mbar.L/s is to be used only for cryogenic components operating in vacuum!

In a general way, leave some freeness to the possible supplier who might propose a smart original solution, especially for cycles.

In a general way, do not hesitate to make contact with possible suppliers to check if there is any possible over specification prior to issuing the technical specification.

A remark of the ISO standard:

ISO standards are of interest on large systems. ISO 10440 (similar to API 619) deals with oil-lubricated screw compressors, and ISO 10438 (API 614) deals with oil systems. It is important that the person who is in charge of the redaction of the specification evaluates which standard is relevant for the project (interest/extra cost). One should not apply the whole API standards without making such a selection: it would lead to a higher unnecessary cost and, sometimes, a lower efficiency. A quote from ISO: "Users of this part of ISO 10440 should be aware that further or differing requirements may be needed for individual applications. This part of ISO 10440 is not intended to inhibit a vendor from offering, or the purchaser from accepting, alternative equipment or engineering solutions for the individual application. This may be particularly appropriate where there is innovative or developing technology. Where an alternative is offered, the vendor should identify any variations from this part of ISO 10440 and provide details".

Obviously, all ISO items that have been mentioned in the technical specification must be checked in the technical offer.

Chapter 14
Commissioning Tests of a Refrigeration-Liquefaction Plant

Abstract The commissioning tests must show that the system reaches or exceeds the required performances but also that all operating procedures run automatically (with no human intervention) and smoothly. Operating conditions during the commissioning tests must be preciously saved: they will be the reference for the life of the system.

14.1 Introduction

The *targets* of the commissioning tests are to be described into the plant specification.

The tests are to be performed in a *logical order*: from the room temperature compression station to the cryogenic limits of supply towards the cryostat.

Each component must be *fully tested* before going to the next stage.

The commissioning tests must show that the system reaches or exceeds the required performances but also that all operating procedures run automatically (with no human intervention) and smoothly. An error calculation should be performed for each performance measurement.

During the tests, monitoring and logging of all parameters must be performed in order to build an "initial values" data base that will be used for comparisons and diagnosis during the whole lifetime of the system. There are the references of the system.

The tests are to be done, if possible, in the most difficult operating conditions. If not, calculate the incidence of parameters that are not at worst conditions (e.g. suction temperature of a compressor).

In most of the cases, commissioning tests necessitate special equipment (test equipment) that can be, at minimum, a sensor but, sometimes, especially at low temperatures, a special equipment (cryostat).

Finally, this chapter describes some possibilities to perform commissioning tests. They should be taken as examples. There are not unique, and they can (or must) be adapted to specific conditions.

G. Gistau Baguer, *Cryogenic Helium Refrigeration for Middle and Large Powers*, International Cryogenics Monograph Series,
https://doi.org/10.1007/978-3-030-51677-2_14

The technical specification can incorporate part of tests description.

14.2 An Example of the Sequence of a Test Programme

The detailed tests are described in the following paragraphs.

Compressor station:

- Leak tightness
- Cycle gas purity
- Vibration measurements
- Performances at various operating conditions (nominal and reduced flow rates)
- Long duration test
- Oil concentration at the discharge of the bulk oil separator
- Efficiency of coalescers

Full flow dryer

Cold box:

- Insulation vacuum
- Leak tightness
- Cycle adsorbers
- Regeneration and switching or by-passing procedure for adsorbers
- Retention capacity of the adsorbers
- Turbine performances after optimisation
- Cryogenic compressor or/and supercritical circulator

Liquid helium storage

14.3 Status of the System Prior to Starting the Final Part of Commissioning Tests

It is assumed that, prior to the final commissioning tests, the supplier has already performed its usual tests to ensure that commissioning tests can be started. Such tests generally include:

- Pressure tests
- Static leak tests (with certificates)
- Electrical tests
- Calibration of all instruments (sensors, valves, analysers, etc.)
- Check of all the safety interlocks
- Check of correct operation of the control procedures in the process control system
- Conditioning of all gas circuits
- Vibration checks on compressors

- Various regenerations
- At least, one cool-down and warm-up followed by a full regeneration of the cold box adsorbers

14.4 Compression Station

The compression station includes the compressor(s), the cycle pressure control valves, the cycle buffer capacity and the oil removal system. As said in Sect. 14.1, the supplier tests have already been performed. Prior to run a global test, which is generally a long duration test during which all parameters are logged, specific tests are to be performed.

- Leak tightness in operation
- Operation at design (nominal and turn-down) conditions of compressors
- Performance of each compressor

14.4.1 Compression Station Leak Tightness in Operation

Configuration of the system
The compression station is operating in a steady state, at nominal conditions.

To be checked
Leak tightness of the shaft seals has to be checked during operation (static leak tests have been already performed). The gas leak rate in operation is to be measured.

Necessary instrumentation
There is generally no need for a specific instrumentation.

Procedure
The compressor is running in a closed loop, and the suction and discharge pressures are controlled by means of the usual valves connected to the cycle buffer capacity. Tuning of the PI parameters of the control loops must be done in order to get perfectly stable pressures. When operating pressures and temperatures are established at nominal values and stable, the cycle buffer capacity is isolated in order to make the system more sensitive to leaks.

During the test that is performed at stable nominal conditions (pressures and temperatures), the pressures and temperatures are monitored. The pressure control system must be operating as shown in Fig. 8.13, left, where LP is controlled by the HP to LP by-pass; therefore, the HP varies according to the total leak. A simple calculation gives the leak mass flow rate. To avoid making temperature corrections due to ambient and cooling water temperature, it is advised to run such a test on a 24-hour module. Remember that such test can also be performed during the time the plant is in operation (remember Sect. 10.4.2).

14.4.2 Cycle Gas Purity

Configuration of the system
The compression station is operating in a steady state, at nominal conditions. The full flow dryer is disconnected.

To be checked
The cycle gas purity must be preserved during a long-term run.

Necessary instrumentation
Suitable analysers:

- O_2, (1 to 1000 ppM V)
- N_2, (1 to 1000 ppM V)
- H_2O, (1 to 1000 ppM V)

Procedure
Such a test can be performed during the leak tightness test and during the long duration test.

Expected results
Check that impurity levels increases are equal or lower than guaranteed.

14.4.3 Vibration Measurements

To be checked
A measurement of the level of vibration is to be performed at locations indicated by the maker and other possible locations for each operating regime, both on the motor and the compressor.

A spectral analysis (remember Sect. 11.5.1.2) is to be performed in order to identify possible sources of vibrations.

Remark Perform vibration tests *also* in reduced operation modes, especially if the slide valve is not at 100%.

Necessary instrumentation
The vibration sensors must be **screwed** to the machines during the measurement (remember Sect. 11.5.1.2.1).

Configuration of the system
The whole compression station is operating at nominal conditions. Such test can be performed during the long run test. It must also be performed when operating at reduced flow.

Procedure
Such a test can be performed during the long duration test.

Expected results
Vibration levels must be in accordance with both industrial standards and supplier specifications.

Remark Results of initial measurements must be saved because they will be a reference during all the lifetime of the machines.

14.4.4 Performances at Nominal Mass Flow Rate for Each Compressor

To be checked
For each compressor: mass flow rate and absorbed power at nominal operating conditions during an *uninterrupted* run of 100 h.

Configuration of the system
The compression station is operating in a steady state, at nominal conditions.

Necessary instrumentation
Parameters to be measured (optimum location of each sensor is to be discussed):

On compressor:

- Helium inlet pressure at first stage PT IC
- Helium inlet temperature at first stage TT IC
- Helium mass flow rate (range) FT
- Absorbed electrical power of the main motors CT M (Fig. 14.1)

These parameters in Table 14.1 are used to calculate the compressor efficiency. The other parameters (remember Table 6.5) are to be saved into the "initial values" data base.

Procedure
According to the compression station structure, establish helium circulation in order to be able to measure separately the performance of each compressor.

Special test equipment is to be installed to control and measure the helium compressor flow.

One must make sure that recycling by-pass valves are closed (and leak tight).

In case of a three-pressure compression station as shown in Fig. 14.2, the situation is a little bit more complicated: FCV LP in series with a mass flowmeter in order to set and measure the LP compressor flows and FCV HP and a flowmeter for the HP compressors are to be installed. Each test by-pass valve should be able to process 120% of the nominal helium mass flow rate at nominal conditions of one compressor.

In order to test a first stage compressor, a second stage compressor must also be running to transfer the LP compressor flow up to the HP where it is measured. Obviously, the HP compressor flow is totally by-passed into its by-pass. The test of

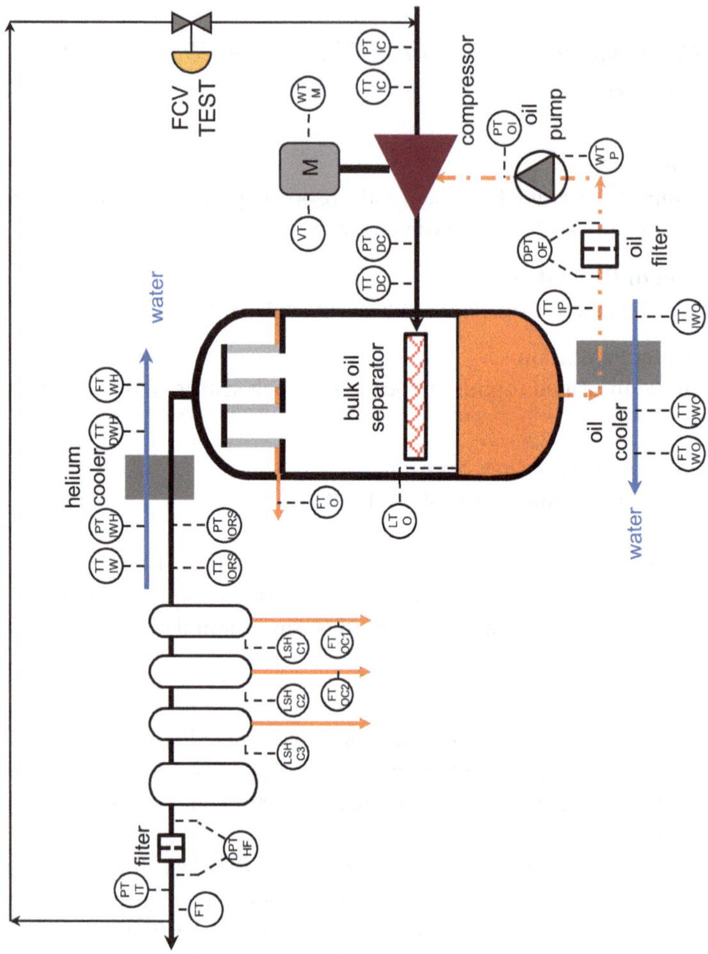

Fig. 14.1 Parameters to be logged for a compressor test

Table 14.1 Parameters to be logged for the test of one stage

Sensor	Duty
PT IC	Helium inlet pressure
TT IC	Helium inlet temperature
PT DC	Helium discharge pressure
TT DC	Helium discharge temperature
FT	Helium mass flow rate (range)
WT M	Absorbed electrical power of the main motor
FT O	Drained oil from BOS
FT OC1	Drained oil from coalescer 1
FT OC2	Drained oil from coalescer 2

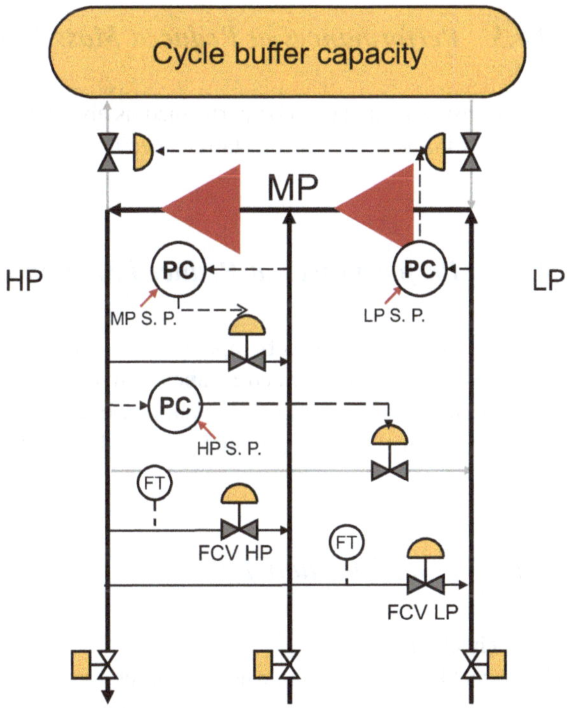

Fig. 14.2 Compressor flow measurement

a second stage compressor necessitates the operation of one first stage compressor in order to control the MP.

Using the above parameters, the operation of oil and helium coolers is to be checked. Each cooler must deliver oil or helium at the specified temperature for the specified water flow at *maximum water specified temperature* when the compression station operates at *nominal conditions*. Pressure drop on the water circuit is also to be checked.

Remark When the suction temperature during the test is higher than the nominal one, a correction on the compressor mass flow rate is to be done (remember Sect. 6.2. 1.1). When it is lower, it is possible, by controlling the water flow rate on the helium coolers, to set the suction temperatures.

All parameters that are monitored during the test must be logged into the initial values data base that will be used for comparisons during the lifetime of the system.

Expected results
When operating at nominal operating conditions (or corrected suction conditions), mass flow rate must be equal or higher than guaranteed; absorbed power must be equal or lower than guaranteed.

14.4.5 Performances at Reduced Mass Flow Rate

Same measurements are to be performed as above 14.4.4 for each operating condition stated in the compression station specification.

14.4.6 Performances at Reduced Suction Pressure

When the compressor is to be operated at a sub-atmospheric pressure, same measurements are to be performed as above 14.4.4 for various suction pressures from atmospheric down to design, for example, every 0.10 bar. It is also interesting to check the minimal suction pressure that can be reached.

14.4.7 Long Duration Test

To be checked
The correct behaviour of the whole compression station during an *uninterrupted* run of 24 hours minimum for each operating condition.

Necessary instrumentation
No extra instrumentation is needed.

Configuration of the system
The compression station is operating in a steady state, at specified conditions.

Procedure
Run the compression station at specified conditions.

Log regularly all the parameters listed in Sect. 14.5.4 and the compressor shaft seal oil flow.

Expected results
When operating at reduced mass flow rate (or corrected suction conditions), mass flow rate must be equal or higher than guaranteed; absorbed power must be equal or lower than guaranteed. The vibrations must be checked, especially at low slide valve positions.

14.4.8 Check of the Oil Removal System Performance

14.4.8.1 The Bulk Oil Separator

To be checked
The oil concentration upstream the oil demister that is included in the bulk oil separator.

Necessary instrumentation
Special valving to measure the oil quantity that is separated or automatic system that measure the oil flow (remember Sect. 6.3.8.5.1.)

Configuration of the system
The compression station is operating in a steady state, at nominal conditions.

Procedure
The correct operation of the bulk oil separator is of paramount importance: the consequence of a bad oil separation is an overloading of the coalescer stages. To measure its efficiency, as explained in Sect. 6.3.8.5, the mass of oil that is separated by the demister devices that are located in the bulk oil separator as shown in Fig. 14.1 plus the mass of oil that is separated at the first coalescer is compared to the helium mass that has been processed during the duration of the test (FTO in Table 14.1).

Remark Such a measurement cannot be performed in case the oil demister is a mesh device (Fig. 14.3).

Expected results
The oil concentration downstream the bulk oil separator must be in accordance with the technical specification.

14.4.8.2 The Coalescers

To be checked
The oil concentration upstream the first coalescer (calculation with FT OC1 and FT OV2 in Table 14.1).

Necessary instrumentation
Special valving to measure the oil quantity that is separated.

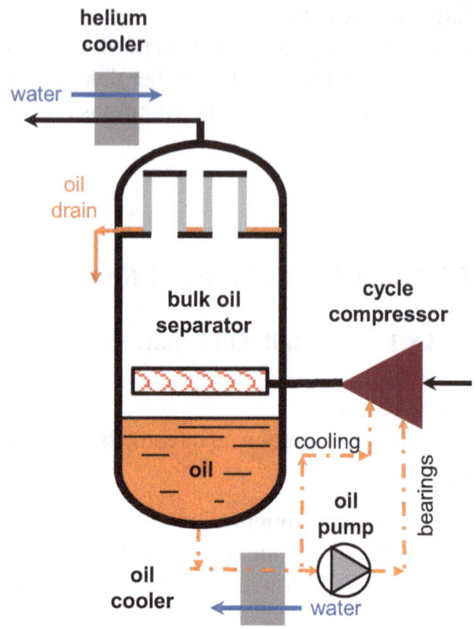

Fig. 14.3 Check of the bulk oil separator

Configuration of the system

The compression station is operating in a steady state, at nominal conditions.

Procedure

During the time of commissioning tests, it is only possible to check the performance of the first stage of coalescers, as explained in Sect. 6.3.8.5.

Remark: One must be sure that the coalescer cartridge is saturated before starting measurements (remember Fig. 6.62!).

Expected results

The oil concentration must be in accordance with the technical specification.

14.4.8.3 The Oil Vapour Adsorber

To be checked

The oil vapour concentration in helium.

Procedure

There is not an easy possibility to check the performance of the oil vapour adsorber during the commissioning tests. Furthermore, with a new activated charcoal load, there are very few chances that oil vapour is released during the rather short time of the test.

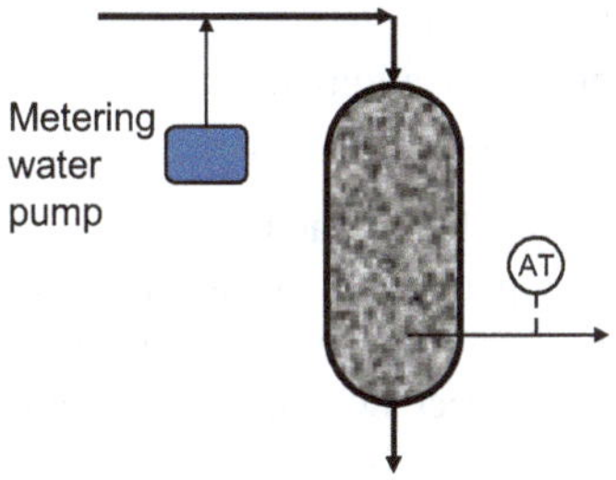

Fig. 14.4 Test of a dryer

14.4.9 *Check of the Full-Flow Dryer*

To be checked

- The purity of the processed helium
- The automatic regeneration procedure
- The regeneration temperature
- Duration
- Nitrogen consumption for a regeneration cycle

Configuration of the system

The helium must be the design operating conditions: pressure, temperature and flow.

Procedure

It is not possible to check the full-flow dryer performance in the very operating conditions for which it has been designed because the real water concentration is, obviously, very low (remember Sect. 9.5.1.2.1) and the time for the test would be unrealistic. However, it is interesting to check the *quantity of water* that can be trapped in different conditions, i.e. at higher concentration, in order to reduce the duration of the test, as shown in Fig. 14.4. A controlled flow water pump injects water upstream the dryer. No analysis is necessary: the water concentration is calculated according to the helium flow. A hygrometer is connected to the analysis discharge port of the dryer (remember Sect. 9.4.3.2.1). The test is run until the hygrometer shows that the moisture level starts to increase. For unsaid reasons, this kind of test, that has been proposed a number of times by the author, has never been performed, even if there is not any danger to pollute the system.

Expected results

Check that the measured quantity of water that has been trapped is in agreement with the guaranteed value.

14.5 The Cold Box

The cold box includes mainly the heat exchangers, the cryogenic expansion turbines, the adsorbers and the phase separator(s), cryogenic compressors or/and supercritical helium circulators and cryogenic and room temperature valves. As said in Sect. 14.1, the supplier tests have already been performed. Prior to run a global test, which is generally a long duration test during which all parameters are logged, specific tests are to be performed.

- Insulation vacuum
- Operating at design (nominal and turn-down) conditions of turbines
- Performance of each turbine
- Correct switching between adsorbers

Remark In the configuration of the system, the wording: "The cold box is operating in a steady state, at nominal conditions" means that cycle pressures, heat loads at various temperature levels and withdrawal of helium are at nominals conditions. Steady state means that all cold box pressure and temperatures are stable for 1 hour.

14.5.1 Insulation Vacuum

To be checked
The pressure in the vacuum enclosure.

Necessary instrumentation
Vacuum gauges that are part of the supply. The secondary gauge is to be located far away from the connection of the secondary pump.

Configuration of the system
The cold box is at room temperature. It has been pumped down with possible help of a mobile primary pump. The vacuum set is operating.

Procedure
Measurement of the vacuum enclosure pressure.

Expected results
When the cold box is at room temperature, the insulation vacuum must be lower than 10^{-3} mbar. Such value depends on the "history" of former operation of the cold box. It will obviously improve during cool-down.

14.5.2 Cold Box Leak Tightness

14.5.2.1 Air Leaks into Vacuum

An air leak is generally not detectable due to cryo-trapping on cold surfaces. It is advised to isolate the vacuum set before warming up the cold box. At room temperature, the cold box vacuum should be almost the same as before cool-down, but it is not granted that all water is desorbed from the so large surface of the multi-layer insulating material.

14.5.2.2 Helium Leaks into Vacuum

A helium leak detector can be connected to the cold box vacuum in case the vacuum level seems to be too high at normal operating temperatures. One should not forget that leaks might have a different behaviour at room and cold temperatures.

14.5.3 Cycle Adsorbers

14.5.3.1 Regeneration and Switching Procedure of the Warm Adsorbers

According to the way of heating, the test procedure can be slightly different.

To be checked
The correct conditions of operation (automatic or manual) of the regeneration and the switching procedure.

Necessary instrumentation
Nitrogen flowmeter.

Configuration of the system
The cold box is operating at nominal conditions. It is not necessary that the adsorber to be regenerated be saturated.

Procedure
Launch the regeneration sequence or implement manually the procedure.

Expected results
Regeneration temperature, duration of the total regeneration sequence and quantity of regeneration nitrogen in accordance with the specification.

No, or acceptable, excursion of the cold box temperatures.

14.5.3.2 Regeneration and By-Pass Procedure of the Cold Adsorber

To be checked
The correct conditions of operation (automatic or manual) of the regeneration procedure and the by-pass switching procedure.

Necessary instrumentation
Nitrogen flowmeter.

Configuration of the system
The cold box is operating at nominal conditions. It is not necessary that the adsorber to be regenerated be saturated.

Procedure
Launch the regeneration sequence or implement manually the procedure.

Expected results
Regeneration temperature, duration of the total regeneration sequence and quantity of regeneration nitrogen in accordance with the specification.

No, or acceptable, excursion of the cold box temperatures.

14.5.3.3 Retention Capacity of the Adsorbers

14.5.3.3.1 Warm Adsorber

To be checked
The retention capacity of air gases at design operating conditions. Such a test is not very often performed.

Necessary instrumentation
Air injection system.

Analysers.

Configuration of the system
The cold box is operating at nominal conditions. The adsorber to be tested has been recently regenerated (Sect. 14.5.3.1).

Procedure
Dry air is injected at the inlet of the cold box. The air concentration can be the same as specified in the technical specification if the operating time is 100 hours; if not, the air concentration should be increased in order that the test duration shows that the quantity of air to be trapped is the same (if the spec requests for 50 ppM V for 200 hours, run the test with 100 ppM V during 100 hours). Either the air flow is monitored and set in accordance with the helium flow or an analyser is connected downstream the air injection. The operation analyser is connected at the analysis port towards the end of the adsorber bed (remember Sect. 9.4.3.3.1).

Expected results
Measure the quantity of air that is trapped until the analyser reaches the switching level.

14.5.3.3.2 Cold Adsorber

To be checked
The retention capacity of hydrogen and neon at design operating conditions. Such a test is very seldom performed.

Necessary instrumentation
Impurity injection system.
Analysers.

Configuration of the system
Same as for the warm adsorber.

Procedure
Same as for the warm adsorber.

Expected results
Same as for the warm adsorber.

14.5.4 Turbine Performance

To be checked
This is not a real performance test. Such a performance test has been performed at the maker's facility with the suitable accuracy measurements. Here, efficiency, flow and power at design inlet pressure and temperature are checked *with the sensors that are available in the cold box*. Such measurements will be part of the "initial values" file.

Necessary instrumentation
The sensors that are installed in the refrigerator are sufficient, provided their accuracy has been correctly selected (remember Sect. 13.2.2.2) (Fig. 14.5).

The parameters in Table 14.2 are used to calculate the turbine efficiency and power. The other parameters (remember Tables 6.14 or 6.15) are to be saved into the "initial values" data base.

Configuration of the system
The turbine optimisations (rotational speed and labyrinth control in case of a gas static bearing system) have been performed.

The cold box is operating in a steady state, at nominal conditions.

Check that the optimisation of both rotational speed (remember Sect. 6.6.3.4) and labyrinth tuning in case of static bearings (remember Sect. 6.6.3.2.1) (Fig. 6.11) is correct.

Check that the labyrinth tuning system behaves correctly during changes of operating conditions.

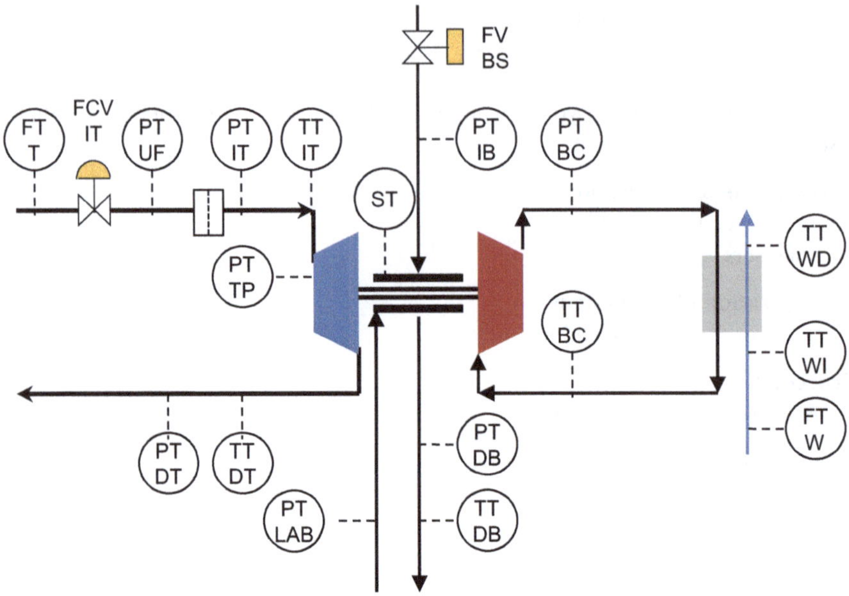

Fig. 14.5 Parameters to be logged during the test of a turbine with static bearings

Table 14.2 Parameters to be logged for the test of one turbine

Sensor	Duty
PT IT	Helium inlet pressure
TT IT	Helium inlet temperature
PT DT	Helium discharge pressure
TT DT	Helium discharge temperature
FT	Helium mass flow rate
ST	Turbine rotational speed

In case of a turbine fitted with dynamic bearings, the parameters to be logged are slightly different, as shown on Fig. 14.6. Of course, no checks to be performed about and labyrinth tuning.

Expected results
Efficiency, flow and consequently power are compared to the factory test if any or the expected values.

14.5.5 Thermal Shields

To be checked
The power that can be absorbed for the design return temperature.

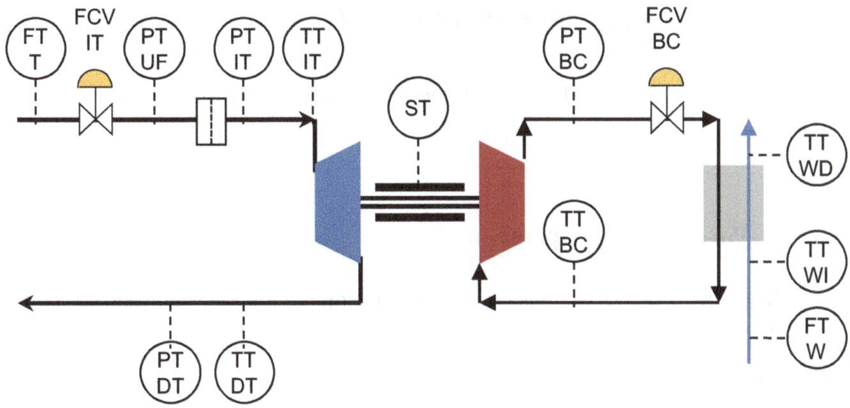

Fig. 14.6 Parameters to be logged during the test of a turbine with dynamic bearings

Fig. 14.7 Measurements to be made to check the available cryogenic power on thermal shields

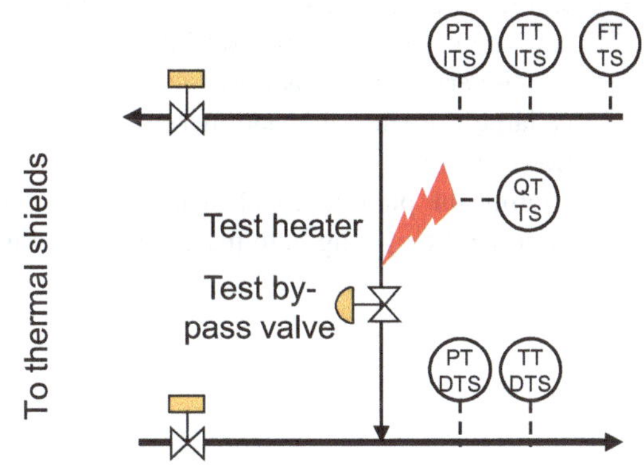

Necessary instrumentation
A by-pass valve and a heater, preferably located in the cold box (see Fig. 14.7).

Configuration of the system
The refrigerator is operating in a steady state, at nominal conditions for all temperature and pressures.

Procedure
The cold box is operating at nominal conditions.

Circulate all the shield flow through the test by-pass valve that is set in order that the pressure difference between upstream and downstream reaches the design value. Inject the design power.

The test is run for 24 hours since the shield temperatures are stable.

Expected results
When the design power is injected, the discharge temperature must be equal or lower than the design value.

14.5.6 *Lowest Temperature Power (> 4.5 K)*

According to the duty to be fulfilled by the plant, there are various ways to check its output.

14.5.6.1 Pure Liquefier Operation

Remember in Sect. 7.7.2, the two possible operations of a pure liquefier: constant or rising level.

To be checked
There are two possibilities to check the liquefaction capacity:

- Operating the liquefier at *rising level* for a small standard liquefier or a very large machine. In case of a small standard liquefier, there are few settings for optimisation because the system is well known as it is standard; therefore the time to fill the stationery dewar is acceptable to get a significant measurement. In case of a very large liquefier, the stationery dewar is large and allows for long operating periods of time.
- Operating the plant at constant level, by vaporising, warming up to room temperature and recycling helium, as shown in Fig. 14.8.

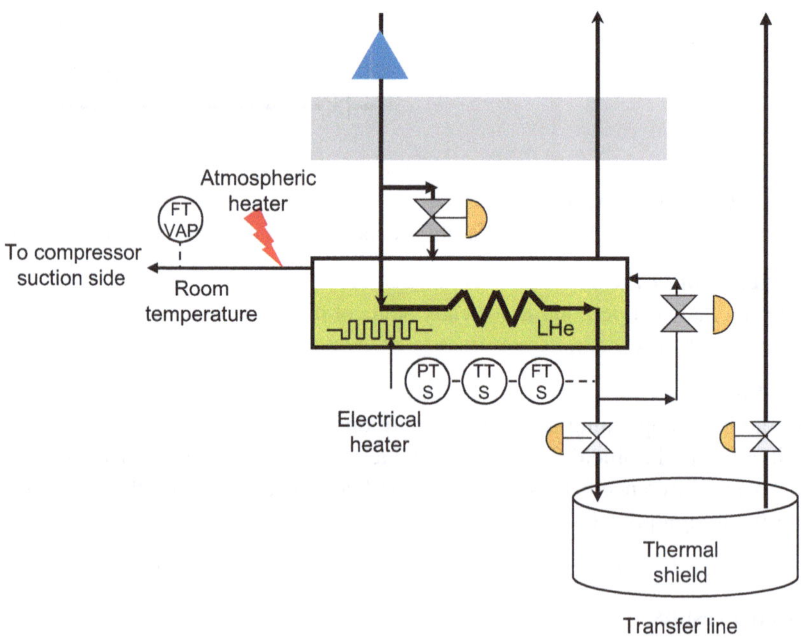

Fig. 14.8 Measuring the refrigerator capacity in a mixed duty operation

Necessary instrumentation
Measurement of the quantity of liquid helium in the storage (either level or scale) or room temperature flowmeter if liquid is vaporised and recycled.

Configuration of the system
The cold box is operating at nominal conditions, liquid helium is stored into the stationary dewar or recycled.

Procedure
The check of the liquid helium output is generally performed by measuring the liquid helium volume that is deposited into the stationary dewar or by weighting it, or measuring, at room temperature, the helium flow that is recycled.

Expected results
The measured output must be equal or higher than the guaranteed value.

14.5.6.2 Mixed Duty Operation

To be checked
The cryogenic power at design pressure, the liquid helium output *at constant level* and the temperature of supercritical helium.

Necessary instrumentation

Heater in the cold box phase separator or in the test cryostat.
Temperature, pressure and flow measurements of super critical helium.
Device to withdraw saturated vapour, with a flowmeter.

Configuration of the system
The cold box is operating at nominal conditions.

Procedure
Here, one has to check the mixed duty refrigeration + liquefaction at constant level or isothermal power and liquid helium output. Each one is to be measured separately. The isothermal power is provided by a simple electrical heater that is dipped into the liquid helium bath of the test device, generally the phase separator that can be a "natural" part of the cold box or a special device added for performing the tests (see Fig. 14.8). The liquid helium output could be withdrawn as liquid, but there are chances that, due to fast vaporisation of liquid and large density variations, flow instabilities happen. It is advised to vaporise some liquid, withdraw saturated vapour instead and warm it, for example, with an atmospheric heater before returning the flow to the suction side of the cycle compressor after flow measurement. To generate such a flow of saturated vapour, one has to inject the corresponding power into the liquid helium bath.

Therefore, to measure 300 W at 4.5 K + 0.5 g/s liquid, one has to inject:

$$300 + 0.5 \times \text{latent heat} = 300 + 0.5 \times 18.59 = 309.29 \text{ W}$$

into the liquid and withdraw 0.5 g/s of saturated vapour.

In order to check the subcooling heat exchanger, the requested flow is to be circulated through the test valve, measured at FT S, and pressure and temperature measured at PT S and TT S.

Expected results

If the refrigerator exceeds the expected performance, it is generally the liquid output that is fixed, the isothermal power being automatically adapted in order to control the phase separator level.

The temperature of the supercritical helium must be equal or lower than the guaranteed value.

14.5.7 Lowest Temperature (< 4.5 K)

This test includes the cryogenic compressor tests that cannot be performed independently. This is a *performance* test because it has generally not been performed at the maker's facility: test loops for such equipment do not exist to the knowledge of the author. Here, efficiency, flow and absorbed power at any specified pressures and temperatures are measured *with the sensors that are available in the cold box* (remember Sect. 13.2.2.2). A control of the stability of the lowest pressure at constant heat load is interesting in case of cooling RF cavities. Such measurements will be part of the "initial values" file.

It is also the occasion to check a few points on the surge line of each compressor.

To be checked

The available power at the correct temperature.
The performances of the cryogenic compressors.
The stability of the suction pressure of the first stage compressor.
The correct evolution of the pump-down procedure.
The correct operation during turn-down situations.
The correct operation during reconnection scenario.

14.5.7.1 Necessary Instrumentation

Instrumentation for measuring the cryogenic power As explained in Sect. 4.3.1.3, the final Joule Thomson heat exchanger is generally not located into the cold box, and, furthermore, it is often split into numerous ones sitting in the various cryostats. To check the cryogenic power without being connected to the cryostats, one needs such a one-off heat exchanger to perform the power measurement. Such a

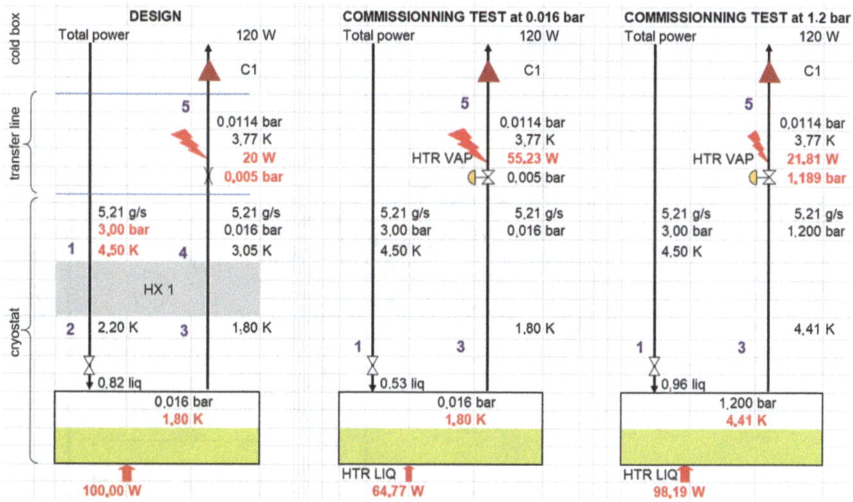

Fig. 14.9 Principle of the test equipment in order to measure the cryogenic power at a temperature lower than 4.5 K

heat exchanger might be rather large (it would take room in the cold box) and would be used very seldom. To avoid the investment of this heat exchanger, one should find conditions where the temperature at the suction side of the first stage compressor is same as in the real arrangement.

Figure 14.9, left, shows the design operating conditions:

- The heat exchanger is fed, e.g. with supercritical helium at 3.00 bar, 4.50 K.
- Expansion down to 0.016 bar gives liquid at 1.80 K.
- In order to absorb 100.00 W at 1.80 K, the cycle mass flow rate is 5.21 g/s.
- Saturated vapour is warmed up to 3.05 K in the heat exchanger.
- The transfer line adds a heat leak of 20.00 W and has a pressure drop of 0.005 bar.
- Helium reaches the suction side of the first compressor at 0.0114 bar, 3.77 K.

In the commissioning phase, where there is no connection to the cryostat, one must find a way to get same conditions at the suction side of the first compressor: 0.0114 bar, 3.77 K.

The arrangement in Fig. 14.9, centre, fulfils the duty:

- A phase separator, operating at 0.016 bar, collects liquid that is produced by expansion of same flow of supercritical helium: 5.21 g/s at 3.00 bar, 4.50 K.
- A heater HTR LIQ vaporises the liquid and controls the level.
- A pressure drop of 0.005 bar is provided by a valve.
- A heater HTR VAP injects power into helium.

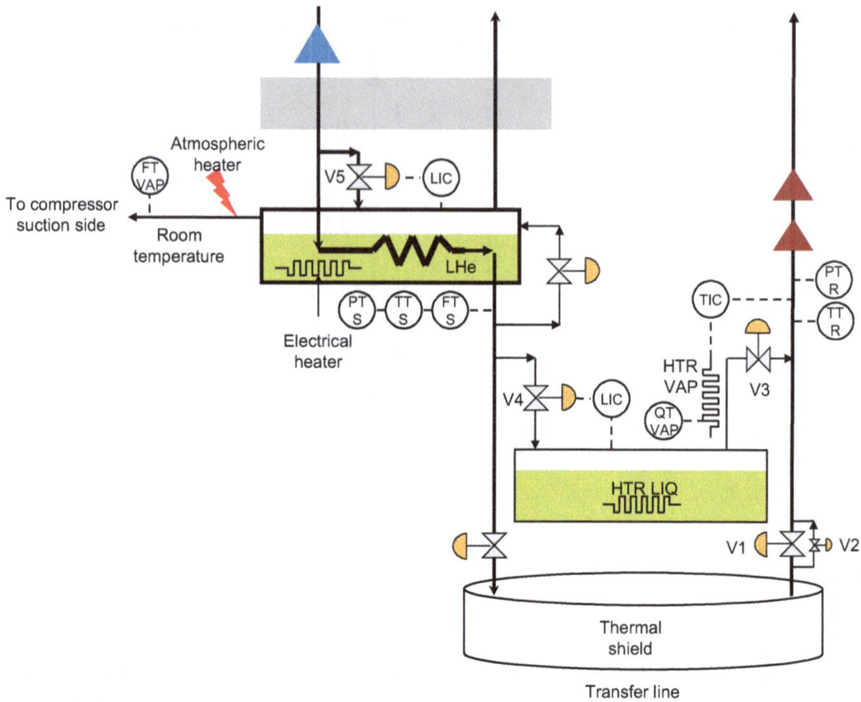

Fig. 14.10 Power measurement at a temperature lower than 4.5 K

One has to get, at point 5 (suction side of the compressor), the design conditions. For that, the valve is set at a position that gives the required pressure of 0.0114 bar, and the heater is energised in order that the temperature reaches 3.77 K.

In the centre part, one sees that, in order to reach same temperature as at the inlet of first stage, 55.23 W must be injected into the saturated vapour issued from the phase separator, while the valve provides a pressure drop of 0.005 bar. Not surprisingly, the total power is equal to 120.00 W.

Remark The phase separator might operate at any other pressure than design, as shown in the right part of Fig. 14.9. Note that the powers dumped into liquid and vapour are different, because the latent heat changes with pressure, but their total is same. However, one must be aware that a test phase separator operating at a higher pressure than design will not allow a pump-down simulation as described here under.

The test equipment is shown in Fig. 14.10, where the phase separator houses the heater HTR LIQ. The saturated vapour is heated by heater HTR VAP. The power injected QT LIQ into HTR LIQ generates the required mass flow; the level is controlled by valve V 4.

The heater HTR VAP allows getting the expected design return temperature measured with TT R (simulation of the cryostat JT heat exchanger warm end temperature difference plus transfer line heat load). The valve V3 allows setting the suction pressure of the first compressor.

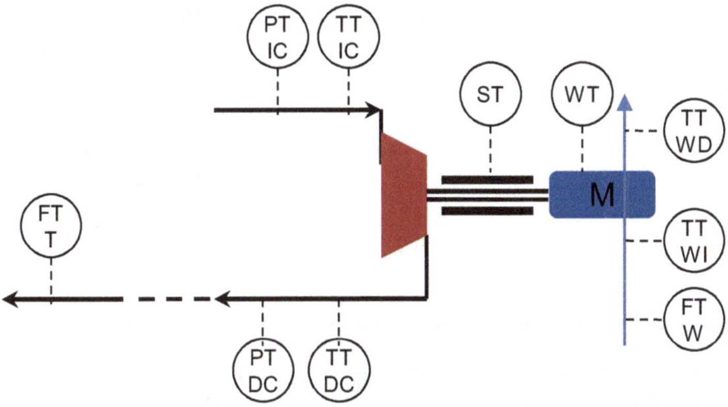

Fig. 14.11 Parameters to be logged during the test of a cryogenic compressor or circulator

By the way, this is the almost same equipment that is suggested in Fig. 6.171 to re-start the system! One has only to add valve V3 if the test phase separator is operating at a higher pressure than design.

Instrumentation for measuring the cryogenic compressor performances The sensors that are installed in the refrigerator are sufficient, provided their accuracy has been correctly selected (remember Sect. 13.2.2.2).

Parameters to be logged for each compressor (see Fig. 14.11):

- Inlet pressure PT IC (PT R in case of first stage)
- Inlet temperature TT IC (TT R in case of first stage)
- Discharge pressure PT DC
- Discharge temperature TT DC
- Rotational speed ST
- Mass flow rate FT
- Motor input power WT

These parameters are used to calculate the compressor efficiency.
Other parameters are to be saved into the "initial values" data base.

- Motor cooling water inlet temperature TT WI
- Motor cooling water discharge temperature TT WD
- Motor cooling water flow FT W (Table 14.3)

14.5.7.2 Test Procedures

There are various tests to be performed:

- Steady state regimes
- Simulation of a pump-down sequence

Table 14.3 Parameters to be logged for the test of a cryogenic compressor

Sensor	Duty
PT IC	Helium inlet pressure
TT IC	Helium inlet temperature
PT DC	Helium discharge pressure
TT DC	Helium discharge temperature
FT	Helium mass flow rate
ST	Turbine rotational speed
WT	Motor power

- Simulation of the beginning of the sequence for reconnection to the transfer line

Steady States The cold box is operating in a steady state at, successively, all required conditions (nominal, operation 1, operation 2, etc.), with liquid in the 4.5 K phase separator.

The first stage suction conditions are set at, successively, all required values. Expected results for each operating condition:

- The pressure measured with PT R must be equal or lower than the design pressure.
- The total power injected into heaters QT LIQ and QT VAP must be equal or higher than the nominal value.
- The suction pressure PT IC must stay within the excursions allowed in the technical specification (plus or minus x mbar within y hours).

Simulation of a Pump-Down
Sequence The target is to simulate the pump-down sequence for an important quantity of liquid helium (therefore, a rather long pumping-down time), as in the actual operation. However, the quantity of liquid that is in the phase separator of the test equipment is very small compared to the actual volume of liquid that is distributed in the whole machine. As a consequence, the time of pump-down would be very short. In order to get a realistic pumping time, one has to provide saturated vapour at a controlled rate.

If the flow processed by the cryogenic compressors that is naturally provided by pressure change on liquid and energy transferred from materials that are cooled down is totally replaced by vapour produced by a heater, the suction pressure of the compressor stays constant.

Therefore, the pump-down simulation sequence is as follows:

- The set point of the CCC train control system is at design pressure.
- The liquid level is kept constant by valve V4.
- V3 is kept full open.
- The inlet temperature to the CCC first stage TT R is kept constant by HTR VAP.
- The bath pressure is controlled by HTR LIQ, as shown in Fig. 14.10, according to a ramp that varies linearly with time.

The CCC control system increases the rotational speed of the CCCs in order to decrease the suction pressure. The heater HTR LIQ provides sufficient flow in order to vary the suction pressure according to a ramp.

Remark The temperature TT R is not exactly constant in the actual operation, due to the behaviour of the JT heat exchanger(s) and the constant heat load of the transfer line. However, it allows to perform a rather significant pump-down simulation. Of course, it is possible to vary TT R according to the operating conditions provided that a simulation calculation taking care of the JT heat exchanger and the transfer line has formerly been performed.

Simulation of the Beginning of the Sequence for Reconnection to the Transfer Line The same procedure as above is to be followed, except that the compressor suction pressure is progressively ramped down to a value that is slightly lower than the pressure upstream valve V1 in Fig. 14.10. The test equipment alone does not allow to go farther in the reconnection procedure.

14.6 Liquid Helium Storage

A liquid helium storage dewar can be part of the helium plant.

14.6.1 A Remark on How to Measure Accurately the Heat Leaks of a Liquid Helium Dewar

The time for temperature stabilisation of the insulating material of a liquid helium dewar might be rather long (at least days, sometimes weeks).

Heat leaks have various origins:

- The heat that is deposited on the surface of the internal reservoir is split between the wetted part where it vaporises liquid and the part that is in contact with the saturated vapour where it warms up the vapour.
- The supporting system that brings a constant heat load. On a vertical dewar, the heat leak goes to the vapour (see Fig. 14.12, right). However, on a horizontal dewar, the heat leak goes to liquid or vapour depending on the liquid level (see Fig. 14.12, centre). Therefore, such heat leak vaporises liquid or warms up saturated vapour.

As there is generally not a temperature measurement on the vapour that gets out of the internal reservoir, it is not possible to measure the amount of heat that goes to the vapour and make an accurate measurement of heat getting into the internal reservoir.

For such reason, it is advised to get the maximum helium mass flow that should be released by the dewar, according to the liquid level, from the supplier.

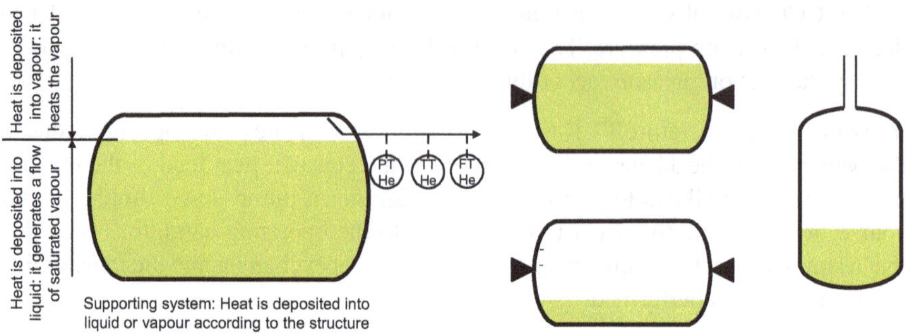

Fig. 14.12 Origin of heat leaks into a liquid helium dewar

14.6.2 Storage with Liquid Nitrogen Cooled Shields

The two tests (thermal shields and inner reservoir) are performed simultaneously.

14.6.2.1 The Heat Leaks on Thermal Shields

To be checked
The heat leaks on the thermal shield or the guaranteed nitrogen mass flow rate that is released.

Necessary instrumentation

- Room temperature gaseous nitrogen flowmeter FT N2, with pressure and temperature corrections if it is not a mass flowmeter (see Fig. 14.13)
- Pressure sensor PT N2

Configuration of the system
The liquid nitrogen volume should be around 75%. The liquid helium level should be about 75%.

Procedure
Measurement of the gaseous nitrogen mass flow rate that is released by the shield circuit. (When dealing with liquid nitrogen, the incidence of level is not so important as for helium, see Sect. 14.6.1.)

The helium storage operating conditions must be steady: the nitrogen and helium flows must be stable for 1 day prior to starting measurements.

The heat leak is not very different from:

$$\dot{Q} = \dot{m} \times L_{N2} \quad (14.1)$$

with $\dot{m}$ nitrogen mass flow rate and L_{N2} nitrogen heat of vaporisation (latent heat).

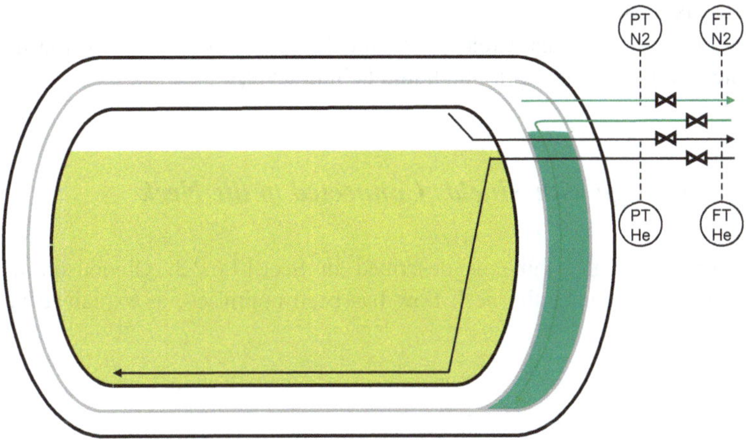

Fig. 14.13 Measurement of the performance of a liquid helium dewar

Expected results
Heat leaks on the shield or nitrogen mass flow rate must be equal or lower than the value stated into the specification.

14.6.2.2 The Heat Leaks on Liquid Helium

To be checked
The guaranteed heat leak of the dewar.

Necessary instrumentation:

- Room temperature gaseous helium flowmeter FT He, with pressure and temperature corrections if it is not a mass flowmeter (see Fig. 14.13) or a scale (this needs to accurately know the nitrogen boil-off)
- Pressure sensor PT He

Configuration of the system
If the thermal shield is cooled with liquid nitrogen, the liquid nitrogen volume should be around 75%. The liquid helium volume should be at least around 75%. If the later condition is not fulfilled, one has to check from the maker the possible consequence on the performance.

Procedure
Measurement of the gaseous helium mass flow rate that is released by the storage in nominal conditions. The inner reservoir *absolute* pressure must be controlled in order that the flow measurement is not affected by atmospheric pressure variations.

The helium storage operating conditions must be steady: the nitrogen (if any) and helium flows must be stable for 1 day prior to starting measurements.

Expected results
The helium mass flow that is released by the dewar must be equal or lower than the guaranteed value, according to the liquid helium level.

14.6.3 Storage with Shields Connected to the Neck

The procedure is the same as described in Sect.14.6.2.2. Obviously, it must be performed only after the neck flow has been optimised, as explained in Sect. 10.4.4.

Chapter 15
The Cryo Tool Box

Abstract The Cryo Tool Box allows building a tool (calculations on spread sheets) that can be used for everyday conventional thermodynamic calculations. A way to use comfortably software as HEPAK and REFPROP is showed. It incorporates calculations of room temperature compressor, free expansion of a fluid, heat exchanger, cryogenic expansion turbine, cryogenic centrifugal compressor and equivalent power.

15.1 Introduction

At the time he discovered cryogenics (1965!), the author was fortunately not involved in process calculations, so he did not suffer too much of the lack of easiness to use the available information on helium thermodynamic properties.

A major event in the world of helium process calculations occurred in 1972 with the publication of the well-known NBS technical note 631: “Thermodynamic properties of helium-4 from 2 to 1500 K with Pressures to 1000 Atmospheres” that was issued by the US National Bureau of Standards (Fig. 15.1). It presented tabular data in increments of pressure and temperature (Fig. 15.2). However, one can understand that there were few chances that, for each process point, the displayed value was the one that was looked for; therefore it gave way to a number of manual interpolations with the obvious introduction of quantity of mistakes (at least for the author).

Later on, when the author started giving lessons on helium refrigeration (1997), he noticed soon that it was of interest to show the participants how to run simple thermodynamic calculations in order that they could make either approximate calculations to get an order of magnitude or even accurate calculations on the process they plan to build or operate.

The target of such a chapter is to run some calculations that are related to typical component operations in spread sheets dealing with topics that have been met along the book: a room temperature compressor, a heat exchanger, an expander and so on. Even if calculations can be performed with a large number of fluids, only helium,

G. Gistau Baguer, *Cryogenic Helium Refrigeration for Middle and Large Powers*, International Cryogenics Monograph Series,
https://doi.org/10.1007/978-3-030-51677-2_15

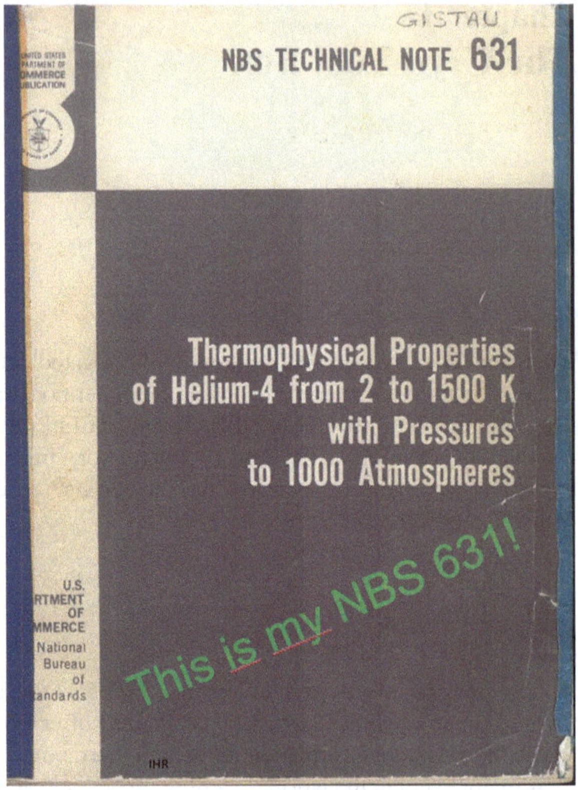

Fig. 15.1 The author's (very) old NBS 631 table

1.2 ATMOSPHERE ISOBAR

THERMODYNAMIC PROPERTIES OF HELIUM 4

TEMPERATURE DEG. K	VOLUME CC/G ** X .01	ISOTHERM DERIVATIVE CC-ATM/G	ISOCHORE DERIVATIVE ATM/K	INTERNAL ENERGY J/G	ENTHALPY J/G	ENTROPY J/G-K
* 2.167	0.06726					
2.5	0.06789	487.0	2.63	3.868	4.694	1.940
3.0	0.06953	407.0	3.46	4.902	5.747	2.326
3.5	0.07211	309.0	3.81	6.180	7.057	2.730
4.0	0.07629	203.0	3.83	7.835	8.762	3.185
* 4.424	0.08281	106.0	3.51	9.790	10.80	3.667
* 4.424	0.4870	29.4	0.469	24.02	29.94	7.993
4.5	0.5142	34.0	0.441	24.50	30.76	8.174
5.0	0.6580	57.1	0.335	27.02	35.03	9.077
5.1	0.6830	60.9	0.321	27.46	35.77	9.223
5.2	0.7073	64.6	0.309	27.88	36.48	9.363
5.3	0.7310	68.1	0.299	28.30	37.19	9.496
5.4	0.7542	71.6	0.289	28.70	37.87	9.625
5.5	0.7769	74.9	0.280	29.10	38.54	9.748

Fig. 15.2 A look inside NBS 631

nitrogen and hydrogen will be used here. Each spread sheet is then saved into a common file "Cryo Tool Box", and later, these calculations can be easily performed again by copy and paste, sparing time and avoiding possible mistakes.

15.2 The HEPAK Software

Today, working with NBS 631 would really be "prehistory"! One can forget any trouble of this kind and run a software that can insert a helium property into calculations as simply as making a multiplication! This is thanks to CRYODATA that developed HEPAK®, a wonderful software for cryogenic people. This software can be run in connection with EXCEL® that allows calling automatically helium properties during the calculation.

"HEPAK®[1] is a computer programme for calculating the thermophysical properties of helium from fundamental state equations of the NBS Technical note 631. The state equation domain has been extended and is valid from 0.8 Kelvin or the melting line to 1500 Kelvin, *including the super fluid region, the lambda line* and liquid-vapour mixtures; as a function of pressure, these equations are valid up to 1000 bar, except in the temperature range of 80–300 where the state equations are valid to 10,000 bar. HEPAK also includes thermal transport properties in both normal and *super fluid*, the dielectric constant, refractive index and fluid surface tension. The user of HEPAK should be knowledgeable in thermophysical concepts as applied to fluid properties. The author of the HEPAK programme is Vincent Arp.

HEPAK is an ActiveX DLL (executable code) that can be interfaced to other applications (including Excel in the Visual Basic for Applications environment). Information on interfacing the software to other applications is described in the User's guide" (this is an excerpt of the HEPAK User's guide).

Important Remark Up to now (2020), HEPAK is the only software that allows calculating in the superfluid region.

15.3 The REFPROP® Software

REFPROP is an acronym for REFerence fluid PROPerties. This programme, developed by the National Institute of Standards and Technology (NIST), calculates the thermodynamic and transport properties of industrially important fluids and their mixtures. These properties can be displayed in tables and plots through the graphical user interface (not used here); they are also accessible through spread

---

[1] **Contacting CRYODATA Inc. Horizon Technologies**

General contact information:

Horizon Technologies	*Phone:* 303.972.2530
7555 S. Webster St.	*Fax:* 303.972.2530
Littleton, Colorado 80128, USA	*E-mail:* services@htess.com

Customer/Technical Support: support@htess.com

sheets or user-written applications accessing the REFPROP dll (used here). REFPROP is based on the most accurate pure fluid and mixture models currently available. It implements three models for the thermodynamic properties of pure fluids: equations of state explicit in Helmholtz energy, the modified Benedict-Webb-Rubin equation of state and an extended corresponding states (ECS) model.

The property formulations and fluid data files were programmed by:

Eric W. Lemmon, Ian H. Bell, Marcia L. Huber, and Mark O. McLinden Applied Chemicals and Materials Division National Institute of Standards and Technology Boulder, CO 80305. *(This is an excerpt of the REFPROP Documentation, Release 10.)*

The latest issue of REFPROP (June 04 2018) is release 10 that is used here.

However, spread sheets written in former issue 9.1, which has different calling sequences, can still be run with the 10 release.

15.4 The GASPAK® Software

CRYODATA issued also GASPAK, a software that deals with a number of fluids.

GASPAK® is an integrated package of thermodynamic equations for properties of fluids. The equations are valid from the triple point and melting line through the compressed liquid and the liquid vapour mixture region to well above room temperature. Pressure wise, they are valid up to the melting line or several hundred atmospheres (upper temperature and pressure limits vary from fluid to fluid). The user of GASPAK should be knowledgeable of thermo-physical concepts as applied to fluid properties. Authors are R.D. McCarty, V. Arp, and J. R. Fox. *(This is an excerpt of the GASPAK User's guide).*

15.5 The Suggested Way to Work with a Fluid Property Software

There are various ways to work with a fluid property software. However, as the author is not a specialist in computer usage, he is used to do that in what he thinks to be a very simple way that is explained here under. The basic principle is: "to carefully type something **once**, and later on, use it again and again, only by **copy and paste**". The reason that leads to this principle is concentrate once, check carefully and, then, forget the way it has been done.

For that, the author suggests to build a spread sheet that is called "gas properties" (detailed explanations are given in Sect. 15.6). In this spread sheet, carefully type the function that calls for properties (calling sequence), each of them referring to two selected inputs. Therefore, one gets a spread sheet containing the usual properties that one is going to use.

To perform a calculation, on the working sheet, every time a property is to be calculated according to two given inputs, just copy from the "gas properties" sheet the cell that performs the corresponding calculation according to the same inputs, and paste it into the correct cell of the working sheet. For some frequently used properties, it is tricky to create special macros (see Sect. 15.8.2.1).

Explanations will be given here under, by making very small steps, in order that anybody and even persons not very familiar with Excel are able to follow.

Remark All calculations that are dealt with in this chapter using REFPROP 10 can also be performed with other software like HEPAK or GASPAK, provided some adaptations are implemented. The interests of using REFPROP are that the syntax is simpler (although very much looking like to HEPAK's) and the same spread sheet can perform calculation with any gas, including more than one gas in a calculation (like for nitrogen pre-cooling for helium) and even gas mixtures. Unfortunately, for helium, REFPROP does not operate at pressures and temperatures that are lower than the lambda point. For calculations at temperatures lower than 3.0 K, it is advised to use HEPAK.

In this chapter, calculations are performed with REFPROP 10.

15.6 Connecting REFPROP or HEPAK and EXCEL

Just follow the instructions that are given in the USER's guide.

REFPROP or HEPAK will not interfere on any other job when it is not use it and one will not notice any delay in usual calculations.

15.7 Building the REFPROP Gas Property Tool

This tool allows to run easily thermodynamic calculations.

A reminder before starting: two parameters (e.g. pressure and temperature) are required to define a thermodynamic state point (e.g. enthalpy).

Remark All procedures that are described here under can also be implemented with HEPAK.

15.7.1 Calculation of a Property

The general calculation format is:

$$= \text{REFPROP}(\text{Output property code, Fluid name, Input property code,}\\ \text{Units, Value of 1st input property, Value of 2nd input property})$$

Parameters are separated by a comma.

The first input in the call to the function is the output property code.

The second input is the pure fluid name: helium, nitrogen or hydrogen.

The third input is the combination of input properties being sent to the function. Thus, for a given pressure and temperature, the second input would be "PT". Other valid inputs include the density (D), energy (E), enthalpy (H), entropy (S) and quality (Q).

Finally, the values of input properties are displayed, according to the unit system.

Let us begin by a simple example: one wants to calculate the enthalpy ("H") of helium ("HELIUM") corresponding to a given input doublet expressed in the MKS system ("MKS") – pressure ("P") and temperature ("T")

$$= \mathrm{REFPROP}(''\mathrm{H}'', ''\mathrm{HELIUM}'', ''\mathrm{PT}'', ''\mathrm{MKS}'', 100, 300)$$

Other properties can be calculated by using any of the following functions: pressure, temperature, enthalpy, entropy, density, isochoric heat capacity (or c_V), isobaric heat capacity (or c_p,), mass quality and so on.

For saturation states, the second input is Tliq or Tvap for a given saturated temperature, Pliq or Pvap for a given saturated pressure or Dliq or Dvap for a saturated density. Equilibrium liquid properties are returned when these inputs contain "liq", and saturated vapour properties are returned when "vap" is used. Of course, only one input is necessary.

$$= \mathrm{REFPROP}(''\mathrm{H}'', ''\mathrm{HELIUM}'', ''\mathrm{Pliq}'', ''\mathrm{MKS}'', 100)$$
$$= \mathrm{REFPROP}(''\mathrm{H}'', ''\mathrm{HELIUM}'', ''\mathrm{Tvap}'', ''\mathrm{MKS}'', 4.5)$$

The third input to the property functions defines the units. Here, "MKS" will be used. The units that correspond to these inputs are kPa, K, kg/m^3, kJ/kg, kJ/kg-K, m/s, μPa-s, W/m-K, mN/m.

For example, the following line would be placed in a cell to calculate the density of argon at 1 kPa and 100 K (the result will be in units of kg/m^3):

$$= \mathrm{REFPROP}(''\mathrm{D}'', ''\mathrm{ARGON}'', ''\mathrm{PT}'', ''\mathrm{MKS}'', 1, 100)$$

Pressure expressed in kPa units are, sometimes, not easily significant. Therefore, the author prefers to use "bar" for pressure, "g/s" for mass flow rates "J/g" for enthalpy and J/g.K for entropy. Consequently, some coefficients are to be introduced to work directly with such units: pressure in bars is to be multiplied by 100 to get kPa, but no coefficient is to be used for enthalpy, entropy and density. Now, the function to calculate enthalpy of helium in J/g at 1.0 bar and 300.0 K comes to:

$$= \mathrm{REFPROP}(''\mathrm{H}'', ''\mathrm{HELIUM}'', ''\mathrm{PT}'', ''\mathrm{MKS}'', 1 * 100, 300.0)$$

In order to run same calculations with various gases, it is interesting to type the fluid name on a cell (here A1) and refer to the cell.

The spread sheet is displayed in Fig. 15.3, top

	A	B	C	D
1	helium	P	T	H
2		Pressure	Temperature	Enthalpy
3		(bar)	(K)	(J/g)
4		1	300	=REFPROP("H",A1,"PT","mks",B4*100,C4)

	A	B	C	D
1	helium	P	T	H
2		Pressure	Temperature	Enthalpy
3		(bar)	(K)	(J/g)
4		1.00	300.00	1563.3194

Fig. 15.3 The enthalpy spread sheet

Remark About Inputs

It is suggested to type inputs **in red bold figures**. Consequently, when looking to a former spread sheet, one immediately sees which and where the inputs are.

Some info:

- In line 1, type the uppercase letter that refers to the property (that allows to check the function).
- In line 2, type the parameter name.
- In line 3, type the units.
- In cells B4 and C4, inputs are **BOLD RED**: easy to see.

When the calculation is performed, it comes to Fig. 15.3, bottom.

Remark If the reader wants to perform same calculations, he/she should use the same cell coordinates (line and column) as in the example, to be in a position to make easier checks if results are different.

In the same way as for enthalpy, build the property tool that contains the properties that one uses generally: with the same inputs (pressure and temperature), calculate, for example, entropy, specific weight and quality.

15.7.2 Calculation of Some Most Common Properties

$$= \mathrm{REFPROP}(''\mathrm{H}'', ''\$\mathrm{A}\$1'', ''\mathrm{PT}'', ''\mathrm{MKS}'', \mathrm{B4} * 100, \mathrm{C4})$$
$$= \mathrm{REFPROP}(''\mathrm{S}'', ''\$\mathrm{A}\$1'', ''\mathrm{PT}'', ''\mathrm{MKS}'', \mathrm{B4} * 100, \mathrm{C4})$$
$$= \mathrm{REFPROP}(''\mathrm{D}'', ''\$\mathrm{A}\$1'', ''\mathrm{PT}'', ''\mathrm{MKS}'', \mathrm{B4} * 100, \mathrm{C4})$$

The only change is the function value.

Now, one can assemble the tool (Fig. 15.4):

	A	B	C	D	E	F
1	helium	P	T	H	S	D
2		Pressure	Temperature	Enthalpy	Entropy	Density
3		(bar)	(K)	(J/g)	(J/g.K)	(kg/m^3)
4		1.00	300.00	1563.3194	28.0075	0.1604

Fig. 15.4 Calculation of some properties

This property tool calculates enthalpy, entropy and density according to the same input doublet: pressure and temperature in cells B4 and C4.

15.7.3 *Other Property Calculations*

But one also needs to calculate the temperature that corresponds to a given input doublet as pressure and enthalpy or other doublets. Use the same function with different inputs:

Calculations according to pressure and enthalpy:

$$=\mathrm{REFPROP}(''\mathrm{T}'', ''\$\mathrm{A}\$1'', ''\mathrm{HELIUM}'', ''\mathrm{PH}'', ''\mathrm{MKS}'', \mathrm{A5}*100, \mathrm{C5})$$
$$=\mathrm{REFPROP}(''\mathrm{S}'', ''\$\mathrm{A}\$1'', ''\mathrm{HELIUM}'', ''\mathrm{PH}'', ''\mathrm{MKS}'', \mathrm{A5}*100, \mathrm{C5})$$
$$=\mathrm{REFPROP}(''\mathrm{D}'', ''\$\mathrm{A}\$1'', ''\mathrm{HELIUM}'', ''\mathrm{PH}'', ''\mathrm{MKS}'', \mathrm{A5}*100, \mathrm{C5})$$
$$=\mathrm{REFPROP}(''\mathrm{Qmass}'', ``\$\mathrm{A}\$1", ''\mathrm{HELIUM}'', ''\mathrm{PH}'', ''\mathrm{MKS}'', \mathrm{A5}*100, \mathrm{C5})$$

Calculations according to pressure and entropy:

$$=\mathrm{REFPROP}(''\mathrm{T}'', ''\$\mathrm{A}\$1'', ''\mathrm{HELIUM}'', ''\mathrm{PS}'', ''\mathrm{MKS}'', \mathrm{A5}*100, \mathrm{C6})$$
$$=\mathrm{REFPROP}(''\mathrm{H}'', ''\$\mathrm{A}\$'', ''\mathrm{HELIUM}'', ''\mathrm{PS}'', ''\mathrm{MKS}'', \mathrm{A5}*100, \mathrm{C6})$$
$$=\mathrm{REFPROP}(''\mathrm{D}'', ''\$\mathrm{A}\$1'', ''\mathrm{HELIUM}'', ''\mathrm{PS}'', ''\mathrm{MKS}'', \mathrm{A5}*100, \mathrm{C6})$$
$$=\mathrm{REFPROP}(''\mathrm{Qmass}'', ''\$\mathrm{A}\$1'', ''\mathrm{HELIUM}'', ''\mathrm{PS}'', ''\mathrm{MKS}'', \mathrm{A5}*100, \mathrm{C6})$$

And, finally, the tool box comes to (see Fig. 15.5):

In line 5, quality is a number between 0 and 1 that shows the ratio of vapour (Fig. 15.5). In other lines, it is said that “state in single phase, quality not calculated”.

15.7.4 *Properties at Saturation*

REFPROP allows also getting helium properties *at saturation*; therefore, the calling sequence deals with *only one parameter* (most usually, the pressure or the temperature).

	A	B	C	D	E	F	G
1	helium	P	T	H	S	D	Q
2		Pressure	Temperature	Enthalpy	Entropy	Density	Quality
3		(bar)	(K)	(J/g)	(J/g.K)	(kg/m^3)	(-)
4		1.00	300.00	1563.3194	28.0075	0.1604	#State is s
5		1.00	4.2098	15.0000	3.5656	21.7606	0.73
6		1.20	6.5170	35.3442	7.4131	9.9341	#State is s
7		1.20	6.4846	35.1496	7.3832	10.0000	#State is s

Fig. 15.5 Calculation of a few properties

$$= \text{REFPROP}(''\text{T}'', ''\$\text{A}\$1'', ''\text{Pvap}'', ''\text{MKS}'', \text{B4} * 100)$$
$$= \text{REFPROP}(''\text{T}'', ''\$\text{A}\$1'', ''\text{Pliq}'', ''\text{MKS}'', \text{B5} * 100)$$
$$= \text{REFPROP}(''\text{P}'', ''\$\text{A}\$1'', ''\text{Tvap}'', ''\text{MKS}'', \text{C4})$$
$$= \text{REFPROP}(''\text{P}'', ''\$\text{A}\$1'', ''\text{Tliq}'', ''\text{MKS}'', \text{C5})$$
$$= \text{REFPROP}(''\text{H}'', ''\$\text{A}\$1'', ''\text{Pvap}'', ''\text{MKS}'', \text{B7} * 100)$$
$$= \text{REFPROP}(''\text{H}'', ''\$\text{A}\$1'', ''\text{Pliq}'', ''\text{MKS}'', \text{B8} * 100)$$

and so on.

In a similar way, one can calculate other properties of saturated vapour and of saturated liquid

$$= \text{REFPROP}(''\text{S}'', ''\$\text{A}\$1'', ''\text{Tvap}'', ''\text{MKS}'', \text{C4})$$

or

$$= \text{REFPROP}(''\text{D}'', ''\$\text{A}\$1'', ''\text{Pliq}'', ''\text{MKS}'', \text{B8} * 100)$$

orand so on.

This new part of the tool box (lines 9 and 10) comes to (Fig. 15.6):

15.7.5 *The Gas Property Sheet*

The basic gas property sheet appears finally like that (Fig. 15.7):

This is a *very basic tool*. It is generally sufficient for most calculations for helium refrigeration, but one can customise it as for most used properties, units, colours and so on.

At least, one is now in a position to use the tool for calculations! It is advised to open a special file in the directory: "Cryo Tool Box". This spread sheet could be named: "Gas properties". Save it preciously in the "Cryo Tool Box" work book, and therefore, it can be used at any time.

Remark It is interesting to protect the sheets in order to avoid some involuntary modification that could bring an error!

	A	B	C	D	E	F	G
1	helium	P	T	H	S	D	Q
2		Pressure	Temperature	Enthalpy	Entropy	Density	Quality
3		(bar)	(K)	(J/g)	(J/g.K)	(kg/m^3)	(-)
4	Sat. vapour	1.00	4.2098	20.5718	4.8892	16.6713	1.00
5	Sat. liquid	1.00	4.2098	-0.0733	-0.0149	124.9442	0.00
6							
7	Sat. vapour	1.1908	4.40	20.3832	4.6035	20.1241	1.00
8	Sat. liquid	1.1908	4.40	0.9848	0.1948	120.9128	0.00

Fig. 15.6 Properties at saturation

	A	B	C	D	E	F	G
1	helium	P	T	H	S	D	Q
2		Pressure	Temperature	Enthalpy	Entropy	Density	Quality
3		(bar)	(K)	(J/g)	(J/g.K)	(kg/m^3)	(-)
4		1.00	300.00	1563.3194	28.0075	0.1604	#State is sinc
5		1.00	4.2098	15.0000	3.5656	21.7606	0.73
6		1.20	6.5170	35.3442	7.4131	9.9341	#State is sinc
7		1.20	6.4846	35.1496	7.3832	10.0000	#State is sinc
8							
9	Sat. vapour	1.00	4.2098	20.5718	4.8892	16.6713	1.00
10	Sat. liquid	1.00	4.2098	-0.0733	-0.0149	124.9442	0.00
11							
12	Sat. vapour	1.1908	4.40	20.3832	4.6035	20.1241	1.00
13	Sat. liquid	1.1908	4.40	0.9848	0.1948	120.9128	0.00

Fig. 15.7 The gas property basic spread sheet

15.7.6 A Faster Way to Get Often Called Properties: Excel Macros

For most often used properties as ($h = f(P, T)$, $T = f(P, h)$, it is interesting to create macros, following Excel instructions. For example, in Fig. 15.8, the macro CTRL +SHIFT+H calculates the enthalpy according to cell A1 for the gas, cell located on the same line, at −2 columns for pressure and cell located on the same line, at −1 column for temperature (this is same disposition as in the gas properties spread sheet).

Similarly, in Fig. 15.9, CTRL+SHIFT+H returns the temperature according to cell A1 for the gas, cell located on the same line, at −1 column for pressure and cell located on the same line, at +1 column for enthalpy

More complicated macros can be written, for example, for calculating a turbine.

```
'
 Sub hRP ( )
'
'
 EnthalpyREFPROP Macro
'
'
 Shortcut key: Ctrl+Shift+H
'
     ActiveCell.FormulaR1C1
="=REFPROP(""H"",R1C1,""PT"",""MKS"",RC[-2]*100,RC[-1])"
End Sub
'
```

Fig. 15.8 A macro for calculating enthalpy from pressure and temperature with REFPROP

```
'
 Sub tRP ( )
'
'
 TemperatureREFPROP Macro
'
'
 Shortcut key: Ctrl+Shift+T
'
     ActiveCell.FormulaR1C1
="=REFPROP(""T"",R1C1,""PT"",""MKS"",RC[-1]*100,RC[1])"
End Sub
'
```

Fig. 15.9 A macro for calculating temperature from pressure and enthalpy with REFPROP

15.8 Building Some Tools

The Way It Is Suggested to Work

Even if the systems that are calculated are simple, it is of interest to have a sketch that shows where the parameters (inputs and outputs) are located.

As it is the first spread sheet that is built here, let us explain the way to proceed:

- At the left of the current sheet, draw a sketch of the system where inputs are fed (**in red bold**).
- Transfer the inputs into the "active part" of the spread sheet, by copy from cell of the sketch to cell of the spread sheet.

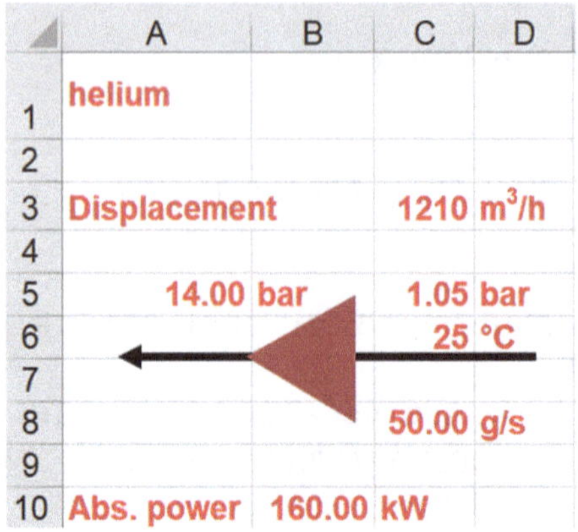

Fig. 15.10 Room temperature compressor sketch

- Calculate.
- Display the results of the calculation on the sketch, by copy from cell of the spread sheet to cell of the sketch. This makes possible to see all parameters in the sketch.

It is advised to use same cell coordinates as in the examples in order to make possible corrections easier.

15.8.1 The Room Temperature Compressor

Draw the Sketch (Fig. 15.10)
Inputs:

- Suction pressure: 1.05 bar
- Suction temperature: 25.0 °C
- Discharge pressure:14.0 bar
- Processed mass flow rate: 50.0 g/s

Type, in bold red, each input into the correct cell:

- Suction pressure in cell A5
- Suction temperature in cell A6
- And so on

Transfer the Inputs from the Sketch to the Spread Sheet
The spread sheet is "fed" by inputs that have been typed into the sketch: in cell F5, copy C5, in cell G5, copy C6 and so on (see Fig. 15.11).

Calculate
The calculation takes place *in the spread sheet only* (columns F and more). The formula to be implemented into cell K5 for isothermal power is (refer to paragraph 6. 2.1.1):

	A	B	C	F	G	H	I	J	K	L	M	N
1	helium			Suct. press.	Suct. Temp.	Disch. press.	Mass flow	Shaft power	Isoth. power	Isoth. eff.	Displacement	Vol. eff.
2				(bar)	(°C)	(bar)	(g/s)	(kW)	(kW)	(-)	(m^3/h)	(-)
3	Displacement		1210 m^3/h									
4												
5		14.00 bar	1.05 bar	=C5	=C6	=A5	=C8	=B10			=C3	
6			25 °C									
7												
8			50.00 g/s									
9												
10	Abs. power	160.00 kW										
11	Isoth. power		kW									
12	Isoth. eff.		isoth.									
13	Volum. eff.		vol.									

Fig. 15.11 Transferring inputs into the spread sheet

	A	B	C	F	G	H	I	J	K	L	M	N
1	nitrogen			Suct. press.	Suct. Temp.	Disch. press.	Mass flow	Shaft power	Isoth. power	Isoth. eff.	Displacement	Vol. eff.
2				(bar)	(°C)	(bar)	(g/s)	(kW)	(kW)	(-)	(m^3/h)	(-)
3	Displacement		200 m^3/h									
4												
5		14.00 bar	1.05 bar	1.05	25.00	14.00	50.00	25.00	11.46	0.458	200.00	0.758
6			25 °C									
7												
8			50.00 g/s									
9												
10	Abs. power	25.00 kW										
11	Isoth. power	11.46 kW										
12	Isoth. eff.	0.458 isoth.										
13	Volum. eff.	0.758 vol.										

Fig. 15.12 Finalised calculation of a room temperature helium compressor

$$\dot{W} = \dot{m} \times T_1 \times R/M \times \mathrm{Ln}(P_2/P_1) \quad (6.4)$$

with:

$R = 8.314472\ \mathrm{J.mol^{-1}.K^{-1}}$
M = molar mass of the gas. For helium, REFPROP calculation of the molar mass is
= REFPROP("M";A1) = 4.0026 g

$$= (\mathrm{I5}/1000) * (273.15 + \mathrm{G5}) * (8.314472/(\ \mathrm{REFPROP}(''\mathrm{M}'', \$\mathrm{A}\$1)) * \mathrm{Ln}(\mathrm{H5}/\mathrm{F5})$$

In cell L5: = K5/J5
In cell N5:

$$= \mathrm{I5} * 3.6/(\mathrm{REFPROP}(''\mathrm{D}'', \$\mathrm{A}\$1, ''\mathrm{PT}'', ''\mathrm{mks}'', \mathrm{F5} * 100, (\mathrm{G5} + 273.15)))/\mathrm{M5}$$

Display the Results
When the calculation is finalised, it appears as in Fig. 15.12. Notice that, in order to have all information on the sketch, one can copy values of Isothermal

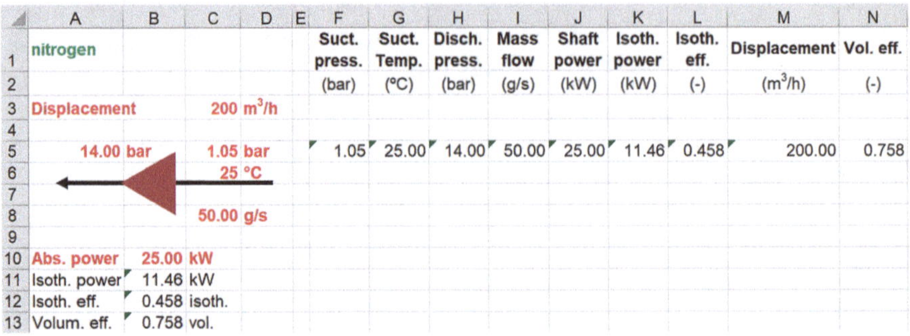

Fig. 15.13 Calculation of a room temperature nitrogen compressor

power, isothermal efficiency and volumetric efficiency in clines 11 to 13. Therefore, all parameters of the calculation appear on the sketch, which makes a comfortable display.

Save this sheet as "Room temperature compressor" (or RTC) into the work book called: "Cryo Tool Box".

The spread sheet can be easily used for a nitrogen compressor, as shown in Fig. 15.13.

15.8.2 The Free Expansion of a Fluid

15.8.2.1 Free Expansion

Let us begin with a very simple exercise:

What is the temperature change when some power (positive or negative) is dumped into a system that is circulated with a helium mass flow rate and, furthermore, having pressure drops?

As in Sect. 15.8.1:

- Draw the sketch with inputs.
- Transfer inputs into the spread sheet.
- Calculate.
- Display results on the sketch.

Draw the Sketch (Fig. 15.14)
Inputs:

- Initial conditions: $P_{\text{in}} = 14.00$ bar, $T_{\text{in}} = 20.00$ K, $\dot{m}$ 50.00 g/s
- Thermal power: $Q = 100$ W
- Discharge condition: $P_{\text{dis}} = 1.00$ bar,

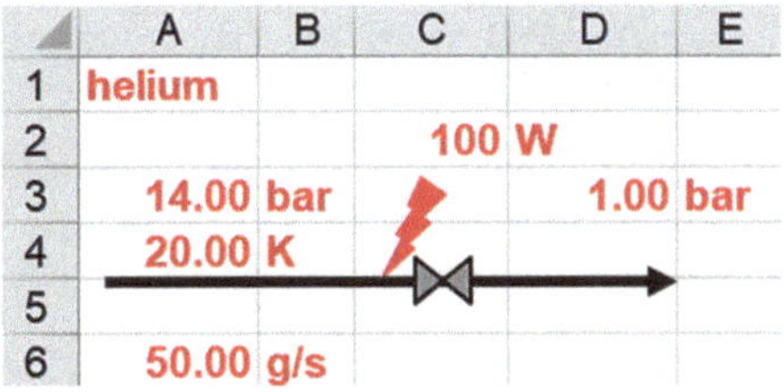

Fig. 15.14 Sketch of the system

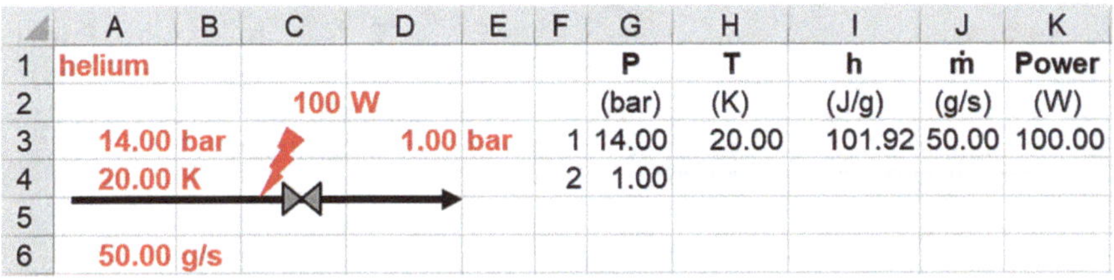

Fig. 15.15 Transferring inputs

Transfer Inputs (Fig. 15.15)

A Trick To insert a dot on top of a letter as $\dot{m}$ or $\dot{Q}$, set the pointer just after the letter, open "insert", "symbol" and insert symbol # 0307, Unicode hexadecimal.

Calculate

1. Cell I3: calculate the inlet enthalpy (copy and paste cell D4 in "He prop", or use CTRL+h).
2. Cell I4: calculate of the enthalpy at discharge:

$$h_{\text{dis}} = h_{\text{in}} + \left(\dot{Q}/\dot{m}\right)$$

3. Cell H4: calculate the discharge temperature according to discharge pressure and enthalpy (copy and paste cell C5 in "gas properties", or use CTRL+t) (Fig. 15.16).

This is the first tool for thermodynamic calculations. This spread sheet could be named: "Free expansion". Save it preciously in the "Cryo Tool Box" work book, and therefore, it can be used at any time when this situation is met.

Did you notice that, since we started this exercise, we have completely forgotten that we are using REFPROP?

Now, one can "play" with this spread sheet: modify anyone (or all) input(s) and see how the discharge conditions are changing.

	A	B	C	D	E	F	G	H	I	J	K
1	helium						P	T	h	ṁ	Power
2			100	W			(bar)	(K)	(J/g)	(g/s)	(W)
3	14.00	bar		1.00	bar	1	14.00	20.00	101.92	50.00	100.00
4	20.00	K		19.16	K	2	1.00	19.16	103.92		
5											
6	50.00	g/s									

Fig. 15.16 Free expansion

	A	B	C	D	E	F	G	H	I	J	K
1	helium						P	T	h	ṁ	Power
2			0	W			(bar)	(K)	(J/g)	(g/s)	(W)
3	14.00	bar		1.00	bar	1	14.00	300.00	1567.56	50.00	0.00
4	300.00	K		300.82	K	2	1.00	300.82	1567.56		
5											
6	50.00	g/s									

Fig. 15.17 Expansion of helium through a valve

	A	B	C	D	E	F	G	H	I	J	K
1	helium						P	T	h	ṁ	Power
2			0	W			(bar)	(K)	(J/g)	(g/s)	(W)
3	14.00	bar		1.00	bar	1	14.00	300.00	1567.56	50.00	0.00
4	300.00	K		300.82	K	2	1.00	300.82	1567.56		
5								-0.82			
6	50.00	g/s									

Fig. 15.18 Adding the temperature difference calculation

15.8.2.2 Expanding Helium at Room Temperature

Let us re-do the very first calculation that was done in Sect. 1.3 (Fig. 15.17).

15.8.2.3 Looking for the Inversion Temperature of a Gas

Remember Sect. 1.4.2.2 where the inversion temperature is the temperature at which an isenthalpic expansion does not change the gas temperature (Fig. 15.18).

In the cell H5 of Fig. 15.17, insert a calculation of the difference between cell H3 and cell H4. In the "What-if analysis"/Goal seek", ask what is the value of cell A6 in order cell I5 = 0. The result in in Fig. 15.19: 43.21 K.

Let us perform the same calculation with nitrogen. It is only necessary to type nitrogen instead of helium in cell A1 and run the "What-if analysis"/" goal seek" (Fig. 15.20).

	A	B	C	D	E	F	G	H	I	J	K
1	helium						P	T	h	$\dot{m}$	Power
2			0	W			(bar)	(K)	(J/g)	(g/s)	(W)
3	14.00	bar		1.00	bar	1	14.00	43.21	229.41	50.00	0.00
4	43.21	K		43.21	K	2	1.00	43.21	229.41		
5								0.00			
6	50.00	g/s									

Fig. 15.19 The inversion temperature of helium

	A	B	C	D	E	F	G	H	I	J	K
1	nitrogen						P	T	h	$\dot{m}$	Power
2			0	W			(bar)	(K)	(J/g)	(g/s)	(W)
3	14.00	bar		1.00	bar	1	14.00	603.54	630.79	50.00	0.00
4	603.54	K		603.54	K	2	1.00	603.54	630.79		
5								0.00			
6	50.00	g/s									

Fig. 15.20 The inversion temperature of nitrogen

	A	B	C	D	E	F	G	H	I	J	K	L
1	helium						P	T	h	$\dot{m}$	Power	Quality
2			0	W			(bar)	(K)	(J/g)	(g/s)	(W)	(-)
3	1.20	bar		0.50	bar	1	1.20	4.41	1.04	50.00	0.00	0.00
4	4.41	K		3.55	K	2	0.50	3.55	1.04			0.17
5												
6	50.00	g/s										

Fig. 15.21 Expansion of pure equilibrium liquid helium

15.8.2.4 Expanding Saturated Liquid

It is interesting to see how saturated liquid that is expanded behaves. This is an opportunity to deal with the "quality" function (Fig. 15.21).

Quality is the mass proportion of vapour; therefore, one can see that expanding pure liquid helium gives a two-phase mixture containing $1 - 0.1703 = 0.83$ liquid only. The physical explanation of this phenomenon is that part of liquid has to vaporise to reach the new equilibrium. To vaporise liquid, the heat that is necessary is provided by the remaining part of the liquid that, consequently, cools down. Cells H2 (Temperature) and I2 (Enthalpy) must deal with saturation properties. Therefore, cell C10 of Table 15.7 must be copied in cell H3 of Table 15.21 and cell H10 in cell I2.

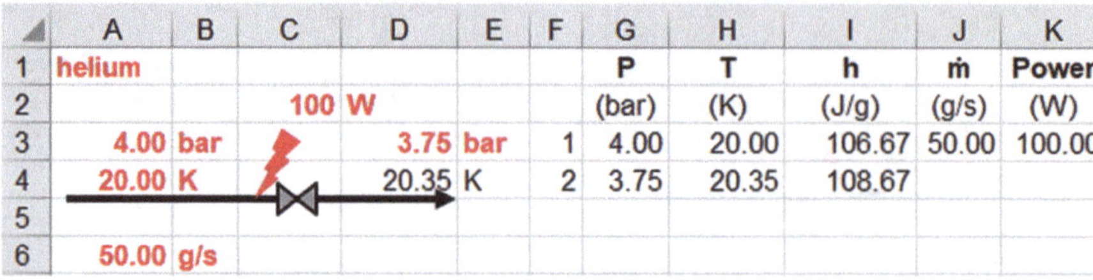

	A	B	C	D	E	F	G	H	I	J	K
1	helium						P	T	h	ṁ	Power
2			100	W			(bar)	(K)	(J/g)	(g/s)	(W)
3	4.00	bar		3.75	bar	1	4.00	20.00	106.67	50.00	100.00
4	20.00	K		20.35	K	2	3.75	20.35	108.67		
5											
6	50.00	g/s									

Fig. 15.22 Filling input values

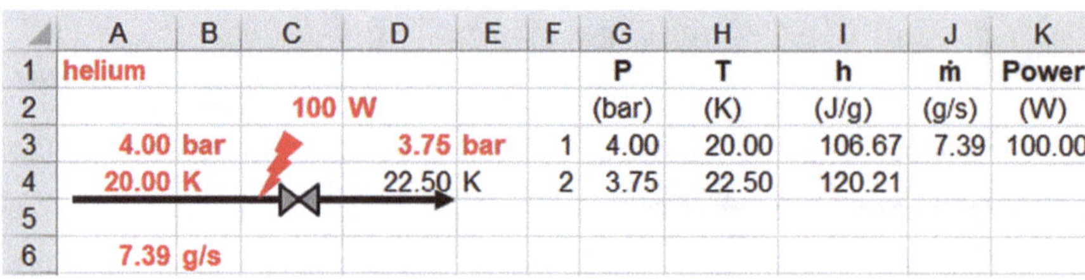

	A	B	C	D	E	F	G	H	I	J	K
1	helium						P	T	h	ṁ	Power
2			100	W			(bar)	(K)	(J/g)	(g/s)	(W)
3	4.00	bar		3.75	bar	1	4.00	20.00	106.67	7.39	100.00
4	20.00	K		22.50	K	2	3.75	22.50	120.21		
5											
6	7.39	g/s									

Fig. 15.23 Calculating the mass flow rate

15.8.2.5 Calculating a Mass Flow

Here, one uses the "What if analysis" tool in another way. What is the helium mass flow rate that is circulated into a circuit, knowing inlet and discharge pressures and temperatures and the power that is dumped into helium?

(This is a usual way to measure a mass flow rate.)

Inputs:

- Inlet conditions: $P_{in} = 4.00$ bar, $T_{in} = 20.00$ K,
- Thermal power: $\dot{Q} = 100$ W
- Discharge condition: $P_{dis} = 3.75$ bar, $T_{in} = 22.50$ K,

Fill-in the new input values (Fig. 15.22):

As the mass flow rate is unknown, leave the former value: 50 g/s. The discharge temperature that is displayed 20.35 K is lower than the one that is measured. One could make manual changes on the value of cell A8 (mass flow) until one gets 22.50 in cell D4; it works, but it is a tedious process. Better use the "Goal Seek" tool of Excel (in What-if analysis)!

In cell D4, type the "Goal Seek" box 22.5 and the box "cell to modify" A8. The target tool returns 7.38 in cell A8 (Fig. 15.23).

By the way, this is a thermal mass flowmeter!

Remark: this way of explaining is somewhat "heavy": steps are small, and one could have gone faster. This way has been selected according to the author's experience when running such calculations sessions with participants of the numerous helium refrigeration schools that he animated: he noticed that not everybody is familiar with thermodynamic calculations and even with Excel.

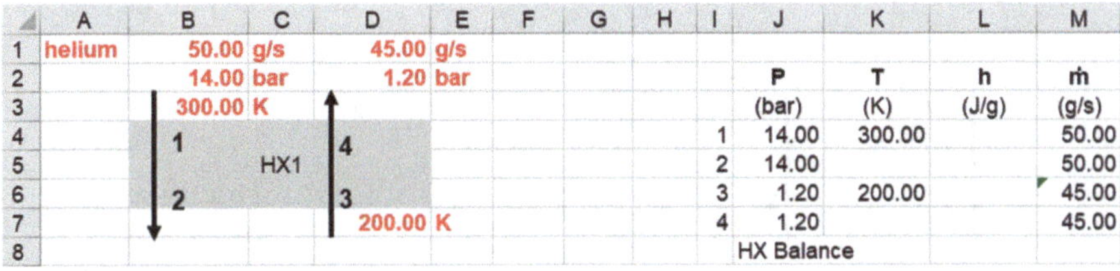

Fig. 15.24 Transferring inputs

15.8.3 The Heat Exchanger

15.8.3.1 A Simple Heat Exchanger (Operating at Temperatures Higher Than 100 K)

Problem: cold helium is used to cool warm helium into a heat exchanger (assumption – no pressure drops, no heat in-leaks). What is the temperature of the warm circuit helium at its outlet (point 2 in the sketch)? By the way, these are the operating conditions of the heat exchanger that is discussed in paragraph 2.3.

Sketch and Inputs

Let us call 1 and 2 the warm gas inlet and outlet and 3 and 4 the cold gas inlet and outlet (keep the same tagging in all heat exchanger calculations) (Fig. 15.24).

Inputs:

- Conditions at 1: $P_1 = 14.00$ bar, $T_1 = 300.00$ K, $\dot{m}_1 = 50.00$ g/s
- Conditions at 2: $P_2 = 14.00$ bar, $\dot{m}_2 = 50.00$ g/s
- Conditions at 3: $P_3 = 1.20$ bar, $T_3 = 200.00$ K, $\dot{m}_3 = 45.00$ g/s
- Conditions at 4: $P_4 = 1.20$ bar, $\dot{m}_4 = 45.00$ g/s

Calculate (Fig. 15.25 top)

There are two unknown values: T_2 and T_4.

Cell D3: select a reasonable temperature difference value (remember it is 1% of the temperature): 3.0 K. Therefore: $T_4 = T_1 - 3.00$.

Cells L4, L6 and L7 are copy and paste of cell D4 (enthalpy = f(P and T)) of "gas properties".

Cell L5: type $h_2 = [\dot{m}_3 \times (h_3 - h_4) + \dot{m}_1 \times h_1]/\dot{m}_1$ Eq. 2.5

Cell K5: copy and paste cell C5 (temperature $= f(P$ and $h)$) of "gas properties".

The spread sheet is ready and runs!

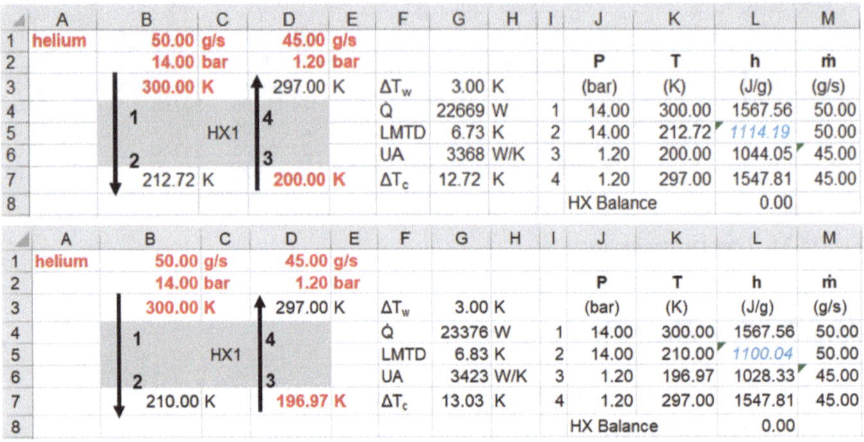

Fig. 15.25 A simple heat exchanger

$$T_2 = 212.72 \text{ K}$$

It is a good practice to check if the heat exchanger mass and thermal balance are correct. In cell L8, type:

$$= \text{M4} * (\text{L4-L5}) + \text{M6} * (\text{L6-L7})$$

This spread sheet can be improved by adding the calculation of the power, the linear mean logarithmic temperature difference and so on in columns G, G and H.

This spread sheet could be named: "Simple heat exchanger". Save it preciously in the "Cryo Tool Box" work book, and therefore, it can be used at any time when this situation is met.

Remark

As it was done with the "Free expansion" spread sheet, one can use the heat exchanger spread sheet *without any change* to calculate in another way.

For example, what should be the warm mass flow rate to get $T_2 = 210$ K?

This is possible by using the "What if analysis/Goal Seek" tool: set cell B7, fill-in the "to value" box 210 and the box "By changing cell" D7. The Goal Seek tool returns 196.97 in cell D7, as shown in Fig. 15.25 bottom.

Warning! Such a tool can only be used for heat exchangers operating above 100 K. For lower temperatures, the heat exchanger must be cut into elementary heat exchangers, assuming that in each elementary heat exchanger, thermo-physical properties are constant (see Sect. 15.8.3.2).

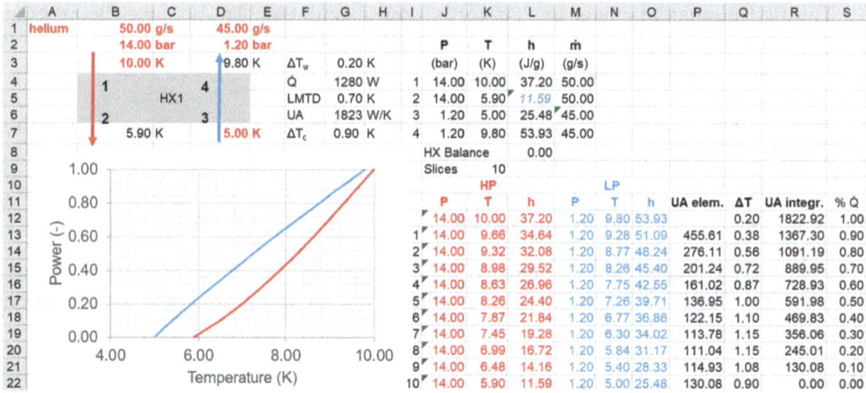

Fig. 15.26 Cold HX

15.8.3.2 A Heat Exchanger in Which Helium Properties Are Not Constant

Remember Sect. 2.4.1.4.

In the area A1 to M8, except F5 to G6, calculations are performed as done in Sect. 15.8.3. (Fig. 15.26).

Remark If, at point 3, it were saturated vapour, enthalpy should be calculated accordingly (REFPROP("H";A1;"Tvap";"mks";J6*100).

In order to calculate the heat exchanger UA, one has to cut the heat exchanger in slices, here, 10 slices.

- Input the number of slices in cell K9,
- Copy column headings in J10 to R11,
- Type the point numbers in cells I13 to I22,
- Copy HP pressure in cells J12 to J22,
- Copy LP pressure in cells M12 to M22,
- Copy K4 and L4 in K12 and L12,
- Copy K7 and L7 in N12 and O12.

The first elementary heat exchanger transfers one / number of slices of the energy, therefore the enthalpy at its HP cold end (cell L13) is: L12-((L4-L5)/K9).

- Copy the formula to calculate temperature = f(P and h) which is in cell C5 of Gas properties (REFPROP("T";A1;"PH";"mks";J13*100;L13) of Fig.15.7 in K13,
- Perform same calculations on the LP side,
- Calculate the elementary UA of this heat exchanger in P13: G$4/K$9)/(((K12-N12)-(K13-N13)) /LN((K12-N12)/(K13-N13))),
- In cell Q12, calculate the temperature difference at the warm end of the heat exchanger,
- In cell Q13, calculate the temperature difference at the cold end of the heat exchanger,

The first elementary heat exchanger is now calculated. Apply same calculation process to the other elementary heat exchangers in cells K14 to R22.

Plotting the heat exchanger diagram: power = f(T)

- Add column S to state that each elementary heat exchanger power is 1/10 of the whole heat exchanger power,
- Select K12 to K22 and S12 to S22 and plot the HP part of the diagram. (red line),
- Select N12 to N22 and S12 to S22 and plot the LP part of the diagram. (blueline).

This spread sheet could be named: "heat exchanger" (or HX). Save it preciously in the "Cryo Tool Box" work book, and therefore, it can be used at any time when this situation is met.

Remark 1 If the LMTD had been calculated, assuming that thermal properties are constant (linear diagram), the result would have been 0.27 k instead of 0.79 K.

Remark 2 If some temperatures are lower than around 3 K, it is mandatory to perform the calculation with the HEPAK software instead of REFPROP.

15.8.3.3 A Heat Exchanger Operating with Two Different Fluids

The fluids might be different, for example, helium that is cooled against nitrogen (Fig. 15.27).

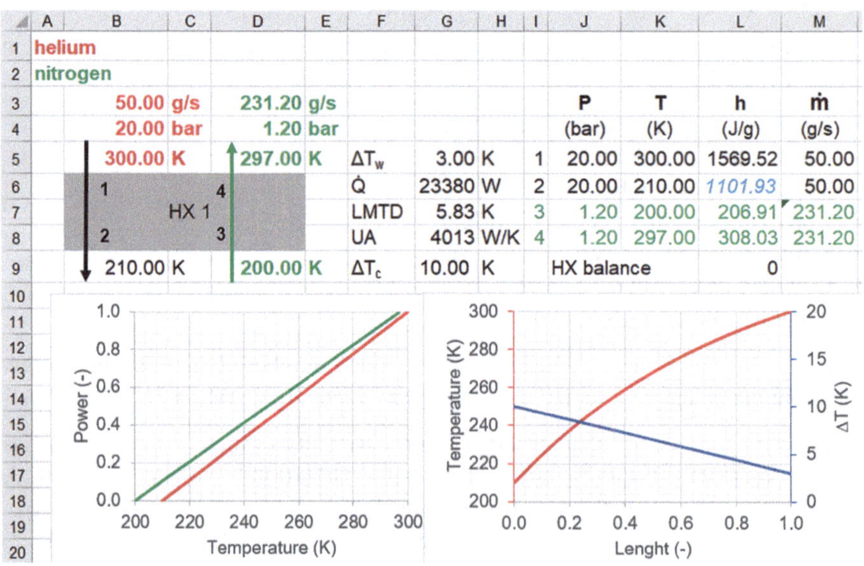

Fig. 15.27 A heat exchanger operating between helium and nitrogen

	A	B	C	D	E	F	G	H	I	J	K	L	M
1	helium					P	T	h	s	$\dot{m}$	$\dot{Q}$	Δh isent	Δh real
2		14.00	bar			(bar)	(K)	(J/g)	(J/g.K)	(g/s)	(W)	(J/g)	(J/g)
3		25.00	K		1	14.00	25.00			50.00			
4		50.00	g/s		η	0.75							
5		0.75	isentr.		is	1.20							
6					2	1.20							
7		1.2	bar										
8			K										

Fig. 15.28 The cryogenic expander 1

	A	B	C	D	E	F	G	H	I	J	K	L	M
1	helium					P	T	h	s	$\dot{m}$	$\dot{Q}$	Δh isent	Δh real
2		14.00	bar			(bar)	(K)	(J/g)	(J/g.K)	(g/s)	(W)	(J/g)	(J/g)
3		25.00	K		1	14.00	25.00	130.49	9.42	50.00	-2982		
4		50.00	g/s		η	0.75							
5		0.75	isentr.		is	1.20		50.98	9.42			-79.51	-59.63
6					2	1.20	12.93	70.85					
7		1.2	bar										
8		12.93	K										

Fig. 15.29 The cryogenic expander 2

This heat exchanger is the same as in Fig. 2.25.

The only change in the calculation is that properties at points 3 and 4 have to be calculated according to the other fluid, here nitrogen. In cell L7:

$$= \text{REFPROP}("\text{H}", \$\text{A}\$2, "\text{PT}", "\text{MKS}", \text{J7} * 100, \text{K7})$$

15.8.4 The Cryogenic Expander

Problem: What is the discharge temperature of a cryogenic expander, knowing its inlet pressure and temperature, efficiency and discharge pressure?

Remember Sect. 6.6.1

Sketch and Inputs (Fig. 15.28)

Inputs:

- Inlet: $P_1 = 14.00$ bar, $T_1 = 25.00$ K, $\dot{m} = 50.00$ g/s
- Discharge: $P_2 = 1.20$ bar,
- Isentropic expander efficiency: $\eta = 0.75$

Line 5 is planned for an intermediary calculation of the ideal expander

Calculation (Fig. 15.29)

- Cell H3: P_1 and T_1 => h_1

- Cell I3: P_1 and T_1 => s_1
- Cell I5: = I3
- Cell H5: P_2 and s_2, => h_{is} (copy cell D6 in the spread sheet, gas properties, and paste it on the working spread sheet)
- Cell L5: isentropic enthalpy difference: $Dh_{is} = h_1 - h_{is}$
- Cell M5: actual enthalpy difference: $Dh_{act} = \eta \times (h_1 - h_{is})$
- Cell H6: actual enthalpy at the discharge side: $h_2 = h_1 - Dh_{act}$
- Cell G6: actual discharge temperature: P_2 and h_2 => T_2

Finally, we get the discharge temperature of our actual expander: 12.93 K.

Improving the Spread Sheet by "Compacting" Calculations

The spread sheet in Fig. 15.29 is large although one only needs two lines: inlet and discharge (lines 2 and 6). Accordingly, we are not too much interested in the values of entropy. Is there a way to "compact" the spread sheet?

In Fig. 15.30 top, let us use only the input cells.

Let's start by compacting cell H6. The formula in the cell in Fig. 15.29 is:

$$\text{H6} = \text{H3} + \text{M6}$$

H3 is replaced by the formula in H3:

$$\text{H3} = \text{REFPROP}(\text{"H"}, \$\text{A}\$1, \text{"PT"}, \text{"MKS"}, \text{F3} * 100, \text{G3})$$
$$\text{H6} = \text{REFPROP}(\text{"H"}, \$\text{A}\$1, \text{"PT"}, \text{"MKS"}, \text{F3} * 100, \text{G3}) - \text{M6}$$

M6 is replaced by the formula in M6:

$$= \text{L5} * \text{F4}$$

In Fig. 15.30 top, H6 comes to:

$$\text{H6} = \text{REFPROP}(\text{"H"}, \$\text{A}\$1, \text{"PT"}, \text{"MKS"}, \text{F3} * 100, \text{G3}) - (\text{L5} * \text{F4})$$

Compacting L5. The formula in the cell in Fig. 15.29 is:

$$\text{L5} = \text{H5} - \text{H3}$$

$$\text{L5} = \text{REFPROP}(\text{"H"}, \$\text{A}\$1, \text{"PS"}, \text{"MKS"}, \text{F5} * 100, \text{I5}) - \text{REFPROP}(\text{"H"}, \$\text{A}\$1, \text{"PT"}, \text{"MKS"}, \text{F3} * 100, \text{G3})$$

I5

$$\text{I5} = \text{I3} = \text{REFPROP}(\text{"S"}, \$\text{A}\$1, \text{"PT"}, \text{"MKS"}, \text{F3} * 100, \text{G3})$$

H6 comes to:

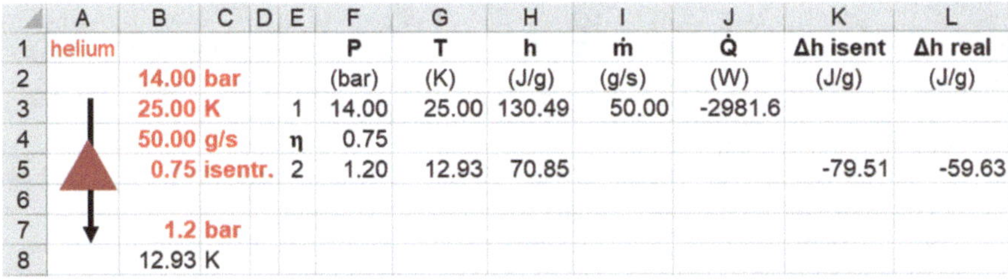

	A	B	C	D	E	F	G	H	I	J	K	L
1	helium					P	T	h	ṁ	Q̇	Δh isent	Δh real
2		14.00	bar			(bar)	(K)	(J/g)	(g/s)	(W)	(J/g)	(J/g)
3		25.00	K		1	14.00	25.00	130.49	50.00	-2981.6		
4		50.00	g/s		η	0.75						
5		0.75	isentr.		2	1.20	12.93	70.85			-79.51	-59.63
6												
7		1.2	bar									
8		12.93	K									

Fig. 15.30 Finalised turbine spread sheet

H6 = (REFPROP("H",A1,"PT","mks",F3*100,G3))
−(((REFPROP("H",A1,"PT","mks",F3*100,G3))
−(REFPROP("H",A1,"SP","mks",(REFPROP("S",A1,"PT","mks",
F3*100,G3)),F5*100)))*F4)

One sees that, now, all parameters that are in the H6 cell formula are the input parameters:

- Inlet pressure: F3
- Inlet temperature: G3
- Efficiency: F4
- Discharge pressure: F6

It can be checked that when cells H3, I3, H5, I5, L5 and M6 are deleted, the result in cell H6 is not affected.

The spread sheet can be now "compacted" as shown in Fig. 15.31 bottom. As one is not interested in the entropy values, the column will be used for mass flow rate in order to calculate the cryogenic power of the expander. It can be even more compacted in order to calculate directly the discharge temperature.

This spread sheet could be named: "Expander". Save it preciously in the "Cryo Tool Box" work book, and therefore, it can be used at any time when this situation is met.

By the way, this is the tool that is displayed in Fig. 6.110.

As explained in Sect. 15.7.6, a macro can be created, that gives directly the discharge temperature of the turbine.

As it has been done with the former spread sheets, one can use the spread sheet *without any change* to calculate in another way. For example, what should be the efficiency of an expander operating in the same conditions: (P_1, T_1, P_2) but which discharge temperature is $T_2 = 13.20$ K?

This is possible by using the "Goal Seek" tool: set cell G5, fill-in the "to value" box = 28.52 and the box "By changing cell" B5. The Goal Seek tool returns: 0.73 in cell F4.

	A	B	C	D	E	F	G	H	I	J	K	L
1	helium					P	T	h	ṁ	Power	ΔH isent	ΔH real
2		0.50	bar			(bar)	(K)	(J/g)	(g/s)	(W)	(J/g)	(J/g)
3		5.00	K		1	0.50	5.00	29.02	400.00			
4		400.00	g/s		η	0.75						
5		0.75	isentr.		2	1.20	7.72	42.37		5336	17.79	13.34
6		1.20	bar									
7		7.72	K									

Fig. 15.31 A cryogenic compressor

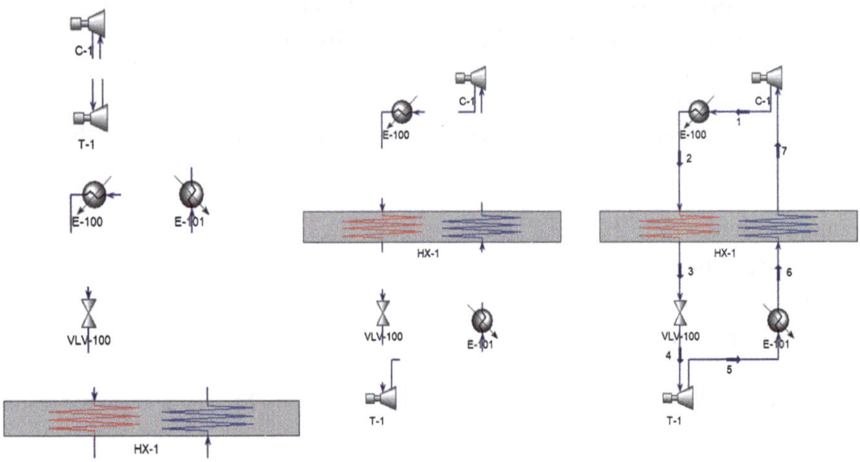

Fig. 15.32 HYSIS flow diagram

15.8.5 The Cryogenic Compressor or Circulator

It is, obviously, a very similar calculation as for an expander. The difference sits in the calculation of the real enthalpy difference: one has to divide the isentropic enthalpy drop by the efficiency to get the real one (Fig. 15.31).

This spread sheet could be named: "Cryo compressor". Save it preciously in the "Cryo Tool Box" work book, and therefore, one can use it any time one meets this situation.

By the way, this is the calculation that is displayed in Fig. 6.140!

15.8.6 Industrial Software to Perform Thermodynamic Calculations

These are very simple information about the HYSIS (Aspen Tech) software that is widely used for thermodynamic calculations.

As a simple example, let us consider a Brayton cycle.

On the left part of Fig. 15.32, a few components of the HYSIS library are displayed:

- A compressor
- An expander
- Two thermal loads (heat exchangers)
- A valve
- A process heat exchanger

In the centre, the various components have been dragged to their location on the flow diagram. Each component and process point is given a marker (C-1, E-100, etc.).

On the right part, they have been connected by lines that are drawn between the inlet and outlet of components to be connected.

As soon as the components are connected, the software establishes the relationship between them.

Inputs are introduced, and the software returns the information that are requested at each process point.

15.8.7 The Equivalent Power

This tool allows calculating the equivalent power of a mixed duty refrigerator. Inputs are written in bold red in Fig. 15.33.

Shields Calculate in D8, the mass flow rate that allows absorbing the cryogenic power (remember Sect. 1.4), according to inlet and outlet conditions. In D8, type:

	A	B	C	D	E	F	G	H
1	**helium**							
2	Room temp.	**300**	(K)	Shields	Supercritical	Isothermal	Liquid	
3	Power		(W)	**3000**	**2000**	**10000**		
4	P in		(bar)	**18.00**	**3.00**		1.20	
5	T in		(K)	**70.00**	**4.50**		300	
6	P out		(bar)	**17.50**	**2.70**	**1.20**	**1.20**	
7	T out		(K)	**80.00**	**4.90**	4.4087	4.4087	
8	Mass flow		(g/s)	57.25	925.92		**13.00**	
9	Carnot power		(W/W)	10055	139615	670479	86643	906792
10	Carnot power at	4.5	K					65.67
11	Equivalent at	**4.5**	K	153	2126	10210	1319	13809
12	Absorbed power		(kW)					**3150**
13	Specific power		(W/W)					228
14	Carnot yield		(-)					0.288
15	Absorbed power		(kW)					3130

Fig. 15.33 Calculation of the equivalent power

$$D3/((\text{REFPROP}(''H'', \$A\$1, ''PT'', ''MKS'', D6 * 100, D7)) \\ -(\text{REFPROP}(''H'', \$A\$1, ''PT'', ''MKS'', D4 * 100, D5)))$$

In D9, calculate the Carnot power (remember Sect. 3.2.2):

$$D8 * ((B2 * ((\text{REFPROP}(''S'', \$A\$1, ''PT'', ''MKS'', D6 * 100, D7)) \\ - (\text{REFPROP}(''S'', \$A\$1, ''PT'', ''MKS'', D4 * 100, D5)))) \\ - ((\text{REFPROP}(''H'', \$A\$1, ''PT'', ''MKS'', D6 * 100, D7)) \\ - (\text{REFPROP}(''H'', \$A\$1, ''PT'', ''MKS'', D4 * 100, D5))))$$

Supercritical It is exactly the same calculation as for the shields. One has only to drag cells D8 and D9 into column E.

Isothermal Calculate the Carnot power in F9, as indicated in Sect. 3.1:

$$F3 * ((B2 - F7)/F7)$$

Liquid Calculate the Carnot power:

$$G8 * ((\$B\$2 * ((\text{REFPROP}(''S'', \$A\$1, ''PT'', ''mks'', G6 * 100, G5)) \\ - (\text{REFPROP}(''S'', \$A\$1, ''Pliq'', ''mks'', G6 * 100)))) \\ - ((\text{REFPROP}(''H'', \$A\$1, ''PT'', ''mks'', G6 * 100, G5)) \\ - (\text{REFPROP}(''H'', \$A\$1, ''Pliq'', ''mks'', G6 * 100))))$$

In cell H9, calculate the sum of Carnot powers.
In line 11, calculate the equivalent powers. For example, in D11:

$$D9/\$H\$10$$

Estimation of the Compressor Power (Remember Sect. 3.5.5):
In cell H15:

$$437 * (0.001 * H11) \wedge 0.75$$

This spread sheet could be named: “Equivalent power”. Save it preciously in the “Cryo Tool Box” work book and therefore, it can be used at any time when this situation is met.

Chapter 16
The Saga of Cryogenic Refrigeration

Abstract It is always interesting and exciting to get some information on the history of an activity. A survey starting from liquefaction of the "permanent" gases up to the present state of cryogenic refrigeration is described.

16.1 The Liquefaction of "Permanent" Gases

In 1845, after M. Faraday succeeded in liquefying various gases by compressing, cooling and both compressing and cooling simultaneously, he called some gases "permanent" gases, because he failed to liquefy them. In 1877, almost simultaneously, L. P. Cailletet, in France, and R. Pictet, in Switzerland, succeeded in liquefying oxygen, by cooling and expansion. In 1898, J. Dewar, in the UK, liquefied hydrogen.

16.2 The First Liquefaction of Helium

Only the latest of many tentative trials for liquefying helium are presented here. After hydrogen that was liquefied on May 10, 1898, by James Dewar, liquefaction of helium was expected soon, and a race started between the most possible winner, J. Dewar, and a challenger, Kamerlingh Onnes, in Leiden, Netherlands (Fig. 16.1). Understanding that the job was difficult, K. Onnes started in a rather "industrial" way: he first set up a glass-blowing and an instrument maker school. He then built a 14 L/h air liquefier and, later, a 4 L/h liquid air pre-cooled hydrogen liquefier too.

Then, he had to get helium gas! He got it out of monazite (Fig. 16.2, left). Due to the alpha decay of thorium and uranium, monazite contains 1–2 cm^3 of helium gas per gram, which can be extracted by heating. Crude helium was purified by passing on CuO, and air impurities were frozen at liquid hydrogen temperature. Finally, pressurised helium was circulated several times through activated charcoal cooled

G. Gistau Baguer, *Cryogenic Helium Refrigeration for Middle and Large Powers*, International Cryogenics Monograph Series,
https://doi.org/10.1007/978-3-030-51677-2_16

Fig. 16.1 Heike Kamerlingh Onnes

at liquid hydrogen temperature. Such a heavy job allowed to obtain an amount of 360 NL of pure helium gas (Fig. 16.2 right)!

To compress helium for the liquefier, he used a Cailletet piston compressor that he modified as shown in Fig. 16.3, left: the part in contact with helium was made out of mercury pushed by the piston; a little bit of glycerine is allowed for lubricant.

The liquefier itself was totally made out of glass (see Fig. 16.4). Thermal insulation was mainly made out of wool.

On the day before the attempt, a reserve of 75 litres of liquid air is produced. The next day as soon as 6:30 A.M., the hydrogen liquefaction begins by the expansion of the gas, previously cooled to about 80 K with liquid air. At 1:30 p.m., 20 litres of liquid hydrogen are produced. The liquefaction of helium can start. Helium gas compressed up to 100 bar (remember that the amount of helium that had Onnes was only 0.36 m^3!) is first cooled by liquid air and then with liquid hydrogen boiling at 0.1 bar approximately, so around 14 K (he knew already of the interest to have the coldest possible temperature at the warm end of the JT heat exchanger. Remember Sect. 3.3.2.6.1), and finally expanded to atmospheric pressure. At 6:30 P.M., on July 10, 1908, while the supply of liquid hydrogen is about to run out, the liquid helium is discovered thanks to somebody who had the idea to apply light behind the glass

Fig. 16.2 Monazite, left; storage of precious gases, right!

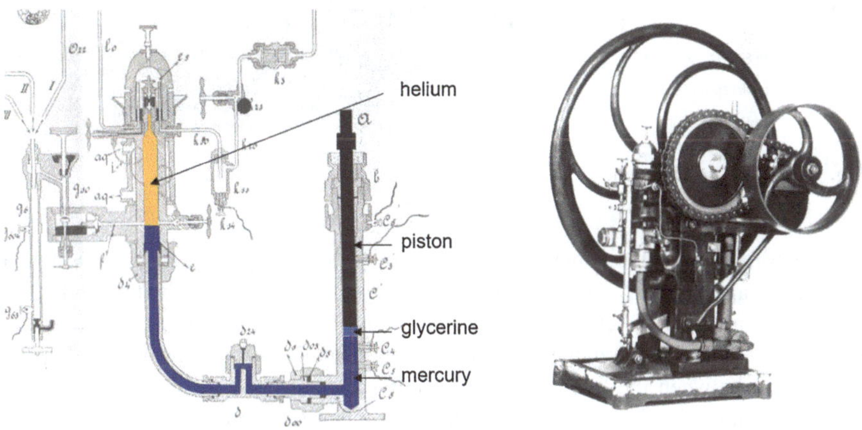

Fig. 16.3 The helium compressor

cryostat, therefore making visible the liquid-vapour interface of a coffee cup volume of liquid helium!

As a scientific person, K. Onnes wanted to take full advantage of having liquid helium, and he started pumping on it in order to get solid helium. But solid helium did not appear. It is only in 1926 that Willem Keesom, a Kamerlingh Onnes student in Leiden, solidified helium by cooling it to less than 2.0 K (−271.15 °C) and subjecting it to a pressure of 25 bar (remember Fig. 1.23).

16.3 The Simon Cryostat

In the beginning of the 1930s, F. Simon, a German scientist, who left Germany in 1933 to the UK, where he, later, was ennobled to become Sir Francis Simon, developed a very smart and cheap way to liquefy modest quantities of helium.

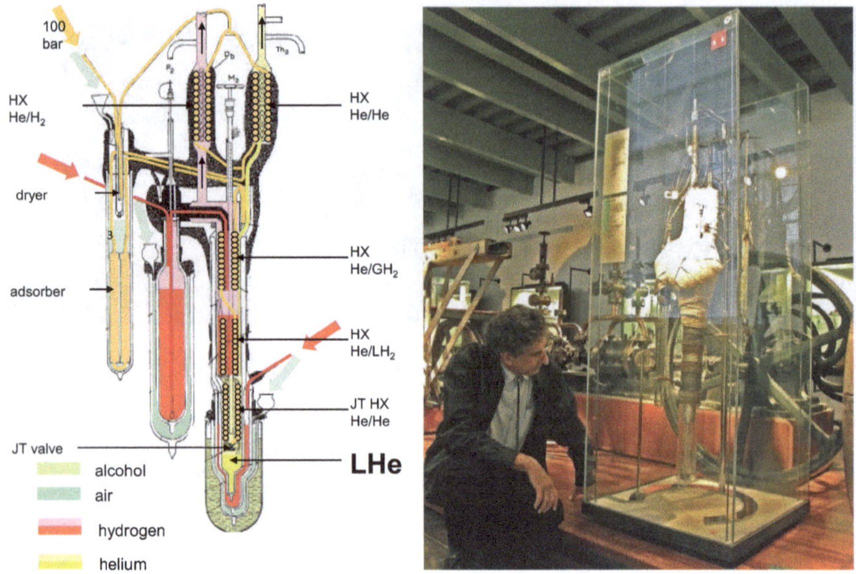

Fig. 16.4 The Onnes liquefier

The Simon cryostat consists in a helium high-pressure capacity to which a hydrogen capacity is in contact (see Fig. 16.5). Both capacities are surrounded by another capacity in which vacuum can be pulled and which is dipped into liquid nitrogen. It is a so-called thermal interrupter that allows or not a good thermal conduction between the hydrogen and helium capacities and the outer wall that is dipped into a liquid nitrogen bath, depending if it is, respectively, filled with helium or pulled to vacuum. Piping and valves are not represented in the sketch.

The helium capacity is filled with helium at 150 bar (Fig. 16.5, 1). The thermal interrupter is filled with atmospheric pressure helium. The helium and hydrogen capacities are firstly cooled down with nitrogen. Helium in the thermal interrupter allows cooling down of both the helium and hydrogen capacities. Vacuum is pulled out of the thermal interrupter. Liquid hydrogen is then allowed in the hydrogen capacity that is in contact with the helium capacity (Fig. 16.5, 2). Later hydrogen is solidified by pumping on the liquid, down to 14.0 K (Fig. 16.5, 3). Finally, helium that is slowly vented expands isentropically, and liquid helium appears in the capacity (Fig. 16.5, 4). Fortunately, as the specific heat of the material of the capacity is negligible, a very small amount of energy is brought in helium during the cooling process.

The density of helium at 150 bar and 14.0 K is 195.01 kg/m^3. Isentropic expansion down to 1 bar turns 70% of helium into liquid. The density of helium at 1.0 bar and 4.2 K is 124.94 kg/m^3. The theoretical volume of liquid that has been generated is 1.1 times the volume of the capacity; therefore, the capacity is miraculously full of liquid!

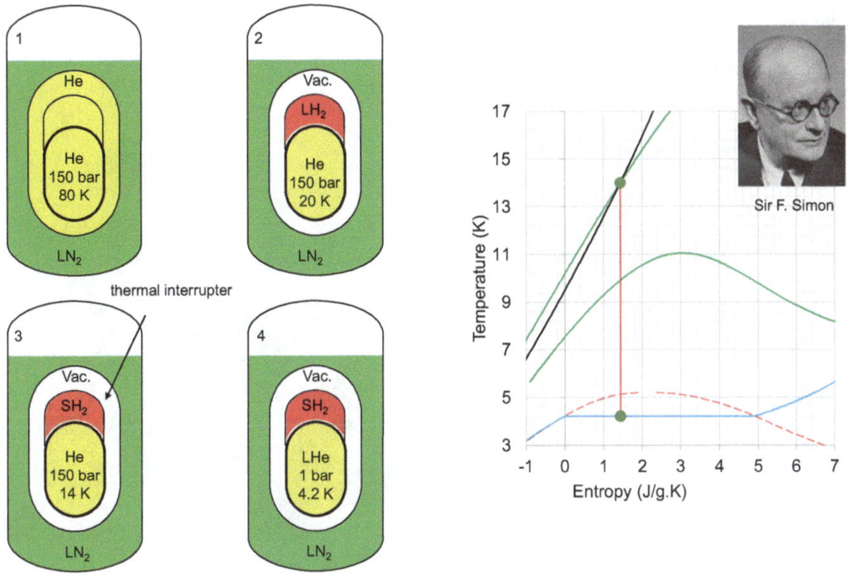

Fig. 16.5 The Simon cryostat operating principle

One can follow the Simon expansion on the T-s diagram, Fig. 16.5, right

16.4 Opening the Door to Liquid Helium Applications

However, without the discovery of superconductivity, cryogenic applications of liquid helium would have remained confidential. In 1910, in order to check two opposing theories that said that at near absolute zero, a metal resistance would decrease to zero or it would grow to infinity, Gilles Holst, a student of Kamerlingh Onnes, cooled different metals to the temperature of liquid helium and found that their resistance decreased greatly. By cooling mercury to 4.18 K, he measured a zero electrical resistance! Intrigued, Kamerlingh Onnes circulated an electric current in a ring of lead immersed in liquid helium (therefore, lead was a superconductor), then put it in short circuit, and found that current continued to flow unabated!

In 1913, Francis Perrin proposed to build a huge coil where the magnetic field could reach 10 Tesla, but as he was aware of the enormous amount of heat that would have been generated by Joule effect, he planned to cool it with liquid air. Kamerlingh Onnes replied that it would be much more interesting by using a superconducting wire with which the Joule effect would be zero. However, it was not until the early 1960s that the MIT (Massachusetts Institute of Technology) in Boston, USA, carried the dream: the first superconducting coil, shortly followed by Sir Martin Wood in the UK (Fig. 16.6).

Fig. 16.6 The first European superconducting coil in the hand of its father Sir Martin Wood, in 1963

Today, one knows that, without superconductivity, a number of machines could not exist, for example, the controlled thermonuclear fusion would spend all the power in Joule heating to reach the high necessary magnetic fields!

16.5 The Pure Joule-Thomson Liquefiers

16.5.1 Air Liquefiers

In 1895 Carl von Linde (1842–1934), a German engineer, developed an air liquefier based on a two-stage Joule-Thomson expansion: from 200 bar down to 40 bar and from 200 bar down to atmospheric pressure, as shown in the sketch, in Fig. 16.7, left.

16.5.2 Helium Liquefiers

In 1950, Ashmead, working in Cambridge, UK, built a 3.8 L/h helium liquefier.

In 1952, Professor Albert Lacaze, working at CNRS in Grenoble, France, developed a series of liquefiers based on pure Joule Thomson process. In other words, like the Ashmead machine, it was a machine that was similar to the one built by K. Onnes, but it was all metallic. A hydrogen liquefier was built first (Fig. 16.8).

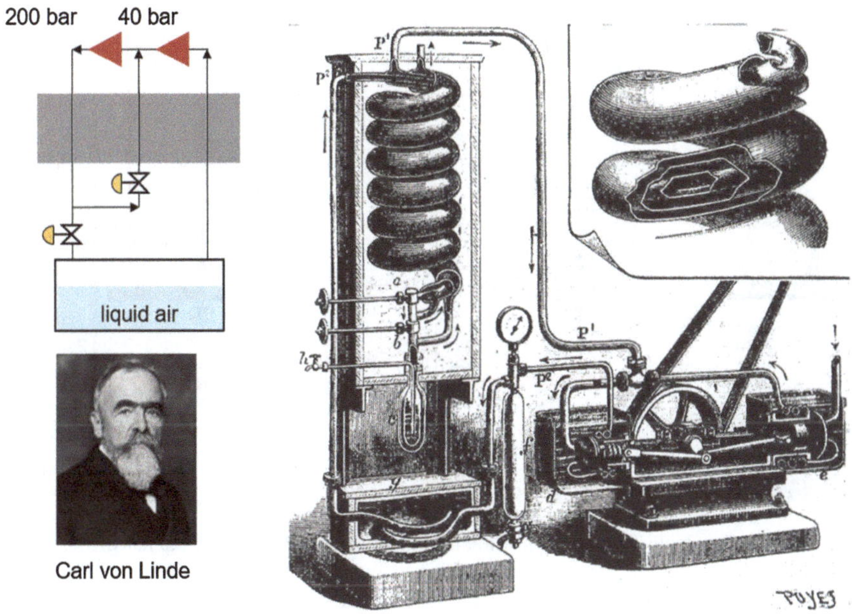

Fig. 16.7 The Linde air liquefier based on pure JT expansion

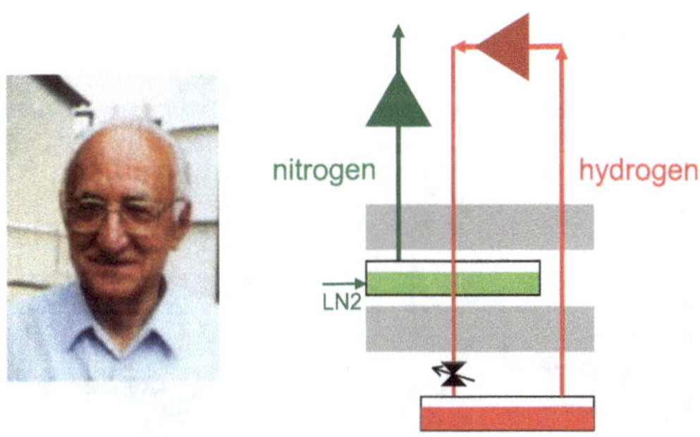

Fig. 16.8 A pure Joule Thomson expansion hydrogen liquefier (CNRS, Grenoble, France), A. Lacaze

Same machine has also been used to liquefy neon in the 1960s.

Later, using the hydrogen cycle for pre-cooling the helium cycle, he developed a 7 L/h liquefier. To get better pre-cooling, both liquid nitrogen and liquid hydrogen baths were pumped, as shown in Fig. 16.9. Later hydrogen was pumped with a warm ejector in order to avoid venting hydrogen (Fig. 16.8, right).

A total of 70 units of this type of liquefier were constructed and sold by an Air Liquide subsidiary: TBT (*Très Basses Temperatures*) (Fig. 16.10).

See the heat exchangers of this liquefier in Fig. 6.68!

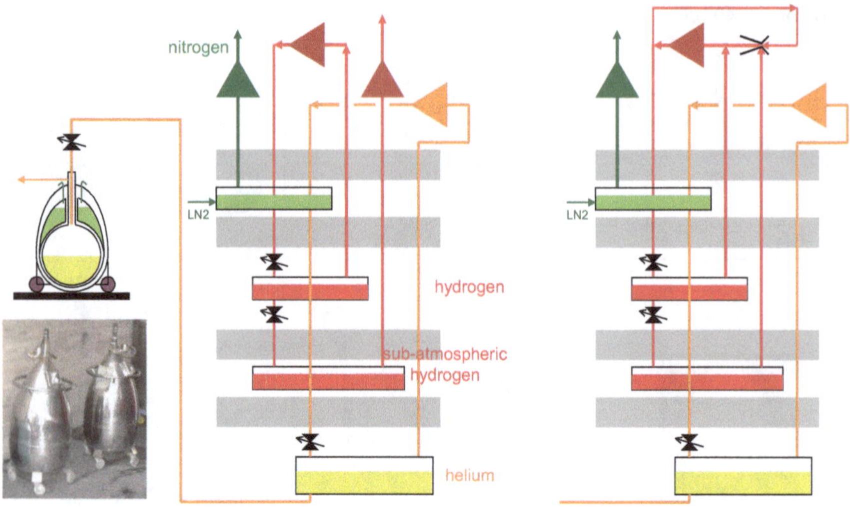

Fig. 16.9 A pure Joule Thomson expansion helium liquefier (TBT, Grenoble, France)

Fig. 16.10 The TBT liquefier

16.6 The Liquefiers with Expansion Machines

16.6.1 The Reciprocating Expander

It seems that the use of an expansion machine for reaching low temperatures was suggested by Carl Wilhelm Siemens, a German engineer, in 1857.

In 1902, Georges Claude, a French engineer, successfully used a reciprocating expansion machine to liquefy air (Fig. 16.11). He used a leather packing as a piston seal without any lubrication except liquid air. With an air pressure of only 40 bar, instead of 200 bar for the JT expansion cycle built by C. Linde, Claude achieved an expansion with external work, resulting in a lower temperature than had before been only possible by simple expansion in a valve. Furthermore, this process was more efficient. This development was the starting point of the Air Liquide company. At the end of the 1960s, the author operated reciprocating expanders, shown in Fig. 16.12, in an air separation plant.

In 1934, while he worked at the Clarendon Laboratory, Cambridge, UK, P. Kapitza built a 1.7 L/h helium liquefier using a Claude cycle with one reciprocating expander and liquid nitrogen pre-cooling.

Fig. 16.11 The Claude air expander. G. Claude at left

Fig. 16.12 An air reciprocating expander from 200 bar to 5 bar

The Collins Liquefier

Remember Sect. 3.5.3.2

The first commercial helium expansion machine has been the reciprocating expander. This equipment, mainly developed by Samuel (Sam) Collins at MIT, USA, in the 1940s, has been and is still very popular for small laboratory machines, mainly liquefiers, in the USA (see Fig. 16.13, left). Its maintenance, which is rather heavy, can anyway be easily performed by the operator. Its rotational speed is limited to 5 Hz (300 rpm). It necessitates a heavy flywheel.

It was incorporated into a very popular small helium liquefier (see Fig. 16.13, right) (Fig. 16.14).

16.6.2 *The Expansion Turbine*

Reciprocating expanders found rapidly limits when going towards larger machines: for higher volumetric flows, it is not possible to reasonably increase both displacement and frequency; therefore, as in many other systems, the technology shifted from reciprocating to rotating machinery, which allows processing higher flow rates within a small volume and a higher reliability.

The idea of a cryogenic expansion turbine was first suggested by Lord Rayleigh in 1898. The first machine has been designed around 1934 by Guido Zerkowitz, an Italian engineer working for the German firm of Linde AG (see Fig. 16.15).

In 1939, the Russian physicist Piotr Kapitza (see Fig. 16.16) perfected the design of cryogenic centripetal turbo expanders for air. His first practical prototype had a

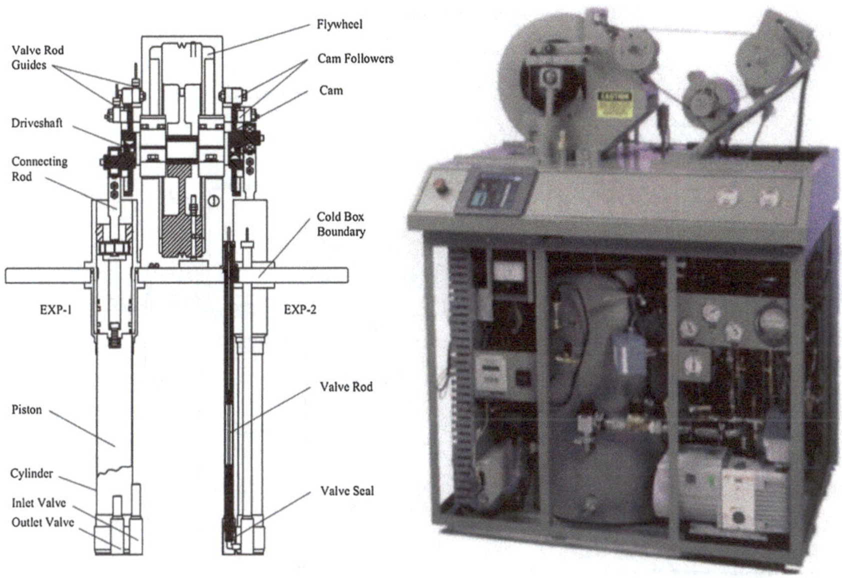

Fig. 16.13 The Collins reciprocating expander, left. A Collins helium liquefier fitted with a reciprocating expander on top left of the box, right. One can see the flywheel of the expander

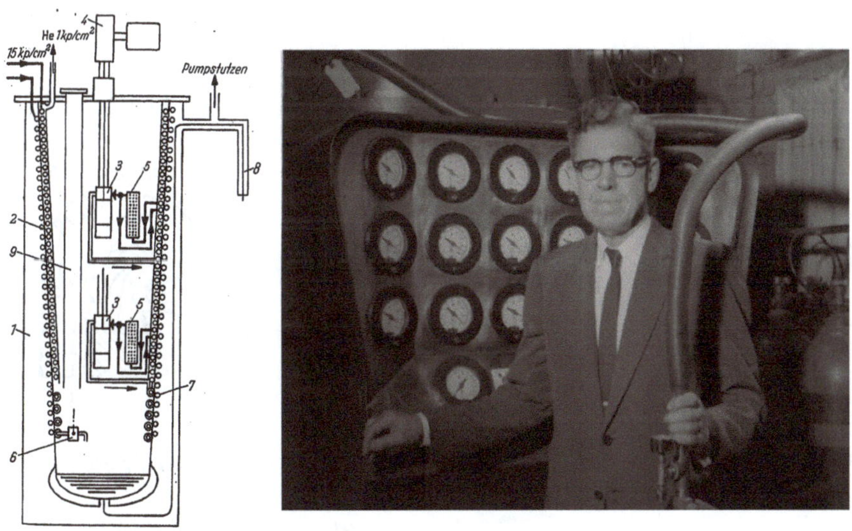

Fig. 16.14 A sketch of the first liquefier, left; Sam Collins, right

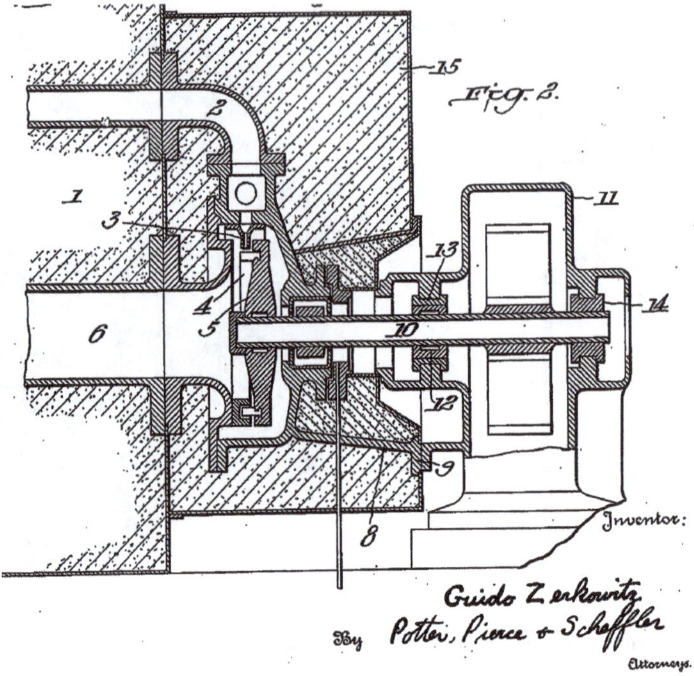

Fig. 16.15 Excerpt of the Zerkowitz patent

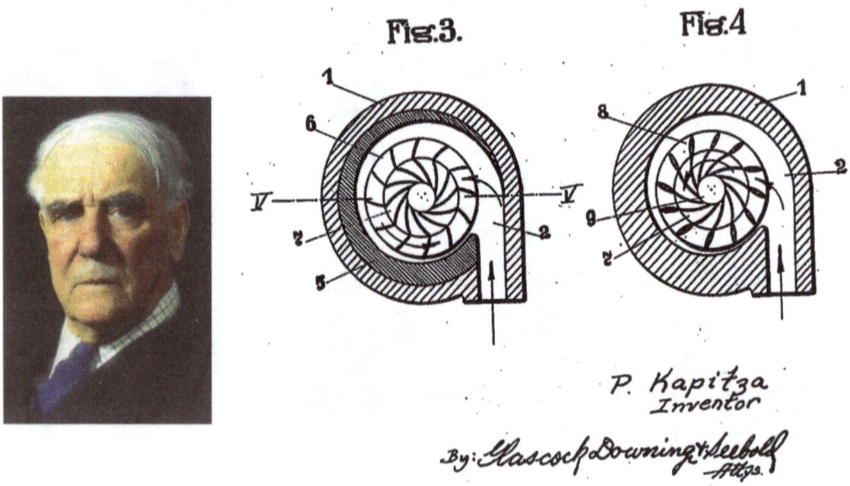

Fig. 16.16 Excerpt of the Kapitza patent

Fig. 16.17 The first Air Liquide turbine liquefier left, J. C. Villard the Air Liquide turbine "father", when younger (courtesy Air Liquide)

diameter of 80 mm, operated at 40,000 rpm, and expanded 1000 cubic metres of air per hour. It used a water pump as a brake and is said to have had an efficiency of around 80%. Most cryogenic turbo expanders in industrial use since then, for industrial gas liquefaction and low-temperature refrigeration, have been based on the Kapitza centripetal wheel design. In 1978, Piotr Kapitza (at last!) was awarded a Nobel Physics Prize in the area of low-temperature physics.

Liquefiers and Refrigerators Operating with Turbines
In 1968, Air Liquide delivered its first turbine liquefier. It was a one-turbine, liquid nitrogen pre-cooled, Claude cycle (see Fig. 16.17). At this time, operating one turbine was a challenge, so it is easy to understand why only one turbine was integrated in this machine! It produced initially 25 L/h and shortly later 35 L/h due to the improvement of the turbine efficiency.

In the same period of time, both British Oxygen Company (BOC), thanks to H. Sixsmith (see Fig. 16.18), and Sulzer issued their turbine liquefiers too.

16.7 The Space Adventure

The space activity has been a powerful incentive for cryogenics.

As early as 1903, C. Tsiolkovsky (1857–1935), a Russian scientist, proposed a rocket engine powered by a combination of liquid oxygen and liquid hydrogen. In 1926, R. Goddard (1882–1945), an American engineer, launched a test rocket using liquid oxygen and gasoline. At the end of WW II, the V2 German rocket was powered by a combination of alcohol and liquid oxygen. In 1945, the US Navy Bureau of Aeronautics started investigating the potential of liquid hydrogen as a

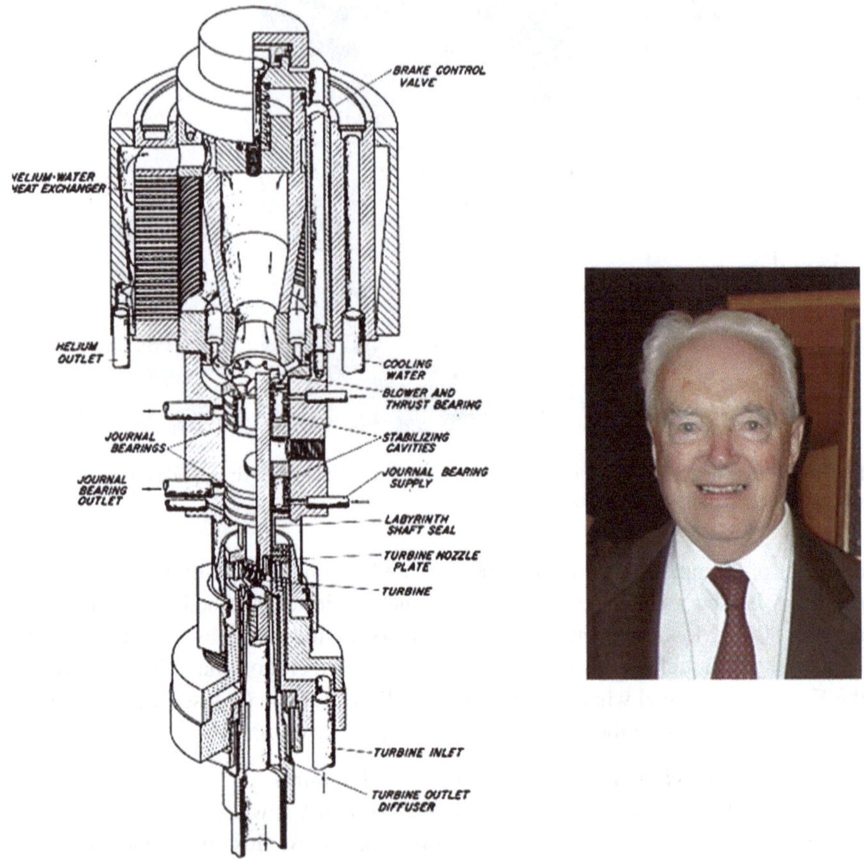

Fig. 16.18 H. Sixsmith the father of the BOC static gas bearing turbine

rocket propellant. And, finally, Centaur, the first US rocket using liquid hydrogen and oxygen, was launched in 1962.

Very large volumes of liquid hydrogen were needed: the technology of hydrogen liquefaction, storage and transfer benefited from major developments. Simulation chambers necessitated cooling systems at temperatures, as low as 20 K for cryopumping shrouds.

16.8 The Cryogenerators and Cryocoolers

Cryogenerators operate on regenerative cycles. Except for the Stirling, they are a rather new technology that developed very fast, mainly boosted by space applications, mainly for cooling detectors, where tiny cryogenic powers with high efficiency

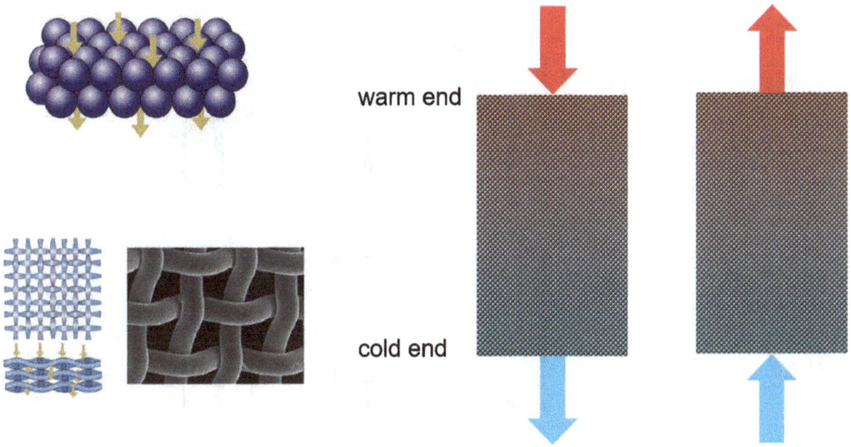

Fig. 16.19 A regenerator, left top small beads, left bottom mesh (50–100 μm diameter wires), right operating principle

and reliability are needed. They are also used for re-condensation or liquefaction of helium. The so-called cryocoolers are small cryogenerators.

16.8.1 A Reminder on Regenerator

A regenerator is a heat exchanger made out of a divided material (mesh, small beads, see Fig. 16.19, left). It is circulated by an alternate flow of fluid: it accumulates heat when it is circulated in one direction by warm gas that cools down (see Fig. 16.19, centre) and then releases the stored heat when it is circulated in the opposite direction by cold gas that warms up (see Fig. 16.19, right).

Similarly, as a heat exchanger, a regenerator has to fulfil difficult duties: it should have a high volumetric thermal capacity (remember Fig. 1.29), low axial and large longitudinal thermal conductivity, large heat transfer coefficient, low flow resistance and small dead volume.

16.8.2 The Stirling Cryogenerator

In the 1960s Jan Kohler, working for Philips, NL, adapted to cryogenics the principle of the hot air engine initially developed by R. Stirling, a Scottish reverend, in 1816.

A piston and a displacer move in a cylinder (see Fig. 16.20) filled with a working gas, most often helium.

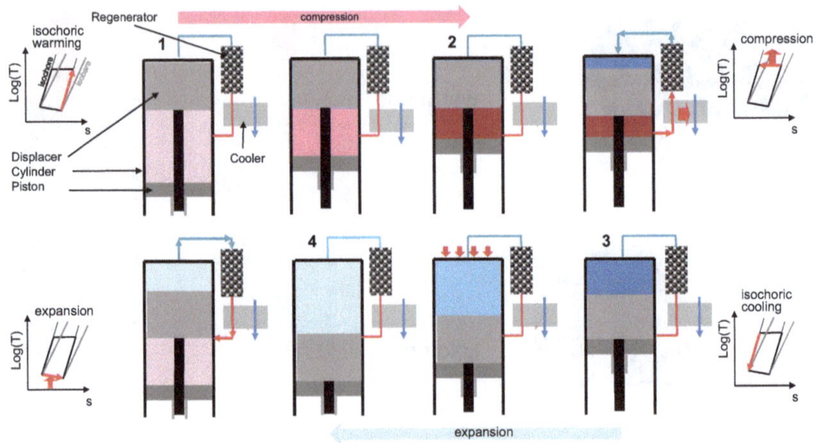

Fig. 16.20 Operating principle of the Stirling cryogenerator. Dark colours mean HP, red means warm, blue means cold

1-2. The displacer is at the cold end of the cylinder (up), the piston moves towards the end of the cylinder (upwards), and the LP working fluid is compressed and warmed up.

2-3. The displacer moves to the warm end of the cylinder (downwards); the HP working fluid circulates first the water cooler where it cools down to room temperature and then the regenerator where it cools down to cold temperature, at constant volume (isochoric cooling).

3-4. The piston and the displacer move downwards, the working fluid is expanded to LP, and it cools down.

4-1. The displacer moves to the cold end of the cylinder (upwards); the working fluid circulates the regenerator where it warms up to room temperature at constant volume (isochoric warming).

The piston and the displacer are actuated by two separate rod-crankshaft systems set at 90° to get the phase shift. The Stirling cryogenerator can be operated at frequencies from 15 to 60 Hz. Its practical operating temperature is around liquid nitrogen's (Fig. 16.21). Its cryogenic power ranges from a few watts to 200 kW. It is used for liquefaction of air gases or pre-cooling of colder boiling gases (see Sect. 16.9).

With a two-stage arrangement, the Stirling cryogenerator can reach around 15 K.

A special type of Stirling cryogenerator incorporates a linear oscillator based on a similar principle as used in loud speakers (see Fig. 16.22, centre). The pistons are generally operating in opposition to reduce vibrations. The regenerator-displacer operates with a phase shift in resonance with the pistons thanks to a spring (Fig. 16.22, left). The pistons and displacer can also be supported by flexure bearings, shown in Fig. 16.22, right, that have both a very high axial flexibility

Fig. 16.21 A one-stage SPC-1 Stirling cryogenerator (Stirling Cryogenics)

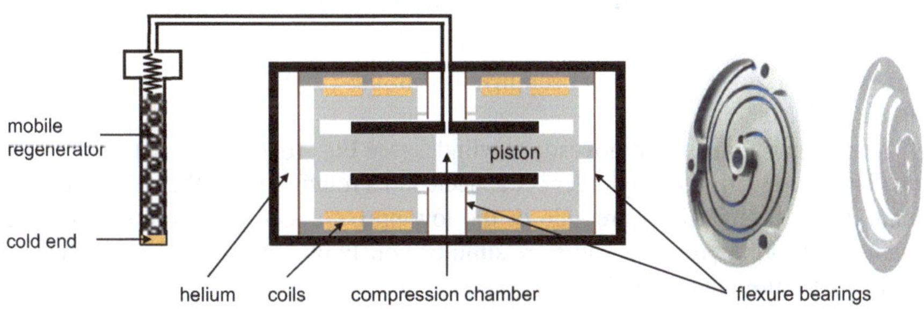

Fig. 16.22 A Stirling cryogenerator actuated by a linear oscillator on flexure bearings

and a very high radial rigidity. The MTBF of such systems is, surprisingly, as high as 50,000 hours.

Such a structure allows locating the cold part at a distance from the oscillator, which is particularly interesting for space applications.

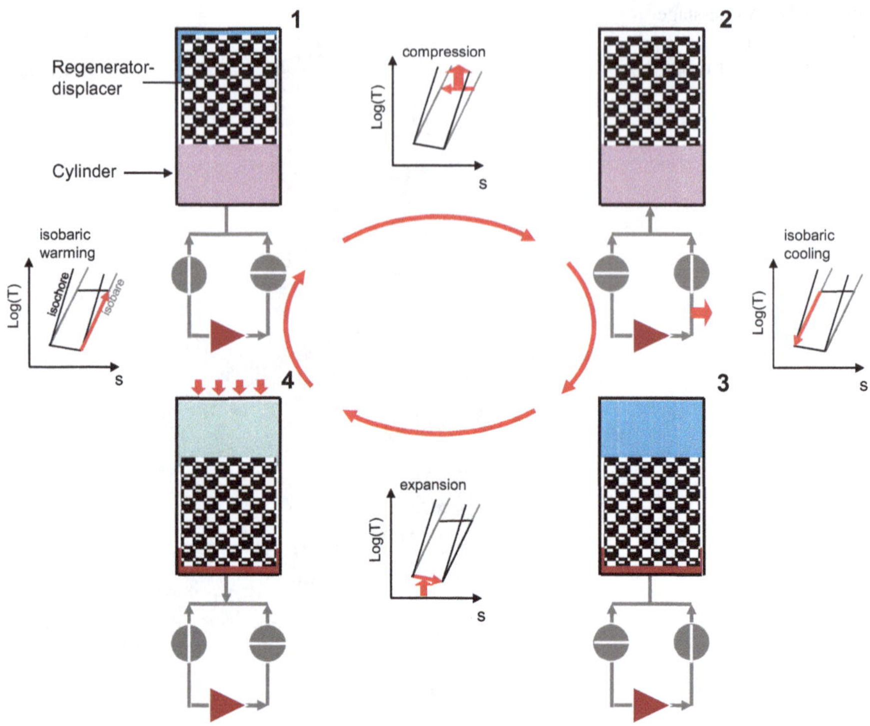

Fig. 16.23 Operating principle of the Gifford-McMahon cryogenerator

16.8.3 The Gifford-McMahon (GM) Cryogenerator

This kind of cryogenerator has been invented in the 1950s by William Gifford and Howard McMahon.

A displacer-regenerator moves in a cylinder (see Fig. 16.23) filled with a working gas. The warm part of the cylinder is connected to a rotating valve that, alternatively, connects to the high-pressure side or the low-pressure side of a compressor. The regenerator and the rotating valve are simultaneously operated by a same dedicated motor (not represented here).

1-2. The cylinder is connected to the HP, the regenerator is up, and the pressure of the working fluid increases.

2-3. The cylinder is still connected to the HP, the regenerator moves downwards, and the HP working fluid cools down at constant pressure (isobaric).

3-4. The cylinder is connected to the LP, and the HP working fluid is expanded: it cools down.

4-1. The cylinder is still connected to the LP, the regenerator moves upwards, and the working fluid warms up.

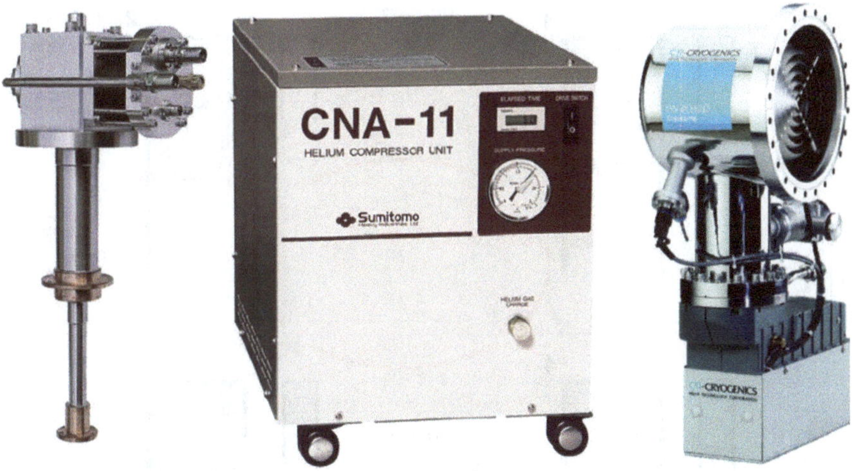

Fig. 16.24 A bare GM cryogenerator with its compressor, left (Sumitomo). A cryopump cooled by a GM cryocooler (courtesy Applied Cryogenics, Inc)

The operating frequency of a Gifford-McMahon cryogenerator is around 1 Hz. Its practical operating temperature is around that of liquid nitrogen.

With a two-stage arrangement, it can reach around 15 K.

Gifford-McMahon cryogenerators are used for cooling vacuum cryopumps at around 15 K that are most used in the semiconductor industry or for the pre-cooling of gases boiling at colder temperatures.

The GM cryogenic power ranges from a few W up to 500 W at 77 K and from a few hundreds of mW up to 100 W at 20 K (Fig. 16.24).

16.8.4 The Pulse Tube (PT) Cryocooler

In the 1960s, William Gifford introduced the concept of the pulse tube cryocooler, based on a mix of principles of both Stirling and GM. Development took a number of years, but today, such a technology is mature. Its main interest is that there is no moving part in the cold end; therefore, no vibrations are generated.

To "ease" the explanation of the operating principle, an intermediary step is to look at a "pedagogic" Stirling as shown in Fig. 16.25. The working fluid is circulated by the displacer that is located in another cylinder.

1-2. The displacer is at the cold end of the cylinder (up), the piston moves towards the end of the cylinder (upwards), the LP working fluid is compressed, and heat is rejected to ambient.

2-3. The displacer moves to the warm end of the cylinder (downwards), and the HP working fluid circulates the regenerator: it cools down.

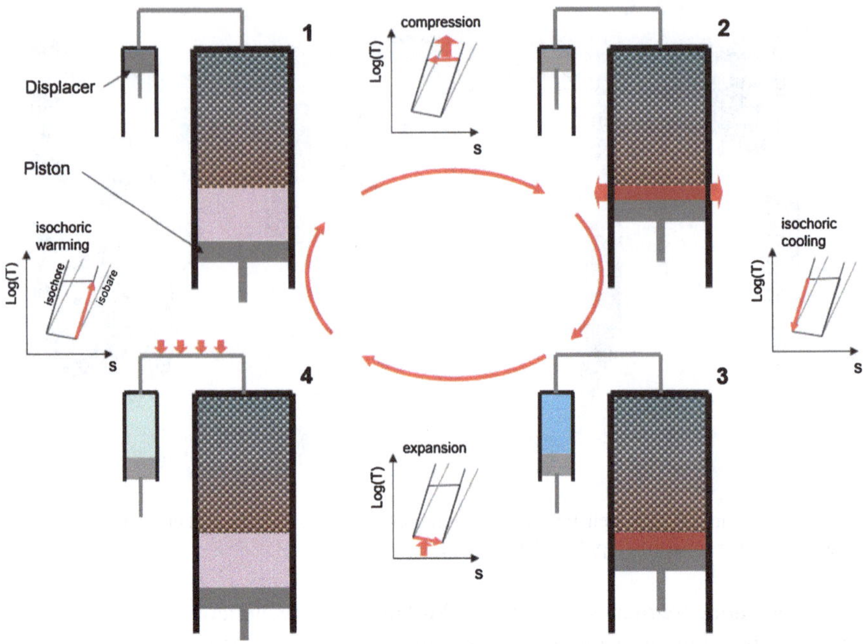

Fig. 16.25 Operating principle of the pulse tube cryogenerator: a pedagogic Stirling

Fig. 16.26 Structure of the pulse tube cryocooler

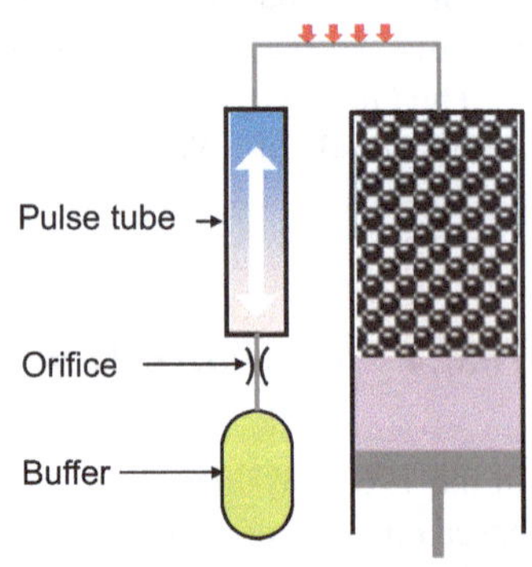

3-4. The piston moves downwards, the working fluid is expanded to LP, and it cools down.
4-1. The displacer moves to the cold end of the cylinder (upwards), and the working fluid circulates the regenerator: it warms up.

In the actual pulse tube cryocooler (see Fig. 16.26), the working fluid that is located in the pulse tube is circulated by pressure waves that perform the same duty as the displacer. The correct phase shift can be, among other arrangements, created by the combination of a buffer capacity and a calibrated orifice.

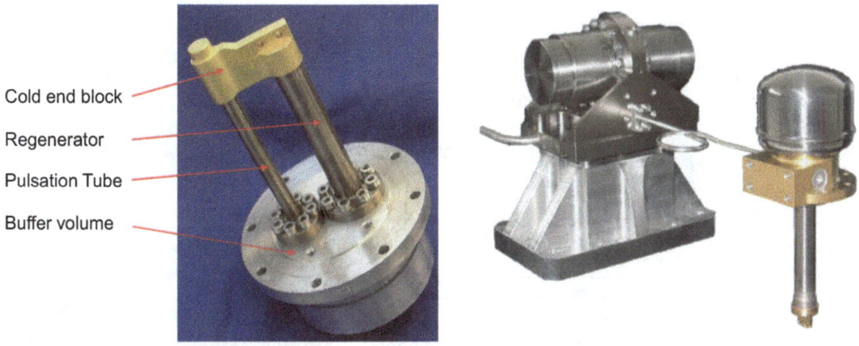

Fig. 16.27 A pulse tube cryogenerator, left. The full assembly with a linear oscillator, right. (courtesy Air Liquide)

The specificity of the pulse tube cryocooler is that there is no moving part in the cold end; therefore, there are no vibrations and the availability is high. For such reasons, it is very interesting for space applications.

A pulse tube cryocooler can also be fed by an oscillator of same type as described in Sect. 16.8.2, which allows operating at higher frequencies, up to 100 Hz.

The cryogenic power of pulse tubes driven by a pressure oscillator (frequency from a few tens of Hz up to 100 Hz, also called Stirling-type pulse tube) ranges from a few hundreds of mW up to about 5 W at 77 K.

Low-frequency pulse tubes (in the range of 1–2 Hz to achieve the 4 K temperature range) have also been developed, where the pressure oscillation is obtained from a GM-type cryocooler compressor with a set of valves. Their cryogenic power ranges from 100 mW up to 2 W at 4 K (Fig. 16.27).

16.8.5 The Brayton Cryocooler

The MELFI (Minus Eighty Degree Laboratory Freezer for ISS) is a rack-sized facility that provides the Space Station with a refrigerated volume for the storage and fast freezing of life science and biological samples. It is a one-shaft Brayton cycle where the compressor and turbine wheels and the motor are on the same shaft, supported by gas dynamic bearings (remember Sect. 6.5.1.2.2.2) allowing the compressor to recover the turbine energy (see Fig. 16.28, left). The cycle-coiled heat exchanger has the shape of a toroid that surrounds the active cartridge (Fig. 16.28, right). It is operated with nitrogen. The MELFI cryocooler can absorb up to 90 W at −80 °C.

Vibration free and high availability are the two main advantages that make it a good choice for space.

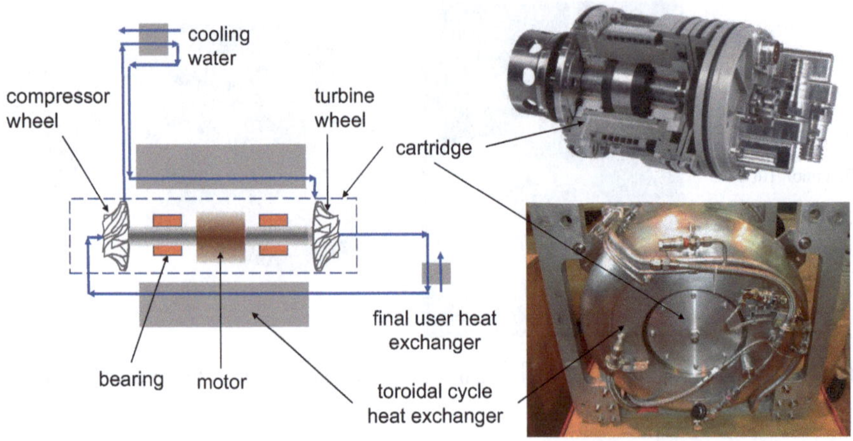

Fig. 16.28 The MELFI cryocooler operating principle, left, the rotating-part cartridge, right top; assembly, right, bottom (copyright P. Avanian, Air Liquide)

16.9 Liquefiers Pre-cooled by Cryogenerators

Such machines are composed of a JT cycle that is pre-cooled by a cryogenerator.

16.9.1 Pre-cooling with a Stirling Cryogenerator

In 1970, Air Liquide designed a small 3 L/h liquefier: the ALP 300 (AL for Air Liquide, P for Philips, 300 L/week)). It incorporated a Philips Stirling two-stage cryogenerator (remember Sect. 16.7) operating simultaneously at temperatures around 80 K and 16 K. A simple JT cycle was pre-cooled by the cryogenerator, thanks to three coiled heat exchangers (see Fig. 16.29, left).

On same principle, a few years later, Philips designed the PLHe 209, a 9 L/h liquefier incorporating two cryogenerators, according to the flow diagram shown in Fig. 16.29, left. This machine had some tricky aspects. The heat exchangers were of the perforated plate type as described in Sect. 6.4.1.4. The compressor was using "an oil-free compressor by virtue of an oil-supported rolling diaphragm which acted as the positive seal of the pistons". Such compressor was expensive; therefore, the designer wanted to reduce its displacement. The solution was to operate the compressor at a rather high suction pressure: 2.5 bar. The compression from atmospheric pressure return from the dewar up to 2.5 bar was performed by a cold ejector (remember Sect. 4.3.1.2.3) (Fig. 16.30).

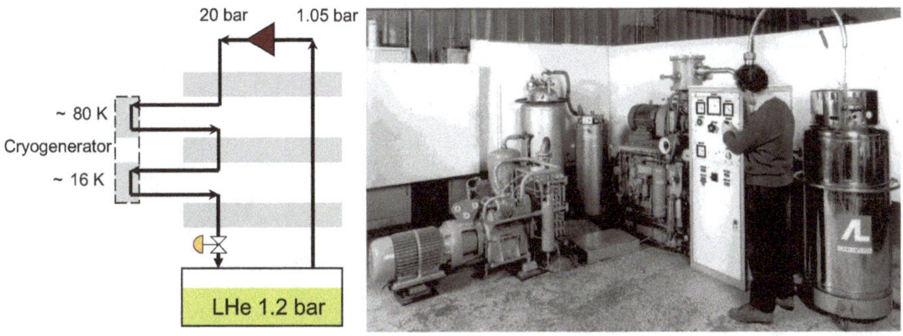

Fig. 16.29 The ALP 300 helium liquefier (Air Liquide DR)

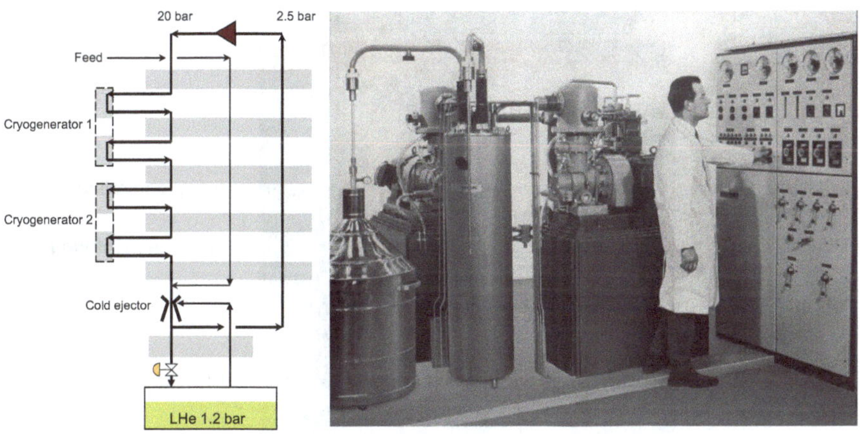

Fig. 16.30 The PLHe 209 (Philips)

16.9.2 Pre-cooling with a Pulse Tube Cryocooler

The cold finger (coldest part) of the cryocooler cools and condenses helium, directly in the dewar (Fig. 16.31).

Such helium liquefiers produce from 15 to 80 L/day.

16.10 Automation of Refrigeration Plants

Up to the end of the 1970s, refrigeration plants were fitted with a scarce number of control loops that were only able to keep the machine in a steady state when no major perturbations occurred (see Sect. 8.5). Commands were manual as well as set points

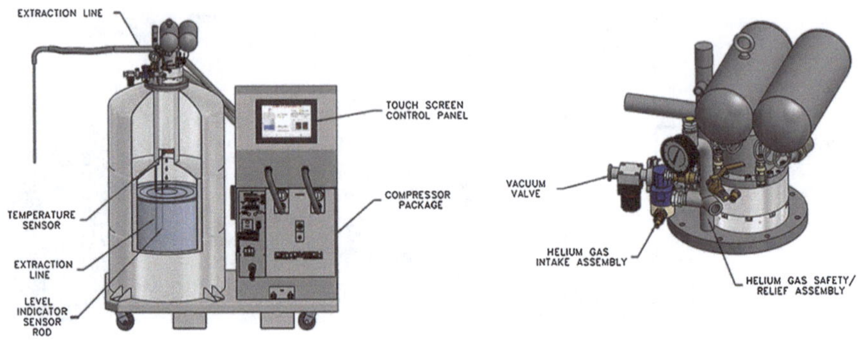

Fig. 16.31 Helium liquefier with PT cryocooler. (Courtesy Cryomech)

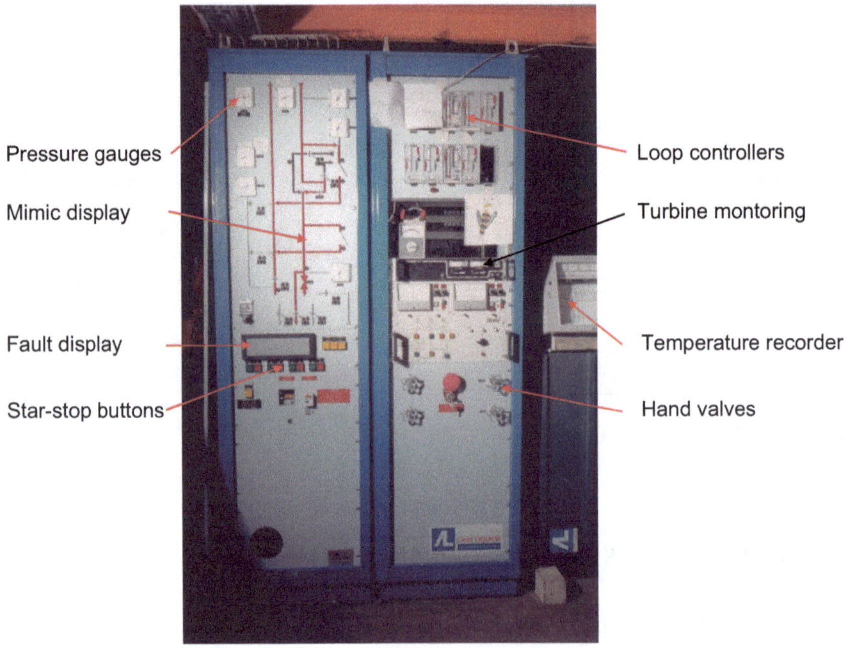

Fig. 16.32 The control panel of the Cello refrigerator (1979) (courtesy Air Liquide)

for control loops; most of information was only displayed (see Fig. 16.32), and logs were filled up by hand. The follow-up of the cool-down of a coil was plotted by hand on graph paper as shown in Fig. 16.33, left.

At this time, technology began making improvements in various fields:

- Sensors providing an electric signal became more easily available and cheaper.

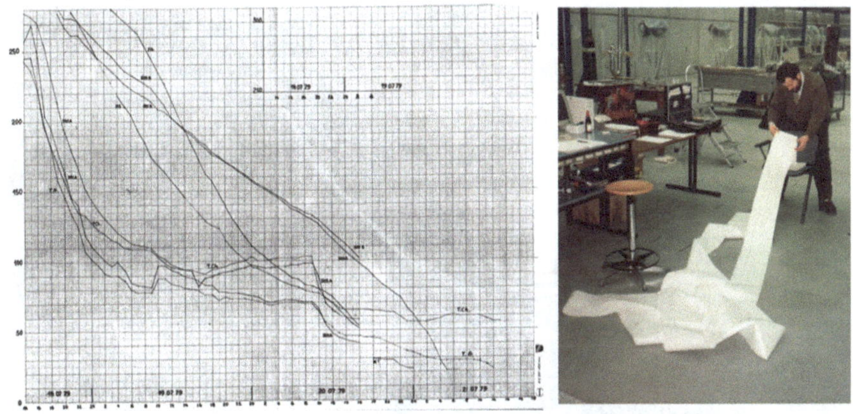

Fig. 16.33 Some antic ways to follow a cool-down (totally manually at CELLO, DESY, Hamburg, Germany, in 1979, with the help of a printer at ALPI, Padova, Italy, in 1989)

- The computer technology developed, and small power machines became available, even if they had a very small memory capacity (16 kilo-octets!) and very poor calculation possibilities.
- Some people realised that a specific language should be developed to ease programming of system controls, forgetting the complicated language of computer guys.

In 1979, the author started developing what he expected to be a semi-automatic system. During the course of this job, the idea came naturally to switch to a full-automatic system. With the help in automation that came from the CEA Saclay, HELIAL was finally started up in 1980 (see Fig. 16.34).

The automation experience acquired with HELIAL allowed to automatise the rather complicated Tore Supra refrigerator a few years later (remember Sect. 12.8.1).

Today, all helium refrigerators or liquefiers are automatic.

16.11 The Oil Lubricated Screw Compressor

Today, almost all helium refrigeration systems are fed by oil lubricated screw compressors. This has been an important step in the 1980s: the availability of the compressor increased substantially when switching from the dry reciprocating compressor to the screw one (in the end of the 1970s, the author spent half of the weekends of one calendar year trying to fix a piston ring compressor at DESY, Hamburg, Germany!) although the efficiency was a little bit decreased.

Fig. 16.34 A 7 L/h HELIAL automatic liquefier (courtesy Air Liquide)

16.12 The Cryogenic Centrifugal Compressor

The rationale of the cryogenic compressor has been explained in Sect. 12.8.1: the size of the room temperature sub-atmospheric station and the quasi-impossibility to build a low-pressure heat exchanger when going towards large cryogenic powers led to the development of cryogenic centrifugal compressors for the Tore Supra refrigerator, in the 1980s. Such technology, which has been recognised as unique, is now implemented in all middle and large cryogenic power refrigerators operating at superfluid helium temperatures.

16.13 The Static Pressurised Superfluid Helium

The static pressurised superfluid helium application has been developed in *Commissariat à l'Energie Atomique* (CEA), Grenoble, France, under the guidance of Gérard Claudet, in the 1970s (remember Sect. 1.4.2.4). It had been first implemented in the Tore Supra machine and is now a natural part of a lot of systems operating at superfluid helium temperatures.

16.14 All-Cryogenic Rotating Machine Refrigerators

In the 2010s, the concept of the MELFI turbo-Brayton (remember Sect. 4.6.1.1) refrigerator was up-sized to large powers operating at around liquid nitrogen temperatures. In 2020, the cryogenic power range of such refrigerator is rather wide (remember Sect. 12.2).

It is a very promising solution for large efficient and reliable refrigerator pre-cooling (remember Sect. 4.6.1.1).

16.15 The Present State of the Art in Cryogenic Helium Refrigeration

16.15.1 Cryogenic Power

Since the 1950s, the power of a refrigeration unit has been increased by a factor of more than 1000: see Figs. 16.35 and 16.36!

Today (2020) operating temperatures as low as 1.8 K are routinely used (remember Sect. 12.8).

16.15.2 Availability

The availability of large refrigeration systems is rather high: as shown in Fig. 16.37, the CERN LHC cryogenic that operates at 1.8 K reaches availability figures as high as 98.6% for a very complicated system[1].

Such a result is, among others, a consequence of a sophisticated maintenance activity.

16.15.3 Efficiency

The present specific power of about 220 W/W at 4.5 K is mainly limited by the efficiency of the compression system. If centrifugal compressors could be used for large helium refrigeration plants, it would drastically increase both efficiency and reliability. For example, in case centrifugal compressors with an isothermal efficiency of 72% were used instead of screw compressors with an isothermal

[1]L. Delprat et al., in LHC availability and helium management from run 1 to run 2, 7th international workshop on cryogenics operation, 2015, Fermi National Accelerator Laboratory

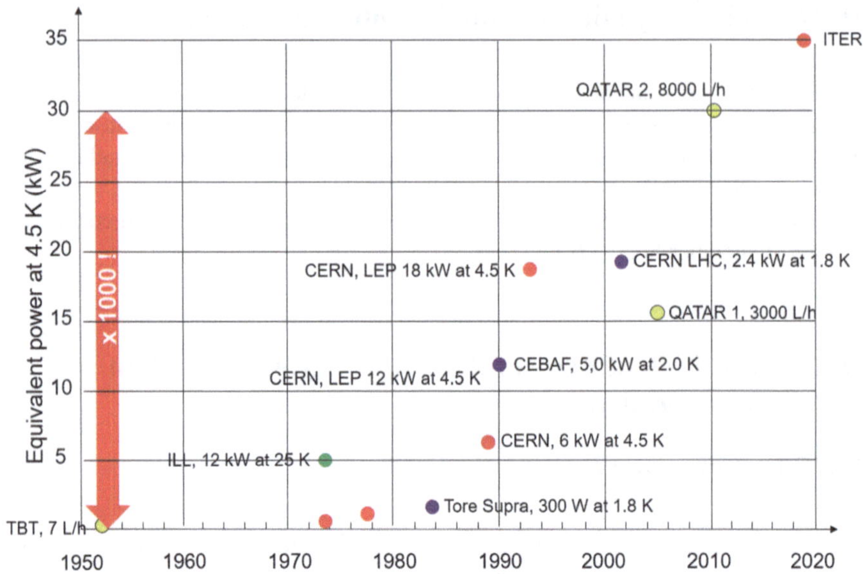

Fig. 16.35 Evolution of the power of a one-off refrigerator unit versus time

Fig. 16.36 The author in the large cold box of the 7000 L/h liquefier (courtesy D. Grillot Air Liquide). To the right, a photo *at same scale* of the heat exchangers of the 7 L/h liquefier he was in charge in 1965

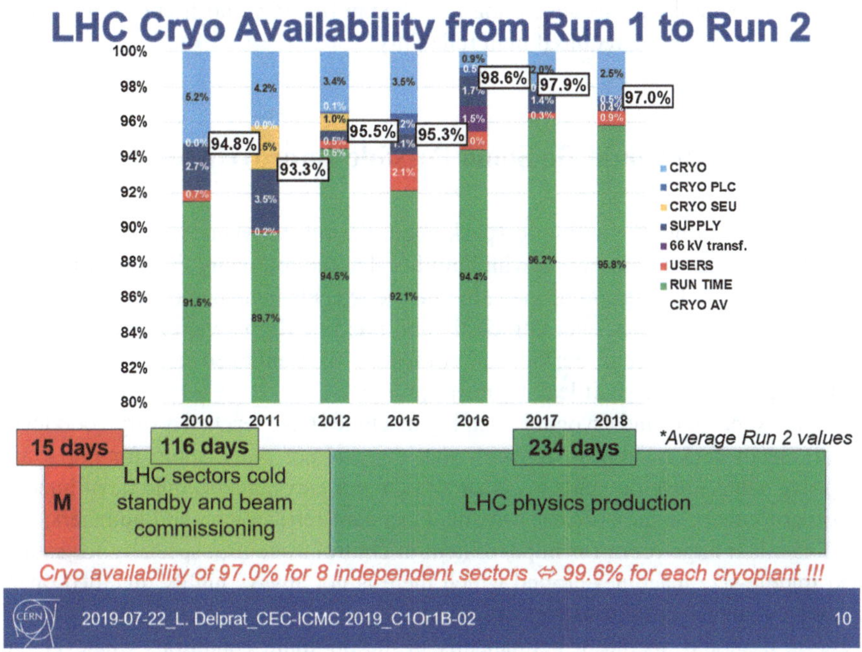

Fig. 16.37 Availability of the LHC cryogenic system

efficiency of 55%, the LEP refrigerator efficiency (remember Sect. 3.5.5) would reach 38% instead of 29% (specific power 170 W/W).

16.15.4 Technology

The present conventional components of a helium refrigerator are the oil lubricated screw compressor, the plate and fin brazed aluminum heat exchanger, the cryogenic expansion turbine and the cryogenic centrifugal compressor.

16.15.5 Space Special Applications

A lot of cryogenics are involved in modern satellites dedicated to Earth observation, scientific missions or meteorological satellites in order to cool detectors for seeing and measuring (bolometers, microcalorimeters, etc.). Such systems must be simultaneously vibration free, highly reliable (no maintenance!), highly efficient and of low volume and weight.

Special technologies as adsorption and flexure bearings (no friction) have been developed to reach the required high reliability.

16.16 Which Could Be Some Possible Improvements?

It is always a difficult exercise to speak about the future. Many wrong predictions have been made; therefore, it is with humility that the following lines are written.

On the technology point of view, the hermetic centrifugal cycle compressor is a very promising solution, avoiding costly and pressure drop generating oil removal systems and also, special purification equipment as mentioned in Sect. 6.2.6.6 in order to get rid of the shaft leak tightness problem.

Heat exchangers and cryogenic rotating machines have reached a very good level of efficiency and reliability. It seems that there is not much room for important improvements. Developments to get small two-phase expanders could be of interest in order to increase the efficiency of small standard refrigerators and liquefiers.

There is a large room for improvement in control systems, especially for small- and middle-size machines around which there is not always highly qualified operating personnel. The control system should not only take care of the operation of the system; it should operate it permanently at the optimum conditions whatever the operating regime is, help much more the operator, perform diagnosis with the help of artificial intelligence and even suggest corrective actions. Maintenance actions should also be remembered in time and acknowledged.

For small standard liquefiers, the control system should incorporate a software that gives information to the operator about maintenance actions and performs the helium inventory management. An automatic mobile dewar filling system would also be very useful and would improve the liquid transfer by running optimised filling procedures.

Chapter 17
A Digest in Thermodynamics for Helium Refrigeration

Abstract Along the book, various thermodynamic formulae are used. Most of them are rather well-known, but some readers might be interested in getting a little bit deeper. This is the aim of this chapter.

17.1 A Refresher in Elementary Engineering Thermodynamics

The first principle of thermodynamics expresses the conservation of energy. For a closed cycle, it can be written:

$$dU + dK = dQ + dW \tag{17.1}$$

U: internal energy
K: kinetic energy
Q: heat exchanged by the system
W: work exchanged by the system

Remember that:

$$dW = -PdV \tag{17.2}$$

Note that $dU + dK = dQ + dW$ Eq. 17.1 is a possible definition of the internal energy U of a system, once the internal energy of a reference state U_0 is known.

Consider now a static fluid occupying a volume V at pressure P:

$$dK = 0 \tag{17.3}$$

If the static fluid undergoes an isochore (constant volume) transformation:

G. Gistau Baguer, *Cryogenic Helium Refrigeration for Middle and Large Powers*, International Cryogenics Monograph Series,
https://doi.org/10.1007/978-3-030-51677-2_17

$$dW = 0 \tag{17.4}$$

therefore, $dU + dK = dQ + dW$ Eq. 17.1 yields:

$$dQ = dU \tag{17.5}$$

Similarly, for an isobar (constant pressure) transformation:

$$dQ = dU + PdV = dU + d(PV) \tag{17.6}$$

The form of $dQ = dU + PdV = dU + d(PV)$ Eq. 17.6 leads to the introduction of a new function of state called *enthalpy H*:

$$H = U + PV \tag{17.7}$$

So that, at constant pressure:

$$dQ = dH \tag{17.8}$$

The mathematical simplicity of $dQ = dU$ Eq. 17.5 and $dQ = dH$ Eq. 17.8 is exploited in engineering calculations: while it is convenient to represent an isochore transformation of a system by its variation in internal energy, enthalpy should preferably be used to describe its isobar evolution (e.g. at atmospheric pressure). As a consequence, the specific heats at constant volume and constant pressure are:

$$c_\mathrm{v} = \frac{dU}{dT} \tag{17.9}$$

$$c_\mathrm{p} = \frac{dH}{dT} \tag{17.10}$$

The latent heat of vaporisation of a fluid at saturation conditions is the enthalpy difference between vapour and liquid:

$$L_\mathrm{v} = H_\mathrm{vapour} - H_\mathrm{liquid} \tag{17.11}$$

Now, consider a fluid steadily flowing across a pressure gradient, in a fixed-wall channel (e.g. a pipe). Although there is no production of mechanical work, the fluid still receives or performs work against the upstream and downstream pressures:

$$dW = -d(PV) \tag{17.12}$$

Hence, the conservation of energy can be written:

$$dU + dK = dQ - d(PV) \tag{17.13}$$

or

$$dH + dK = dQ \tag{17.14}$$

If the flow is adiabatic, with small variations in kinetic energy, e.g. in a thermally insulated pipe, $dH + dK = dQ$ Eq. 17.14 reduces to:

$$dH = 0 \tag{17.15}$$

and the fluid undergoes an isenthalpic transformation.

For an adiabatic flow with large change in velocity, e.g. accelerating through a nozzle:

$$dH = -dK \tag{17.16}$$

Thus, the enthalpy drop is used to accelerate the fluid.

As:

$$K = \frac{v^2}{2} \tag{17.17}$$

Starting from stagnation, the fluid final velocity is:

$$v = \sqrt{2}\Delta H \tag{17.18}$$

The second principle of thermodynamics (Clausius' formulation) states that for a closed system exchanging heat quantities Q_i at temperatures T_i:

$$\sum_i \frac{dQ_i}{T_i} \geq 0 \tag{17.19}$$

This inequality reduces to equality in the case of reversible processes.

The form of $\sum_i \frac{dQ_i}{T_i} \geq 0$ Eq. 17.19 leads to the introduction of another function of state called *entropy* S, defined by:

$$dQ = TdS \tag{17.20}$$

A useful consequence of equation is the representation of thermodynamics processes in the temperature vs entropy diagram: the heat quantity reversibility exchanged in a given transformation is proportional to the area below the transformation line on the diagram (see Sect. 3.2.4). The interest of such a representation clearly appears in the discussion of refrigeration cycles.

17.2 Logarithmic Mean Temperature Difference (LMTD)

From Wikipedia, the free encyclopaedia

The logarithmic mean temperature difference (also known as log mean temperature difference, LMTD) is used to determine the temperature driving force for heat transfer in flow systems, most notably in heat exchangers. The LMTD is a logarithmic average of the temperature difference between the hot and cold feeds at each end of the double pipe exchanger. For a given heat exchanger with constant area and heat transfer coefficient, the larger the LMTD, the more heat is transferred. The use of the LMTD arises straightforwardly from the analysis of a heat exchanger with constant flow rate and fluid thermal properties.

17.2.1 *Definition*

We assume that a generic heat exchanger has two ends (which we call "*A*" and "*B*") at which the hot and cold streams enter or exit on either side; then, the LMTD is defined by the logarithmic mean as follows:

$$\mathrm{LMTD} = \frac{\Delta T_A - \Delta T_B}{\mathrm{In}\left(\frac{\Delta T_A}{\Delta T_B}\right)} = \frac{\Delta T_A - \Delta T_B}{\mathrm{In}\Delta T_A - \mathrm{In}\Delta T_B} \tag{17.21}$$

where ΔT_A is the temperature difference between the two streams at end *A* and ΔT_B is the temperature difference between the two streams at end *B*. With this definition, the LMTD can be used to find the exchanged heat in a heat exchanger:

$$Q = U \times Ar \times \mathrm{LMTD} \tag{17.22}$$

where Q is the exchanged heat duty (in watts), U is the heat transfer coefficient (in watts per kelvin per square meter) and Ar is the exchange area. Note that estimating the heat transfer coefficient may be quite complicated.

This holds both for co-current flow, where the streams enter from the same end, and for countercurrent flow, where they enter from different ends.

In a cross-flow, in which one system, usually the heat sink, has the same nominal temperature at all points on the heat transfer surface, a similar relation between exchanged heat and LMTD holds, but with a correction factor. A correction factor is also required for other more complex geometries, such as a shell and tube exchanger with baffles.

17.2.2 Derivation

Assume heat transfer [1] is occurring in a heat exchanger along an axis z, from generic coordinate A to B, between two fluids, identified as 1 and 2, whose temperatures along z are $T_1(z)$ and $T_2(z)$.

The local exchanged heat flux at z is proportional to the temperature difference:

$$q(z) = U(T_2(z) - T_1(z))/D = U(\Delta T(z))/D, \tag{17.23}$$

where D is the distance between the two fluids.

The heat that leaves the fluids causes a temperature gradient according to Fourier's law:

$$\frac{dT_1}{dz} = k_a(T_1(z) - T_2(z)) = -k_a\Delta T(z) \tag{17.24}$$

$$\frac{dT_2}{dz} = k_b(T_2(z) - T_1(z)) = -k_b\Delta T(z) \tag{17.25}$$

where k_a and k_b are the thermal conductivities of the intervening material at points A and B, respectively. Summed together, this becomes

$$\frac{d\Delta T}{dz} = \frac{d(T_2 - T_1)}{dz} = \frac{dT_2}{dz} - \frac{dT_1}{dz} = K\Delta T(z) \tag{17.26}$$

where $K = k_a + k_b$.

The total exchanged energy is found by integrating the local heat transfer q from A to B:

$$Q = \int_A^B q(z)dz = \frac{U}{D}\int_A^B \Delta T(z)dz = \frac{U}{D}\int_A^B \Delta T dz \tag{17.27}$$

Use the fact that the heat exchanger area Ar is the pipe length B–A multiplied by the inter-pipe distance D:

$$Q = \frac{UAr}{(B - A)}\int_A^B \Delta T dz = \frac{UAr\int_A^B \Delta T dz}{\int_A^B dz} \tag{17.28}$$

In both integrals, make a change of variables from z to ΔT:

$$Q = \frac{UAr\int_{\Delta T(A)}^{\Delta T(B)} \Delta T \frac{dz}{d\Delta T} d(\Delta T)}{\int_{\Delta T(A)}^{\Delta T(B)} \frac{dz}{d\Delta T} d(\Delta T)} \tag{17.29}$$

With the relation for ΔT found above, this becomes:

$$Q = \frac{UAr \int_{\Delta T(A)}^{\Delta T(B)} \frac{1}{K} d(\Delta T)}{\int_{\Delta T(A)}^{\Delta T(B)} \frac{1}{K \Delta T} d(\Delta T)} \tag{17.30}$$

Integration at this point is trivial and finally gives:

$$Q = U \times Ar \times \frac{\Delta T(B) - \Delta T(A)}{\ln [\Delta T(B)/\Delta T(A)]}, \tag{17.31}$$

from which the definition of LMTD follows.

17.3 Exergy

17.3.1 Definition of Exergy

The general definition of exergy is: "The maximum useful work which can be extracted from a system as it reversibly comes into equilibrium with its environment".

Exergy can be defined for all processes; its definition is rather simple for a thermal process of a single fluid that one considers in cryogenics.

For thermodynamics of fluids, exergy is defined as:

$$\Delta \dot{E} = \dot{m} \times \Delta e \tag{17.32}$$

Δe: specific exergy difference

$$\Delta e = \Delta h - T_0 \times \Delta s \tag{17.33}$$

Δh: specific enthalpy difference
Δs: specific entropy difference
T_0: reference temperature

Similar to enthalpy and entropy, *exergy only makes sense as exergy difference between two states.* The specific exergy difference of a certain state is therefore defined as:

$$\Delta e = (h - h_0) - T_0 \times (s - s_0) \tag{17.34}$$

h: specific enthalpy
s: specific entropy
T_0: reference temperature

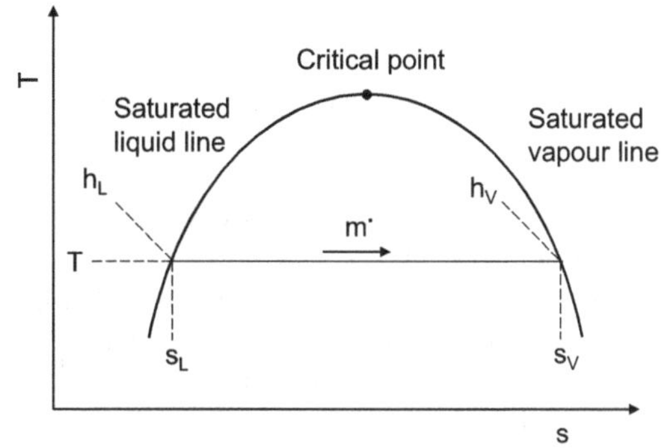

Fig. 17.1 A Carnot engine

h_0: specific enthalpy at reference temperature and pressure
s_0: specific entropy at reference temperature and pressure

One may express the specific exergy difference as the difference of two specific exergy values:

$$\Delta e = e - e_0 \tag{17.35}$$

$$e = h - T_0 \times s \tag{17.36}$$

$$e_0 = h_0 - T_0 \times s_0; \;\; e_0 = \text{constant} \tag{17.37}$$

Due to the combination of the functions of state, enthalpy and entropy, one can call the specific exergy a "derived function of state".

17.3.2 Thermodynamic Significance of Exergy

17.3.2.1 The Carnot Factor

In order to understand exergy, one needs to understand the significance of the Carnot factor, which is mathematically derived from the first and second law of thermodynamics. Figure 17.1 represents the general sketch of a Carnot engine.

The first law of thermodynamics postulates conservation of energy.

$$\dot{Q} = \dot{Q}_0 + W \tag{17.38}$$

The second law of thermodynamics postulates that the entropy in any process is either constant or increases.

$$\frac{\dot{Q}}{T} \leq \frac{\dot{Q}_0}{T_0} \quad \text{or} \quad \frac{\dot{Q} \times T_0}{T} \leq \dot{Q}_0 \tag{17.39}$$

Combining the two equations, we get:

$$W \leq \dot{Q} \times \left(1 - \frac{T_0}{T}\right); \text{ with: } \left(1 - \frac{T_0}{T}\right) \text{ being the Carnot factor.} \tag{17.40}$$

For an ideal, isentropic process W equals the right side of equation above; for non-ideal cycles, W is smaller.

The maximum work that can be extracted from a thermal process is such that the heat flow from the high temperature multiplied by the Carnot factor.

For cryogenics, the process sketched in Fig. 17.1 can be used without any modification. For temperatures T smaller than T_0, the Carnot factor becomes negative. Work will consequently not be generated but must be added to the process.

Remark, this is simply one other way of expressing that heat will never flow by itself from a lower temperature to a higher.

For temperatures T much higher than T_0, the Carnot factor approaches a maximum value of one. The maximum work that can be extracted from an ideal process consequently approaches the heat flow at temperature T. For temperatures T approaching zero, the Carnot factor approaches $-\infty$. The minimum work to be added to a process at low temperatures is consequently a multiple of the heat extracted.

For a process between 4.41 K (LHe boiling at 1.2 bar) and an ambient temperature of 300 K, the work to be added according to the Carnot factor would be:

$$W \leq \dot{Q} \times -67.03$$

The minimum work to be injected into an ideal process removing heat at 4.41 K is 67.03 times the amount of heat removed.

17.3.2.2 Derivation of Exergy

In order to more easily understand the origin of the formula for exergy, let us consider two cases, an isothermal phase transition at constant pressure and a non-isothermal state transition with pressure change.

17.3.2.2.1 Isothermal Phase Transition at Constant Pressure

Isothermal phase transition is one of the rare processes that can be considered as fully reversible. The heat injected for evaporation can be fully recovered by condensation.

In helium cryogenics, isothermal phase transition is generally referred to as "pure refrigeration". Figure 17.2 shows a sketch of the two-phase area in a T-s diagram.

According to the Carnot factor, the minimum work to be injected into a thermal process is:

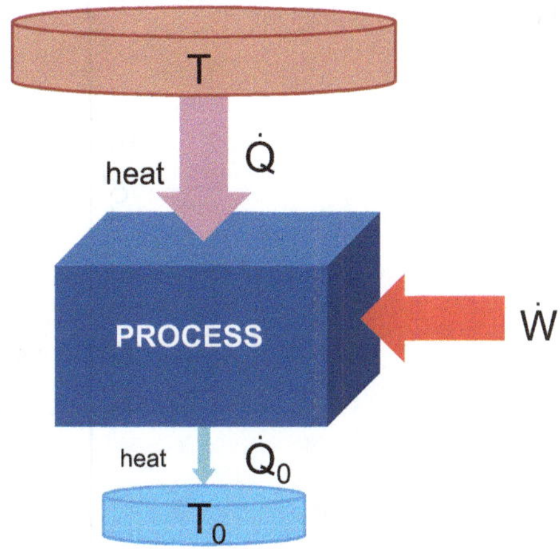

Fig. 17.2 A liquid to vapour transition in the two-phase area of a T-s diagram

$$W \leq \dot{Q} \times \left(1 - \frac{T_0}{T}\right) \tag{17.41}$$

The heat of evaporation is:

$$\dot{Q} = \dot{m} \times \Delta h \tag{17.42}$$

Δh: specific enthalpy difference; $h_V - h_L$

For a reversible isothermal process is valid:

$$\Delta h = T \times \Delta s \tag{17.43}$$

Δs: specific entropy difference; $s_V - s_L$

For minimum work to be injected to the process one such gets:

$$W_{\text{min}} = \dot{m} \times \Delta h \times \left(1 - T_0 \times \frac{\Delta s}{\Delta h}\right) \tag{17.44}$$

or

$$W_{\text{min}} = \dot{m} \times (\Delta h - T_0 \times \Delta s) \tag{17.45}$$

W_{min} describes the minimum work to be injected into the process, which equals exergy, resulting in:

$$\Delta \dot{E} = \dot{m} \times (\Delta h - T_0 \times \Delta s) \tag{17.46}$$

or

Fig. 17.3 Log T-s diagram sketch

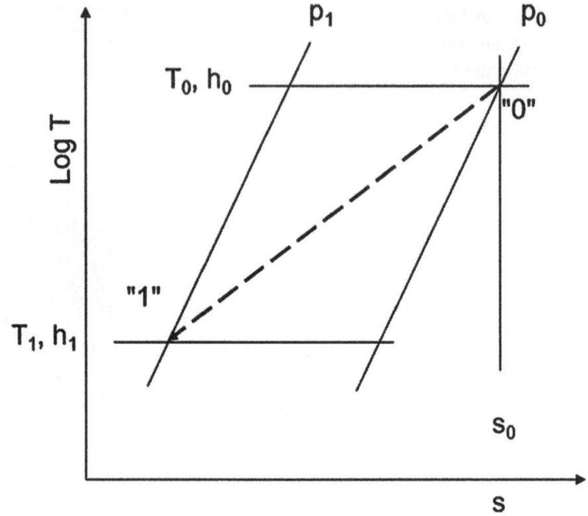

$$\Delta e = \Delta h - T_0 \times \Delta s; \ \Delta e = e_V - e_L \tag{17.47}$$

The specific exergy difference for the isothermal phase transition equals the Carnot formula in a different form.

17.3.2.2.2 Non-isothermal and Non-isobaric Process for an Ideal Gas

Figure 17.3 shows the sketch of a Log T-s diagram for an ideal gas.

For an ideal gas, the following formulae for entropy and enthalpy difference are valid:

$$\Delta h_{0\to1} = c_p \times (T_1 - T_0) \tag{17.48}$$

$$\Delta s_{0\to1} = c_p \times Ln\frac{T_1}{T_0} - R \times Ln\frac{p_1}{p_0} \tag{17.49}$$

For the purely thermal, i.e. isobaric part of the process "moving" a mass flow from property point "0" to property point "1", we can refer to the Carnot formula.

$$W \leq \dot{Q} \times \left(1 - \frac{T_0}{T}\right) \tag{17.50}$$

As the process is not isothermal, one uses the integrated mean value of the Carnot factor over T.

$$W_{\text{th min } 0\to 1} = \dot{m} \times \Delta h_{0\to 1} \times \frac{\int_0^1 \left(1 - \frac{T_0}{T}\right) dT}{(T_1 - T_0)} \tag{17.51}$$

$$W_{\text{th min } 0\to 1} = \dot{m} \times c_p \times (T_1 - T_0) \times \frac{[T - T_0 \times LnT]_{T_0}^{T_1}}{(T_1 - T_0)} \tag{17.52}$$

$$W_{\text{th min } 0\to 1} = \dot{m} \times c_p \times (T_1 - T_0) - \dot{m} \times c_p \times T_0 \times Ln\frac{T_1}{T_0} \tag{17.53}$$

For the isothermal part of the process "moving" a mass flow from property point "0" to property point "1", one can refer to the isothermal compression work.

$$W_{\text{cmp min } 0\to 1} = \dot{m} \times R \times T_0 \times Ln\frac{p_1}{p_0} \tag{17.54}$$

Adding the minimum work for the thermal process to the minimum compression work, one gets:

$$W_{\text{min } 0\to 1} = W_{\text{th min } 0\to 1} + W_{\text{cmp min } 0\to 1} \tag{17.55}$$

$$W_{\text{min } 0\to 1} = \dot{m} \times c_p \times (T_1 - T_0) - \dot{m} \times c_p \times T_0 \times Ln\frac{T_1}{T_0} + \dot{m} \times R \times T_0 \times Ln\frac{p_1}{p_0}$$

$$W_{\text{min } 0\to 1} = \dot{m} \times (\Delta h_{0\to 1} - T_0 \times \Delta s_{0\to 1}) \tag{17.56}$$

$W_{\text{min } 0\to 1}$ again describes the minimum effort to be injected in the process to achieve the change of property for the gas flow, which one defined as exergy.

$$\Delta \dot{E}_{0\to 1} = \dot{m} \times (\Delta h_{0\to 1} - T_0 \times \Delta s_{0\to 1}) \tag{17.57}$$

$$\Delta e = \Delta h - T_0 \times \Delta s \tag{17.58}$$

The specific exergy difference for the general process follows simply from the integration of the Carnot factor and the isothermal compression between the reference point "0" and any process point.

Due to the close relation of exergy and the Carnot factor, the terms "exergetic efficiency" and "Carnot efficiency" or "exergetic capacity" and "Carnot capacity" are often used synonymously in cryogenics.

One can now understand, and probably explain to others, the origin of the formula:

$$\Delta e = \Delta h - T_0 \times \Delta s. \tag{17.59}$$

This definition of exergy is limited to thermodynamic processes.

17.3.2.3 The Definition of the Reference State

Logically, from the definition of exergy given at the top of this chapter, the reference state T_0, p_0 should be ambient temperature and pressure. The question is often to which ambient to refer, as one can refer to standard conditions, average conditions or actual conditions.

For cryogenics, handling property states far off ambient, the difference has only a small influence. As one very much like simple figure values, a commonly used reference point for cryogenics is 300 K, 1.0 bar.

References

1. A. Ravex article Les techniques de l'ingénieur
2. Gistau G.M., Claudet G. (1986) The TORE SUPRA 300 W – 1.75 K Refrigerator. In: Fast R.W. (eds) Advances in cryogenic engineering, vol 31. Springer, Boston, MA
3. H. Quack Mendelssohn award talk ICEC 23/ICMC
4. http://gperinic.web.cern.ch/gperinic/bibliography.htm
5. Iarocci, M.A. Brown, D. Sondericker, J. Wu, King-Chuen, Benson, J. Farah, Yaqub Lac, C. Morgillo, A. Nicoletti, A. Quimby, E. Rank, J. Rehak, M. Werner, A.. (2003). RHIC cryogenics. Nuclear Instruments and Methods in Physics Research Section A: Accelerators, Spectrometers, Detectors and Associated Equipment. 499. 264-279. https://doi.org/10.1016/S0168-9002(02)01939-3.
6. E. Monneret and et al. ITER Cryoplant Final Design and Construction, IOP Conf. Series: Materials Science and Engineering 171 (2017) 012031 doi:https://doi.org/10.1088/1757-899X/171/1/012031
7. Parish H., Gistau G., Hood C., Kreinbrink K., Appleton W. (1991) Design of a Large 2.0 K Refrigerator for CEBAF. In: Nonte J. (eds) Supercollider 3. Springer, Boston, MA
8. R. G. Scurlock, *History and Origins of Cryogenics*, (Clarendon Press, Oxford, 1992)
9. D. van Delft. *Freezing physics, Heike Kamerlingh Onnes and the quest for cold,* (Royal Netherlands Academy of Arts and Science, 2007)

G. Gistau Baguer, *Cryogenic Helium Refrigeration for Middle and Large Powers*, International Cryogenics Monograph Series,
https://doi.org/10.1007/978-3-030-51677-2

Index

G. Gistau Baguer, *Cryogenic Helium Refrigeration for Middle and Large Powers*, International Cryogenics Monograph Series,
https://doi.org/10.1007/978-3-030-51677-2

C